ELECTRONIC DEVICES AND CIRCUITS

■ MICHAEL HASSUL

CALIFORNIA STATE UNIVERSITY,
LONG BEACH

■ DONALD E. ZIMMERMAN

CALIFORNIA STATE UNIVERSITY,
LONG BEACH

Prentice Hall

Upper Saddle River, New Jersey   *Columbus, Ohio*

Library of Congress Cataloging-in-Publication Data
Hassul, Michael.
    Electronic devices and circuits / Michael Hassul, Donald E. Zimmerman.
       p.   cm.
    Includes index.
    ISBN 0-13-500869-7
     1. Electronic circuits.   2. Electronic apparatus and appliances.
  I. Zimmerman, Donald E.   II. Title.
  TK7867.H367   1997
  621.3815—dc20   96-30563
                                            CIP

Cover Photo: Telegraph Color Library/FPG Int'l
Editor: Linda Ludewig
Production Editor: Mary M. Irvin
Design Coordinator: Julia Zonneveld Van Hook
Cover Designer: Russ Maselli
Production Manager: Pamela D. Bennett
Marketing Manager: Debbie Yarnell
Production Coordinator: Proof Positive/Farrowlyne Associates, Inc.

This book was set in Times Roman by The Clarinda Company and was printed and bound by Von Hoffmann Press, Inc. The cover was printed by Von Hoffman Press, Inc.

 © 1997 by Prentice-Hall, Inc.
Simon & Schuster/A Viacom Company
Upper Saddle River, New Jersey 07458

Printed in the United States of America

10 9 8 7 6 5 4 3 2 1

ISBN: 0-13-500869-7

Prentice-Hall International (UK) Limited, *London*
Prentice-Hall of Australia Pty. Limited, *Sydney*
Prentice-Hall of Canada, Inc., *Toronto*
Prentice-Hall Hispanoamericana, S.A., *Mexico*
Prentice-Hall of India Private Limited, *New Delhi*
Prentice-Hall of Japan, Inc., *Tokyo*
Simon & Schuster Asia Pte. Ltd., *Singapore*
Editora Prentice-Hall do Brasil, Ltda., *Rio de Janeiro*

PREFACE

There are several good introductory electronic textbooks on the market, so why write another one? The answer lies in our combined 30 plus years of teaching electronics. Those years have taught us—and studies have confirmed—that students learn new material best when their texts (and classes) meet the following general conditions:

☐ Present a manageable amount of material to be learned in a given time frame.

☐ Provide for self-review of material presented.

☐ Repeat learned material before new material is presented.

☐ Build more complex material on the basis of learned material.

☐ Use a just-in-time approach in which background material (for example, physics) is presented only when it is going to be used immediately.

We have also learned additional teaching methods that apply more specifically to the instruction of electronics:

☐ Use everyday metaphors to explain the basic behavior of electronic devices.

☐ Use recognizable real-life situations in which a particular electronic device or circuit is employed.

☐ Use circuits that are in use in actual practice—avoid strictly "academic" circuits.

☐ Provide enough fully-worked examples to demonstrate each basic point and some common variations.

☐ Recognize the widespread use of computer-aided analysis and design programs.

Our goal was to write a textbook that meets these criteria. The result is *Electronic Devices and Circuits.*

■ CHAPTER OVERVIEW

Our first step was to write a textbook that uses 30 chapters to cover the material that most books cover with much fewer chapters. Because the overall length of our book is the same as the others, we have presented the material in smaller "bites." This means that each chapter contains a less intimidating array of objectives, allows for quicker review of a smaller amount of material, provides for more worked examples per topic, and provides for more homework problems per topic.

Circuits-Based Instruction

Most textbooks give a full description of a device—physics, data sheets, detailed models—before showing the device in a circuit. At times, there are several chapters between the introduction of a device and its actual use. We believe that the device and its use in a circuit should be introduced at the same time. This is done by using the simplest reasonable model for the device. We introduce additional device features only after describing the basic circuits.

Emphasis on Troubleshooting

Troubleshooting electronic circuits is one of the most important aspects of a technologist's career. Therefore, we emphasize troubleshooting by including troubleshooting examples in most chapters as well as devoting a stand-alone chapter on systematic troubleshooting of transistor circuits.

Computer Applications

We include computer problems at the end of most chapters to give students a chance to use their available computer package to analyze the circuits they have already analyzed by hand and to examine the results of failures of various elements in the circuit. Most of the circuits in the textbook have already been simulated for both PSpice and Electronic Workbench. A disk containing these circuit files is available from your instructor.

■ TEXT ORGANIZATION

We also took particular care when we organized the material. *Electronic Devices and Circuits* is divided into five parts.

Part 1 *Diodes:* Chapters 2 to 6

Diodes are first introduced as ideal on-off switches in Chapter 2, where we describe the basic diode circuits. Chapter 3 presents a more realistic model for the diode, discusses practical diodes, and reviews the circuits of Chapter 2. Chapter 3 also introduces troubleshooting diode circuits. Chapter 4 reviews the behavior of the capacitor and inductor and shows how diodes can be combined with these elements. Chapter 5 is devoted to the Zener diode, which provides a stable voltage. Chapter 6 discusses some special diodes used in communication and optical circuits. (Diode physics is presented in Appendix A.)

Part 2 *Bipolar Junction Transistors (BJT):* Chapters 7 to 19

Chapter 7 introduces, with minimal physics, the basic building block of analog electronic circuits, the NPN BJT. (BJT physics are described in Appendix B.) Chapter 8 details a systematic approach to troubleshooting BJT circuits. Chapter 9 describes the DC behavior of the BJT and introduces the general concept of amplification. Chapter 10 describes the AC behavior of the BJT and formally defines amplification. Chapter 11 describes a way to analyze complex circuits by combining simple models of individual subcircuits. Chapter 12 describes the buffer amplifier and how to use it to electrically isolate a source and load. Chapter 13 is presented for those who want a more detailed explanation of BJT behavior and modeling.

Chapter 14 describes another BJT, the PNP BJT. Most textbooks teach the NPN and the PNP at the same time. The problem with this approach is that in the real world, PNP circuits do not mirror NPN circuits. That is, PNP bias circuits that are used to provide the required DC currents and voltages are different than NPN bias circuits. To avoid confusion, the PNP is covered in a separate chapter. This chapter also gives the student a chance to review the basic BJT amplifier circuits.

Chapters 15 to 18 contain transistor circuits that provide different biasing schemes, good high-frequency response, differential amplification, and constant current sources.

Chapter 19 discusses the various power output stages that can be built with BJTs. Because failures in electronic devices often occur in the output stage, we present a reasonably complete discussion of these stages and appropriate troubleshooting techniques.

Part 3 *Field Effect Transistors (FET):* Chapters 20 to 22

These chapters introduce an alternative transistor to the BJT. The FET has different DC behavior from the BJT but the same AC behavior. Therefore, once you have learned how BJT circuits behave, you already know how FET circuits behave. As a result, the FET material can be covered in just three chapters.

Part 4 *The Operational Amplifier (Op-Amp):* Chapters 23 to 25

The operational amplifier is introduced in Chapter 23, which discusses its ideal and real behavior and introduces the two basic Op-Amp circuits. Chapter 24 describes additional Op-Amp circuits, and Chapter 25 describes how to use the Op-Amp to provide precision rectification or to produce trigger signals that indicate that an event has occurred.

Part 5 *Advanced Topics* Chapters 26 to 30

Although earlier chapters briefly described the frequency behavior of some electronic devices, Chapter 26 formally introduces this topic. The frequency behavior of the simple resistor-capacitor circuit is first reviewed and then extended to BJT, FET, and Op-Amp circuits. Chapter 27 continues the discussion of frequency behavior by describing how we build electronic filters (bass control, treble control, and equalizers).

Chapter 28 introduces the concept of feedback and then applies that information to electronic amplifiers that use feedback. Chapter 29 describes circuits that can produce sine waves, square waves, and triangle waves. These wave shapes are used in communications, testing, and measurement. Finally, Chapter 30 describes a class of electronic devices used to switch electronic signals on and off under a variety of circumstances. These devices are useful in applications such as motor controllers and light dimmers.

Supplements and Ancillaries

Available with this text are the Instructor's Solution Manual, written by Narottam Shrestha; a Test Item File, written by Steve Rice; and a Test Manager, Transparencies, and a PSpice and Electronic Workbench data disk. In addition, a Laboratory Manual, written by Donald Zimmerman, is available.

■ ACKNOWLEDGMENTS

No textbook can be written without the cooperation of a large number of people. Michael Hassul would like to thank his first teacher of electronics, his father, Leo J. Hassul. His father was a hobbyist who built his own hi-fi electronics and helped his son build his first amateur radio transmitter. The most influential teacher in his career (as well as his first university teacher of electronics) was Prof. D. Schilling, then of the Polytechnic Institute of Brooklyn, now of City University of New York. Prof. Schilling showed me that you can teach with humor, that learning can be fun, and that it isn't necessary to sacrifice accuracy when material is presented in a relaxed manner, minimizing extraneous, and at times, intimidating, background information.

Hassul would also like to thank colleagues and students for their input to this project. Among colleagues, Profs. Narottam Shrestha (who wrote the solutions manual) and Ken James (who reviewed parts of the manuscript) stand out. Also influential have been Profs. Paul Neudorfer, Hassan Babaie, Gordon Carpenter, and Bahram Shahian. Thanks to all of my students who have taught the teacher and, in particular, to Gerald Bruce and Roger Savoie, Jr., who beta tested the manuscript and caught many errors. Thanks also to Reena Pradhanang for typesetting the solutions manual.

Also due thanks are the Manhattan Beach Public Library, which served as an excellent source of technical reference material, and the folks at The Hungry Mind coffee house-bookstore for providing a friendly atmosphere where an author could bang away for hours at the keyboard of his laptop computer.

Finally, Hassul wants to acknowledge the contributions to the text and his sanity of his wife Laurie J. Spector. Perhaps only married authors can truly appreciate the support and sacrifices of their spouse in such an undertaking. In addition to her general support, Laurie also acted as the sounding board for much of the humor (and made her own contributions to same) and the everyday similes and metaphors used to describe the behavior of electronic devices.

Don Zimmerman acknowledges the contributions of Matthew Fichtenbaum, a graduate student and instructor at the Massachussetts Institute of Technology, who taught him that there was a way to look at electronic devices without the overly complex mathematical models that had been presented in his previous courses. That meant that it was possible to develop an intuitive, or "gut level," understanding of how electronic devices work. Since then, Zimmerman has learned that the troubleshooting of electronic circuits depends on this "gut level" understanding.

Mr. Fichtenbaum was also responsible for introducing a young, eager, but confused kid to the wonderful world of electronics, facilitating the creation of a lifetime of positive self-esteem. He opened the door to a vast number of truly gratifying career opportunities that competence in electronics has made possible.

Both authors wish to express their gratitude to the excellent reviewers of this text. Most of their recommendations were incorporated into the text, resulting in a much better book. We thank Stephen Cheshier, Southern College of Technology, Georgia; Herbert D. Daugherty, Ivy Tech State College, Muncie, Indiana; Susan Drake, ITT Technical Institute, Portland, Oregon; Doug Fuller, Humber College, Toronto, Ontario; Jan Jellema, Eastern Michigan University; John Price, DeVry Institute, Columbus, Ohio; Bob Silva, Pima Community College, Arizona; Tim Staley, DeVry Institute, Irving, Texas; and Peter Tampas, Michigan Technological University.

We wish to thank all of the editors and their assistants at Prentice Hall for the work they have done in bringing this project to fruition. Thanks to Holly Hodder, who encouraged our vision of the project. Also, great thanks are owed to Gail Savage of Proof Positive/Farrowlyne Associates, Inc., whose official job description is project editor, but in fact acted as the copy editor, art editor, development editor, and production manager, and provided aid and comfort to bedraggled authors. Thanks also go to Eileen Schmidke, the person responsible for managing design, text, and art for this text. Her hard work and attention to detail helped wrestle a huge manuscript and art program into a carefully presented completed text.

CONTENTS

4

Diode—Reactive Circuits   86

12

13

21

**The MOSFET
682**

27

**Active Filters
898**

CHAPTER

1

INTRODUCTION TO ELECTRONICS

Chapter Objectives

To understand the uses of electronics in today's world.

To learn how to use the techniques and exercises throughout this text.

To understand the difference between electron flow and conventional flow electronics.

To appreciate the applications of special computer programs for studying electronics.

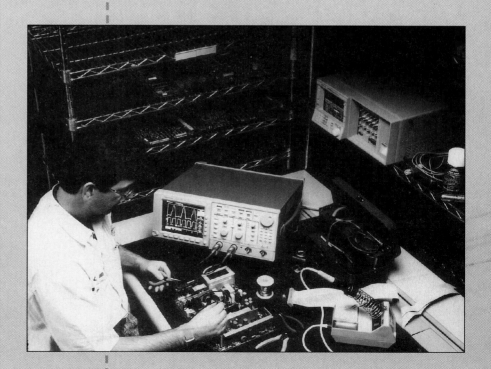

1–1 ■ ELECTRONICS TODAY

We live at the end of the first *Electronics Century* (OK, we made that term up, but it fits). From the moment our clock radios wake us in the morning, to our last waking moment in the glow of late night TV, electronics is an integral part of our daily life.

As Figure 1–1 shows, our Electronics Century began with the invention of the first vacuum tube. The vacuum tube allowed us to amplify weak telephone signals, to construct radio transmitters and receivers, and to build the first computers. The vacuum tube was later replaced with the smaller and more energy-efficient transistor. Today, we can place thousands of transistors on a single small chip to produce an integrated circuit.

FIGURE 1–1
Electronics timeline

1890s	Cathode Ray Tube (oscilloscope)
1900s	Vacuum Tube Diode (radio detection)
	Vacuum Tube Triode (amplification)
1910s	Superheterodyne Reciver
1920s	Commercial AM Radio
	Kinescope (TV cathode ray tube)
	FM Radio
1930s	Commercial FM Radio
	Radar
	Electron Microscope
	Radio Telescope
	Analog Computer
1940s	Inertial Navigation
	Commercial TV
	Commercial Stereo
	Digital Computer
	Transistor
1950s	Video Recorder
	Color TV
	Industrial Robot
	Transistorized Hearing Aid
	Transistorized Computer (Mainframe)
	Compact Pacemaker
	Sputnik
1960s	Minicomputer
	Music Synthesizer
	Light Emitting Diode (LED)
	Laser
	Integrated Circuit
	Communication Satellite
	Electronic Watches
1970s	Microcomputer
	Personal Computer
	Pocket Calculator
	Fiber Optics
	Microchip
	Video Games
1980s	Laser Printer
	VHS and BETA VCRs
	Portable Video Camera
	Home Video Games
	CD Player
	Satellite TV

Tubes

Transistors

Integrated circuits

μ processors

FIGURE 1–2

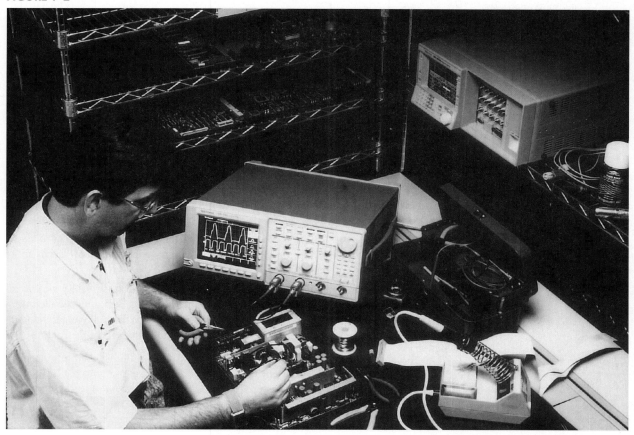

The field of electronics is so broad that almost all human endeavors involve some form of electronics application (Figure 1–2). Unless we as a species self-destruct, leaving the world to the cockroaches, electronics will be with us for a long long time. Consider just a sample of the areas that are now dependent on electronics:

☐ Home: radio, TV, VCR, CD player, remote control, video games, microwave oven, telephone and answering machine, home security, personal computer, light dimmers, appliances, any toy that requires batteries.

☐ Transportation: automobiles, motorcycles, airliners, trains and rapid transit, ships at sea, space shuttle, air and auto traffic control, navigation, radar, communication.

☐ Business: data processing, cellular telephone, fax machine, manufacturing, security, computer networks, commercial satellites, robots.

☐ Entertainment: music (synthesizing; recording; enhancement, e.g., Dolby; mixing; CD-ROMs; amplifying and special effects); music concert and stage special effects; movies; satellite TV.

☐ Medicine: diagnostics (electrocardiogram, CAT and MRI scans); patient monitoring, treatment, and therapy (lasers, ultrasound, even electronic acupuncture); rehabilitation (pacemakers, drug delivery, physical therapy, bio-feedback); patient aids (artificial limbs, hearing aids).

☐ Scientific Research: data acquisition and analysis, computer modeling and numerical problem solving.

☐ Environmental Protection: efficient electronic homes, environmental monitoring, the electric car, environmental control of office buildings, waste treatment.

☐ Athletics: optimization of performance through electronic analysis of motion, control of exercise equipment.

☐ Military: everything, right down to the dogface soldier on the ground using walkie-talkies, satellite communications, and the global positioning satellite.

Because electronics are so pervasive, your future is bright, whether you have enrolled in a program of electronics study because you like to fool around with electronics, or you want to learn electronics to ensure your economic future, or you have a passion, such as music, that can be enhanced with electronics.

Industrial work in electronics is typically divided into three categories: engineer, technologist, and technician. The duties of these different job categories are described here in general detail:

☐ Technician: build circuits from schematic diagrams supplied by engineers. Troubleshoot prototype circuits and repair faulty equipment. Usually follow a predetermined protocol or instruction in checking, verifying, and fixing circuits and equipment.

☐ Technologist: all the work of the technician plus additional responsibilities in management, training, decision making, and documentation. The technologist often has more business training than the technician.

☐ Engineer: primarily responsible for design and analysis. Education is usually very theoretical. The engineer is analogous to the architect who designs a building; the technologist is similar to the contractor that takes the plans from paper to the real world; the technician is similar to the carpenter, electrician, and plumber who does the construction.

Often the above divisions in the work of technician, technologist, and engineer are not clear cut. For example, in a small entrepreneurial company, a technologist with a gift for design (design is as much art as science) might do the design, as well as the construction and troubleshooting.

1–2 ■ THE STUDY OF ELECTRONICS

Some of you have been playing with electronics as a hobby. Some of you have a natural instinct for electronics. Most of you have never seen a diode or transistor before nor are lucky enough to have an "electronic gene." Common to all of you, however, is your choice to be an active participant in the electronics field. You must, therefore, take this course very seriously.

Unlike the general education course that is intended to broaden your perspective of life (do not sell these courses short—in the long haul, these courses are extremely important to your development as members of society), your work future depends on this introductory electronics course. You cannot "skate" through this course, then expect to learn the material in your more advanced electronics courses.

Passive participation, where you simply regurgitate what your professor has told you, may work in some courses but will not work here. If electronics is to be your career, you must actively participate in your own education.

Multiple Small Exposures

Many of you can benefit from a study pattern that we call **multiple small exposures.** That is, you learn by repeated exposures to small amounts of material. There are many details to learn in electronics. Multiple small exposures can prevent these details from overwhelming you.

When the authors were young and broke (no longer young, but still broke), we could only afford cheap transportation, such as an old car or motorcycle. Either because of interest or lack of funds for a real mechanic, we learned to troubleshoot and repair our vehicles. We didn't do this by getting the shop manual and reading it from cover to cover, trying to become instantaneous expert mechanics. Rather, we would go to the section in the manual that applied to our specific problem.

For example, carburetor problems were very common in older vehicles. Figure 1–3 shows how we taught ourselves to fix these problems. First, we got the shop manual for the motorcycle. Then we would look at the pictures of the carburetor. Next we would read the text. We would repeat this process until we had a fairly clear idea of how the carburetor fit into the fuel system and how the carburetor itself is adjusted. Finally, we would go to the bike and try to fix or adjust the carburetor. When we didn't succeed at first, we would go back to the manual and repeat the process until we succeeded.

FIGURE 1–3
Multiple small exposures

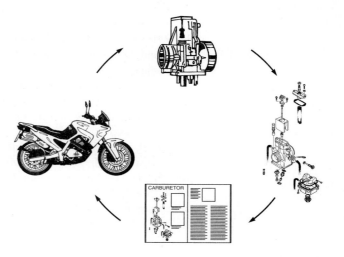

In this way, by looking at a small section of the manual, we learned bit by bit how to fix most anything on a motorcycle or car. By taking bite-sized chunks of the manual, we did not get discouraged. We learned.

In the application of multiple small exposures to the study of the electronics, do not worry about immediately memorizing details. Rather, proceed as follows when you begin a new chapter.

☐ Look at the figures in the chapter. If there is a graph, pay attention to the variables in the graph, but your goal should be to just get an appreciation of the general shape of the plot.

☐ Read the section headings. These headings will give you an idea of the material covered in the chapter. At this point, concentrate on one section at a time until you understand the material there.

☐ Look at the boldface terms. New concepts and names are often printed in bold. This is an easy way to learn the buzz words in electronics.

☐ Look at key equations, usually set off in boxes. These are the bottom-line equations needed to determine the values of circuit variables. Other equations represent intermediate steps that can be skipped on first reading.

☐ Skim the examples. Most of what you need to learn is contained in the examples.

☐ Read the text in the section.

☐ Repeat the process until you have mastered the material in a given section. Test your knowledge with the section review questions.

Study Techniques

A relatively painless method for applying the multiple small exposure technique is to set aside 15 minutes twice a day when you begin the study of a new chapter. In the first 15-minute session, you might just look at the figures and their captions. In the second 15-minute session, you might look at the section headings and the boldface words in a given section. The next day, look at the key equations in the first session and the text in the second session.

As you apply this technique, you will naturally find that you are spending more than 15 minutes per session. The trick is to avoid cramming; don't force yourself to put in hours of study. Instead of reading a magazine on the can, take the book in with you and skim a chapter. You want to avoid forming negative feelings about your chosen career. Multiple small exposures can help you enjoy solving the challenges inherent in the study of electronics. It's similar to solving puzzles.

The final process in learning electronics is to take your book and class learning into the lab. Putting the actual electronic hardware together and measuring circuit responses is the best way to really understand these devices. In the lab, you will construct circuits, test them, and determine the cause of any problems that occur. This is your apprenticeship in electronics.

A final word about your studies. As we showed in Figure 1–1, the devices we use today are much faster, smaller, and more energy efficient than the devices used years ago. It is also true that we can do much, much more with today's electronics than could be done 50 years ago.

Still, if Rip Van Electron fell asleep in 1940 and woke today, he might not recognize the transistor, but he would recognize many of the circuits. For example, Figure 1–4 shows the similarity between a vacuum tube cathode follower circuit from a 1946 reference book and a FET follower circuit from Chapter 22 of this textbook. The basic circuitry is the same in each.

FIGURE 1–4
Similar vacuum tubes and transistor circuits

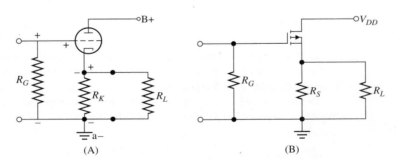

(A)   (B)

Users of electronic devices, whether tubes or integrated circuits, do not need to be experts on the device's physics. Rather, we are interested in the relations between input and output voltages and currents. The devices in this book are known as **solid state** devices and are constructed with materials known as **semiconductors.** Even if these devices are replaced by new devices (optical, plastic, organic, or even biological) with similar input-output properties, what you learn now about such things as AC-to-DC converters, amplifiers, filters, and oscillators will last you a lifetime.

1–3 ▪ A WORD ABOUT CURRENT

You know from your AC/DC courses that electrical current is a measure of the movement of electrical charges. You are also aware that current has a direction. What you might not know is that there are two different definitions for the current direction: **electron flow** and **conventional flow.**

In electron flow, we define the current direction as the direction taken by electrons (negative charges) in the circuit. Figure 1–5A shows a circuit in which the electrons move counterclockwise. As you can see, the current is defined in the direction of the electron flow. Ohm's law for electron flow is shown in Figure 1–5B. Note that electron flow current goes from the negative to the positive terminal of the resistor.

We can also define the current direction as the direction of flow of *positive* charges. Figure 1–6A shows the direction a positive charge would take around the circuit, and the current direction for this convention. Figure 1–6B shows that Ohm's law is the same as before but now the current goes from the positive to the negative terminal of the resistor.

FIGURE 1–5
Electron flow

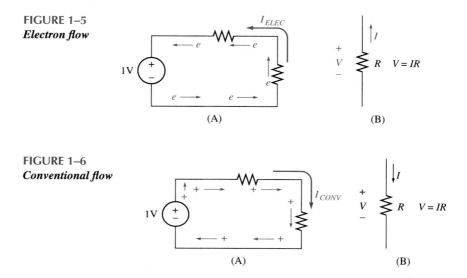

(A)　　　　　　　　　　　　　　　　(B)

FIGURE 1–6
Conventional flow

(A)　　　　　　　　　　　　　　　　(B)

Interestingly, the conventional flow definition is based on work by Benjamin Franklin, who assumed that electricity flowed from an excess of some sort of electrical substance to a deficit of this substance. Hence, electricity flows from positive to negative.

Approximately one-half of the technology institutes in this country use electron flow. Conventional flow is used by the other half of the technology institutes. Conventional flow is also used by *all* electrical engineers. Because engineers determine circuit symbols, the arrows on electronic devices are based on the conventional flow definition.

Figure 1–7 shows the diode that will be discussed in the next chapter. In the first figure, electron flow is used to define the diode current. In the second figure, conventional flow is used. You can see that the conventional flow current is in the direction pointed to by the triangle in the diode symbol.

This textbook is written with the electron flow definition for current. A companion text is written with the conventional flow definition for current. Be aware of this difference in the definition of current direction when you discuss circuits with someone educated at a different school.

FIGURE 1–7
Electron flow and conventional flow in a diode

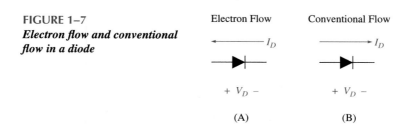

Electron Flow　　　　　Conventional Flow

(A)　　　　　　　　　　　　(B)

Also be aware that the only real difference between electron flow and conventional flow is the direction of the current arrow shown in a circuit diagram. Voltage definitions do not change, and all of the equations in the two versions of this text are identical.

1–4 ■ COMPUTER SIMULATION

There are many ways computer simulations can help you practice with and understand electronic circuits. For example, your instructor might have you write programs that will solve many of the equations in this textbook in BASIC, FORTRAN, or one of the newer C languages. You can also use a spreadsheet program or a programmable calculator to set up and solve these equations. Many circuit analysis programs are now available. The most popular of these are based on SPICE (Simulation Program with Integrated Circuit Emphasis.) This program was first designed in the early 1970s at the University of California, Berkeley.

The following discussion is not intended to teach you how to use particular computer programs. Rather, we want to show you the way some of the more common programs simulate circuits. Your instructor will explain the specific programs available to you.

FIGURE 1–8
Transistor amplifier in PSPICE

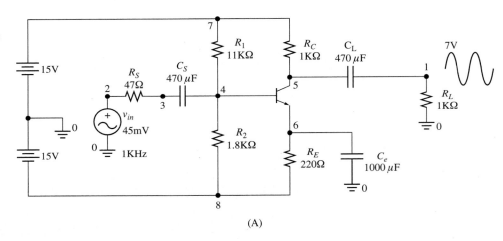

(A)

```
Transistor amplifier
*circuit description
vin    2    0     sin (0 45M 1000)
Vcc    7    0     15
Vee    0    8     15
RS     2    3     47
R1     4    7     11K
R2     4    8     1.8K
RC     5    7     1K
RE     6    8     220
RL     1    0     1K
CS     3    4     470U
CL     5    1     470U
CE     6    0     1000U
*transistor description
Q1     5    4     6 qmod
.model  qmod  npn
*analysis and output requests
.tran .05M 5M
.plot tran V(1) V(2)
.probe
.end
```

(B)

Sample Circuit Simulation Programs

In the basic SPICE program, you create a description of the circuit that tells the computer what elements the circuit contains, how they are connected, and what models you use for electronic devices such as diodes and transistors. You also tell the computer what type of analysis you want to perform. Consider the circuit shown in Figure 1–8A, which is an amplifier that takes a small input signal (45mV) and produces a much larger output signal (approximately 7V).

Figure 1–8B shows the program you would write to produce this simulation. To use the standard SPICE simulation, you first number the nodes, ground always being numbered 0. The program is then written as follows:

Title: First line of program is the title for this simulation.

Comments: Any line beginning with an * is treated as a comment and is not acted on by the computer.

FIGURE 1–9
Transistor amplifier in MicroSim

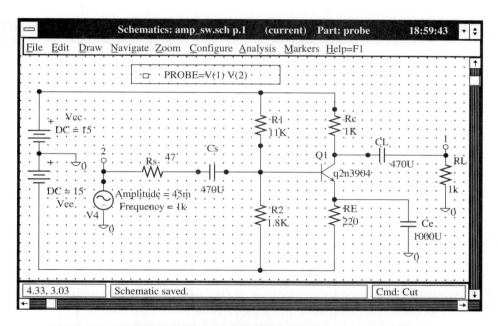

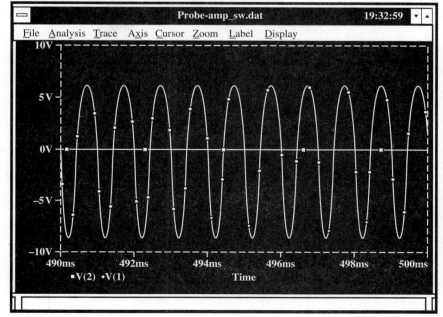

Circuit Elements: Voltage sources (Vxxxxxxx), resistors (Rxxxxxxxx), capacitors (Cxxxxxxx), and inductors (Lxxxxxxx) are described by telling the computer which two nodes the element is connected to and telling the computer the values of these elements. Active devices such as the diode (Dxxxxxxxx) and transistor (Qxxxxxxx) also are described by the nodes to which they are connected. In addition, the active device is given a model name (qmod for Q1), and the model to be used is described on a separate line (.qmod). The description of the circuit elements is known as the *NETLIST.*

Analysis Options and Output Requests: These final lines of the program tell the computer what type of analysis to perform (DC, AC, transient) and what outputs are to be printed and plotted.

There are now a number of programs on the market that allow you to actually draw the circuit on the screen. This is known as **schematic capture.** One such program is available from MicroSim.[1] Figure 1–9A (p. 11) shows what the screen looks like, and Figure 1–9B shows the input and output plots for the amplifier circuit. Note that because the output is so much larger than the input (7V compared to 45mV), the input voltage shows up as an almost straight line. All of the extensive analysis options found in SPICE are available in this program.

Another popular schematic capture program, known as Electronics Workbench,[2] is available from Interactive Image Technologies LTD. As you can see from Figure 1–10, this program allows you to simply drag circuit elements from the parts bin into the circuit you are building. Measurement is performed by actually connecting icons of multimeters and oscilloscopes to the circuit. This program is very user friendly.

FIGURE 1–10

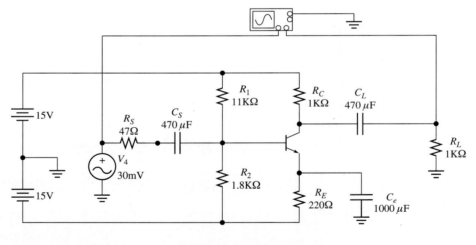

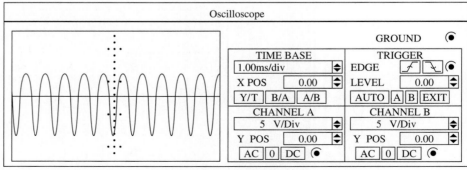

[1]Footnote copy TK
[2]Footnote copy TK

Using Computer Programs

Whatever computer program you have available, use it to your advantage. Most chapters have homework problems that specifically ask you to simulate circuits. Beyond homework problems, you can also simulate the circuits discussed in the text to see how the circuit works. More important, you can easily vary element values (for example, change a resistor value) and see how that change affects the circuit response.

Equally important, you can use computer simulations to help you learn to troubleshoot a circuit. For example, as we describe in Chapter 8, problems in most transistor circuits can be found by looking at the expected and measured DC voltages. At DC, all capacitors are opened and all AC sources are removed. Consider, for example, the amplifier circuit with all capacitors and AC sources removed (Figure 1–11).

Figure 1–11A shows that we are measuring the DC voltage at the top of the transistor and get 697mV. To see what would happen if resistor R_2 has burned out, we remove R_2 from the circuit. Figure 1–11B shows that we now get a large negative voltage at the transistor. To see what would happen if resistor R_2 fails by shorting out, we replace R_2 with a short (Figure 1–11C). We now get a large positive voltage. Therefore, if our circuit is not

FIGURE 1–11
Troubleshooting

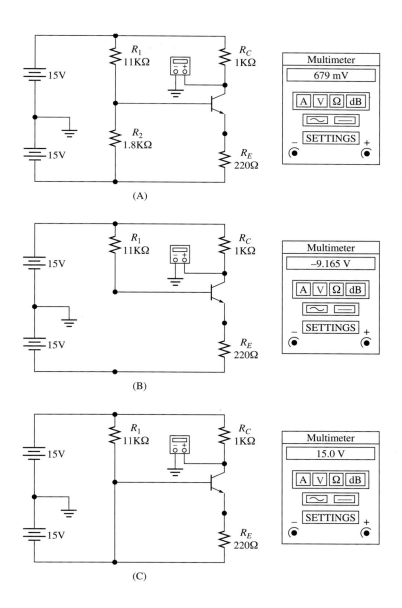

working and we get a large negative voltage at the top of the transistor, we can suspect that R_2 has failed open. If we get a large positive voltage, we can suspect that R_2 has failed short.

To ease your way into computer simulations, we have already simulated most of the circuits in this text (excluding homework circuits—we want to leave you something to do). Disks are available for use with PSPICE, MicroSim, and Electronic Workbench.

PART

DIODES

1

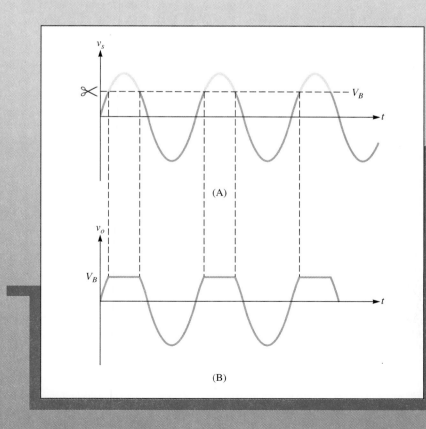

CHAPTER

2

THE IDEAL DIODE

CHAPTER OBJECTIVES

To determine the **ON-OFF** characteristics of the ideal diode.

To determine the state of a diode in a circuit. Is the diode **ON** or **OFF?**

To determine voltage and currents in DC diode circuits.

To determine the output of the half-wave rectifier circuit.

To determine the output of the full-wave rectifier circuit.

To determine the output of clipper circuits.

2-1 ■ INTRODUCTION

The semiconductor devices we discuss in this textbook behave very differently from the resistors, capacitors, and inductors that are familiar to you. Semiconductors all regulate, in some manner, the amount of current that can pass through them in a given direction. The first device we consider is the **diode.** The diode, like the resistor, is a two-terminal device. The circuit symbol for the diode is shown in Figure 2–1A. The triangle is known as the **anode** of the diode. The straight line in the diode symbol is called the **cathode** of the diode.

Because of the physics of the diode, discussed in detail in Appendix A, current can pass only in one direction through the device. That is, the diode is a one-way street for current (Figure 2–1B). Current can pass only from the cathode to the anode of the diode. No current can exist in the diode in the opposite direction.

FIGURE 2–1
The Diode

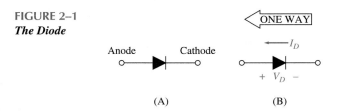

(A) (B)

Figure 2–2 shows, in general form, some important uses for the basic diode property of one-way conduction:

☐ Traffic control of electrical current

☐ AC to DC conversion

☐ Voltage limiting

We can use diodes to direct current to different paths in a circuit. The example shown in Figure 2–2A demonstrates what can happen if you take off on your motorcycle while the kickstand is still down. You may go splat on the ground—very embarrassing (don't ask us how we know). You can use diode circuits to direct current away from the ignition system if you accidentally leave the kickstand down.

The most common use, by far, of the diode is to convert an AC voltage to a DC voltage. You probably have at least one AC adapter for your radio, tape player, or calculator. The adapter converts the AC voltage from the wall socket to the DC voltage needed for your electronic device (Figure 2–2B).

Diodes are also used to limit the voltage swing of a signal. This is often done to protect delicate electronics from too large a voltage. Limiting is also used by guitarists to chop off part of the signal produced by their electric guitars. This introduces a scratchy sound to their music. Guitarists call this limiting device a "fuzz box" (Figure 2–2C).

FIGURE 2–2
Diode Circuits

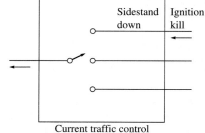

Current traffic control

Oops! Left his sidestand down.

(A)

FIGURE 2–2 *(continued)*

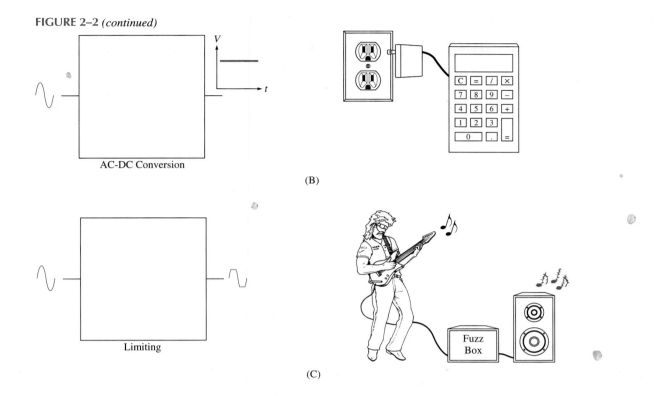

(B)

(C)

2–2 ■ THE IDEAL DIODE

To understand the behavior of any electrical device, you must know how the device current depends on the device voltage. Although the exact dependence of these variables is quite complicated, we can use simple models to understand and analyze diode circuits.

The simplest model is the **ideal diode.** The ideal diode acts as a switch. It is either closed or open. We call these two diode states **ON** or **OFF.** The diode is **ON** when current passes from the cathode to the anode. When the diode is **ON,** we can replace it with a short circuit.

The diode is **OFF** when cathode voltage is greater than anode voltage. When the diode is **OFF,** we can replace it with an open circuit. These cases are shown in Figure 2–3 (p. 20). Note: Even though there is no current in the **OFF** diode, we still label its current.

Unlike polarity definitions for the resistor, capacitor, and inductor, the polarity of V_D and the direction of I_D must be assigned as shown. Diode voltage is defined as positive at the anode and diode current is always assigned in the **ON** direction. With these definitions, we can summarize ideal diode **ON-OFF** behavior.

$$\text{diode } \textbf{ON: } I_D > 0 \qquad V_D = 0 \rightarrow V_1 = V_2$$
$$\text{diode } \textbf{OFF: } I_D = 0 \qquad V_D < 0 \rightarrow V_1 < V_2$$

We often refer to an **ON** diode as being **forward biased** and an **OFF** diode as being **reverse biased.** Diode circuits are built to take advantage of the fact that the diode can be turned **ON** or **OFF** by adjusting the voltage across the diode or the current through the diode.

FIGURE 2–3
The ON and OFF Diode

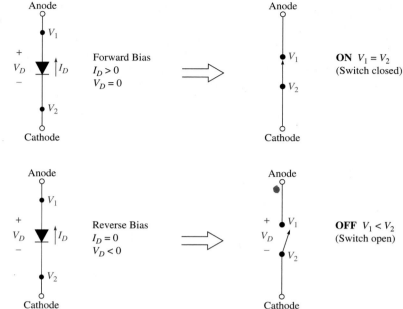

REVIEW QUESTIONS

1. In what two states does the ideal diode operate?
2. When the ideal diode is **ON,** with what can it be replaced?
3. When the ideal diode is **OFF,** with what can it be replaced?
4. What are V_D and I_D when the diode is **ON?**
5. What are V_D and I_D when the diode is **OFF?**

2–3 ■ DIODE-RESISTOR CIRCUITS

We start by taking a look at the simple diode-resistor circuit shown in Figure 2–4A. In circuits with a single voltage source, current always travels to the positive terminal of the source. In this circuit, therefore, current goes from ground to the 10V source. Because this is the allowed direction for current through a diode, the diode is **ON** and acts as a short circuit (Figure 2–4B). In this circuit, we see that because

$$V_1 = V_2 = 0V$$

the voltage across the resistor is

$$V_R = 10 - V_1 = 10V$$

From Ohm's Law we find the resistor current to be

$$I_R = \frac{V_R}{R} = \frac{10}{1K} = 10mA$$

Because the diode is in series with the resistor, their currents must be the same:

$$I_D = I_R = 10mA$$

You can see that when the diode is **ON** in this circuit, the diode voltage is 0V and the diode current is positive.

Things get interesting when we reverse the diode (Figure 2–5A). The circuit current still wants to go from ground to the positive voltage. Now this current is trying to pass through the diode from its anode to its cathode. Because this is not possible, the diode must be **OFF** and, so, acts as an open circuit (Figure 2–5B).

It might seem confusing to define two different currents in Figure 2–5. This happens because the current direction for the diode is fixed by convention. When the diode is reversed, the diode current direction must also be reversed. In Figure 2–5, $I_R = -I_D$.

FIGURE 2–4

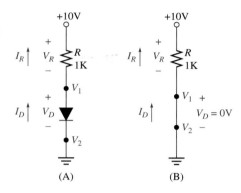

(A)   (B)

FIGURE 2–5

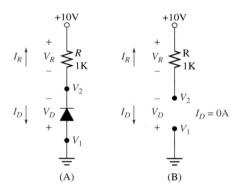

(A)   (B)

Because of the open circuit, the current in the resistor is

$$I_R = 0A$$

From Ohm's Law

$$V_R = RI_R = 0V$$

Because the anode of the diode is grounded

$$V_1 = 0V$$

The voltage at the cathode of the diode is equal to the source voltage less the voltage drop across the resistor, which is zero in this case because there is no resistor current. That is

$$V_2 = 10 - 1KI_R = 10V$$

The diode voltage is found from

$$V_D = V_1 - V_2 = 0 - 10 = -10V$$

Because $V_D < 0V$ our assumption that the diode is **OFF** is correct.

EXAMPLE 2–1
Simple Diode-Resistor Circuits

For the diode circuits in Figure 2–6A and B, determine if the diode is **ON** or **OFF.** Find V_R, I_R, V_D, and I_D.

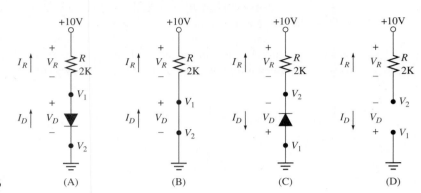

FIGURE 2–6 (A) (B) (C) (D)

Solution

(a) In Figure 2–6A, the diode current goes from ground to the source. Because this is in the allowed direction (i.e., $I_D > 0$), the diode is **ON.** Therefore, we replace the diode with a short circuit (Figure 2–6C). Because the cathode is grounded

$$V_1 = V_2 = 0V$$

so the voltage drop across the resistor is

$$V_R = 10 - V_1 = 10 - 0 = 10V$$

and the resistor current is

$$I_R = \frac{V_R}{R} = \frac{10}{2K} = 5mA$$

Finally, because the diode and resistor have the same current

$$I_D = I_R = 5mA$$

(b) In Figure 2–6B, the current still wants to flow from ground to the positive source. Now, however, this is the wrong direction for the diode, and the diode is **OFF.** Therefore, we replace the diode with an open circuit (Figure 2–6D). Because of the open circuit, all currents are zero:

$$I_D = 0A \text{ and } I_R = 0A$$

From Ohm's Law

$$V_R = RI_R = 0V$$

Because the anode is grounded

$$V_1 = 0V$$

Because there is no voltage drop across the resistor

$$V_2 = 10 - V_R = 10V$$

Therefore, the diode voltage is

$$V_D = V_1 - V_2 = 0 - 10 = -10V$$

Redraw the circuit in Figure 2–6A, changing the voltage source to 5V and the resistor to 10KΩ. Find all voltages and currents.

Answer $V_1 = V_2 = V_D = 0$ V, $I_R = I_D = 0.5$mA

EXAMPLE 2–2
Single Loop Circuit

Find V_o and I_D for the single loop circuit shown in Figure 2–7A.

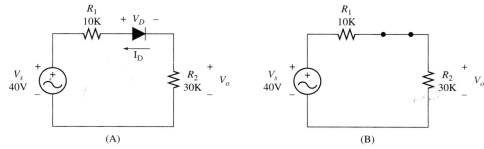

(A)   (B)

FIGURE 2–7

Solution   In a single loop circuit, there is only one current. Because current leaves from the negative terminal of the battery and returns to the positive terminal, the current in this circuit passes from the cathode to the anode. Therefore, the diode is **ON,** and we can replace it with a short circuit (Figure 2–7B). We now have a voltage divider circuit, where

$$V_o = \frac{R_2}{R_1 + R_2} V_s$$

so

$$V_o = \frac{R_2}{R_1 + R_2} V_s = \frac{30K}{10K + 30K} 40 = 30V$$

Because $I_D = I_{R2}$

$$I_D = I_{R_2} = \frac{V_s}{R_1 + R_2} = \frac{40}{10 \text{ K} + 30K} = 1\text{mA}$$

Redraw the circuit in Figure 2–7A, changing both resistors to 5KΩ. Find V_o and I_D.

Answer $V_o = 20$V and $I_D = 4$mA

A more complicated diode circuit is shown in Figure 2–8 (p. 24). Before we can analyze this circuit, we must determine whether the diode is **ON** or **OFF.** Most single-source circuits inspection reveals what paths the currents in the circuit will take. By following these current paths, we can tell which diodes are **ON** or **OFF** as follows:

☐ For a positive source, follow the current as it leaves ground and heads for the source. If this path leads through a diode, compare your current direction with the diode **ON** direction. If the current passes through the diode in the **ON** direction, replace the diode

FIGURE 2–8

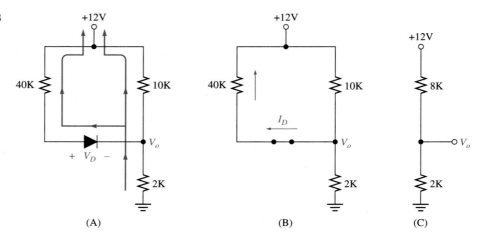

(A)   (B)   (C)

with a short circuit. If your path takes you in the **OFF** direction of the diode, replace the diode with an open circuit.

☐ For a negative source, current goes from the source to ground. Again, if the current goes through a diode in the **ON** direction, replace it with a short circuit. If the current path tries to go through the diode in the **OFF** direction, replace it with an open circuit.

Consider the circuit in Figure 2–8A. We have drawn the two paths from ground to the positive source. Because current passes through the diode from cathode to anode, the diode is **ON.**

In Figure 2–8B we replace the diode with a short circuit and note that the 40KΩ and 10KΩ resistors are in parallel. The equivalent parallel resistance of 8KΩ and the 2KΩ resistor form a voltage divider (see Figure 2–8C). We find the output voltage from

$$V_o = \frac{R_{2K}}{R_{2K} + R_{8K}} \, 12 = \frac{2K}{2K + 8K} \, 12 = 2.4V$$

Now that we have found V_o, we can return to Figure 2–8B to find I_D. You can see that the diode current is also the current in the 40KΩ resistor. The voltage across this resistor is

$$V_{40K} = 12 - V_D = 12 - V_o = 12 - 2.4 = 9.6V$$

Therefore

$$I_D = I_{40K} = \frac{V_{40K}}{40K} = \frac{9.6}{40K} = 0.24mA$$

EXAMPLE 2–3
Multiple Resistor
Circuit

Find the currents and voltages in Figure 2–9A. Note that this is the same circuit as Figure 2–8A, but the diode is reversed.

Solution   As shown in Figure 2–9A, current in this circuit tries to flow from ground to the source. The diode, therefore, is reverse biased and is replaced with an open circuit (Figure 2–9B). We see that the 10KΩ and 2KΩ resistors form a voltage divider, while the 40KΩ resistor is hanging in midair. We find V_o from the voltage divider:

$$V_o = \frac{R_{2K}}{R_{2K} + R_{10K}} \, 12 = \frac{2K}{2K + 10K} \, 12 = 2V$$

Now

$$V_1 = V_o = 2V$$

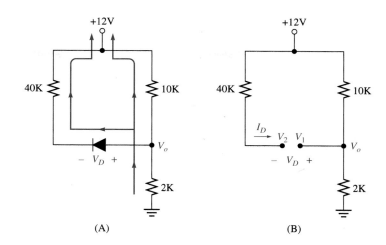

FIGURE 2–9   (A)   (B)

Because there is no current in the 40KΩ resistor

$$V_2 = 12V$$

The diode voltage is

$$V_D = V_1 - V_2 = 2 - 12 = -10V$$

The negative diode voltage confirms that the diode is **OFF.**

DRILL EXERCISE   Redraw the circuit in Figure 2–9, changing the supply voltage to 6V. Find V_o and V_D.

Answer $V_o = 1V$, $V_D = -5V$

REVIEW QUESTIONS

1. What direction does current flow in a grounded circuit driven by a positive voltage source?
2. What direction does current flow in a grounded circuit driven by a negative voltage source?
3. How can you tell if a diode is **ON** or **OFF** in a circuit driven by a single source?
4. Draw a single loop circuit with a voltage source and two resistors. What is the voltage divider formula for your circuit?

2–4 ■ IS A DIODE ON OR OFF?—TWO APPROACHES

The diode circuits you will most commonly deal with are not much more difficult than those we have presented; however, try not to be intimidated by larger circuits. The critical question you must answer in any diode circuit is which diodes are **ON** and which diodes are **OFF.** Often you can determine the state of a diode by inspecting the circuit to see in which direction the current is going. This is the process we demonstrated with the previous diode circuits. In fact, it is the only technique needed for most of the diode circuits discussed in the following sections and chapters.

You cannot always depend on simple circuit inspection to determine the state of a diode. We can develop a general procedure by reconsidering the behavior of the ideal diode, summarized in Figure 2–10. If a diode is **ON,** I_D is positive; if a diode is **OFF,** V_D is negative. This leads to two approaches to determining the state of a diode.

FIGURE 2–10

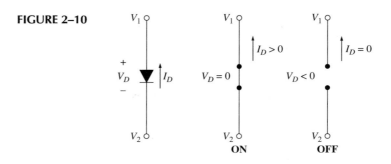

Current Test

Assume that the diode under consideration is **ON,** and replace the diode with a short circuit. Analyze the circuit to find I_D. If I_D is positive, your initial assumption was correct and the diode is **ON.** If I_D turns out to be negative, your assumption was wrong and the diode must be **OFF** ($I_D = 0$A).

Voltage Test

Assume that the diode under consideration is **OFF** and replace it with an open circuit. Analyze the circuit to find the anode voltage (V_1) and the cathode voltage (V_2). If V_1 is less than V_2, the diode is reverse biased ($V_D < 0$) and your initial assumption is correct. The diode is **OFF.** If V_1 is greater than V_2, the diode is forward biased and your initial assumption is wrong. The diode is **ON** ($V_D = 0$V).

Which test is easier to use? The answer depends on your personal preference (in the classroom, your personal preference usually means your instructor's preference), or on the particular diode circuit you are examining. Consider the circuit shown in Figure 2–11.

In Figure 2–11B, you have applied the current test. You assume the diode is **ON** and replace it with a short circuit; however, it is not immediately clear to you how to solve for I_D.

In Figure 2–11C, you apply the voltage test. You assume the diode is **OFF** and replace it with an open circuit. You now note that because the 10V source supplies the series combination of R_1 and R_2, these two resistors form a voltage divider. The same is true for R_3 and R_4. Therefore, you find the anode and cathode voltages from

$$V_1 = \frac{R_2}{R_1 + R_2} \, 10 = \frac{2\text{K}}{2\text{K} + 2\text{K}} \, 10 = 5\text{V}$$

FIGURE 2–11

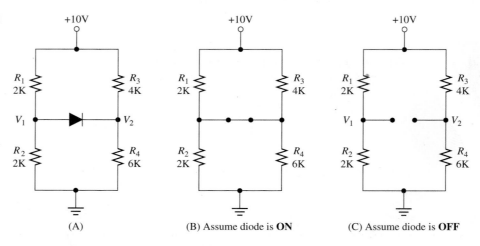

(A)   (B) Assume diode is **ON**   (C) Assume diode is **OFF**

and

$$V_2 = \frac{R_4}{R_3 + R_4} \, 10 = \frac{6K}{4K + 6K} \, 10 = 6V$$

You can see that in this circuit, the anode voltage is less than the cathode voltage, so the diode voltage

$$V_D = V_1 - V_2 = 5 - 6 = -1V < 0$$

is negative. The diode, therefore, is reverse biased and your assumption is correct. The diode is **OFF.**

EXAMPLE 2–4
Diode Test

Determine if the diode in Figure 2–12 is **ON** or **OFF.**

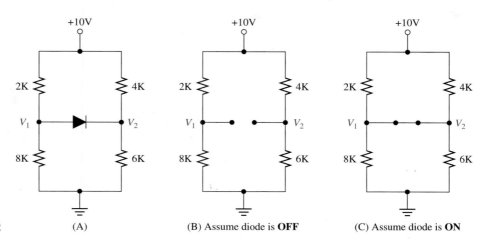

FIGURE 2–12   (A)   (B) Assume diode is **OFF**   (C) Assume diode is **ON**

Solution   We assume that the diode is **OFF** and use Figure 2–12B:

$$V_1 = \frac{8K}{2K + 8K} \, 10 = 8V \text{ and } V_2 = \frac{6K}{4K + 6K} \, 10 = 6V$$

Therefore

$$V_D = V_1 - V_2 = 8 - 6 = 2V > 0$$

This answer shows that there is enough voltage across the diode to make it forward biased. The diode is **ON;** therefore the initial assumption was wrong. This means that the values found for V_1 and V_2 are wrong. The correct answers would come from the analysis of the circuit shown in Figure 2–12C. Note that because of the short circuit, $V_1 = V_2$.

This circuit is a bit messy to work with. Although we do not expect you to find V_1 and V_2, we give you the result for completeness. The two top resistors are in parallel, so

$$2K \, \| 4K = 1.33K\Omega$$

Likewise, the two bottom resistors are in parallel, so

$$8K \, \| 6K = 3.43K\Omega$$

The two equivalent resistors form a voltage divider, so

$$V_1 = V_2 = \frac{3.43K}{3.43K + 1.33K} \, 10 = 7.2V$$

DRILL EXERCISE Redraw the circuit in Figure 2–12 (p. 27), changing the 8KΩ resistor to 2KΩ. Find V_1, V_2, and V_D. What is the state of the diode?

Answer $V_1 = 5V$, $V_2 = 6V$, $V_D = -1V$. Diode is **OFF**.

REVIEW QUESTIONS
1. How do you use the current test to determine the state of a diode?
2. How do you use the voltage test to determine the state of a diode?

2–5 ▪ CURRENT TRAFFIC CONTROL—DIODE LOGIC

As we mentioned earlier, diodes can be used to control the direction of current flow in a circuit. This control of current flow is also called **current steering.** Consider the circuit shown in Figure 2–13A. The cathode voltage of D1 is 6V. Because of the voltage drop across R_S, the anode voltage of the diode must be less than 5V. So, the cathode voltage of D1 is greater than the anode voltage. Therefore, D1 must be **OFF** and there is no current in the diode. In this case, then, all of the source current is directed, or steered, through R_L.

FIGURE 2–13

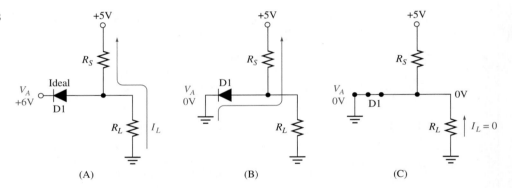

(A)   (B)   (C)

Now consider the same circuit in Figure 2–13B, in which V_A is set to 0V. In this case, D1 is **ON,** creating a short circuit path to ground for the source current. Therefore, all of the source current is diverted through the diode, so that $I_L = 0A$.

You may be thinking that there should still be some current in R_L. To see that this is not the case, examine Figure 2–13C. You can see that when D1 is **ON,** 0V is applied across R_L. Therefore, the current in R_L is

$$I_L = \frac{V_o}{R_L} = 0A$$

Another way to think about this circuit is to remember that current follows the path of least resistance. Because there is a short circuit path through D1 to ground, all of the source current takes this path.

A Diode Logic Circuit

In Figure 2–14A, we have added another diode to the circuit just analyzed. Now we have two inputs in addition to the 5V source. If V_A is set to 0V, D1 becomes a short circuit and diverts current away from R_L (Figure 2–14B). If V_B is set to 0V, D2 becomes a short circuit and, once again, current is diverted away from R_L.

FIGURE 2–14
Diode AND Logic

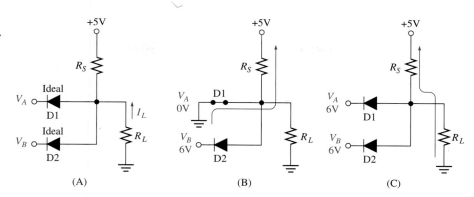

(A) (B) (C)

If both of the inputs are set to 6V, both diodes are **OFF** and current is steered to the load resistor (Figure 2–14C).

In other words, there is an output voltage only if both V_A and V_B are set to 6V. This type of diode arrangement forms a logical **AND** circuit. The motorcycle example discussed at the beginning of this chapter can illustrate the use of the diode **AND** circuit. In this example, you have tried to ride off with your sidestand down. Your sidestand catches something on the ground, and you fall. Even if you don't hurt yourself, you will definitely look uncool to your friends.

We can use the diode **AND** circuit of Figure 2–14 to save you some pain and embarrassment. Place sensors to measure voltages at the kickstand (V_A) and the gearbox (V_B). If the kickstand is up, $V_A = 0V$; if the kickstand is down, $V_A = 6V$. When the motorcycle is at rest, $V_B = 0V$; if the motorcycle starts to move, $V_B = 6V$.

Now, if the motorcycle starts to move and the kickstand is down, both

$$V_A = 6V \text{ and } V_B = 6V$$

Both diodes will be **OFF,** and current will be directed to the load (Figure 2–14C). The voltage created across the load can be used to trigger a relay that will kill the ignition. If the kickstand is up, $V_A = 0V$ and D1 will turn **ON.** The resulting short circuit will direct current away from the ignition kill circuit (Figure 2–14B). You ride off, looking very cool.

DRILL EXERCISE

What happens to the motorcycle ignition if the motorcycle is not moving?

Answer With the motorcycle at rest, $V_B = 0V$, and D2 will turn **ON.** The resulting short circuit will direct current away from the ignition kill circuit. With the motorcycle at rest, the motor will continue to run, even if the kickstand is down.

REVIEW QUESTIONS

1. What is another name for current traffic control?
2. What happens to the load current if either diode in Figure 2–14A is **ON?**
3. Under what condition will current be steered to the load in Figure 2–14A?
4. What kind of logic does the circuit in Figure 2–14 perform?
5. What will happen to you if you attempt to ride off with your kickstand down?

2–6 ■ AC to DC Conversion—The Half-Wave Rectifier

You are taking a test and your calculator goes dead! It's panic time. But you're smart—you remembered to bring your AC adapter. You plug your adapter into a wall socket and ace the test. Have you ever wondered about what is in the adapter? The answer is that your adapter contains circuitry that converts AC to DC voltage. The primary element in this circuitry is the diode.

Converting AC to DC is the most common use of the diode. We will introduce this topic here and continue to develop the circuitry in later chapters. The circuit shown in Figure 2–15A is similar to the single diode circuits discussed in the previous section. The difference here is that the driving voltage is not a constant. The input voltage varies with time. For example, consider the input voltage wave form shown in Figure 2–15B. The voltage, with respect to ground, is a constant 4V for the 1st second, goes to a constant –4V for the 2nd second, returns to 4V for the 3rd second, and so on. This type of wave shape is known as a **square wave.**

FIGURE 2–15
Half-Wave Rectifier

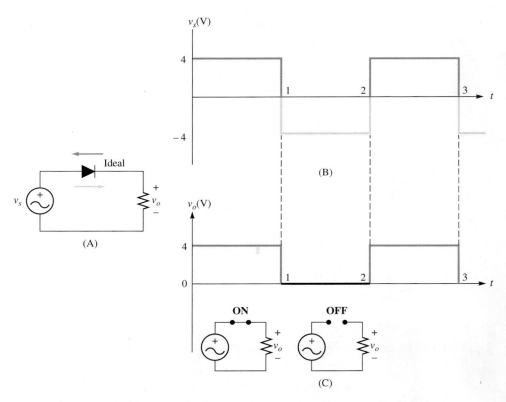

We can analyze this circuit by taking snapshots in time. Just as a regular camera takes photographs of activities at specific points in time, we freeze the circuit and measure voltages and currents at specific points in time. At each point in time, the input acts as DC voltage. At each point in time, therefore, we analyze the circuit using the techniques of the previous section.

In practice, we usually consider a range of times in a single analysis. In this circuit, for example, we know that the input is positive for t between 0 and 1 second and negative for t between 1 and 2 second, and so on. In this circuit the diode is **ON** when the source is positive and **OFF** when the source is negative. When the diode is **ON,** we replace it with a short circuit. When the diode is **OFF,** we replace it with an open circuit. This results in two possible outputs:

$$\text{Diode } \mathbf{ON}: v_o = v_s = 4 \text{ V}$$

$$\text{Diode } \mathbf{OFF}: v_o = 0 \text{ V}$$

Note that by convention we use lower case letters for AC currents and voltages. The output voltage for this circuit is shown in Figure 2–15C.

You can see that we have chopped off the negative half of the input signal. This chopping process is known as **rectification.** More specifically, because we are throwing away one-half of the input signal, it is known as **half-wave rectification.** We can also use rectification to chop off the the positive part of a signal (see Example 2–5). Rectification is the primary use of the diode. So much so, in fact, that many manufacturers use the term **rectifier** instead of diode.

EXAMPLE 2–5
Half-Wave
Rectifier Examples

Sketch v_o for the rectifier circuits shown in Figure 2–16A and B (p. 32). Although these wave forms are not sinusoidal, the rectifier behaves the same. The diode is **ON** whenever current passes from the cathode to the anode.

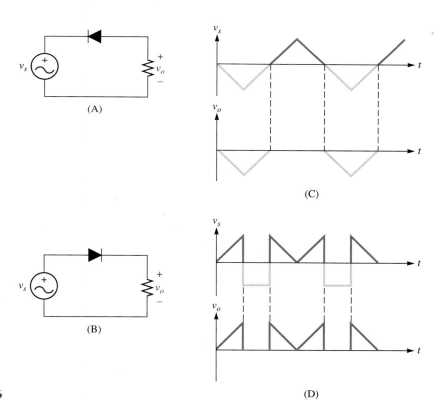

FIGURE 2–16

(C)

(D)

Solution
(a) The diode is **OFF** when the input is positive, and $v_o = 0$V. The diode is **ON** when the input is negative, and $v_o = v_s$ (see Figure 2–16C).
(b) The diode is **ON** when the input is positive, and $v_o = v_s$. The diode is **OFF** when the input is negative, and $v_o = 0$V (see Figure 2–16D).

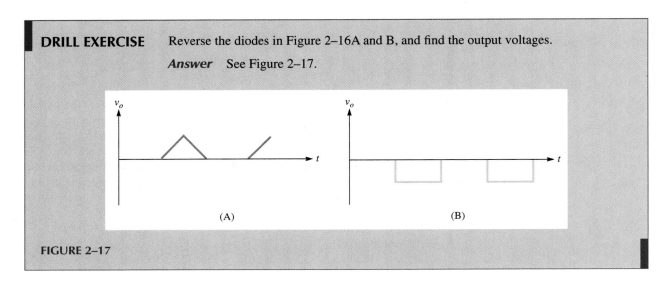

DRILL EXERCISE   Reverse the diodes in Figure 2–16A and B, and find the output voltages.

Answer   See Figure 2–17.

(A)   (B)

FIGURE 2–17

AC-DC Conversion

As we've already mentioned, the conversion of AC power to DC power is probably the most common use of the diode. Most of the electronic gear that you own—products such as CD players, amplifiers, and computers, are designed to work with DC power. When we plug these devices into wall sockets, internal circuitry converts the AC voltage to the DC voltage that they need. Most adapters, or converters, that you plug into the wall when the batteries run low on your calculator or on your "Game Boy" contain a rectification circuit.

If we drive the rectifier shown in Figure 2–18 with a sinusoidal input voltage, the output voltage will consist of only the positive halves of the input signal. We started out by calling rectification a process for converting an AC signal to a DC signal; however, the output voltage does not look very much like a constant. How, then, does rectification produce DC?

The answer is that there is DC, and then again, there is DC. By this we mean that while we would like our DC signals to be pure constants, we often can settle for something

FIGURE 2–18
Average Value of Half-Wave Rectifier Voltage

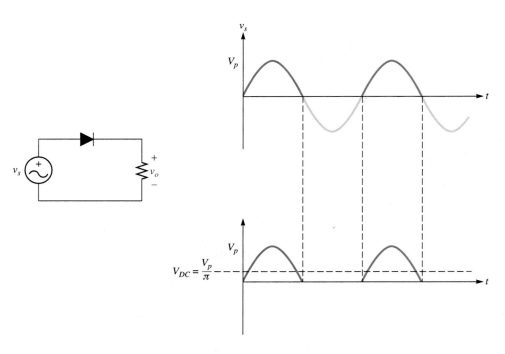

less. We can settle for a signal that has a DC component, as well as an AC component. Finding the exact DC component of a signal requires the use of calculus. That is, the DC value of a signal is equal to the average value of the signal. Don't panic! Because the average value of a signal is equal to the area under the signal, often we can approximate the DC value of a signal with some basic geometry.

We have indicated in Figure 2–18 the area contained by both the input sinusoid and the output half-wave rectified signal. Note that we consider only a single cycle of each signal. The area above the horizontal axis is considered positive (colored in dark blue), and the area below this axis is negative (colored in light blue). You can see that for each cycle of the sinusoid, there is as much positive area as negative area. Therefore, the total area is 0. A pure sinusoid has no DC component.

The half-wave rectified signal, on the other hand, has only positive area. This area gives us the average value, which also gives us the DC value, of the output voltage. The DC value of a half-wave rectified sinusoid is given by

$$\text{Half-Wave Rectified Signal } V_{DC} = \frac{V_p}{\pi}$$

where V_p is the peak value of the sine wave (Figure 2–18).[1]

Reminder: AC voltage is often given as an rms value. Therefore, to find the DC value of a half-wave rectified signal that is measured in true rms volts, convert the rms value to peak value as follows:

$$V_{rms} = \frac{V_p}{\sqrt{2}} \text{ and } V_p = \sqrt{2}\, V_{rms}$$

EXAMPLE 2–6
Finding DC Values

Find the DC values for a half-wave rectified voltage if
(a) $V_p = 110V$ **(b)** $V_{rms} = 115V$

Solution
(a) For $V_p = 110V$

$$V_{DC} = \frac{V_p}{\pi} = \frac{110}{\pi} = 35V$$

(b) When the voltage is given in rms, we first find the peak value. For $V_{rms} = 115V$

$$V_p = \sqrt{2}\, V_{rms} = \sqrt{2} \times 115 = 163V$$

The DC value can now be found

$$V_{DC} = \frac{V_p}{\pi} = \frac{163}{\pi} = 51.9V$$

DRILL EXERCISE

What is the peak value and average (DC) value of a half-wave rectified AC voltage with an *rms* value of 120V?

Answer $V_p = 170V$ and $V_{DC} = 54V$

---

[1]For the adventuresome reader, the DC value of the half-wave rectified signal is obtained by integrating a sine wave over a half period. That is,

$$V_{DC} = V_{average} = \frac{1}{T} \int_0^{T/2} V_p \sin \frac{2\pi}{T} t\, dt = \frac{V_p}{2\pi} (\cos 0 - \cos \pi) = \frac{V_p}{\pi}$$

REVIEW QUESTIONS

1. Draw a half-wave rectifier circuit so the diode turns **ON** when the input goes positive.
2. What does the output of your rectifier look like for a sine wave input?
3. Draw a half-wave rectifier circuit so the diode turns **ON** when the input goes negative.
4. What does the output of this rectifier look like for the sine wave input?
5. What is another name for the DC value of a signal?
6. What is the DC value of a half-wave rectified signal?
7. How do you find the peak value of a sinusoidal voltage if you know the rms value?

2–7 ■ AC to DC Conversion—The Full-Wave Rectifier

The Diode Bridge Circuit

It may have occurred to you that the half-wave rectifier is not very efficient. We are tossing out one-half of the input voltage. This is a lot of lost energy. Full-wave rectifiers make use of the entire signal. The **diode bridge,** shown in Figure 2–19, is the most common realization of a full-wave rectifier. To understand the operation of the bridge circuit, follow the source current.

When the source is positive, current leaves from the bottom of the source and travels the route shown in Figure 2–20A. Diodes D3 and D2 are **ON.** Diodes D1 and D4 are **OFF.** The current path enters at the negative side of the resistor, which means the output voltage is positive. This gives us the first half-cycle (Figure 2–20B).

When the source is negative, current leaves from the top of the source and travels through D1 and D4, as shown in Figure 2–20C. Note how the current still enters the resistor on the negative side, resulting once again in a positive output voltage (Figure 2–20D). The total signal is the sum of the two half-wave signals. We get a rectified output with positive going half-cycles during both positive and negative halves of the input signal (Figure 2–20E). This is the **full-wave rectified signal.**

FIGURE 2–19
Full-Wave Rectifier

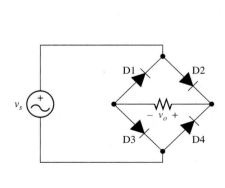

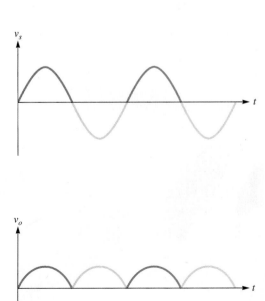

FIGURE 2–20

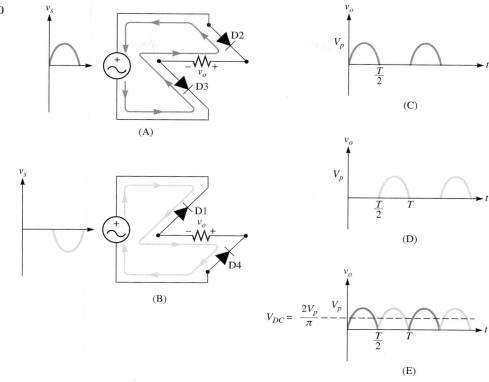

Because both halves of the input cycle are used, the full-wave rectified signal has twice the area as the half-wave signal. Therefore, the full-wave signal provides twice the DC content of the half-wave signal. That is

$$\text{Full-Wave Rectified Signal} \quad V_{DC} = \frac{2V_p}{\pi}$$

In Chapter 4, we show you how to build a full-wave rectifier with a center-tapped transformer and just two diodes.

DRILL EXERCISE

What is the peak value and DC (average) value of a full-wave rectified AC voltage with an rms value of 110 V?

Answer $V_p = 156\text{V}$ and $V_{DC} = 99.3\text{V}$

REVIEW QUESTIONS

1. Draw a full-wave rectifier circuit.
2. What does the output of your rectifier look like for a sine wave input?
3. What is the DC value of a full-wave rectified signal?

2–8 ■ DIODE LIMITERS (CLIPPERS)

We begin by again analyzing the resistor-diode circuits shown in Figure 2–21 (p. 36). In the first circuit the diode is **ON** when the input in positive. When the diode turns **ON,** the output is shorted to ground, so the output is 0V. When the diode is **OFF** (input is negative), we replace it with an open circuit and see that $v_o = v_s$.

We have reversed the diode in Figure 2–21B. Now the diode is **OFF** when the input is positive and **ON** when the input is negative. In this case we preserve the positive half of the input cycle. This output may look familiar. It is the same response we got from the rectifier circuits of Section 2–6. In fact, the two circuits considered here are alternative implementations of a rectifier.

Top Clipping

The **diode limiter** is a modification of these half-wave rectifier circuits. Consider the circuit shown in Figure 2–22, where we have added a battery to the rectifier circuit of Figure 2–21A.

When is the diode **ON** in this circuit? That is, when does current flow from the battery to the input source? The only point you need to remember is that current flows from the most negative (or smallest) voltage to the most positive (or largest) voltage. In this circuit then, current flows only in the **ON** direction for the diode when the input wave form is larger than $V_B = 10V$ (blue region in Figure 2–22A). Figure 2–22B shows the circuit when the diode is **ON:**

$$\text{Diode ON} \quad v_o = V_B = 10V$$

When the input signal is less than $V_B = 10V$, the diode is **OFF** (Figure 2–22C). During this time, the battery is disconnected from the circuit and

$$\text{Diode OFF} \quad v_o = v_s$$

FIGURE 2–21

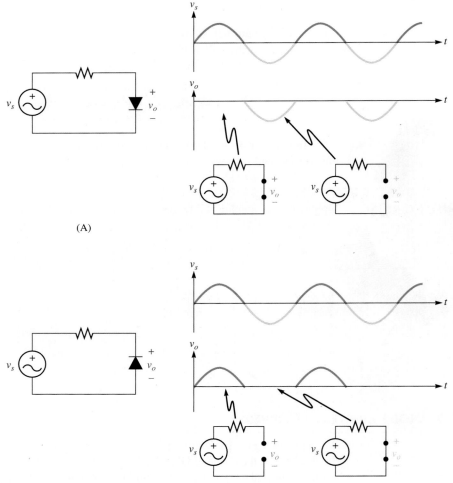

(A)

(B)

FIGURE 2–22
Top Clipping

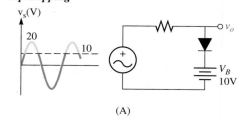

(A)

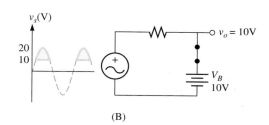

(B)

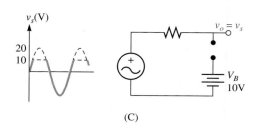

(C)

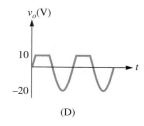

(D)

You can see that the output voltage in this circuit can never be larger than +10V. As soon as the output tries to exceed +10 V, the diode turns **ON** and *limits* the output to +10V. The total response looks similar to the input signal with the top clipped off (Figure 2–22D). For this reason, this circuit is also known as a **diode clipper.**

We see from the rectifier circuit of Figure 2–21A, that rectification is a special case of clipping. Imagine that the battery in Figure 2–22 is set to 0V, which gives us the same circuit as Figure 2–21A. This means that we are limiting the output to a maximum positive voltage of 0V, and we get the output shown in Figure 2–21A. Therefore, we have clipped the input at 0V.

Figure 2–23 shows you a *scissors cutting technique* that provides a simple way to think of clipping. Draw a dotted line through the input signal at the level of V_B. Cut along this line and discard everything *above* the cut.

FIGURE 2–23

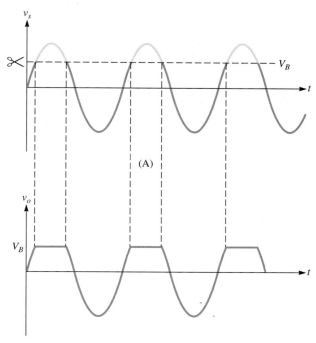

(A)

(B)

EXAMPLE 2–7
Diode Limiters—
Top Clipping

Find the outputs for the two circuits shown in Figure 2–24.

Solution
(a) We draw a dotted line through the input signal at 5V. Clipping along this line results in the output signal shown in Figure 2–24C.
(b) When we draw the line corresponding to 5V, we note that the entire input signal is below this line. The result, therefore, is that the output signal is the entire input signal (Figure 2–24D). Note the interesting result here. Because the input voltage never reaches 5V, the diode never turns on.

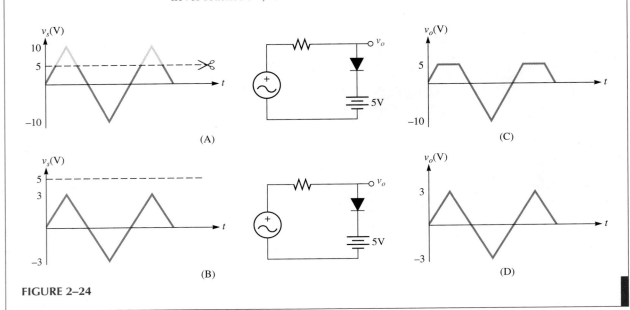

FIGURE 2–24

DRILL EXERCISE Change the reference voltages in Figure 2–24A to $V_B = 2V$ and find the output voltages.

Answer See Figure 2–25.

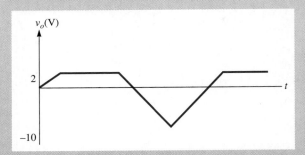

FIGURE 2–25

Bottom Clipping

What happens if we reverse the diode in the previous clipping circuits? We still cut the input signal along the line set to the battery voltage. Only now we keep the part of the signal above the cut and discard the signal *below* the cut. To see this, examine the circuit of Figure 2–26. Note that we have also reversed the reference battery.

FIGURE 2–26
Bottom Clipping

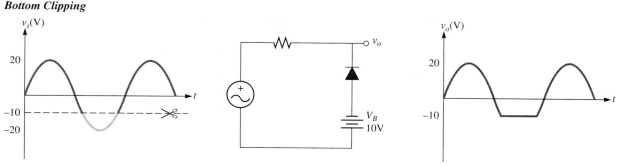

The diode is now **ON** whenever the input is less than –10V and is off whenever the input is greater than –10V. This results in an output that is clipped at –10V, as shown in the figure. Again, we can visualize this output by clipping the input signal along a line drawn at –10V and discarding everything *below* this line.

EXAMPLE 2–8
Diode Limiters—
Bottom Clipping

Find the outputs for each of the circuits shown in Figure 2–27.

Solution   Both signals are clipped at –5V (Figure 2–27C and D).

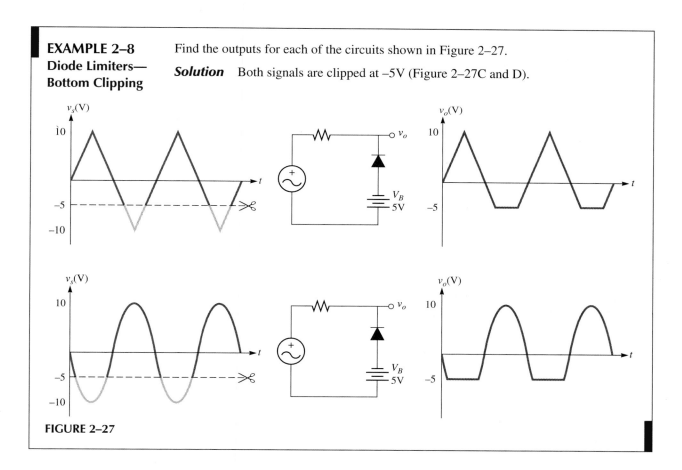

FIGURE 2–27

Symmetric Clipping

We can simultaneously clip the top and bottom part of a signal by using the circuit shown in Figure 2–28 (p. 40). Although it is possible to use two different reference voltages in this circuit, most commonly we use the same voltages. In this case, this circuit will produce symmetrical clipping. That is, we will chop off the same amount from the top and bottom of the input signal.

FIGURE 2–28
Symmetrical Clipping

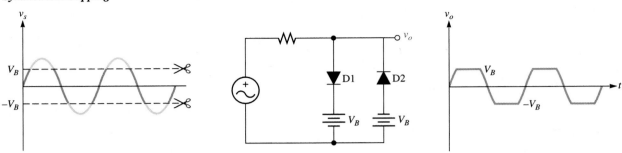

When the input is greater than V_B, diode D1 is **ON** and diode D2 is **OFF.** In this case, the output voltage is fixed at V_B. When the input is less than $-V_B$, diode D2 turns **ON,** while D1 is **OFF.** Now, the output is fixed at $-V_B$.

The key to analyzing this circuit is to understand that only one of the diodes can be **ON** at any one time. Try this simple exercise: In Figure 2–28 assume both diodes are **ON,** and replace them with short circuits. Do you see any problem? That's right. We end up with two different voltage sources in parallel. As you recall, it is impossible to put in parallel two ideal voltage sources of different value. Therefore, both diodes cannot be **ON** at the same time. We can summarize the behavior of this general clipping circuit as

$$v_s > V_B \rightarrow \text{D1 is ON, D2 is OFF} \rightarrow v_o = V_B$$
$$v_s < -V_B \rightarrow \text{D2 is ON, D1 is OFF} \rightarrow v_o = -V_B$$
$$-V_B < v_s < V_B \rightarrow \text{D1 and D2 are OFF} \rightarrow v_o = v_s$$

EXAMPLE 2–9
Symmetric
Clipping Circuits

Find the output for each of the clipping circuits shown in Figure 2–29.

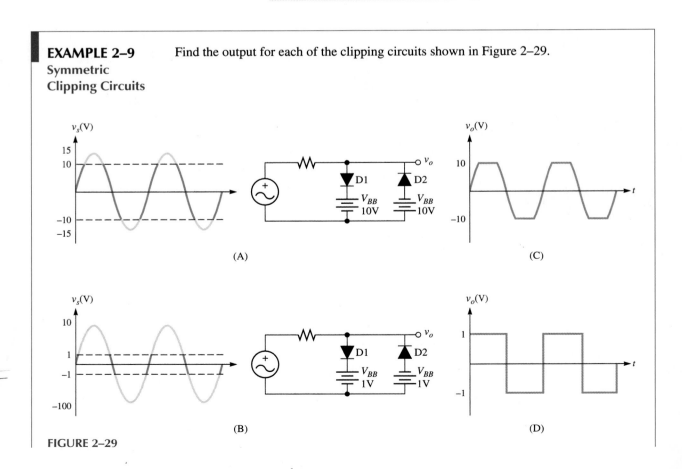

FIGURE 2–29

Solution

(a) We can use the scissors cutting technique to find the output. This circuit clips off the signal above 10V and below –10V (Figure 2–29C).

(b) This is a straightforward clipping circuit that limits the output to +1 and –1V (Figure 2–29D). Note that we have greatly expanded the output scale so that you can see the output wave form.

DRILL EXERCISE   In Figure 2–29A (p. 41) change the reference batteries to 20V and find v_o.

Answer   The input signal never gets large enough to turn either diode **ON.** Therefore, both diodes remain **OFF,** and v_o is the same as v_s.

Why do we use limiters? In some cases to protect delicate electronics from damage, we simply do not want a voltage to get too large, either positive going or negative going. The third circuit in Example 2–9 (p. 25) shows another use. The output in this circuit looks remarkably like a square wave. In fact, clipping can be used to create a square wave from a sine wave.

Clipping distorts the input signal. Although this is often undesirable, especially in classical music, rock guitarists purposely use electronics to distort sounds. The fuzz box is a clipping circuit that produces a rocking, raunchy sound that is pleasing to the ear of the heavy metal fan, and even to some professors.

REVIEW QUESTIONS

1. What term is synonymous with clipping?
2. Draw a top clipping circuit.
3. Describe the scissors cutting technique for the top clipping circuit.
4. Draw a bottom clipping circuit.
5. Describe the scissors cutting technique for the bottom clipping circuit.
6. Draw the symmetrical clipping circuit.
7. Describe the scissors cutting technique for the symmetrical clipping circuit.

SUMMARY

- The ideal diode is **ON** if I_D is positive.
- The ideal diode is **OFF** if V_D is negative.
- The **ON** diode is replaced with a short circuit.
- The **OFF** diode is replaced with an open circuit.
- For the current test, assume the diode is on and find I_D. If I_D is positive, the diode is **ON.** If I_D is negative, the diode is **OFF** and the true value of the current is $I_D = 0$A.
- For the voltage test, assume the diode is **OFF** and find V_D. If V_D is negative, the diode is **OFF.** If V_D is positive, the diode is **ON** and the true value of the voltage is $V_D = 0$V.
- The half-wave rectifier chops off half of the input sinusoid, producing a signal with a DC level of $\dfrac{V_p}{\pi}$.
- The full-wave rectifier uses both parts of the input cycle to create a signal with a DC level of $\dfrac{2\,V_p}{\pi}$.
- The top clipping circuit eliminates the portion of the input signal that is greater than the reference voltage.

■ The bottom clipping circuit eliminates the portion of the input signal that is below the reference voltage.

■ The symmetrical clipping circuit eliminates the portions of the input signal that are above and below the reference voltages.

SECTION 2–2   The Ideal Diode

1. For the diode shown in Figure 2–30
   (a) Label V_D and I_D.   **(b)** If $I_D = 20$mA, what is V_D?   **(c)** Is the diode **ON** or **OFF?**

SECTION 2–3   Diode-Resistor Circuits

2. For the diode shown in Figure 2–30
   (a) Label V_D and I_D.   **(b)** If $V_1 = 3$V and $V_2 = 4$V, what are V_D and I_D?
   (c) Is the diode **ON** or **OFF?**

FIGURE 2–30

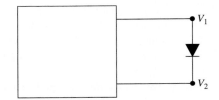

SECTION 2–3   Diode Resistor Circuits

FIGURE 2–31

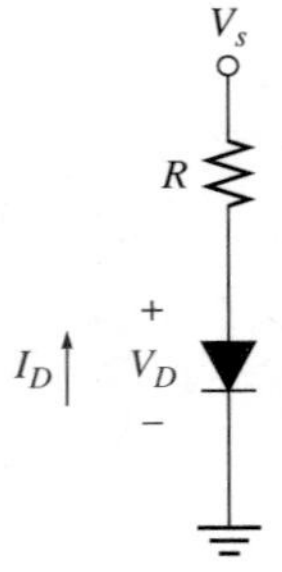

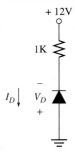

3. For the diode circuit shown in Figure 2–31, $V_S = +10$V and $R = 1$KΩ:
   (a) Find V_D.   **(b)** Find I_D.
4. For the diode circuit shown in Figure 2–31, $V_S = +5$V and $R = 2$KΩ:
   (a) Find V_D.   **(b)** Find I_D.
5. For the diode circuit shown in Figure 2–32:
   (a) Find V_D.   **(b)** Find I_D.
6. For the diode circuit shown in Figure 2–33, $V_S = +15$V, $R_1 = 10$KΩ, $R_2 = 5$KΩ:
   (a) Find V_D.   **(b)** Find I_D.   **(c)** Find V_o.
7. For the diode circuit shown in Figure 2–33, $V_S = +10$V, $R_1 = 2$KΩ, $R_2 = 3$KΩ:
   (a) Find V_D.   **(b)** Find I_D.   **(c)** Find V_o.
8. Redraw the diode circuit shown in Figure 2–33, reversing the direction of the diode. Now, if
   $V_S = +15$V, $R_1 = 10$KΩ, $R_2 = 5$KΩ
   (a) Find V_D.   **(b)** Find I_D.   **(c)** Find V_o.
9. Redraw the diode circuit shown in Figure 2–33, reversing the direction of the diode. Now, if
   $V_S = +10$V, $R_1 = 2$KΩ, $R_2 = 4$KΩ
   (a) Find V_D.   **(b)** Find I_D.   **(c)** Find V_o.

FIGURE 2–32

FIGURE 2–33

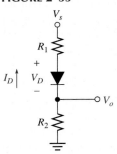

FIGURE 2–34

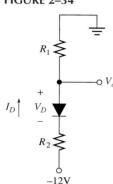

10. For the diode circuit shown in Figure 2–34, $R_1 = R_2 = 1\text{K}\Omega$:

 (a) Find V_D. (b) Find I_D. (c) Find V_o.

11. For the diode circuit shown in Figure 2–34, $R_1 = 2\text{K}\Omega$ and $R_2 = 4\text{K}\Omega$:

 (a) Find V_D. (b) Find I_D. (c) Find V_o.

12. Redraw the circuit of Figure 2–34, reversing the direction of the diode. If $R_1 = R_2 = 1\text{K}\Omega$

 (a) Find V_D. (b) Find I_D. (c) Find V_o.

13. Reverse the diode in Figure 2–34. If $R_1 = 2\text{K}\Omega$ and $R_2 = 4\text{K}\Omega$

 (a) Find V_D. (b) Find I_D. (c) Find V_o.

SECTION 2–4   Is a Diode ON or OFF?

14. For the circuit shown in Figure 2–35:

 (a) Is the diode **ON** or **OFF**? (b) Find V_D. (c) Find I_D.

15. Redraw the circuit shown in Figure 2–35 with the diode reversed.

 (a) Is the diode **ON** or **OFF**? (b) Find V_D. (c) Find I_D.

16. You know two tests for determining whether a diode is **ON** or **OFF**.

 (a) Describe how you would apply the Current Test to the circuit in Figure 2–35.

 (b) Describe how you would apply the Voltage Test to the circuit in Figure 2–35.

 (c) Which test seems easiest for you to apply to this circuit? Why?

17. For the circuit in Problem 15

 (a) Describe how you would apply the Current Test to this circuit.

 (b) Describe how you would apply the Voltage Test to this circuit.

 (c) Which test seems easiest for you to apply to this circuit? Why?

FIGURE 2–35

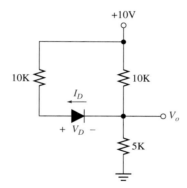

SECTION 2–5   Current Traffic Control—Diode Logic

18. For the circuit in Figure 2–14 (p. 29), let $R_S = R_L = 1\text{K}\Omega$. Find I_L if

 (a) $V_A = +6\text{V}$. (b) $V_A = 0\text{V}$.

19. For the circuit in Figure 2–13 (p. 28), let $R_S = R_L = 1\text{K}\Omega$. Find I_L if

 (a) $V_A = +6\text{V}$ and $V_B = +6\text{V}$. (b) $V_A = +6\text{V}$ and $V_B = 0\text{V}$.

 (c) $V_A = 0\text{V}$ and $V_B = +6\text{V}$. (d) $V_A = 0\text{V}$ and $V_B = 0\text{V}$.

SECTION 2–6   AC to DC Conversion—The Half-Wave Rectifier

20. For the rectifier circuit shown in Figure 2–36 (p. 44), find the following when v_s is positive:

 (a) I_D. (b) V_D. (c) v_o.

21. For the rectifier circuit shown in Figure 2–36 (p. 44), find the following when v_s is negative:

 (a) I_D. (b) V_D. (c) v_o.

22. Reverse the direction of the diode in Figure 2–36 (p. 44), and redraw the circuit. Do not change v_s. Find the following when v_s is positive:

 (a) I_D. (b) V_D. (c) v_o.

FIGURE 2–36

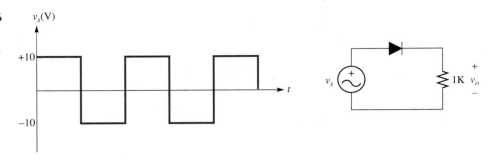

23. For the circuit of Problem 22, find the following when v_s is negative:

    (a) I_D.   (b) V_D.   (c) v_o.

24. The rectifier circuit in Figure 2–37 has the sinusoidal input voltage as shown.

    (a) Draw v_s and right below it draw v_o, using the same time scales. Show maximum and minimum values.

    (b) What is the greatest value of I_D?   (c) What is the smallest value of I_D?

    (d) What is the greatest value of V_D?   (e) What is the smallest value of V_D?

25. For the input voltage in Figure 2–37

    (a) What is the average value of this signal?

    (b) What is the rms value of this signal?

26. A half-wave rectified signal has a peak value of 70V. What is the average value of this signal?

27. You want to produce a half-wave rectified signal that has an average value of 25V. What peak value is required of the input voltage?

28. Reverse the diode in Figure 2–37 and repeat Problem 24.

FIGURE 2–37

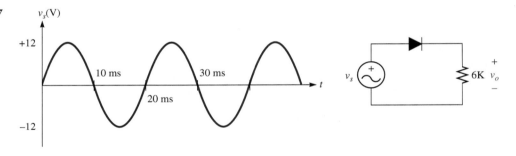

SECTION 2–7 AC to DC Conversion—The Full-Wave Rectifier

29. Figure 2–38 shows a full-wave bridge rectifier.

    (a) Redraw the circuit with the diodes that are **ON** when the input is positive.

    (b) Redraw the circuit with the diodes that are **ON** when the input is negative.

    (c) Draw v_s and v_o on the same scale.

FIGURE 2–38

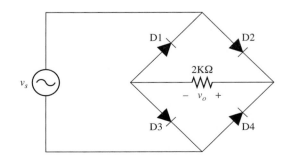

30. A full-wave rectified signal has a peak value of 100V. What is the average value of this signal?

SECTION 2–8   Diode Limiters (Clippers)

31. Draw the output voltage for the clipping circuit of Figure 2–39.

32. Draw the output voltage for the clipping circuit of Figure 2–40.

FIGURE 2–39   **FIGURE 2–40**

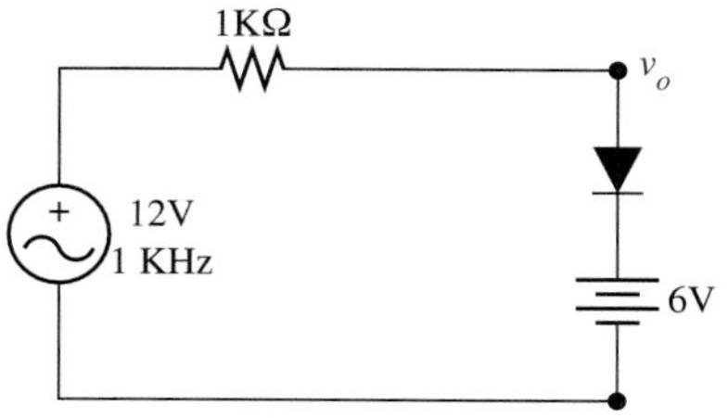

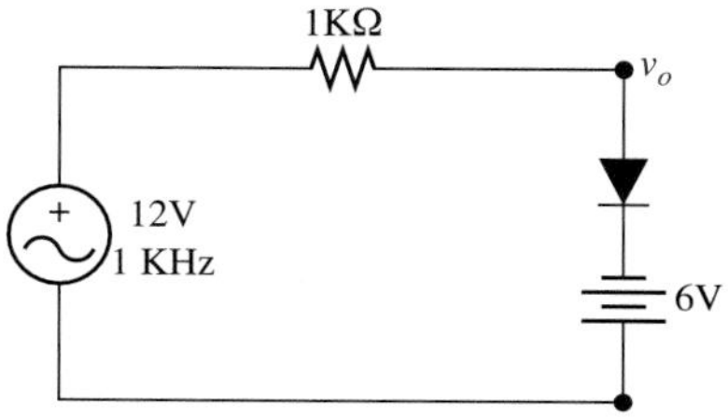

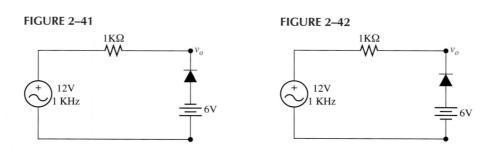

33. Draw the output voltage for the clipping circuit of Figure 2–41.

34. Draw the output voltage for the clipping circuit of Figure 2–42.

FIGURE 2–41   **FIGURE 2–42**

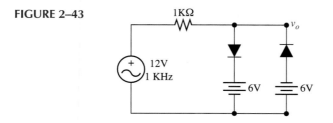

35. Draw the output voltage for the clipping circuit of Figure 2–43.

FIGURE 2–43

36. Draw a symmetrical clipping circuit with $V_B = 5V$. Draw the output of this circuit if

    (a) V_s is a 10V sinusoid.   **(b)** V_s is a 50V sinusoid.   **(c)** V_s is a 1V sinusoid.

CHAPTER

3

THE REAL DIODE

CHAPTER OBJECTIVES

To determine the I-V characteristics of the real diode.

To determine when the real diode is **ON** or **OFF.**

To analyze rectifiers with the real diode model.

To analyze limiters with the real diode model.

To describe the diode breakdown region and to find the *PIV* for a diode.

To select a commercially available diode for use in a circuit.

To read diode data sheets.

To troubleshoot diode circuits.

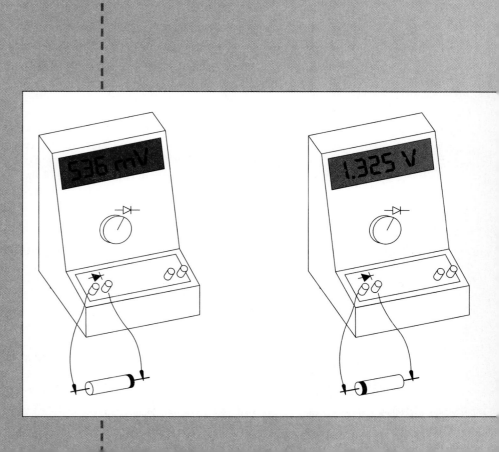

3–1 ■ INTRODUCTION

In Chapter 2 we described the behavior of several diode circuits. In those circuits we assumed that the diode was an ideal device, which was either a short circuit if **ON** or an open circuit if **OFF.** Real diodes behave somewhat differently.

Please note that we are not saying that there are two types of diodes, ideal and real. When we discuss the ideal diode, we are referring to an ideal model for the diode. In this chapter we introduce more realistic models. Even the models we use in this chapter are approximations to the actual I-V behavior of the diode. So, when we use the term "real diode," we are referring to an improved diode model, not a different type of diode.

Several commercially available diodes are shown in Figure 3–1. Because diodes are usually used for rectification, many manufacturers use the term **rectifier** instead of diode. Note that diode bridge rectifiers are available as single devices. The original diodes were constructed from vacuum tubes. In the 1940s and 1950s, engineers at Bell Laboratories found that they could construct diodes from crystalline material, primarily made from silicon or germanium. These materials have conductive properties between good conductors (metals, for example) and insulators. This is where the term **semiconductor** comes from. Most modern electronics are built with semiconductors.

FIGURE 3–1

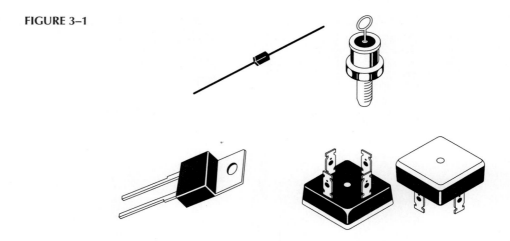

Most modern diodes are made from the semiconductor material silicon. For this reason, we limit the discussion in this chapter and the next two to the silicon diode. In Chapter 6 we discuss additional diode types, including the germanium diode.

In this chapter, we will show you the current-voltage characteristic for the semiconductor diode. Because this relation is quite complex, we introduce much simpler models for the diode and then reexamine some of the circuits of Chapter 2. You will find that the ideal analysis found in the preceding chapters is only slightly modified in this chapter. We end with a discussion of troubleshooting diode circuits.

3–2 ■ DIODE CURRENT AND VOLTAGE (THE I-V CURVE)

There are two ways to understand the behavior of any electrical device. We can get down to the molecular level to determine the physics of the device. From the physics we can construct a model for the device. Or, we can determine the device behavior by experimentally finding the dependency of the device current on the device voltage. In general, if you are constructing the device, you need to know the device physics very well. If you are using the device in a circuit, however, the current-voltage (I-V) characteristic is all you need to know. Because most of us use electronic devices rather than fabricate them, we will use the current-voltage approach. A description of diode physics can be found in Appendix A.

We begin our discussion by taking another look at the resistor. In Figure 3–2A we have set up an experiment where we apply different voltages to a resistor and measure the current that results. In Figure 3–2B we plot the measured current versus the applied voltage. This plot is known as the device **I-V characteristic,** or **curve.** The data show several important points:

□ When no voltage is applied, there is no current.

□ A positive voltage results in a positive current.

□ A negative voltage results in a negative current.

□ The current and voltage are linearly related. That is, the data points can be connected with a straight line.

FIGURE 3–2

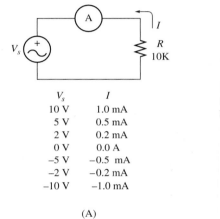

V_s	I
10 V	1.0 mA
5 V	0.5 mA
2 V	0.2 mA
0 V	0.0 A
–5 V	–0.5 mA
–2 V	–0.2 mA
–10 V	–1.0 mA

(A)

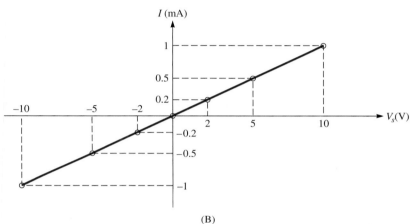

(B)

Because resistor current and voltage are related by a straight line, we can write

$$I = \frac{1}{R} V$$

where $\frac{1}{R}$ is the slope of the straight line. You all recognize this equation as a variation of Ohm's Law

$$V = RI$$

We present current versus voltage rather than the more familiar version of Ohm's Law because this is the standard approach in electronics.

In Figure 3–3A (p. 50) we have redrawn the I-V curve that results when $R = 10K\Omega$, along with the curves for $R = 5K\Omega$ and $R = 20K\Omega$. You can see that as the resistance gets smaller, we get more current for the same voltage. As the resistance gets larger, we get less current for the same voltage. Also note that as R gets smaller, the line gets steeper (increased slope); as R gets bigger, the line becomes less steep (decreased slope).

We now examine two special cases, the short circuit and the open circuit. The short circuit is shown in Figure 3–3B. The important fact about a short circuit is that no matter how much current (positive or negative) is in the short circuit, the voltage across the short circuit is 0V. This means that the I-V plot for the short circuit will be the vertical axis itself, as shown in the figure. In Figure 3–3A, we showed you that the I-V curve gets steeper as the resistance gets smaller. The short circuit has a resistance of 0Ω, leading to the steepest possible slope—a vertical line.

The open circuit is shown in Figure 3–3C. Here, no matter how much voltage (positive or negative) is applied across an open circuit, the current is 0A. This means that the

FIGURE 3–3

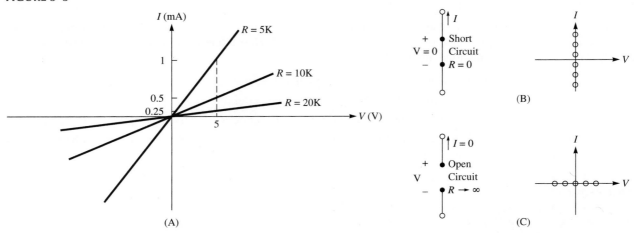

(A)   (B)   (C)

I-V curve for an open circuit is the horizontal axis, as shown in the figure. The resistance of an open circuit is ∞, which leads to a slope of 0 on the I-V curve.

We are now in a position to show you the I-V curve for the ideal diode. Remember that the ideal diode is a short circuit when it is forward biased, or **ON,** and an open circuit when it is reverse biased, or **OFF.** Because the ideal diode is **OFF** when V_D is negative, and **ON** when I_D is positive, we get the I-V curve shown in Figure 3–4.

The ideal diode I-V curve is not a simple straight line. When a device I-V curve is not a straight line, we say that the device is **non-linear.** With all non-linear devices, we must be sure we know the region of the I-V curve in which we are operating.

FIGURE 3–4

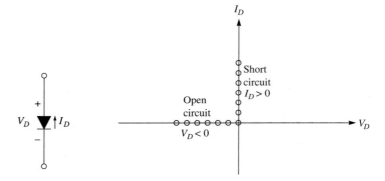

Before we turn to the real diode, we want to examine one last I-V curve: the curve for the ideal battery. A 5V battery and its I-V curve are shown in Figure 3–5. We note that no matter what current is in the battery, the voltage across the battery is 5V. Once again, this gives us a vertical line for the I-V curve. If we compare the short circuit I-V curve (Figure 3–3B) with the battery I-V curve, we learn the following:

☐ An ideal battery has zero internal resistance. The battery I-V curve has the same slope as the short circuit.

☐ As the battery voltage decreases to 0V, the battery becomes a short circuit. A battery with 0V is described with a vertical line at $V = 0V$.

Note that all real batteries have some internal resistance. The I-V curve for a real battery, therefore, will not be truly vertical. In most cases, we do not lose much by assuming that a real battery is ideal.

FIGURE 3–5

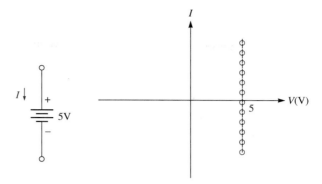

1. Draw an I-V curve for a linear resistor.
2. What kind of curve do you get for the linear resistor?
3. What is the I-V curve for a short circuit?
4. What is the I-V curve for an open circuit?
5. Draw the I-V curve for a 10V battery.
6. What is the internal resistance of an ideal battery?

3–3 ■ THE REAL DIODE

To find the I-V curve for a real diode, we set up the circuit shown in Figure 3–6A (p. 52) and measured diode voltage and current as we varied the input voltage source. The results of this experiment are tabulated below.

V_D (V)	I_D (mA)
-4	-1×10^{-11}
-2	-1×10^{-11}
0	0
0.2	3×10^{-8}
0.3	1.6×10^{-6}
0.4	8.9×10^{-5}
0.5	4.8×10^{-3}
0.6	0.26
0.7	14.0
0.72	32.0
0.74	72.0
0.76	160.0
0.78	350.0

This experiment shows several interesting results. When the diode voltage is negative, diode current is negative and very small. We see that the reverse-biased real diode is not a perfect open circuit, which would have no current. The current that flows in the reverse-biased diode is often called the **leakage current.** The leakage current is so small, we usually assume that even the reverse-biased real diode is an open circuit.

Note a major difference between the ideal and real diode models for V_D greater than 0 V. The ideal diode becomes a short circuit ($V_D = 0$) as soon as we get a positive diode current. The real diode, on the other hand, does not become a short circuit at this point. As

FIGURE 3–6

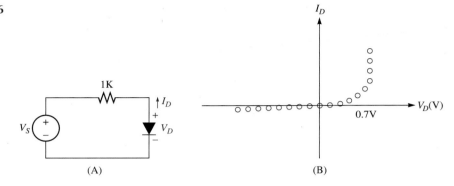

(A)   (B)

the voltage across the real diode becomes more positive, the diode current increases slowly. In fact, for $V_D < 0.5V$, the diode current is still very small.

When the diode voltage reaches approximately 0.7V, we suddenly get a rush of current. The current then increases very rapidly with small additional increases in diode voltage. For this reason, we say that a real diode turns **ON** when the voltage across it reaches approximately 0.7V. For diode voltages less than this threshold voltage, we can assume diode current is 0A; the diode is **OFF**. For the real diode, then, we have the following approximations:

> **Real Diode ON-OFF Behavior**
> **ON:** $V_D = 0.7V$ $I_D > 0$
> **OFF:** $V_D < 0.7V$ $I_D = 0$

Figure 3–7 shows the plots of the I-V curve of the ideal and the real diode. You can see that the main difference is that the ideal diode turns **ON** at 0V, while the real diode turns **ON** at approximately 0.7V. We note again that 0.7V is only an approximation. The turn-on voltage for commercially available diodes varies from 0.6 to 1V.

FIGURE 3–7
The ideal and real diode
I-V curves

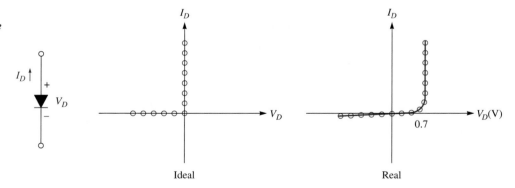

Ideal   Real

The I-V curve for the real diode can be modeled with the following *exponential* equation:

$$I_D = I_R(e^{V_D/V_T} - 1)$$

where I_R is the reverse-bias current and V_T is a device voltage parameter that depends on temperature. For the diode in Figure 3–6, we used the typical values:

$$I_R = 10^{-11}mA \quad \text{and} \quad V_T = 26mV \text{ (at room temperature)}$$

The diode current-voltage formula can be solved with a calculator, so such problems should not scare you. However, if the diode is part of a larger circuit, where both diode current and voltage are unknown, this equation cannot be easily used. In fact, a more accurate analysis of diode circuits requires the use of computer programs. Fortunately, we do not have to use the diode formula to get reasonably accurate answers in diode circuits.

The Improved Diode Model

If you compare the real diode I-V curve of Figure 3–7 to the battery I-V curve of Figure 3–5 (p. 51), you can see a striking similarity for positive current. In both cases, the curve is a nearly vertical line. That is, for positive I_D, the diode acts as a battery of 0.7V. This is the major difference between the ideal and real diode. Both real and ideal diodes approximate open circuits when they are **OFF;** the ideal diode is a short circuit when **ON,** while the real diode can be approximated with a 0.7V battery when **ON.** These models are shown in Figure 3–8.

FIGURE 3–8
Diode models

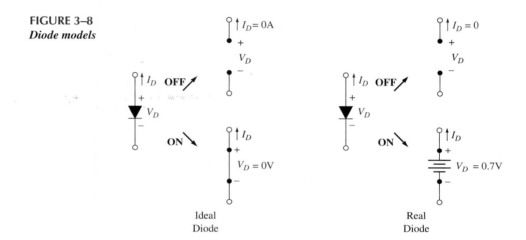

Ideal Diode

Real Diode

An Important Note

Do not confuse the diode with a battery. Unlike a battery, the diode (except for some diodes discussed in Chapter 6) is not a source of energy. It might be helpful to think of the diode as a device in which the resistance decreases as the current through it increases. The voltage across the **ON** diode, therefore, remains at a constant 0.7V. It is this constant 0.7V that we model with a battery.

The use of the 0.7V battery to represent the diode turn-on voltage is an improvement over the simple on-off model of the previous chapter. However, a more accurate model of the real diode also includes some resistance. (In fact, some authors refer to the model used here as the ideal model. In a later section, we will show you a more detailed model for the diode.) For the most part, the improved model presented here is all you will ever need. Therefore, unless we tell you otherwise, this is the model we will use for the real diode. We must modify our testing procedures when we want to determine if a real diode is **ON** or **OFF.**

Current Test

We assume that the diode under consideration is **ON** and replace it with a 0.7V battery. The battery has the same polarity as V_D. We now analyze the circuit to find I_D. If I_D is positive, our initial assumption was correct and the diode is **ON.** If I_D turns out to be negative, our assumption was wrong, and the diode must be **OFF.** The diode should be replaced with an open circuit and the circuit reanalyzed.

EXAMPLE 3–1
The Current Test I

Determine the state of the diode for the circuit in Figure 3–9A.

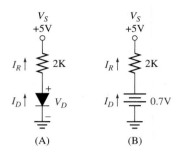

FIGURE 3–9   (A)   (B)

Solution   Assume the diode is **ON** and replace it with a 0.7V battery (Figure 3–9B). The diode current is the same as the resistor current and is found from

$$I_D = I_R = \frac{5 - 0.7}{2K} = \frac{4.3}{2K} = 2.15mA$$

Because the diode current is positive, the diode is indeed **ON.**

DRILL EXERCISE

Redraw the circuit in Figure 3–9, changing the voltage source to 1V. Find I_D and determine the state of the diode.

Answer $I_D = 0.15mA$. Diode is **ON.**

EXAMPLE 3–2
The Current Test II

Determine the state of the diode for the circuit in Figure 3–10A, and find I_D and V_D.

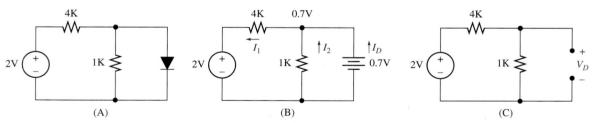

(A)   (B)   (C)

FIGURE 3–10

Solution   Assume that the diode is **ON** and replace it with a 0.7V battery (Figure 3–10B). First note that

$$I_1 = I_2 + I_D$$

So

$$I_D = I_1 - I_2$$

Now,

$$I_1 = \frac{2 - 0.7}{4K} = \frac{1.3}{4K} = 0.325mA$$

and

$$I_2 = \frac{0.7}{1K} = 0.7\text{mA}$$

This leads to

$$I_D = I_1 - I_2 = 0.325 - 0.7 = -0.375\text{mA}$$

Because the diode current is negative, the diode must be **OFF** and the true value of the diode current is $I_D = 0A$. To properly analyze the circuit, replace the diode with an open circuit. The circuit is now a voltage divider (Figure 3–10C) and the diode voltage is found from

$$V_D = \frac{1K}{1K + 4K} 2 = \frac{1}{5} 2 = 0.4V$$

Because it takes 0.7V to turn a diode **ON,** this answer confirms that the diode is **OFF.**

DRILL EXERCISE   Redraw the circuit in Figure 3–10, changing the input voltage to 2.5V. Determine the state of the diode, and find I_D and V_D.

Answer   Diode is **OFF.** $I_D = 0A$, $V_D = 0.5V$

Voltage Test

As with the ideal diode, to apply the voltage test to a real diode, we assume the diode is **OFF** and replace it with an open circuit. We then analyze the circuit to find V_D. Now, if $V_D < 0.7$, the diode is **OFF.** If $V_D > 0.7V$, our assumption was wrong, and the diode is **ON.** Replace it with a 0.7V battery.

EXAMPLE 3–3
The Voltage Test

Determine the state of the diode in Figure 3–11A. Find I_D and V_D.

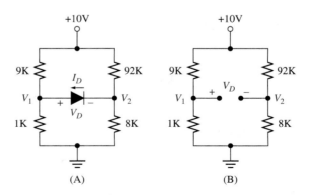

FIGURE 3–11   (A)   (B)

Solution   We assume that the diode is **OFF** and replace it with an open circuit (Figure 3–11B). We now analyze the two voltage dividers to get

$$V_1 = \frac{1K}{1K + 9K} 10 = 1V$$

and

$$V_2 = \frac{8K}{8K + 92K} \, 10 = 0.8V$$

So, the diode voltage is

$$V_D = V_1 - V_2 = 1 - 0.8 = 0.2V$$

Because $V_D < 0.7V$, our assumption was correct and the diode is **OFF.** Therefore, the value we found for V_D is correct. Because the diode is **OFF**

$$I_D = 0A$$

DRILL EXERCISE   Redraw the circuit in Figure 3–11 (p. 55), changing the input to 50V. Determine the state of the diode.

Answer   Diode is **ON.**

REVIEW QUESTIONS

1. Sketch the I-V curve for a real diode.
2. When does the real diode turn **ON?**
3. What is the improved model for the diode?
4. How do you use the current test for the real diode?
5. How do you use the voltage test for the real diode?

3–4 ■ DIODE CIRCUITS WITH REAL DIODES

We have already showed you some common diode circuits. How do these circuits behave when we consider that they are constructed of real, rather than ideal, diodes? The simplest approach is to recognize that the real diode can be modeled as the ideal diode in series with the battery that represents the turn-on voltage (Figure 3–12A). Note that V_D is measured across the ideal diode-battery combination. Also note that because the two elements are in series, the battery could also be placed at the anode of the ideal diode. Figures 3–12B and C show that the ideal diode-battery combination give us the correct values for I_D and V_D for the **ON** and **OFF** real diode.

We will now review several of the diode circuits of the previous chapter. Each of the real diodes in these circuits is first replaced with the ideal diode-battery combination and then analyzed.

The Half-Wave Rectifier Circuit

Consider the half-wave rectifier circuit shown in Figure 3–13A. We first replace the real diode with the ideal diode–battery combination (Figure 3–13B). As you can see, the battery representing the diode is in series with the source. The simplest procedure here is to combine the diode battery with the source.

This means that the ideal diode circuit is driven by $v_s - 0.7$. As shown in Figure 3–13C, the 2V sine wave is shifted down by 0.7V, now oscillating between −2.7V and

FIGURE 3–12
*The improved
diode model*

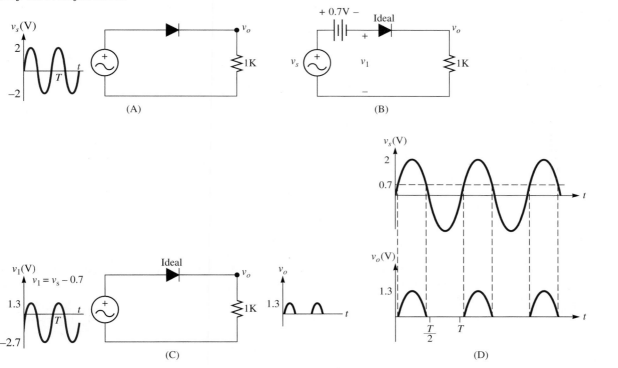

FIGURE 3–13
Half-wave rectifier circuit

1.3V. Using the results of the preceding chapter, we see that the output is a rectified signal with a peak value of 1.3V. For comparison, in Figure 3–13D we show the actual input and the output voltages on the same plot. Note that the output signal does not begin at 0s. This is because the real diode does not turn **ON** until the input signal reaches 0.7V.

| **EXAMPLE 3–4**
The Half-Wave
Rectifier | Find the output voltage for the half-wave rectifier shown in Figure 3–14A (p. 58). Note that we have reversed the diode.

Solution   We first replace the diode with the battery-ideal diode (Figure 3–14B). Note the polarity of the 0.7V battery, and its placement. The voltages at the sources are added together (+ terminal of signal source is connected to − terminal of 0.7V |

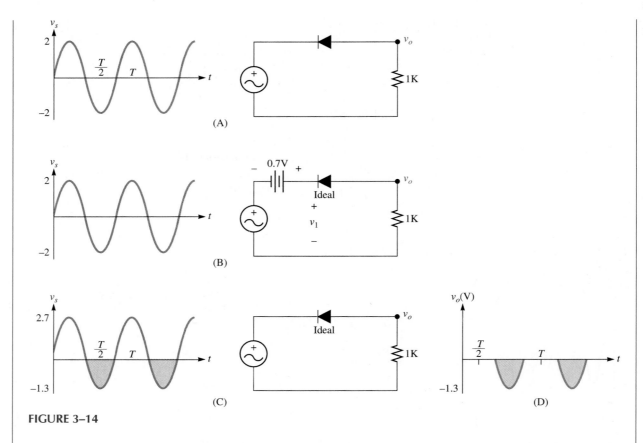

FIGURE 3–14

battery). The resulting signal is a sine wave that goes from 2.7V to −1.3V (Figure 3–14C).

Because the diode turns **ON** when the input goes negative, we get the result shown in Figure 3–14D.

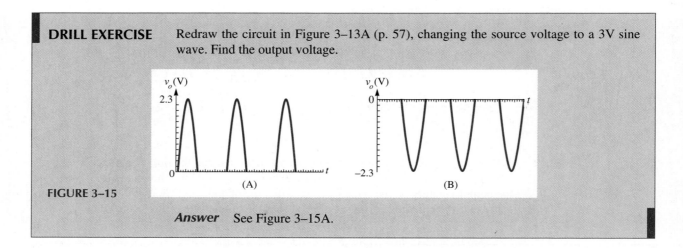

DRILL EXERCISE   Redraw the circuit in Figure 3–13A (p. 57), changing the source voltage to a 3V sine wave. Find the output voltage.

FIGURE 3–15

Answer   See Figure 3–15A.

DRILL EXERCISE   Redraw the circuit in Figure 3–14A, changing the source voltage to a 3V sine wave. Find the output voltage.

Answer   See Figure 3–15B.

The Full-Wave Rectifier

The full-wave bridge rectifier is shown in Figure 3–16A. This circuit is often built with commercially available diode bridges (see Figure 3–1). Whether built with individual diodes or the diode bridge, the analysis is the same. We will analyze this circuit only for the positive part of the input cycle, when D2 and D3 are **ON.** The same results will hold for the negative part of the input cycle.

In Figure 3–16B, we have replaced the real diodes, D2 and D3, with the ideal diode-battery combination. As strange as this circuit might look, you should note that the two **ON** diode batteries, the load, and the source are all in series. This means that we can move the batteries around the circuit until we can explicitly show that they are in series with the source. This is done in Figure 3–16C, which indicates that the ideal diode rectifier bridge is driven by $v_s - 1.4$V.

FIGURE 3–16
Full-wave rectifier circuit

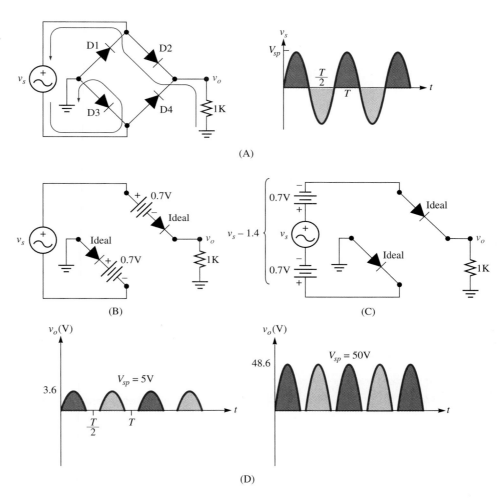

Figure 3–16D shows the output of the rectifier for two different input voltage levels. Note that we have included the contribution from the negative part of the input cycle. (Convince yourselves that we have done this properly.) When the input voltage is small (5V), we lose some peak voltage, which is now 3.6V. As the input voltage increases (50V), the loss in peak value of the output voltage is small.

There is also a significant dead time between the output pulses when the input is 5V. The dead-time distortion is caused by the fact that input voltage must reach 1.4V before the diodes turn **ON.** When the peak input voltage is only 5V, it takes a while for the voltage to reach 1.4V. As the input voltage increases (50V), 1.4V is reached sooner, so the dead time between pulses becomes negligible.

EXAMPLE 3–5
The Full-Wave Rectifier

Find the total output of the full-wave rectifier shown in Figure 3–16A (p. 59) for a 5V sinusoidal input.

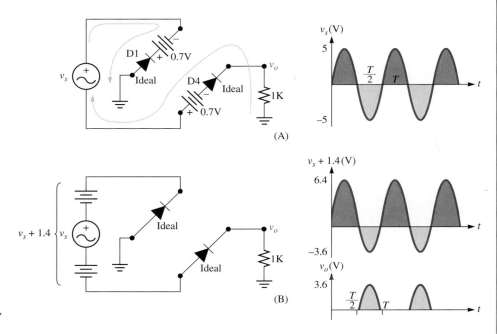

FIGURE 3–17

Solution We already have the output for the positive part of the input cycle. When the input swings negative, diodes D1 and D4 turn **ON** (Figure 3–17A). We add the three sources together to get a signal that swings from −3.6V to 6.4V (Figure 3–17B). The resulting output is also shown in Figure 3–17B.

On the positive half-cycle of the input, D2 and D3 are **ON** and we get the rest of the output shown in Figure 3–16.

DRILL EXERCISE Find the total output of the full-wave rectifier for a 100V input sinusoid.

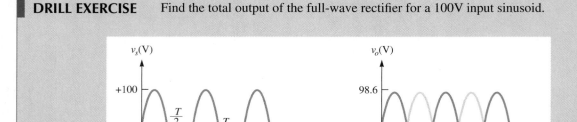

FIGURE 3–18

Answer See Figure 3–18.

The Clipping Circuit

Figure 3–19A shows a simple clipper (limiter) circuit. In Figure 3–19B we replace the real diode with the ideal diode-battery combination. Here we place the diode battery so that it is clearly in series with the limiter reference battery. We add the two batteries to get the ideal diode clipper circuit, shown in Figure 3–19C. We now have a circuit that clips at $V_B + 0.7V$ (Figure 3–19D).

FIGURE 3–19
Top clipping

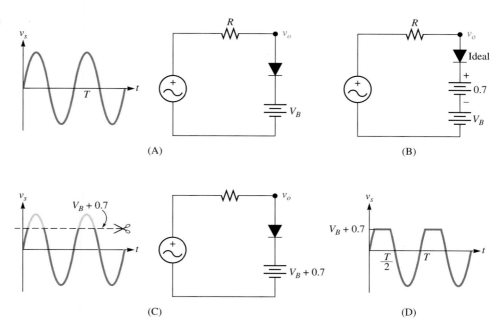

EXAMPLE 3–6
Two-Sided Limiting with Real Diodes

Find the output for the limiter circuit of Figure 3–20A.

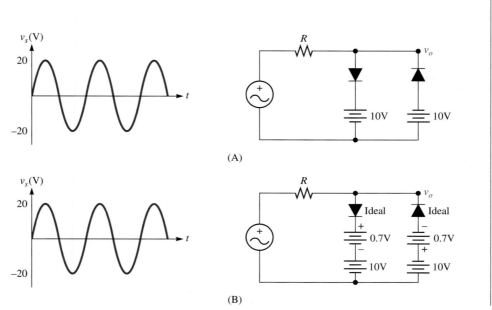

FIGURE 3–20
Symmetrical clipping

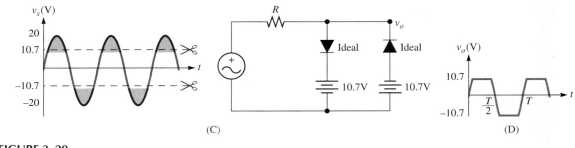

FIGURE 3–20
(continued)

Solution   We first replace each real diode with the ideal diode-battery combination (Figure 3–20B, p. 61), then add the series connected batteries (Figure 3–20C). This gives us a positive limiter reference of $10 + 0.7 = 10.7V$, and a negative limiter reference of $-(10 + 0.7) = -10.7V$. The resultant wave form is shown in Figure 3–20D.

DRILL EXERCISE

Change the input sinusoid in Figure 3–20A (p. 61) to a 10V sine wave and find the output.

Answer   The diodes never turn **ON,** so the output voltage equals the input voltage. That is, the output voltage is also a 10V sine wave.

REVIEW QUESTIONS

1. Draw the improved model for the real diode with the battery at the anode.
2. Draw the improved model for the real diode with the battery at the cathode.
3. Draw a half-wave rectifier with the improved model. How does the 0.7V battery combine with the signal source?
4. Draw a full-wave rectifier with the improved model. How do the 0.7V batteries combine with the signal source?
5. Draw a symmetrical clipper with the improved model. How do the 0.7V batteries combine with the reference batteries?

3–5 ■ DIODE RESISTANCE

Real diodes do not behave exactly as ideal batteries when they are turned **ON.** At moderate current levels, real diodes have a small forward-biased resistance of 10 to 100Ω. This can be seen in the diode exponential curve, which is repeated in Figure 3–21A. The forward resistance causes the rising curve to have a finite slope. Remember, any slope on an I-V curve represents conductance. The forward resistance of the diode, therefore, is given by the inverse of the slope in the I-V curve. The graph in Figure 3–21B approximates the exponential rise of the actual diode I-V curve with a straight line that has the slope $1/R_F$. At this point we have assumed that the reverse-biased current is small enough to ignore.

To incorporate the diode forward resistance into your analysis, you can use the diode model shown in Figure 3–22A. When the real diode is **OFF,** you still replace it with an open circuit (Figure 3–22B). When the diode is **ON,** you now replace it with a series connection of the diode turn-on battery and the diode forward resistance (Figure 3–22C). Again, because the model contains a series connection of elements, we can place the individual elements in any order.

FIGURE 3–21

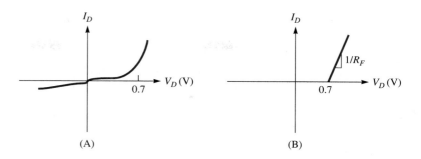

(A)   (B)

FIGURE 3–22
Diode forward resistance

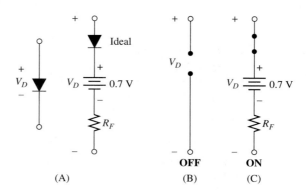

(A)   (B)   (C)

To see how diode forward resistance can affect circuit behavior, let us examine the rectifier circuit of Figure 3–23A. We replace the real diode with the model shown in Figure 3–22 and arrange the model components as shown in Figure 3–23B. This places the two voltage sources in series and creates a simple voltage divider when the diode is **ON** (Figure 3–23C).

The output voltage is given by

$$v_o = \frac{R_L}{R_L + R_F} (v_s - 0.7)$$

FIGURE 3–23

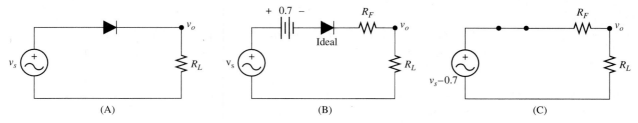

(A)   (B)   (C)

If the load resistor (R_L) is about the same size as R_F, then we must take R_F into consideration. If $R_L > 100R_F$, we can safely ignore the diode forward resistance; there will be only a 1 percent error introduced in this case.

REVIEW QUESTIONS

1. Sketch the I-V curve for the diode.
2. Draw a circuit model for the real diode that takes into account its forward resistance.
3. Draw all the possible combinations of ideal diode, turn-on battery, and forward resistance that can be used to model the real diode.
4. When can we ignore diode resistance?

3–6 ■ DIODE RATINGS—THE BREAKDOWN REGION

The ON Diode

Diodes, like resistors, will burn up if they dissipate more power than they are designed to handle. As with all electrical devices, diode average power is found from

$$P = VI$$

Because the **ON** diode has an approximately constant voltage of 0.7V, we see that the forward-biased diode dissipates the average power

$$P_D = 0.7 \times I_D$$

where I_D is the average diode current and is determined by the rest of the circuit.

Most manufacturers specify the maximum forward-biased current that a diode can handle, rather than the maximum power the diode can dissipate. This maximum current rating is for the average (DC) level of the diode current. In pure DC circuits, this current is fairly easy to find. Consider the circuit in Figure 3–24. We replace the real diode with the ideal diode-battery model (Figure 3–24B), and then add the two batteries together (Figure 3–24C). The diode current is found as

$$I_D = \frac{4.3}{10K} = 0.43mA$$

The power dissipated by the diode is

$$P_D = 0.7I_D = 0.7 \times 0.43mA = 0.3mW$$

FIGURE 3–24

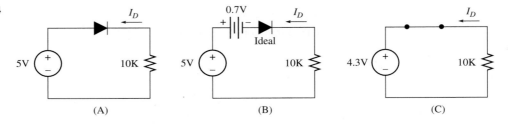

(A)   (B)   (C)

For safety, we would choose a diode with a 0.8mA or greater current rating or a diode with a power rating greater than 0.6mW.

Finding the average diode current in an AC circuit can be very daunting. Consider, for example, a sinusoidal input to the same rectifier circuit we just analyzed (Figure 3–25). We first replace the real diode with the ideal diode-battery combination, then add the two sources together, resulting in a sinewave that oscillates between $-2.7V$ and $1.3V$ (Figure 3–25B). The final output is shown in Figure 3–25C, along with the diode current

$$I_D = I_R = v_o/10K$$

The peak diode current is 0.13mA, but what is the average value of this signal? Rectified signals can have some very strange-looking wave shapes. Unless you are familiar with integral calculus, there is no way to find the exact average value of diode currents; however, there is a way to make a simple approximation. In Figure 3–25C, we have dotted in a square pulse around a typical rectified diode current pulse. The average value of a square pulse train is given by

$$\text{square pulse average: } I_{Daverage} \approx \frac{WI_p}{T}$$

FIGURE 3–25

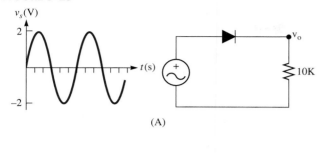

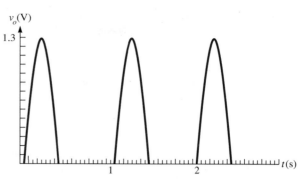

(A)

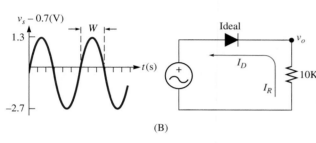

(B)

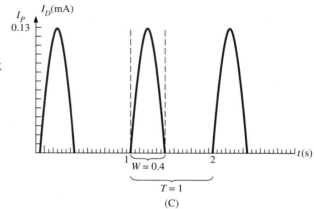

(C)

where I_p is the maximum diode current, W is the width of the current pulse, and T is the period of the input sinusoid (1s in this figure). For Figure 3–25C, we get

$$I_{Daverage} \approx \frac{WI_p}{T} = \frac{0.4 \times 0.13\text{mA}}{1} = 0.052\text{mA}$$

The power dissipated by the diode is

$$P_D = 0.7 \times I_{Daverage} = 0.7 \times 0.052\text{mA} = 0.0364\text{mW}$$

Note that this method of calculating average diode current and power will always overestimate the value. For example, a half-wave rectified current has an actual average value of 0.318 I_p. The square pulse approximation (where $W = T/2$) gives a value of 0.5 I_p. However, we find the average current and power to ensure we don't use a diode that will burn out, which would be very bad for your continued job prospects. There is, therefore, nothing wrong with overestimating the power requirements.

EXAMPLE 3–7
Approximating Power

Approximate $I_{Daverage}$ and P_D for the half-wave rectifier in Figure 3–26 (p. 66).

Solution Figure 3–26B shows the rectified output voltage that has a peak value of 49.3V (remember the diode voltage drop). The peak diode current is, therefore

$$I_p = \frac{49.3}{1\text{K}} = 49.3\text{mA}$$

The approximate average diode current is

$$I_{Daverage} = \frac{WI_p}{T} = \frac{0.01 \times 49.3\text{mA}}{0.02} = 24.7\text{mA}$$

and the average power dissipated is

$$P_D = 0.7 \times I_{Daverage} = 0.7 \times 24.7\text{mA} = 17.3\text{mW}$$

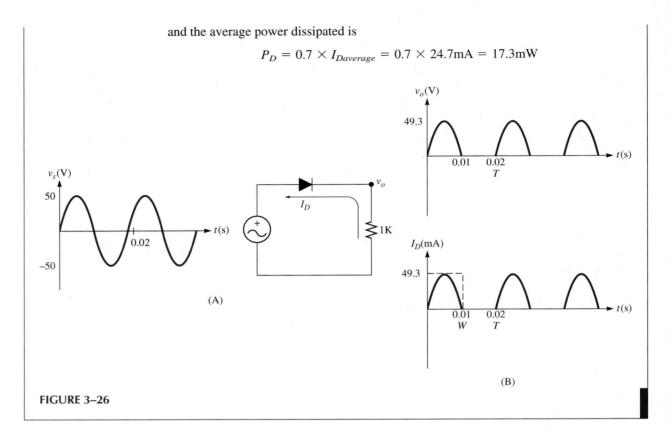

FIGURE 3–26

The Breakdown Region

In Figure 3–6 (p. 52) we showed you the diode I-V curve; however, we did not show you the complete curve. As you know, when V_D is negative, the diode is reverse biased and carries almost no current. If we increase the reverse bias by making V_D more and more negative, we reach a point where the diode goes into **breakdown**.

When enough reverse-biased voltage is applied to a diode, the diode junction breaks down electrically, not physically. That is, the internal forces created by the voltage applied to the reverse-biased diode are great enough to break free the electrons that are normally bound to the atoms in the crystalline structure. These free electrons create a rush of current through the reverse-biased junction. We say that in breakdown, reverse current *burns* through the junction.

If we include the breakdown region in the diode I-V curve, we get the picture shown in Figure 3–27. Once again, as the diode is driven deeper into the reverse-biased region, we reach a voltage where the diode goes into breakdown. This voltage is variously known as the **peak inverse voltage** (*PIV*), the **breakdown voltage** (*BV*), or the **maximum reverse**

FIGURE 3–27

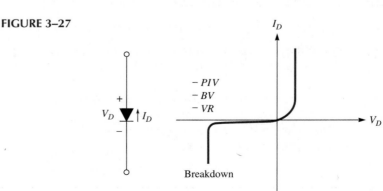

voltage (*VR*). Note that by convention, *PIV* is given as a positive number. This is why the plot shows the value −*PIV* at breakdown. This also applies to *BV* and *VR*.

For the type of diode circuits we have examined so far, we must avoid the breakdown region. The correct behavior of these circuits requires the diode to be an open circuit when it is reverse biased. That is, reverse-biased diode current must be close to 0A. In breakdown, reverse-biased diode currents are not 0A, but, in fact, can be very large. (This excessive current can permanently damage a diode.)

To find the required *PIV* for a diode in a given circuit, we must find the largest reverse-biased voltage that the diode will sustain. Consider the rectifier circuit in Figure 3–28. When the diode is reverse biased, we replace it with an open circuit (Figure 3–28B). We first note that

$$V_D = v_s - v_o$$

FIGURE 3–28

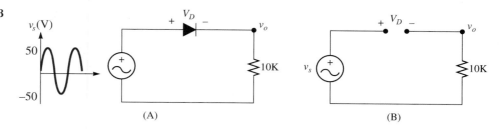

(A)   (B)

The diode is **OFF** when the input goes negative. During this time $v_o = 0$V, and we get the maximum expected reverse-biased diode voltage:

$$V_{Dreverse} = +50\text{V} \text{ (note that } V_{Dreverse} = -V_D)$$

For safety, we would choose a diode with a *PIV* of 100V or greater.

EXAMPLE 3–8
Breakdown

Find the *PIV* required for the diode in the limiter circuit shown in Figure 3–29. Use a safety factor of 2:1, that is, double the maximum expected reverse-bias voltage.

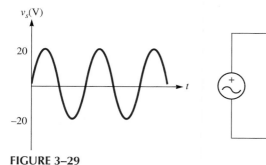

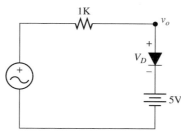

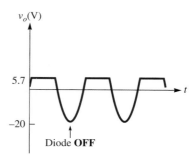

FIGURE 3–29

Solution As shown in the figure, the diode is reverse biased when the input drops below 5.7V (don't forget the diode turn-on voltage). Because the cathode is fixed at 5V

$$V_D = v_o - 5$$

The *PIV* is found when the input falls to −20V

$$V_{Dreverse} = 20 + 5 = 25\text{V}$$

so we would use a diode with

$$PIV = 50\text{V}$$

FIGURE 3–30

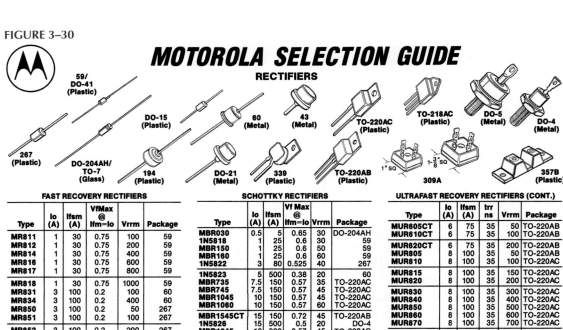

MOTOROLA SELECTION GUIDE
RECTIFIERS

FAST RECOVERY RECTIFIERS

Type	Io (A)	Ifsm (A)	VfMax @ Ifm=Io	Vrrm	Package
MR811	1	30	0.75	100	59
MR812	1	30	0.75	200	59
MR814	1	30	0.75	400	59
MR816	1	30	0.75	600	59
MR817	1	30	0.75	800	59
MR818	1	30	0.75	1000	59
MR831	3	100	0.2	100	60
MR834	3	100	0.2	400	60
MR850	3	100	0.2	50	267
MR851	3	100	0.2	100	267
MR852	3	100	0.2	200	267
MR820	5	300	0.2	50	194
MR821	5	300	0.2	100	194
MR826	5	300	0.2	600	194
IN3879	6	150	0.2	50	DO-4
1N3880	6	150	0.2	100	DO-4
1N3881	6	150	0.2	200	DO-4
1N3883	6	150	0.2	400	DO-4
MR1366	6	150	0.2	600	DO-4
1N3889	12	200	0.2	50	DO-4
1N3890	12	200	0.2	100	DO-4
1N3891	12	200	0.2	200	DO-4
1N3893	12	200	0.2	400	DO-4
MR1376	12	200	0.2	600	DO-4
1N3899	20	250	0.2	50	DO-5
1N3900	20	250	0.2	100	DO-5
1N3901	20	250	0.2	200	DO-5
1N3903	20	250	0.2	400	DO-5
MR1386	20	250	0.2	600	DO-5
1N3909	30	300	0.2	50	DO-5
1N3910	30	300	0.2	100	DO-5
1N3913	30	300	0.2	400	DO-5
1N4933	1	30	0.2	50	59
1N4934	1	30	0.2	100	59
1N4935	1	30	0.2	200	59
1N4936	1	30	0.2	400	59
MR810	1	30	0.75	50	59

R Suffix: Reverse Polarity

Type	Io (A)	Ifsm (A)	Vrrm	Package
1N4001	1	30	50	DO-41
1N4002	1	30	100	DO-41
1N4003	1	30	200	DO-41
1N4004	1	30	400	DO-41
1N4005	1	30	600	DO-41
1N4006	1	30	800	DO-41
1N4007	1	30	1000	DO-41
1N5392	1.5	50	100	DO-15
1N5393	1.5	50	200	DO-15
1N4719	3	300	50	60
1N4720	3	300	100	60
1N4721	3	300	200	60
1N4722	3	300	400	60
1N4723	3	300	600	60
1N4724	3	300	800	60
1N4725	3	300	1000	60
MR500	3	100	50	267
MR501	3	100	100	267
MR502	3	100	200	267
MR504	3	100	400	267
MR506	3	100	600	267
MR508	3	100	800	267
MR510	3	100	1000	267
1N5406	3	200	600	267
MR750	6	400	50	194

SCHOTTKY RECTIFIERS

Type	Io (A)	Ifsm (A)	Vf Max @ Ifm=Io	Vrrm	Package
MBR030	0.5	5	0.65	30	DO-204AH
1N5818	1	25	0.6	30	59
MBR150	1	25	0.6	50	59
MBR160	1	25	0.6	60	59
1N5822	3	80	0.525	40	267
1N5823	5	500	0.38	20	60
MBR735	7.5	150	0.57	35	TO-220AC
MBR745	7.5	150	0.57	45	TO-220AC
MBR1045	10	150	0.57	45	TO-220AC
MBR1060	10	150	0.57	60	TO-220AC
MBR1545CT	15	150	0.72	45	TO-220AB
1N5826	15	500	0.5	20	DO-4
MBR1645	16	300	0.57	45	TO-220AC
MBR2045CT	20	150	0.72	45	TO-220AB
1N5829	25	800	0.48	20	DO-4
MBR3035CT	30	400	0.72	35	TO-3
MBR2545CT	30	400	0.73	45	TO-220AB
MBR3035PT	30	400	0.72	35	TO-218AC
MBR3045PT	30	400	0.72	45	TO-218AC
MBR3535	35	600	0.55	35	DO-4
MBR3545	35	600	0.55	45	DO-4
1N5832	40	800	0.59	20	DO-5
MBR6035	60	800	0.6	35	DO-5
SD51	60	800	0.6	45	DO-5
MBR6045	60	800	0.6	45	DO-5
MBR12035C	120	1500	0.68	35	357B
MBR12045C	120	1500	0.68	45	357B
MBR12050C	120	1500	0.68	50	357B
MBR12060C	120	1500	0.68	60	357B
MBR20035C	200	1500	0.71	35	357B
MBR20045C	200	1500	0.71	45	357B
MBR20050C	200	1500	0.71	50	357B
MBR20060C	200	1500	0.71	60	357B
MBR30035C	300	2500	0.64	35	357B
MBR30045C	300	2500	0.64	45	357B

ULTRAFAST RECOVERY RECTIFIERS

Type	Io (A)	Ifsm (A)	trr ns	Vrrm	Package
MUR105	1	35	25	50	DO-41
MUR110	1	35	25	100	DO-41
MUR115	1	35	25	150	DO-41

GENERAL PURPOSE RECTIFIERS

Type	Io (A)	Ifsm (A)	Vrrm	Package
MR751	6	400	100	194
MR754	6	400	400	194
MR756	6	400	600	194
MR758	6	400	800	194
MR760	6	400	1000	194
MR1120	12	300	50	DO-4
1N1199A	12	300	50	DO-4
MR1121	12	300	100	DO-4
MR1121R	12	300	100	DO-4
1N1200	12	300	100	DO-4
N1200A	12	300	100	DO-4
MR1122	12	300	200	DO-4
1N1202	12	300	200	DO-4
1N1202A	12	300	200	DO-4
MR1124	12	300	400	DO-4
1N1204	12	300	400	DO-4
1N1204A	12	300	400	DO-4
1N1204RA	12	300	400	DO-4
MR1126	12	300	600	DO-4
MR1126R	12	300	600	DO-4
1N1206	12	300	600	DO-4
1N1206A	12	300	600	DO-4
1N1206R	12	300	600	DO-4
1N1206RA	12	300	600	DO-4
MR1128	12	300	800	DO-4

ULTRAFAST RECOVERY RECTIFIERS (CONT.)

Type	Io (A)	Ifsm (A)	trr ns	Vrrm	Package
MUR605CT	6	75	35	50	TO-220AB
MUR610CT	6	75	35	100	TO-220AB
MUR620CT	6	75	35	200	TO-220AB
MUR805	8	100	35	50	TO-220AC
MUR810	8	100	35	100	TO-220AC
MUR815	8	100	35	150	TO-220AC
MUR820	8	100	35	200	TO-220AC
MUR830	8	100	35	300	TO-220AC
MUR840	8	100	35	400	TO-220AC
MUR850	8	100	35	500	TO-220AC
MUR860	8	100	35	600	TO-220AC
MUR870	8	100	35	700	TO-220AC
MUR880	8	100	35	800	TO-220AC
MUR890	8	100	35	900	TO-220AC
MUR8100	8	100	35	1000	TO-220AC
MUR1520	15	200	35	200	TO-220AC
MUR1540	15	200	35	400	TO-220AC
MUR1550	15	200	35	500	TO-220AC
MUR1615	16	100	35	150	TO-220AB
R711XPT	30	150	100	100	TO-218AC
R712XPT	30	150	100	200	TO-218AC
MUR3010PT	30	400	35	100	TO-218AC
MUR3015PT	30	400	35	150	TO-218AC
MUR3020PT	30	400	35	200	TO-218AC

RECTIFIER BRIDGES

Type	Io (A)	Ifsm (A)	Vrrm	Package
MDA2501	25	400	100	309A
MDA2502	25	400	200	309A
MDA2504	25	400	400	309A
MDA2506	25	400	600	309A
MDA3500	35	400	50	309A
MDA3501	35	400	100	309A
MDA3502	35	400	200	309A
MDA3504	35	400	400	309A
MDA3506	35	400	600	309A
MDA3508	35	400	800	309A
MDA3510	35	400	1000	309A

Type	Io (A)	Ifsm (A)	Vrrm	Package
MR1130	12	300	1000	DO-4
MR1130R	12	300	1000	DO-4
MR2404	24	400	400	339
1N3491	30	300	50	DO-21
1N3491R	30	300	50	DO-21
1N3492	30	300	100	DO-21
1N3492R	30	300	100	DO-21
1N3493	30	300	200	DO-21
1N3493R	30	300	200	DO-21
1N3495R	30	300	400	DO-21
MR328	30	300	600	DO-21
MR330	30	300	800	DO-21
MR331	30	300	1000	DO-21
1N3660	30	400	100	DO-21
1N3660R	30	400	100	DO-21
1N3661	30	400	200	DO-21
1N3668	30	400	400	DO-21
1N3663R	30	400	400	DO-21
1N1183A	40	800	50	DO-5
1N1184A	40	800	100	DO-5
1N1186A	40	800	200	DO-5
1N1188A	40	800	400	DO-5
1N1190A	40	800	600	DO-5
MR5040	50	600	400	43

DRILL EXERCISE

Redraw the the limiter circuit of Figure 3–29 (p. 67), changing the input to a 10V sine wave, reversing the diode, and reversing the battery. Using a 2:1 safety factor, find the *PIV* you would use for this diode

Answer   *PIV* = 30V.

REVIEW QUESTIONS

1. What is the average power dissipated by the diode?
2. How can you approximate the average diode current for rectified signals?
3. What causes a diode to go into breakdown, and what happens to diode current in breakdown?
4. How do you find the *PIV* for a diode?

3–7 ◼ COMMERCIALLY AVAILABLE DIODES— THE DIODE DATA SHEET

Catalog Data

Figure 3–30 shows a page from a catalog of an electronic component supplier that describes commercially available diodes. Note that Motorola uses the term *rectifier* instead of *diode*. There are two basic types of diodes that are used, fast (or ultrafast) recovery diodes and general purpose diodes. You can also see listed Schottky rectifiers and rectifier bridges. Schottky diodes are a special type of fast-recovery diode and will be discussed in Chapter 6. The rectifier bridge provides on a single chip the four diodes required for full-wave rectification.

Fast-recovery diodes are used in high-speed and high-frequency applications, such as AM demodulation and the high-speed switching circuits found in computers. The general purpose diode is used in low-frequency applications, such as those discussed in the previous chapter. Let's examine the data given for these diodes (note that not all data is given for all diodes).

☐ *Io*, the average forward-biased current. This is the average current, $I_{Daverage}$, that we calculated in the previous section.

☐ *Ifsm*, the maximum surge current. This is the current that the diode can withstand for a short time.

☐ *Vfmax*, the forward-biased diode turn-on voltage.

☐ *Vrrm*, the peak maximum reverse voltage (also known as peak inverse voltage, *PIV*).

☐ *trr*, the recovery time. This tells us how fast the diode can be turned on or off.

EXAMPLE 3–9
Diode Selection

Select a general purpose diode that can be used in the circuit of Figure 3–31A (p. 70).

Solution   We first note that I_D is the current in the 10Ω resistor, so $I_D = v_o/10$, and is shown in Figure 3–31B. The peak diode current is, therefore

$$I_{Dp} = \frac{v_{op}}{10} = \frac{160}{10} = 16A$$

In Chapter 2, we showed you that for a half-wave rectified signal, the average value is given by peak value divided by π; therefore

$$I_{Daverage} = \frac{I_{Dp}}{\pi} = \frac{16}{\pi} = 5.1A$$

When the diode is **OFF** the cathode voltage is at 0V, while the anode voltage drops to $-160V$. The maximum reverse-biased voltage is, therefore, 160V.

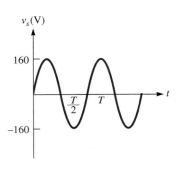

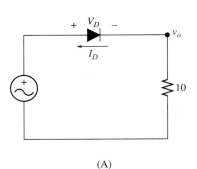

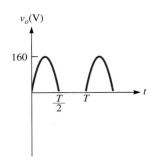

(A)

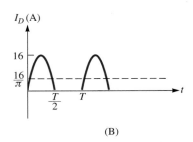

(B)

FIGURE 3–31

If we use a safety factor of 2:1 for these two values, we want a diode that has an *Io* of approximately 10A and a *Vrrm* of approximately 320V. Scanning the selection guide shown in Figure 3–30, we see that the MR1124 or one of the 1N1204s will fit the bill.

DRILL EXERCISE   Redraw the circuit in Figure 3–31A, changing the 10Ω load to 5Ω and select an appropriate diode for this circuit. Use a 2:1 safety factor.

Answer   *Vrrm* = 320V and *Io* = 20A. Diode MR2404 can be used here.

Diode Data Sheets

Semiconductor manufacturers provide data sheets that are more detailed than the information in vendor catalogs. Figure 3–32 shows Motorola's long-form data sheet for the general purpose diodes 1N4001 through 1N4007. Figure 3–33 (p. 72) shows National Semiconductors' short-form data sheet for a family of high-speed diodes. We will consider the Motorola data sheet first.

The first set of data in Figure 3–32 gives the maximum ratings for the 1N400x family.

FIGURE 3–32

MOTOROLA
SEMICONDUCTOR TECHNICAL DATA

Order this document
by 1N4001/D

Axial-Lead
Standard Recovery Rectifiers

This data sheet provides information on subminiature size, axial lead mounted rectifiers for general–purpose low–power applications.

Mechanical Characteristics
- Case: Epoxy, Molded
- Weight: 0.4 gram (approximately)
- Finish: All External Surfaces Corrosion Resistant and Terminal Leads are Readily Solderable
- Lead and Mounting Surface Temperature for Soldering Purposes: 220°C Max. for 10 Seconds, 1/16″ from case
- Shipped in plastic bags, 1000 per bag.
- Available Tape and Reeled, 5000 per reel, by adding a "RL" suffix to the part number
- Polarity: Cathode Indicated by Polarity Band
- Marking: 1N4001, 1N4002, 1N4003, 1N4004, 1N4005, 1N4006, 1N4007

**1N4001
thru
1N4007**

1N4004 and 1N4007 are Motorola
Preferred Devices

**LEAD MOUNTED
RECTIFIERS
50–1000 VOLTS
DIFFUSED JUNCTION**

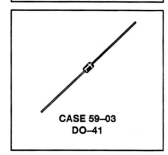

**CASE 59–03
DO–41**

MAXIMUM RATINGS

Rating	Symbol	1N4001	1N4002	1N4003	1N4004	1N4005	1N4006	1N4007	Unit
*Peak Repetitive Reverse Voltage Working Peak Reverse Voltage DC Blocking Voltage	V_{RRM} V_{RWM} V_R	50	100	200	400	600	800	1000	Volts
*Non–Repetitive Peak Reverse Voltage (halfwave, single phase, 60 Hz)	V_{RSM}	60	120	240	480	720	1000	1200	Volts
*RMS Reverse Voltage	$V_{R(RMS)}$	35	70	140	280	420	560	700	Volts
*Average Rectified Forward Current (single phase, resistive load, 60 Hz, see Figure 8, T_A = 75°C)	I_O	1.0							Amp
*Non–Repetitive Peak Surge Current (surge applied at rated load conditions, see Figure 2)	I_{FSM}	30 (for 1 cycle)							Amp
Operating and Storage Junction Temperature Range	T_J T_{stg}	– 65 to +175							°C

ELECTRICAL CHARACTERISTICS*

Rating	Symbol	Typ	Max	Unit
Maximum Instantaneous Forward Voltage Drop (i_F = 1.0 Amp, T_J = 25°C) Figure 1	v_F	0.93	1.1	Volts
Maximum Full–Cycle Average Forward Voltage Drop (I_O = 1.0 Amp, T_L = 75°C, 1 inch leads)	$V_{F(AV)}$	—	0.8	Volts
Maximum Reverse Current (rated dc voltage) (T_J = 25°C) (T_J = 100°C)	I_R	0.05 1.0	10 50	μA
Maximum Full–Cycle Average Reverse Current (I_O = 1.0 Amp, T_L = 75°C, 1 inch leads)	$I_{R(AV)}$	—	30	μA

*Indicates JEDEC Registered Data

Preferred devices are Motorola recommended choices for future use and best overall value.

Rev 5

 **MOTOROLA**

FIGURE 3–33

National Semiconductor

Diode Data

Computer Diodes (Glass Package)

Device No.	Package No.	V_{RRM} V Min	I_R nA Max	@ V_R V	V_F V Min	V_F V Max	@ I_F mA	C pF Max	t_{rr} ns Max	Test Cond.	Proc. No.
1N625	DO-35	30	1000	20		1.5	4		1000	(Note 1)	D4
1N914	DO-35	100	25 / 5000	20 / 75		1.0	10		4	(Note 2)	D4
1N914A	DO-35	100	25 / 5000	20 / 75		1.0	20		4	(Note 2)	D4
1N914B	DO-35	100	25 / 5000	20 / 75		0.72 / 1.0	5 / 100		4	(Note 2)	D4
1N916	DO-35	100	25 / 5000	20 / 75		1.0	10		4	(Note 2)	D4
1N916A	DO-35	100	25 / 5000	20 / 75		1.0	20		4	(Note 2)	D4
1N916B	DO-35	100	25 / 5000	20 / 75		0.73 / 1.0	5 / 30		4	(Note 2)	D4
1N3064	DO-35	75	100	50		0.575 / 0.650 / 0.710 / 1.0	0.250 / 1.0 / 2.0 / 10.0	2	4	(Note 3)	D4
1N3600	DO-35	75	100	50	0.54 / 0.66 / 0.76 / 0.82 / 0.87	0.62 / 0.74 / 0.86 / 0.92 / 1.0	1.0 / 10.0 / 50.0 / 100.0 / 200.0	2.5	4	(Note 4)	D4
1N4009	DO-35	35	100	25		1.0	30	4	2	(Note 2)	D4
1N4146	DO-35	See Data for 1N914A/914B									
1N4147	DO-35	See Data for 1N914A/914B									
1N4148	DO-35	See Data for 1N914									
1N4149	DO-35	See Data for 1N916									
1N4150	DO-35	See Data for 1N3600									
1N4151	DO-35	75	50	50		1.0	50	4	2	(Note 2)	D4
1N4152	DO-35	40	50	30	0.49 / 0.53 / 0.59 / 0.62 / 0.70 / 0.74	0.55 / 0.59 / 0.67 / 0.70 / 0.81 / 0.88	0.1 / 0.25 / 1.0 / 2.0 / 10.0 / 20.0	4	2	(Note 2)	D4
1N4153	DO-35	75	50	50	See 1N4152			4	2	(Note 2)	D4
1N4154	DO-35	35	100	25		1.0	30	4	2	(Note 2)	D4

Row 1: The peak inverse voltages for these diodes ranges from 50 to 1000V.

Row 2: Diodes can handle a slightly larger one-time (nonrepetitive) peak reverse voltage. These diodes can handle single-peak inverse voltage that ranges from 60 to 1200V.

Row 3: Data from Row 1 is presented here as rms voltage (rms = $\sqrt{2}$ peak).

Row 4: The diode can handle this maximum average forward-biased current. Notice that it is the same for the entire family.

Row 5: The diode can tolerate a single-pulse current of 30A. The surge current of 30A is much greater than the average current-handling capability of 1A.

Row 6: The operating temperature range of the diode is measured in degrees centigrade. In Fahrenheit, this temperature range is $-85°$ to $347°F$.

The next set of data gives the electrical characteristics of the diode. These numbers are the ones we use when we model the diode.

Rows 7 and 8: These two lines report on diode turn-on voltage under slightly different test conditions. The values range from 0.8 to 1.1V. This is somewhat larger than the 0.7V that we have used in this chapter. The discrepancy is that the manufacturer gives the turn-on voltage for a diode current of 1.0A. This is a typical voltage for a current of 1.0A. As diode currents drop into the milli-ampere range, the turn-on voltage also drops.

Rows 9 and 10: These two lines report on diode reverse-biased current under slightly different test conditions and temperatures. The information shows that the largest level of the reverse-biased current is 50μA. Typically, the reverse current is 50nA at room temperature ($T_j = 25°C$) and increases as temperature increases. We have always assumed that the reverse-biased diode is an open circuit and, so, would have no current in it. The actual reverse-biased current is so small that our approximation is valid. One more point about this current. This is *not* the current value used in the exponential formula given for the diode I-V curve in Section 3–3. An explanation of this discrepancy follows.

The Diode Reverse Current

The diode exponential equation given in Section 3–3 is repeated here:

$$I_D = I_R(e^{V_D/V_T} - 1)$$

where V_T is known as the **thermal voltage** and is approximately 26mV at room temperature, and I_R is the semiconductor reverse current. If we know the diode forward voltage at a given diode forward current, we can solve for I_R as follows:

$$I_R = \frac{I_D}{e^{V_D/V_T} - 1}$$

We can get the data we need from Row 7 of the data sheet shown in Figure 3–32 (p. 71). We see that the diode voltage is approximately 1.0V when the diode current is 1.0A. We find the reverse diode current as

$$I_R = \frac{I_D}{e^{V_D/V_T} - 1} = \frac{1.0}{e^{1.0/.026} - 1} = \frac{1.0}{5.1 \times 10^{16}} = 2 \times 10^{-17}A$$

This current is much, much smaller than the reverse current given in Rows 9 and 10. Note that 0.05μA is equal to 5×10^{-8} A.

Why the difference? The answer is that in a real diode, current can flow along the surface of the device and the diode casing, as well as through the semiconductor material of the diode itself. The total current in a diode, therefore, is the sum of the semiconductor current given by the diode exponential equation plus the surface-case current.

The current that flows along the diode casing is small enough that it can be ignored when the diode is forward biased. When the diode is reverse biased, however, this surface-case current is actually larger than the semiconductor reverse-bias current and dominates the total diode current.

Figure 3–33 (p. 72) shows the short-form data sheet for another family of diodes. These diodes are specifically designed to be used in high-speed circuits. Although presented in a different manner from the previous data sheet, peak inverse voltage, forward currents, reverse current, and forward voltage are all given here. In addition, a parameter known as the **reverse recovery time** (t_{rr}) is also given.

Reverse Recovery Time

In this and the previous chapters, we have not discussed the fact that it takes some time to move enough charge into a diode to overcome the reverse bias. At low signal frequencies this time factor is not a problem. As the signal frequency increases, we must be able to move charges into and out of the diode at increasingly faster rates.

Consider, for example, a rectifier that has a reverse recovery time of 2.0μs. We will try to use this diode in a power circuit with a frequency of 60Hz and in a computer circuit that operates at 1.0MHz. Remembering that time and frequency are related by $T = 1/f$, the diode in the power circuit must turn **ON** every

$$T = 1/(60\text{Hz}) = 16.7\text{ms}$$

The diode in the computer circuit must turn **ON** every

$$T = 1/(1\text{MHz}) = 1.0\mu\text{s}$$

You can see that the given recovery time of 2.0μs is fast enough for use in the power circuit but is not fast enough for use in the computer circuit.

EXAMPLE 3–10
Using the Data Sheet

For the 1N914B diode (Figure 3–33, p. 72):

(a) Find the reverse-bias semiconductor current.
(b) What is the maximum frequency at which this diode can operate?

Solution
(a) We first note that the forward voltage for the 1N914B is 0.72V for a forward current of 5.0mA. Using the diode exponential equation given previously, we find the reverse-bias semiconductor current is

$$I_R = \frac{I_D}{e^{V_D/V_T} - 1} = \frac{5 \times 10^{-3}}{e^{0.72/.026} - 1} = \frac{5 \times 10^{-3}}{1.1 \times 10^{12}} = 4.55 \times 10^{-15}\text{A}$$

Note once again that the semiconductor reverse current is much smaller than the total reverse current of 25mA = 2.5×10^{-10}A.

(b) The reverse recovery time for this diode is 4.0ns. This corresponds to a maximum operating frequency of

$$f = 1/T = 1/(4 \times 10^{-9}) = 250 \times 10^6 = 250\text{MHz}$$

REVIEW QUESTIONS

1. List several diode parameters that can be found in a company's catalog data.
2. How can you use this data to select a diode for use in a circuit?
3. List several diode parameters that can be found in a manufacturer's data sheet.
4. Why is the reverse current given in a diode's data sheet much greater than predicted from the diode I-V curve?

3–8 ■ DIODE TESTING

As a working electronics technologist, you will face two situations in which diodes must be tested. In the first instance, you have been given a diode circuit design that you must build. In this case, it would be prudent to test any diode you plan to use before you solder it into the circuit.

In the second instance, you are presented with an already-built diode circuit that does not appear to be working. In this case, you must put on your sleuthing cap to determine if the diode might be at fault. You would first try to test the diode while it is still in the circuit. If this testing confirms that the diode might be bad, you would then disconnect it from the circuit and test it in isolation.

Diodes can be easily tested by using commonly available multimeters, such as the analog volt-ohm meter (VOM) and the digital multimeter (DMM) shown in Figure 3–34. The analog VOM displays its results in the form of a moving needle, whereas the digital DMM displays its results with a calculator-like read-out. Figure 3–34 shows the results of measuring the resistance of a 1KΩ resistor.

FIGURE 3–34

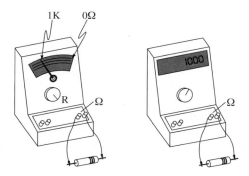

As many textbooks recommend, a diode can be tested by measuring its forward and reverse resistance. The forward resistance should be small, and the reverse resistance should be large. In Figure 3–35A we use an analog VOM to measure the resistance of the forward-biased diode (+ connected to the anode) and see that the resistance is small. We then reverse the diode direction (+ to cathode, Figure 3–35B) and measure a large reversed-biased resistance. This diode is good. Note that we set the VOM to the KΩ range to prevent excessive current in the diode.

FIGURE 3–35

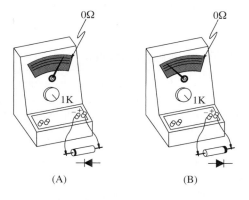

(A)   (B)

A Bad Diode (Using the VOM)

If a diode is bad, you will measure the same resistance in both directions. Most diodes fail by opening up. In this case, you will measure a high resistance in both directions. If the diode happens to fail by internally shorting out, you will measure a small resistance in both directions. In either case, if you do not see a measurable difference in resistance when you measure in both directions, the diode is bad.

A Bad Diode (Using the DMM)

The resistance scale of the DMM generally produces a current that is too small to turn a diode **ON**. The solution to this dilemma is provided by the DMM manufacturers. Because the resistance scale cannot be used to test a diode, the DMM is usually supplied with a special diode test scale and diode test terminals. This is shown in Figure 3–36. With this scale the DMM supplies an approximately constant current to the diode and measures the voltage developed across the diode.

FIGURE 3–36

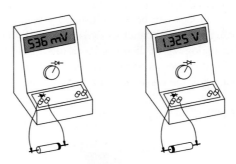

Because the diode resistance is low in the forward direction, the voltage developed across the forward-biased diode is low, approximately 500 to 800mV. The reverse-biased diode has a very high resistance, so the voltage developed across it is large. If the diode is good, you will measure a large difference in its forward- and reverse-direction voltages.

If a diode is bad, you will measure the same voltage in both directions. Most diodes fail by opening up. In this case, you will measure a high voltage in both directions. If the diode happens to fail by internally shorting out, you will measure a small voltage in both directions. In either case, if you do not see a measurable difference in voltage when you measure in both directions, the diode is bad.

**REVIEW
QUESTIONS**

1. How can you test a diode with an analog VOM?
2. How can you test a diode with a DMM?

3–9 ■ DIODE CIRCUIT TROUBLESHOOTING

Suppose you have an electronic gismo that isn't working, and you have to fix it. Electronic circuit failure can manifest itself in several ways (Figure 3–37). You switch on the device and it smokes and smells bad; you switch on the device and it just sits there with no output; you switch on the device and you get an output but not the correct output, e.g., the fuzz box doesn't fuzz.

The first step in troubleshooting is visual inspection. Are any leads broken; do any solder joints look bad; is a circuit element obviously burned? If the circuit looks OK, then you proceed with electrical tests.

The surest way to test a diode circuit is with an oscilloscope. Diode circuits are designed to change the shape of the input wave. Because the look of the wave shape is critical, you need a device that will show the exact diode circuit output. This is what the oscilloscope will do. Consider the half-wave rectifier shown in Figure 3–38A. If the circuit is working, then you will see the sinusoidal input (v_i) on Channel 1 and the half-wave rectified signal (v_L) on Channel 2.

FIGURE 3–37

FIGURE 3–38

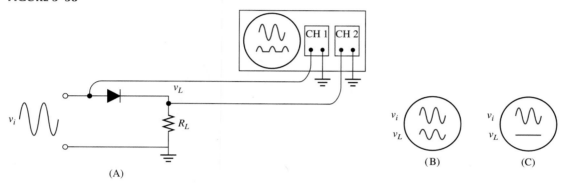

Figures 3–38B and C show some possible outputs if failure occurs. In Figure 3–38B, the output voltage is the same as the input voltage. Because the input voltage is not being rectified, the diode must have failed. In fact, the diode must be shorted out. This is not a typical failure mode for a diode—they usually fail by opening up. In Figure 3–38C, the output is 0V; that is, there is no output. In this case, there is no clear-cut determination of which circuit element is bad.

The general circuit shown in Figure 3–39 explains the dilemma. The output is zero if the series element is open *or* if the shunt element is shorted. This illustrates that the output shown in Figure 3–38C could be caused by an opened diode or a shorted load. Because electrical elements most often fail open, the problem is probably with the diode. The only way to tell for sure is to remove the diode and test it as described in the previous section.

FIGURE 3–39

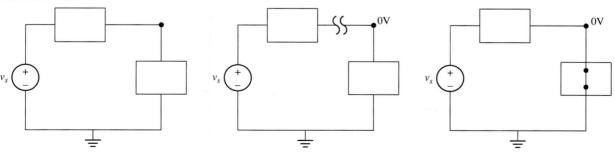

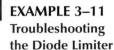

EXAMPLE 3–11
Troubleshooting the Diode Limiter

Figure 3–40A shows a diode limiter circuit.

(a) Find the expected output.
(b) Figure 3–40B shows the measured output. Determine the likely cause of failure.
(c) Figure 3–40C shows the measured output. Determine the likely cause of failure.

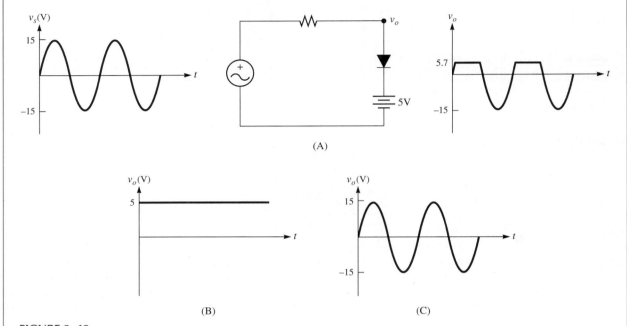

FIGURE 3–40

Solution

(a) The expected output is shown in Figure 3–40A.
(b) The only way the output can always be 5V is if the diode has shorted out.
(c) If the diode has opened up, there will be no current draw from the source and, therefore, no voltage drop across the resistor. The output will equal the input.

Multimeter Testing

Can you troubleshoot a diode circuit if you don't have an oscilloscope? The answer is yes. You can use a multimeter, either an analog VOM or a digital DMM. Use the resistance scale (set to 1KΩ range to prevent excess current in the diode) on the VOM or the diode test scale on the DMM. **CAUTION:** Be sure to shut off the source before you use either of these meters. If you do not, you will get spurious readings and you may damage your meter.

In the previous section, you saw how to test a diode by measuring across it in both directions. If the diode is good, the two measurements will differ significantly. If the diode is bad, the two measurements will be similar.

You can apply the same procedure to testing a diode that is in a circuit. Measure across it in both directions. If the measurements are different, the diode is good. If the measurements are similar, the diode is probably bad. The other circuit elements will affect your readings, but the general principle remains the same. If necessary, remove the diode from the circuit and test it separately.

Figure 3–41 shows an example of this procedure. In Figure 3–41A we turn off the source and measure across the diode in both directions. Because the voltages are significantly different, the diode is good. Note that the resistor is now in parallel with the diode,

FIGURE 3–41

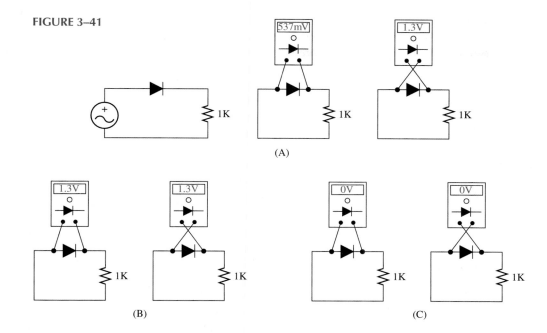

(A)

(B)   (C)

so the voltage measured by the DMM will be affected by the resistor. In particular, when the diode is **OFF,** the DMM does not see the very high reverse resistance of the diode. The DMM is now measuring the voltage across the 1KΩ resistor (this voltage is created by the output current of the DMM). Compare the reverse voltage seen here with the reverse voltage when the diode was tested separately in Figure 3–36.

In Figure 3–41B, you measure the same high voltage in both directions across the diode. You correctly conclude that the diode has failed open. Your meter is measuring the voltage across the 1KΩ resistor.

In Figure 3–41C, you recognize that you are measuring across a short circuit. The problem here is that either the diode or the resistor may be shorted. In this case, you must remove the diode and test it separately.

REVIEW QUESTIONS

1. What device is needed to measure AC signals in a circuit?
2. How will a failed short diode affect the output of a rectifier?
3. How will a failed open diode affect the output of a rectifier?
4. How will a failed short diode affect the output of a limiter?
5. How will a failed open diode affect the output of a limiter?
6. How can you use a multimeter to troubleshoot a diode circuit?

SUMMARY

▪ The real diode I-V curve is given by $I_D = I_R (e^{V_D/V_T} - 1)$, where typical values for the device parameters are: $I_R = 10^{-11}$mA and $V_T = 26$mV.

▪ When the real diode is **ON,** $V_D = 0.7$V. It can be approximated with a 0.7V battery.

▪ The real diode is **OFF** if $V_D < 0.7$V. Replace the **OFF** diode with an open circuit.

▪ The diode current test proceeds as follows: Assume the diode is **ON** and replace it with a 0.7V battery. Solve for I_D. If $I_D > 0$, the diode is **ON.** If $I_D < 0$, the diode is **OFF,** and you replace it with an open circuit.

- The diode voltage test proceeds as follows: Assume the diode is **OFF** and replace it with an open circuit. Solve for V_D. If $V_D < 0.7V$, the diode is **OFF**. If $V_D > 0.7V$, the diode is **ON**, and you replace it with a 0.7V battery.

- To analyze a real diode circuit, replace the **ON** diode with a 0.7V battery and add that battery to any other battery with which it is in series. Analyze the remaining circuit as an ideal diode circuit.

- The diode breakdown voltage is the reverse-bias voltage that causes a large reverse current to flow. This voltage has many labels: *PIV, BV, VR*.

- Catalog data typically provides information on the forward-bias turn-on voltage, the breakdown voltage, and the maximum forward currents the diode can handle.

- Manufacturer's data sheets provide more detailed information than catalog data.

- Diodes can be tested on the resistance scale of analog volt-ohmmeters (VOM) or on the diode testing scale of digital multimeters (DMM). In either case, measure across the diode in both directions. The diode is good if the two measurements are significantly different. The diode is bad if the two measurements are similar.

- Troubleshooting a diode circuit is easiest if you use an oscilloscope to measure output voltage. A comparison of the actual output with the expected output voltage will usually indicate if the diode is bad.

- Diodes can be tested in the circuit by measuring across them with multimeters. Similar measurements in both directions indicate that the diode is bad. Don't forget that other circuit elements can influence these measurements.

PROBLEMS

SECTION 3–3   The Real Diode

1. The forward- and reverse-bias behavior of the diode is described by the equation

$$I_D = I_R \left(e^{V_D/V_T} - 1\right)$$

   where for a given diode, $I_R = 1.5 \times 10^{-12}$ mA and $V_T = 26$ mV.

   (a) Use your calculator to find the diode current in mA for the following diode voltages:

   $$V_D = -5, -3, -1, 0, 0.1, 0.3, 0.5, 0.7, 0.71, 0.75, 0.8$$

   (b) Use your results to plot I_D vs. V_D.

2. For Figure 3–42A label the diode voltage and current, and apply the current test to the real diode.

   (a) What would the current be if the diode is **ON**?

   (b) Is the diode **ON** or **OFF**?

   (c) If the diode is **OFF**, what are V_D and I_D?

FIGURE 3–42

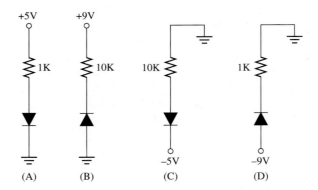

(A)   (B)   (C)   (D)

3. Repeat Problem 2 for Figure 3–42B.

4. Repeat Problem 2 for Figure 3–43C.

5. Repeat Problem 2 for Figure 3–42D.

6. For Figure 3–42A, label the diode voltage and current. Apply the voltage test to the real diode.

   (a) What would the current be if the diode is **OFF**?   (b) Is the diode **ON** or **OFF**?

   (c) If the diode is **ON**, what are V_D and I_D?

7. Repeat Problem 6 for Figure 3–42B.

8. Repeat Problem 6 for Figure 3–42C.

9. Repeat Problem 6 for Figure 3–42D.

10. For Figure 3–43A label the diode voltage and current, and apply the current test to the real diode.

   (a) What would the diode current be if the diode is **ON?**

   (b) Is the diode **ON** or **OFF?**   **(c)** If the diode is **ON,** what are V_D and I_D?

FIGURE 3–43

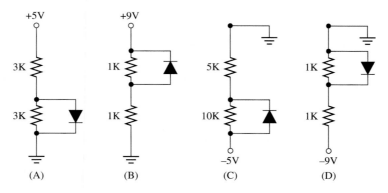

(A)   (B)   (C)   (D)

11. Repeat Problem 10 for Figure 3–43B.

12. Repeat Problem 10 for Figure 3–43C.

13. Repeat Problem 10 for Figure 3–43D.

14. For Figure 3–43A, label the diode voltage and current. Apply the voltage test to the real diode.

   (a) What would the current be if the diode is **OFF?**   **(b)** Is the diode **ON** or **OFF?**

   (c) If the diode is **OFF,** what are V_D and I_D?

FIGURE 3–44

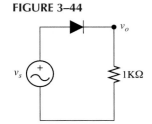

15. Repeat Problem 14 for Figure 3–43B.

16. Repeat Problem 14 for Figure 3–43C.

17. Repeat Problem 14 for Figure 3–43D.

SECTION 3–4   Diode Circuits with Real Diodes

18. The input to the half-wave rectifier in Figure 3–44 is a 1.5V 10Hz sine wave. On the same plot, draw the input and output wave forms.

19. The input to the half-wave rectifier in Figure 3–44 is a 75V 60Hz sine wave. On the same plot, draw the input and output wave forms.

FIGURE 3–45

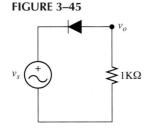

20. The input to the half-wave rectifier in Figure 3–45 is a 3V 1KHz sine wave. On the same plot, draw the input and output wave forms.

21. The input to the half-wave rectifier in Figure 3–45 is a 30V 400Hz sine wave. On the same plot, draw the input and output wave forms.

22. The source for the full-wave rectifier in Figure 3–46 is a 170V 60Hz sine wave.

   (a) Redraw the circuit and indicate which diodes are **ON** when $v_s = +170$V

   (b) What is V_D for the diodes that are **ON?** What is I_D?

   (c) For the diodes that are **OFF,** what are I_D and V_D?

FIGURE 3–46

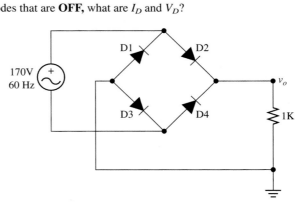

(d) Draw the sine wave input v_s and, right below it, draw v_o on the same time scale.

(e) What is the maximum value of v_o?   (f) What is the minimum value of v_o?

(g) Which diodes are **ON** when $v_s = -170\text{V}$?

23. Let the source in Figure 3–46 (p. 81) be a 2V 60Hz sine wave.

(a) Repeat Problem 22.

(b) Compare the output of the bridge rectifier for the 2V and 170V inputs.

FIGURE 3–47

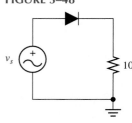

24. Figure 3–47 shows a limiter circuit.

(a) Replace the diode with the improved model, and label the diode current and voltage.

(b) For what range of v_s is the diode **ON**?   (c) For what range of v_s is the diode **OFF?**

(d) Let $v_s = +12\text{V}$. What are V_D and I_D?   (e) Let $v_s = -12\text{V}$. What are V_D and I_D?

(f) Let v_s be a 12V 60Hz sine wave. Draw v_s and right below it, properly aligned, draw v_o. What is the maximum value of v_o? Where are the two signals equal?

25. The half-wave rectifier shown in Figure 3–48 is driven with an 80V sine wave.

(a) What is the maximum current that will flow in the diode?

(b) What is the average forward current ($I_{Daverage}$) that will flow in the diode?

FIGURE 3–48

(c) What is the maximum reverse-biased voltage for the diode?

(d) Using a 2:1 safety factor, what values of Io and PIV should we insist the diode have so that the diode could be safely used in the circuit?

(e) Choose a diode from the selection guide in Figure 3–30 that meets the above requirements.

SECTION 3–5   Diode Resistance

26. The diode in Figure 3–23 (p. 63) has a forward resistance of $R_F = 100\Omega$. If this circuit is driven by $V_S = +10\text{V}$. Find the output voltage for the following load resistances

(a) $R_L = 1\text{M}\Omega$ (b) $R_L = 10\text{K}\Omega$

(c) $R_L = 1\text{K}\Omega$ (d) $R_L = 100\Omega$

SECTION 3–6   Diode Ratings—The Breakdown Region

27. Figure 3–49 shows a half-wave rectified diode current.

(a) What is the approximate average value of the diode current?

(b) What is the approximate power dissipated by the diode?

FIGURE 3–49

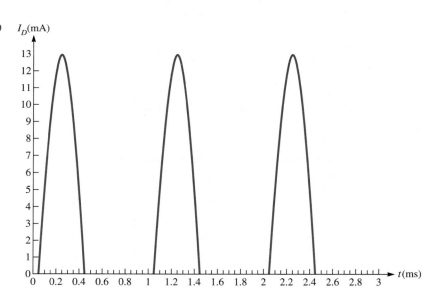

FIGURE 3–50

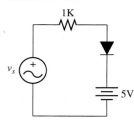

28. The limiter circuit in Figure 3–50 is driven by a 10V sine wave.

(a) What is the maximum reverse-bias voltage the diode will see?

(b) Using a 2:1 safety factor, what PIV do you need for this diode?

FIGURE 3–51

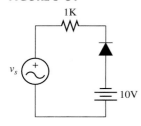

1K

v_s

10V

29. The limiter circuit in Figure 3–51 is driven by a 25V sine wave.

   (a) What is the maximum reverse-bias voltage the diode will see?

   (b) Using a 2:1 safety factor, what *PIV* do you need for this diode?

SECTION 3–7   Commercially Available Diodes—The Diode Data Sheet

30. Refer to Figure 3–32 (p. 71), and list the data given for the 1N4001 diode. What information is given by each parameter?

31. We need a diode with the following: breakdown voltage greater than 100V; maximum average rectified current greater than 10A; diode must operate up to 20MHz. Use Figure 3–30 (p. 68) to select a diode or diodes that would meet our needs.

SECTION 3–9   Troubleshooting Problems

32. Using a VOM set to measure resistance, you test a diode. In one direction you get a reading of 723Ω. Reversing the leads, you get a similar reading.

   (a) Is the diode good or bad?

   (b) If the diode is bad, what is wrong with it?

   (c) What readings should you get with a good diode? (Try this in the lab.)

33. You are testing a diode with a DMM set on the diode test scale. In one direction, you get a reading of 535mV. You reverse the diode leads and get an "Off-Scale" reading (usually a 1 followed by blank digits).

   (a) Is this diode good or bad?   **(b)** What do each of the readings mean?

34. You are testing a diode with a DMM set on the diode test scale. In one direction, you get a reading of 347mV. You reverse the diode leads and a reading of 297mV.

   (a) Is this diode good or bad?   **(b)** What do each of the readings mean?

35. You are testing a diode with a DMM set on the diode test scale. In both directions, you get an "Off-Scale" reading.

   (a) Is this diode good or bad?   **(b)** What do these readings mean?

36. The diode in Figure 3–42A (p. 80) has a measured voltage of 5V. What is wrong with the diode?

37. The diode in Figure 3–42B has a measured voltage of 0V. What is wrong with the diode?

38. The diode in Figure 3–43A (p. 81) has a measured voltage of 2.5V. What is wrong with the diode?

39. The diode in Figure 3–43B has a measured voltage of 4.5V. What is wrong with the diode?

40. The diode in Figure 3–43C has a measured voltage of 0V. What is wrong with the diode?

41. You have built the circuit shown in 3–46 (p. 81). On a dual channel oscilloscope, you look at v_s and v_o simultaneously. When v_s is positive, v_o looks the same as v_s. However, v_o = 0V when v_s is negative.

   (a) Which two diodes could be bad?   **(b)** Has the bad diode failed open or short?

42. You are testing the half-wave rectifier circuit shown in Figure 3–48. The output voltage, v_o, is zero at all times. What is wrong with the diode?

43. You are testing the circuit of Problem 18 (p. 81) with a dual trace oscilloscope, simultaneously observing v_s and v_o. The two voltages are identical. What could be wrong with the diode?

44. The limiter circuit shown in Figure 3–47 has a defect. On your dual trace oscilloscope, the input and output voltages are identical. What could be wrong with the diode?

45. For the circuit as in Figure 3–47, your oscilloscope shows that v_o = 5V. What could be wrong with the diode to cause this DC output?

46. You have built the circuit of Figure 3–28 (p. 67) with a diode that has a *PIV* of 25V. After applying power to the circuit, you find that you have no voltage at the output.

   (a) What is the likely cause of this problem?

   (b) If the diode failed, explain why it was destroyed.

47. The bridge rectifier shown in Figure 3–52 (p. 84) was constructed with a single bridge rectifier unit (see Figure 3–1D) rated at Io = 25A and *PIV* = 50V.

   (a) Switch S was closed when v_s = 0V and rising to its maximum positive value. Which diode will be first to fail? Why did it burn up?

   (b) Switch S was closed when v_s = 0V and falling to its maximum negative peak voltage. Which diode is the first to fail? Why did it burn up?

FIGURE 3–52

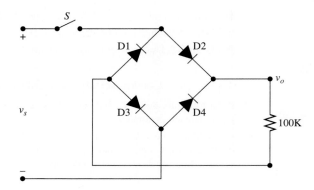

Computer Problems

For these problems use whatever computer simulation package is available to you. Whenever you simulate a sine wave, set the frequency to 60Hz.

48. List the diode models available in your computer program. For each diode model, what is the reverse-biased current and the breakdown voltage? Note that your library might have a diode model that is labeled IDEAL. This is *not* the simple **ON-OFF** model that we presented as ideal in Chapter 2.

49. Simulate the circuit shown in Figure 3–6 (p. 52). Vary the input voltage from −4V to 1V, and find the diode current and voltage. Plot the diode current vs. the diode voltage.

50. Simulate the circuit shown in Figure 3–11 (p. 55).
  (a) Is the diode **ON** or **OFF?** How can you tell from the computer output?
  (b) What are V_D and I_D?

51. Simulate the half-wave rectifier of Figure 3–13 (p. 57).
  (a) Create simultaneous plots of v_s and v_o.
  (b) Increase the input to 50V and repeat.

52. Simulate the full-wave rectifier of Figure 3–16 (p. 59).
  (a) Create simultaneous plots of v_s and v_o when the input is a 2 V sine wave.
  (b) Increase the input to 50V and repeat.

53. Simulate the limiter circuit of Figure 3–20 (p. 61). Create simultaneous plots of v_s and v_o.

54. If your program allows, set the diode breakdown voltage to 25V and repeat Problem 49, increasing the input voltage range to −50V to 1V.

CHAPTER

4

DIODE—REACTIVE CIRCUITS

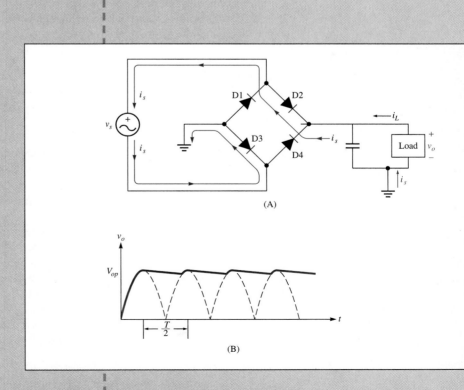

C HAPTER OBJECTIVES

To review capacitor behavior.

To determine the output of filtered rectifiers.

To determine the ripple for filtered rectifiers.

To determine the use of transformers in AC-to-DC conversion.

To determine required diode current ratings.

To analyze the peak detector circuit.

To analyze the AM demodulator circuit.

To analyze the clamping circuit.

To determine the use of the diode in preventing inductor flyback.

To troubleshoot diode-capacitor circuits.

4–1 ■ INTRODUCTION

In previous chapters we showed you how diode-resistor circuits can be used to build recti-fiers and limiters. In this chapter we will show you that by combining diodes with resistors and capacitors you can build even more useful circuits. For example, AC-to-DC conver-sion (Figure 4–1A) can be greatly improved by adding a capacitor to the rectifier circuits. Another example is the use of diode-capacitor circuits in AM radio reception (Figure 4–1B).

FIGURE 4–1

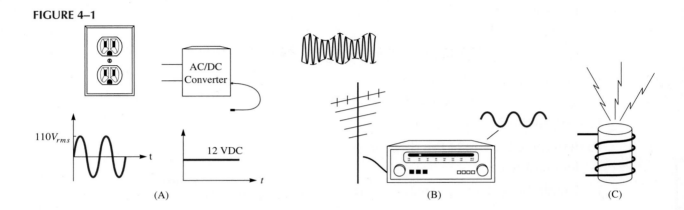

(A)   (B)   (C)

Diodes can also be used to prevent electrical surges. Electrical surges are a common problem in inductive circuits, such as motors and relays. Relays are electromechanical switches that are actuated with electromagnetic coils. If the current driving the coil is acci-dentally interrupted, the magnetic field will collapse, causing a very large induced voltage. This induced voltage can literally burn the relay up (Figure 4–1C). Diodes can be used to prevent these electrical surges from damaging inductive devices.

In this chapter we will describe the behavior and uses of these diode-capacitor and diode-inductor circuits. We begin with a review of capacitor properties and then discuss rectification with filtering, peak detection, AM demodulation, clamping, and voltage dou-bling. We finish the chapter with a review of inductor behavior and the use of diodes in in-ductor circuits.

4–2 ■ THE CAPACITOR AS BATTERY

The essential characteristic of the capacitor is that it stores charge. That is

$$Q = CV$$

where V is the applied voltage; C is the capacitance, measured in farads (F); and Q is the stored charge, measured in coulombs (C). This charge can then be used to provide an elec-trical current to a load. In TV sets, capacitors are charged to very high voltages. The un-wary technician who unplugs the set and immediately sticks his or her fingers into the chassis could have a hair-raising experience if he or she touched these capacitors. The large charge on the capacitor would discharge through the technician, causing pain, con-sternation, and the uttering of very colorful language.

Consider the circuit shown in Figure 4–2A, where we have charged the capacitor to 1 V. Theoretically, if the capacitor is ideal, it will hold this charge forever. In Figure 4–2B, we close the switch and connect the capacitor to the load. The load now draws current from the capacitor; that is, the capacitor discharges through the load. As the capacitor dis-charges, the capacitor voltage decreases. This process continues until the capacitor is com-pletely discharged—its voltage falls to 0V.

FIGURE 4–2

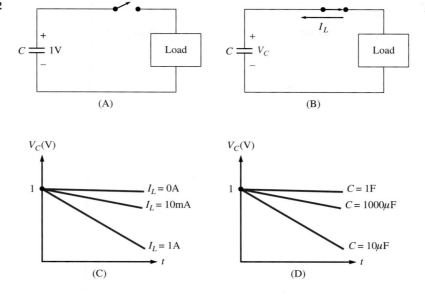

(A)

(B)

(C)

(D)

How fast does the capacitor discharge? The answer depends on the amount of current that is drawn off by the load. The current that the capacitor must supply is given by

$$I_L = \frac{\Delta Q}{\Delta t} = \frac{\Delta(CV)}{\Delta t} = C\frac{\Delta V}{\Delta t}$$

where ΔV is the decrease in the capacitor voltage and Δt is the discharge time. We can find the change in capacitor voltage as

$$\Delta V = \frac{I_L\,\Delta t}{C}$$

While it is not true in reality, for simplicity we will assume the load current is constant. In this case capacitor voltage will decrease linearly. That is, a plot of capacitor voltage versus time will be a straight line. The formula shows that a large load current will cause a large drop in capacitor voltage; a small load current will cause very little voltage drop. This is shown in Figure 4–2C. The size of the capacitor also determines how much the capacitor voltage drops. The larger the capacitor, the less the capacitor voltage drops (Figure 4–2D).

You can see in Figures 4–2C and D that for small load currents and large capacitors the change in capacitor voltage is small compared to the initial voltage on the capacitor. That is, capacitor voltage is approximately constant. In this case, the capacitor acts as a battery. Once charged, the capacitor will hold its voltage. This is the property that we use in diode-capacitor circuits.

Before we turn to the circuits of interest, we remind you that the values we have used for the load currents and capacitances are not realistic. Nonetheless, even though currents in electronic circuits are typically in the milliampere range and capacitors are in the microfarad range, the analysis we present is still valid.

EXAMPLE 4–1

Capacitor Discharge

In the circuit in Figure 4–2, the capacitor is now charged to 10V. Assume the capacitor supplies a constant load current of 1.0mA. Find C so that after 0.5s, the capacitor voltage falls to only 9.9V.

Solution We begin with the voltage equation and rearrange it to solve for C:

$$\Delta V = \frac{I_L \, \Delta t}{C}$$

so

$$C = \frac{I_L \Delta t}{\Delta V}$$

In this problem, the desired voltage drop is

$$\Delta V = 10 - 9.9 = 0.1 \text{V}$$

For $I_L = 1$mA and $\Delta t = 0.5$s

$$C = \frac{I_L \, \Delta t}{\Delta V} = \frac{1 \times 10^{-3} \times 0.5}{0.1} = 5 \times 10^{-3} = 5000 \mu\text{F}$$

DRILL EXERCISE   Find C if the load currents in Example 4–2 (p. 89) are 2mA and 10mA.

Answer $C = 10{,}000\mu\text{F}$, $C = 50{,}000\mu\text{F}$

REVIEW QUESTIONS

1. How does the voltage across a capacitor depend on the charge stored?
2. How does capacitor current depend on capacitor voltage?
3. How do you find the capacitor required to achieve a desired voltage drop in a given time frame?

4–3 ■ HALF-WAVE RECTIFIER WITH CAPACITOR

In previous chapters, we showed how rectification can be used to create a signal with some DC component. The problem with those rectifiers is that, in addition to the DC component, the output signal still has a significant AC component. We can greatly improve the AC-to-DC conversion process by placing a capacitor in parallel with the load.

The capacitor shown in Figure 4–3A helps to filter out the AC component of the rectified signal. To understand how this filter works, remember that a capacitor acts as open circuit for DC and short circuit for AC. As the rectified signal, which contains both DC and AC components, approaches the capacitor, the AC component "sees" a short circuit to ground. The DC component "sees" an open circuit, so it continues on its way. The total output of the filtered rectified signal is, therefore, a pretty good DC signal.

Ripple

Of course, the capacitor does not become a true short for AC signals. To determine how close we have come to a pure DC signal, you must examine the circuit of Figure 4–3 in more detail. This analysis will also provide you with some clues as to how C is chosen. This analysis depends on the fact that capacitors act as batteries for a short time, holding their stored voltage at a constant level.

During the initial part of the positive cycle, the diode is **ON** and the capacitor voltage charges to V_{op} (Figure 4–3B). For a real diode, the capacitor voltage will be a diode drop lower than the peak input voltage. That is

$$V_{op} = V_{sp} - 0.7$$

Note that the load voltage in this circuit is the same as the capacitor voltage.

Now it gets a bit tricky. The input voltage reaches its peak and starts to decrease. At the same time, the capacitor is still holding its maximum charge. That is, capacitor voltage stays at V_{op}. Because the capacitor voltage is now greater than the source voltage, the cathode of the diode is at a higher voltage than the anode. The diode, therefore, opens, and we are left with a simple capacitor discharging circuit, in which the capacitor supplies voltage to the load (Figure 4–3B).

FIGURE 4–3
Half-wave rectifier with filter

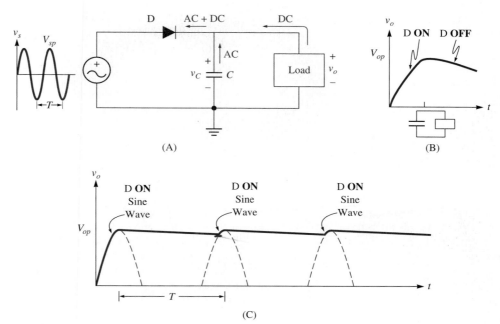

Left on its own, the capacitor will eventually discharge completely. However, during this time the source voltage will increase again until it is larger than the capacitor discharge voltage. Once again, the anode to cathode voltage is positive, turning the diode **ON.** The capacitor voltage will recharge to V_{op}. This process repeats during each cycle, producing the wave form shown in Figure 4–3C.

The final output is much closer to a pure DC signal than the unfiltered rectified signal. The DC value is approximately V_{op}. Note that we still have some variation in the output voltage. This variation is the AC component of the signal and is known as the **ripple.** The amount of ripple allowed in a rectified voltage is a key design specification. For example, a rectified signal intended to drive a DC motor can have a fair amount of ripple. A rectified signal intended to supply the DC power to a low-voltage digital device must have very little ripple.

Finding a formula for ripple can be quite daunting when you consider that we have to analyze the wave form of Figure 4–3C. To simplify the derivation, we assume that the load current is constant. This allows us to assume that the capacitor discharge is a straight line, as discussed in the previous section. Furthermore, in a well-designed rectifier the ripple is small, so we can ignore the recharging time. These two assumptions lead to the simple straight-line approximation of the ripple that is shown in Figure 4–4 (p.92).

The capacitor discharge voltage is given approximately by

$$\Delta V \approx \frac{I_L \, \Delta t}{C} = \frac{I_L T}{C}$$

where $\Delta t = T$, the period of the input signal. The period of a signal is the inverse of the frequency (f) of the signal, where frequency is given in Hz; that is, $T = 1/f$. If the load is a resistor (Figure 4–4B), we can use Ohm's Law to find the load current. While the actual load current will vary slightly, we assume that it is constant, which gives us

$$I_L \approx \frac{V_{op}}{R}$$

where V_{op} is the approximate DC value of the rectified signal. Substituting for the load current in the ripple equation, we get

$$\text{ripple} \approx \Delta V = \left(\frac{V_{op}}{R}\right)\left(\frac{T}{C}\right) = \frac{V_{op}T}{RC}$$

FIGURE 4–4

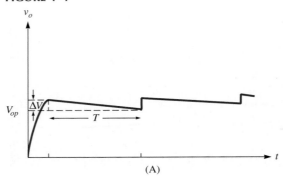

(A)

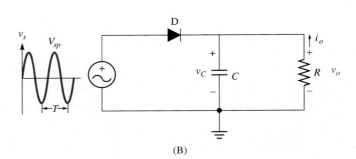

(B)

Ripple is most often expressed as a percentage of the maximum value because the absolute value of the ripple is not very important. After all, a 1V ripple is insignificant if the maximum voltage is 100V, but it is very significant if the maximum voltage is 2V. As a percentage, ripple is given by

$$\text{ripple (\%)} = 100\,\frac{\Delta V}{V_{op}}\% = 100\,\frac{\dfrac{V_{op}T}{RC}}{V_{op}}\% = 100\,\frac{T}{RC}\%$$

Consider a filtered half-wave rectifier (see Figure 4–4) with the following parameters: $V_{op} = 160V$, $T = 16.7\text{ms}$, $R = 100\Omega$, and $C = 1000\mu\text{F}$. Because the input voltage is very large, we ignore the 0.7V diode drop. The ripple in the rectified signal is

Reminder:

$$V_{rms} = \frac{V_p}{\sqrt{2}} \text{ for a}$$

sine wave.

$$\text{ripple (\%)} = 100\,\frac{0.0167}{100 \times 1000 \times 10^{-6}} = 16.67\%$$

To decrease the ripple, we would normally use a larger capacitor.

EXAMPLE 4–2
Half-Wave
Rectifier with
Capacitor I

Referring to Figure 4–4B, sketch the output and find the ripple given the following:

$$V_s = 110\text{VRMS } 60\text{Hz } (T = 16.7\text{ms}), R = 1\text{K}\Omega, C = 1000\mu\text{F}$$

You may ignore the diode voltage drop.

Solution   The output voltage peak equals the input voltage peak and is given by

$$V_{op} = \sqrt{2}\ V_{rms} = \sqrt{2} \times 110 = 156\text{V}$$

The ripple and ripple (%) are

$$\text{ripple} = \Delta V = \frac{V_p\ T}{RC} = \frac{156 \times 0.0167}{1\text{K} \times 1000 \times 10^{-6}} = 2.61\text{V}$$

$$\text{ripple (\%)} = 100\ \frac{T}{RC} = 100 \times \frac{0.0167}{1\text{K} \times 1000 \times 10^{-6}} = 1.67\%$$

EXAMPLE 4–3
**Half-Wave
Rectifier with
Capacitor II**

Find a new capacitor to reduce the ripple in Example 4–2 to 0.5%.

Solution   To find a capacitor to achieve a given ripple, rearrange the ripple equation to solve for *C*.

$$C = 100\ \frac{T}{\text{ripple (\%)}\ R} = 100\ \frac{0.0167}{0.5 \times 1\text{K}} = 0.00334\text{F} = 3340\mu\text{F}$$

Note that capacitor values are given in μF.

DRILL EXERCISE

Choose a capacitor to reduce the ripple in the previous example to 0.1%. Find the ripple voltage (ΔV).

Answer $C = 16700\mu\text{F}$, $\Delta V = 0.156\text{V}$

rms Ripple Factor

Ripple is very often defined by comparing the rms value of the voltage ripple to the DC value of the voltage. The rms value of ripple is given approximately by

$$\Delta V_{rms} = \frac{\Delta V}{\sqrt{3}}$$

(Because we divide by $\sqrt{3}$, instead of $\sqrt{2}$, the ripple is best modeled with a triangular rather than a sine wave.) The rms ripple factor is given by

$$\text{rms ripple factor} = \frac{ripple}{\sqrt{3}} = \frac{V_{op}\ T}{\sqrt{3}RC}$$

**REVIEW
QUESTIONS**

1. How does a capacitor behave for DC?
2. How does a capacitor behave for AC?
3. How can we use these properties to improve AC-to-DC conversion?
4. What is ripple?
5. How do you find ripple?
6. How do you find ripple (%)? Why would you want to find it?
7. How do you choose a capacitor to meet a given ripple specification?
8. How is the rms ripple factor defined? Why do we divide by $\sqrt{3}$?

4-4 ■ THE FULL-WAVE RECTIFIER WITH CAPACITOR

Figure 4–5A shows the most common circuit used to implement a filtered full-wave rectifier. Remember, the full-wave rectifier makes use of both halves of the input cycle. Although it looks different from the bridge circuit shown in Chapter 3, it is in fact the same. On the positive input cycle, the source current goes through D3 to ground. The return path to the source is through the load, in the positive direction, and D2 (Figure 4–5A). That is, D2 and D3 are **ON**, and D1 and D4 are **OFF.**

FIGURE 4–5
Full-wave rectifier with filter

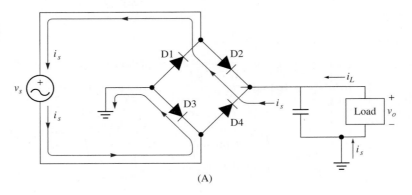

(A)

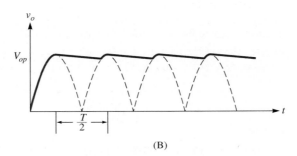

(B)

On the negative input cycle, the source current path goes through D1 to ground. Again, on the return to the source through D4, the path enters the load in the positive direction. During this half of the input cycle, D1 and D4 are **ON,** and D2 and D3 are **OFF.**

The filtered signal shown in Figure 4–5B looks very much like the output of the filtered half-wave rectifier. In fact, the only difference is that the time between successive peaks is now T/2. This means, that all else being equal, the ripple from a full-wave rectifier has one-half the ripple from a half-wave rectifier.

$$\text{ripple} = \Delta V = \frac{I_L T}{2C}$$

We remind you that T is the period of the input signal. Because we get two positive half-cycles during this time, the period of the full-wave rectified output is $T/2$. For a resistive load, the ripple and ripple (%) are given by

$$\text{ripple} = \Delta V = \frac{V_{op} T}{2\,RC}$$

and

$$\text{ripple (\%)} = 100\,\frac{T}{2\,RC}$$

EXAMPLE 4–4
Full-Wave Rectifier with Capacitor

Referring to Figure 4–5, find a capacitor that will produce a filtered signal with a 1% ripple, given the following: $v_s = 308\text{VAC}$ with a period of 0.02s and $R = 50\Omega$.

Solution Solving the ripple equation for C gives us

$$C = 100\,\frac{T}{\text{ripple (\%)}\,2R} = 100\,\frac{0.02}{1 \times 2 \times 50} = 0.02\ \text{F} = 20{,}000\mu\text{F}$$

You can see that even with the full-wave rectifier, we need a very large capacitor to achieve a small ripple when the load resistance is small.

Rectifier with Transformer

Up to this point, the rectifiers we have discussed provide a DC output voltage that is close to the peak value of the AC input voltage. Typically, this AC voltage is supplied from the wall socket in your home or at work. This voltage is approximately 110VRMS (AC voltages are usually specified in rms). The peak AC voltage that is available, therefore, is 156V. This means that the rectifiers discussed will provide approximately 156VDC.

While large DC motors might require such a large DC voltage, most modern electronics would be fried by 156VDC. Many of these devices require between 5 and 50VDC for operation. The best way to reduce the large DC voltage produced by the basic rectifier is to reduce the input voltage. We can use transformers to reduce the AC input voltage to a usable level.

A transformer is shown in Figure 4–6A. As you might recall from your AC circuits course, the output voltage of the transformer is given by

$$v_2 = \frac{v_1}{n}$$

where $n{:}1$ is the turns ratio of the transformer. If the output voltage of a transformer is smaller than the input voltage, the transformer is known as a **step-down transformer.** The output of the step-down transformer is now connected to the rectifier (Figure 4–6B). For example, if you want to produce a 5VDC output, you first need to reduce the 156V peak voltage to 5V peak. To find the turns ratio you would need for your step-down transformer, you solve the transformer equation for n:

$$n = \frac{v_1}{v_2} = \frac{156}{5} \approx 31$$

With a 31:1 turns ratio step-down transformer, we can produce the required 5VDC with the bridge rectifier shown in Figure 4–6B.

FIGURE 4–6

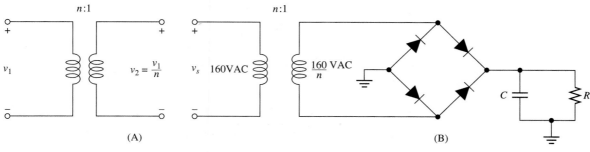

(A)   (B)

Note that in the following exercises we ignore the internal resistance of the transformer. Transformer internal resistance, as well as diode forward resistance, can have a significant impact on the ripple of the rectified voltage.

EXAMPLE 4–5
Bridge Rectifier with Transformer

Given a driving voltage of 160VAC at 60Hz (T = 16.7ms), and a load resistor of 100Ω, build a rectifier that will provide approximately 8VDC with 5% ripple.

Solution Use the circuit shown in Figure 4–6 (p. 95). The first step in this design is to choose the proper step-down transformer. To achieve the desired output of 8V, you need to reduce the driving voltage of 160V. To find the proper transformer turns ratio (n:1), divide the driving voltage by the desired voltage:

$$n = \frac{160}{8} = 20$$

The final part of this design is to choose the capacitor in the rectifier. To find the required capacitor, use the full-wave rectifier ripple formula:

$$C = 100\ \frac{T}{\text{ripple (\%)}\ 2\ R} = 100\ \frac{0.0167}{5 \times 2 \times 100} = 0.00167 = 1670\mu\text{F}$$

DRILL EXERCISE

Given the transformer circuit in Figure 4–6 and the following information:

$$V_s = 115\text{RMS at 60Hz},\ n = 10,\ C = 1000\mu\text{F},\ R = 1\text{K}\Omega$$

find the peak AC input voltage, the DC output voltage, and ripple (%).

Answer V_{sp} = 163V, V_o = 16.3VDC, ripple (%) = 0.835%

The Center-Tapped Transformer

As long as we are going to use a transformer, we can build a different full-wave rectifier, one that uses only two diodes. This circuit uses a center-tapped transformer (Figure 4–7). The center-tapped transformer acts as though it has a single primary and two secondaries. The coils are wound so that when the primary is energized, the currents in both secondaries travel in the same direction.

Returning to the circuit in Figure 4–7, let us assume that on the positive half-cycle current (when current goes from bottom to top in the primary) goes from top to bottom in both secondary coils. This means that on the positive half of the input cycle, diode D1 is **ON** and D2 is **OFF**. The output current path is through D1 and into the bottom of the load, creating a positive output half-cycle. When the input reverses direction, the secondary coil

FIGURE 4–7
Full-wave rectifier with center-tapped transformer

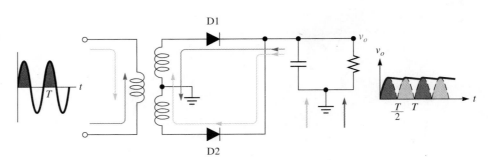

currents also reverse; output current is now moving from bottom to top in the secondary coils. In this case, D2 is **ON** and D1 is **OFF.** The output current path is through D2 and, again, into the bottom of the load, creating the other positive output half-cycle.

Voltage level control is achieved by choosing the proper turns ratio between the primary and secondary coils. Because the output uses only one secondary coil at a time, if we need a ten-to-one ratio, each secondary coil has a 10:1 turns ratio with the primary. The filtering properties of the capacitor in this circuit give the same results as in the bridge rectifier.

EXAMPLE 4–6
Center-Tapped
Transformer
Rectifier

Repeat Example 4–5, using a center-tapped transformer.

Solution   We now use the circuit of Figure 4–7. Since the design specifications have not changed, the solution in Example 4–4 still applies. That is

$$n = 20 \quad \text{and} \quad C = 1670\mu F$$

The difference in this circuit is that each half of the transformer secondary must have a 20:1 turns ratio with the primary.

REVIEW QUESTIONS

1. What are the ripple and ripple (%) formulas for the full-wave rectifier?
2. How does the output voltage of a transformer depend on the input voltage?
3. If the output voltage is smaller than the input voltage, what do we call the transformer?
4. Given an input voltage and a desired output voltage, how do you find the required turns ratio of a transformer?
5. Draw the transformer-driven, filtered bridge rectifier.
6. What is the advantage of the center-tapped transformer full-wave rectifier?

4–5 ■ DIODE CURRENT RATINGS

To choose the right diode for a rectifier, you must know what currents the diode will have to handle.

Half-Wave Rectifier

The load voltage and diode current for a half-wave rectifier are shown in Figure 4–8 (p. 98). You can see that there is an initial surge of diode current as the capacitor charges. After that a repetitive train of diode current pulses correspond to the on-time of the diode. We are interested in three current measures:

1. the average diode current ($I_{D_{ave}}$),
2. the nonrepetitive peak surge current ($I_{pk_{nr}}$),
3. the repetitive peak surge current (I_{pk_r}).

The easiest way to find the average diode current is to realize that this current is approximately equal to the load current. That is

$$I_{D_{ave}} \approx I_L$$

FIGURE 4–8

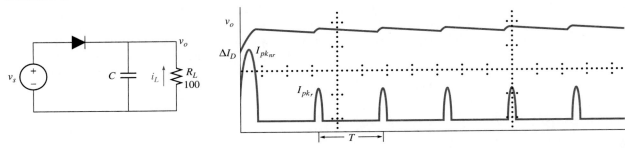

We want to choose a diode that can handle this average current. Then, to provide a safety factor, we actually would choose a diode that can handle an even larger average current.

The nonrepetitive peak surge current occurs during the initial charging of the capacitor. Because the capacitor is initially uncharged, the voltage across it is initially 0V. Therefore, initially the diode is effectively connected to ground. The current in the diode is then determined by the internal resistance of the source (R_s) and the forward resistance of the diode (R_F).

$$I_{pk_{nr}} \approx \frac{V_{sp} - 0.7}{R_s + R_F}$$

where V_{sp} is the peak value of the driving voltage and 0.7V is the voltage drop across the diode.

If we ignore the initial current surge, the average value of the diode current is the average value of the repetitive current surges. The average value of the current surges is approximately

$$I_{D_{ave}} = I_L \approx \frac{I_{pk_r}}{2} \frac{t_{on}}{T}$$

where t_{on} is the time during which the diode conducts (see Figure 4–8). We can rearrange this formula to find

$$I_{pk_r} \approx 2 I_L \frac{T}{t_{on}}$$

If the capacitor is larger, the on-time of the diode decreases and the repetitive peak surge current increases.

EXAMPLE 4–7
Diode Currents

The drive signal in Figure 4–8 is a 50VAC 60Hz voltage. Find the average diode current, the peak nonrepetitive surge current, and the peak repetitive surge current. The on-time for the diode is $t_{on} = 1.55$ms. The source and diode resistances are $R_S = 1\Omega$ and $R_F = 2\Omega$.

Solution
The average (DC) value of the load current is given approximately by

$$I_L \approx \frac{V_{op}}{R_L} = \frac{V_{sp} - 0.7}{R_L} = \frac{49.3}{100} = 0.493\text{A}$$

The nonrepetitive peak current is given approximately by

$$I_{pk_{nr}} = \frac{V_S - 0.7}{R_s + R_F} = \frac{49.3}{1 + 2} = 16.4\text{A}$$

Because the period of the driving voltage is $T = \dfrac{1}{60} = 16.7\text{ms}$, the repetitive peak current is given by

$$I_{pk_r} = 2\,I_L\,\frac{T}{t_{on}} = 2 \times 0.493 \times \frac{16.7\text{ ms}}{1.55\text{ ms}} = 10.6\text{A}$$

The conduction time for the diode (t_{on}) can be found with the help of your calculator from

$$t_{on} = \frac{T}{4} - \frac{T}{2\,\pi}\,\sin^{-1}\left(\frac{V_{op} - \Delta V}{V_{op}}\right)$$

Note: Be sure that when you evaluate the inverse sine, your calculator is set in radian mode.

EXAMPLE 4–8
Determining the Diode On-Time

A half-wave rectifier with a 60 Hz input produces an output with a peak value of 75V and a ripple of 3V. What is the on-time for the diode?

Solution The rectifier output has a period of

$$T = \frac{1}{f} = \frac{1}{60} = 16.7\text{ms}$$

The on-time is

$$t_{on} = \frac{T}{4} - \frac{T}{2\,\pi}\,\sin^{-1}\left(\frac{V_{op} - \Delta V}{V_{op}}\right) = \frac{16.7\text{ ms}}{4} - \frac{16.7\text{ ms}}{2\,\pi}\,\sin^{-1}\frac{72}{75} = 0.754\text{ ms}$$

Full-Wave Rectifiers

Figure 4–9 (p. 100) shows the center-tapped transformer and bridge full-wave rectifiers. In the transformer rectifier each diode supplies one-half of the total average current to the load. In the bridge rectifier each pair of **ON** diodes supplies one-half of the average load current.

Also note that the frequency of the full-wave rectified voltage is twice the frequency of the half-wave rectified voltage. The period of the full-wave rectified voltage is $T/2$.

The diode currents for the **ON** diodes in a full-wave rectifier are

	Transformer Rectifier	**Bridge Rectifier**
$I_{D_{ave}}$	$\dfrac{I_L}{2}$	$\dfrac{I_L}{2}$
$I_{pk_{nr}}$	$\dfrac{V_{sp} - 0.7}{R_S + R_F}$	$\dfrac{V_{sp} - 1.4}{R_S + 2R_F}$
I_{pk_r}	$I_L\dfrac{T}{t_{on}}$	$I_L\dfrac{T}{t_{on}}$

FIGURE 4–9

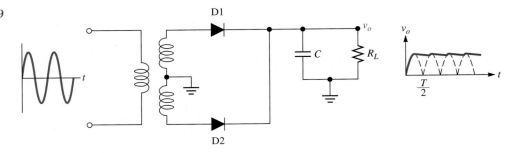

FIGURE 4–10

GENERAL PURPOSE RECTIFIERS

Type	Io (A)	Ifsm (A)	Vrrm	Package
MR751	6	400	100	194
MR754	6	400	400	194
MR756	6	400	600	194
MR758	6	400	800	194
MR760	6	400	1000	194
MR1120	12	300	50	DO–4
1N1199A	12	300	50	DO–4
MR1121	12	300	100	DO–4
MR1121R	12	300	100	DO–4
1N1200	12	300	100	DO–4
N1200A	12	300	100	DO–4
MR1122	12	300	200	DO–4
1N1202	12	300	200	DO–4
1N1202A	12	300	200	DO–4
MR1124	12	300	400	DO–4
1N1204	12	300	400	DO–4
1N1204A	12	300	400	DO–4
1N1204RA	12	300	400	DO–4
MR1126	12	300	600	DO–4
MR1126R	12	300	600	DO–4
1N1206	12	300	600	DO–4
1N1206A	12	300	600	DO–4
1N1206R	12	300	600	DO–4
1N1206RA	12	300	600	DO–4
MR1126	12	300	600	DO–4

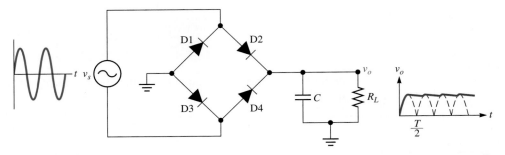

Because there are two **ON** diodes in the bridge rectifier, we subtract two diode voltage drops from the input voltage and include two diode forward resistances. For the center-tapped transformer rectifier, V_{s_p} is the transformer secondary peak voltage. The source resistance (R_s) is the resistance of the transformer secondary.

Diode Current Ratings

Figure 4–10 shows the data sheet for diodes used in rectifier circuits. Typically, only two maximum current ratings are given: Average Rectified Forward Current (I_o) and Nonrepetitive Peak Surge Current (I_{FSM}).

To use this data, we would apply a safety factor of, for example, 20 percent to these maximum values. We can use the 1N1204 if

$$I_{D_{ave}} < \frac{I_o}{1.2} = \frac{12}{1.2} = 10A$$

and

$$I_{pk_{nr}} < \frac{I_{FSM}}{1.2} = \frac{300}{1.2} = 250A$$

REVIEW QUESTIONS

1. What three diode currents are of interest in rectifier circuits?
2. How do you find the average diode current in the half-wave rectifier?
3. What causes the nonrepetitive peak surge current? What is its value for the half-wave rectifier?
4. How do you calculate the repetitive peak surge current?
5. How do you find the diode on-time?
6. How are the three diode currents defined for the center-tapped transformer full-wave rectifier?
7. How are the three diode currents defined for the bridge full-wave rectifier?
8. What diode maximum current ratings are typically given by a manufacturer?

4–6 ■ THE PEAK DETECTOR

A circuit that is similar to the rectifier-filter circuit is the **peak detector.** Consider the following problem. We want to determine the maximum temperature reached by a patient during the day. Into the patient we have inserted a temperature probe that provides an electrical output that is proportional to temperature. We will continuously measure the temperature and every 15 minutes record the maximum temperature that occurred during the 15-minute interval. A typical 15-minute temperature measurement is shown in Figure 4–11A.

The measurement system shown in Figure 4–11B consists of a diode, capacitor, and recording device. The recording device is represented with a resistor (R_m). This circuit is identical to the half-wave rectifier with a capacitor. It functions in the same manner.

FIGURE 4–11
Peak detector

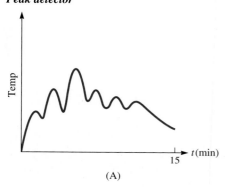

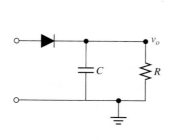

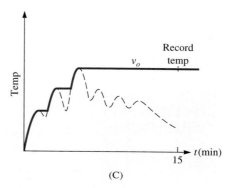

(A)   (B)   (C)

When the input initially rises, the diode turns **ON** and the capacitor charges to the first peak. When the input decreases, the diode turns **OFF** and the capacitor holds the first peak value. As the input rises again, the diode will eventually turn **ON** again and charge the capacitor to the new peak value. The capacitor will charge to the largest peak value, at which time the diode will permanently turn **OFF.** The last value recorded will thus be the maximum value of the input signal. The only practical difference between this circuit and the half-wave rectifier circuit is that the capacitor in the peak detector must hold its voltage for a longer time. The capacitor in the half-wave rectifier only needs to hold its charge for one period of the input signal (usually 0.0167s), while the capacitor in the peak detector must hold its value for the duration of the measurement (Figure 4–11C).

EXAMPLE 4–9
Peak Detector

(a) Draw the output of a peak detector if it is measuring the signal shown in Figure 4–12A.
(b) The recording device has a resistance of $R_m = 1\text{M}\Omega$ and must lose less than 1% of the peak signal per hour. Find the required capacitance.

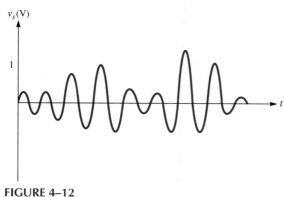

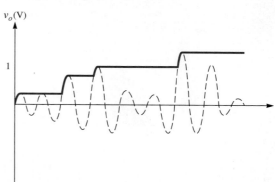

FIGURE 4–12

Solution

(a) See Figure 4–12B (p. 101).

(b) Because the detector and the filtered half-wave rectifier perform the same basic function, we can use the ripple formula of the half-wave rectifier to find the required capacitance.

$$C = 100 \frac{T}{ripple \ (\%) \ R_m}$$

where the ripple (%) represents the allowable signal loss and T represents the time the detector must hold the signal nearly constant. Because a 1% loss is allowed, and because the detector is required to hold the measurement for 1 hour, or 3600 seconds,

$$C = 100 \frac{T}{ripple \ (\%) \ R_m} = 100 \frac{3600}{1 \times 1 \times 10^6} = 0.36\text{F}$$

This system requires a very large capacitor.

REVIEW QUESTIONS

1. What is the purpose of the peak detector?

2. Draw a peak detector circuit.

3. What circuit does the peak detector resemble?

4. What is the major difference between the peak detector and the half-wave rectifier?

4–7 ■ THE DIODE AM DEMODULATOR

A circuit that is related to the peak detector can be used to build a simple amplitude modulation (AM) radio receiver. Long before many of you were born, kids were building simple crystal radios that had a tuned antenna and a circuit to separate the audio signal from the carrier. This separation process is known as AM **detection,** or AM **demodulation.**

At the AM broadcast station the audio signal is combined with a carrier wave to produce an AM signal (Figure 4–13). We see that the AM signal is composed of a high-frequency carrier wave that is modulated by the lower-frequency audio signal. To emphasize the low-frequency content of this wave, we have drawn lines that connect the peaks to each other, and the valleys to each other. These lines form what is known as the **envelope** of the wave.

FIGURE 4–13

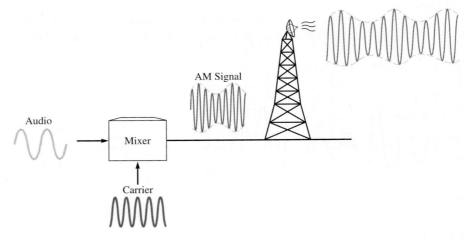

You can see that either the top or the bottom envelope of the wave represents the original signal. The detector has two jobs: (1) eliminate the redundant bottom half of the AM signal; and (2) remove the high-frequency carrier from the remaining wave. The detector output is then a replica of the transmitted audio signal.

The diode circuit shown in Figure 4–14 is a simple detector that rectifies the AM signal, cutting off the bottom half. The capacitor acts as a filter that looks like a short circuit when frequency is high and looks like an open circuit when frequency is low. Therefore, the high-frequency carrier component of the wave is shunted to ground, while the lower-frequency audio signal passes to the output.

FIGURE 4–14
AM detector

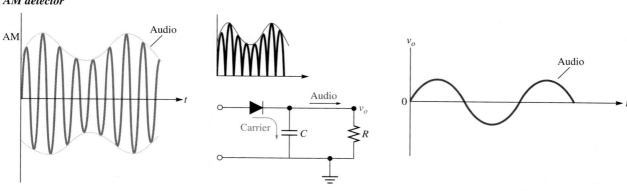

You may have noticed that this circuit is the same as the rectifier-filter circuit used in AC-DC conversion (see Figure 4–3, p. 91). In fact, the detector circuit works the same way the half-wave rectifier-filter circuit works.

To see how the detector works, we show an expanded version of the AM signal (including envelope) in Figure 4–15A. During the initial positive cycle of the AM signal, the diode in Figure 4–14 turns **ON** and the capacitor charges to the first peak value. As soon as the AM signal decreases, the diode is reversed biased and turns **OFF.** The capacitor now discharges through the resistor. The capacitor voltage continues to decrease until the AM signal catches up to it. At this point, the diode turns **ON** again, the capacitor charges to the new peak, and the process repeats.

It may seem that we want a very large capacitor so that the capacitor voltage will fall very slowly. However, in Figure 4–15B we show that this is not true. All is well at the be-

FIGURE 4–15

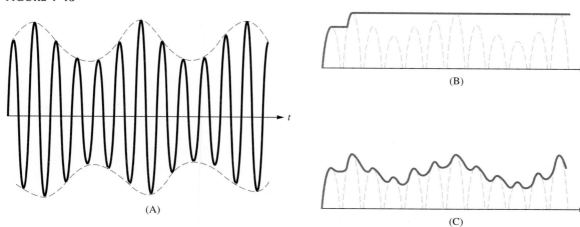

ginning; the capacitor voltage tracks the rise in the AM signal until the peak voltage is reached. However, if the capacitor is large and holds this voltage indefinitely, we get the result shown. We never track the downward movement of the AM signal.

The correct design criteria for the detector is to choose a capacitor so that the RC circuit will discharge fast enough to follow the falling slope of the audio envelope (Figure 4–15C, p. 103). We assume that the audio signal is a pure tone; that is, the audio signal is a sine wave with a period of T_a. The maximum slope of the envelope representing the audio signal (m_a) is given by

$$m_a = \frac{2\pi V_p}{T_a}$$

where V_p is the peak value of the envelope representing the audio signal.

This is the slope that the capacitor discharge must match. The slope of the discharging capacitor voltage (m_{exp}) is given by

$$m_{exp} = \frac{V_p}{RC}$$

Equating the two slopes gives us

$$\frac{V_p}{RC} = \frac{2\pi V_p}{T_a}$$

Solving for C gives us

$$C = \frac{T_a}{2\pi R}$$

The audio signal that is the highest moves the fastest; therefore, we use this frequency to select C.

EXAMPLE 4–10
AM Demodulation

Find the capacitor needed to demodulate an AM signal that has a maximum frequency of 10KHz. The demodulator has a load resistance of 1 KΩ.

Solution   Use the circuit of Figure 4–14 (p. 103). First find the period of the audio signal:

$$T_a = \frac{1}{f_a} = \frac{1}{10K} = 1 \times 10^{-4}\text{s}$$

Now, find C:

$$C = \frac{T_a}{2\pi R} = \frac{1 \times 10^{-4}}{2\pi \times 1K} = 0.0159\mu\text{F}$$

DRILL EXERCISE

KDOG is broadcasting a special program for man's best friend. The audio signal now has a frequency of 50KHz. Repeat the above example for this frequency.

Answer $T_a = 0.2 \times 10^{-4}\text{s}$, $C = 0.00318\mu\text{F}$

1. What is AM detection?

2. What is another name for "detection"?

3. Draw a simple AM detection circuit.

4. What other circuit does the detector circuit resemble?

5. What is the maximum slope of an audio signal?

6. What is the slope of an RC discharge?

7. How do you select a capacitor for the AM demodulator?

4–8 ■ DIODE CLAMPING CIRCUITS

The clipping circuits discussed in the previous chapter can prevent a signal from exceeding a given voltage. The problem with these circuits is that they change the input wave form. A **clamping circuit** also prevents a signal from exceeding a desired voltage, but it does not change the wave shape. Instead of clipping off part of the signal, the clamper works by shifting the entire signal up or down as required.

The circuit shown in Figure 4–16 gives you a conceptual idea of how the clamper works. The problem: Without changing the shape of the input wave (10V sine wave), prevent the output voltage from going above 6V. The solution: Connect a 4V battery in series with the source. The maximum output is now $10 - 4 = 6V$. The only problem with this circuit is that if the input increases to 11V, the output will increase to 7V. What we need is a circuit that automatically adjusts the battery voltage to ensure the output will never rise above 6V.

FIGURE 4–16

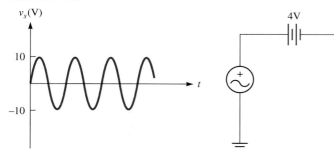
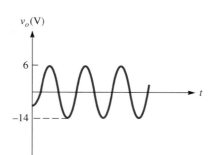

The circuit shown in Figure 4–17 (p. 106) is a diode clamper circuit in which the capacitor serves as the adjustable "battery." As you know, capacitors tend to hold the voltage to which they have been charged. Can you see that the output voltage of the clamper can never be greater than V_B? If the output tries to get larger than V_B (ignoring the 0.7V diode turn-on voltage), current will flow from the battery to the source, turning the diode **ON.** When the diode is **ON,** it becomes a short circuit, limiting the output to V_B. That is, the output voltage can never be higher than V_B volts. The battery, therefore, determines the clamping level of the circuit.

Remember: We are ignoring the 0.7V diode voltage drop.

To simplify the analysis of this circuit, we use the square wave input shown in Figure 4–17A. The diode turns **ON** during the first positive half of the input cycle (Figure 4–17B). The output voltage is now at its upper limit (V_B). The capacitor will quickly charge to the DC value:

$$V_C = V_{sp} - V_B$$

FIGURE 4–17
Clamping circuit

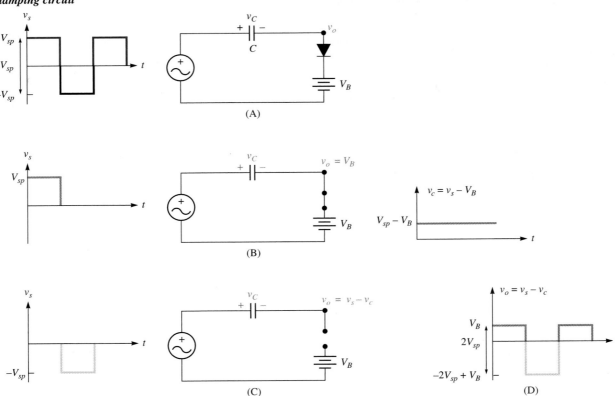

where V_{sp} is the maximum value of the input voltage. During the negative half-cycle, the diode is reversed biased and becomes an open circuit (Figure 4–17C). At this time, the output voltage is found from

$$v_o = v_s - V_C$$

Once the capacitor is charged to V_C, it will tend to hold this value, acting as a battery. This means that the clamper circuit produces an output that is simply a DC shift of the input. The result is shown in Figure 4–17D. The maximum output voltage can be found from

$$V_{op} = V_{sp} - V_C$$

Using the value we previously found for V_C, we find

$$V_{op} = V_{sp} - (V_{sp} - V_B) = V_B$$

So

$$\boxed{V_{op} = V_B}$$

We see that the upper limit of the output signal is equal to the battery voltage. This is the same result we got when we studied the clamper circuit.

Although we used a square wave input to describe the clamper, this circuit will work with all inputs of the appropriate frequency. Because the circuit in Figure 4–17 subtracts a DC level from the input circuit, we call it **negative clamping.**

If the diode and the battery are reversed, we will add a DC level to the input signal. This is known as **positive clamping.** In many electronic circuits, the DC level of a signal is lost as the signal is processed. Clamping can be used to restore the lost DC level. In this case, we call the clamping circuit a **DC restorer.** DC restorers are commonly used in television circuitry.

EXAMPLE 4–11
Clamping Circuits

Find the output wave form, given the input wave forms and the reference battery voltages in Figure 4–18A–D. (Ignore the 0.7V diode turn-on voltage.)

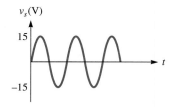

(A)

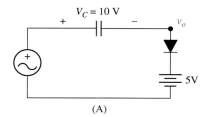

(A)

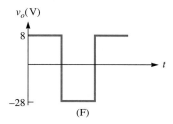

(E)

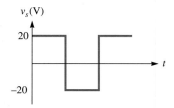

(B)

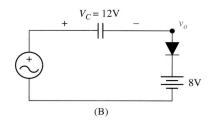

(B)

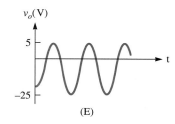

(F)

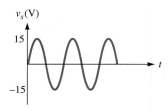

(C)

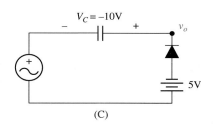

(C)

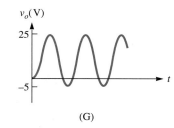

(G)

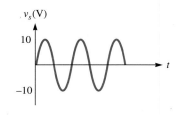

(D)

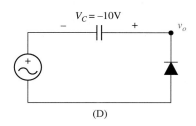

(D)

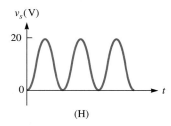

(H)

FIGURE 4–18

Solution

(a) In this negative clamping circuit, the input signal is shifted down by 15 − 5 = 10V (Figure 4–18E).

(b) In this negative clamping circuit, the input signal is shifted down by 20 − 8 = 12V (Figure 4–18F).

(c) In this positive clamping circuit, the input signal is shifted up by 15 − 5 = 10V (Figure 4–18G).

(d) In this positive clamping circuit, the input signal is shifted up by 10 − 0 = 10V (Figure 4–18H).

DRILL EXERCISE   Sketch the output for each of the clamper circuits in Figure 4–19.

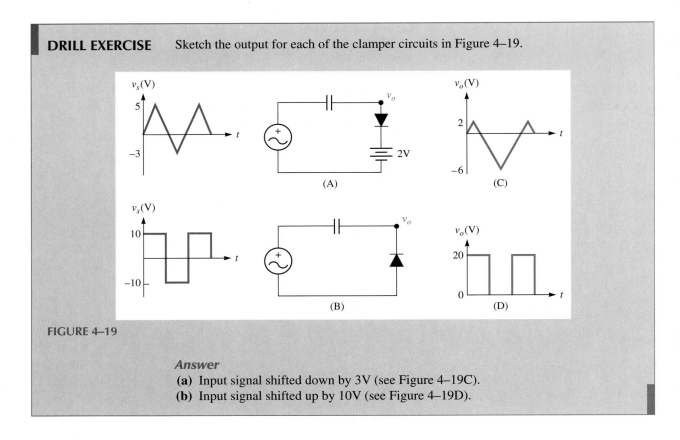

FIGURE 4–19

Answer

(a) Input signal shifted down by 3V (see Figure 4–19C).

(b) Input signal shifted up by 10V (see Figure 4–19D).

REVIEW QUESTIONS

1. What do clipping and clamping circuits have in common?
2. What are the differences between clipping and clamping circuits?
3. Draw a clamping circuit.
4. What purpose does the capacitor serve?
5. What voltage do we see across the capacitor?
6. What is the output of a clamper?

4–9 ■ THE VOLTAGE DOUBLER

With a diode and a capacitor, we can build a rectifier to create an approximate DC voltage from an AC voltage. The DC level we can create will be equal to the peak value of the AC voltage. We can use additional capacitors and diodes to build a circuit that will create a DC voltage that is a multiple of the peak AC input voltage. The basic idea in these circuits is to charge each capacitor to the peak input voltage and to arrange the capacitors so that their stored voltages will add.

The simplest of these circuits, shown in Figure 4–20A, is the **voltage doubler.** Note that the output voltage is taken across the second capacitor. This circuit is too complicated for an exact analysis; however, we can give you a basic idea of how this circuit works.

On the negative half-cycle, D1 is **ON** and D2 is **OFF.** The first capacitor (C1) will charge to

$$V_{C1} = V_{sp}$$

(note the polarity of V_{C1}). During the positive half-cycle, D1 turns **OFF** and D2 turns **ON**. The second capacitor (C2) now charges to

$$V_{C2} = V_{C1} + V_{sp} = 2V_{sp}$$

That is, the DC level of C1 adds to the input signal. The output voltage is now twice the input peak voltage.

We can use an alternative analysis if we rearrange the voltage doubler circuit as shown in Figure 4–20B. You can see now that the voltage doubler is actually a combination of a positive clamper and a rectifier. The positive clamper raises the input signal to a peak value of $2V_{sp}$ (see Figure 4–18D in Example 4–11). The rectifier converts the $2V_{sp}$ AC signal into a DC voltage of $2V_{sp}$.

FIGURE 4–20
Voltage doubler

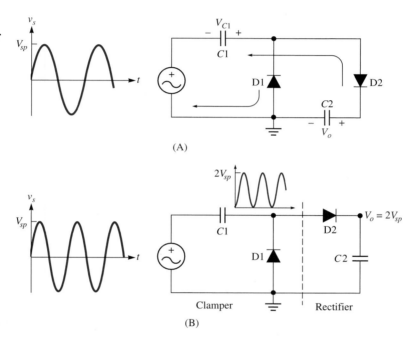

(A)

(B)

4–10 ■ DIODE-INDUCTOR CIRCUIT (FLYBACK)

The inductor is a device that stores energy in its magnetic field (Figure 4–21, p. 110). As current is passed through the inductor, a magnetic field is created. This magnetic field cuts through the windings of the inductor, inducing a voltage. If the current is suddenly removed, the magnetic field abruptly collapses, inducing a very large voltage in the inductor. This is known as **flyback** because the induced voltage is produced in the opposite direction to the original voltage.

To understand flyback, first review the voltage-current equation for the inductor:

$$v = L\frac{\Delta i}{\Delta t}$$

FIGURE 4–21
Inductor flyback

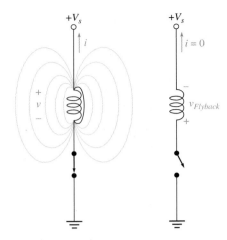

We can use this equation to study the inductor circuit shown in Figure 4–22. The current in an inductor is found from

$$\Delta i = \frac{v\,\Delta t}{L}$$

In this circuit, we applied a constant 12V to the inductor (Figure 4–22A). As the graph shows, the constant voltage creates a current that increases with time. You see that the current grows linearly, which means that the magnetic field also grows linearly.

We now open the switch (Figure 4–22B), which causes the current to drop immediately to 0A. That is, the current drops from I_o to 0A, in 0s. Returning to the original inductor voltage-current equation, we see that the induced voltage at the switch is

$$v_{SW} = L\frac{\Delta i}{\Delta t} = L\frac{I_o}{0} \to \infty$$

When the switch is opened, the magnetic field collapses instantaneously. This rapid collapse of the magnetic field induces a huge voltage across the coil.

In theory, abruptly shutting off the current leads to an infinite spike in voltage (Figure 4–22C). In reality, an inductor is a complicated device that has resistance and capaci-

FIGURE 4–22

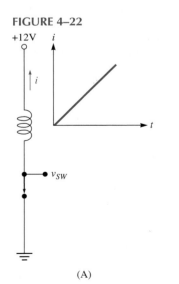

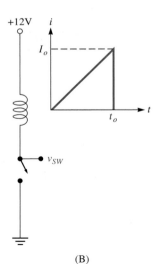

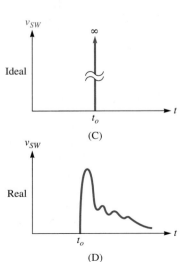

(A)   (B)   (D)

tance, as well as inductance. The actual flyback voltage will reach a large but finite value (Figure 4–22D).

If you drove to school today, you made extensive use of flyback voltages. The ignition coil (a transformer) in your car creates a 20 to 40KV drive to your spark plugs. In this case, flyback voltage is created by switching off the current to the primary windings of the ignition coil. The flyback voltage is then increased through the use of coupled coils; that is, the ignition coil is a step-up transformer.

Flyback voltage can also be used to create very large DC voltages. We simply connect the flyback circuit to a rectifier circuit (Figure 4–23A). In this circuit we periodically open and close the switch to produce a series of flyback pulses. As each pulse peaks, the diode turns **ON** and the capacitor is charged (Figure 4–23B). We need to provide the series of pulses because each pulse doesn't last long enough to fully charge the capacitor. Figure 4–23C shows that it takes several pulses to fully charge the capacitor to its final value. This is a common application in television sets, which require large DC voltages to operate the picture tube.

FIGURE 4–23

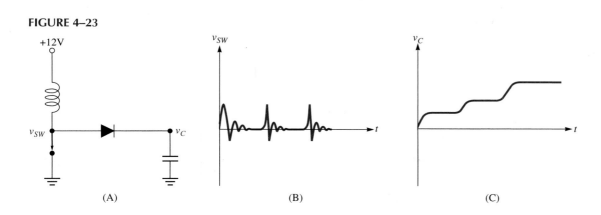

(A) (B) (C)

Flyback Prevention

Flyback voltage is present in all circuits containing inductive devices or properties. For example, motors driven with pulsed inputs will exhibit flyback. In these cases, the large voltage produced by flyback can damage the device. We can prevent flyback and protect inductive devices by connecting a diode across the inductor (Figure 4–24A). In this circuit when the switch opens, the flyback voltage turns the diode **ON** and provides a short circuit path for the inductor current (Figure 4–24B). This prevents the current from falling abruptly to 0A, preventing the flyback voltage from getting large.

FIGURE 4–24

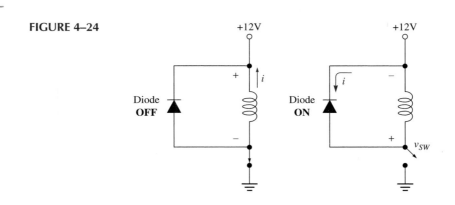

4–11 ■ TROUBLESHOOTING DIODE-CAPACITOR CIRCUITS

All the troubleshooting techniques we discussed in Chapter 3 apply to troubleshooting the diode circuits in this chapter. Here we also have to be concerned with failures of the capacitor in the diode circuit.

The Capacitor

Capacitors can fail open or short. They can fail open because of broken leads within the capacitor or because the electrolytic insulator can dry out. A bad connection to the circuit can also produce the same result as a failed open capacitor.

Capacitors can fail short if there is an internal short between the plates. In a circuit if the capacitor leads are touching, or if an electrolytic capacitor is inserted backwards, the effect will be the same as a failed short capacitor.

If you measure a good capacitor with an ohmmeter, you will get a very large resistance reading. Therefore, a failed short capacitor can be determined with an ohmmeter. Simply connect the ohmmeter leads to the capacitor (observe correct polarity for electrolytic capacitors). If the measured resistance is very low, the capacitor has an internal short. Failed open capacitors are much harder to detect with an ohmmeter. Many DMMs come with the capability to directly test a capacitor.

The best way to determine if a capacitor in a diode circuit has failed is to use the oscilloscope, just as we did in the previous chapter. Knowing how the output of a particular circuit is supposed to look will usually clue you in to what has failed. In the following examples, we will assume that all resistors are good.

EXAMPLE 4–12
Troubleshooting I

Consider the half-wave rectifier and its input-output measurements shown in Figure 4–25. Which device is most likely to have failed? How has it failed?

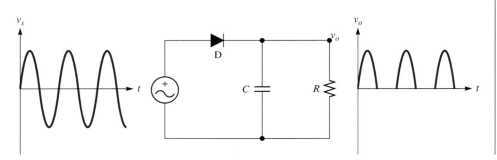

FIGURE 4–25

Solution Because the output signal is rectified, the diode must be good; however, the output voltage is not being held constant. That is, the circuit behaves as though the capacitor is not there. Therefore, the capacitor must have failed open. Replace the capacitor.

EXAMPLE 4–13
Troubleshooting II

Consider the half-wave rectifier and its input-output measurements shown in Figure 4–26. Which device is most likely to have failed? How has it failed?

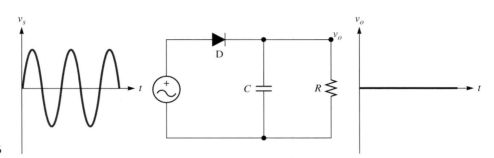

FIGURE 4–26

Solution This circuit has no output. This could be caused either by (a) a failed open diode, in which case no signal gets to the output; or (b) a failed short capacitor, which ties the output to ground. We can tell which device is defective by disconnecting the input source and testing the diode with a meter. **Always remove all voltage sources before using an ohmmeter in a circuit.** If the diode is good, check the capacitor for an internal short and replace it, if necessary.

SUMMARY

- The capacitor stores voltage and acts as a battery for a short period of time.
- AC-to-DC conversion is greatly improved if a capacitor is placed across the load.
- The peak value of a sinusoid is given by $V_p = \sqrt{2}VRMS$.
- The output DC voltage of a rectifier with a capacitor (VDC) approximately equals V_p of the input.
- The ripple (%) of a rectified signal is given by $100\,\dfrac{\Delta V}{VDC}\,\%$.
- The ripple (%) of a half-wave rectified voltage is given by $100\,\dfrac{T}{RC}\,\%$.
- The ripple (%) of a full-wave rectified voltage is given by $100\,\dfrac{T}{2RC}\,\%$.
- A step-down transformer is used to reduce the input VAC.
- Diode currents of interest in a rectifier are the average diode current, the nonrepetitive peak surge current, and the repetitive peak surge current.
- The peak detector holds the maximum value that a signal reaches.
- The AM demodulator extracts the audio signal from the carrier signal.
- The capacitor in the AM demodulator is given by $C = \dfrac{T_a}{2\pi R}$, where T_a is the smallest period of the audio signal.
- Clamping circuits add or subtract a DC level to the input signal.
- The voltage doubler provides a DC level that is twice the input VAC.
- Flyback voltage is the large voltage induced in an inductor when the current is interrupted.
- A diode placed across an inductor can minimize the effect of flyback voltage.

PROBLEMS   *Note:* Except as noted, you may ignore the diode voltage drops in the following problems.

SECTION 4–2   The Capacitor as Battery

1. In Figure 4–27, let $R_1 = 20$ KΩ and assume the capacitor is charged to 20 V.

(a) The switch (S) is closed for 1ms and then reopened. How much does the voltage on the capacitor drop; that is, what is ΔV? Assume the discharge current is constant.

(b) The switch (S) is closed for 5ms and then reopened. How much does the voltage on the capacitor drop; that is, what is ΔV? Assume the discharge current is constant.

2. In Figure 4–27, let $R_1 = 10$KΩ and assume the capacitor is charged to 100V.

(a) The switch (S) is closed for 1ms and then reopened. How much does the voltage on the capacitor drop; that is, what is ΔV? Assume the discharge current is constant.

(b) The switch (S) is closed for 5ms and then reopened. How much does the voltage on the capacitor drop; that is, what is ΔV? Assume the discharge current is constant.

3. In Figure 4–27, the capacitor is charged to 10V. Find R_1 so that the capacitor voltage will drop to 9.9V 2ms after the switch is closed.

4. Let $R_1 = 10$KΩ in Figure 4–27. Find the capacitor required so that 10ms after the switch has closed, the output drops from 50V to 48V.

SECTION 4–3   Half-Wave Rectifier with Capacitor

5. In Figure 4–28, v_s is a 50VAC 60Hz sine wave.

(a) What is V_{op}?   **(b)** What is T?

(c) What is the ripple ΔV?   **(d)** What is the ripple (%)?

6. In Figure 4–28, v_s is a 50VAC 60Hz sine wave. What is the load current?

7. In Figure 4–28, v_s is a 50VRMS 120Hz sine wave.

(a) What is V_{op}?   **(b)** What is T?

(c) What is the ripple ΔV?   **(d)** What is the ripple (%)?

8. In Figure 4–28, v_s is a 50VRMS 120Hz sine wave. Find the load current.

9. In Figure 4–28, v_s is a 100VRMS 60Hz sine wave. Select C to get a ripple (%) of 3%.

10. In Figure 4–28, v_s is a 70VAC 60Hz sine wave. Select C for an rms ripple factor of 5V.

SECTION 4–4   Full-Wave Rectifier with Capacitor

11. Referring to Figure 4–5 (p. 94), draw v_o on the vertical axis and time on the horizontal axis.

(a) When are diodes D2 and D3 **ON?**   **(b)** When are diodes D1 and D4 **ON?**

(c) When are diodes D2 and D3 **OFF?**   **(d)** When are diodes D1 and D4 **OFF?**

(e) During the time when D2 and D3 are **OFF,** what is the maximum reverse-bias voltage that appears across each diode?

FIGURE 4–27

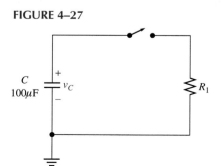

FIGURE 4–28

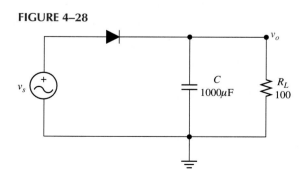

12. For the full-wave rectifier circuit in Figure 4–29, v_s is an 80VAC 60Hz sine wave.

    (a) How often is the capacitor recharged?   **(b)** What is V_{op}?

    (c) What is the approximate I_L?   **(d)** What is ΔV?

    (e) What is the ripple (%)?

    (f) Draw a diagram that shows v_s on the vertical scale and time on the horizontal scale. Label the zero crossing times of v_s on the time scales.

    (g) Draw a diagram of v_o immediately below your drawing of v_s. Be sure the time scales match in both diagrams. On your diagram of v_o show the following: V_{op}, ΔV, T.

    (h) If you consider the ripple on v_o, what is the maximum value of I_L? What is the minimum value of I_L?

FIGURE 4–29

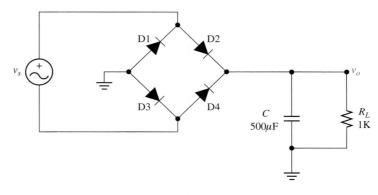

13. For the full-wave rectifier circuit in Figure 4–29, v_s is a 110VRMS 60Hz sine wave.

    (a) How often is the capacitor recharged?   **(b)** What is V_{op}?

    (c) What is the approximate I_L?   **(d)** what is ΔV?

    (e) What is the ripple (%)?

    (f) Draw a diagram that shows v_s on the vertical scale and time on the horizontal scale. Label the zero crossing times of v_s on the time scales.

    (g) Draw a diagram of v_o immediately below your drawing of v_s. Be sure the time scales match in both diagrams. On your diagram of v_o show the following: V_{op}, ΔV, T.

    (h) If you consider the ripple on v_o, what is the maximum value of I_L? What is the minimum value of I_L?

14. In Figure 4–29, v_s is a 110RMS 60Hz sine wave. Choose a capacitor so that ripple (%) = 2%.

15. In Figure 4–29, v_s is a 110RMS 60Hz sine wave. Choose a resistor so that ripple (%) = 5%.

16. For the full-wave rectifier shown in Figure 4–30, the input voltage is a 170V 60Hz sine wave.

    (a) What is the amplitude of v_s? What is the frequency of v_s?

    (b) What is V_{op}?   **(c)** What is T?   **(d)** What is ΔV?

    (e) What is the approximate load current, (I_L)?   **(f)** What is the ripple (%)?

    (g) Draw a diagram of v_o vs. time, observing all requirements given previously for such a figure.

    (h) When are diodes D2 and D3 **ON?**   **(i)** When are diodes D2 and D3 **OFF?**

17. Choose a new capacitor in Figure 4–30 so that ripple (%) = 2%.

FIGURE 4–30

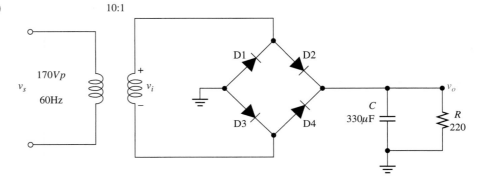

18. For the center-tapped transformer rectifier circuit shown in Figure 4–31, v_s is a 170VAC 60Hz sine wave.

  (a) What is the maximum value attained by v_i?

  (b) What are the maximum values of v_{Ts1} and v_{Ts2} relative to the grounded center tap of the secondary winding of the transformer?

  (c) What is V_{op}?

  (d) What is the rate at which the capacitor is charged from the transformer?

  (e) What is the average value of I_L? **(f)** Draw a diagram of v_o showing V_{op}, ΔV, and T.

FIGURE 4–31

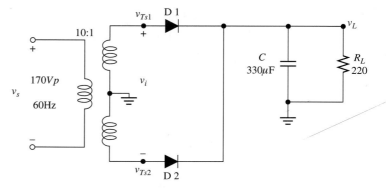

SECTION 4–5 Diode Current Ratings

For Problems 19–23, include the diode voltage drops in your calculations.

19. In Figure 4–28 (p. 114), v_s is a 60VAC 60Hz sine wave.

  (a) What is V_L? **(b)** What is I_L?

  (c) What is $I_{D_{ave}}$? **(d)** Using a 20% safety factor, what I_o is needed here?

  (e) If the source and diode resistances are $R_S = 1\Omega$ and $R_F = 3\Omega$ respectively, what is $I_{pk_{nr}}$?

  (f) Using a 20% safety factor, what I_{FSM} is needed here?

20. A half-wave rectifier with a 60Hz input produces an output that has a peak value of 50V and ripple (%) = 5%. What is the diode on-time?

21. In this problem, you are going to derive the diode on-time formula. As shown in Figure 4–32, when the diode is **ON,** the output voltage is a sine wave

$$\text{diode ON: } v_o = V_{op} \sin \frac{2\pi}{T}t$$

The diode on-time is found from

$$t_{on} = \frac{T}{4} - t_c$$

  (a) Find t_c. (Hint: @ $t = t_c$, $v_o = V_{op} - \Delta V$.) **(b)** Find t_{on}.

FIGURE 4–32

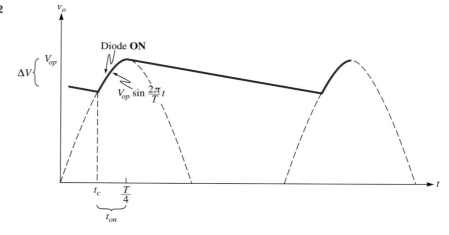

22. A center-tapped transformer rectifier has the following data: $V_{secondary}$ is a 50VAC 60Hz sine wave, $R_L = 200\Omega$, $R_s = 0.5\Omega$, $R_F = 2\Omega$, $t_{on} = 0.5$ms.

   (a) Find $I_{D_{ave}}$.   **(b)** Find $I_{pk_{nr}}$.   **(c)** Find I_{pk_r}.

   (d) Using 20% safety factors, what I_o and I_{SFM} are required?

23. A bridge rectifier has the following data: v_s is a 110RMS 60Hz sine wave, $R_L = 100\Omega$, $R_s = 1\Omega$, $R_F = 2\Omega$, $t_{on} = 0.25$ms.

   (a) Find $I_{D_{ave}}$.   **(b)** Find $I_{pk_{nr}}$.

   (c) Find I_{pk_r}.   **(d)** Using 20% safety factors, what I_o and I_{SFM} are required?

SECTION 4–6   **The Peak Detector**

24. For the peak detector circuit of Figure 4–33, the input is a 1KHz sine wave with a changing amplitude. The amplitude changes as follows:

$$v_{sp} = 0\text{V at first}$$
$$v_{sp} = 1\text{V for 3 cycles}$$
$$v_{sp} = 3\text{V for 3 cycles}$$
$$v_{sp} = 1\text{V for 3 cycles}$$
$$v_{sp} = 0\text{V for the rest of the time}$$

   (a) Draw a diagram of v_s vs. time. Divide the time axis into 1ms intervals.

   (b) Immediately below this diagram, show when D1 is **ON,** thereby charging the capacitor.

   (c) How many times does v_o change to a new voltage?

   (d) After v_o has reached its highest value, how long does it take for v_o to fall by 37%? 95%? 99%?

FIGURE 4–33

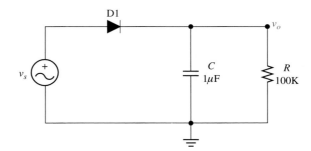

SECTION 4–7   **The Diode AM Demodulator**

25. For the AM demodulator shown in Figure 4–34, the AM signal is represented by v_s. Let the audio envelope be a 1V 1KHz sine wave. Let the carrier signal be a 3V 1MHz sine wave.

   (a) What is the period T_a of the audio envelope?

   (b) What is the maximum slope of the audio signal?

   (c) What is the maximum slope of the carrier signal?

FIGURE 4–34

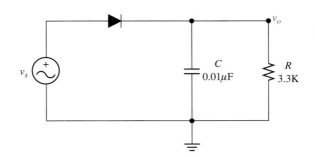

SECTION 4–8 Diode Clamping Circuits

26. For the clamper circuit shown in Figure 4–35, the input signal is a 10V 1KHz sine wave.

    (a) Draw a diagram of v_s vs. time.   **(b)** What is the period (T) of the input signal?

    (c) Draw a diagram of v_o immediately below v_s.

    (d) Draw a diagram of i_L immediately below v_o.

FIGURE 4–35

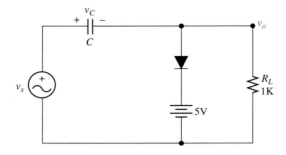

27. Draw a diode clamping circuit that will turn a 10V square wave into a +20V pulse train.

28. Draw a diode clamping circuit that will turn a 5V square wave into a −10V pulse train.

SECTION 4–11 Troubleshooting Diode-Capacitor Circuits

29. Figure 4–36 shows AC measurements on a filtered half-wave rectifier. What is the likely problem in this circuit?

FIGURE 4–36

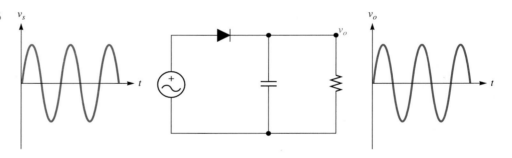

30. Figure 4–37 shows AC measurements on a filtered half-wave rectifier. What is the likely problem in this circuit?

FIGURE 4–37

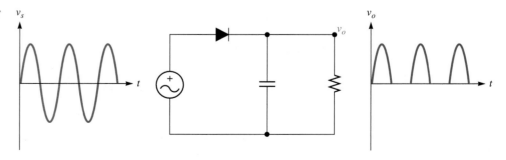

31. Figure 4–38 shows AC measurements on a filtered half-wave rectifier. What is the likely problem in this circuit?

FIGURE 4–38

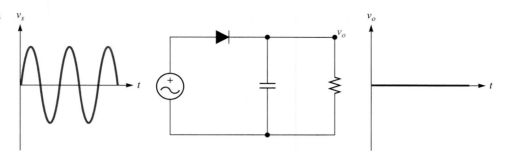

32. Figure 4–39 shows AC measurements on a filtered full-wave rectifier. What is the likely problem in this circuit?

FIGURE 4–39

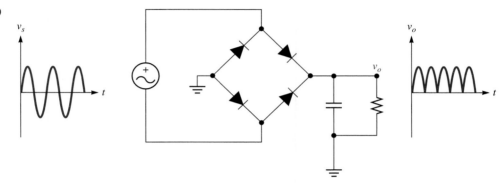

33. Figure 4–40 shows AC measurements on a clamping circuit. What is the likely problem in this circuit?

FIGURE 4–40

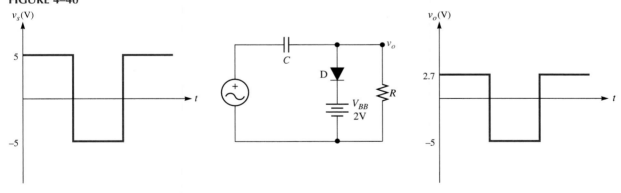

Computer Problems

34. Simulate the circuit of Problem 5 (p. 114).
    (a) Try to fit straight lines to the output voltage.   (b) Find the peak output voltage.
    (c) Find the ripple and ripple (%).
35. Simulate the circuit of Problem 12 (p. 115).
    (a) Try to fit straight lines to the output signal.   (b) Find the ripple and ripple (%).
36. Simulate the circuit of Problem 16 (p. 115).
    (a) Try to fit straight lines to the output voltage.   (b) Find the ripple and ripple (%).
37. Simulate the clamper circuit of Problem 26.
    (a) Plot v_o and V_C.   (b) Are these plots what you expected? Explain.

Chapter

5

The Zener Diode—Voltage Regulation

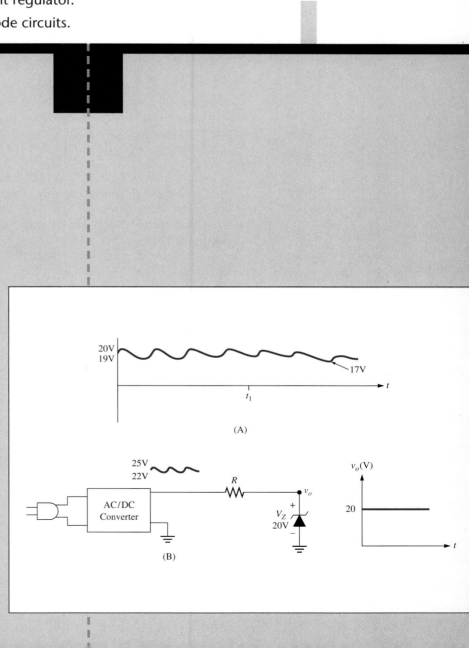

C HAPTER OBJECTIVES

To determine the I-V characteristic and circuit connection of the Zener diode.

To determine when a Zener diode comes out of breakdown.

To build a symmetrical clipping circuit with Zener diodes.

To determine the proper operating conditions for Zener regulators.

To use an integrated circuit regulator.

To troubleshoot Zener diode circuits.

5–1 ■ INTRODUCTION

In previous chapters we described several rectifier circuits that are used to convert an AC voltage to a DC voltage. These rectifiers produce a DC signal that contains some AC ripple. In some applications this small amount of ripple is not acceptable. Even in applications where this ripple is acceptable, it can still cause you trouble.

Let's say that you have built a precision rectifier that has 5% ripple and is driven by a 20V AC source. The output of the rectifier should vary, therefore, from 19V to 20V. When you use the circuit, you find that the output actually varies from 17V to 20V, a 15% variation. What has happened? The supply voltage has changed on you, from 20V to 18V (Figure 5–1).

FIGURE 5–1

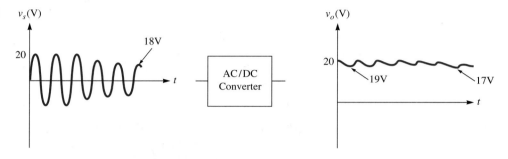

Supply-voltage variation can cause some nasty problems. The alternator in your automobile provides an increasing amount of AC voltage as you speed up. The battery in your car is charged by the rectified alternator output, so without protection you may destroy the battery when you decide to ignore the speed limit and pretend you are at the Indy 500.

Consider another situation. You sit down at your computer to write the great American novel. You are happily banging away at your keyboard when your neighbor decides to turn on every power tool in her garage. This causes the lights in the entire neighborhood to dim. At the same time, your computer screen goes blank (Figure 5–2). You have just lost hours of work. The neighbor-induced decrease in the line voltage to your house causes a disastrous decrease in your computer rectifier output, and a disastrous increase in your blood pressure.

FIGURE 5–2

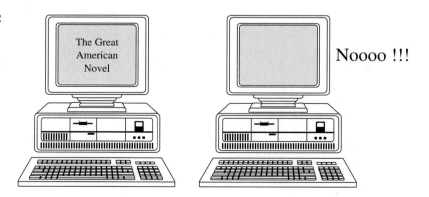

Fortunately, electronic designers are aware of these supply-voltage variations and have provided us with **voltage regulators.** These devices produce a nearly constant output voltage that is independent of reasonable supply or load variations. The simplest voltage regulator is the Zener diode. Better regulation can be obtained with integrated circuits that are specifically designed for this purpose. We will first discuss the Zener diode, then finish the chapter with a brief description of integrated circuit voltage regulators.

5–2 ▪ THE ZENER DIODE

In the previous chapter we showed you the I-V characteristic of a real diode. This characteristic was separated into three regions: forward bias, reverse bias, and breakdown (Figure 5–3). The relatively constant forward voltage of approximately 0.7V is sometimes used to provide voltages to other circuit elements.

Consider the diode tree shown in Figure 5–4A. This tree can provide output voltages of approximately 0.7V, 1.4V, and 2.1V. The diode tree in Figure 5–4B can provide voltages of approximately 9.3V, 8.6V, and 7.9V. Note that diodes are not specifically constructed to provide a highly regulated forward-voltage drop.

FIGURE 5–3

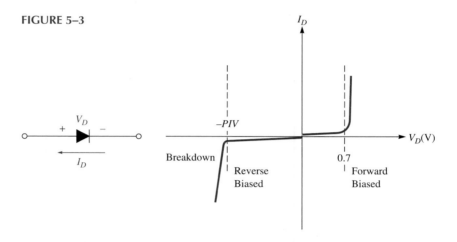

FIGURE 5–4

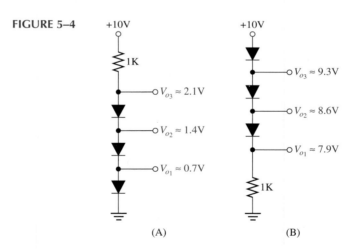

(A)   (B)

Zener Diode Basics

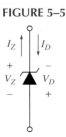

FIGURE 5–5

Look back to the breakdown region in Figure 5–3. With special fabrication techniques we can build diodes that have specific, extremely consistent breakdown voltages from device to device. These diodes are known as Zener diodes, and the breakdown voltage in these devices is known as the Zener voltage, V_Z.

Figure 5–5 shows the special circuit symbol used for the Zener diode. Because the Zener diode is meant to be used in the reverse-bias direction, we define the Zener voltage and current (V_Z and I_Z) in the opposite direction to V_D and I_D. That is, for the Zener diode

$$V_D = -V_Z \quad \text{and} \quad I_D = -I_Z$$

It is extremely important to remember that the Zener diode is still a diode. The I-V curve of the Zener looks similar to the I-V characteristic of a regular diode. The only difference is that a regular diode is manufactured to work well in the forward region while the Zener diode is manufactured to work well in the breakdown region.

What is the advantage of the Zener? The answer lies in the detail of the I-V curve in the reverse-bias region. In Figure 5–6 we show the reverse-bias region for a regular diode (Figure 5–6A) and a Zener diode (Figure 5–6B). Note that we have turned the standard diode I-V plot upside down and inside out to emphasize that our primary interest is in the reverse-bias region. Representing the 3rd quadrant in this manner (I and V negative) is also how manufacturers provide such information on data sheets.

FIGURE 5–6

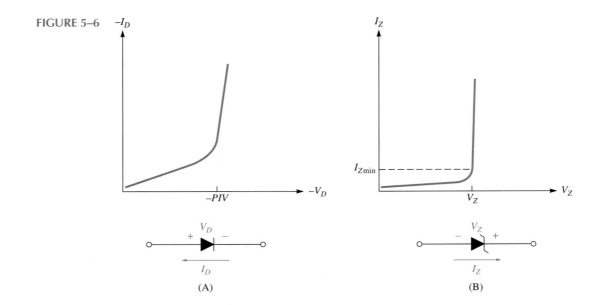

(A)   (B)

Comparing the two plots in Figure 5–6, you can see that the Zener diode curve in breakdown is approximately a vertical line, indicating that in breakdown the Zener diode produces nearly constant voltage. This is not true for the regular diode. In breakdown, therefore, the Zener diode acts as a battery with a voltage of V_Z (Figure 5–7A). Note that the polarity of the battery is the same as the polarity of V_Z.

Because the Zener diode is still a diode, the reverse-bias current must be large enough ($I_Z > I_{Z_{min}}$, Figure 5–6) to put it into breakdown. If the reverse-bias current is not large enough to put the Zener diode into breakdown, then the Zener acts as a regular diode that is **OFF.** If the Zener diode is not in breakdown, therefore, it acts as an open circuit (Figure 5–7B). We say that enough current must *burn* through the Zener to keep it in breakdown.

FIGURE 5–7

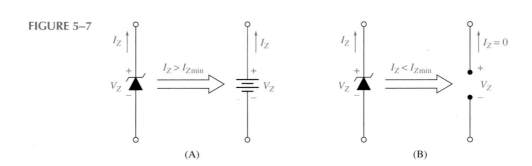

(A)   (B)

The Real Zener Diode

Figure 5–8 shows a typical manufacturer's short form data sheet, which lists common Zener diodes and some device parameters.

First column: the device number you look for when you go to the store.

Second column: the Zener voltage for the device. Note the rather odd voltage values for Zener diodes. Above 10V, Zeners are commonly available in integer values up to 200V. Zener diodes are available in 1%, 2%, 10%, and 20% tolerances. For example, a Zener diode with a nominal voltage of 6V and a tolerance of 10% will have a true voltage somewhere between 5.4V and 6.6V.

Third column: the minimum reverse current that is needed for the diode to Zener (usually labeled I_R by the manufacturer).

Fourth column: power rating for the device. Because the voltage of a Zener diode is fixed, the power dissipated by the Zener diode is

$$P_D = V_Z I_Z$$

FIGURE 5–8
Zener diodes data sheet

Device No.	V_Z (V)	$I_{Z_{min}}$ (μA)	$P_{D\,(mW)}$	$R_{Z(\Omega)}$
1N5226	3.3	25	500	28
1N4729	3.6	100	1000	10
1N4730	3.9	50	1000	9
1N5229	4.3	5	500	22
1N4732	4.7	10	1000	8
1N4733	5.1	10	1000	7
1N5232	5.6	5	500	11
1N5233	6.0	5	500	7
1N4735	6.2	10	1000	2
1N957	6.8	150	500	4.5
1N958	7.5	75	500	5.5
1N5237	8.2	3	500	8
1N5238	8.7	3	500	8
1N960	9.1	25	500	10
1N961	10.0	10	500	8.5

where I_Z is determined by the circuit to which the Zener is connected. Sometimes, rather than giving the power rating for the Zener, manufacturers will give the maximum reverse bias current that the diode can tolerate ($I_{Z_{max}}$).

Fifth column: internal resistance. The real Zener diode is not perfectly modeled with just a battery. The Zener curve in breakdown (Figure 5–6B) is not perfectly vertical because the diode has a small internal resistance that will cause the actual voltage across the Zener to vary somewhat. We ignore this resistance in this textbook.

1. Draw the I-V curve for a diode and label the forward, reverse, and breakdown regions.
2. In what region does the Zener diode operate?
3. Draw the circuit symbol for the Zener diode and label its voltage and current. How do V_Z and I_Z compare to V_D and I_D?
4. What is the difference in the reverse-bias I-V curve for the regular and Zener diodes?
5. In breakdown, how is the Zener diode modeled?
6. What is $I_{Z_{min}}$?
7. If $I_Z < I_{Z_{min}}$, in what state is the diode operating, and how is the diode modeled?
8. How do you find the power dissipated by a Zener diode?

5–3 ■ ZENER DIODE CIRCUITS

FIGURE 5–9

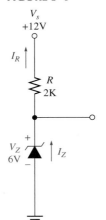

Figure 5–9 shows the basic Zener diode circuit. Note that the Zener is connected so that it is reverse biased. **Zener diodes are not normally connected in the forward bias mode.** To analyze a Zener diode circuit, assume that the diode is in breakdown, treat it as a battery with a value of V_Z, and solve for the Zener current (I_Z). The final step is to make sure that I_Z is greater than the minimum current needed to keep the diode in breakdown.

For the Zener diode in Figure 5–9, the device parameters are given as

$$V_Z = 6V$$

and

$$I_{Z_{min}} = 100\mu A = 0.1mA$$

As long as the reverse-biased current of the Zener diode is greater than 0.1mA, the Zener diode will be in breakdown and will act as a 6V battery. Note that the supply voltage must always be greater than the Zener voltage. The Zener current is found from

$$I_Z = I_R = \frac{12 - V_Z}{2K} = \frac{12 - 6}{2K} = 3mA$$

Because $I_Z > I_{Z_{min}}$, the diode is in breakdown and acts as a 6V battery. The power dissipated by the Zener is given by

$$P_D = V_Z I_Z = 6 \times 3mA = 18mW$$

Now, let's connect our Zener circuit to a load, as shown in Figure 5–10A. Note that in this circuit the Zener current and the current in the resistor are not the same. The Zener current is now

$$I_Z = I_R - I_L$$

If the Zener is in breakdown, it acts as a 6V battery and the resistor current is still

$$I_R = \frac{12 - 6}{2K} = 3mA$$

The Zener current is

$$I_Z = I_R - I_L = 3mA - I_L$$

In Figure 5–10B, we show the Zener characteristic curve for this diode. Points a–e correspond to the Zener current levels for different amounts of load currents. *Point a* gives us the Zener diode current when there is no load ($I_Z = I_{Z_{no\text{-}load}}$)

point	I_L (mA)	I_Z (mA)
a	0	3 ($I_{Z_{no\text{-}load}}$)
b	1	2
c	2	1
d	2.8	0.2
e	**2.95**	**0.05** ← wrong answer

Note that the final value is boldfaced; this answer is *wrong*. As the load draws more current, the current in the Zener diode decreases. As the current in the Zener diode decreases, we move down the I-V curve towards the knee of the curve. If we try to draw 2.95 mA of current from the Zener, we end up on the other side of the knee ($I_Z < I_{Z_{min}}$).

At this point the Zener diode is not in breakdown. It has become just another reverse-biased diode. That is, the Zener diode opens up, and we get a circuit we did not intend to build (Figure 5–10C). To keep the Zener diode in breakdown, the maximum current that the load can draw is given by

$$I_{L_{max}} = I_{Z_{no\text{-}load}} - I_{Z_{min}}$$

FIGURE 5–10

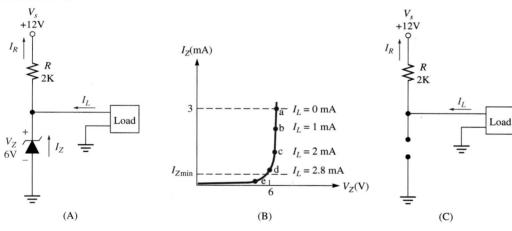

(A)   (B)   (C)

**EXAMPLE 5–1
Zener Circuit**

Find the maximum current the load can draw in the Zener circuit of Figure 5–11. Assume that for this Zener diode, $I_{Z_{min}} = 0.1$mA.

FIGURE 5–11

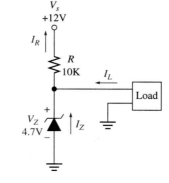

Solution When the load draws no current

$$I_{Z_{no\text{-}load}} = I_R$$

so

$$I_{Z_{no\text{-}load}} = I_R = \frac{12 - V_Z}{10\text{K}} = \frac{12 - 4.7}{10\text{K}} = 0.73\text{mA}$$

and the maximum available load current is

$$I_{L_{max}} = I_{Z_{no\text{-}load}} - I_{Z_{min}} = 0.63\text{mA} - 0.1\text{mA} = 0.63\text{mA}$$

Therefore, the load can draw 0.63mA before the Zener comes out of breakdown.

DRILL EXERCISE   Repeat Example 5–1 (p. 127) with a supply voltage of 6V and $I_{Z_{min}} = 0.1$mA.

Answer $I_{L_{max}} = 0.03$mA

Zener Diode–Resistor Load Circuit

In Figure 5–12 a resistor is the load. If the diode is in breakdown

$$V_L = V_Z$$

and the currents in the source and load resistors are

$$I_R = \frac{12 - V_Z}{2\text{K}} = \frac{12 - 6}{2\text{K}} = 3\text{mA}$$

and

$$I_L = \frac{V_Z}{6\text{K}} = \frac{6}{6\text{K}} = 1\text{mA}$$

The Zener diode current is, therefore

$$I_Z = I_R - I_L = 3\text{mA} - 1\text{mA} = 2\text{mA}$$

So, the Zener diode is in breakdown and acts as a 6V battery. The power dissipated in the diode is

$$P_D = V_Z I_Z = 6 \times 2\text{mA} = 12\text{mW}$$

FIGURE 5–12

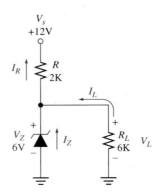

Is the Zener Diode in Breakdown?—An Alternate Approach

There is an alternative method for determining whether a Zener diode is in breakdown. Assume that the Zener diode is **OFF**, and replace it with an open circuit. Find the voltage across the Zener diode. If this voltage is less than V_Z, the Zener diode is indeed **OFF**. If the voltage is greater than V_Z, your assumption was wrong and the Zener diode is in breakdown. In this case, the true voltage across the Zener diode is given by V_Z.

For example, if we assume the Zener diode in Figure 5–12 is open, we can find the diode voltage (V_L) from the voltage divider

$$V_L = \frac{R_L}{R_L + R} V_S = \frac{6K}{6K + 2K} 12 = 9V$$

Because this voltage is greater than the Zener voltage, the Zener diode must be in breakdown and the load voltage is actually

$$V_L = V_Z = 6V$$

EXAMPLE 5–2
Zener Diode–Resistor Circuit I

For the circuit of Figure 5–12, find I_Z for the following load resistors. Then find the power dissipated by the Zener. Be sure to check if the Zener diode is in breakdown.

(a) $R_L = 12K\Omega$
(b) $R_L = 3K\Omega$
(c) $R_L = 1K\Omega$

Solution In all cases, if the diode is in breakdown, $V_Z = 6V$ and

$$I_R = \frac{12 - V_Z}{2K} = \frac{12 - 6}{2K} = 3mA$$

(a) For $R_L = 12K\Omega$

$$I_L = \frac{V_Z}{12K} = \frac{6}{12K} = 0.5mA$$

so

$$I_Z = I_R - I_L = 3mA - 0.5mA = 2.5mA$$
$$P_D = V_Z I_Z = 6 \times 2.5mA = 15mW$$

As the load resistor gets larger, the load current decreases. This results in even more current burning through the Zener, putting it deeper into breakdown. It also results in a greater power dissipation.

(b) For $R_L = 3K\Omega$

$$I_L = \frac{V_Z}{3K} = \frac{6}{3K} = 2mA$$

so

$$I_Z = I_R - I_L = 3mA - 2mA = 1mA$$
$$P_D = V_Z I_Z = 6 \times 1mA = 6mW$$

As the load resistor gets smaller, the load current increases, drawing more current away from the Zener diode. This results in less current burning through the Zener, bringing it closer to the knee of its I-V curve (Figure 5–10B, p. 127). The power dissipated by the Zener also decreases.

(c) For $R_L = 1\text{K}\Omega$

$$I_L = \frac{V_Z}{1\text{K}} = \frac{6}{1\text{K}} = 6\text{mA}$$

STOP! The no-load current for the Zener diode is only 3mA. The Zener cannot supply 6 mA to the load. The Zener diode is no longer in breakdown, it is now **OFF** and

$$I_Z \approx 0\text{A}$$

Note that if you just merrily went along plugging numbers into your calculator, you would have found that $I_Z = -3\text{mA}$. The Zener current can never be negative.

DRILL EXERCISE
Redraw the circuit in Figure 5–12 (p. 128), changing the supply voltage to 10V. Find I_Z for no-load and for $R_L = 12\text{K}\Omega$, $6\text{K}\Omega$, and $3\text{K}\Omega$.

Answer $I_Z = 1.5\text{mA}, 1.0\text{mA}, 0\text{A}$

Minimum Load Resistance

The load current in Figure 5–11 (p. 127) is given by

$$I_L = \frac{V_Z}{R_L}$$

For a given Zener diode, the maximum load current occurs when the load resistance is smallest:

$$I_{L_{\text{max}}} = \frac{V_Z}{R_{L_{\text{min}}}}$$

The minimum load resistor that can be used with a given Zener diode is

$$R_{L_{\text{min}}} = \frac{V_Z}{I_{L_{\text{max}}}}$$

EXAMPLE 5–3
Zener Diode–Resistor Circuit II

In Figure 5–12 (p. 128) find the smallest load resistor that can be used and still have the Zener diode operate properly. Assume that $I_{Z_{min}} = 0.1\text{mA}$.

Solution The maximum current the Zener diode can supply is

$$I_{Z_{\text{max}}} = I_{Z_{no\text{-}load}} - I_{Z_{\text{min}}} = \frac{12 - 6}{2\text{K}} - 0.1\text{mA} = 2.9\text{mA}$$

The minimum allowable load resistor is found from

$$R_{L_{\text{min}}} = \frac{V_Z}{I_{Z_{\text{max}}}} = \frac{6}{2.9\text{mA}} = 2.07\text{K}\Omega$$

Any load resistor smaller than 2.07KΩ will bring the Zener out of breakdown.

DRILL EXERCISE Redraw the circuit in Figure 5–12 (p. 128), changing the supply voltage to 10V. Find the smallest load resistor that can be used in this circuit.

Answer $R_{L_{min}} = 3.16 \text{K}\Omega$

REVIEW QUESTIONS

1. How is a Zener diode always connected in a circuit? Why?
2. Can the supply voltage ever be lower than the Zener voltage? Why?
3. Draw the Zener diode reverse-bias I-V curve.
4. On the I-V curve, indicate what happens as the Zener diode supplies current to a load.
5. Draw a Zener diode circuit with a load resistor.
6. What is the no-load current? Write a formula for it.
7. Assume that the Zener is in breakdown. Write a formula for the load current.
8. What is the maximum current available to the load?
9. When is the load current maximum?
10. What is the minimum load resistor that can be used with a Zener diode?
11. How do you know if your Zener diode is actually in breakdown?

5–4 ■ ZENER CLIPPER CIRCUITS

Figure 5–13 shows a symmetrical clipping circuit similar to those studied in Chapters 2 and 3. The top and bottom portion of the input sine wave are clipped at +5.7 and –5.7V (we include the 0.7V diode drop).

FIGURE 5–13

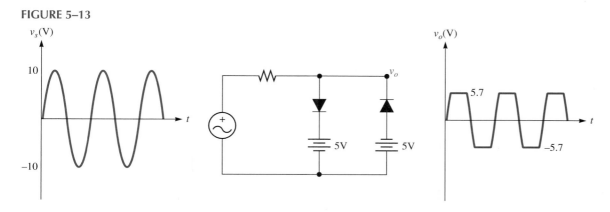

We can use Zener diodes to create the same circuit, eliminating the need for the batteries. Because batteries are not perfect—for instance, they are bulky, run down in time, and can leak nasty chemicals—anytime we can replace a battery with a Zener diode we improve the circuit.

Consider first the Zener circuit shown in Figure 5–14A (p. 132). On the positive half-cycle of the input, the Zener is reverse biased. Until the input reaches 5V, the Zener does not go into breakdown and, therefore, acts as a regular reverse-biased diode. That is, the Zener acts as an open circuit (Figure 5–14B) and

$$v_o = v_s$$

FIGURE 5–14

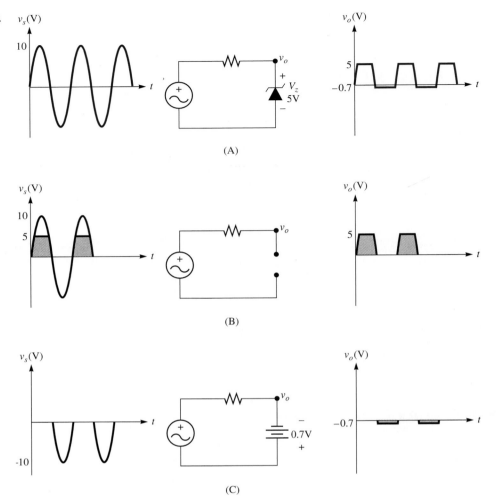

(A)

(B)

(C)

When the input exceeds 5V, the Zener goes into breakdown and acts as a battery. At this time

$$v_o = V_Z = 5V$$

That is, the output is clipped at 5V. When the input, still on the positive cycle, drops below 5V, the Zener once again opens.

When the input signal goes negative, the Zener diode becomes forward biased, and as with all forward-biased diodes, acts as a 0.7V battery. Therefore, on the negative half-cycle of the input, the output is fixed at –0.7V (Figure 5–14C). The total result is shown in Figure 5–14A.

Symmetrical clipping can be achieved with the Zener diode circuit of Figure 5–15A. To see how this circuit works, we look at the output for the positive half-cycle (Figure 5–15B). During the positive half-cycle, current is moving in the **ON** direction for Z1 and the reverse bias direction for Z2. When the input is positive and less than 5.7V (it takes 0.7V to turn **ON** Z1 and another 5V to put Z2 into breakdown), both diodes are **OFF,** and the output equals the input.

When the input voltage tries to increase above 5.7V, Z1 turns **ON**, acting as a 0.7V battery, and Z2 goes into breakdown, acting as a 5V battery. The output now is fixed at 5.7V.

When the positive input drops below 5.7V, both diodes are again **OFF** and the output equals the input. On the negative half-cycle, the current now moves in the direction to turn

FIGURE 5–15

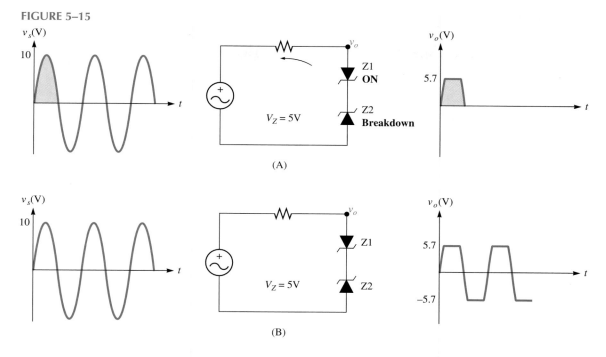

(A)

(B)

Z2 **ON** and put Z1 into breakdown. The negative half-cycle, therefore, produces an output that is the mirror image of the output during the positive half-cycle (Figure 5–15B).

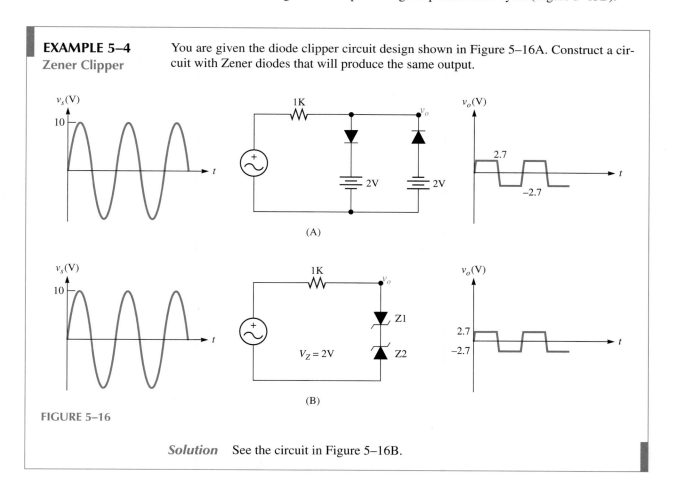

EXAMPLE 5–4
Zener Clipper

You are given the diode clipper circuit design shown in Figure 5–16A. Construct a circuit with Zener diodes that will produce the same output.

(A)

(B)

FIGURE 5–16

Solution See the circuit in Figure 5–16B.

REVIEW QUESTIONS

1. Why is it a good idea to limit the number of batteries used in a circuit?

2. Explain the operation of the Zener diode symmetrical clipper.

3. What two voltages determine the limiting voltage in a Zener diode clipping circuit?

5–5 ■ ZENER VOLTAGE REGULATOR CIRCUIT

We now return to the problem of voltage regulation discussed in the introduction. We have seen in Chapter 3 that even with a transformer, rectifier, and capacitor filter, the DC signal we produce still has a small AC ripple on it. Furthermore, any variation in the line voltage supplying the rectifier will increase the variation of the rectifier output voltage (Figure 5–17A).

FIGURE 5–17

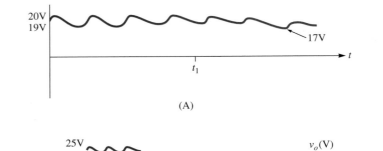

(A)

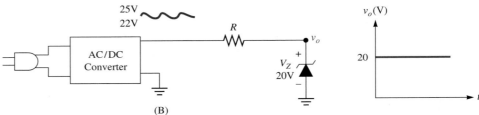

(B)

We can produce much better DC voltage regulation by using the Zener circuit shown in Figure 5–17B. Note that to produce a 20V output, we use a 20V Zener and increase the rectifier output to approximately 25V. The output is now a very well-regulated 20V. The only requirement is that the rectifier output be large enough to put the Zener into breakdown.

In any real application, we must take into account the current drawn by the load. Figure 5–18 shows the more realistic situation of a Zener voltage regulator that drives a load resistor. For comparison take a peek back at Figure 5–12 (p. 128). The only difference in the circuit in Figure 5–18 is that the supply signal is not a constant 12V. The supply voltage now varies between 22V and 25V.

You already know that we need to keep the Zener diode in the breakdown region. This means that, ignoring $I_{Z_{min}}$, the load current is limited to the no-load current in the Zener. In this case, the no-load current varies as the input voltage varies. That is, when the input signal is 25V, the no-load current is

$$I_Z = I_R = \frac{25 - 20}{2K} = 2.5mA$$

When the input signal is 22V, the no-load current is

$$I_Z = I_R = \frac{22 - 20}{2K} = 1mA$$

How much current is available to drive the load? Because the input can drop to 22V at any time, we must assume that only 1mA is available to the load.

As long as the load draws less than 1mA, the Zener diode will remain in breakdown and we will produce a very well-regulated 20V output.

EXAMPLE 5–5
Zener Regulator—
Minimum Load

As the load resistance decreases in Figure 5–18, more current is drawn from the Zener regulator. How small can R_L get before we lose regulation (ignore $I_{Z_{min}}$)?

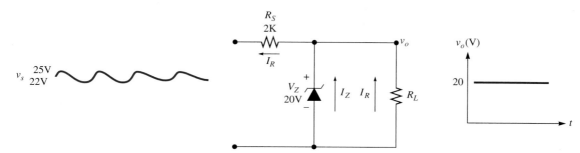

FIGURE 5–18

Solution   The current drawn by the load is given by

$$I_L = \frac{V_Z}{R_L}$$

Because the 20V Zener in this circuit can supply a maximum current of 1mA

$$R_{L_{min}} = \frac{V_Z}{I_{L_{max}}} = \frac{20}{1\text{mA}} = 20\text{K}\Omega$$

As long as the load resistance is greater than 20KΩ (allowing for no safety margin), this regulator will work as advertised.

DRILL EXERCISE

Redraw the Zener regulator circuit of Figure 5–18 so that the input varies from 28V to 32V. Find the maximum current available to the load, and the smallest load resistance that can be used.

Answer $I_{L_{max}} = 4\text{mA}, R_{L_{min}} = 5\text{K}\Omega$

This drill example illustrates that we can increase the current available to the load by increasing the supply voltage. We can also increase the available load current by decreasing the regulator source resistance.

EXAMPLE 5–6
Zener Regulator—
Regulator Resistor

Reduce the regulator source resistor in Figure 5–18 to

$$R_s = 1\text{K}\Omega$$

Find the maximum current available to the load and the smallest resistance that can be used (ignore $I_{Z_{min}}$).

Solution   When the input drops to 22V, the no-load current is

$$I_Z = \frac{22 - 20}{1K} = 2\text{mA}$$

and we have doubled the available load current

$$I_{L_{max}} = 2\text{mA}$$

The minimum load resistor is found from

$$R_{L_{min}} = \frac{V_Z}{I_{L_{max}}} = \frac{20}{2\text{mA}} = 10\text{K}\Omega$$

DRILL EXERCISE

Redraw the Zener regulator circuit of Figure 5–18 (p. 135), changing the regulator source resistor to $R_s = 1\text{K}\Omega$. Let the input vary from 28V to 32V. Find the maximum current available to the load and the smallest load resistance that can be used.

Answer $I_{L_{max}} = 8\text{mA}$; $R_{L_{min}} = 2.5\text{K}\Omega$

REVIEW QUESTIONS

1. You have built a filtered rectifier designed to have a 5% ripple. Instead, the output changes by more than 10%. Why?

2. Draw a simple Zener circuit, without a load, that will produce a nearly constant DC output.

3. For the Zener regulator to work, what must be true of the supply voltage?

4. Now add a load resistor to your Zener regulator.

   (a) How much current is available to the load?

   (b) How much current will the load actually draw?

   (c) What will happen if the load draws too much current?

   (d) What are two ways to increase the available current to the load? Explain your answers.

5–6 ■ INTEGRATED CIRCUIT VOLTAGE REGULATORS

Very often you will find that functions such as voltage regulation can be accomplished by using an integrated circuit, or chip, instead of a discrete device. You have already seen one example of this, the diode bridge of Chapter 3 (see Figure 3–1, p. 48). Some integrated circuit voltage regulators can produce a fixed or a variable output voltage. One advantage of these devices is that properties such as thermal compensation, short circuit protection, and surge protection can be built into the device.

Consider the LM2931 series of voltage regulators available from Nation Semiconductor. This series comes as a three-terminal fixed voltage device or a five-terminal adjustable voltage device (Figure 5–19A). We don't want to scare you by showing you the circuit of the LM2931 (you can look it up in a data book); however, the regulator is built using five Zener diodes and a lot of other goodies.

FIGURE 5–19

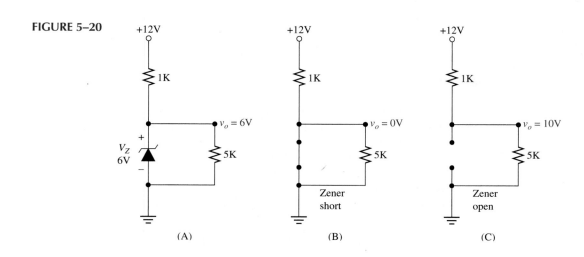

Fixed Voltage

Adjustable Voltage
(A)

v_s
6–26V

LM2931

v_o
5V

(B)

v_s

LM2931

v_o
5V + V_Z

$+$
V_Z
$-$

(C)

Figure 5–19B shows how easy it is to use the three-terminal fixed voltage device. You supply a DC (or approximate DC), voltage that can range from 6V to 26V. The output voltage is regulated to within 5% of 5V. The capacitors are sometimes used to shunt to ground any AC signal that might be present at the input or output terminals.

Often integrated circuits are combined with discrete devices for customized application. Figure 5–19C shows how a Zener diode is combined with the LM2931 to produce the output voltage

$$V_o = 5 + V_Z$$

5–7 ■ TROUBLESHOOTING ZENER DIODES AND CIRCUITS

Zener diodes cannot be individually tested with a multimeter. This is because multimeters usually don't have enough output voltage to put the Zener into breakdown. Even if your meter could put a low-voltage Zener into breakdown, Zeners will produce a low-resistance measurement in both the reverse breakdown region and the forward **ON** region.

A bad Zener can only be found by troubleshooting the Zener circuit. Consider the basic Zener circuit shown in Figure 5–20. If the Zener diode is working properly, you will measure approximately 6V (remember the Zener voltage tolerance) across the load. If the Zener has failed short, you will measure 0V (Figure 5–20B). This same problem could also be caused by a shorted load resistor (R_L) or an opened source resistor (R_s). The only way to

FIGURE 5–20

+12V

1K

V_Z
6V
$+$
$-$

$v_o = 6V$

5K

(A)

+12V

1K

$v_o = 0V$

5K

Zener
short

(B)

+12V

1K

$v_o = 10V$

5K

Zener
open

(C)

tell which device has failed is to remove the resistors and check them with an ohmmeter. If the resistors are good, then the diode is bad.

If the diode has failed open in this circuit, you will measure 10V (Figure 5–20C, p. 137). Essentially, any measurement across the load besides 6V indicates the Zener diode has probably failed.

You will need an oscilloscope to troubleshoot Zener diode circuits that have AC inputs. Consider the symmetrical clipping circuit and its expected output shown in Figure 5–21A. If either Zener fails open, you will see the input signal at the output (Figure 5–21B). You will not know which Zener is bad, so you will have to replace them one at a time until the circuit works.

FIGURE 5–21

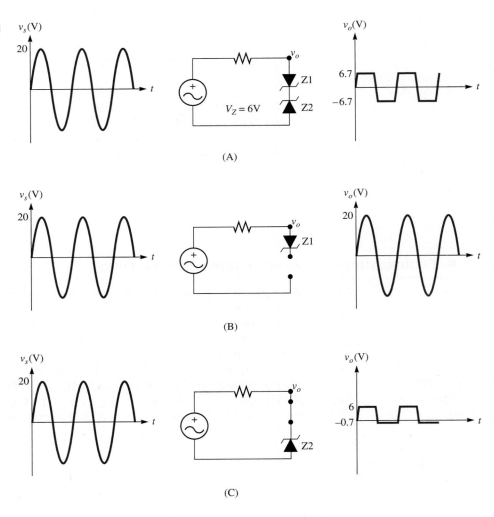

If diode Z1 fails short, you will get the wave form shown in Figure 5–21C. The top of the output signal is clipped at 6V, while most of the bottom of the output signal is lost. If Z2 fails short, you will lose most of the top part of the output signal.

Finally, Figure 5–22A shows a properly working Zener regulator circuit that provides a constant 20V output. If the Zener diode fails open, you will see the same ripple in the output signal as you see in the input signal (Figure 5–22B). The voltage level of the ripple signal will depend on the voltage divider formed by R_L and R_s.

FIGURE 5–22

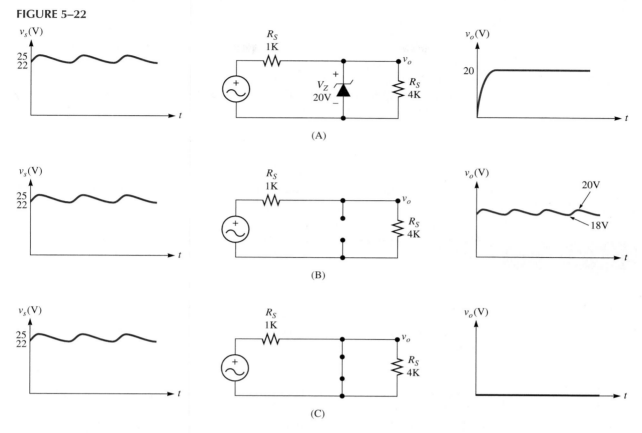

(A)

(B)

(C)

If the Zener has failed short, you will see 0V at the output (Figure 5–22C). Again, this problem could be caused by a shorted load resistor or an opened source resistor. The only way to tell is to remove the resistors and check them with an ohmmeter. If the resistors are good, then the diode is bad.

EXAMPLE 5–7
Troubleshooting a Zener Diode Circuit

Study the Zener circuit shown in Figure 5–23. Several possible output measurements are shown. For each measurement, determine if the diode is working or if it may be bad. If the diode is bad, how did it fail?

A. $v_o = 10.5V$
B. $v_o = 16V$
C. $v_o = 0V$

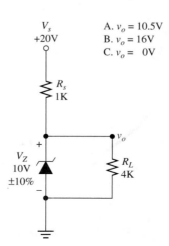

FIGURE 5–23

Solution

(a) Because of the diode tolerance, we can expect between 9 and 11 VDC at the output. This measured output falls within this range, so the Zener diode is functioning properly.

(b) Because the output voltage is very different from the Zener voltage, we suspect that the Zener has failed open. In fact, if the Zener is removed, the remaining voltage divider would give us

$$v_o = \frac{R_L}{R_L + R_s} \, 20 = \frac{4K}{4K + 1K} \, 20 = 16V$$

which is what we get at the output.

(c) Because the output is 0V, either the diode is shorted, R_L is shorted, or R_s is open.

SUMMARY

- The Zener diode is usually in the breakdown region, where it behaves as a battery with a constant voltage (V_Z).
- The Zener diode is always connected in a circuit so that it is in reverse bias.
- If too little reverse-bias current burns through the Zener, it will come out of breakdown and act as a regular reverse-biased diode; that is, it will become an open circuit.
- Zener diodes are rated for maximum power dissipation ($P = V_Z I_Z$) or for maximum reverse-bias current ($I_{Z_{max}}$).
- The maximum current available to a load is approximately the no-load current in the Zener. If the load tries to draw more current, the Zener will come out of breakdown.
- The maximum current available to a load can be increased by increasing the supply voltage or decreasing the supply resistance.
- The regular diode-battery combination in a clipping circuit can be replaced with a Zener diode.
- Zener diodes can be used to regulate the rippled output voltage of a filtered rectifier to produce a very constant DC voltage.
- Zener diodes cannot be tested with a multimeter; a bad Zener must be found by troubleshooting the Zener circuit.
- If the voltage across the Zener load is anything but V_Z (within its tolerance rating), the Zener has probably failed.

PROBLEMS

SECTION 5–2   The Zener Diode

1. Draw a typical diode I-V characteristic curve.
   (a) Indicate the regions in which the regular diode operates.
   (b) Indicate the region in which the Zener diode operates.

2. You have a 9V battery, a resistor, and a number of regular silicon diodes. Draw a circuit that will provide outputs of 0.7V, 1.4V, 2.1V, and 2.8V.

3. You have a 9V battery, a resistor, and a number of regular silicon diodes. Draw a circuit that will provide outputs of 8.3V, 7.6V, 6.9V, and 6.2V.

4. Figure 5–24 shows Zener diode data provided by an electronics retailer. What is the maximum current that the following Zener diodes can handle?
   (a) 1N746   (b) 1N4731   (c) 1N4764

5. Choose the appropriate Zener diode from Figure 5–24 to match the following requirements.
   (a) $V_Z = 12V$, $P_{D_{max}} = 300mW$ (b) $V_Z = 12V$, $P_{D_{max}} = 750mW$
   (c) $V_Z = 4.3V$, $I_{Z_{max}} = 0.2A$

SECTION 5–3   Zener Diode Circuits

6. Redraw the Zener diode circuit shown in Figure 5–25, and label V_Z and I_Z. The nominal voltage of the Zener diode is 4V. Find the Zener diode voltage and current if $V_s = 12V$.

FIGURE 5–24
Zener diodes

Part#	Voltage	Watt
1N746	3.3	250mW
1N4739	9.1	250mW
1N963	12	400mW
1N5256	30	400mW
1N5229	4.3	500mW
1N5235	6.8	500mW
1N4728	3.3	1W
1N4729	3.6	1W
1N4731	4.3	1W
1N4733	5.1	1W
1N4735	6.2	1W
1N4740	10	1W
1N4747	20	1W
1N4764	100	1W

FIGURE 5–25

7. Redraw the Zener diode circuit shown in Figure 5–25, and label V_Z and I_Z. The nominal voltage of the Zener diode is 4V. Find the Zener diode voltage and current if $V_s = 6V$.

8. Redraw the Zener diode circuit shown in Figure 5–25, and label V_Z and I_Z. The nominal voltage of the Zener diode is 4V. Find the Zener diode voltage and current if $V_s = 3V$.

9. For the circuit in Figure 5–25, the Zener diode has a nominal voltage of 4V with a 5% tolerance. For $V_s = 12V$,

    (a) Find the maximum current in the Zener diode.

    (b) Find the minimum current in the Zener diode.

10. Redraw the Zener diode circuit shown in Figure 5–26, and label V_Z and I_Z. The nominal value of the Zener diode is 4V. Find the Zener diode voltage and current if $V_s = -12V$.

11. Redraw the Zener diode circuit shown in Figure 5–26, and label V_Z and I_Z. The nominal value of the Zener diode is 4V. Find the Zener diode voltage and current if $V_s = +12V$.

12. Redraw the circuit shown in Figure 5–27, and label V_Z and I_Z. Find the Zener diode voltage and current.

13. Redraw the circuit shown in Figure 5–28, and label V_Z and I_Z. Find the Zener diode voltage and current.

14. Redraw the circuit shown in Figure 5–29, and label the Zener current, source resistor current, and load resistor current. Find I_S, I_L, and I_Z if the load resistor is removed (no-load). Is the Zener diode in breakdown? If the Zener diode is not in breakdown, do not solve for the currents.

FIGURE 5–26

FIGURE 5–27

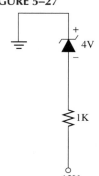

FIGURE 5–28

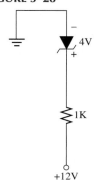

FIGURE 5–29

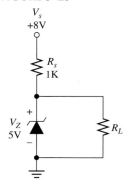

15. Redraw the circuit shown in Figure 5–29 (p. 141), and label the Zener current, source resistor current, and load resistor current. Find I_S, I_L, and I_Z if R_L = 5KΩ. Is the Zener diode in breakdown? If the Zener diode is not in breakdown, do not solve for the currents.

16. Redraw the circuit shown in Figure 5–29, and label the Zener current, source resistor current, and load resistor current. Find I_S, I_L, and I_Z if R_L = 2.5KΩ. Is the Zener diode in breakdown? If the Zener diode is not in breakdown, do not solve for the currents.

17. Redraw the circuit shown in Figure 5–29, and label the Zener current, source resistor current, and load resistor current. Find I_S, I_L, and I_Z if R_L = 1KΩ. Is the Zener diode in breakdown? If the Zener diode is not in breakdown, do not solve for the currents.

18. For the circuit shown in Figure 5–29, what is the maximum current available to the load?

19. For the circuit shown in Figure 5–29, increase the supply voltage so that you can double the available load current.

20. For the circuit shown in Figure 5–29, choose a new R_s in order to double the available load current.

SECTION 5–4   Zener Clipping Circuits

21. Replace the clipping circuit of Figure 5–30 with an equivalent Zener diode circuit. Be sure to include the turn-on voltage of the regular diodes. Select the closest available Zener diodes in Figure 5–8 (p. 125).

FIGURE 5–30

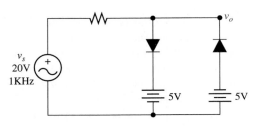

FIGURE 5–31

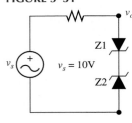

22. For the Zener diode clipping circuit of Figure 5–31, the input is a 20V 1Hz sine wave. Draw the input sine wave with its magnitude on the vertical axis and time on the horizontal axis. Draw several cycles, and label the points on the time axis where the input sine wave peaks and where the input sine wave crosses the time axis. Immediately under the input sine wave, using the same amplitude and time scale, draw the output wave form.

23. For the Zener diode clipping circuit of Figure 5–31, the input is a 100V 1Hz sine wave. Draw the input sine wave with its magnitude on the vertical axis and time on the horizontal axis. Draw several cycles, and label the points on the time axis where the input sine wave peaks and where the input sine wave crosses the time axis. Immediately under the input sine wave, using the same amplitude and time scale, draw the output wave form. What wave form does this output represent?

24. For the Zener diode clipping circuit of Figure 5–31, the input is a 5V 1Hz sine wave. Draw the input sine wave with its magnitude on the vertical axis and time on the horizontal axis. Draw several cycles, and label the points on the time axis where the input sine wave peaks and where the input sine wave crosses the time axis. Immediately under the input sine wave, using the same amplitude and time scale, draw the output wave form.

SECTION 5–5   Zener Voltage Regulator Circuit

25. The Zener circuit shown in Figure 5–32 is driven by a rectifier that produces a 50V output with a 5V ripple.
    (a) What is the maximum source resistor current?
    (b) What is the minimum source resistor current?
    (c) How much current is available to drive a load?

26. Redraw the circuit shown in Figure 5–32 and connect a load resistor across the Zener diode. If R_L = 200KΩ, find the source resistor current, load resistor current, and Zener current.

FIGURE 5–32

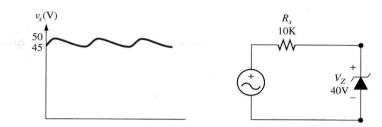

27. Using the results of Problem 25, find the smallest load resistor that can be used in the Zener circuit of Figure 5–32.

SECTION 5–6 Integrated Circuit Voltage Regulators

28. An integrated circuit voltage regulator produces an output that is 20V regulated to within ±5%. What range of output voltages can you expect from this device?

29. Using the LM2931, what Zener diode from Figure 5–24 (p. 141) do you need to produce a combined output voltage of 8.6V?

SECTION 5–7 Troubleshooting Zener Diodes and Circuits

30. For the circuit shown in Figure 5–33, the output voltage is measured as 0V. Is the Zener diode good or bad? If it is bad, how has it failed? Assume all resistors are good.

31. For the circuit shown in Figure 5–33, the output voltage is measured as 4.1V. Is the Zener diode good or bad? If it is bad, how has it failed? Assume all resistors are good.

32. For the circuit shown in Figure 5–33, the output voltage is measured as 6.7V. Is the Zener diode good or bad? If it is bad, how has it failed? Assume all resistors are good.

FIGURE 5–33

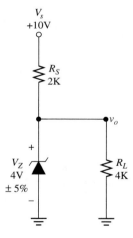

33. The Zener diodes in the clipping circuit shown in Figure 5–34 (p. 144) have a nominal voltage of $V_Z = 10V$. Four possible output measurements are given. For each output, determine if one or both diodes have failed. If a diode is bad, has it failed open or short?

34. The Zener regulator circuit shown in Figure 5–35 (p. 145) has the three possible measured outputs given in the figure. For each output measurement, determine whether the Zener is good or bad. If the Zener is bad, how has it failed? Assume the resistors are good.

35. A 2V integrated circuit voltage regulator is connected as shown in Figure 5–36 (p. 145). What can you say about this circuit if the output is measured as

(a) $V_o = 7.1V$?   (b) $V_o = 2V$?

FIGURE 5–34

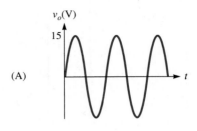

(A)

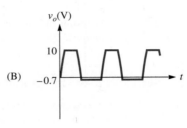

(B)

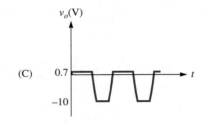

(C)

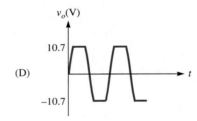

(D)

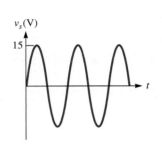

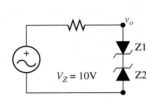

Computer Problems

36. Simulate the circuit of Figure 5–4A (p. 123), and find the output voltages. Repeat for each diode model in your computer's library.

37. Simulate the circuit of Figure 5–4B and find the output voltages. Repeat for each diode model in your computer's library.

38. Simulate the circuit shown in Figure 5–9 (p. 126). Vary the supply voltage (V_s) from 12V down to 0V in 0.1V steps, and find V_Z and I_Z. Plot I_Z vs. V_Z.

39. Simulate the circuit shown in Figure 5–12 (p. 128). Vary the supply voltage from 12 V down to 0 V, and find the point at which the Zener comes out of breakdown (i.e., $V_Z \neq 6$ V).

40. Simulate the circuit shown in Figure 5–12. Set $V_s = 12$V. Vary the supply resistor (R_s) from 2KΩ to 10KΩ, and find the point at which the Zener comes out of breakdown.

41. Simulate the circuit shown in Figure 5–12. Set $R_s = 2$KΩ. Vary the load resistor (R_L) from 6KΩ to 1KΩ, and find the point at which the Zener comes out of breakdown.

42. In this problem, you are going to combine a rectifier circuit from Chapter 4 with a Zener regulator circuit.

    (a) Simulate the half-wave rectifier shown in Figure 4–28 (p. 114) in Chapter 4. Plot the output voltage vs. time.

FIGURE 5–35

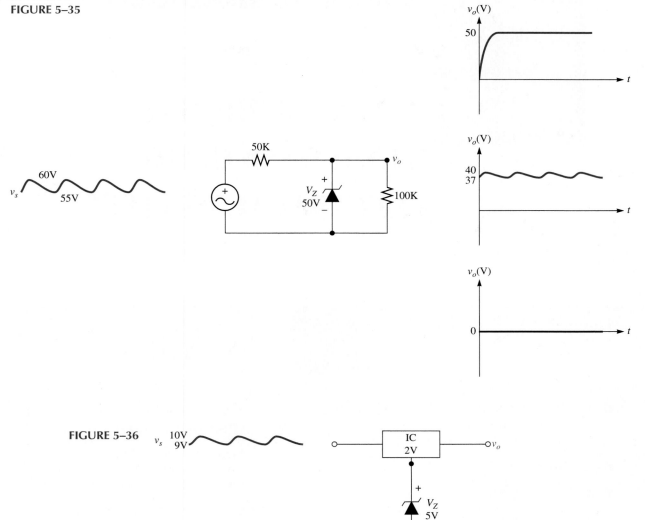

FIGURE 5–36

(b) Remove R_L from the rectifier circuit. Connect the output of the rectifier so that it is now the input supply voltage of a Zener regulator with the following parameters:

$$R_s = 10\Omega \text{ (Note how small this resistor is.)}$$
$$V_Z = 40V$$
$$R_L = 100\Omega \text{ the original load.}$$

Find and plot the load voltage and the new output of the rectifier. Compare these results to Part a.

(c) Change R_s to 100Ω and repeat part b. Explain your results.

Chapter

6

Special-Purpose Diodes and Opto-Electrical Devices

CHAPTER OBJECTIVES

To determine the properties and uses of the germanium
diode.

To determine the properties and uses of the Schottky
diode.

To determine the properties and uses of the tunnel diode.

To determine the properties and uses of the varactor
diode.

To determine the properties and uses of the photo diode in
the photoconductive mode.

To determine the uses of the photo diode in the photovoltaic mode.

To determine the properties and uses of the photoresistor.

To determine the properties and uses of the light-emitting
diode.

To use the opto-isolator to couple parts of an electrical
circuit.

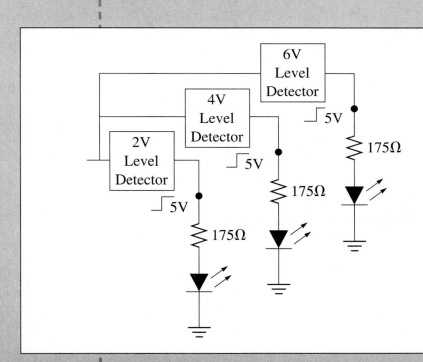

6–1 ■ INTRODUCTION

Specialized Diodes

In Chapters 2–4, we described the standard silicon diode and its circuit applications. In this chapter we will discuss the following specialized diodes (Figure 6–1):

☐ the germanium (Ge) diode

☐ the Schottky diode

☐ the tunnel diode

☐ the varactor

☐ the photo diode

☐ the light-emitting diode (LED)

FIGURE 6–1
Specialized diodes

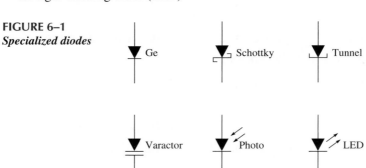

The germanium and Schottky diodes are used in many of the same types of circuits as silicon diodes. The remaining diodes have properties that make them very useful in communication, sensor, and display applications.

Also related to the photo diode and the LED are these opto-electrical devices:

☐ the photoresistor

☐ the photo transistor

☐ the opto-isolator

In the subsequent sections we will give a very brief description of each of these special purpose devices. The intent here is to familiarize you with these devices and their uses, not to turn you into, for example, a tunnel diode expert.

Troubleshooting Specialized Diodes

Some of the diodes listed above are made with different materials than silicon, some have very special electrical behavior, and some convert light energy to electrical energy. Still, all of them are diodes and, therefore, can be tested with the same techniques used for regular silicon diodes (Chapter 3, Section 9). For this reason, we present troubleshooting only in the homework problems.

6–2 ■ THE GERMANIUM DIODE

The first diodes were **germanium diodes,** which were made (logically) of the semiconductor material germanium (not geranium—flower power never got past the 1960s). The I-V characteristics of germanium and silicon diodes are shown in Figure 6–2. You can see that although the general shape of the germanium characteristic is similar to silicon, there are significant differences. While the turn-on voltage for silicon diodes is approximately 0.7 V, the turn-on voltage for germanium diodes is approximately 0.3 V:

Silicon: $V_{D(ON)} \approx 0.7V$

Germanium: $V_{D(ON)} \approx 0.3$

The turn-on voltage for the germanium diode is approximately one-half the turn-on voltage of the silicon diode. This is the major advantage of the germanium diode.

We can also see disadvantages in the germanium I-V curve. Its breakdown voltage is smaller than that of the silicon diode. Silicon diodes have breakdown voltages that range from 50 to 1000V. Germanium diodes typically have breakdown voltages less than 100V.

FIGURE 6–2

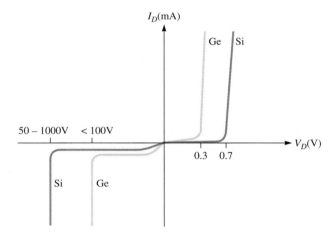

Another disadvantage of the germanium diode is its larger reverse-bias current. Remember, when we design and analyze diode circuits, we assume that the reverse-bias current of the diode is 0A. The larger the actual reverse-bias current, the less accurate our approximate analysis.

EXAMPLE 6–1

The Germanium Diode

The rectifier circuit of Figure 6–3A uses a germanium diode.

(a) Sketch the output if $V_{D(ON)} = 0.3V$.

(b) What is the maximum reverse-bias voltage in this circuit? What *PIV* do we need? Remember, *PIV* is the *peak inverse voltage* and is equivalent to a diode's breakdown voltage.

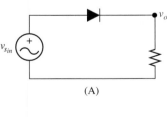

(A)

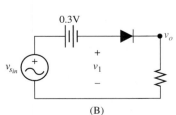

(B)

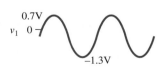

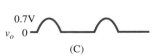

(C)

FIGURE 6–3

Solution

(a) We represent the diode drop with a 0.3V battery (Figure 6–3B, p. 149). The input signal is shifted down by 0.3V, as shown in the second plot of Figure 6–3C. The rectified output is shown in the last plot in Figure 6–3C.

(b) When the input goes negative, the diode opens and the output goes to 0V. The reverse-bias voltage is found from

$$v_D = v_{in} - v_o = v_{in}$$

The maximum reverse-bias voltage occurs when the input reaches its negative peak:

$$V_{D_{max}} = V_{in_{max}} = -1\text{V}$$

For safety, we would choose a diode with a *PIV* of at least 2V.

REVIEW QUESTIONS

1. Sketch the I-V characteristic of the germanium diode. Be sure to label the turn-on voltage.

2. What is the typical turn-on voltage of the germanium diode?

3. What are disadvantages of the germanium diode?

6–3 ■ THE SCHOTTKY DIODE

The germanium and standard silicon diodes are built with P- and N-type material, but we can also construct a diode by bonding a metal (aluminum or platinum) to, for example, N-type silicon. That is, we form the diode's anode with metal rather than P-type material (Figure 6–4A). This diode is known as a **Schottky diode.** Schottky diodes can also be constructed with a metal, P-type junction.

The circuit symbol and I-V curve for the Schottky diode are shown in Figure 6–4B. You can see that the I-V characteristic of the Schottky diode is similar to the I-V characteristic of the germanium diode. For the Schottky diode, as for the germanium diode

$$\text{Schottky: } V_{D(ON)} \approx 0.3\text{V}$$

The Schottky diode also has the same range of breakdown voltages as the germanium diode.

What is not shown on the I-V curve is the fact that the Schottky diode can switch **ON** and **OFF** much faster than the PN junction diodes. Also, the Schottky diode produces less unwanted noise than either the silicon or germanium diode. These two characteristics

FIGURE 6–4
Schottky diode

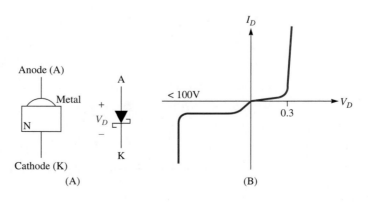

make the Schottky diode very useful, for example, in high-speed switching power circuits. And while the Schottky diode and the germanium diode both have a low turn-on voltage, the Schottky diode has the additional advantage of speed. As a result, you rarely see a germanium diode used today.

EXAMPLE 6–2
The Schottky Diode

Figure 6–5A shows a full-wave rectifier. Figure 6–5B shows the essential components of this circuit and the 5V AC secondary voltage of the transformer. Assume that

$$V_{D(ON)} = 0.3V$$

(a) When the secondary voltage is positive, which diode is **ON** and what is the output voltage?

(b) When the secondary voltage is negative, which diode is **ON** and what is the output voltage?

(c) What is the total output voltage?

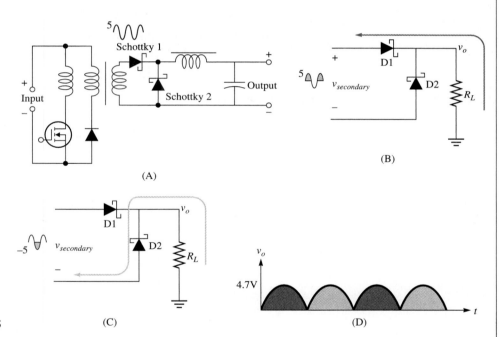

FIGURE 6–5

(A)

(B)

(C)

(D)

Solution

(a) Secondary voltage is positive. Figure 6–5B shows the current flow, indicating that Schottky 1 is **ON** and Schottky 2 is **OFF.** The current passes through the load such that the peak output voltage is positive and given by

$$V_{op} = V_{secondary} - V_{D(on)} = 5 - 0.3 = 4.7V$$

(b) Secondary voltage is negative. Figure 6–5C shows the current flow, indicating that Schottky 1 is **OFF** and Schottky 2 is **ON.** The current passes through the load such that the peak output voltage is still positive and given by

$$V_{op} = -V_{secondary} - V_{D(on)} = -(-5) - 0.3 = 4.7V$$

(c) The total output voltage is the full-wave rectified voltage shown in Figure 6–5D.

6–4 ■ THE TUNNEL DIODE

Figure 6–6A shows a very interesting I-V curve. In the middle of the curve (region II) the current actually decreases as the voltage increases. This is the I-V characteristic of a special device known as a **tunnel diode;** its circuit symbol is shown in Figure 6–6B.

Let's find the slopes in the three regions of the example tunnel diode I-V curve shown in Figure 6–7. In region I ($v_D < V_p$)

$$m_I = \frac{\Delta i}{\Delta v} = \frac{0.6\text{mA} - 0.4\text{mA}}{0.11\text{V} - 0.09\text{V}} = 10\text{mS}$$

This corresponds to an internal diode resistance of

$$R_I = \frac{1}{m_I} = \frac{1}{10\text{mS}} = 100\Omega$$

Let's jump to region III ($v_D > V_v$):

$$m_{III} = \frac{\Delta i}{\Delta v} = \frac{0.6\text{mA} - 0.4\text{mA}}{0.82\text{V} - 0.78\text{V}} = 5\text{mS}$$

FIGURE 6–6

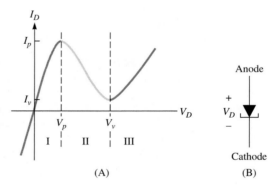

(A)   (B)

FIGURE 6–7

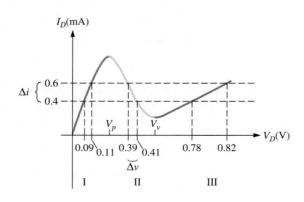

This corresponds to an internal diode resistance of

$$R_{III} = \frac{1}{m_{III}} = \frac{1}{5\text{mS}} = 200\Omega$$

Now, in region II ($V_p < v_D < V_v$)

$$m_{II} = \frac{\Delta i}{\Delta v} = \frac{0.6\text{mA} - 0.4\text{mA}}{0.39\text{V} - 0.41\text{V}} = -10\text{mS}$$

This corresponds to an internal diode resistance of

$$R_{II} = \frac{1}{m_I} = \frac{1}{-10\text{mS}} = -100\Omega$$

That's right, in region II the tunnel diode acts as a negative resistance! As a result, as the applied voltage is increased, the resulting current decreases. Negative resistances are useful in oscillator construction, and this is where you will sometimes find a tunnel diode.

Note: The negative resistance of the tunnel diode only manifests itself for AC signals. We found the resistance by taking the ratio of the *change* in v_D to the *change* in i_D. For DC, a positive voltage will produce a positive current.

EXAMPLE 6–3
The Tunnel Diode Circuit

Is the tunnel diode in Figure 6–8A operating in its negative resistance region, as shown in color in Figure 6–8B?

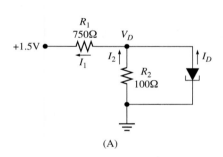

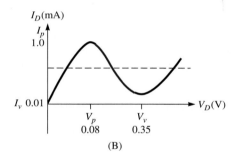

FIGURE 6–8 (A) (B)

Solution   Because there is no simple mathematical relation between the tunnel diode current and tunnel diode voltage, we cannot easily get an exact answer here. In fact, looking at the I-V curve in Figure 6–8B, you can see that the same current can be produced by three different voltages.

To get an approximate answer, we will first find the current at several different assumed diode voltages. Then we will try to find a diode voltage-current pair that is consistent with the diode I-V curve.

We see that the diode current is the difference in the two resistor currents:

$$I_D = I_1 - I_2$$

where the resistor currents are given by

$$I_1 = \frac{V_{in} - V_D}{R_1} = \frac{1.5 - V_D}{750}$$

and

$$I_2 = \frac{V_D}{R_2} = \frac{V_D}{100}$$

The diode current is, therefore

$$I_D = I_1 - I_2 = \frac{1.5 - V_D}{750} - \frac{V_D}{100}$$

Let's find the diode current for a voltage lower than V_p, a voltage slightly higher than V_p, a voltage close to V_v and a voltage greater than V_v:

V_D	I_D
0.05V	1.4mA
0.1V	**0.87mA**
0.3V	−1.4mA
0.4V	−2.5mA

Only the second result gives us both a diode voltage and a diode current that are approximately consistent with the diode I-V curve. The diode, therefore, is operating in its negative resistance region and, what is more, is operating near its peak current.

Figure 6–9 shows a piezo-electric crystal-controlled (C_R) circuit that converts a 28MHz input signal to a 1MHz output signal. While the circuit looks very complicated, finding DC values is not too difficult. Remember, for DC, capacitors act as open circuits and inductors act as short circuits. The crystal is also an open circuit for DC. For DC, therefore, the circuit reduces to the components shown in color in the figure. If you compare this piece of this complex circuit to Figure 6–8 (p. 153), you will see that we have already performed the DC analysis of it.

FIGURE 6–9

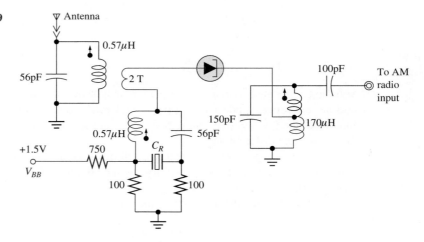

REVIEW QUESTIONS

1. Sketch a tunnel diode I-V curve. What is unusual about this curve?
2. On your I-V curve, show how can you find the internal resistance of the tunnel diode in the middle region of the curve.
3. What can you say about the resistance of the tunnel diode in the middle region of its I-V curve?

6–5 ■ THE VARACTOR DIODE

In Appendix A (p. A-1), we show you that at the PN junction of a diode, the physics of the device results in a negative charge on the P side and positive charge on the N side (Figure 6–10A). The region between these positive and negative charges, known as the *depletion region*, does not contain any mobile charges.

The result of PN-junction physics is that we end up with charges separated by an insulator. This is exactly how we describe a capacitor. In fact, all PN junctions have an associated capacitance (C_j). When voltage is applied to the diode, the depletion region decreases (forward bias) or increases (reverse bias), changing the value of the PN-junction capacitance. That is, the junction capacitance will increase with reverse bias and decrease with forward bias.

FIGURE 6–10
Varactor

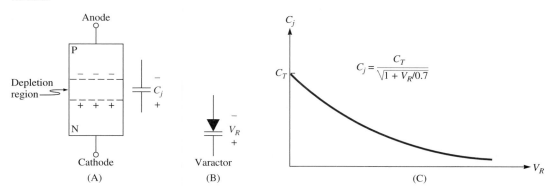

This capacitance is usually an annoyance that slows down the response of the diode. In integrated circuits, however, we take advantage of this property to build capacitors using reverse-biased PN junctions. Integrated circuit PN-junction capacitances are usually limited to less than 100pF.

Some diodes are specially manufactured so that the PN-junction capacitance has a known and controllable relation to the applied diode voltage. This produces a device known as a **varactor,** which has a voltage-controlled capacitance. Other names for the varactor are *tuning capacitor diode* and *varicap diode*.

Figure 6–10B shows the circuit symbol for a varactor, and Figure 6–10C shows how a varactor's capacitance depends on the applied reverse-bias voltage. Note that as the reverse-bias voltage increases, the varactor capacitance decreases. The value C_T is varactor capacitance when there is no reverse-bias voltage. The relation between reverse-bias voltage (V_R) and varactor capacitance is

$$C = \frac{C_T}{\sqrt{1 + V_R/0.7}}$$

Note: For some varactors the denominator is a cube root rather than a square root. The useful range of the varactor is approximately 1/2 to 1/3 of the zero bias value (C_T).

Figure 6–11 (p. 156) shows a data sheet for some commercially available varactors.

FIGURE 6–11

P_{TOT} mW	V_{BR} V	Nominal Capacitance (pF)
330	25	10
330	25	15
330	25	22
330	25	33
330	25	47
330	25	68
330	25	100

P_{TOT} mW	V_{BR} V	I_R μA	C_d(pF) Min.	Max.	V/√Hz
330	30	0.05	1.8	2.8	25/1
330	30	0.02	26	32	3/1
330	30	0.1	20	25	3/1

P_{TOT} mW	V_R V	I_F mA	C_d(pF) Nom.	Max.
330	30	20	6.8	7.5
330	30	20	8.2	9
330	30	20	10	11
330	30	20	12	13.2
330	30	20	15	16.5
330	30	20	18	19.8
330	30	20	22	24.2
330	30	20	27	29.7
330	30	20	33	36.3

EXAMPLE 6–4
The Varactor Diode

The circuit shown in Figure 6–12A uses the varactor ZC833ACT-ND, listed in Figure 6–11. Find the capacitance of the varactor.

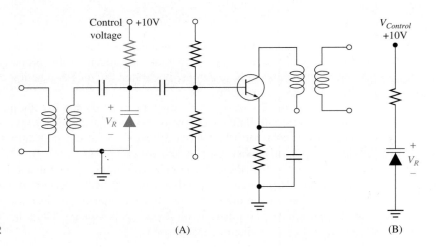

FIGURE 6–12   (A)   (B)

Solution   Because we are looking for the DC value across the varactor, we can open the coupling capacitors to get the circuit shown in Figure 6–12B. Now, because the diode is reverse biased, there is no current in the resistor and

$$V_R = V_{control} = 10V$$

This varactor has a nominal capacitance of 33pF. The capacitance of the varactor is, therefore

$$C = \frac{C_T}{\sqrt{1 + V_R/0.7}} = \frac{33pF}{\sqrt{1 + 10/0.7}} = 8.44pF$$

DRILL EXERCISE
Find the capacitance of the varactor in Figure 6–12 if the control voltage is 5V.

Answer $C = 11.6pF$

REVIEW QUESTIONS

1. Why is there a capacitance at a PN junction?
2. How does junction capacitance depend on the diode voltage?
3. What is the typical range of values for a PN-junction capacitance?
4. What is a varactor?
5. How does the capacitance of the varactor vary with the applied reverse-bias voltage? What is the formula?

6–6 ■ THE PHOTO DIODE

We can use specialized diodes known as **photo diodes** to convert light energy into electrical energy. Devices that convert one form of energy into another are known as *transducers, sensors,* or *detectors.* In fact, the photo diode is also known as a *photo detector.*

You know that the reverse-bias current of a diode is very small (Figure 6–13A). If a diode is constructed so that light (photons) can reach the junction, then the energy imparted by the photons to the atoms in the junction will create more free electrons (and more holes).

FIGURE 6–13
Photo diode

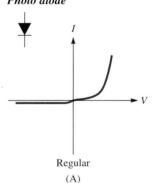

Regular
(A)

Increasing

Photo diode
(B)

(C)

These additional free electrons will create a larger reverse-bias current. As the incident light intensity increases, the reverse-bias current increases (Figure 6–13B). Also note that the entire I-V curve shifts to the right as light intensity increases. The circuit symbol for the photo diode is shown in Figure 6–13C (the arrows represent the incident light).

If we use the photo diode in the reverse-bias region, we have a device that has a current output that is dependent on incident light intensity. Used in this fashion, the photo diode is operating in the *photoconductive mode*. If we use the photo diode in the forward-bias region, we have a device that produces an output voltage in response to light illumination; used in this manner, the photo diode is operating in the *photovoltaic mode*.

The Photoconductive Mode

To use a photo diode in the photoconductive mode, it is operated strictly in the **OFF** region. To emphasize this point, we define a reverse-diode current and voltage (I_R and V_R), then plot the family of I_R-V_R curves that result from increases in the incident light illuminance (E in Figure 6–14A). Figure 6–14B shows a plot of the reverse current versus the incident light illuminance for a given photo diode. You can see that this plot is pretty close to a straight line, allowing us to write

$$I_R = mE$$

where m is the slope of the straight line in Figure 6–14B and is known as the *sensitivity* of the photo diode. For Figure 6–14B

$$I_R \approx 10 \ \frac{\mu A}{mW/cm^2} \ E$$

Note that most manufacturers label the reverse current when the diode is in the dark as *dark current* (I_D), and label the reverse current when the diode is in the light as *short circuit current* (I_{SC}). That is

$$I_R = I_D \ \text{dark}$$

$$I_R = I_{SC} \ \text{light}$$

Note: In this text, we have already used the symbol I_D to represent forward diode current. Here, manufacturers have chosen to use this same symbol to represent the *reverse* current. Using the same symbol to represent different variables and parameters can be confusing. Unfortunately, there are a limited number of letters in the English and Greek alphabet (we seem to have stopped at using just two alphabets), so this situation is often unavoidable. So, as with any language, in the language of electronics, you must be aware of the context in which the symbol appears.

FIGURE 6–14

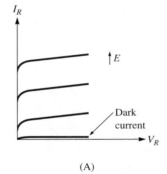 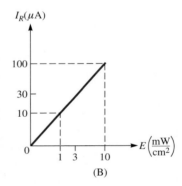

(A) (B)

EXAMPLE 6–5
The Photoconductive Cell

For the photo diode circuit shown in Figure 6–15:

(a) Find the voltage (V_o) across the photo diode in the dark. Is the diode reverse biased?
(b) The diode is now exposed to a pulse train of light with intensity of 3mW/cm². Use the graph given in Figure 6–14B to determine the current due to the light.
(c) Use the results of part B to find the output voltage.

Solution
(a) Because the 12V source is trying to pull current though the diode in the wrong direction, the diode is reverse biased. Therefore, the diode is operating in the photoconductive mode. In the dark we can assume that the reverse-bias current is negligible, so

$$\text{dark: } I_R \approx 0A$$

And because

$$V_o = 12 - I_R \ 10K$$

$$\text{dark: } V_o = 12V$$

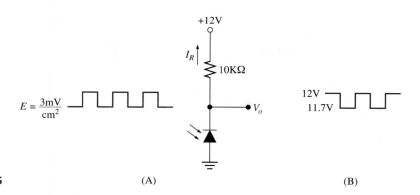

FIGURE 6–15 (A) (B)

(b) When the light is turned on, we use Figure 6–14B to find that for $E = 3\text{mW/cm}^2$:

$$\text{light: } I_R = 30\mu A$$

(c) In the light, the output voltage drops to

$$\text{light: } V_o = 12 - 10KI_R = 12 - 10K \times 30\mu A = 12 - 0.3 = 11.7V$$

The output voltage train is shown in Figure 6–15B.

Note that the photoconductive output voltage in the previous example only varied 300mV between the light and dark conditions. This is due to the fact that a photo diode only produces small currents. Most photo diode circuits, therefore, feed the detector output into an amplifier (Figure 6–16).

FIGURE 6–16

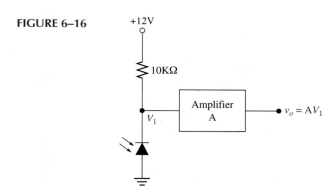

The Light Spectrum

When described as a wave, light is an electromagnetic signal, so it exhibits the same sinusoidal behavior as radio waves. That is, light waves are also described by their frequency (f(Hz)). Radio waves occupy the spectrum from the KHz range (AM radio) to the GHz range (satellite communication); light waves occupy the spectrum from the infrared (30THz) to the ultraviolet (3PHz). Visible light ranges from 400THz (red) to 800THz (violet). The electromagnetic spectrum is shown in Figure 6–17A (p. 160).

Figure 6–17B shows the spectrum in terms of the wavelength of the signal. The wavelength (λ) of an electromagnetic signal is the physical distance between the peaks of the sinusoidal wave. The wavelength and frequency of a signal are related by

$$\lambda = \frac{c}{f}$$

where c is the speed of light: $c = 3 \times 10^{10}\text{cm/s}$.

FIGURE 6–17

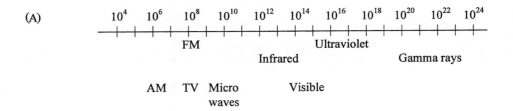

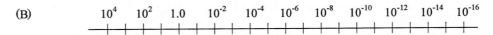

EXAMPLE 6–6
Frequency and Wavelength

Find the wavelength of the following signals:

(a) AM: $f = 640$KHz
(b) FM: $f = 2$MHz
(c) infrared: $f = 100$THz
(d) visible light: $f = 600$THz
(e) ultraviolet: $f = 1$PHz

Solution

(a) The wavelength for the AM signal is

$$\lambda = \frac{c}{f} = \frac{3 \times 10^{10}\text{cm/s}}{640\text{KHz}} = 4.68 \times 10^{4}\text{cm}$$

(b) The wavelength for the FM signal is

$$\lambda = \frac{c}{f} = \frac{3 \times 10^{10}\text{cm/s}}{2\text{MHz}} = 1.5 \times 10^{4}\text{cm}$$

(c) The wavelength for the infrared signal is

$$\lambda = \frac{c}{f} = \frac{3 \times 10^{10}\text{cm/s}}{100\text{THz}} = 3 \times 10^{-4}\text{cm}$$

(d) The wavelength for the visible light signal is

$$\lambda = \frac{c}{f} = \frac{3 \times 10^{10}\text{cm/s}}{600\text{THz}} = 5 \times 10^{-5}\text{cm}$$

(e) The wavelength for the ultraviolet signal is

$$\lambda = \frac{c}{f} = \frac{3 \times 10^{10}\text{cm/s}}{1\text{PHz}} = 3 \times 10^{-5}\text{cm}$$

Photo diodes are made from several different materials, such as silicon, germanium, and selenium, because each of these materials responds only to a specific subset of the light spectrum (Figure 6–18). Note that selenium has its peak response in the visible spectrum, silicon peaks in the near infrared region, and germanium peaks in the far infrared spectrum. By incorporating a specific substance, a photo diode can meet very specific needs.

FIGURE 6–18

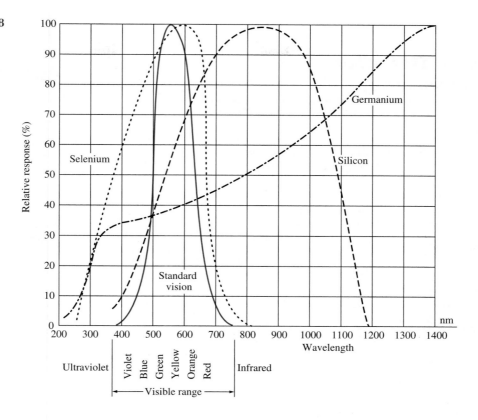

Commercial Photo Diodes Data Sheets

$$1 \text{ lux} = 1.46 \times 10^{-3} \frac{\text{W}}{\text{M}^2}$$

Figure 6–19 (p. 162) shows the data sheets for a commercial photo diode. The opto-electrical characteristics for this diode give us the following information:

☐ *Dark current* (I_D): the reverse current that exists when the diode is in the dark. For this device, the typical dark current is 3nA.

☐ *Short circuit current* (I_{SC}): reverse current in the diode at the indicated incident light intensity; 50 for a light intensity of 1000 lux. This gives us a device sensitivity of

$$0.05 \frac{\text{mA}}{\text{lux}}$$

☐ *Peak sensitivity wavelength* (λ): the light wavelength to which the diode is most sensitive. This device is most receptive to 940nm light, which is in the infrared range. This diode is not meant to detect visible light; it is an infrared detector.

Figure 6–20A, B, and C (p. 162) show three data graphs for this diode. The first graph shows the relationship between the the diode short circuit current and the incident light intensity, showing that

$$I_{SC} \text{ (\textmu A)} \approx 0.05E \text{ (lux)}$$

which agrees with the range calculated from the device's short-circuit current data.

The second graph shows the spectral response of the diode, or the wavelengths to which the diode is sensitive. The peak response occurs at a wavelength of 940nm, which is the value we got from the opto-electrical properties.

The third graph shows the directional sensitivity of this particular diode. Photo diodes are often constructed with internal lenses that allow light to enter the diode only

FIGURE 6–19

Photo Detector
Diode Output

This device is designed for infrared remote control and other sensing applications, and can be used in conjunction with the MLED81 infrared emitting diode.

Features:
- Low Cost
- Designed for Automated Handling and Accurate Positioning
- Sensitive Throughout the Near Infrared Spectral Range
- Infrared Filter for Rejection of Visible Light
- High Speed

Applications:
- Remote Controls in Conjunction with MLED81
- Other High Speed Optical Sensing Applications

MRD821

Motorola Preferred Device

**PHOTO DETECTOR
DIODE OUTPUT**

**CASE 381-01
STYLE 1**

MAXIMUM RATINGS

Rating	Symbol	Value	Unit
Reverse Voltage	V_R	35	Volts
Forward Current — Continuous	I_F	100	mA
Total Power Dissipation @ T_A = 25°C Derate above 25°C	P_D	150 3.3	mW mW/°C
Ambient Operating Temperature Range	T_A	− 30 to + 70	°C
Storage Temperature	T_{stg}	− 40 to + 80	°C
Lead Soldering Temperature, 5 seconds max, 1/16 inch from case	—	260	°C

ELECTRICAL CHARACTERISTICS (T_A = 25°C unless otherwise noted)

Characteristic	Symbol	Min	Typ	Max	Unit
Dark Current (V_R = 10 V)	I_D	—	3	30	nA
Capacitance (f = 1 MHz, V = 0)	C_J	—	175	—	pF

OPTICAL CHARACTERISTICS (T_A = 25°C unless otherwise noted)

Characteristic	Symbol	Min	Typ	Max	Unit
Wavelength of Maximum Sensitivity	λ_{max}	—	940	—	nm
Spectral Range	$\Delta\lambda$	—	170	—	nm
Sensitivity (λ = 940 nm, V_R = 20 V)	S	—	50	—	$\mu A/mW/cm^2$
Temperature Coefficient of Sensitivity	ΔS	—	0.18	—	%/K
Acceptance Half-Angle	φ	—	± 70	—	°
Short Circuit Current (Ev = 1000 lux[1])	I_S	—	50	—	μA
Open Circuit Voltage (Ev = 1000 lux[1])	V_L	—	0.3	—	V

NOTE 1. Ev is the illumination from an unfiltered tungsten filament source, having a color temperature of 2856K (standard light A, in accordance with DIN5030 and IEC publication 306-1).

FIGURE 6–20

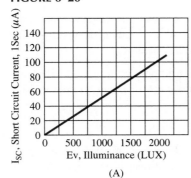

(A)

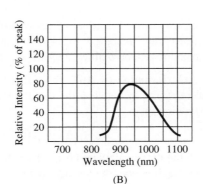

(B)

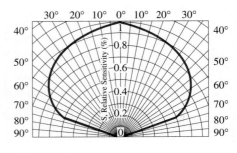

(C)

from specific directions. The diode will respond with varying sensitivity to light anywhere within the indicated contour but will not respond to light that lies outside this contour.

Figure 6–21 explains how to interpret the graph.

☐ Figure 6–21A: The diode has maximum response (100%) to a light source located directly in front of it (0°).

☐ Figure 6–21B: If the light source is placed at a 50° angle to the diode, we follow a line from the 50° mark to the angular response contour. Read the relative sensitivity from the circular line that passes through this point on the angular response contour. The relative sensitivity here is approximately 0.8. This means that the diode response to a light located at an angle of 50° is 80% of its maximum response.

Figure 6–22 shows some examples of the photo diodes available from an electronic parts retailer. The *acceptance angle* gives us the directional range (but not the directional sensitivity); $I_{C(on)}$ is their label for I_{SC}. (You can see that there is little standardization in the labeling of device parameters.)

FIGURE 6–21

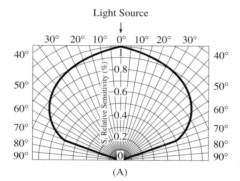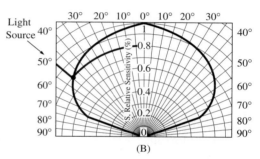

(A)   (B)

FIGURE 6–22

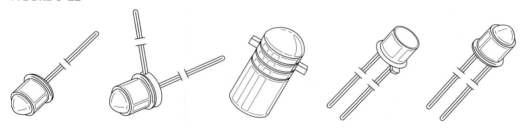

Photodiode Detectors

Type	Fig.	Acceptance Angle	min I_C (on) @ $V_R = 20V$ μA	H mW/cm²
SD1420-002	1	24°	5.0	20
SD1420-002L	2	24°	5.0	20
SD2420-002	3	48°	7.0	20
SD3421-002	4	90°	10.0	5
SD5421-002	5	18°	40.0	5

The Photovoltaic Mode

Look back at Figure 6–13B (p. 157) and you will see another striking fact about the family of curves for the photo diode. In the dark the photo diode behaves just like a regular diode; when there is no current, there is no voltage (the I-V curve goes through the origin). However, as the incident light intensity increases, even when there is no diode current, there is a diode voltage. This zero current voltage increases as light intensity increases.

This means that if you just sit a photo diode with no external voltage sources on the table and shine the right kind of light on it, the diode will produce a voltage. Used in this manner, the diode is operating in the *photovoltaic mode*. Because the relation between the diode voltage and incident light is not linear, we usually don't try to use the photovoltaic mode to give us the exact light intensity. Rather, we use the voltage output to tell us whether or not light is present and to give us a qualitative idea of light intensity.

Standard photo diodes can produce voltages up to approximately 0.5V and can provide currents up to approximately 1mA, but most photo diodes produce considerably less current than that. Because the output is small, we usually connect the diode output to a voltage amplifier. Figure 6–23 shows how we can use a photodiode and voltage amplifier to trip a relay when the sun comes out (an electronic rooster).

Photo diode output voltage and current can be increased by using a series and parallel array of these devices. In fact, the **solar cell** that you are probably aware of is a photovoltaic device that can output up to 1mA of current per cell.

FIGURE 6–23

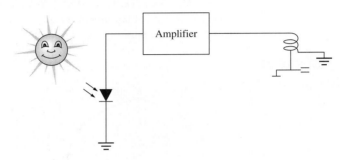

EXAMPLE 6–7
The Photovoltaic Mode

When illuminated, each photo diode in the array in Figure 6–24 produces 0.5V and a maximum current of 0.1mA.

(a) Find the total output voltage of the array and the maximum available current.
(b) If a diode fails open, what happens to the output voltage and available current?
(c) How can the output voltage be increased?
(d) How can the output current be increased?

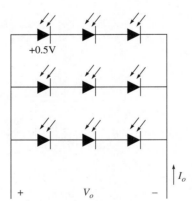

FIGURE 6–24

Solution

(a) Each row of the array contains three photo diodes in series. Because each diode produces 0.5V, the total voltage across a row (the rows are in parallel, so each row produces the same voltage) is the array output voltage:

$$V_o = 3 \times 0.5 = 1.5\text{V}$$

Because the diodes in a given row are in series, each diode must carry the same current, which is a maximum of 0.1mA. Each row can contribute 0.1mA, so the total available output current is

$$I_o = 3 \times 0.1\text{mA} = 0.3\text{mA}$$

(b) If a diode fails open, then one row is effectively removed from the array. Because the other two rows still have three diodes in series, the output voltage is still

$$V_o = 1.5\text{V}$$

The loss of a row, however, results in the loss of its current. With just two rows providing current, the available current is now

$$I_o = 0.2\text{mA}$$

(c) To increase the output voltage, increase the number of diodes in each row. Note that all rows must contain the same number of diodes.

(d) To increase the output current, increase the number of rows in the array. For example, if a fourth row of three diodes is added to the array, the available current will increase by 0.1mA; we will now have 0.4mA available.

DRILL EXERCISE   Using the photo diode in the example, design an array that will produce a 9V output with 1mA of current available.

Answer   18 photo diodes in each row; 10 rows in the array.

REVIEW QUESTIONS

1. Where on the I-V curve do we operate a photo diode in the photoconductive mode?
2. Sketch the relation between a photo diode current and the incident light intensity.
3. What measures of light intensity do we use? What are the relations between these measures?
4. What is the wavelength of a light wave? What is the relation between wavelength and frequency?
5. What is the range of frequencies for radio waves? For infrared light? For visible light? For ultraviolet light?
6. What is the range of wavelengths for radio waves? For infrared light? For visible light? For ultraviolet light?
7. What three materials are used in the construction of photo diodes?
8. Where does the peak response of silicon occur? Germanium? Selenium?
9. What are the typical parameters given for commercial photo diodes?
10. What happens to the zero current voltage of a photo diode as incident light intensity increases? Is this relation linear?
11. What is the operating mode of a photo diode when it is used without external voltage sources?
12. What are the typical ranges for photovoltaic output voltage and current?
13. In a photo diode array, what determines the output voltage? The available output current?

6–7 ■ THE PHOTORESISTOR (PHOTOCONDUCTIVE CELL)

The **photoresistor** is a cousin of the photo diode; it also converts light into an electrical signal. Photoresistors are also known as *photoconductive cells*. Do not confuse a photoconductive cell with the photoconductive mode of the photo diode (sorry about the confusion—we didn't invent the terminology). The photoresistor is not a diode with a PN junction. Instead it is simply a hunk of N-type material. The resistance of this material is determined by the number of free electrons in the material. As the material absorbs light energy, the number of free electrons increases, so its resistance decreases.

Figure 6–25A shows a typical photoresistor and its circuit schematic. Figure 6–25B shows data for several commercial photoresistors. This data indicates the device resistance in the dark and in the light.

Figure 6–26 shows a simple voltage divider circuit, in which the voltage measured across the photoresistor is a measure of light intensity. That is

FIGURE 6–25

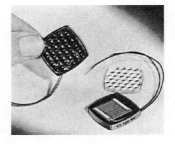

(A)

PHOTORESISTORS

TYPE	Voltage (Max.)	2 fc Resistance in KΩ	Maximum Dark Resistance (Ω) 5 sec. after 2 fc.
VT-201	100	4.3	500K
VT-202	100	8	750K
VT-204	300	110	20M
VT-211	100	2.3	500K
VT-211H	100	3.5	1M
VT-212L	100	7.5	5M
VT-214L	300	64	100M
VT-214	300	140	200M
VT-223H	300	65	5M
VT-232	10	100	50M
VT-241	100	2	500K
VT-242	100	15	10M
VT-242H	200	30	10M
VT-301	200	3	200K
VT-304	300	144	50M
VT-311L	200	2.1	500K
VT-314	300	140	200M
VT-322L	300	15.5	5M
VT-333L	200	14	5M
VT-333	200	25	10M
VT-34L	200	0.7	100K
VT-343	300	28	10M
VT-50L	200	1.3	100K
VT-501	200	2.5	200K
VT-502L	300	8	1M
VT-801	200	3	250K
VT-833	300	33	15M
VT-841L	100	0.9	500K

(B)

FIGURE 6–26
Photoresistor

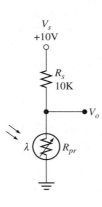

$$V_o = \frac{R_{pr}}{R_{pr} + R_s} V_s$$

Note: Because R_{pr} appears in both the numerator and denominator of this equation, the relation between light intensity and V_o is not linear.

EXAMPLE 6–8
The Photoresistor

The circuit of Figure 6–26 uses the VT-241, described in Figure 6–25B. Find the output voltage when the photoresistor is

(a) In the dark.
(b) In the light ($E = 2$fc); (1fc $= 1.36 \times 10^{-4}$ W/M^2).

Solution   The output voltage is given by

$$V_o = \frac{R_{pr}}{R_{pr} + R_s} V_s = \frac{R_{pr}}{R_{pr} + 10K} 10$$

(a) In the dark this photoresistor has a resistance of 500KΩ. Therefore, in the dark, the output voltage is

$$V_o = \frac{R_{pr}}{R_{pr} + 10K} 10 = \frac{500K}{500K + 10K} 10 = 9.80V$$

(b) In the light this photoresistor has a resistance of 2KΩ. The output in the light is

$$V_o = \frac{R_{pr}}{R_{pr} + 10K} 10 = \frac{2K}{2K + 10K} 10 = 1.67V$$

REVIEW QUESTIONS

1. What is a photoresistor?
2. How does light change the resistance of a photoresistor?
3. As light intensity increases, what happens to the resistance of a photoresistor?
4. What simple circuit can we use to convert light intensity into a voltage? Is the voltage linearly related to light intensity? Why?

6–8 ▪ LIGHT-EMITTING DIODES (LED)

LED Basics

The light-emitting diode (LED) and its circuit symbol is shown in Figure 6–27. Note that the arrows now leave the diode, showing that this diode uses energy to produce light. The relation between light energy and electrical energy that we discussed for the photo diode is still true here, only in reverse. In the LED, electrical energy applied to the junction raises

FIGURE 6–27

Case 173-01
Plastic

Case 209-02
Convex Lens
Metal

Case 349-01
Plastic

Case 349C-01
Plastic

FIGURE 6–28

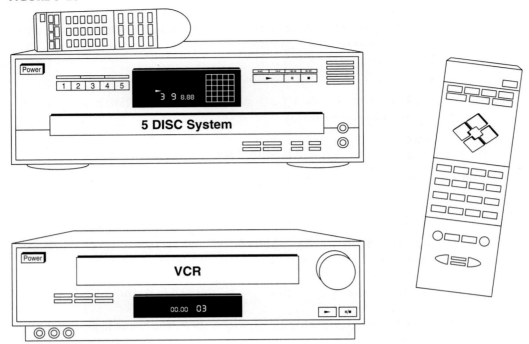

the energy level of the electrons. As electrons return to their original energy level, they emit photons and, so, give off light.

The LED is one of today's most common electro-optical devices, forming the displays on such devices as clocks, VCRs, calculators, and (the couch potato's favorite) the remote control (Figure 6–28). Too many of us are familiar with the flashing LED display that seems to mock us when we try to program some of today's electronics.

LEDs light up when they are forward biased. The typical forward voltage (V_F) and current (I_F) for LEDs are

V_F: 1.5 → 2.5V

I_F: <20mAfor AC operation and <10mA for DC (battery) operation

We limit the current in DC operation to reduce the drain on the battery.
CAUTION! LEDs typically have reverse breakdown voltages on the order of 5V. This means you must be careful not to use LEDs with a high level of reverse bias.

The color of the light given off by an LED depends on the material used in its construction and on the construction technique. Figure 6–29 shows the LED colors generally available (including infrared). LEDs are also constructed so they emit light only in given directions.

LEDs are commonly used as indicators that something has happened. For example, in the circuit in Figure 6–30, we will know when the switch has closed because the LED will light up. For the device in this circuit we assume that

$$V_F = 2V$$

If the LED is **ON** then

$$I_F = \frac{12 - V_F}{R_S} = \frac{12 - 2}{2K} = 5mA$$

FIGURE 6–29

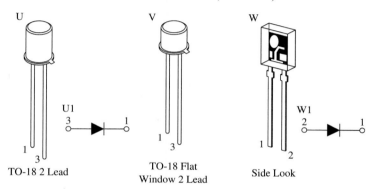

Package	Lens Size	Color	Electro-Optical Characteristics (Ta = 25°C)					Dimensions		Lead Spacing
			P_D (mW)	I_F (mA)	V_R (V)	I_O (mcd)	λ_P (nm)	Lens Height	Lead Length	
	φ2.0	Red Diffused	70	20	4	1.0	700	8.0	14, 15.5	2.54
	φ2.0	Green Diffused	90	20	4	1.0	565	8.0	14, 15.5	2.54
	φ2.0	Amber Diffused	90	20	4	2.5	590	8.0	14, 15.5	2.54
	φ2.8	† Red Clear	70	25	4	0.6	700	5.5	15.5, 17.0	2.54
	φ2.8	Green Clear	90	30	4	0.8	565	5.5	15.5, 17.0	2.54
	φ2.8	Amber Clear	90	30	4	0.8	590	5.5	15.5, 17.0	2.54
	φ2.8	Orange Clear	90	30	3	1.0	630	5.5	15.5, 17.0	2.54
	φ3.5	Red Diffused	70	25	4	0.7	700	2.3	12.5, 12.5	2.54
	φ3.5	Green Diffused	90	30	4	1.2	565	2.3	12.5, 12.5	2.54
	φ3.5	Amber Diffused	90	30	4	0.8	590	2.3	12.5, 12.5	2.54
	φ3.5	Orange Diffused	90	30	3	1.5	630	2.3	12.5, 12.5	2.54

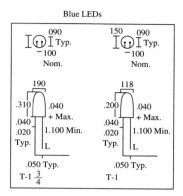

Blue LEDs

Semiconductor Infrared Emitters (IRED LEDs)

Dia-gram	Circuit	Forward Voltage (V)	Max. Forward Current (mA)	Max. Power Output (mW)	Beam Angle +/−	Peak Wave Length (nm)
U	U1	1.7	100	6.0	10	940
V	U1	1.7	100	6.0	40	940
U	U1	1.7	100	25.0	10	940
U	U1	1.7	100	5.4	8	940
V	U1	1.7	100	5.4	40	940
U	U1	1.7	100	12.0	10	880
U	U1	1.7	100	9.0	10	880
U	U1	1.7	100	10.5	10	880
V	U1	1.7	100	12.0	40	880
V	U1	1.7	100	9.0	40	880
V	U1	1.7	100	10.5	40	880
W	W1	1.7	60	0.28	30	940
W	W1	1.35	50	0.6	30	880

FIGURE 6–30
Light emitting diode (LED)

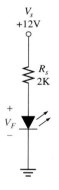

EXAMPLE 6–9
A Bad LED Design

We can often learn from bad designs. The LED circuit in Figure 6–31A (p. 170) is meant to indicate the voltage level of the input signal. This is what is supposed to happen: As the voltage increases beyond 2V, the first LED lights up; as the voltage increases beyond 4V, the second LED lights up; as the voltage increases beyond 6V, the third LED lights up.

What problems make it impossible for this circuit to work? Each LED has a turn-on voltage of 1.5V and a maximum safe operating current of 20mA.

Solution Problem #1: The design does not take into account the voltage drop across the LEDs. For example, LED1 will not light up until the input voltage reaches

$$V_{in} = V_{LED1} + V_{Z1} = 1.5 + 2 = 3.5V$$

A proper design would use Zener diodes of 0.5, 2.5, and 4.5V.

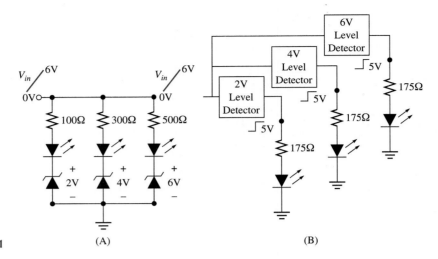

FIGURE 6–31

(A)   (B)

Problem #2: Even if we correct Problem #1, the current in each LED will vary with the input voltage. The LED current is given by

$$I_{LED} = \frac{V_{in} - (V_{LED} + V_Z)}{R}$$

To correct Problem #1, we would use a 0.5V Zener for the first LED. The current for this LED is now

$$I_{LED1} = \frac{V_{in} - (V_{LED} - V_Z)}{R} = \frac{V_{in} - 2}{100}$$

As the input voltage hits 2V, the LED will begin to light up. However, at this point there is very little current, so the LED will barely glow. As the input voltage continues to increase, the LED will grow brighter. The maximum current in this LED is

$$I_{LED1_{max}} = \frac{6 - 2}{100} = 40mA$$

which exceeds the maximum allowable LED current. To reduce the maximum LED current, we can increase the resistor to 200Ω; however, this will result in an even dimmer LED output.

Figure 6–31B shows a better indicator design. The level detectors output a constant voltage (5V) as the input exceeds a set reference voltage. That is, when the input passes 2V, the first level detector provides a 5V output to drive the first LED; the second level detector provides a 5V output when the input passes 4V; the third detector produces a 5V output when the input passes 6V. When the input voltage is less than the reference voltage, the level detectors output 0V, and the associated LED is dark.

With this circuit, each LED is always driven by 5V; when lit, the current in each LED is then

$$I_{LED} = \frac{5 - 1.5}{175} = 20mA$$

The Seven-Segment Display

The seven-segment display shown in Figure 6–32A is probably the most common application of LED technology. Each segment in this display is an LED constructed to produce a bar of light. By turning on the appropriate LED segments, the device displays every letter

FIGURE 6–32

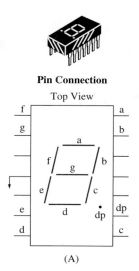

Pin Connection

Top View

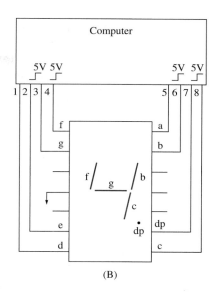

in the alphabet and every number from 0 to 9. Figure 6–32B shows a typical method for controlling the seven-segment display. When the voltage on a computer output line goes high (5V), the associated LED bar lights up. For example, to display the number 4, computer output lines 3, 4, 6, and 8 would go high.

Data sheets for some commercial seven-segment displays are shown in Figure 6–33. The data for the individual LEDs in the segment are similar to those for stand-alone LEDs.

FIGURE 6–33

SEVEN SEGMENT DIGITAL DISPLAY

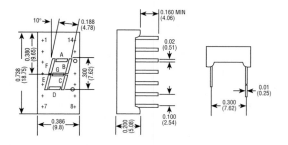

COMMON ANODE

PIN #	FUNCTION
1.	CATHODE A
2.	CATHODE F
3.	COMMON ANODE
4.	NO PIN
5.	NO CONNECTION
6.	NO PIN
7.	CATHODE E
8.	CATHODE D
9.	CATHODE DP
10.	CATHODE C
11.	CATHODE G
12.	NO PIN
13.	CATHODE B
14.	COMMON ANODE

COMMON CATHODE

PIN #	FUNCTION
1.	ANODE F
2.	ANODE G
3.	NO PIN
4.	COMMON CATHODE
5.	NO PIN
6.	ANODE E
7.	ANODE D
8.	ANODE C
9.	ANODE DP
10.	NO PIN
11.	NO PIN
12.	COMMON CATHODE
13.	ANODE B
14.	ANODE A

0.3" DIGIT SIZE

PART NO.	PEAK WAVE-LENGTH λP(nm)	EMITTED COLOR	MAXIMUM RATINGS				OPTO-ELECTRICAL CHARACTERISTICS						
			I_F	V_R	P_D	Topr/Tstg	$V_F(V)$		I_F	$I_R(\mu A)$	V_R	$I_V(\mu cd)$	I_F
			(mA)	(V)	(mW)	(°C)	typ.	max.	(mA)	max.	(V)	typ.	(mA)
MTN1130-ASR	655	Red	25	3	50	–40 ~ 85	1.7	2.0	20	100	3	300	10
MTN2130-AG	567	Green	20	5	60	–40 ~ 85	2.2	3.0	10	100	5	1200	10
MTN4130-AO	635	Organge	20	5	60	–40 ~ 85	2.1	3.0	10	100	5	1300	10
MTN4130-AHR	635	Hi-Effi Red	20	5	60	–40 ~ 85	2.1	3.0	10	100	5	1300	10
MTN1130-CSR	655	Red	25	3	50	–40 ~ 85	1.7	2.0	20	100	3	300	10
MTN2130-CG	567	Green	20	5	60	–40 ~ 85	2.2	3.0	10	100	5	1200	10
MTN4130-CO	635	Orange	20	5	60	–40 ~ 85	2.1	3.0	10	100	5	1300	10
MTN4130-CHR	635	Hi-Effi Red	20	5	60	–40 ~ 85	2.1	3.0	10	100	5	1300	10

6–9 ■ THE PHOTO TRANSISTOR AND OPTO-ISOLATOR

The Photo Transistor

Similar to the photo diode, the **photo transistor** (Figure 6–34A) produces an output current in response to incident light. However, due to a different construction, the photo transistor produces significantly more current than the photo diode. The data for photo diodes and photo transistors in Figure 6–34B shows that the photo transistor produces 10 to 20mA. Photo diodes typically produce less than 1mA.

From a practical point of view, the difference in current output is the only difference between the photo transistor and the photo diode. All of our previous discussions about the photo diode applies to the photo transistor. This is why we present the photo transistor now, before we have discussed transistor properties.

FIGURE 6–34

(A)

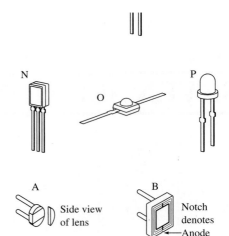

NPN Photo Transistor

Diagram	V_{CEO}	I_C	Peak Sensitivity Wavelength	tr/tt
N	20V	10mA	900nm	8μs
O	20V	20mA	800nm	2.5/3.5μs
K	20V	20mA	800nm	4μs
P	30V	20mA	800nm	4μs

Photodiodes

Diagram	Short Circuit Current $I_{SC}(\mu A)$	Dark Current I_D(nA)	Peak Sensitivity Wavelength (nm)	Breakdown Voltage V_{BR}(V)	Active Area (mm sq.)
				Blue Enhanced Photodiodes	
A	155	1.0	950	10	17
A	70	5	950	15	7.75
B	85	2	950	10	100

The Opto-Isolator (Opto-Coupler)

The *opto-isolator,* also known as the *opto-coupler,* is a device that combines an LED and a photo transistor in a single package. Figure 6–35A shows several opto-isolators and their circuit schematics. Figure 6–35B shows the basic operating principle of the opto-isolator. When an electrical signal activates the LED, the LED light will be modulated in the same manner as the input. For example, a sinusoidal variation in the LED input current will produce an output light that varies sinusoidally in intensity. The LED light shines on the photo transistor, producing an output current that has the same wave shape as the input.

It might occur to you that a simple piece of wire could accomplish the same task as this fancy bit of semiconductor electronics. To see why the opto-isolator is used, consider the situation shown in Figure 6–36A. We have connected electrical equipment to record

FIGURE 6–35
Photo transistor

Dia-gram	Isolation Voltage	Current Transfer Ratio - Typ.	BVCEO (Min.)	Typ. ton/toff (μsec.)
A1	1775	20%	30	3.0/3.0
A1	2500	20%	30	2.5/2.6
A1	2500	20%	30	2.0/2.0
A1	2500	20%	30	3.0/3.0
A1	1500	20%	30	2.0/2.0
A1	2500	10%	30	3.0/3.0
A1	1500	10%	30	2.0/2.0
A1	2500	10%	30	2.5/2.6
A1	500	10%	30	2.0/2.0
B1	2500	100%	30	5.0/40
B1	2500	100%	30	5.0/40
B1	2500	100%	30	5.0/40
B1	2500	100%	30	5.0/40
B1	1500	100%	30	5.0/40
B1	2500	50%	30	5.0/40
B1	1500	50%	30	5.0/40
B1	2500	500%	30	5.0/100
B1	2500	500%	30	5.0/100
B1	2500	500%	30	5.0/100
B1	2500	500%	30	5.0/100

A1 — Photo-Transistor Output

A2 — Dual Photo Transistor Output

B1 — Photo-Darlington Output

(A)

(B)

some vital signs of our patient. We have done this by simply wiring our recorder directly to the sensors on the patient. We start recording data and look forward to lunch.

Suddenly, there is a short in the measuring equipment, causing a large electrical surge to travel down the wires to the sensors, and, unfortunately, to the patient. Instead of resting comfortably on the table, the patient is face down on the floor and turning blue (Figure 6–36B). Lunch is off. If we had used an opto-isolator to couple the sensor signal into the recorder (Figure 6–36C), the electrical surge would blow out the opto-isolator, not the patient.

FIGURE 6–36

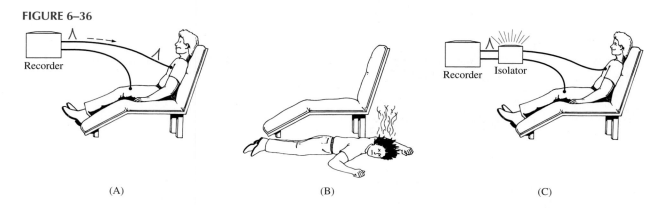

(A)   (B)   (C)

Opto-isolators are used whenever we want to electrically isolate two parts of a circuit. Figure 6–37 (p. 174) shows the following data for some commercially available opto-isolators:

☐ *Isolation voltage:* These devices can withstand 2500V and maintain input-output isolation.

☐ *Current transfer ratio:* These devices produce an output current that is 0.5 to 3 times the input current.

☐ *ton/toff:* How fast the opto-isolator can be turned on and off; these devices turn on in 3μs and turn off in 5μsec.

FIGURE 6–37

Photo-Transistor Output Optoisolators

Dia-gram	Isolation Voltage	Current Transfer Ratio - Typ.	BVCEO (Min.)	Typ. ton/toff (μsec.)
K	2500	50-300%	40	3/5
L	2500	50-300%	40	3/5
M	2500	50-300%	40	3/5

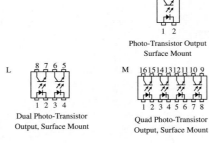

Photo-Transistor Output
Surface Mount

Dual Photo-Transistor
Output, Surface Mount

Quad Photo-Transistor
Output, Surface Mount

FIGURE 6–38
Fiber optic components

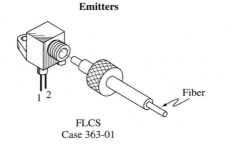

Emitters

FLCS
Case 363-01

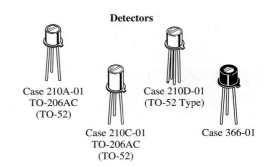

Detectors

Case 210A-01
TO-206AC
(TO-52)

Case 210C-01
TO-206AC
(TO-52)

Case 210D-01
(TO-52 Type)

Case 366-01

Fiber Optics

You are probably aware that many of our telephone lines, computer data lines, and other communication systems use fiber optics to transmit signals. The basic components of a fiber optic system are an emitter, detector, and the fiber optic wire (Figure 6–38). The emitter and detector are actually the LED and the photo transistor discussed in this chapter.

REVIEW QUESTIONS

1. What is the primary difference between the photo transistor and the photo diode?
2. How is a opto-isolator constructed?
3. What is another name for an opto-isolator?
4. Why do we use opto-isolators?

SUMMARY

■ The germanium diode has a turn-on voltage of approximately 0.3V, a breakdown voltage of less than 100V, and a larger reverse-bias current than the silicon diode.

■ The Schottky barrier diode is constructed with a metal-silicon junction. It has a turn-on voltage of approximately 0.3V and a breakdown voltage of less than 100V. The Schottky diode is faster than either of the PN-junction diodes.

■ The tunnel diode has a negative AC resistance region.

■ Varactor diodes have a junction capacitance that is set by their reverse-bias voltage. The junction capacitance decreases from its nominal value (C_T) according to the following:

$$C = \frac{C_T}{\sqrt{1 + V_R/0.7}}$$

- The photo diode has an I-V curve that depends on the light intensity incident upon it. The spectral response of this diode depends on its materials and construction.
- When reverse biased (in the photoconductive mode), the photo diode has a reverse-biased current that is linearly dependent on the incident light intensity.
- When not reverse biased (in the photovoltaic mode), the photo diode has a zero current voltage that depends on the incident light intensity.
- The photoresistor (also known as the photoconductive cell) has a resistance that decreases when it is exposed to light.
- The light-emitting diode (LED) produces a light when it is energized by approximately 2V. The color of the light emitted depends on the material used to construct the diode.
- The photo transistor has the same general behavior as the photo diode; however, the photo transistor produces more current.
- Opto-isolators (opto-couplers) are composed of an LED and a photo transistor. They are used to electrically isolate parts of a circuit.

PROBLEMS

SECTION 6–2   The Germanium Diode

1. Sketch the I-V curve for a germanium diode. Indicate the turn-on voltage and the breakdown voltage.
2. For the germanium diode circuit in Figure 6–39
   (a) Is the diode **ON** or **OFF?**   **(b)** What is V_D?
   (c) Find I_D.   **(d)** Find V_o.
3. Draw the output voltage of the germanium diode rectifier in Figure 6–40.

FIGURE 6–39

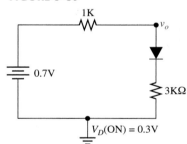

FIGURE 6–40

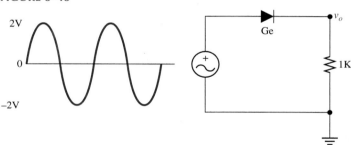

4. For the circuit of Figure 6–41
   (a) Find the peak inverse voltage the diode will experience.
   (b) Can a germanium diode be used in this circuit? Why?

FIGURE 6–41

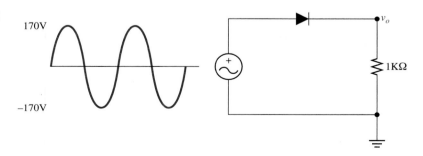

SECTION 6–3   The Schottky Diode

5. How does the I-V curve of the Schottky diode compare to the I-V curve of the germanium diode?

6. What advantages does the Schottky diode have over the germanium diode?
7. For the Schottky diode circuit in Figure 6–42
   (a) If $R_1 = 1K\Omega$, is the diode **ON** or **OFF?** Find V_o.
   (b) If $R_1 = 5K\Omega$, is the diode **ON** or **OFF?** Find V_o.

FIGURE 6–42

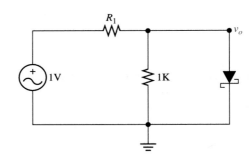

SECTION 6–4   The Tunnel Diode

8. For the tunnel diode I-V curve in Figure 6–43
   (a) Find the device resistance at point 1.
   (b) Find the device resistance at point 2.
   (c) Find the device resistance at point 3.

FIGURE 6–43

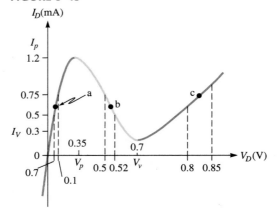

FIGURE 6–44

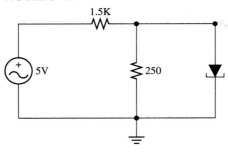

FIGURE 6–45

9. The tunnel diode in Figure 6–44 has the I-V curve shown in Figure 6–43. Find the diode current for the diode voltages given below. Which voltage-current pair is consistent with the I-V curve?

$$V_D = 0.3, 0.4, 0.65, 0.75V$$

SECTION 6–5   The Varactor Diode

10. The varactor in Figure 6–45 has a nominal capacitance of $C_T = 20pF$. Find the varactor capacitance.
11. The varactor in Figure 6–46 (p. 177) has a nominal capacitance of $C_T = 40pF$. Find the varactor capacitance.

FIGURE 6–46

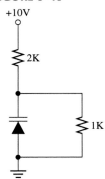

+10V

2K

1K

12. You are using a varactor that has $C_T = 30pF$. You want to set its capacitance to 20pF. Find the necessary reverse-bias voltage.

SECTION 6–6   The Photo Diode

13. Figure 6–47 shows a photo diode circuit, where the diode has a sensitivity of

$$\emptyset.1 \; \frac{\mu A}{lux}$$

  (a) Find V_o in the dark.
  (b) Find V_o if light incident on the diode has the illuminance $E = 2000$ lux.
14. Find the wavelength for each of the following frequencies:
  (a) $f = 5MHz$ **(b)** $f = 200THz$ **(c)** $f = 740THz$
15. Find the frequency corresponding to the following wavelengths:
  (a) $\lambda = 1 \times 10^6 cm$ **(b)** $\lambda = 2 \times 10^{-4} cm$ **(c)** $\lambda = 5 \times 10^{-6} cm$
16. A photo diode has the spectral response shown in Figure 6–20B (p. 162).
  (a) What is the wavelength of input that produces the peak response?
  (b) What is the frequency of input that produces the peak response?
  (c) What is the range of wavelengths that produce at least 50% of the peak response?
17. A photo diode has the angular response shown in Figure 6–20C.
  (a) What is the relative sensitivity of the diode to a light source that is located at a 40° angle to the diode?
  (b) What is the relative sensitivity of the diode to a light source that is located at a 60° angle to the diode?
18. Each diode in the photo diode array in Figure 6–48 produces a voltage of 0.75V and can supply up to 0.2mA. Find the total array voltage and the maximum available current.

FIGURE 6–47

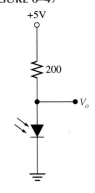

+5V

200

V_o

FIGURE 6–48

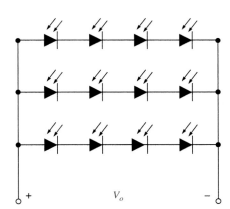

+ V_o −

FIGURE 6–49

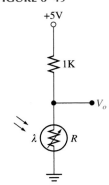

+5V

1K

V_o

λ R

19. A photo diode produces 0.5V with a maximum available current of 0.5mA. Design a photo diode array that will produce 4V @ 2mA.

SECTION 6–7   The Photoresistor

20. Figure 6–49 shows a photoresistor circuit. If the photo resistor is a VT-204 (Figure 6–25B, p. 166), find the output voltage:
  (a) In the dark.   **(b)** In the light.
21. You need a photoresistor that will produce 2 to 3V in the light. Which photoresistors in Figure 6–25B will satisfy your requirements?

FIGURE 6–50

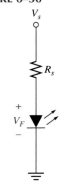

SECTION 6–8 Light-Emitting Diodes (LED)

22. In Figure 6–50, $V_S = +5V$, $V_F = 1.5V$, $R_s = 1K\Omega$. What is I_F?

23. In Figure 6–51, $V_S = +5V$, $V_F = 1.5V$. What is the smallest R_S that can be used if we want the diode current to remain below 15mA?

24. In Figure 6–51, $V_S = +5V$, $V_F = 1.5V$, $R_s = 1K\Omega$, $V_Z = 2V$. What is I_F?

25. In Figure 6–51, $V_S = +5V$, $V_F = 1.5V$, $R_s = 1K\Omega$, $V_Z = 4V$. What is I_F?

26. Referring to Figure 6–32, which lines must go high to produce

    (a) The number 5?   **(b)** The letter A?

TROUBLESHOOTING PROBLEMS

Note: For diode troubleshooting techniques, please refer to Chapter 3, Section 9.

27. The germanium diode rectifier circuit in Figure 6–52 is not working. What is a likely cause of the following observed outputs:

    (a) $v_o = 0V$

    (b) The output is the same as the input.

28. Refer to Figure 6–8 in Example 6–3, p. 153. The measured voltage across the tunnel diode is 0.18V.

    (a) What are the two resistor currents and the tunnel diode current?

    (b) What is the likely cause of these results?

FIGURE 6–51

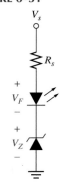

29. How can you tell if a photo diode has failed?

30. How can you tell that an LED has failed?

31. Figure 6–53 is a circuit that is supposed to indicate when the input jumps to +5V.

    (a) Why won't this circuit work?

    (b) You have changed the Zener diode to +3V. The LED gets very bright and then goes dead. What has happened?

FIGURE 6–52

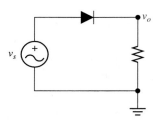

FIGURE 6–53

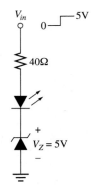

BIPOLAR JUNCTION TRANSISTORS (BJT)

PART

2

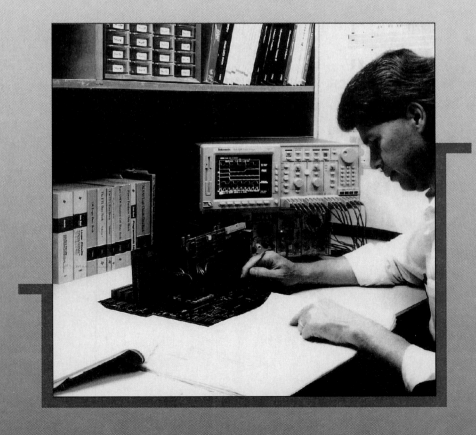

CHAPTER

7

THE NPN BIPOLAR JUNCTION TRANSISTOR

To determine the currents and voltages of the NPN Bipolar Junction Transistor in the **CUT-OFF, ACTIVE,** and **SATURATION** states.

To determine the maximum power dissipation of the transistor.

To find the DC values for the transistor currents and voltages for transistors operating in the **ACTIVE** state.

To find the DC values for the transistor currents and voltages for transistors operating in the **CUT-OFF** and **SATURATION** states.

To determine the operating state of a transistor.

To extract relevant information from a transistor data sheet.

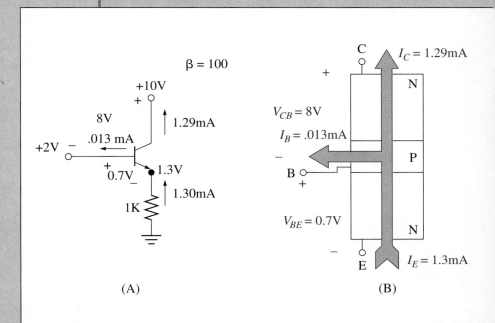

(A)   (B)

7–1 ■ INTRODUCTION

In the previous chapters we showed you how P-type and N-type material are combined to produce rectifiers and Zener diodes. These PN devices are used primarily to produce the DC voltages required to operate today's electronics—electronics that provide us with entertainment, industrial processes, transportation, education, communication, computers, defense, and, most importantly, jobs. Although these electronic components have different functions and uses, they all share common circuits that serve as building blocks. The primary component of these building block circuits is the transistor.

In the 1950s engineers at Bell Laboratories constructed the first useful transistors by sandwiching P-type material between two N-type materials (see Figure 7–1A). The actual process they used, known as "double diffusion," resulted in the transistor shown more accurately in Figure 7–1B. This process starts with N-type material; then P-type material is diffused into it. Additional N-type material is then diffused into the P region. This three-terminal device is the NPN *B*ipolar *J*unction *T*ransistor, or NPN BJT. The circuit symbol for the NPN BJT is shown in Figure 7–1C, where the terminals are labeled the base (B), the collector (C), and the emitter (E).

FIGURE 7–1
Representations of a transistor.

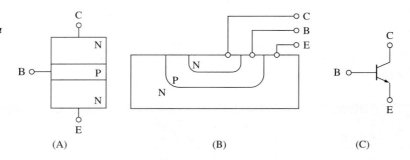

(A)　　　　　　　(B)　　　　　　　(C)

The BJT actually comes in two "flavors." It can also be built by sandwiching N-type material between two P-type materials to produce the PNP BJT. The current and voltage polarities of the PNP are opposite to the NPN's polarities. Otherwise, both transistor types behave alike, although the NPN is the most commonly used BJT. Here we will introduce the basic transistor circuits with the NPN transistor. We will discuss the PNP BJT in Chapter 14.

It is helpful to think of the transistor as an electronic valve. In fact, the British term for the vacuum tube, the device that preceded the transistor, is *valve*. Consider an example out of the childhood of one of the authors. Growing up in New York City before air-conditioning, we kids use to cool off in the hot muggy summers by turning on the fire hydrants—at least until someone called the cops. The point here is that the hydrant can be thought of as a three-terminal device: The input comes from the city's water main, the output gushes forth from the hydrant outlet, and the valve stem at the top of the hydrant controls the flow of the water. (Figure 7–2A).

FIGURE 7–2

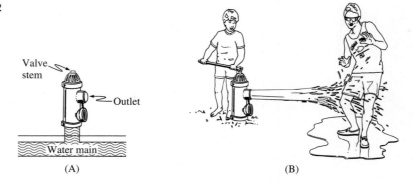

(A)　　　　　　　　　　　　　　(B)

The fire hydrant could be put in one of three states. The first—the hydrant turned off—was preferred by the police, firemen, and most of the adult population on the block. This was, of course, unacceptable to us hot, sweaty little hooligans. We usually preferred the second state, with the hydrant turned on to some moderate level of output, which we could modulate by increasing or decreasing the valve opening. Finally, when we were feeling even more mischievous than usual, which happened more often than I care to remember, we would move to the third state by cranking the valve fully open to get maximum water flow, a raging torrent which we could use to knock down unsuspecting pedestrians.

This analogy stresses two points: First, a three-terminal device, hydrant or transistor, can operate in three states: (1) **CUT-OFF,** in which the hydrant has no water flow and the transistor has no current output; (2) **ACTIVE,** in which the hydrant output is controlled by the direction and amount of force applied to the valve stem, and the transistor output collector current is controlled by base current; and (3) **SATURATED,** in which the hydrant valve stem is turned on as much as possible to obtain maximum water output, and the transistor base current is increased to obtain maximum collector current.

Second, both devices produce very large outputs with very small inputs at the controlling terminal. With the hydrant, the force applied to the valve stem by a ten-year-old boy was amplified many times to produce an output with enough force to bowl over grown men. Similarly, with the transistor, the output current is much larger than the controlling base current. That is, the transistor is a current amplifier.

In the following sections, as we discuss the transistor, you may find it helpful to keep the hydrant-transistor analogy in mind. The emitter is similar to the water main, the collector is similar to the hydrant outlet, and the base is similar to the valve stem, controlling the flow of electrons from the emitter to the collector.

7–2 ■ BJT STATES OF OPERATION

To understand the behavior of the transistor, you must know the relations between the transistor voltages and currents. Two-terminal devices, such as resistors, can be described with one voltage and one current. Three-terminal devices, however, are described with three voltages and three currents. These voltages and currents are defined for the BJT in Figure 7–3, where we have redrawn the NPN sandwich of Figure 7–1A in a more readable form. Figure 7–3B shows the BJT's circuit schematic, along with its associated voltages and currents.

The BJT's construction gives the device the appearance of two back-to-back diodes (Figure 7–3A), one from the base to the emitter and another from the base to the collector. Although the BJT's behavior cannot be completely described this way—you cannot build a BJT by connecting two diodes back-to-back—it is a convenient model to keep in mind when you try to determine the state of the BJT.

FIGURE 7–3

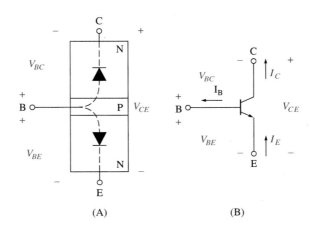

(A)   (B)

Overview

You know that it takes approximately 0.7V to turn a diode **ON**. This is also true of the PN junctions in the BJT. The base-emitter junction, which we will call the **emitter diode,** turns on when the base-to-emitter voltage (V_{BE}) reaches approximately 0.7V. The base-collector junction, which we will call the **collector diode,** actually turns on when the base-to-collector voltage (V_{BC}) reaches a slightly smaller voltage, approximately 0.5V.

The states of the emitter and collector diodes completely determine the state of the BJT. If both V_{BE} and V_{BC} are less than their turn-on voltages, both diodes are **OFF** and the transistor itself is **CUT-OFF.** If both V_{BE} and V_{BC} equal approximately 0.7V and 0.5V respectively, both diodes are **ON.** This is similar to completely opening a valve; that is, the transistor is as turned on as it is going to get, and we get maximum possible current at the collector. In this case, we say the transistor is **SATURATED.** We will discuss these two states in more detail in a later section.

For this textbook the most interesting state occurs when the emitter diode is **ON** and the collector diode is **OFF;** that is, V_{BE} equals 0.7V and V_{BC} is less than 0.5V. In this case we say the transistor is in the **ACTIVE** state in which the output current of the transistor is modulated by base current and base voltage.

In summary, the state of a transistor is determined as follows:

CUT-OFF:	Emitter diode and collector diode are **OFF.**
SATURATED:	Emitter diode and collector diode are **ON.**
ACTIVE:	Emitter diode is **ON**—collector diode is **OFF.**

The Active State

During the **ACTIVE** state of the BJT, the modulation of the output current by the base current allows us to construct circuits to make such things as audio amplifiers and televisions. The physics of how this current modulation works is quite complicated and can be found in Appendix A (p. A-8). Here, we only give a very simple explanation.

In the **ACTIVE** state, the emitter diode of the transistor is **ON** and the collector diode is **OFF,** but the transistor does *not* act like two back-to-back diodes. To get a feel for how the **ACTIVE** transistor behaves, consider the NPN BJT shown in Figures 7–4 and 7–5.

We first assume that the emitter diode is **ON** (that is, $V_{BE} = 0.7$V) and that the collector terminal is not connected to anything (Figure 7–4A). In this case what we actually have is a diode between the base and the emitter; the collector is simply a lump of silicon. Current flows only from the emitter to the base. That is, when the emitter diode is **ON,** the electrons in the emitter cross the junction and enter the base. These electrons are then pulled out of the base, creating the base current (Figure 7–4B).

FIGURE 7–4

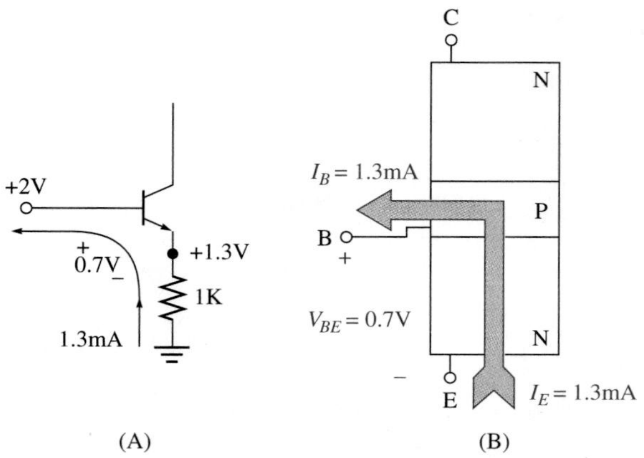

(A)   (B)

FIGURE 7–5

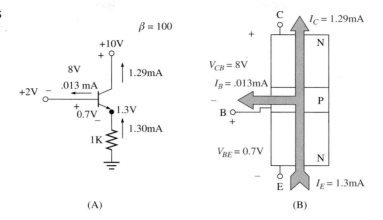

(A)   (B)

We now connect a positive voltage at the collector (Figure 7–5). Because the collector is more positive than the base, the collector diode is reverse biased, so it is **OFF.** (Note that we have changed the definition of the collector-base voltage. We do this to be consistent with the fact that in the active mode, the collector is positive.) It might seem that since the collector diode is reverse biased, there can be no current into the collector. This is where common sense and physics conflict. In fact, there is a current, a fairly large one, that flows to the collector.

As explained in Appendix B, proper transistor construction allows the large positive voltage at the collector to attract most of the electrons that flow from the emitter into the base. Typically, 99% or more of the electrons that cross the base-emitter junction are pulled out of the collector. This means that only 1% of these electrons are pulled out at the base (see Figure 7–5). We can express these relations as

$$I_C = 0.99I_E$$

and

$$I_B = 0.01I_E$$

Note that because in most transistors the collector current is more than 99% of the emitter current, we usually set the collector current approximately equal to the emitter current. That is

$$I_C \approx I_E$$

Most commonly, we express the collector current as a function of the base current. To find this relation, we divide the preceeding equations. That is

$$\frac{I_C}{I_B} = \frac{0.99I_E}{0.01I_E} = \frac{0.99}{0.01} = 99$$

Therefore

$$I_C = 99I_B$$

In other words, for each electron that is pulled out of the base, 99 electrons are pulled out of the collector. The base current, therefore, acts as the valve wrench that controls the flow of the collector current. In general, we write

$$I_C = \beta I_B$$

where β is known as the **current gain** of the transistor. Typically, β varies from 50 to 500; that is, collector current is 50-to-500 times the base current in the typical transistor.

Figure 7–6 shows a model often used to describe the behavior of the transistor in the **ACTIVE** state. As Figure 7–6B shows, a diode represents the base-emitter junction, and a controlled current source represents the relation between the collector current and the base current (Figure 7–6B). While we can often determine the behavior of a transistor circuit without resorting to this model, it is useful in determining how we treat the transistor. For example, for DC voltages and currents we know that we can accurately model an **ON** diode as a 0.7V battery (Figure 7–6C). As we shall see, for AC we can model the diode as a resistor.

FIGURE 7–6

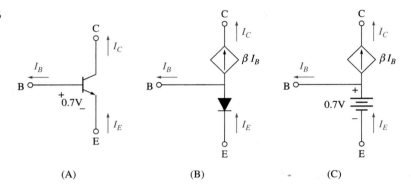

(A)   (B)   (C)

EXAMPLE 7–1
Transistor Current Gain, β

Find β for the following emitter current–collector current relations:

(a) $I_C = 0.90I_E$ **(b)** $I_C = 0.98I_E$ **(c)** $I_C = 0.995I_E$

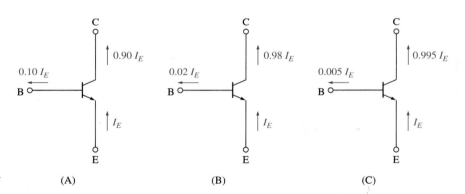

FIGURE 7–7   (A)   (B)   (C)

Solution

(a) Because 90% of I_E goes to the collector, 10% of I_E must go to the base, so

$$I_B = 0.10I_E \text{ (Figure 7–7A)}$$

and

$$\beta = \frac{I_C}{I_B} = \frac{0.9I_E}{0.1I_E} = \frac{0.9}{0.1} = 9$$

(b) Because 98% of I_E goes to the collector, 2% of I_E must go to the base (Figure 7–7B), so

$$I_B = 0.02I_E$$

and

$$\beta = \frac{I_C}{I_B} = \frac{0.98 I_E}{0.02 I_E} = \frac{0.98}{0.02} = 49$$

(c) Because 99.5% of I_E goes to the collector, 0.5% of I_E must go to the base (Figure 7–7C, p. 186), so

$$I_B = 0.005 I_E$$

and

$$\beta = \frac{I_C}{I_B} = \frac{0.995 I_E}{0.005 I_E} = \frac{0.995}{0.005} = 199$$

DRILL EXERCISE   Find β if $I_C = 0.99 I_E$ and $I_C = 0.998 I_E$.

Answer $\beta = 99$, $\beta = 499$

EXAMPLE 7–2
Emitter Current-
Collector Current
Relationships

Find the relation between emitter and collector currents for the following current gains (β):

(a) $\beta = 50$ **(b)** $\beta = 150$ **(c)** $\beta = 250$

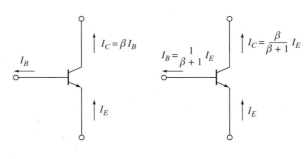

FIGURE 7–8   (A)   (B)

Solution   Because the emitter current supplies both the base and collector currents, we get the following (which is actually an expression of Kirchhoff's Current Law):

$$I_E = I_B + I_C$$

We know that $I_C = \beta I_B$ (see Figure 7–8A), so

$$I_E = I_B + I_C = I_B + \beta I_B = (1 + \beta) I_B$$

Therefore, the base current is given by

$$I_B = \frac{1}{1 + \beta} I_E$$

We again use the fact that the collector current is β times the base current to get the relation between the collector and emitter currents (Figure 7–8B):

$$I_C = \beta I_B = \frac{\beta}{1 + \beta} I_E$$

Using this relation we can now solve the problems in this example:

(a) $\beta = 50$ so $I_C = \dfrac{50}{1 + 50}I_E = 0.98I_E$

(b) $\beta = 150$ so $I_C = \dfrac{150}{1 + 150}I_E = 0.993I_E$

(c) $\beta = 250$ so $I_C = \dfrac{250}{1 + 250}I_E = 0.996I_E$

DRILL EXERCISE   Find the relation between I_C and I_E for $\beta = 125$, $\beta = 500$

Answer $I_C = 0.992I_E$, $I_C = 0.998I_E$

The results of Example 7–2 show that for typical transistors, in which β ranges from 50 to 250, the collector current is 98 to 99.6% of the emitter current. This confirms that for typical transistors the collector current is so close to the emitter current that we can simply say

$$I_C \approx I_E$$

As a result, when we analyze transistor circuits, we always set the collector current equal to the emitter current.

REVIEW QUESTIONS

1. BJT is an abbreviation for what?
2. Draw the circuit schematic for the NPN BJT; label the base, collector, and emitter terminals; label all of its currents and voltages.
3. What are the three possible states of the transistor?
4. Describe the transistor states in terms of the two transistor PN junctions: the emitter and collector diodes.
5. What does the parameter β describe, and what are its typical values?
6. What is the approximate relation between the collector and emitter currents?

7–3 ■ TRANSISTOR RATINGS

Transistor Power

Resistors dissipate power in the form of heat. If a 0.25W resistor is used in a circuit where it must dissipate 1 W, the resistor will burn up. Likewise, transistors dissipate power in the form of heat. Therefore, we must be sure to use a transistor with the correct power rating to avoid destroying the transistor.

You determine how much power a resistor can dissipate by simply multiplying the voltage across the resistor times the current in the resistor, i.e., $P = VI$ (Figure 7–9A). The transistor, however, is a three-terminal device. How do we find the power dissipated in a device that has three voltages and three currents (Figure 7–9B)?

When the transistor is **ACTIVE,** as here, the emitter diode is **ON** and the collector diode is **OFF.** The power dissipated due to the base current is $V_{BE}I_B$. Because $V_{BE} = 0.7$V

FIGURE 7–9

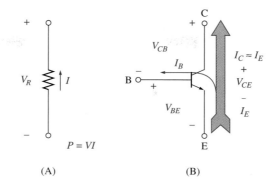

(A)   (B)

and I_B is very small (one hundredth of I_C), the power dissipated at the base is negligible and can be safely ignored.

As represented in Figure 7–9B, the largest current in the transistor is the emitter-to-collector current, which we consider to be I_C. It is this current that can damage the transistor. Therefore, power dissipated in the transistor is primarily due to collector current and collector-to-emitter voltage. That is

$$P = V_{CE}I_C$$

Typically, V_{CE} is in the 1-to-25V range and I_C is in the milliampere range. This means that most transistors will dissipate power in the milliwatt range.

When we analyze transistor circuits, we find transistor power after we have determined the collector current and the collector-to-emitter voltage.

Base Collector Breakdown (V_{CBO})

There is another consideration with a transistor. In the **ACTIVE** mode the collector diode is reverse biased. Recall from your study of the diode that if the reverse-bias voltage of the diode gets too large, the diode will break down. Similarly, if the reverse-bias voltage across the collector diode (V_{CB}) gets too large, the collector-base junction will break down, and the transistor will cease to operate properly.

For this reason, transistors are rated for the maximum allowable reverse-biased collector-base voltage. This maximum value is labeled V_{CBO}. In addition to finding the power that a transistor will dissipate in a given circuit, you should also find the maximum value expected for V_{CB}. This value should be less than the rated V_{CBO} for your transistor.

REVIEW QUESTIONS

1. Why is power dissipation in the emitter diode negligible?
2. How is transistor power dissipation defined?
3. Why is it important to know the power dissipated by a transistor?
4. Show the reverse-biased collector-base voltage (V_{CB}) on a transistor diagram.
5. Describe reverse-biased PN-junction breakdown, and explain why it must be avoided.

7–4 ■ SIMPLE DC TRANSISTOR CIRCUITS (I_B Control)

Now we're ready to show you some simple transistor circuits. Figure 7–10A (p. 190) shows a very basic circuit that includes a base supply voltage (V_{BB}), a collector supply voltage (V_{CC}) and a base resistor (R_B). A voltage with a double subscript, e.g., V_{BB}, repre-

FIGURE 7–10
I_B control

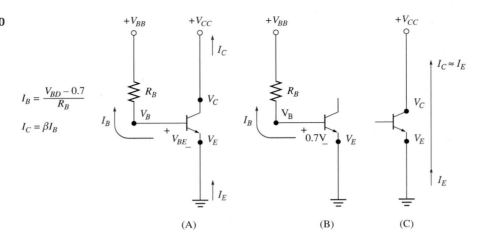

$$I_B = \frac{V_{BD} - 0.7}{R_B}$$

$$I_C = \beta I_B$$

(A)   (B)   (C)

sents a DC supply voltage. The DC voltages and currents in a transistor circuit are known as the transistor's **DC bias** values.

Note that although the labeled currents are the same as in our other transistor diagrams, we have changed the voltage notation. We no longer show V_{BE}, V_{CB}, and V_{CE}. Rather, we now show the voltages that can be measured at the base, collector, and emitter nodes of the transistor. We remind you that voltage with a single subscript is node voltage and is measured between the node and ground.

The following relations hold:

$$V_{BE} = V_B - V_E = 0.7\text{V (if the emitter diode is **ON**)}$$
$$V_{CB} = V_C - V_B$$
$$V_{CE} = V_C - V_E$$

We have found it helpful to think of the transistor circuit as two separate circuits: the **base circuit** and the **collector circuit.** The base circuit is shown in Figure 7–10B and consists of the base supply voltage, the base resistor, the base-to-emitter junction, and, in this case, ground. The base circuit alone determines base current and voltage.

The collector circuit is composed of the collector supply voltage, the collector voltage, the emitter voltage, and any resistors connected to the collector or emitter. For now, all we need to know about the collector circuit is this: If the transistor is in the **ACTIVE** state, the collector current is β times the base current.

When we analyze these transistor circuits, we always find the base current first. This is because if the transistor is **ACTIVE,** the base current controls the collector current ($I_C = \beta I_B$). Also, although not technically true, it is helpful to think that the collector current controls the emitter current ($I_E \approx I_C$). In general, transistors are **ACTIVE** whenever V_{BB} is greater than the emitter diode turn-on voltage of 0.7V.

Consider the circuit in Figure 7–11A. In Figure 7–11B, we show the 0.7V drop that exists across the emitter diode when it is **ON**. In this circuit the base current is the current in the 10KΩ resistor and is given by

$$I_B = \frac{2 - V_B}{10\text{K}}$$

We find V_B by noting that the emitter is grounded

$$V_E = 0\text{V}$$

and that the base voltage is 0.7V greater than the emitter voltage. That is

$$V_B = V_E + 0.7 = 0 + 0.7 = 0.7\text{V}$$

FIGURE 7–11

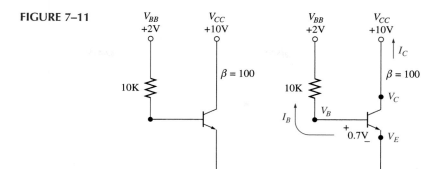

(A)   (B)

Using this value for base voltage in the base current equation, we get

$$I_B = \frac{2 - V_B}{10K} = \frac{2 - 0.7}{10K} = \frac{1.3}{10K} = 0.13\text{mA}$$

Now if the transistor is **ACTIVE,** the base current controls the collector current:

$$I_C = \beta I_B = 100 \times 0.13\text{mA} = 13\text{mA}$$

Because the base current is assumed to be very small compared to the collector current, we always assume that the emitter current is the same as the collector current:

$$I_E \approx I_C = 13\text{mA}$$

We have found that it is a good idea to list both transistor currents and node voltages. This will enable you easily to determine the state of the transistor, the power dissipated by the transistor, and the collector-to-base reverse-bias voltage (V_{CB}), and to provide a basis for troubleshooting transistor circuits. We call this list the **DC Bias List.** For this circuit, the **DC Bias List** is

I_B	0.13mA
I_C	13mA
I_E	13mA
V_B	0.7V
V_C	10V (collector is connected 10V source)
V_E	0V (emitter is grounded)

From the DC Bias List of transistor currents and voltages, we see that

$$V_{CB} = V_C - V_B = 10 - 0.7 = 9.3\text{V}$$
$$V_{CE} = V_C - V_E = 10 - 0 = 10\text{V}$$

and

$$P = V_{CE}I_C = 10 \times 13\text{mA} = 130\text{mW}$$

These results indicate that we need to find a transistor that has a V_{CBO} greater than 9.3V and is rated for greater than 130mW.

EXAMPLE 7–3
I_B **Control with**
Base Resistor

Find I_B, I_C, I_E, V_B, V_E, V_C, V_{CB}, V_{CE}, and P for the transistor circuit shown in Figure 7–12, using the following base resistors. Assume the current gain (β) is 200.

(a) $R_B = 1K\Omega$ **(b)** $R_B = 10K\Omega$ **(c)** $R_B = 100K\Omega$

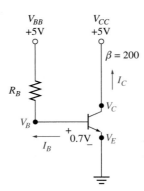

FIGURE 7–12

Solution   Because the emitter is grounded, $V_E = 0$. If the emitter diode is **ON,** the base voltage must be 0.7V above the emitter voltage; therefore, $V_B = 0.7V$. From Ohm's Law, we find the base current to be

$$I_B = \frac{5 - V_B}{R_B} = \frac{5 - 0.7}{R_B} = \frac{4.3}{R_B}$$

The collector current is β times the base current, so

$$I_C = \beta I_B = 200 I_B$$

(a) If the base resistor is $R_B = 1\ K\Omega$

$$I_B = \frac{4.3}{R_B} = \frac{4.3}{1K} = 4.3\text{mA}$$

and

$$I_C = \beta I_B = 200 I_B = 200 \times 4.3\text{mA} = 860\text{mA}$$

The **DC Bias List** is

I_B	4.3mA
I_C	860mA
I_E	860mA
V_B	0.7V
V_C	5V (collector is connected to the 5V source)
V_E	0V (emitter is grounded)

$$V_{CB} = V_C - V_B = 5 - 0.7 = 4.3V$$
$$V_{CE} = V_C - V_E = 5 - 0 = 5V$$
$$P = V_{CE}I_C = 5 \times 860\text{mA} = 4300\text{mW} = 4.3\text{W}$$

(b) If the base resistor is increased to $R_B = 10K\Omega$, the base and collector currents are

$$I_B = \frac{4.3}{10K} = 0.43\text{mA}$$

and

$$I_C = 200 I_B = 86\text{mA}$$

The **DC Bias List** is

I_B	0.43mA
I_C	86mA
I_E	86mA
V_B	0.7V
V_C	5V (collector is connected to the 5V source)
V_E	0V (emitter is grounded)

$$V_{CB} = V_C - V_B = 5 - 0.7 = 4.3V$$
$$V_{CE} = V_C - V_E = 5 - 0 = 5V$$
$$P = V_{CE}I_C = 5 \times 86\text{mA} = 430\text{mW}$$

(c) Finally, we increase the base resistance to $R_B = 100\text{K}\Omega$ and find the base and collector currents:

$$I_B = \frac{4.3}{100\text{K}} = 0.043\text{mA}$$

and

$$I_C = 200I_B = 8.6\text{mA}$$

The **DC Bias List** is

I_B	0.043mA
I_C	8.6mA
I_E	8.6mA
V_B	0.7V
V_C	5V (collector is connected to the 5V source)
V_E	0V (emitter is grounded)

$$V_{CB} = V_C - V_B = 5 - 0.7 = 4.3V$$
$$V_{CE} = V_C - V_E = 5 - 0 = 5V$$
$$P = V_{CE}I_C = 5 \times 8.6\text{mA} = 43\text{mW}$$

You can see from this example that as you increase the base resistance, there is a decrease in base current and, therefore, collector current.

DRILL EXERCISE

Redraw the circuit shown in Figure 7–12 with $R_B = 50\text{K}\Omega$. Find the transistor currents, voltages, power, and V_{CB}.

Answer $I_B = 0.086\text{mA}, I_C = I_E = 17.2\text{mA}$
$V_B = 0.7\text{V}, V_C = 5\text{V}, V_E = 0\text{V}$
$P = 86\text{mW}, V_{CB} = 4.3\text{V}$

EXAMPLE 7–4
I_B **Control With Base Supply Voltage**

In the transistor circuit of Figure 7–13, (p. 194) assume $\beta = 100$. Find the transistor currents, voltages, collector-to-base reverse-bias voltage, collector-to-emitter voltage, and power dissipated for the following base supply voltages:

(a) $V_{BB} = 5\text{V}$ **(b)** $V_{BB} = 3\text{V}$ **(c)** $V_{BB} = 1\text{V}$

Note the use of the double subscript for the voltage. Double subscripts are always used to indicate supply voltages. Do not confuse this voltage with the base node voltage (V_B), which must always be found by analyzing the circuit.

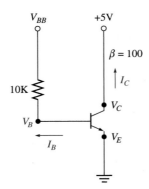

FIGURE 7–13

Solution For this circuit $V_E = 0V$ and $V_B = 0.7V$, so the base current is

$$I_B = \frac{V_{BB} - 0.7}{10K}$$

and

$$I_C = 100I_B$$

(a) If the base supply voltage is $V_{BB} = 5V$, the base and collector currents are

$$I_B = \frac{5 - 0.7}{10K} = \frac{4.3}{10K} = 0.43mA$$

and

$$I_C = 100 \times 0.43mA = 43mA$$

The **DC Bias List** is

I_B 0.43mA

I_C 43mA

I_E 43mA

V_B 0.7V

V_C 5V (collector is connected to the 5V source)

V_E 0V (emitter is grounded)

$V_{CB} = V_C - V_B = 5 - 0.7 = 4.3V$

$V_{CE} = V_C - V_E = 5 - 0 = 5V$

$P = V_{CE}I_C = 5 \times 43mA = 215mW$

(b) We now lower the base supply voltage to $V_{BB} = 3V$.
The base and collector currents are

$$I_B = \frac{3 - 0.7}{10K\Omega} = 0.23mA$$

and

$$I_C = 100 \times 0.23mA = 23mA$$

The **DC Bias List** is

I_B	0.23mA
I_C	23mA
I_E	23mA
V_B	0.7V
V_C	5V (collector is connected to the 5V source)
V_E	0V (emitter is grounded)

$$V_{CB} = V_C - V_B = 5 - 0.7 = 4.3V$$
$$V_{CE} = V_C - V_E = 5 - 0 = 5V$$
$$P = V_{CE}I_C = 5 \times 23mA = 115mW$$

(c) Finally, we lower the base supply voltage to $V_{BB} = 1V$
The base and collector currents are

$$I_B = \frac{1 - 0.7}{10K} = 0.03mA$$

and

$$I_C = 100 \times 0.03mA = 3mA$$

The **DC Bias List** is

I_B	0.03mA
I_C	3mA
I_E	3mA
V_B	0.7V
V_C	5V (collector is connected to the 5V source)
V_E	0V (emitter is grounded)

$$V_{CB} = V_C - V_B = 5 - 0.7 = 4.3V$$
$$V_{CE} = V_C - V_E = 5 - 0 = 5V$$
$$P = V_{CE}I_C = 5 \times 3mA = 15mW$$

As the base supply voltage decreases, the base and collector currents decrease.

DRILL EXERCISE   Redraw the circuit in Figure 7–13 with $V_{BB} = 4V$; find the transistor currents, voltages, power, and V_{CB}.

Answer $I_B = 0.33mA$, $I_C = I_E = 33mA$
$V_B = 0.7V$, $V_C = 5V$, $V_E = 0V$
$P = 165mW$, $V_{CB} = 4.3V$

7–5 ■ THE COLLECTOR RESISTOR

The transistor is usually used with a resistor connected between the collector and its supply voltage (Figure 7–14A). The collector resistor serves two main purposes: it allows us to control the voltage at the collector; it also protects the transistor from excessive collector current and, therefore, from excessive power dissipation.

FIGURE 7–14

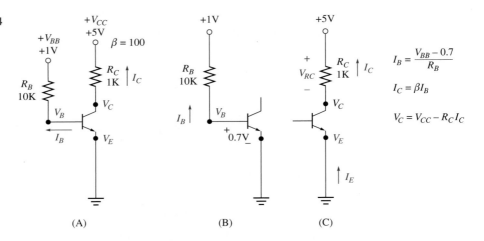

(A)   (B)   (C)

We begin the analysis of this transistor circuit in the same manner as we analyzed the transistor circuits of the previous section. Consider the base circuit in Figure 7–14B. Because the emitter is grounded

$$V_E = 0$$

and because we assume the transistor is **ACTIVE**

$$V_B = V_E + 0.7 = 0 + 0.7 = 0.7V$$

The base current is found from

$$I_B = \frac{V_{BB} - 0.7}{R_B} = \frac{1 - 0.7}{10K} = \frac{0.3}{10K} = 0.03mA$$

Note that the collector resistor was not used in these calculations. The base of the transistor communicates with the emitter but does not communicate with the collector. As previously stated, in the **ACTIVE** state the base current and voltage depend only on the base circuit and are always independent of collector voltage and collector resistance.

Because we assume the transistor is in the **ACTIVE** state, the collector current is

$$I_C = \beta I_B = 100 \times 0.03mA = 3mA$$

We complete the analysis by finding the collector voltage from the collector circuit shown in Figure 7–14C. The voltage drop across the collector resistor is given by Ohm's Law

$$V_{RC} = R_C I_C$$

The collector voltage is found by subtracting the voltage drop across the collector resistor from the supply voltage; that is

$$V_C = V_{CC} - R_C I_C$$

For this circuit

$$V_C = V_{CC} - R_C I_C = 5 - 1K \times 3mA = 2V$$

The DC Bias List for this circuit is

DC Bias List

I_B 0.03mA

I_C 3mA

I_E 3mA

V_B 0.7V

V_C 2V

V_E 0V (emitter is grounded)

$V_{CB} = V_C - V_B = 2 - 0.7 = 1.3V$

$V_{CE} = V_C - V_E = 2 - 0 = 2V$

$P = V_{CE} I_C = 2 \times 3mA = 6mW$

If you compare this circuit to the last example in the previous section (Example 7–4C), you can see that the addition of the collector resistor reduces the collector voltage, collector-to-base reverse-biased voltage, and the power dissipated by the transistor. When the collector voltage decreases, the collector diode is reverse biased by a smaller amount. (We will see that there is a limit to how small the collector voltage can get before the collector diode turns **ON.**)

From now on, we will show the base and collector circuits only if they help us understand transistor circuit behavior. Even though we do not show the base and collector circuits explicitly, you can always visualize them when you look at a transistor circuit.

EXAMPLE 7–5
Collector Resistor

For the circuit in Figure 7–15, find the transistor currents, voltages, collector-to-base reverse-biased voltage, and power for these collector resistors. Assume $\beta = 100$.

(a) $R_C = 1K\Omega$ **(b)** $R_C = 2K\Omega$

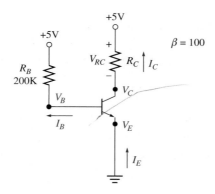

FIGURE 7–15

Solution Because the transistor base current does not depend on R_C, we will find I_B first. The emitter is grounded, so $V_E = 0$, and $V_B = 0.7$V. The base current is found from

$$I_B = \frac{5 - 0.7}{200\text{K}} = 0.0215\text{mA}$$

and the collector current is

$$I_C = \beta I_B = 100 \times 0.0215\text{mA} = 2.15\text{mA}$$

The collector voltage is found by subtracting the voltage across the collector resistor from the supply voltage. That is

$$V_C = V_{CC} - R_C I_C$$

(a) For the collector resistor $R_C = 1\text{K}\Omega$

$$V_C = V_{CC} - R_C I_C = 5 - 1\text{K} \times 2.15\text{mA} = 2.85\text{V}$$

The complete **DC Bias List** for this part is

I_B	0.0215mA
I_C	2.15mA
I_E	2.15mA
V_B	0.7V
V_C	2.85V
V_E	0V (emitter is grounded)

$V_{CB} = V_C - V_B = 2.85 - 0.7 = 2.15\text{V}$
$V_{CE} = V_C - V_E = 2.85 - 0 = 2.85\text{V}$
$P = V_{CE} I_C = 2.85 \times 2.15\text{mA} = 6.13\text{mW}$

(b) We now increase the collector resistor to $R_C = 2\text{K}\Omega$:

$$V_C = V_{CC} - R_C I_C = 5 - 2\text{K} \times 2.15\text{mA} = 0.7\text{V}$$

The complete **DC Bias List** for this part is

I_B	0.0215mA
I_C	2.15mA
I_E	2.15mA
V_B	0.7V
V_C	0.7V
V_E	0V (emitter is grounded)

$V_{CB} = V_C - V_B = 0.7 - 0.7 = 0\text{V}$
$V_{CE} = V_C - V_E = 0.7 - 0 = 0.7\text{V}$
$P = V_{CE} I_C = 0.7 \times 2.15\text{mA} = 1.51\text{mW}$

Note that in this example, $V_{CB} = 0.7 - 0.7 = 0$V. This means that the collector diode is no longer reverse biased. In fact, if V_C gets much smaller, the base-to-collector voltage will increase enough to turn the collector diode **ON.** In this case the transistor will no longer be in the **ACTIVE** mode but, in fact, will transition into the **SATU-RATED** state.

Redraw the transistor circuit shown in Figure 7–15 (p. 197) using the following values:

$$V_{BB} = 10V, \quad V_{CC} = 10V, \quad R_B = 200K\Omega, \quad \text{and} \quad R_C = 1K\Omega.$$

Find the transistor currents, voltages, power, and V_{CB}.

Answer $I_B = 0.0465mA, I_C = I_E = 4.65mA$
   $V_B = 0.7V, V_C = 5.35V, V_E = 0V$
   $P = 24.9mW, V_{CB} = 4.65V$

REVIEW QUESTIONS

1. Draw a transistor circuit with base and collector supply voltages and base and collector resistors.
2. Find the base current in terms of V_{BB} and R_B. Find the collector current.
3. Find the voltage drop across the collector resistor. Find V_C.

7–6 ▪ THE CUT-OFF AND SATURATION STATES

CUT-OFF

Let's find the base current for the transistor circuit shown in Figure 7–16, proceeding as in the previous section. From the base circuit in Figure 7–16B we get

$$V_E = 0 \qquad \text{so} \qquad V_B = 0.7V$$

The base current is now found:

$$I_B = \frac{0 - 0.7}{10K} = \frac{-0.7}{10K} = -0.07mA$$

But wait a minute, the base current is negative! This is simply not possible. Remember, the emitter diode is a PN junction in which current can only pass in one direction. Base current *must* be positive. What is happening here?

The answer is that the emitter diode is **OFF**. This means that the base current is 0A and the transistor is in the **CUT-OFF** state. When the transistor is in **CUT-OFF**, all transistor currents are 0A.

To see how transistor voltages determine whether a transistor is in **CUT-OFF**, reconsider the base circuit in Figure 7–16B. The base supply voltage (V_{BB}) must be large

FIGURE 7–16
The collector resistor

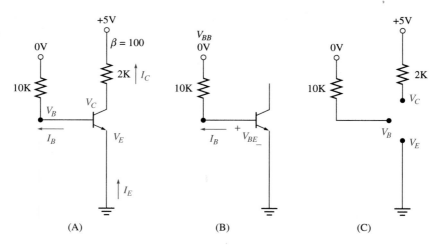

(A)   (B)   (C)

enough to turn the emitter diode **ON.** In addition, V_{BB} must be large enough to provide a voltage across the base resistor to produce a base current. This means that if $V_{BB} < 0.7V$, the typical transistor will be in **CUT-OFF.**

$$\text{CUT-OFF:} \quad V_{BB} < 0.7V$$

To analyze a transistor circuit in which the transistor is **CUT-OFF,** remove the transistor, setting all transistor currents to 0A (Figure 7–16C, p. 199). In this circuit

$$V_B = 0, \qquad V_E = 0, \qquad \text{and} \qquad V_C = 5V$$

Because there is no transistor current, the power dissipated by a **CUT-OFF** transistor is 0W. The reverse-biased collector-base voltage is

$$V_{CB} = V_C - V_B = 5V$$

Maximum reverse-biased collector-base voltage always occurs when the transistor is **CUT-OFF.**

SATURATION

Figure 7–17 shows a transistor circuit that is similar to Figure 7–11 (p. 191). In this case the base resistor is reduced to 100KΩ. Assuming the transistor is **ACTIVE**

$$V_E = 0 \qquad \text{and} \qquad V_B = 0.7V$$

The base current is, therefore

$$I_B = \frac{5 - 0.7}{100\,K} = \frac{4.3}{100K} = 0.043\text{mA}$$

and

$$I_C = 100I_B = 100 \times 0.043\text{mA} = 4.3\text{mA}$$

To find the collector voltage, we subtract the collector resistor voltage from the supply voltage:

$$V_C = V_{CC} - R_C I_C = 5 - 2K\Omega \times 4.3\text{mA} = -3.6V$$

The fact that the collector voltage is negative should give you pause. We have only 5V between the collector supply battery and ground. It is impossible, therefore, for a negative collector voltage to show up. Whatever state this transistor is in, V_C must be greater than 0V and less than 5V. Before considering the ramifications of this analysis, it is instructive

FIGURE 7–17

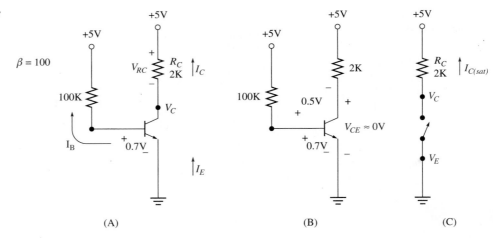

(A)   (B)   (C)

to examine the DC Bias List for this circuit. Remember, we have assumed that the transistor is in the **ACTIVE** state, resulting in this **DC Bias List:**

I_B 0.043mA

I_C 4.3mA

I_E 4.3mA

V_B 0.7V

V_C −3.6V

V_E 0V (emitter is grounded)

$V_{CB} = V_C − V_B = −3.6 − 0.7 = −4.3V$

$V_{CE} = V_C − V_E = −3.6 − 0 = −3.6V$

Note that both V_{CB} and V_{CE} are negative. This cannot be; these two voltages can never be negative. What is going on here? The answer is that the transistor is *not* in the **ACTIVE** region, so this DC Bias List is wrong (indicated by the X through the list).

If the collector-to-base voltage is negative, then the collector diode must be **ON.** That is, if

$$V_{CB} = −4.3V$$

then

$$V_{BC} = −V_{CB} = 4.3V$$

and the collector diode, as well as the emitter diode, is forward biased. In this case the transistor is in **SATURATION.**

When the collector-base PN junction is forward biased, the voltage across the junction is fixed at approximately 0.5V. Therefore, as soon as the collector diode turns on, its voltage is set to 0.5V, and the analysis we performed is invalid. Because the collector diode is **ON,** the actual base collector voltage is

> **SATURATION:** $V_{BC} \approx 0.5V$

The actual collector-emitter voltage can be found from Figure 7–17B:

$$V_{CE(sat)} = V_{BE} − V_{BC} = 0.7 − 0.5 = 0.2V$$

For simplicity, we usually ignore this small voltage and assume that in **SATURATION** the collector-emitter voltage is 0V. In other words, when a transistor is in **SATURATION,** the collector voltage approximately equals the emitter voltage.

> **SATURATION:** $V_{CE(sat)} \approx 0V \rightarrow V_C \approx V_E$

Taken together, this information tells us that in **SATURATION** the collector circuit of the transistor behaves as though a switch has been closed between the collector and emitter (Figure 7–17C). Therefore, in **SATURATION** collector voltage equals emitter voltage.

It is very important that you remember that when a transistor is in **SATURATION,** the transistor currents lose their special relation. That is, in **SATURATION**

$$I_C \neq \beta I_B \qquad \text{and} \qquad I_C \neq I_E$$

If the transistor is in **SATURATION,** we must use the collector circuit to find I_C. Because the base current is not affected by changes in the collector circuit, the value we found for I_B is still correct. From Figure 7–17C we see that the collector current can be found from

$$I_C = I_{C(sat)} = \frac{5 − V_C}{2K} = \frac{5 − 0}{2K} = 2.5mA$$

where, because the transistor is in **SATURATION**, $V_C = V_E = 0$ (the emitter is grounded).

In **SATURATION** the emitter current is found from

$$I_E = I_B + I_C = 0.043\text{mA} + 2.5\text{mA} = 2.54\text{mA}$$

The actual **DC Bias List** for this circuit is

I_B 0.043mA (not affected by **SATURATION**)

I_C 2.5mA $\left(I_C = I_{C(sat)} = \dfrac{V_{CC}}{R_C}\right)$

I_E 2.54mA ($I_E = I_B + I_C$)

V_B 0.7V

V_C 0V (transistor is **SATURATED**, so $V_C = V_E$)

V_E 0V (emitter is grounded)

$V_{CB} = -0.5\text{V}$

$V_{CE} = 0\text{V}$

$P = V_{CE}\,I_C = 0\text{W}$

In summary, when a transistor is in **SATURATION,** you find the approximate collector current and collector voltage by placing a closed switch between the collector and the emitter. That is, in **SATURATION** $V_C \approx V_E$. The base current and voltage are found from the base circuit and are, therefore, not affected by the collector-emitter switch.

It is important to note that because V_{CE} is very small, very little power is dissipated in a **SATURATED** transistor. It acts as though there is a closed switch, a piece of wire, between the collector and emitter. Because the voltage across a piece of wire is approximately zero, the power dissipated is zero, no matter how large the current is in a piece of wire.

As the next several examples show, the value of the base supply voltage, collector supply voltage, base resistor, or collector resistor can put a transistor into **SATURATION**.

EXAMPLE 7–6
Base Supply Voltage and Saturation I

The circuit in Figure 7–18A replicates the circuit in Figure 7–14 with the base supply voltage increased to $V_{BB} = +2\text{V}$ ($\beta = 100$). Find the state of the transistor, I_B, I_C, I_E, and V_{CE}.

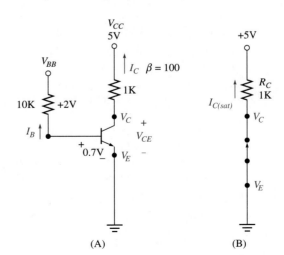

FIGURE 7–18   (A)   (B)

Solution   First we assume that the emitter diode is **ACTIVE** and solve for I_B:

$$I_B = \frac{V_{BB} - 0.7}{10\text{K}} = \frac{2 - 0.7}{10\text{K}} = 0.13\text{mA}$$

We then assume that the transistor is **ACTIVE** and find I_C:

$$I_C = \beta I_B = 13\text{mA}$$

Finally, we find V_C:

$$V_C = V_{CC} - R_C I_C = 5 - 1\text{K} \times 13\text{mA} = -8\text{V}$$

Because V_C is negative, the transistor is **SATURATED,** so we must find the collector current from the **SATURATED** collector circuit, shown in Figure 7–18B:

$$V_C \approx V_E = 0\text{V}$$

and

$$I_C = I_{C(sat)} = \frac{V_{CC}}{1\text{K}} = \frac{5}{1\text{K}} = 5\text{mA}$$

The emitter current is found from

$$I_E = I_B + I_C = 0.13 + 5 = 5.13\text{mA}$$

The results for this **SATURATED** transistor are

$$I_B = 0.13\text{mA}, \ I_C = I_{C(sat)} = 5\text{mA}, \ I_E = 5.13\text{mA}$$

and

$$V_C \approx V_E = 0\text{V}$$

DRILL EXERCISE   Redraw the circuit in Figure 7–18 with $V_{BB} = 2.5$V; find all transistor currents and voltages.

Answer $I_B = 0.18$mA, $I_C = I_{C(sat)} = 5$mA, $I_E = 5.18$mA
$V_B = 0.7$V, $V_C \approx V_E = 0$V

In summary, as the base supply voltage increases, base current and collector current increase. The transistor moves toward **SATURATION** and, for a sufficiently large base voltage, will enter **SATURATION.**

Note: In a **SATURATED** transistor the base current and the collector current are independent of each other. The base current is still (and always is) found only from the base circuit. The collector current is found approximately by closing the imaginary switch between the collector and the emitter in the collector circuit.

EXAMPLE 7–7
Base Supply Voltage and Saturation II

For the circuit in Figure 7–19 (p. 204), find the base supply voltage (V_{BB}) that just puts the transistor into **SATURATION** (assume $\beta = 200$).

Solution   When a transistor first goes into **SATURATION,** we can assume that the collector shorts to the emitter but the collector current is still β times the base current. That is, at the edge of **SATURATION**

$$I_C = \beta I_B = I_{C(sat)}$$

Therefore, at the edge of **SATURATION**

$$I_B = \frac{I_{C(sat)}}{\beta}$$

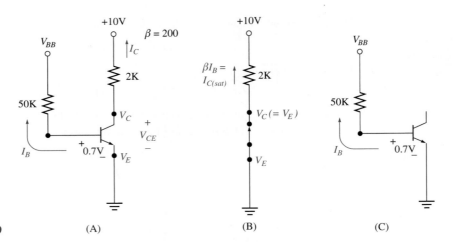

FIGURE 7–19
(A) (B) (C)

From the **SATURATED** collector circuit shown in Figure 7–19B, we find

$$I_{C(sat)} = \frac{10}{2K} = 5mA$$

The base current we want is

$$I_B = \frac{I_{C(sat)}}{\beta} = \frac{5mA}{200} = 0.025mA$$

We now turn our attention to the base circuit in Figure 7–19C to find the base current:

$$I_B = \frac{V_{BB} - 0.7}{50K}$$

Multiplying both sides of this equation by 50K gives us

$$50KI_B = V_{BB} - 0.7$$

Finally, solving for the base supply voltage results in

$$V_{BB} = 0.7 + 50KI_B$$

That is, the base supply voltage must equal the voltage drop across the emitter diode plus the voltage drop across the base resistor. To drive a transistor into **SATURATION**, V_{BB} must be large enough to supply the voltage needed to turn on the emitter diode plus the voltage needed to produce sufficient base current.

From the base current previously found, the base supply voltage needed to just put the transistor into **SATURATION** is

$$V_{BB} = 0.7 + 50KI_B = 0.7 + 50K \times 0.025mA = 0.7 + 1.25 = 1.95V$$

That is, for

$$V_{BB} \geq 1.95V$$

the transistor will be in **SATURATION**.

DRILL EXERCISE   Redraw the circuit in Figure 7–19 with $R_C = 5K\Omega$; find $I_{C(sat)}$ and the base supply voltage that just puts the transistor into **SATURATION**.

Answer $I_{C(sat)} = 2mA$, $V_{BB} \geq 1.2V$

EXAMPLE 7–8
Base Resistor and Saturation

The circuit in Figure 7–20A replicates the circuit in Figure 7–14 with the base resistor decreased to 5KΩ (β = 100). Find the state of the transistor, I_B, I_C, I_E, and V_{CE}.

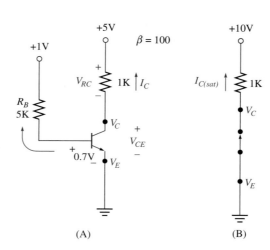

FIGURE 7–20   (A)   (B)

Solution   We first assume that the emitter diode is **ON** and solve for I_B:

$$I_B = \frac{V_{BB} - 0.7}{R_B} = \frac{1 - 0.7}{5K} = 0.06\text{mA}$$

We then assume that the transistor is **ACTIVE** and find I_C:

$$I_C = \beta I_B = 6\text{mA}$$

Finally, we find V_C:

$$V_C = V_{CC} - R_C I_C = 5 - 1K \times 6\text{mA} = -1\text{V}$$

Because V_C is negative, the transistor is **SATURATED,** so we must find the collector current from the **SATURATED** collector circuit (Figure 7–20B):

$$V_C \approx V_E = 0\text{V}$$

and

$$I_C = I_{C(sat)} = \frac{V_{CC}}{2K} = \frac{5}{1K} = 5\text{mA}$$

The emitter current is

$$I_E = I_B + I_C = 5.06\text{mA}$$

The results for this **SATURATED** transistor are

$$I_B = 0.06\text{mA}, \ I_C = I_{C(sat)} = 5\text{mA}, \ I_E = 5.06\text{mA, and } V_C \approx V_E = 0\text{V}$$

DRILL EXERCISE

Redraw the circuit in Figure 7–20 with $R_B = 3K\Omega$; find the transistor currents and voltages.

Answer $I_B = 0.1\text{mA}$, $I_C = I_{C(sat)} = 5\text{mA}$, $I_E = 5.1\text{mA}$
$V_B = 0.7\text{V}$, $V_C \approx V_E = 0\text{V}$

As the base resistor decreases, both base current and collector current increase and the transistor may **SATURATE.**

EXAMPLE 7–9
Base Resistor and Saturation II

For the circuit in Figure 7–21, find the base resistor (R_B) that just puts the transistor into **SATURATION** (assume $\beta = 200$).

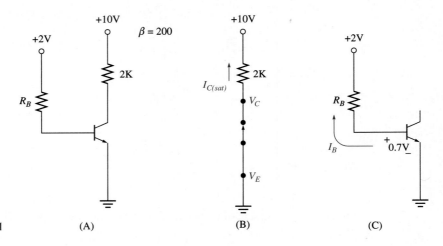

FIGURE 7–21 (A) (B) (C)

Solution When a transistor first goes into **SATURATION,** we assume the collector shorts to the emitter, but the collector current is still β times the base current. That is, at the edge of **SATURATION**

$$I_C = \beta I_B = I_{C(sat)}$$

The base current is, therefore

$$I_B = \frac{I_{C(sat)}}{\beta}$$

From the **SATURATED** collector circuit shown in Figure 7–21B, we find

$$I_{C(sat)} = \frac{10}{2K} = 5mA$$

The base current we want is

$$I_B = \frac{I_{C(sat)}}{\beta} = \frac{5mA}{200} = 0.025mA$$

We now turn our attention to the base circuit (Figure 7–21C), to find the base current:

$$I_B = \frac{V_{BB} - 0.7}{R_B}$$

Solving for the base resistor gives us

$$R_B = \frac{V_{BB} - 0.7}{I_B}$$

Therefore

$$R_B = \frac{2 - 0.7}{0.025mA} = \frac{1.3}{0.025mA} = 52K\Omega$$

That is, for

$$R_B \leq 52K\Omega$$

the transistor will be in **SATURATION.**

DRILL EXERCISE

Redraw the circuit in Figure 7–21 with $V_{BB} = 10V$; find $I_{C(sat)}$ and the base resistor that puts the transistor in **SATURATION.**

Answer $I_{C(sat)} = 5mA$, $R_B \leq 372K\Omega$

EXAMPLE 7–10
Collector Resistor and Saturation I

The circuit in Figure 7–22A replicates the circuit in Figure 7–14 with the collector resistor increased to $2K\Omega$ ($\beta = 100$). Find the state of the transistor, I_B, I_C, I_E, and V_{CE}.

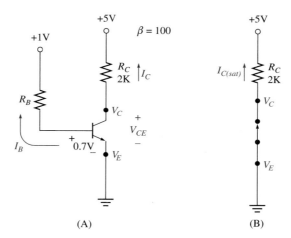

FIGURE 7–22   (A)   (B)

Solution   First we assume that the emitter diode is **ON** and solve for I_B.

$$I_B = \frac{V_{BB} - 0.7}{R_B} = \frac{1 - 0.7}{10K} = 0.03mA$$

Then we assume that the transistor is **ACTIVE** and find I_C:

$$I_C = \beta I_B = 100 \times 0.03mA = 3mA$$

Finally, we find V_C:

$$V_C = V_{CC} - R_C I_C = 5 - 2K \times 3mA = -1V$$

Because V_C is negative, the transistor is **SATURATED,** and we must find the collector current from the **SATURATED** collector circuit (Figure 7–22B), where

$$V_C \approx V_E = 0V$$

and

$$I_C = I_{C(sat)} = \frac{V_{CC}}{R_C} = \frac{5}{2K} = 2.5mA$$

The emitter current is

$$I_E = I_B + I_C = 0.03 + 2.5 = 2.53mA$$

The results for this **SATURATED** transistor are

$$I_B = 0.03\text{mA}, I_C = I_{C(sat)} = 2.5\text{mA}, I_E = 2.53\text{mA},$$

and

$$V_C \approx V_E = 0\text{V}$$

DRILL EXERCISE   Redraw the circuit in Figure 7–22 (p. 207) with $R_C = 3\text{K}\Omega$; find the transistor currents and voltages.

Answer $I_B = 0.03\text{mA}, I_C = I_{C(sat)} = 1.67\text{mA}, I_E = 1.7\text{mA}$
$V_B = 0.7\text{V}, V_C \approx V_E = 0\text{V}$

As the collector resistor increases, the collector voltage decreases. As the collector voltage approaches the emitter voltage, the transistor approaches **SATURATION**.

EXAMPLE 7–11
Collector Resistor and Saturation II

For the circuit in Figure 7–23 find the collector resistor (R_C), that just puts the transistor into **SATURATION** (assume $\beta = 100$).

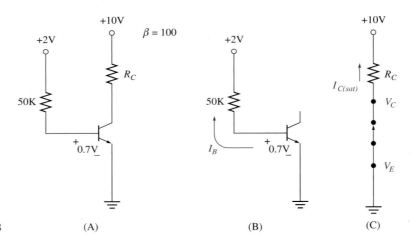

FIGURE 7–23   (A)   (B)   (C)

Solution   When a transistor first goes into **SATURATION,** the collector shorts to the emitter, but the collector current is still β times the base current. That is, at the edge of **SATURATION**

$$I_C = \beta I_B = I_{C(sat)}$$

In this example the base resistor is given, so find the base current first. From the base circuit of Figure 7–23B

$$I_B = \frac{V_{BB} - 0.7}{R_B} = \frac{2 - 0.7}{50\text{K}} = 0.026\text{mA}$$

We now want the collector current to equal the **SATURATION** current:

$$I_{C(sat)} = \beta I_B = 100 \times 0.026\text{mA} = 2.6\text{mA}$$

In **SATURATION** (see Figure 7–23C)

$$V_C \approx V_E = 0V$$

This means that the voltage drop across the collector resistor is V_{CC} and the current in the collector resistor is $I_{C(sat)}$. From Ohm's Law, therefore

$$R_C = \frac{V_{CC}}{I_{C(sat)}} = \frac{10}{2.6mA} = 3.85K\Omega$$

The transistor will be in **SATURATION** for

$$R_C \geq 3.85K\Omega$$

DRILL EXERCISE   Redraw the circuit in Figure 7–23 with $V_{BB} = 10V$; find the collector resistor that will just put the transistor into **SATURATION.**

Answer $R_C \geq 538\Omega, I_{C(sat)} = 18.6mA$

EXAMPLE 7–12
Collector Supply
Voltage and
Saturation I

The circuit in Figure 7–24A replicates the circuit in Figure 7–14 with the collector supply voltage decreased to $2V(\beta = 100)$. Find the state of the transistor, I_B, I_C, I_E, and V_{CE}.

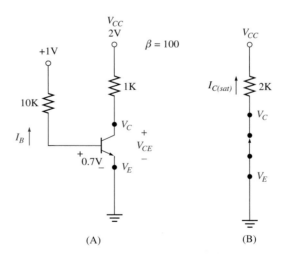

FIGURE 7–24   (A)   (B)

Solution   First we assume that the emitter diode is **ON** and solve for I_B:

$$I_B = \frac{1 - 0.7}{10K} = 0.03mA$$

Again, the base current does not depend on either the collector resistor or the collector supply voltage. We now assume that the transistor is **ACTIVE** and find I_C:

$$I_C = \beta I_B = 100 \times 0.03mA = 3mA$$

Finally, we find V_C:

$$V_C = V_{CC} - R_C I_C = 2 - 1K \times 3mA = -1V$$

Because V_C is negative, the transistor is **SATURATED** and

$$V_C \approx V_E = 0V$$

So

$$I_C = I_{C(sat)} = \frac{V_{CC}}{R_C} = \frac{2}{1K} = 2mA$$

The emitter current is

$$I_E = I_B + I_C = 2.03mA$$

The results for this **SATURATED** transistor are

$$I_B = 0.03mA, \; I_C = I_{C(sat)} = 2mA, \; I_E = 2.03mA,$$

and

$$V_C \approx V_E = 0V$$

As the collector supply voltage decreases, less collector current is available to the transistor, so the transistor is more likely to **SATURATE**.

DRILL EXERCISE   Redraw the circuit in Figure 7–24 (p. 209) with $V_{CC} = 3V$; find the transistor currents and voltages.

Answer $I_B = 0.03mA, \; I_C = I_{C(sat)} = 3mA, \; I_E = 3.03mA$
$V_B = 0.7V, \; V_C \approx V_E = 0V$

EXAMPLE 7–13
Collector Supply Voltage and Saturation II

For the circuit in Figure 7–25, find the collector supply voltage (V_{CC}) that just puts the transistor into **SATURATION** (assume $\beta = 100$).

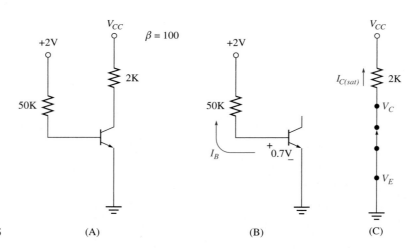

FIGURE 7–25   (A)   (B)   (C)

Solution   When a transistor first goes into **SATURATION,** the collector shorts to the emitter, but the collector current is still β times the base current. That is, at the edge of **SATURATION**

$$I_C = \beta I_B = I_{C(sat)}$$

In this example, the base resistor is given, so the base current can be found first. From the base circuit of Figure 7–25B

$$I_B = \frac{V_{BB} - 0.7}{R_B} = \frac{2 - 0.7}{50K} = 0.026\text{mA}$$

We now want the collector current to equal the **SATURATION** current:

$$I_{C(sat)} = \beta I_B = 100 \times 0.026\text{mA} = 2.6\text{mA}$$

In **SATURATION** (see Figure 7–23C, p. 208)

$$V_C \approx V_E = 0\text{V}$$

This means that the voltage drop across the collector resistor is V_{CC} and the current in the collector resistor is $I_{C(sat)}$. From Ohm's Law, therefore

$$V_{CC} = I_{C(sat)}R_C = 2.6\text{mA} \times 2\text{K}\Omega = 5.2\text{V}$$

The transistor will be in **SATURATION** for

$$V_{CC} \leq 5.2\text{V}$$

DRILL EXERCISE   Redraw the circuit of Figure 7–25 with $\beta = 200$; find the collector supply voltage that just puts the transistor into **SATURATION**.

Answer $V_{CC} \leq 10.4\text{V}, I_{C(sat)} = 5.2\text{mA}$

The last set of examples and drill exercises show how the supply voltages and the base and collector resistors affect the state of a transistor. We see that a transistor heads toward **SATURATION** as

V_{BB} increases

R_B decreases

R_C increases

V_{CC} decreases

β increases

REVIEW QUESTIONS

1. Define **CUT-OFF,** and draw the circuit model for a transistor in **CUT-OFF.**
2. For what range of values of the base supply voltage is a transistor in **CUT-OFF?**
3. Define **SATURATION,** and draw the circuit model for a transistor in **SATURATION.**
4. How do you find $I_{C(sat)}$? Is $I_{C(sat)}$ the maximum possible collector current?
5. Can βI_B be greater than $I_{C(sat)}$?
6. How does the base supply voltage affect the **SATURATION** of a transistor?
7. How does the base resistor affect the **SATURATION** of a transistor?
8. How does the collector supply voltage affect the **SATURATION** of a transistor?
9. How does the collector resistor affect the **SATURATION** of a transistor?
10. How does the current gain affect the **SATURATION** of a transistor?

7–7 ■ DETERMINING THE STATE OF A TRANSISTOR

As we discussed earlier in this chapter, the state of a transistor is determined by the state of its emitter and collector diodes. In **CUT-OFF** both junctions are **OFF**; in the **ACTIVE** state the emitter diode is **ON** and the collector diode is **OFF**; in **SATURATION** both diodes are **ON.** The transistor voltages and currents are summarized below for each state:

CUT-OFF

$$V_{BE} < 0.7V$$

$$V_{BC} < 0.5V$$

$$I_B = I_C = I_E = 0A; \text{ all transistor currents are } 0A.$$

ACTIVE

$$V_{BE} = 0.7V$$

$$V_{BC} < 0.5V$$

$$V_C > V_E$$

$$I_B, I_C, I_E > 0; \text{ all transistor currents are positive.}$$

$$I_C = \beta I_B$$

SATURATION

$$V_{BE} = 0.7V$$

$$V_{BC} = 0.5V$$

$$V_C \approx V_E$$

$$I_C = I_{C(sat)}$$

This summary leads to the following procedure for determining the state of a transistor:

☐ If $V_{BB} \leq 0.7V$, transistor is **CUT-OFF;** set all currents to 0A.

☐ If $V_{BB} > 0.7V$, assume the transistor is in the **ACTIVE** state.

☐ Set $V_{BE} = 0.7V$ and find I_B.

☐ If $I_B > 0$, $I_C = \beta I_B$.

☐ Find V_C.

☐ If $V_C > 0$, transistor is **ACTIVE;** finish analysis.

☐ If $V_C < 0$, transistor is **SATURATED;** set $V_C \approx V_E$ and $I_C = I_{C(sat)}$.

EXAMPLE 7–14
Determining the
State of a
Transistor I

For the circuit of Figure 7–26, determine the state of the transistor and give the DC Bias List.

Solution   Because the base supply voltage is negative, the transistor must be **CUT-OFF.** Recall that the base supply voltage must be greater than 0.7V to turn the emitter diode **ON.** Even if you did not recognize by inspection that this transistor is **CUT-OFF,** you could determine its state by following this procedure.

In Figure 7–26B, if we assume the transistor is **ACTIVE** and solve for the base current, we get

$$I_B = \frac{V_{BB} - 0.7}{20K} = \frac{-5 - 0.7}{20K} = -0.285 \text{mA}$$

The fact that I_B is negative confirms our conclusion that the transistor is **CUT-OFF.** All of the currents in a **CUT-OFF** transistor are 0A, so we can remove the transistor from the circuit (Figure 7–26C). Because there are no currents in the base and collector resistors, we get

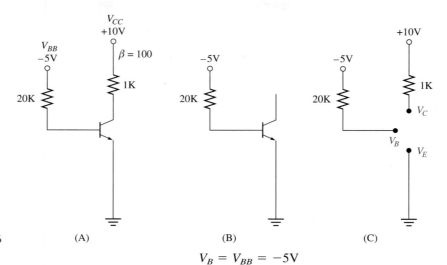

FIGURE 7–26   (A)   (B)   (C)

$$V_B = V_{BB} = -5V$$

and

$$V_C = V_{CC} = 10V$$

The complete **DC Bias List** is

I_B	0A
I_C	0A
I_E	0A
V_B	$-5V$
V_C	10V
V_E	0V (emitter is grounded)

$$V_{CB} = V_C - V_B = 10 - (-5) = 15V$$
$$V_{CE} = V_C - V_E = 10 - 0 = 10V$$
$$P = V_{CE}I_C = 0W$$

Note: In the **CUT-OFF** transistor, the collector-to-base reverse-biased voltage can be very large, 15V in this example. For a given transistor you must not exceed the maximum allowable collector-base voltage. To do so could damage the device. Because transistor currents are 0A, no power is dissipated in the **CUT-OFF** transistor.

DRILL EXERCISE

Redraw Figure 7–26 with $V_{BB} = 0$ (grounded base), and determine the state of the transistor, the transistor currents and voltages, and V_{CB}.

Answer   **CUT-OFF**
$I_B = I_C = I_E = 0A$
$V_B = 0V$, $V_C = 10V$, $V_E = 0V$, $V_{CB} = 10V$

EXAMPLE 7–15
Determining the State of a Transistor II

For the circuit of Figure 7–27 (p. 214), determine the state of the transistor and give the DC Bias List (assume $\beta = 200$).

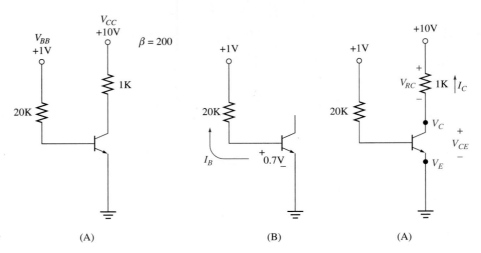

FIGURE 7–27

Solution   We assume the transistor is **ACTIVE** and find the base current from the base circuit in Figure 7–27B:

$$I_B = \frac{V_{BB} - 0.7}{20\text{K}} = \frac{1 - 0.7}{20\text{K}} = 0.015\text{mA}$$

Because I_B is positive, we find I_C from

$$I_C = \beta I_B = 200 \times 0.015\text{mA} = 3\text{mA}$$

We now find V_C and V_{CE}:

$$V_C = V_{CC} - R_C I_C = 10 - 1\text{K} \times 3\text{mA} = 7\text{V}$$

Because V_C is positive, our assumption that the transistor is **ACTIVE** is correct. The complete **DC Bias List** for this circuit is

I_B	0.015mA
I_C	3mA
I_E	3mA
V_B	0.7V
V_C	7V
V_E	0V (emitter is grounded)

$$V_{CB} = V_C - V_B = 7 - 0.7 = 6.3\text{V}$$
$$V_{CE} = V_C - V_E = 7 - 0 = 7\text{V}$$
$$P = V_{CE}I_C = 21\text{mW}$$

DRILL EXERCISE

Redraw the circuit in Figure 7–27, with $V_{BB} = 1.5$V; determine the state of the transistor, all transistor currents and voltages, V_{CB}, and power.

Answer   **ACTIVE**
$I_B = 0.04$mA, $I_C = I_E = 8$mA
$V_B = 0.7$V, $V_C = 2$V, $V_E = 0$V
$V_{CB} = 1.3$V, $P = 16$mW

EXAMPLE 7–16
Determining the State of a Transistor III

For the circuit of Figure 7–28, determine the state of the transistor and give the DC Bias List (assume β = 100).

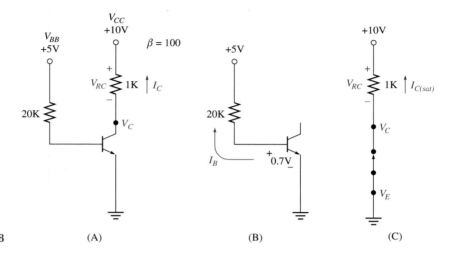

FIGURE 7–28

(A) (B) (C)

Solution We assume the transistor is **ACTIVE** and find the base current from the base circuit shown in Figure 7–28B:

$$I_B = \frac{V_{BB} - 0.7}{20K} = \frac{5 - 0.7}{20K} = 0.215mA$$

Because I_B is positive, we find I_C from

$$I_C = \beta I_B = 100 \times 0.215mA = 21.5mA$$

We now find V_C:

$$V_C = V_{CC} - R_C I_C = 10 - 1K \times 21.5mA = -11.5V$$

Because V_C is negative, our assumption that the transistor is **ACTIVE** is incorrect. This transistor is **SATURATED,** so we must use the **SATURATED** collector circuit shown in Figure 7–28C to find the collector current.

$$I_{C(sat)} = \frac{V_{CC}}{1K} = \frac{10}{1K} = 10mA$$

The complete **DC Bias List** for this **SATURATED** transistor is

I_B 0.215mA
I_C 10mA
I_E 10.2mA
V_B 0.7V
V_C 0V
V_E 0V (emitter is grounded)

$V_{CB} = -0.5V$ (the collector diode is **ON**)
$V_{CE} = 0V$
$P = V_{CE}I_C = 0W$

DRILL EXERCISE   Redraw the circuit in Figure 7–28 (p. 215) with $R_C = 500\Omega$; determine the state of the transistor, all transistor currents and voltages, and V_{CB}.

Answer **SATURATION**
$I_B = 0.215\text{mA}, I_C = I_{C(sat)} = 20\text{mA}, I_E = 20.2\text{mA}$
$V_B = 0.7\text{V}, V_C = V_E = 0\text{V}, V_{CB} = -0.5\text{V}$

REVIEW QUESTIONS

1. What are the voltage and current conditions for a transistor in **CUT-OFF?**
2. What are the voltage and current conditions for an **ACTIVE** transistor?
3. What are the voltage and current conditions for a transistor in **SATURATION?**
4. Summarize the steps used to determine the state of a transistor.
5. What is the power dissipated in a transistor in **CUT-OFF?** Why?
6. What is the power dissipation in a transistor in **SATURATION?** Why?
7. In which state is V_{CB} largest? Why?

7–8 ■ DC TRANSISTOR DATA SHEET

Figure 7–29 shows an abbreviated transistor data sheet for the 2N3904, which includes only DC characteristics of the transistor. A brief discussion of the ELECTRICAL CHARACTERISTICS entries follows.

Line 1: V_{CB} is the voltage from the collector to the base. This voltage is the reverse-biased voltage across the collector diode (Figure 7–4). Recall that if a diode has too much reverse-biased voltage across it, the junction will break down. The reverse-biased collector diode also breaks down at high voltage. This limit is labeled V_{CBO} and is 60V for this transistor.

FIGURE 7–29

ABSOLUTE MAXIMUM RATINGS: 2N3904
Power Dissipation
Total dissipation at 25° C   600mW

Electrical characteristics

SYMBOL	CHARACTERISTIC	MIN	MAX	UNITS	TEST CONDITIONS
BV_{CBO}	Collector-to-emitter Breakdown voltage	60		V	$I_C = 10\mu\text{A}, I_E = 0$
BV_{EBO}	Emitter-to-base Breakdown voltage	6.0		V	$I_E = 10\mu\text{A}, I_C = 0$
h_{FE}	DC Current Gain	40			$I_C = 0.1\text{mA}, V_{CE} = 5.0\text{V}$
		90			$I_C = 1.0\text{mA}, V_{CE} = 5.0\text{V}$
		60	360		$I_C = 10\text{mA}, V_{CE} = 5.0\text{V}$
		40			$I_C = 50\text{mA}, V_{CE} = 5.0\text{V}$
$V_{BE(sat)}$	Base-to-Emitter Saturation voltage		0.85	V	$I_C = 50\text{mA}, I_B = 5.0\text{mA}$
$V_{CE(sat)}$	Collector-to-Emitter Saturation voltage		0.25	V	$I_C = 50\text{mA}, I_B = 5.0\text{mA}$

Line 2: V_{EB} is the reverse-biased voltage across the emitter diode. The breakdown voltage for this junction is labeled V_{EBO}, and is 6V.

Line 3: DC Current Gain is what we have called β. (Most data sheets use the term h_{FE} for this gain. The terms are interchangeable, so we will continue to use β.) Note that for any transistor, the actual DC current gain depends on the size of collector current. For this transistor, β can vary from 40 to 360. For simplicity we ignore these variations and assume a fixed number, usually 100, for transistor current gain.

Line 4: $V_{BE(sat)}$ is the emitter diode turn-on voltage and has a maximum value of 0.85V.

Line 5: $V_{CE(sat)}$ is the collector-emitter voltage when the transistor is in **SATURA-TION.** For this device $V_{CE(sat)}$ has a maximum value of 0.25V. For simplicity, we usually assume that $V_{CE(sat)} \approx 0V$.

SUMMARY

- The NPN BJT is constructed by sandwiching P-type material between two layers of N-type material, resulting in one PN junction between the base and emitter (the emitter diode) and another between the base and collector (the collector diode).
- The transistor is in **CUT-OFF** if both the emitter and collector diodes are **OFF.**
- The transistor is in the **ACTIVE** state if the emitter diode is **ON** and the collector diode is **OFF.**
- The transistor is in the **SATURATED** state if both the emitter and collector diodes are **ON.**
- In the **CUT-OFF** state, all transistor currents are 0A.
- In the **ACTIVE** state, $V_{BE} = 0.7V$, $I_C = \beta I_B$ and $I_E \approx I_C$.
- In the **SATURATED** state, $V_{BE} = 0.7V$, $V_C \approx V_E$, $I_C = I_{C(sat)}$, and $I_E = I_B + I_C$.
- In the **ACTIVE** state, for I_B control

$$I_B = \frac{V_{BB} - 0.7}{R_B}, \ I_C = \beta I_B, \text{ and } V_C = V_{CC} - R_C I_C$$

- In the **SATURATED** state, for I_B control

$$I_B = \frac{V_{BB} - 0.7}{R_B}, \ V_C \approx V_E, \text{ and } I_C = I_{C(sat)} = \frac{V_{CC}}{R_C}$$

- To determine the state of a transistor

    If $V_{BB} \leq 0.7V$, transistor is **CUT-OFF,** set all currents to 0A.

    If $V_{BB} > 0.7V$, assume the transistor is in the **ACTIVE** state.

    Set $V_{BE} = 0.7V$, and find I_B.

    If $I_B > 0$, $I_C = \beta I_B$.

    Find V_C.

    If $V_C > 0$, transistor is **ACTIVE,** finish analysis.

    If $V_C < 0$, transistor is **SATURATED,** set $V_C \approx V_E$ and $I_C = I_{C(sat)}$.

PROBLEMS

SECTION 7–2   BJT States of Operation

1. For the transistor in Figure 7–30 (p. 218):

    (a) Write down the following current values

    $I_E =$

    $I_B =$

    $I_C =$

    (b) What is the current gain (β) for this transistor?

FIGURE 7–30

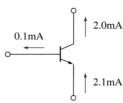

FIGURE 7–31

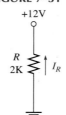

FIGURE 7–32

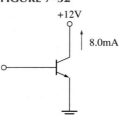

FIGURE 7–33

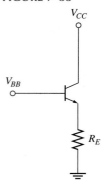

2. For the resistor circuit in Figure 7–31

   (a) How much voltage (V_R) is across the resistor?

   (b) How much current (I_R) is there in the resistor?

   (c) What is the power dissipated by the resistor?

3. For the transistor in Figure 7–32

   (a) What is V_{CC}?   **(b)** What is V_{CE}?

   (c) What is I_C?

4. For the **ACTIVE** transistor in Figure 7–32:

   (a) If $\beta = 5$, what are I_B and I_E?   **(b)** If $\beta = 50$, what are I_B and I_E?

   (c) If $\beta = 200$, what are I_B and I_E?

5. A transistor in the **ACTIVE** state has a base current of $I_B = 0.025$mA.

   (a) Find I_C and I_E if $\beta = 50$.   **(b)** Find I_C and I_E if $\beta = 100$.

   (c) Find I_C and I_E if $\beta = 500$.

6. A transistor in the **ACTIVE** state has an emitter current of $I_E = 1$mA.

   (a) Find I_C and I_B if $\beta = 100$.   **(b)** Find I_C and I_B if $\beta = 250$. ,

   (c) Find I_C and I_B if $\beta = 500$.

7. Redraw Figure 7–33 with the following parameters:

$$V_{BB} = +1\text{V}, V_{CC} = +10\text{V}, R_E = 1\text{K}\Omega, \beta = 100$$

   (a) Find I_B, I_C, I_E.   **(b)** Find V_B, V_C, V_E.

8. Redraw Figure 7–33 with the following parameters:

$$V_{BB} = +2\text{V}, V_{CC} = +10\text{V}, R_E = 1\text{K}\Omega, \beta = 200$$

   (a) Find I_B, I_C, I_E.   **(b)** Find V_B, V_C, V_E.

9. Redraw Figure 7–33 with the following parameters:

$$V_{BB} = +2\text{V}, V_{CC} = +10\text{V}, R_E = 2\text{K}\Omega, \beta = 100$$

   (a) Find I_B, I_C, I_E.   **(b)** Find V_B, V_C, V_E.

SECTION 7–3   Transistor Ratings

10. What is the power dissipated in the transistor in Figure 7–32?

11. Ground the base of the transistor in Figure 7–32. What is the minimum V_{CBO} required for this transistor?

12. What power is dissipated in the transistor of Problem 7? What is the minimum V_{CBO} required for this transistor?

13. What power is dissipated in the transistor of Problem 8? What is the minimum V_{CBO} required for this transistor?

FIGURE 7–34

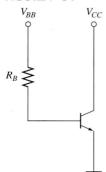

14. What power is dissipated in the transistor of Problem 9? What is the minimum V_{CBO} required for this transistor?

SECTION 7–4   Simple DC Transistor Circuits (I_B control)

15. For the transistor circuit in Figure 7–34

$$V_{CC} = +20V, V_{BB} = +20V, R_B = 2M\Omega, \beta = 100$$

  (a) Find I_B.
  (b) Find I_C. What assumption have you made about the state of operation of the transistor?
  (c) Find I_E.   (d) What is V_B?
  (e) What is V_E?   (f) What is V_C?
  (g) Is this transistor **ACTIVE?** Why?

16. For the transistor circuit in Figure 7–34

$$V_{CC} = +20V, V_{BB} = +20V, R_B = 2M\Omega, \beta = 100$$

  (a) What is V_{CE}?   (b) What is V_{CB}?
  (c) What is V_{BE}?
  (d) What is the power dissipated by the transistor? Give your answer in milliwatts and watts.

17. For the transistor circuit in Figure 7–34

$$V_{CC} = +20V, V_{BB} = +20V, R_B = 2M\Omega, \beta = 200$$

Repeat Problem 15.

18. For the transistor circuit in Figure 7–34

$$V_{CC} = +20V, V_{BB} = +20V, R_B = 2M\Omega, \beta = 200$$

Repeat Problem 16.

19. For the transistor circuit in Figure 7–34

$$V_{CC} = +10V, V_{BB} = +2V, R_B = 200K\Omega, \beta = 100$$

Repeat Problem 15.

20. For the transistor circuit in Figure 7–34

$$V_{CC} = +10V, V_{BB} = +2V, R_B = 200K\Omega, \beta = 100$$

Repeat Problem 16.

21. For the transistor circuit shown in Figure 7–34

$$V_{CC} = +10V, V_{BB} = +0.3V, R_B = 200K\Omega, \beta = 100$$

Repeat Problem 15.

22. For the transistor circuit shown in Figure 7–34

$$V_{CC} = +10V, V_{BB} = +0.3V, R_B = 200K\Omega, \beta = 100$$

Repeat Problem 16.

SECTION 7–5   The Collector Resistor

23. For the transistor circuit shown in Figure 7–35 (p. 220), the DC bias values of the voltages and currents at the leads of the transistor are determined by I_B control. If $\beta = 100$
  (a) Find I_B, I_C, and I_E (assuming the transistor is **ACTIVE**).
  (b) Find V_B, V_E, V_{RC}, and V_C.
  (c) Is the transistor **ACTIVE?** Why?
  (d) If the transistor is **ACTIVE,** are the voltages and currents you calculated in this problem correct?

24. Repeat problem 23 if $\beta = 50$.

25. Repeat problem 23 if $\beta = 200$.

26. For the transistor circuit shown in Figure 7–36:

    (a) What is I_B?

    (b) What is V_B?

    (c) What would I_C be if the transistor were **ACTIVE?**

    (d) What would be the voltage across the collector resistor (V_{RC}) for the value of I_C you have calculated? Is this possible?

    (e) Is the transistor **ACTIVE** or **SATURATED?**

    (f) What is the actual value of I_C? Is this $I_{C(sat)}$?

    (g) When a transistor is **SATURATED,** what two circuit parameters determine the actual value of I_C?

    (h) Write the DC Bias List for this transistor. Use the same order given in the text.

FIGURE 7–35   **FIGURE 7–36**

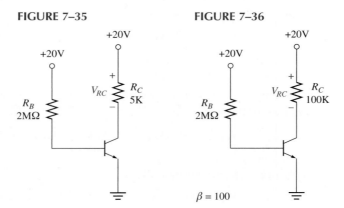

SECTION 7–6   The CUT-OFF and SATURATION Regions

27. For the transistor shown in Figure 7–37, $V_{BB} = 0V$.

    (a) Write the DC Bias List for this circuit.

    (b) Is the transistor **CUT-OFF** or **SATURATED?**

    (c) Explain the state of operation of the transistor based on the value you found for I_B.

28. For the transistor shown in Figure 7–37, $V_{BB} = +15V$.

    (a) Write the DC Bias List for this circuit.

    (b) Is the transistor **CUT-OFF** or **SATURATED?**

    (c) Explain the state of operation of the transistor based on the value you found for I_B.

29. For the transistor circuit shown in Figure 7–38, $V_{BB} = 0V$.

    (a) What is V_C?

    (b) What is I_C?

    (c) If the transistor is acting as a switch, is the switch open or closed?

FIGURE 7–37   **FIGURE 7–38**

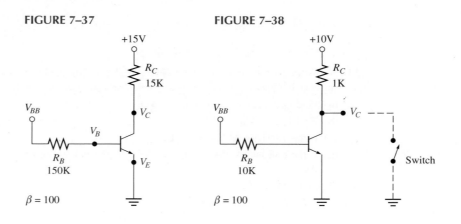

30. For the transistor circuit shown in Figure 7–38, $V_{BB} = +10V$.

(a) What is V_C?

(b) What is I_C?

(c) If the transistor is acting as a switch, is the switch open or closed?

SECTION 7–7 Determining the State of the Transistor

31. For what range of values of V_{BB} is the transistor in Figure 7–39 in **CUT-OFF?**

FIGURE 7–39

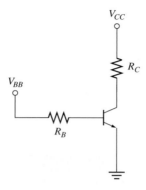

32. In Figure 7–39, $V_{BB} = +2V$, $V_{CC} = +10V$, $R_C = 1K\Omega$, $\beta = 100$.
What value of R_B puts the transistor into **SATURATION?**

33. In Figure 7–39, $V_{BB} = +2V$, $V_{CC} = +10V$, $R_B = 10K\Omega$, $\beta = 100$.
What value of R_C puts the transistor into **SATURATION?**

34. In Figure 7–39, $V_{CC} = +10V$, $R_C = 1K\Omega$, $R_B = 10K\Omega$, $\beta = 100$. What value of V_{BB} puts the transistor into **SATURATION?**

35. In Figure 7–39, $V_{BB} = +2V$, $R_C = 1K\Omega$, $R_B = 10K\Omega$, $\beta = 100$. What value of V_{CC} puts the transistor into **SATURATION?**

36. In Figure 7–39, $V_{BB} = +2V$, $V_{CC} = +10V$, $R_C = 1K\Omega$, $R_B = 10K\Omega$. What value of β puts the transistor into **SATURATION?**

COMPUTER PROBLEMS

37. List the NPN BJT models contained in your computer programs library. What values of current gain are used for these devices?

38. Simulate the circuit of Figure 7–14 (p. 196). Find V_B and V_C. What is the state of the transistor?

39. Simulate the circuit of Figure 7–14. Let

$$V_{BB} = 0V, 0.3V, 0.5V, 1V, 1.2V, \text{ and } 1.5V$$

For each value of V_{BB}, find V_B and V_C. What is the state of the transistor?

40. Simulate the circuit of Figure 7–14. Let

$$R_{BB} = 20K\Omega, 10K\Omega, 6K\Omega, \text{ and } 3K\Omega$$

For each value of R_{BB}, find V_B and V_C. What is the state of the transistor?

41. Simulate the circuit of Figure 7–14. Let

$$V_{CC} = 5V, 4V, 3V, \text{ and } 2V$$

For each value of V_{CC}, find V_B and V_C. What is the state of the transistor?

CHAPTER

8

TROUBLESHOOTING

CHAPTER OBJECTIVES

To use DC Troubleshooting to determine circuit problems.

To record and compare Estimated and Measured Voltages.

To locate circuit problems.

To determine how component failures affect circuit behavior.

To build the Universal Transistor Test Circuit.

To use a voltmeter or oscilloscope to perform DC Troubleshooting.

8–1 ■ INTRODUCTION

As a student of electronics, you have built (or will soon build) many electronic circuits. Typically, you will first learn certain principles of electronic circuit behavior in the classroom, then build circuits in the lab to verify what you have learned. On the job you will build commercial electronic products. Whether you build a circuit in the lab or on the job, you will often find that the circuit does not work properly. Also, on the job you will often be given a circuit that may have once worked but no longer does.

Troubleshooting is the process of determining what is wrong with a circuit: Is this transistor bad? Is that a bad solder joint? In fact, troubleshooting may be one of the most important functions of your career. In this chapter we will describe a troubleshooting approach that we believe will provide a systematic procedure that will greatly enhance your ability to correct faulty electronic circuits.

We have found that we can locate problems in most electronic circuits by measuring just the DC bias voltages and comparing these measurements to estimated values. It is interesting to note that even though most electronic circuits are meant to process AC signals, we do not need to make AC measurements to troubleshoot. DC measurements are usually enough. For this reason, we call our approach **DC Troubleshooting.**

There are three likely situations you will encounter during your education and career when you may be required to troubleshoot.

Temporary Breadboard Building

Breadboarding a circuit is usually the first step in transferring a circuit from paper to reality. This process is a way to build a temporary version of the circuit so we can verify the paper design. As a student, most (if not all) of the circuits you build in the laboratory will be breadboarded. The breadboard itself is a foundation on which you insert elements such as resistors and transistors into holes that are grouped together for making common connections between components. Figure 8–1A shows a white breadboard of the type which is typically used in school laboratories. All too often, we have seen student groups make a mistake in connecting circuit components on a breadboard, then waste an entire lab period making measurements and observations on a circuit that is not functioning properly. As the software people say, "garbage in—garbage out." Nothing of value is learned from such a lab experience. DC Troubleshooting can prevent this waste of time and effort.

Prototype Building

Your job in industry may be one in which you are given a schematic diagram from the engineering department (or even from a book or magazine article) and are told to build a working prototype circuit (Figure 8–1B). Very often a new circuit design does not work. Perhaps someone made a few mistakes when building the circuit, or perhaps the technician followed the published schematic exactly, but the schematic included errors.

Repairing a Circuit That Once Worked

When you are asked to repair the circuit, you really have a chance to look good, demonstrating your skills in electronics. Your boss comes to you with a circuit board and says, "This is our Mighty Microphone Amplifier. It doesn't work. Find out what's wrong and fix it."

Many students, as well as industry workers, start their troubleshooting by praying that they can see something obviously wrong, such as a burned-out component or a broken printed-circuit board trace (connection). Beyond that, they have no other systematic procedure to resort to, so they usually fail to find out what is wrong with the circuit. However, using DC Troubleshooting to analyze the circuit schematic diagram and the actual circuit will nearly always make the problem obvious. You will impress your lab instructor or your boss with your electronics skills.

FIGURE 8–1
Breadboard circuit

Power supply

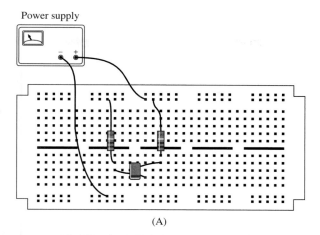

(A)

(B)

8–2 ■ DC TROUBLESHOOTING, BRIEFLY

You can apply DC Troubleshooting to electronic circuits of all complexities, from a single transistor circuit to circuits with many hundreds of transistors. For simplicity, we will discuss only single transistor circuits at this time and assume that you have built such a circuit on a temporary bread-board (Figure 8–1A).

The heart of DC Troubleshooting requires the following steps:

☐ **Estimated Voltages** You make rough calculations to estimate the voltages you expect to see at each of the three transistor leads (you don't need exact values carried out to several decimal places here). Because you make most voltage measurements with ground as a common reference point, you should calculate the Estimated Voltages: V_C, V_B, and V_E. At each lead of the transistor, write down on the circuit schematic diagram the voltages you have estimated for that lead.

☐ **Measured Voltages** Now turn on the power supply, and measure the DC voltages at each lead of your transistor. Write down these Measured Voltages next to the Estimated Voltages on your schematic.

☐ **Assessment** Now for the payoff. A quick glance at the Estimated and Measured Voltages that are written on the schematic will tell you one of two things:

1. If the Measured Voltages are close to the Estimated Voltages, the circuit, in all likelihood, is built correctly and will work.

2. If the Measured Voltages are not close to the Estimated Voltages, it is *guaranteed* that the circuit will not work. It is now your job to find the problem.

Recording the Estimated and Measured Voltages

It is important to get into the habit of writing down the Estimated and Measured Voltages at the leads of the transistor in the same way every time. If you don't use the same method each time you do troubleshooting, you will only confuse yourself. For instance, you should always write both the Estimated and Measured Voltage values at each lead of the transistor. The worst habit is to write in only one number at each lead. You will easily become confused about whether you meant Estimated or Measured Voltage. This is especially true when more than one person is working on the same circuit and each one is writing numbers on the schematic diagram. It's best to establish a simple, straight-forward method for recording the Estimated and Measured Voltages. Use it on all the circuits that you build. Even the method of notation should become a fixed routine.

Estimated Voltages   Figure 8–2A shows a typical I_B control transistor circuit of the type described in Chapter 7. By now you can calculate the transistor currents and voltages. If we neglect the base-emitter voltage drop (remember, we just need rough estimates for DC Troubleshooting), the base current (I_B) is approximately 0.12mA. The voltages we expect at each lead of the transistor are

$$V_B = 0.7V$$
$$V_C = 12 - 1KI_C = 12 - 1K \times 12mA = 0V \ (I_C = 100I_B)$$
$$V_E = 0V$$

These are the *Estimated Voltages* for the circuit when it is working properly. Note that the circuit is just in **SATURATION.** The I_B control transistor is usually operated between **CUT-OFF** and **SATURATION.**

FIGURE 8–2

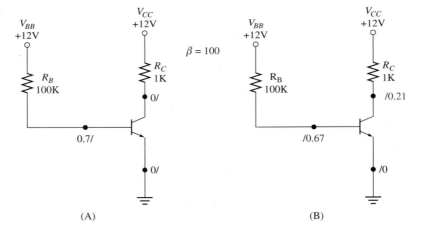

(A)   (B)

To record the Estimated Voltages at each of the lead of the transistor, draw a slash at each transistor lead on the schematic diagram. Write the Estimated Voltages *above* the slash (Figure 8–2A).

Measured Voltages   Having recorded the Estimated Voltages, you now build the circuit and measure the actual voltages at each lead of the transistor. Record these *Measured Voltages* on the circuit diagram by writing them *below* the appropriate slash (Figure 8–2B). This method will tell you which voltage values are estimated and which are measured.

Figure 8–3 shows the circuit with both the Estimated Voltages and the Measured Voltages recorded. This is what your circuit schematic diagram should look like when you have finished your DC Troubleshooting procedure. In this case the two sets of values are in good agreement, so you can safely say that the transistor is working as it should.

FIGURE 8–3

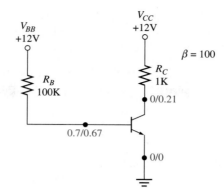

EXAMPLE 8–1
Recording Estimated Voltages

Redraw the circuit in Figure 8–4A, and find and record the Estimated Voltages at the base, collector, and emitter.

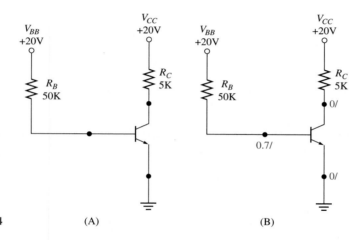

FIGURE 8–4   (A)   (B)

Solution   To estimate the base current, we can neglect the emitter diode voltage drop to find base control current:

$$I_B = \frac{V_{BB}}{R_{BB}} = \frac{20}{50\text{K}} = 0.4\text{mA}$$

If the transistor is in the **ACTIVE** state, the collector current and collector voltages would be

$$I_C = \beta I_B = 40\text{mA}$$

and

$$V_C = V_{CC} - R_C\,I_C = 20 - 5\text{K} \times 40\text{mA} = -180\text{V}$$

Because V_C is negative, the transistor must be **SATURATED**. In the **SATURATED** transistor, collector voltage equals emitter voltage. Therefore, the base, collector, and emitter Estimated Voltages are

$$V_B = 0.7\text{V},\ V_C = V_E = 0\text{V}$$

The Estimated Voltages are recorded in Figure 8–4B.

EXAMPLE 8–2
Recording the Measured Voltages

Measurements made on the circuit in Figure 8–5A give us the following Measured Voltages at the base, collector, and emitter:

$$V_B = 0.73\text{V}, V_C = 0.08\text{V}, \text{ and } V_E = 0\text{V}$$

Redraw the circuit, and record both the Estimated Voltages found in Example 8–1 and the Measured Voltages given here.

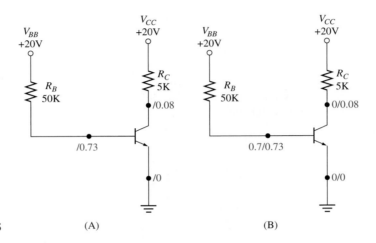

FIGURE 8–5   (A)   (B)

Solution   The Estimated Voltages and Measured Voltages are shown in Figure 8–5B.

DRILL EXERCISE

In Figure 8–3 (p. 227) change the base and collector power supply voltages to 10V. Redraw the circuit, and find the Estimated Voltages at the base, collector, and emitter voltages. The Measured Voltages are

$$V_B = 0.6\text{V}, V_C = 0.2\text{V}, \text{ and } V_E = 0\text{V}.$$

Record the Estimated Voltages and the Measured Voltages on your circuit diagram.

Answer   The Estimated Voltages are $V_B = 0.7\text{V}$, and $V_C = V_E = 0\text{V}$.

REVIEW QUESTIONS

1. What is DC Troubleshooting?
2. What are the Estimated Voltages?
3. How do you record the Estimated Voltages?
4. What are the Measured Voltages?
5. How do you record the Measured Voltages?
6. How do you know if there is a problem with the circuit?

8–3 ■ THE COMPLETE DC TROUBLESHOOTING PROCEDURE

Here we will consider circuits using a single transistor. These circuits include the simple transistor circuits from Chapter 7 and single transistor amplifier circuits, which you will study soon. We assume that the circuit is built on the white breadboards usually used in student labs. In the following discussion we will refer to the schematic of the circuit (Figure 8–6A) and a white board construction in Figure 8–6B. Note that this circuit is biased deep into **SATURATION.**

FIGURE 8–6

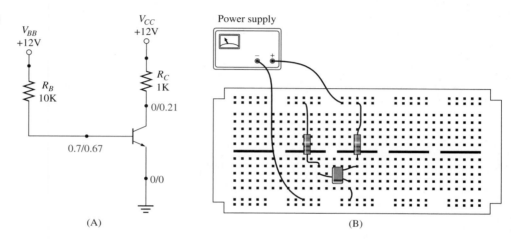

(A)   (B)

Troubleshoot the Circuit

The steps in DC Troubleshooting the circuit in Figure 8–6 are presented here:

☐ **Calculate** the Estimated Voltages at the collector, base, and emitter leads.

☐ At each transistor lead **write** down the Estimated Voltages *above* the slashes on the circuit schematic diagram.

☐ Build the circuit. Check your circuit visually.

☐ Check the **Power Supply Connections.** At each lead of the components that connect to the power supply, measure the power supply voltage. (If more than one power supply voltage connects to the circuit, check each one separately.) This step can catch several problems: failing to wire all power supply points together; broken power and ground leads going to the power supply; even forgetting to turn on the power supply!

☐ Check the **Ground Connections.** Be sure to verify that there is a reading of 0V on each group of component leads that are connected to ground. It is a common error for students to neglect to connect together all the ground points shown in the circuit schematic.

☐ **Measure** the voltage at each lead of the transistor, one at a time. Remember that all voltage measurements are made with the voltmeter's negative lead connected to the circuit's ground connection.

☐ **Write** down the appropriate Measured Voltage *below* the slashes at each transistor lead on the circuit schematic.

☐ **Compare** the Measured Voltage values to the Estimated Voltage values. If these values are close, your circuit is functioning properly. You may now connect up your external input and output. If the Measured Voltage values are not close to the Estimated Voltage values, the circuit will not work.

☐ **Use** your knowledge of transistor and resistor behavior to locate the problem.

Locate the Problem

The first step in locating the problem in a circuit is to examine the connections at the nodes where voltage discrepancies were found. The simplest way to do this is to measure the voltage at both the transistor lead and at the lead of the resistor that is connected to that transistor lead. We can use an expanded **connection diagram** (Figure 8–7B) to record these two measurements. If these two Measured Voltages are the same, as here, the connection between the transistor and the resistor is good. If these measurements are not the same, you have a bad connection.

FIGURE 8–7
Connection diagram

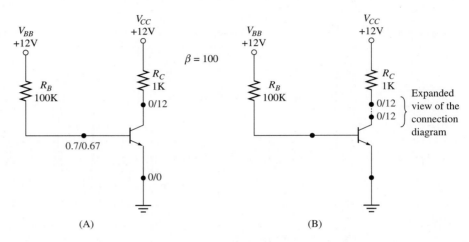

(A)   (B)

If the recorded voltages show that the connection between the transistor and the resistor is bad, fix the connection, and remeasure the voltages at all transistor leads. On a new copy of the schematic, record the new Measured Voltages and Estimated Voltages. If these voltages are now similar, you have found the problem.

If there are still discrepancies between the Estimated and Measured Voltages, the transistor is probably bad. Remove the transistor and test it (see Section 8–4). If your test shows that the transistor is bad, discard it and insert a new transistor into your circuit.

In summary:

☐ If there is a discrepancy between Estimated Voltage and Measured Voltage at a particular lead, measure voltages at the transistor lead and resistor lead.

☐ Record these voltages on an expanded connection diagram.

☐ Compare Measured Voltages on the connection diagram.

☐ If the connection diagram shows discrepancies in Measured Voltages, repair the connection.

☐ If connection diagram does not show discrepancies in Measured Voltages, remove transistor and test it.

EXAMPLE 8–3
DC
Troubleshooting I

Figure 8–8A shows a set of Estimated Voltage and Measured Voltage values on the leads of the **SATURATED** transistor circuit from Figure 8–2. The base and emitter lead voltages are all right, but the collector shows a mismatch. The connection diagram at the collector (Figure 8–8B) shows the voltages measured directly at the transistor collector lead and at the collector resistor lead. What is the likely problem with this circuit?

Solution   Because the connection diagram shows a discrepancy between the voltages measured at the transistor and resistor leads, the most likely problem is that R_C is not connected to the collector lead (Figure 8–8C). There is no current flowing through R_C, so the power supply voltage shows up on the lower lead of R_C. If this is the case and the transistor is all right, then the transistor would still be **SATURATED.** The voltage on the collector would be close to zero volts.

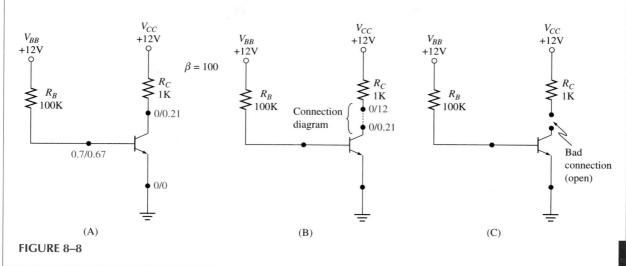

FIGURE 8–8

EXAMPLE 8–4
DC
Troubleshooting II

Figure 8–9A shows a set of Estimated and Measured Voltage values on the leads of the **SATURATED** transistor circuit from Figure 8–2. The base and emitter lead voltages are all right, but the collector shows a big mismatch. The connection diagram at the collector (Figure 8–9B) shows the voltages measured directly at the transistor collector lead and at the collector resistor lead. What is the likely problem with this circuit?

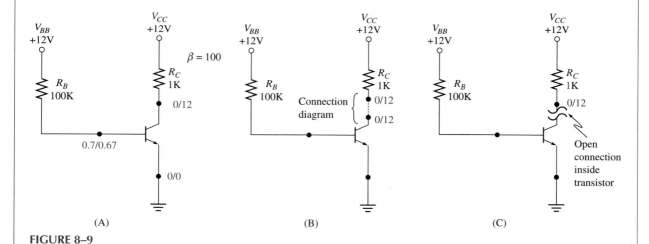

FIGURE 8–9

Solution Because the voltages at both the transistor and resistor leads are the same, there is probably a problem with the transistor. The collector lead might be disconnected from the actual transistor chip inside the package (Figure 8–9C), or the transistor chip might be burnt out, resulting in an open circuit between the collector and the emitter.

Note the circle around the transistor in Figure 8–9C. We use the circle whenever we want to show you what is happening inside the transistor, an internal open in this case. Everything inside the circle is part of the transistor itself.

EXAMPLE 8–5
DC
Troubleshooting III

Figure 8–10A shows voltage values on the leads of a transistor that has a mismatch not only at the collector but also at the base. The connection diagram at the collector and the base is shown in Figure 8–10B. What is the likely problem with this circuit?

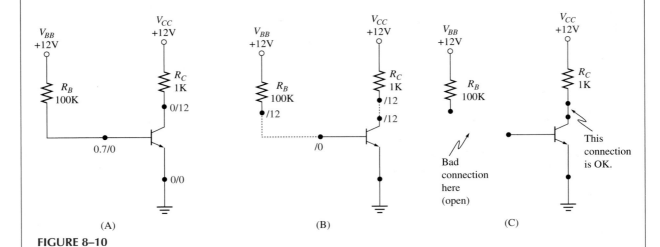

(A)   (B)   (C)

FIGURE 8–10

Solution Remember that this circuit is an example of I_B control. This means that I_B has to be present and has be large enough for the transistor to work and to go into **SATURATION.** Here the voltage at the base is zero and does not show the usual forward-bias voltage of the emitter diode. The emitter diode is not **ON;** therefore, the transistor is in **CUT-OFF.** This explains why the collector voltage is 12V. In **CUT-OFF** there is no collector current and, therefore, no voltage drop across the collector resistor.

Because Figure 8–10B shows a discrepancy between the transistor and resistor voltages at the base connection, there is probably a bad connection between the lower lead of R_B and the base. Figure 8–10C shows that with this bad connection the transistor has no base current. The base has no voltage because the base-emitter diode has no current and is, therefore, **OFF.** When we measure the voltage on the lower lead of R_B, we find 12V because R_B has no current.

EXAMPLE 8–6
DC
Troubleshooting IV

Figure 8–11A shows voltage values on the leads of a transistor that has a mismatch at not only the collector but also at the base. Figure 8–11B shows the connection diagram at the collector and the base. What is the likely problem with this circuit?

Solution Because the connection diagram does not show any voltage discrepancies between the transistor leads and the resistor leads, the most likely problem is with the transistor.

The fact that the base and emitter voltages are the same indicates that there could be a short circuit between the base and emitter of the transistor (Figure 8–11C). This short circuit could be on the circuit board or inside the transistor itself.

Because of the short, current travels directly between ground and the base power supply, bypassing the base-emitter junction. This means that the emitter diode is turned **OFF,** and, therefore, the transistor will be in **CUT-OFF.** All transistor currents will be 0, and the collector voltage will be 12V.

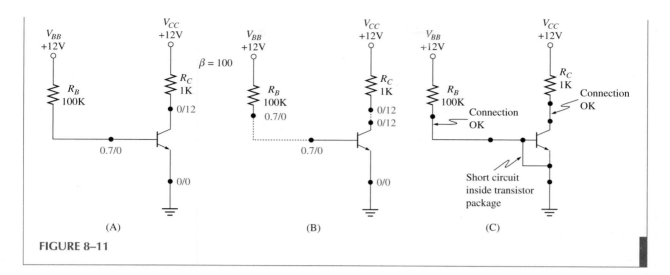

FIGURE 8–11

EXAMPLE 8–7
DC
Troubleshooting V

Figure 8–12A shows a set of Measured Voltage values that are off at every lead of the transistor. They are all 12V. Figure 8–12B shows the connection diagram, which includes all transistor leads. What is the likely problem in this circuit?

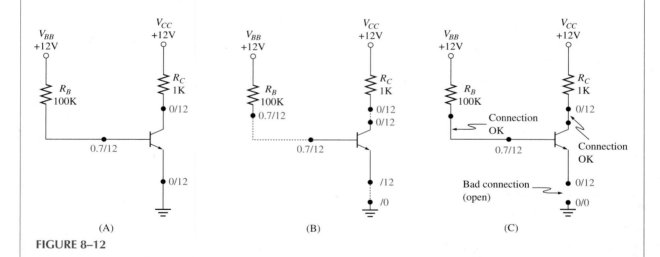

FIGURE 8–12

Solution Because the connection diagram shows a discrepancy at the emitter, the most likely explanation here is that the transistor's emitter is not connected to ground (Figure 8–12C). In this case the entire circuit is just hanging from the positive power supply lead.

If the emitter is not connected to ground, there can be no emitter current and, hence, no base or collector current. Because all currents are 0A, there is no voltage drop across R_B or R_C. Therefore, the voltages at the base and the collector are both 12V. Also, because there is no current in the emitter diode, there is no voltage drop across the base-emitter junction, so the emitter voltage is 12V.

Summary of Examples

In the preceding five examples, we have shown you how to use recorded Estimated and Measured Voltages to determine if the transistor circuit will work as predicted. We then showed you how to use the connection diagram to determine whether connections between components are bad or the transistor itself is bad.

In each example where the transistor was most likely the problem, we gave you a brief explanation of what might be causing the transistor to fail. In the next section we will give you a more complete description of typical component failures. Remember, however, that practical troubleshooting involves just determining *whether* a transistor is bad, not *why* it is bad.

DRILL EXERCISE Figure 8–13 shows a transistor circuit with a discrepancy between the Estimated Voltage and the Measured Voltage at the collector. Two possible connection diagrams are shown in A and B. Which of these connection diagrams corresponds to the following problems:

1. Bad connection at the collector.

2. Bad transistor.

3. Collector lead broken inside the transistor.

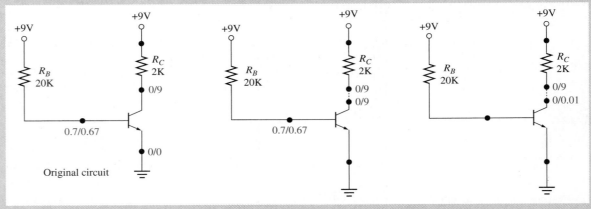

FIGURE 8–13

Answer 1—B, 2—A, 3—A

REVIEW QUESTIONS

1. What are the steps in DC Troubleshooting?

2. What is the first step in locating the problem in a circuit?

3. What is the connection diagram?

4. How does the connection diagram show that a connection is good?

5. How does the connection diagram show that a connection is bad?

6. How do you know if the likely problem is a failure in the transistor?

7. What are some likely causes of transistor failures?

8–4 ■ TYPES OF COMPONENT FAILURES

In the previous section we looked at some problems dealing with the connections between components in a circuit. Now we want to look at some typical types of failures of the components of electronic circuits. So far our transistor circuits have been constructed with resistors and transistors only, so these two components are the only ones we will be talking about here. In later chapters we will add to this list of types of component failures.

Resistor Failures

A resistor fails either by burning out or by breaking. Both are obvious for the most part, but some comments are worth making for the future.

A broken resistor is easy to spot when using the temporary breadboard. On printed circuit boards, however, where the leads are fixed by solder, the broken pieces can be held in perfect alignment, so you need to look closely to spot the problem. Sometimes one of the leads gets pulled out from the resistor body, and this also may not be so easy to see.

A burnt-out resistor usually acts like a burnt-out fuse, i.e., it becomes an open circuit. When this happens, the resistor is usually darkened and easy to spot. The major exception is with wire-wound power resistors, which commonly look the same, even though the wire is an open circuit inside. These resistors are not common in student labs, however.

Most of the later chapters in this book will include a section on troubleshooting problems. Because resistor failures seldom occur, and because they are easy to spot, we will not make them a major part of our list of possible problems that cause a circuit to fail to work.

Transistor Failures

In the student lab we most commonly use transistors known as "low-power amplifier" or "low-power switching" transistors, which usually come in small, black, epoxy enclosures known as TO-92 packages (Figure 8–14). These transistors can burn out if they must dissipate too much power, more than 500mW to 1000mW. Low-power transistors can also burn out if the maximum collector current exceeds 200mA to 500mA. The same problems can occur in high-power transistors; it just takes more current to cause the damage.

When a transistor burns out, any of its three leads can become an open circuit relative to the transistor chip inside the package. Excess heat in a transistor can also cause shorts because transistor chips can melt, causing a short circuit between any two leads inside the package. (It is not worth dealing with multiple combinations of any of these catastrophes in any one transistor, but it can happen!) So you must keep on the lookout for two types of transistor failures: open circuit failures and short circuit failures.

FIGURE 8–14

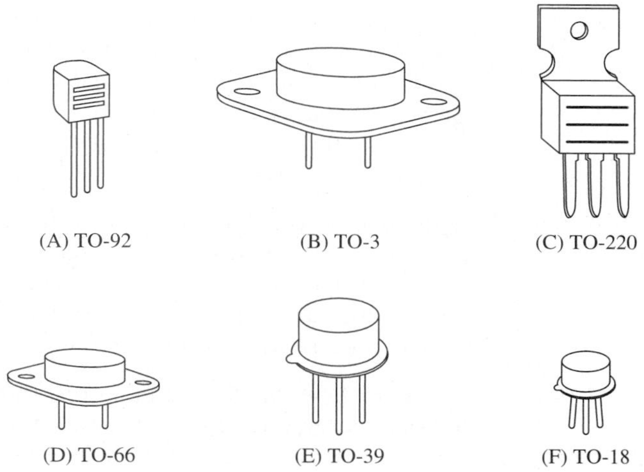

(A) TO-92   (B) TO-3   (C) TO-220

(D) TO-66   (E) TO-39   (F) TO-18

Open Circuit Failures in Transistors

Open Emitter   Figure 8–15A shows the circuit of Figure 8–2 with an open emitter failure. Because the collector diode is not forward biased, it is **OFF** and there can be neither collector nor base current. Therefore, there will be no voltage drops across either resistor, and the voltage at the base and at the collector leads of the transistor will be 12V.

Open Base   Figure 8–15B shows the circuit of Figure 8–2 with an open base failure. Because of the open at the base, there can be no base current, so the transistor is in **CUT-OFF** and all transistor currents are 0A. In this case, the base and collector voltages will both be at 12V.

You can see that an open failure at either the base or the emitter will produce similar results.

Open Collector   Figure 8–15C shows the transistor with an open circuit at the collector. In this case the emitter diode is still **ON,** so we expect to see 0.7V at the base. At the collector, however, we will see 12V because there is no collector current.

FIGURE 8–15
Open failures

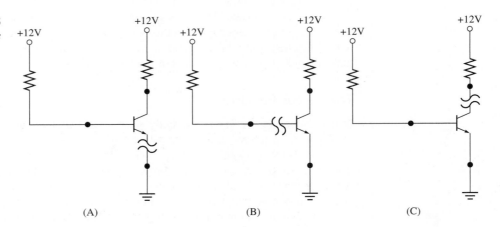

(A)   (B)   (C)

Short Circuit Failures in Transistors

Collector-Emitter Short   Figure 8–16A shows the circuit of Figure 8–2 with a short failure between the collector and the emitter. (The color bracket indicates the short.) The emitter diode is still forward biased, so we expect to see 0.7V at the base. Because the collector is shorted to the emitter, $V_C = V_E = 0V$.

Note that this is also true for the transistor in **SATURATION.** Therefore, a **SATURATED** transistor circuit like this one may not show a discrepancy between Estimated and Measured Voltages, even though the transistor is bad. You will know the transistor is bad when you connect the input signal to the transistor and learn that the collector voltage never changes from 0V.

Base-Emitter Short   Figure 8–16B shows our circuit with a short between the base and the emitter. In this case the base is connected directly to ground, so $V_B = 0V$. Because the current through R_B is diverted to ground, there is no current to forward bias the emitter diode, so the transistor will be **CUT-OFF.** With the transistor in **CUT-OFF,** there is no collector current, so we expect the collector voltage to be 12V.

Collector-Base Short   Figure 8–16C shows the transistor circuit with a short between the base and the collector. In this case the emitter diode is still forward biased, so $V_B = 0.7V$. Now, however, because the collector is shorted to the base, $V_C = V_B = 0.7V$.

FIGURE 8–16
Short failures

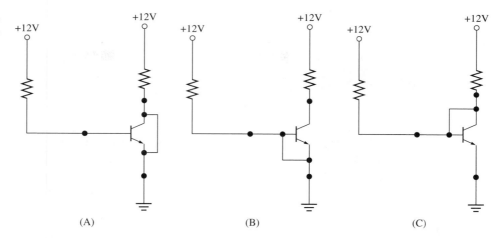

(A)   (B)   (C)

The collector-emitter short is probably the most common type of short circuit failure in transistors. This is because the collector current and the voltage between the collector and the emitter are responsible for the major part of the power dissipation in the transistor. Remember from Chapter 7 that power dissipation in a transistor is mainly due to collector current and the collector-to-emitter voltage, or $P = V_{CE} I_C$. Hence, the transistor chip between the collector and the emitter is most likely to melt first.

REVIEW QUESTIONS

1. What are typical ways that transistors fail?
2. When a resistor burns out or breaks, what does it become?
3. What are the three types of open circuit failures that occur in a transistor?
4. What are the three types of short circuit failures that occur in a transistor?
5. Describe what happens in a circuit for each of the three types of open circuit transistor failures.
6. Describe what happens in a circuit for each of the three types of short circuit transistor failures.

8–5 ■ THE UNIVERSAL TRANSISTOR TEST CIRCUIT (UTTC)

If you suspect that a transistor is defective, it is handy to know how to test it quickly to reassure yourself that it is all right. Several of today's multimeters allow you to find the current gain (β, or h_{FE}) of your transistor. If your transistor is good, the meter will show the actual value of β (Figure 8–17A, p. 238). If your transistor is bad, the reading will be off-scale (Figure 8–17B).

If you do not have such a meter available, you can build the **universal transistor test circuit** (UTTC) shown in Figure 8–18 (p. 238). You can also use this circuit to measure the current gain of the transistor. The UTTC is built using a transistor socket. To test a transistor, insert it into the socket, turn on the supply, and measure the collector voltage.

If we neglect the diode-drop voltage at the base, the UTTC creates an I_B of approximately 10μA. That is

$$I_B \approx \frac{10}{1M} = 10\mu A$$

FIGURE 8–17

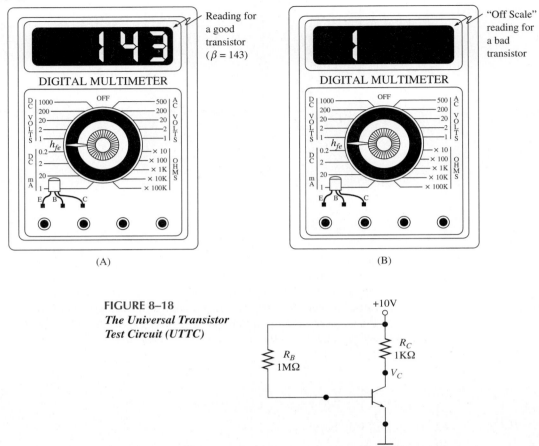

(A)

(B)

FIGURE 8–18
The Universal Transistor Test Circuit (UTTC)

(A)

The collector resistor (R_C) is too small to allow the transistor to **SATURATE,** so the transistor will be in the **ACTIVE** state. When the transistor is active and creating collector current ($I_C = \beta I_B$), R_C is large enough so that you can observe a voltage drop across it. That is

$$V_{RC} = I_C R_C = \beta I_B R_C = \beta 10\mu A \times 1K\Omega = 0.01\beta V$$

A β of 100 will cause I_C to be 1mA, and the voltage across R_C will be 1V. The collector voltage (V_C) will then be 9V. Likewise, a β of 200 will cause V_C to be 8V; and a β of 300 will cause V_C to be 7V. Therefore, assuming the transistor is good, for any reasonable current gain this circuit will be in the **ACTIVE** state.

Determining if a Transistor Is Good or Bad with the UTTC

If the transistor is open at the collector, there will be no I_C, and V_C will be 10V, or the same as the power supply voltage. The same result will also be seen if the transistor is open at the base or even if the transistor is open at the emitter.

If the transistor is shorted between the collector and the emitter, V_C will be 0V. The current gain (β) would have to be 1000 for a good transistor to be **SATURATED** and to show a V_C of 0V. This event is so unlikely that we don't have to consider it as a possibility. All of the other types of transistor failures mentioned earlier will give a V_C of either 10V or 0V.

In summary, if V_C is found to be anywhere between 9.5V and 6.5V (corresponding to a β of 50 to 350), we can safely assume that the transistor is in the **ACTIVE** state and working just fine.

Finding β with the UTTC

We have seen that the value of β affects collector voltage. Therefore, we can use measured collector voltage to find β. We can use the test circuit to find the transistor's DC current gain as follows. From the measured voltages at the base and the collector, we find the base and collector currents:

$$I_B = \frac{10 - V_B}{10M}$$

$$I_C = \frac{10 - V_C}{1K}$$

The DC current gain is given by

$$\beta = \frac{I_C}{I_B}$$

<table>
<tr><td>REVIEW
QUESTIONS</td><td>

1. Describe how some multimeters can be used to test transistors.
2. Draw the Universal Transistor Test Circuit (UTTC).
3. Find the base current in the UTTC.
4. A good transistor will be in what operating state in the UTTC? Why?
5. What will V_C be if there is an open collector, open emitter, or open base in the transistor?
6. What will V_C be if there is a short between the base-emitter, base-collector, or collector-emitter junctions in the transistor?
7. For what range of values of V_C can you assume the transistor is good?
8. How can the UTTC be used to determine β?

</td></tr>
</table>

8–6 ▪ DC TROUBLESHOOTING EXAMPLES

Before we show you a final set of DC Troubleshooting examples, we will review a "bare-bones" list of the steps in DC Troubleshooting. We have left out the steps that quickly become automatic as you become familiar with the process.

☐ Write down the estimated transistor voltages.

☐ Check the power supply.

☐ Check the ground.

☐ Measure and record the actual transistor voltages.

☐ Measure and record voltages on resistor leads.

The last item on this list is necessary only if the Estimated and Measured Voltages on the transistor are not close in value.

EXAMPLE 8–8 **DC** **Troubleshooting VI**	Figure 8–19A (p. 240) shows DC Troubleshooting findings from the transistor circuit we have seen in the earlier examples. Figure 8–19B shows the connection diagram. What is the likely problem in this circuit?

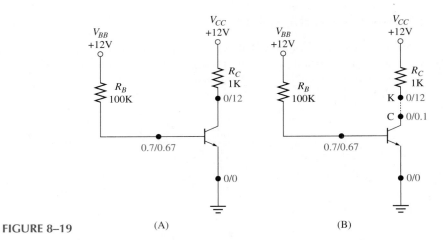

FIGURE 8–19   (A)   (B)

Solution   If you look at the Measured Voltages at connection C-K on the Connection Diagram, the problem becomes obvious. The voltage on the resistor lead (point K) is 12V, while the voltage on the collector of the transistor (point C) is 0.1V. Connection C–K is bad, because the Measured Voltages do not agree. The findings show that R_C is open at this connection; hence, we see power supply voltage at point K.

Even though the circuit is open at the collector, the transistor is still in **SATURATION!** As we showed you in the last chapter, as collector resistor increases, collector current decreases. This means that collector current will eventually become smaller than βI_B and the transistor will **SATURATE**. Therefore, if the transistor is **SATURATED** with a collector resistor of 1KΩ, it will still be **SATURATED** as the collector resistor increases. An open circuit at the collector is equivalent to a collector resistor of infinite ohms. This is very large resistance indeed, so the transistor is still **SATURATED.** This is why we see a normal **SATURATION** voltage at point C.

EXAMPLE 8–9
DC
Troubleshooting
VII

Figure 8–20A shows the DC Troubleshooting findings from the previous example. Again, it shows the mismatch at the collector connection. Figure 8–20B shows the Connection Diagram. What is the likely problem?

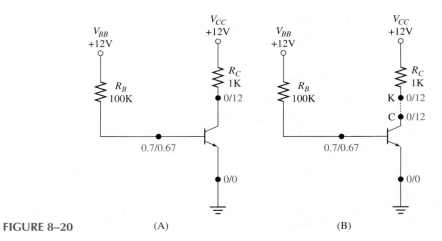

FIGURE 8–20   (A)   (B)

Solution In this example the Measured Voltages at connection C–K agree. This means that this connection is all right. Because the transistor should show a **SATU-RATED** voltage at the collector, we have to conclude that the transistor is bad and that it probably has an open collector lead inside the transistor package.

EXAMPLE 8–10
DC Troubleshooting VIII

Figure 8–21A and the connection diagram in Figure 8–21B show the results of DC Troubleshooting our sample circuit. What is the likely problem?

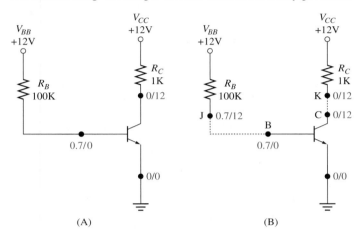

FIGURE 8–21
(A)
(B)

Solution Here we see a mismatch at both the transistor's collector and base connections. Can there be a single problem with the transistor that can cause both discrepancies? If you look at the C–K connection first, you can see that the connection between the collector and the collector resistor is good.

The connection diagram shows that the Measured Voltages at connection J-B do not agree. This connection is bad, and the transistor is not getting any base current. Because there is no base current, the base voltage (V_B) is zero. Therefore, the base resistor carries no current, and we see the power supply voltage at point J.

Also, because there is no base current, the emitter diode is not turned **ON** and the transistor is in the **CUT-OFF** state. In **CUT-OFF** all transistor currents are zero. Therefore, even though the connection at C-K is good, there is no collector current, so we see the supply voltage at the collector. This explains how the single problem at the base causes discrepancies between the Estimated Voltages and Measured Voltages at both the base and the collector.

EXAMPLE 8–11
DC Troubleshooting IX

Figure 8–22A (p. 242) and the connection diagram in Figure 8–22B show the results of DC Troubleshooting. What is the likely problem?

Solution This time the connection diagram shows us that the Measured Voltages at connection J-B are in agreement, so this connection is good.

There is current in R_B because the full power supply voltage is across it. The current pathway connects directly to ground, and the current does not flow through the base of the transistor. Connection J–B is shorted to ground. This short might be inside the transistor, i.e., a base-emitter short.

If the short is not in the transistor, then it exists because of the way the circuit is built. The cause of the short could be a wire from ground that goes to the wrong hole

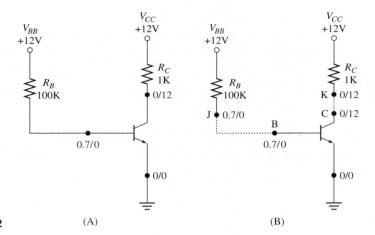

FIGURE 8–22   (A)   (B)

(i.e., a hole that is part of connection J–B). Another possibility is that the leads of two components are touching each other above the surface of the circuit board, and one of these leads is grounded. For example, the lower lead of R_B might be touching the emitter lead of the transistor.

**EXAMPLE 8–12
DC
Troubleshooting X**

Figure 8–23A and the connection diagram in Figure 8–23B shows DC Troubleshooting findings. What is the likely problem?

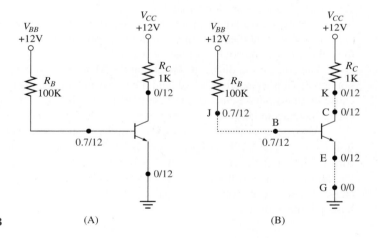

FIGURE 8–23   (A)   (B)

Solution   The voltages show a mismatch at all three leads of the transistor. In our earlier discussion we explained that a bad connection between the emitter and ground could explain the problem.

The connection diagram shows that this is exactly the case. The Measured Voltages at connection E–G show that there is a bad connection between the emitter (point E) and ground (point G). Notice that the Measured Voltages at connection J–B and at connection C–K are in agreement, indicating that these connections are good.

We see the power supply voltage at all the leads of the components because only the positive lead of the power supply is connected to the circuit. The ground bus of the circuit is not connected to the transistor's emitter. Because there is no current in the emitter diode, the transistor is **CUT-OFF** and there are no currents flowing through any of the components of this circuit.

DRILL EXERCISE   A transistor circuit and five possible DC Troubleshooting findings are shown in Figure 8–24. Match the following problems with the appropriate DC Troubleshooting finding.

1. A bad connection between the transistor and base resistor (R_B).

2. A short circuit between the base and the emitter.

3. An open in the emitter lead inside the transistor.

4. An open in the collector lead inside the transistor.

5. A bad connection between the transistor and the resistor at the collector (R_C).

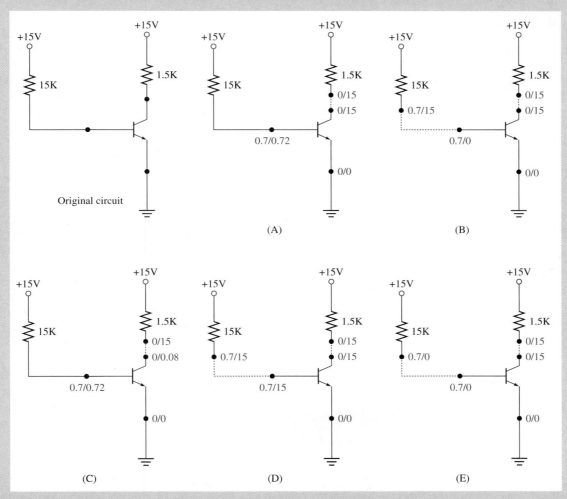

FIGURE 8–24

Answer   1—B, 2—E, 3—D, 4—A, 5—C

8–7 ■ TROUBLESHOOTING IN THE LABORATORY

As a student you will most likely build the circuits that are described in your lab manual. The circuit diagram there will serve as the schematic on which you will write down all Estimated Voltages and Measured Voltages, using the format described in Section 8–2. Do not hesitate to mark up your circuit diagram (or a copy, if you prefer). These written voltages will show your instructor that you are using the systematic approach of DC Troubleshooting. Your instructor may also use the written voltages to help you fix a circuit that you cannot make work.

The primary tool you will use in DC troubleshooting is a voltmeter. First, you will use it to check your power supply and ground connections. Next you measure the voltages at the transistor leads. You will want to observe the following practices common to all voltage measurements.

Voltage measurements should be made relative to ground. As Figure 8–25A shows, you always connect the negative lead of the voltmeter to ground. Make all measurements by moving only the positive lead of the voltmeter to the point in the circuit under observation.

You should avoid making voltage measurements across individual components because you would then have to move both leads of the voltmeter (see Figure 8–25B). Two problems arise with moving both leads: (1) it will be harder for you to systematically determine all required voltages; and (2) you have introduced the possibility of measurement error as you move both hands around the circuit.

FIGURE 8–25(A)

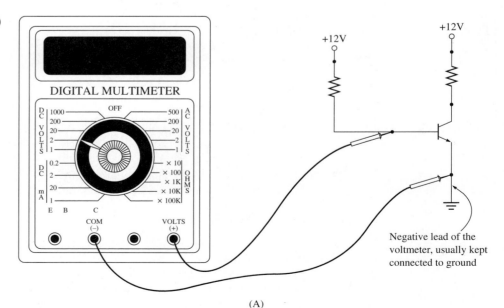

Negative lead of the voltmeter, usually kept connected to ground

(A)

FIGURE 8–25(B)

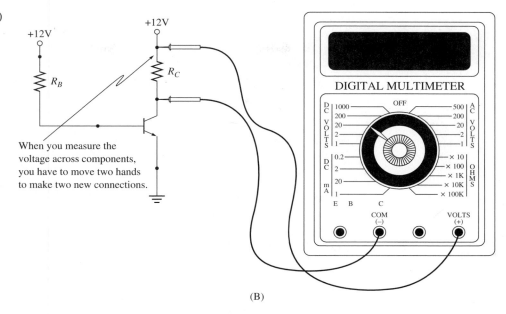

When you measure the voltage across components, you have to move two hands to make two new connections.

(B)

There is another reason to keep one hand in your pocket. You do not want to create a short circuit across your chest accidentally as you fumble around the circuit. This is very distracting for your fellow students and your lab instructor.

If you get used to making all voltage measurements relative to ground when using a voltmeter, you will quickly adapt to using a more advanced voltage measuring device, the oscilloscope. To make DC measurements with an oscilloscope, make sure the input channel is set to "DC," then connect the oscilloscope's ground probe to the circuit ground. You can now make DC voltage measurements by touching the active probe of the oscilloscope to the point of interest in the circuit (Figure 8–26). The advantage of an oscilloscope over a voltmeter is that you can also use an oscilloscope to observe AC signals.

FIGURE 8–26

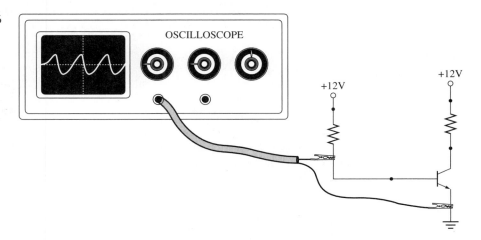

Why Not Current Measurements?

Most digital voltmeters are also capable of measuring current. Current measurements are not part of DC Troubleshooting, however, for the following reasons.

Current measurements are made with the meter in series with the components of the circuit. This means that you have to break the circuit, or undo the circuit connections, to make a current measurement. You then have to reestablish the connections. This can take too much time and can introduce additional problems if you reconnect the wires incorrectly.

Printed circuit boards pose even more problems. You have to unsolder components to put a current meter in series with the current path. Besides being a nuisance, the heat generated from unsoldering and resoldering can damage heat-sensitive electronic components.

Because currents nearly always show their presence by creating voltage drops across a resistor, diode, or transistor terminals, voltage measurement is all that is needed. When you use a voltmeter or oscilloscope, you are measuring across a circuit component (in parallel with the component), so you don't need to interrupt the circuit.

Tips on Using the Temporary Breadboard in the Lab

I. Become Familiar With How the Holes Are Connected Together   Figure 8–27 shows a reproduction of the white temporary breadboard typically used in labs, along with which groups of holes are connected together. Do not hesitate to ask your instructor if you are not sure about which holes are connected. Most connected holes are arranged in vertical groups of five, as shown in the diagram. The horizontal rows of holes are connected together for the purpose of making power supply connections and ground connections, usually called **power** and **ground busses.**

FIGURE 8–27
Breadboard

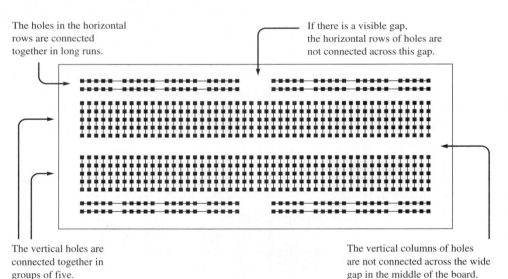

The holes in the horizontal rows are connected together in long runs.

If there is a visible gap, the horizontal rows of holes are not connected across this gap.

The vertical holes are connected together in groups of five.

The vertical columns of holes are not connected across the wide gap in the middle of the board.

The only problem that commonly arises in using the horizontal rows of holes is that on some breadboards, the horizontal rows are separated in the middle. This means that the left half of the row is not connected to the right half. When this is true, there is an obvious wide gap in the middle of the horizontal row. If you want the entire row to be connected together you have to use a wire jumper to connect the two halves of the rows together.

If you are not sure about which groups of holes are connected together, you can always use an Ohmmeter to measure the resistance between holes. Insert a piece of wire into each of two groups of holes, and see if there is a very low resistance (or approximately 0Ω) between the two groups of holes (Figure 8–28A). If so, then the groups are connected together. If the holes are not connected, you will get the off-scale reading of an open circuit (Figure 8–28B). Again, if you are not sure, ask your lab instructor.

II. Build the Circuit Pictorially   You should always build your circuit "pictorially." This means that you should lay out the actual circuit as shown in the circuit diagram (Figure 8–29, p. 249). Place components on the left side of the diagram on the left side of the breadboard in the actual circuit. Do the same for all structures in the circuit—components

FIGURE 8–28(A)

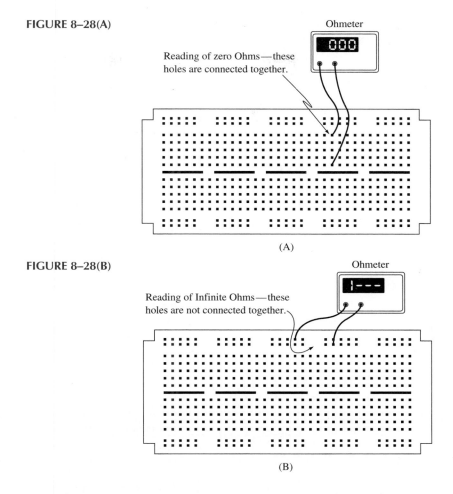

Ohmeter

Reading of zero Ohms—these
holes are connected together.

(A)

FIGURE 8–28(B)

Ohmeter

Reading of Infinite Ohms—these
holes are not connected together.

(B)

on the upper half, lower half of the diagram, and so on. If the actual layout of the circuit is the same as shown in the circuit diagram, you will save yourself a lot of time in troubleshooting the circuit, and in using it.

There is no advantage to deviating from the "picture" of the circuit shown in the diagram when you actually lay out the components as you build the circuit. If you do, you are only inviting trouble. You will especially appreciate this habit when you build circuits containing more than one amplifier stage.

One example of pictorial building has to do with power and ground busses or horizontal row of connections. If the diagram shows the positive power supply on top of the diagram, make sure that the positive power supply bus is on the top of all of your component connections in your actual circuit. Likewise, if the ground connections are in the lowest part of the diagram, then use a horizontal row of holes on the lower part of the breadboard for your ground bus.

If the diagram shows that the circuit has a negative power supply, it is usually drawn in the lowest part of the diagram. If so, make sure that you choose a horizontal row of holes for the negative power supply bus that is below the rest of the components and also below the ground bus.

Common Problems in Using the Temporary Breadboards

Wrong-Hole Alignment **Wrong-hole alignment** occurs when you put a component lead in the vertical group of holes next to the group to which you thought you were connecting. This is easy to do, especially when component leads are bent and kinked or when components lie on top of the holes, hiding them from you.

FIGURE 8–29

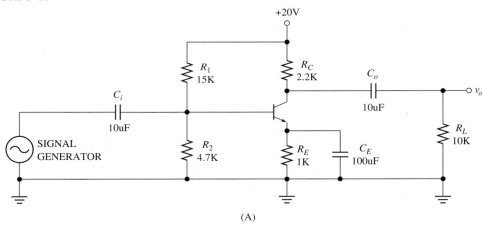

(A)

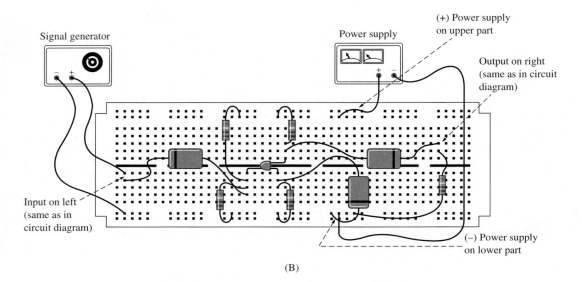

(B)

You can have wrong-hole alignment in the horizontal row of holes also, because there are usually two rows at the top of the breadboard, and two rows at the bottom. Remember that you should use these horizontal rows for the power and ground connections. Wrong-hole alignment here means that some parts of your circuit will not be properly attached to the power supply or will not be properly grounded.

Bad Hole Connection   Underneath each hole are two small leaf springs, which squeeze together to make contact with the component lead. Sometimes the contact surfaces of these springs corrode so that they don't make a good electrical connection, producing a **bad hole connection.** Sometimes the two springs are forced too far apart, and they don't come back together close enough to touch the component lead inserted into the hole. This is usually due to the fact that someone tried to force a lead or wire that was too large in diameter into the hole. In general, don't use wires or leads larger than #22 gauge wire.

If you want to use the board with components that have large leads (such as power resistors or power transistors), first solder #22 gauge wires to the component leads, then insert the smaller wires into the holes in the board. Another method is to put properly-sized wire pieces into the holes on the breadboard and to use clip leads to connect to the component with the large leads.

Sometimes the two small leaf springs get pushed together to the same side of a hole and are bent over permanently because someone pushed a component lead into the hole at an angle instead of straight down. When this happens, the hole is permanently unusable. You can usually see if this is the case, if you look closely enough.

When a bad hole connection is permanent, you may want to mark the hole (with a felt tip pen, for example), to remind you not to use it. This can save you and others a lot of time the next time someone uses the board. However, be sure that the hole is actually bad before you mark it.

Horizontal Row of Holes Not Connected in the Middle The horizontal rows of holes in the uppermost part of the temporary breadboard are usually connected together. These are meant to be power supply busses. The lowermost horizontal rows are also connected together and are meant to be used as ground. Always use these rows for their intended purposes.

Sometimes the horizontal row of holes is interrupted (not connected in the middle). If you want the horizontal row of holes to be continuous, you may have to use a jumper wire across the gap in the middle of the row. All hole problems result in a component not being properly connected to the rest of the circuit, as shown in the circuit diagram. In other words, the component is left open circuited at one end.

REVIEW QUESTIONS

1. What is the primary tool used in DC troubleshooting?
2. Why should all voltage measurements be made with one lead of the voltmeter always connected to ground?
3. How are DC measurements made with an oscilloscope?
4. What is the advantage of using an oscilloscope?
5. Why don't we make current measurements in DC Troubleshooting?
6. Describe the purpose of the horizontal rows of holes found at the top and bottom of a temporary breadboard.
7. What do we mean when we say "build the circuit pictorially"?
8. What are some common problems encountered when using the temporary breadboard?

8–8 ■ DC TROUBLESHOOTING ON THE JOB

Obtaining the Circuit Diagram

When you are building a prototype of a new design on the job, you will already have a diagram of the circuit that you are to build. When you are trying to fix a circuit that once worked, always make every attempt to get the circuit diagram. Usually you will find such a diagram in the manual that comes packaged with the circuit or the equipment that contains the circuit. If not, you will have to call the circuit manufacturer to get a diagram. Usually they can fax or mail you a copy. It is always worth the effort. You can waste a lot of time trying to fix a circuit when you don't have a diagram for it. At the very least you will not be able to apply DC Troubleshooting—your major tool.

In rare instances, you won't be able to obtain the circuit diagram. Then you will have to trace out the diagram from the actual circuit. Start with a lot of paper. Make a drawing of the components and their connections. The first drawing usually will not make sense. Then make another drawing from the first one, trying to group the components around each transistor, or integrated circuit, in individual stages. Create a picture of groups of components that look familiar on a stage-by-stage basis. Keep in mind that most circuits are collections of smaller, single-stage circuits, each of which centers around a single transistor, operational amplifier, or integrated circuit.

Be prepared to draw and redraw the entire circuit several times before it makes sense in its entirety. You can stop only when you can identify all the stages of the circuit as something you have seen before. The whole process takes a lot of time, paper, and patience. You can see why it is preferable to do everything you can to get a proper copy of the diagram from the manual or the manufacturer.

Make Enough Copies of the Circuit Diagram

Whenever you are working on a circuit, do not hesitate to make several photocopies of the circuit diagram. Start with one copy. Write the Estimated Voltages at the required semiconductor leads; then write your Measured Voltages. You will find yourself making a lot of corrections on the diagrams, especially as you begin to gain a more complete understanding of what each stage does and what the DC voltages should be.

Work with only one copy of the circuit diagram at a time—the one which is the most recent. This copy will become filled with such things as notes, and arrows, and erasures. Start again with a fresh copy when the one in use becomes too messy. Begin by inserting the Estimated Voltages and Measured Voltages on the fresh copy.

You will find that when several people are working on the troubleshooting process, the copy in use can get messy and unreadable very rapidly, as each person draws out his or her own idea or theory of how each stage works or what the problem is with the circuit. Troubleshooting often invites a lot of opinions from a lot of people, each hoping to be the one who contributes the key that identifies the problem.

REVIEW QUESTIONS

1. If you don't have the circuit diagram for the circuit you are repairing, what should you do to obtain it?
2. If you can't get the needed diagram, what is your next step?
3. Why is it important to have many copies of the circuit diagram?

SUMMARY

■ DC Troubleshooting is composed of the following steps.

Estimate and record the transistor collector, base, and emitter voltages above the slashes drawn on the circuit diagram.

Measure and record the transistor collector, base, and emitter voltages below the slashes drawn on the circuit diagram.

If the Estimated Voltages and Measured Voltages are similar, the circuit will probably work.

If the Estimated Voltages and Measured Voltages are different, the circuit will not work.

■ If there is a discrepancy between Estimated Voltages and Measured Voltages, then proceed as follows.

Measure and record the voltages at the resistor leads connected to the transistor where a problem exists (connection diagram).

If the connection diagram shows a discrepancy between the resistor and transistor lead voltages, correct the connection.

If the connection diagram shows no discrepancy between the resistor and transistor lead voltages, the transistor is probably bad.

Remove and test the transistor with a multimeter or the Universal Transistor Test Circuit (UTTC).

■ In the laboratory check the connections and the holes on the temporary breadboard.
■ On the job obtain or draw the circuit diagram before attempting DC Troubleshooting.

PROBLEMS

SECTION 8–2 DC Troubleshooting

1. Draw the circuit diagram shown in Figure 8–30.
   (a) What is the Estimated V_B?
   (b) What is the approximate value of I_B (ignore the emitter-diode voltage drop)?
   (c) What is the Estimated V_C?
   (d) What is the expected value of I_C?
   (e) What is the voltage across the collector diode?
   (f) What is the voltage across the emitter diode?
   (g) Are the voltages across the two diodes consistent with a **SATURATED** transistor?
   (h) Write on your circuit diagram the Estimated Voltages at the leads of the transistor. Be sure to write the Estimated Voltages above slash lines, as shown in the text.

2. Repeat Problem 1 for the circuit of Figure 8–31.

3. The transistor shown in Figure 8–32 is **ACTIVE.**
   (a) What is the Estimated V_B?
   (b) What is the approximate value of I_B (ignore the emitter diode voltage drop)?
   (c) What is the Estimated V_C?
   (d) What is the expected value of I_C?
   (e) What is the voltage across the collector diode?
   (f) What is the voltage across the emitter diode?
   (g) Are the voltages across the two diodes consistent with an **ACTIVE** transistor?
   (h) Write on your circuit diagram the Estimated Voltages at the leads of the transistor. Be sure to write the Estimated Voltages above slash lines, as shown in the text.

FIGURE 8–30

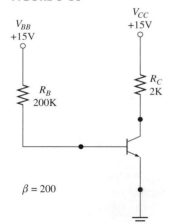

FIGURE 8–31

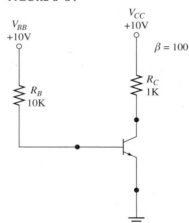

FIGURE 8–32

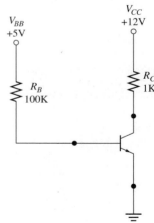

SECTION 8–3 The Complete DC Troubleshooting Procedure

4. Look at the results of DC Troubleshooting and the connection diagram in Figure 8–33 (p. 252). Is there a problem with the circuit connections or the transistor? Explain your answer.

5. Look at the results of DC Troubleshooting and the connection diagram in Figure 8–34 (p. 252). Is there a problem with the circuit connections or the transistor? Explain your answer.

6. Look at the results of DC Troubleshooting and the connection diagram in Figure 8–35 (p. 252). Is there a problem with the circuit connections or the transistor? Explain your answer.

FIGURE 8–33 **FIGURE 8–34** **FIGURE 8–35**

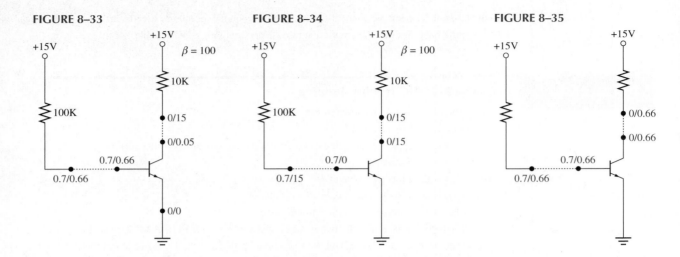

SECTION 8–4 Types of Component Failures

7. For the transistor circuit shown in Figure 8–36, assume all connections are good.

   (a) Find the Estimated Voltages at the three leads of the transistor. Redraw the circuit diagram, and use the proper format to record these voltages.

   (b) The Measured Voltages are

   $$V_B = 0.6V, \ V_C = 24V, \text{ and } V_E = 0V$$

   Using the proper format, record the Measured Voltages, along with the Estimated Voltages.

   (c) Identify a problem with the circuit that could result in the Estimated Voltages and Measured Voltages from parts a and b.

8. For the transistor circuit shown in Figure 8–36, assume all connections are good.

   (a) Find the Estimated Voltages at the three leads of the transistor. Redraw the circuit diagram, and use the proper format to record these voltages.

   (b) The Measured Voltages are

   $$V_B = 24V, \ V_C = 24V, \text{ and } V_E = 0V$$

   Using the proper format, record the Measured Voltages, along with the Estimated Voltages.

   (c) Identify a problem with the circuit that could result in the Estimated Voltages and Measured Voltages from parts a and b.

9. For the transistor circuit shown in Figure 8–36, assume all connections are good.

   (a) Find the Estimated Voltages at the three leads of the transistor. Redraw the circuit diagram, and use the proper format to record these voltages.

   (b) The Measured Voltages are

   $$V_B = 0V, \ V_C = 24V, \text{ and } V_E = 0V$$

   Using the proper format, record the Measured Voltages, along with the Estimated Voltages.

   (c) Identify a problem with the circuit that could result in the Estimated Voltages and Measured Voltages from parts a and b.

10. For the transistor circuit shown in Figure 8–36, assume all connections are good.

   (a) Find the Estimated Voltages at the three leads of the transistor. Redraw the circuit diagram, and use the proper format to record these voltages.

   (b) The Measured Voltages are

   $$V_B = 0.62V, \ V_C = 0.62V, \text{ and } V_E = 0V$$

   Using the proper format, record the Measured Voltages, along with the Estimated Voltages.

   (c) Identify a problem with the circuit that could result in the Estimated Voltages and Measured Voltages from parts a and b.

FIGURE 8–36

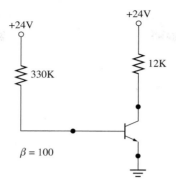

11. For the transistor circuit shown in Figure 8–37:
    (a) Calculate the Estimated Voltages at the three transistor leads.
    (b) Redraw the circuit, and, using the proper format, record the Estimated Voltages.
    (c) Briefly explain why the transistor should be operating in the **SATURATED** state.

FIGURE 8–37

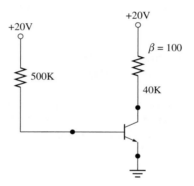

12. The Estimated Voltages for the transistor in Figure 8–37 are

$$V_B = 0.7V, \ V_C = 0V, \ V_E = 0V$$

Redraw the circuit, and show the Measured Voltages if transistor base has failed open.

13. The Estimated Voltages for the transistor in Figure 8–37 are

$$V_B = 0.7V, \ V_C = 0V, \ V_E = 0V$$

Redraw the circuit, and show the Measured Voltages if transistor collector has failed open.

14. The Estimated Voltages for the transistor in Figure 8–37 are

$$V_B = 0.7V, \ V_C = 0V, \ V_E = 0V$$

Redraw the circuit, and show the Measured Voltages if transistor emitter has failed open.

15. The Estimated Voltages for the transistor in Figure 8–37 are

$$V_B = 0.7V, \ V_C = 0V, \ V_E = 0V$$

Redraw the circuit, and show the Measured Voltages if transistor base has shorted to the collector.

16. The Estimated Voltages for the transistor in Figure 8–37 are

$$V_B = 0.7V, \ V_C = 0V, \ V_E = 0V$$

Redraw the circuit, and show the Measured Voltages if transistor base has shorted to the emitter.

17. The Estimated Voltages for the transistor in Figure 8–37 (p. 253) are

$$V_B = 0.7\text{V}, \; V_C = 0\text{V}, \; V_E = 0\text{V}$$

Redraw the circuit, and show the Measured Voltages if transistor collector has shorted to the emitter.

18. The Estimated Voltages for the transistor in Figure 8–37 (p. 253) are

$$V_B = 0.7\text{V}, \; V_C = 0\text{V}, \; V_E = 0\text{V}$$

Redraw the circuit, and show the Measured Voltages if R_C has failed open.

19. The Estimated Voltages for the transistor in Figure 8–37 (p. 253) are

$$V_B = 0.7\text{V}, \; V_C = 0\text{V}, \; V_E = 0\text{V}$$

Redraw the circuit, and show the Measured Voltages if R_C has failed short.

20. The Estimated Voltages for the transistor in Figure 8–37 (p. 253) are

$$V_B = 0.7\text{V}, \; V_C = 0\text{V}, \; V_E = 0\text{V}$$

Redraw the circuit, and show the Measured Voltages if R_B has failed open.

21. The Estimated Voltages for the transistor in Figure 8–37 (p. 253) are

$$V_B = 0.7\text{V}, \; V_C = 0\text{V}, \; V_E = 0\text{V}$$

Redraw the circuit, and show the Measured Voltages if R_B has failed short.

22. Figure 8–38A shows the same transistor circuit used in Figure 8–37 (p. 253). This figure shows a set of values for both the Estimated Voltages and Measured Voltages.

   (a) To see why this circuit is not working properly, which transistor leads and connections would you want to take a close look at?

   (b) Figure 8–38B shows a connection diagram for the circuit. Describe a problem with the circuit that could explain these results.

FIGURE 8–38

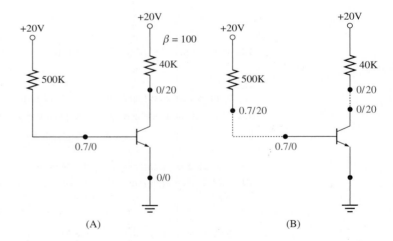

(A)   (B)

SECTION 8–5 The Universal Transistor Test Circuit (UTTC)

23. Figure 8–39 shows the simple circuit for testing a transistor, the Universal Transistor Test Circuit, being used to test a particular transistor.

   (a) The Measured Voltage at the collector is shown. Does this transistor appear to be working properly? Explain your answer.

   (b) Use the results shown in the figure to calculate β for this transistor. Write down each step in your solution.

FIGURE 8–39

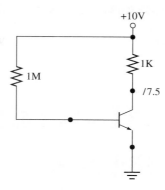

24. Figure 8–40A shows the same circuit used in Figure 8–39. Now, however, there is a different Measured Voltage at the collector.

    (a) Does this transistor appear to be working properly, or is it defective? Explain your answer.

    (b) Figure 8–40B shows the connection diagram for this transistor circuit. Do these findings fit with one of the types of transistor failures discussed in the text? If so, which one and why?

FIGURE 8–40

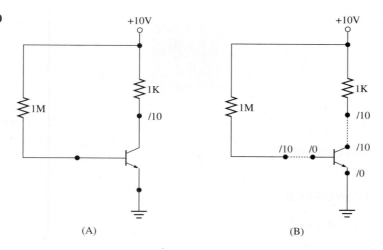

(A) (B)

Computer Problems

Computer simulation is a very useful aid in determining the effects of various circuit failures.

25. Simulate the circuit of Figure 8–35 and record the voltages at the base, collector, and emitter.

26. Open the base of the transistor and repeat problem 25. Also record the voltage at the base side of R_B. (Note: if your computer program will not work with an open at the base, connect a 100MΩ resistor between R_B and the base.)

27. Open the collector of the transistor and repeat problem 25. Also record the voltage at the collector side of R_C. (Note: if your computer program will not work with an open at the collector, connect a 100MΩ resistor between R_C and the collector.)

28. Open the emitter of the transistor and repeat problem 25. (Note: if your computer program will not work with an open at the emitter, connect a 100MΩ resistor between the emitter and ground.)

29. Place a short between the base and the collector and repeat problem 25.

30. Place a short between the base and the emitter and repeat problem 25.

31. Place a short between the collector and the emitter and repeat problem 25.

Chapter

9

Biasing for Linear Applications

CHAPTER OBJECTIVES

To find the currents and voltages in a V_B control bias circuit.

To determine the state of the transistor in a V_B control bias circuit.

To bias a V_B control circuit with a single supply voltage.

To check the validity of the voltage divider approximation.

To determine the voltage gain between the base and the emitter.

To determine the voltage gain between the base and the collector.

To troubleshoot V_B control bias circuits.

(A)   (B)   (C)

9–1 ■ INTRODUCTION

FIGURE 9–1

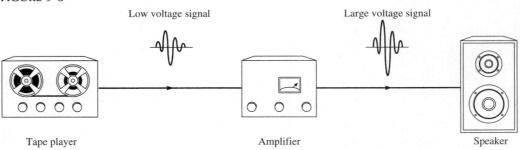

In Chapter 7 we described how the transistor can act in one of three states: **CUT-OFF,** **SATURATED,** and **ACTIVE.** The state of a transistor is entirely determined by the states of the emitter diode and the collector diode (Figure 9–1). The relations between the diode states and the transistor states are

> **CUT-OFF:** Emitter diode and collector diode are **OFF.**
> **ACTIVE:** Emitter diode is **ON**—collector diode is **OFF.**
> **SATURATED:** Emitter diode and collector diode are **ON.**

In the **ACTIVE** transistor (Figure 9–2A), collector current is β times the base current ($I_C = \beta I_B$). If the transistor is **CUT-OFF,** there is no base current, so there is no collector or emitter current. That is, the collector-emitter pathway is open (Figure 9–2B). In **SATURATION** the collector and emitter are, in effect, shorted together. That is, the transistor behaves as though a switch has closed between the collector and emitter (see Figure 9–2C).

FIGURE 9–2
States of the transistor

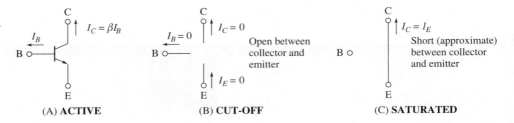

(A) **ACTIVE**   (B) **CUT-OFF**   (C) **SATURATED**

If you operate a transistor between the **CUT-OFF** and **SATURATED** states, you have built an electronic switch. All digital computers are constructed from such electronic switches. Other electronic devices, such as an automobile's electronic ignition, also use transistors as switches.

The current-controlled circuits of Chapter 7 are well suited for use as switches. This is because base current determines whether a transistor is in **CUT-OFF** (an open switch) or in **SATURATION** (a closed switch).

In the **ACTIVE** state the collector current is proportional to the base current. That is, in the **ACTIVE** state a transistor amplifies current (Figure 9–2A). Typically, however, we are interested in amplifying voltage rather than current. For example, the heads on a tape player (or VCR) pick up the magnetic traces on a tape and provide a small voltage output. But the speakers you use to convert the electrical signal to an acoustic (sound) signal need large voltages to operate. The stereo amplifier, with tape head input and speaker output, provides the voltage amplification necessary (Figure 9–3); that is, the output voltage from the amplifier is greater than the input voltage to the amplifier.

Amplification is an example of what we term **linear applications.** In fact, amplifiers are the most common linear devices. We usually amplify signals that vary with time, that

FIGURE 9–3

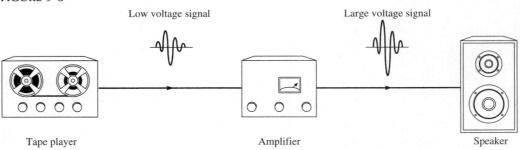

Tape player   Amplifier   Speaker

is, AC signals, rather than DC signals. Although our real interest in a linear device is how much and how well it amplifies AC signals, proper operation of these devices depends on its DC bias voltages and currents.

In the Chapter 7 we showed you some DC (or bias) transistor circuits in which the base current controls the output current. In most amplifiers we build transistor circuits in which the base voltage controls the collector current (V_B control). In this chapter we will show you how to build and analyze V_B control circuits. We will also show you how to convert output current to output voltage.

Because we are now interested in linear amplifiers, and because the transistors in linear amplifiers are usually operated in the **ACTIVE** state, if a transistor in this chapter is either in **CUT-OFF** or in **SATURATION,** there is a problem with the circuit's design or construction. For this reason we will not fully analyze transistors in **CUT-OFF** or in **SATURATION.**

REVIEW QUESTIONS	**1.** When the emitter and collector diodes are **OFF,** what is the state of the transistor?
	2. When the emitter diode is **ON** and the collector diode is **OFF,** what is the state of the transistor?
	3. When both diodes are **ON,** what is the state of the transistor?
	4. When the transistor is used as a switch, in what states does it operate?
	5. When the transistor is used in a linear amplifier, in what state does it operate?

9–2 ■ BASE VOLTAGE BIASING (V_B CONTROL)

When we analyze a transistor circuit, we always assume initially that the transistor is in the **ACTIVE** state. We *must* then always verify our assumption by showing that it is not in either the **CUT-OFF** or the **SATURATED** states.

As shown in Figure 9–1 the three transistor states are determined by the states of the emitter and collector diodes. If the emitter diode is **OFF,** that is, if

$$V_{BE} < 0.7\text{V}$$

then the transistor is **CUT-OFF.** In practice, this means that if $V_B > 0.7\text{V}$, the emitter diode will be **ON** and the transistor will be **ACTIVE** or, perhaps, **SATURATED.**

To be **SATURATED,** the collector diode must turn **ON.** The collector diode will turn **ON** when

$$V_{BC} \approx 0.5\text{V}$$

This means that if the voltage across the collector diode is less than 0.5V, the collector diode will be **OFF** and the transistor will be **ACTIVE.** Rather than look at the collector diode voltage, it is easier to determine that a transistor is *not* **SATURATED** by considering the transistor's collector voltage. Remember, in **SATURATION** the collector approximately shorts to the emitter, that is $V_C \approx V_E$. Therefore, if

$$V_C > V_E$$

the transistor will be **ACTIVE.** (Actually, in **SATURATION,** the collector voltage is still 0.1 to 0.2V above the emitter voltage. However, the assumption that in **SATURATION** the collector voltage is equal to the emitter voltage is accurate enough for our needs.)

Figure 9–4A (p. 260) shows the simplest V_B control circuit. In this circuit, the base voltage is applied directly to the base (there is no base resistor). We have inserted an emit-

FIGURE 9–4

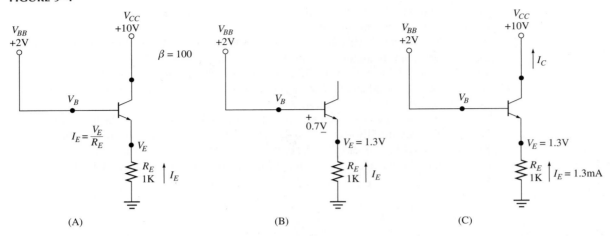

(A) (B) (C)

ter resistor (R_E) between the emitter and ground. Because the base supply voltage is connected directly to the base

$$V_B = V_{BB}$$

We assume the transistor is **ACTIVE** and analyze the base circuit (Figure 9–4B). The emitter voltage (V_E) is equal to the base voltage minus the voltage drop across the base emitter junction:

$$V_E = V_B - 0.7$$

That is, the emitter voltage is set by the voltage applied at the base. We now find I_E from Ohm's law:

$$I_E = \frac{V_E}{R_E} = \frac{V_B - 0.7}{R_E}$$

Finally, because we assume the transistor is **ACTIVE** ($I_C \approx I_E$), the collector current is

$$I_C = \frac{V_B - 0.7}{R_E}$$

In this circuit the collector current depends on the base voltage, not the base current. If the transistor is **ACTIVE,** the base current is found from

$$I_B = \frac{I_C}{\beta}$$

Finally, we note that the collector is directly connected to the collector supply voltage:

$$V_C = V_{CC}$$

Note that in V_B control (Figure 9–4C) we first find the collector current and then the base current. For the values given in this circuit

$$V_E = V_B - 0.7 = 2 - 0.7 = 1.3\text{V}$$

and

$$I_E = \frac{V_E}{R_E} = \frac{1.3}{1\text{K}} = 1.3\text{mA}$$

If the transistor is **ACTIVE,** the collector and base currents are ($\beta = 100$):

$$I_C = I_E = 1.3\text{mA}$$

$$I_B = \frac{I_C}{\beta} = 0.013\text{mA}$$

Finally, the collector voltage is

$$V_C = V_{CC} = 10\text{V}$$

Because the collector voltage is greater than the emitter voltage, our assumption that the transistor is **ACTIVE** is correct.

The **DC Bias List** for this circuit is

$V_B = 2\text{V}$ ($V_B = V_{BB}$ in this circuit)

$V_E = 1.3\text{V}$

$V_C = 10\text{V}$

$I_B = 0.013\text{mA}$

$I_E = 1.3\text{mA}$

$I_C = 1.3\text{mA}$

$V_{CE} = V_C - V_E = 10 - 1.3 = 8.7\text{V}$

$P = V_{CE}I_C = 8.7 \times 1.3\text{mA} = 11.3\text{mW}$

Note that this DC Bias List is presented in a different order than in Chapter 7. The order used here reflects the fact that in V_B control we first find the base voltage.

We can summarize the steps in determining the bias levels in V_B control circuits:

☐ Find V_B:

$V_B = V_{BB}$

V_B controls V_E

$V_E = V_B - 0.7$

V_E controls I_E

$I_E = V_E/R_E$

I_E controls I_C

if **ACTIVE,** $I_C = I_E$

I_C controls I_B

if **ACTIVE,** $I_B = I_C/\beta$

☐ Check to see if transistor is **ACTIVE** by finding V_C.

☐ $V_C = V_{CC}$
If $V_C > V_E$
the transistor is **ACTIVE** and all of the variables found above are correct.
If $V_C < V_E$
STOP! Transistor is **SATURATED.** Put down your pencil. Values you have calculated are wrong.

When a transistor is in **SATURATION,** the current relations we use for **ACTIVE** transistors are not valid. That is, in **SATURATION**

$$I_C \neq I_E$$

and

$$I_B \neq \frac{I_C}{\beta}$$

This is why values calculated assuming an **ACTIVE** transistor are not correct if the transistor is **SATURATED.**

When we build linear amplifiers, we bias transistors in the **ACTIVE** region. For this reason, in this chapter it is not necessary to calculate the correct currents and voltages for a transistor in **SATURATION.** If it is in **SATURATION,** there is a problem with the design.

EXAMPLE 9–1
V_B **Control**

For the circuit in Figure 9–5, find all transistor currents and voltages and the transistor power.

FIGURE 9–5

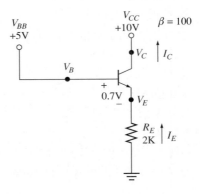

Solution

From the circuit, we find the base voltage:

$$V_B = V_{BB} = 5V$$

Now, V_B controls V_E:

$$V_E = V_B - 0.7 = 5 - 0.7 = 4.3V$$

and V_E controls the emitter current:

$$I_E = \frac{V_E}{R_E} = \frac{4.3}{2K} = 2.15mA$$

Assuming the transistor is **ACTIVE,** the emitter current controls the collector current, which controls the base current:

$$I_C = I_E = 2.15mA$$

and

$$I_B = \frac{I_C}{\beta} = 0.0215mA$$

The collector voltage is

$$V_C = V_{CC} = 10V$$

Because the collector voltage is greater than the emitter voltage, our assumption that the transistor is **ACTIVE** is correct.
The collector-emitter voltage is

$$V_{CE} = V_C - V_E = 10 - 4.3 = 5.7V$$

Finally, the power dissipated is

$$P = V_{CE}I_C = 5.7V \times 2.15mA = 12.3mW$$

DRILL EXERCISE   Change the base supply voltage in Figure 9–5 to $V_{BB} = 8V$. Redraw the circuit, and find the transistor voltages and currents, in the proper order. Find the power dissipated by the transistor. Is the transistor **ACTIVE?**

Answer $V_E = 7.3V$, $I_E = 3.65mA$, $I_C = 3.65mA$, $I_B = 0.0365mA$
$V_C = 10V$, $P = 9.86mW$

ACTIVE

**REVIEW
QUESTIONS**

1. What are the steps in determining the bias levels in V_B control?
2. If the transistor is **ACTIVE,** what are the relations between the transistor currents?
3. In **SATURATION,** what transistor current relations are no longer valid?
4. How do you know that a transistor is **ACTIVE?**
5. What type of application uses V_B control of transistors?
6. In this chapter, why are we *not* interested in finding transistor voltages and currents when the transistor is **SATURATED?**

9–3 ■ THE COLLECTOR RESISTOR

We now add a collector resistor to the V_B control circuit, (Figure 9–6). That means that we must now take into account the voltage drop across the collector resistor, especially how it affects **SATURATION** of the transistor. In the previous section we described the analysis steps for a V_B control circuit. These steps still apply. The only difference now is that we have to account for the voltage drop across the collector resistor. That is,

$$V_C = V_{CC} - R_C I_C$$

FIGURE 9–6

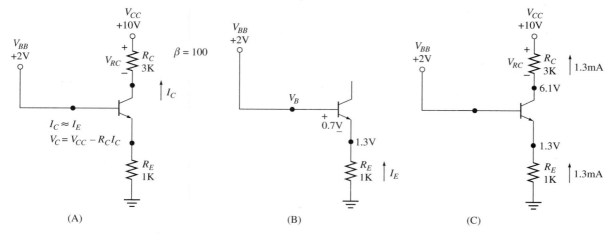

Again, here the base voltage is applied directly to the base (there is no base resistor), so

$$V_B = V_{BB} = 2\text{V}$$

We assume the transistor is **ACTIVE** and analyze the base circuit, shown in Figure 9–6B (p. 263). The base voltage controls the emitter voltage.

$$V_E = V_B - 0.7 = 2 - 0.7 = 1.3\text{V}$$

Now V_E controls I_E:

$$I_E = \frac{V_E}{R_E} = \frac{1.3}{1\text{K}} = 1.3\text{mA}$$

If the transistor is **ACTIVE,** the emitter current controls the collector current, which controls the base current:

$$I_C = I_E$$

and

$$I_B = \frac{I_C}{\beta} = 0.013\text{mA}$$

To complete the analysis, we find V_C:

$$V_C = V_{CC} - V_{RC} = V_{CC} - R_C I_C = 10 - 3\text{K} \times 1.3\text{mA} = 6.1\text{V}$$

We now check to make sure the transistor is not in **SATURATION** by comparing the collector and emitter voltages. Because V_C at 6.1V is greater than V_E at 1.3V, the transistor is **ACTIVE.**

We see here that collector voltage is no longer fixed at the collector supply voltage. Now collector voltage is equal to the collector supply voltage minus the voltage drop across the collector resistor.

The **DC Bias List** for this circuit is

$V_B = 2\text{V}$ ($V_B = V_{BB}$ in this circuit)

$V_E = 1.3\text{V}$

$V_C = 6.1\text{V}$

$I_B = 0.013\text{mA}$

$I_E = 1.3\text{mA}$

$I_C = 1.3\text{mA}$

The power dissipated by the transistor is found from

$V_{CE} = V_C - V_E = 6.1 - 1.3 = 4.8\text{V}$

$P = V_{CE}I_C = 4.8 \times 1.3\text{mA} = 6.24\text{mW}$

EXAMPLE 9–2
V_B **Control with Collector Resistor**

For the circuit in Figure 9–7, find the transistor currents and voltages and the transistor power.

Solution

The base voltage controls the emitter voltage:

$$V_E = V_B - 0.7 = V_{BB} - 0.7 = 5 - 0.7 = 4.3\text{V}$$

Next, the emitter voltage controls the emitter current:

$$I_E = \frac{V_E}{R_E} = \frac{4.3}{2\text{K}} = 2.15\text{mA}$$

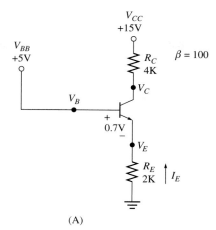

FIGURE 9–7   (A)

If the transistor is **ACTIVE,** the emitter current controls the collector current, which controls the base current:

$$I_C = I_E = 2.15\text{mA}$$

and

$$I_B = \frac{I_C}{\beta} = 0.0215\text{mA}$$

Finally, the collector voltage is found from

$$V_C = V_{CC} - V_{RC} = 15 - 4\text{K} \times 2.15\text{mA} = 6.4\text{V}$$

Because the collector voltage is greater than the emitter voltage, the transistor is **ACTIVE.**

The power is found from

$$V_{CE} = V_C - V_E = 6.4 - 4.3 = 2.1\text{V}$$
$$P = V_{CE}I_C = 2.1 \times 2.15\text{mA} = 4.52\text{mW}$$

DRILL EXERCISE   Change the collector resistor in Figure 9–7 to $R_C = 2\text{K}\Omega$, and redraw the circuit. Find the transistor voltages and currents in the proper order. Find the power dissipated by the transistor. Is the transistor **ACTIVE?**

Answer $V_E = 4.3\text{V}, I_E = 2.15\text{mA}, I_C = 2.15\text{mA}, I_B = 0.0215\text{mA}$
$V_C = 10.7\text{V}, P = 13.8\text{mV}$

ACTIVE

What are the major differences between I_B control discussed in Chapter 7 (Figure 9–8A) and this chapter's V_B control (Figure 9–8B p. 266)? Transistor currents in both of these circuits are similar. Also, in both circuits in Figure 9–8 the collector voltage is 3V. The difference between I_B control and V_B control is seen in the base and emitter voltages.

$$I_B \text{ control (Figure 9–8A) } V_B = 0.7\text{V} \quad \text{and} \quad V_E = 0\text{V}$$
$$V_B \text{ control (Figure 9–8B) } V_B = 1.2\text{V} \quad \text{and} \quad V_E = 0.5\text{V}$$

Because the emitter in I_B control is grounded, the emitter voltage in V_B control is always larger. The larger emitter voltage of V_B control transistor circuit decreases the collector-

FIGURE 9–8

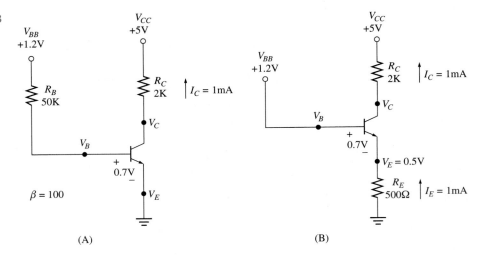

(A)   (B)

emitter voltage (V_{CE}). Smaller collector emitter voltages have two consequences: they reduce transistor power dissipation ($P = V_{CE}I_C$) and the transistor will go into **SATURATION** sooner (as V_C approaches V_E, the transistor **SATURATES**).

EXAMPLE 9–3

Comparison of I_B and V_B Control

Find the transistor voltages and currents for

(a) the I_B control circuit of Figure 9–8A.
(b) the V_B control circuit of Figure 9–8B.

Solution
(a) In the I_B control circuit, the base current is

$$I_B = \frac{V_{BB} - 0.7}{R_B} = \frac{1.2 - 0.7}{50K} = 0.01\text{mA}$$

The collector current is, therefore

$$I_C = \beta I_B = 1\text{mA}$$

The transistor voltages are

$$V_E = 0\text{V}, \; V_B = 0.7\text{V}$$

and

$$V_C = V_{CC} - R_C I_C = 5 - 2K \times 1\text{mA} = 3\text{V}$$

(b) In the V_B control circuit, the base voltage controls the emitter voltage:

$$V_B = V_{BB} = 1.2\text{V}$$

and

$$V_E = V_B - 0.7 = 1.2 - 0.7 = 0.5\text{V}$$

The emitter voltage controls I_E:

$$I_E = \frac{V_E}{R_E} = \frac{0.5}{500} = 1\text{mA}$$

If the transistor is **ACTIVE,** the emitter current controls the collector current, which controls the base current

$$I_C = I_E = 1\text{mA}$$

$$I_B = \frac{I_C}{\beta} = \frac{1}{100} = 0.01\text{mA}$$

Finally, we find the collector voltage from

$$V_C = V_{CC} - R_C I_C = 5 - 2\text{K} \times 1\text{mA} = 3\text{V}$$

EXAMPLE 9–4

V_B **Control;** R_C
Variations

Determine the state of the transistor in Figure 9–9, and find all transistor currents and voltages for the following collector resistors.

(a) $R_C = 2\text{K}\Omega$. **(b)** $R_C = 4\text{K}\Omega$. **(c)** $R_C = 8\text{K}\Omega$.

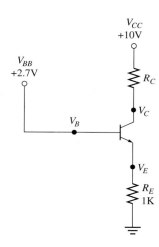

FIGURE 9–9

Solution
Because I_E does not depend on the value of the collector resistor, the emitter current is the same for all three parts. The base voltage controls the emitter voltage, which controls the emitter current:

$$V_E = V_B - 0.7 = 2.7 - 0.7 = 2\text{V}$$

and

$$I_E = \frac{V_E}{R_E} = \frac{2}{1\text{K}} = 2\text{mA}$$

We now find the other transistor variables by assuming that the transistor is **ACTIVE.** We must then confirm this is true by comparing V_C to the V_E we found above.

(a) $R_C = 2\text{K}\Omega$

Assuming the transistor is **ACTIVE,** we find the collector and base currents from

$$I_C = I_E = 2\text{mA}$$

and

$$I_B = \frac{I_C}{\beta} = 0.02\text{mA}$$

We now find the collector voltage:

$$V_C = V_{CC} - R_C I_C = 10 - 2\text{K} \times 2\text{mA} = 6\text{V} > V_E$$

Because $V_C > V_E$, the transistor is **ACTIVE,** so our results are correct.

(b) $R_C = 4\text{K}\Omega$

Assuming the transistor is **ACTIVE,** we find the collector and base currents from

$$I_C = I_E = 2\text{mA}$$

and

$$I_B = \frac{I_C}{\beta} = 0.02\text{mA}$$

We now find the collector voltage:

$$V_C = V_{CC} - R_C I_C = 10 - 4\text{K} \times 2\text{mA} = 2\text{V} = V_E$$

Because $V_C = V_E$, the transistor is just at the edge of **SATURATION.** As we discussed in Chapter 7, at the edge of **SATURATION** the relation between the transistor currents is the same as in the **ACTIVE** state. Therefore, our answers are still correct.

(c) $R_C = 8\text{K}\Omega$

Assuming the transistor is **ACTIVE,** we find the collector and base currents from

$$I_C = I_E = 2\text{mA}$$

and

$$I_B = \frac{I_C}{\beta} = 0.02\text{mA}$$

We now find the collector voltage:

$$V_C = V_{CC} - R_C I_C = 10 - 8\text{K} \times 2\text{mA} = -6\text{V} < V_E$$

Because $V_C < V_E$, the transistor is **SATURATED,** so our results are not correct.

DRILL EXERCISE   Change the emitter resistor in Figure 9–9 (p. 267) to $R_E = 2\text{K}\Omega$, and redraw the circuit. If the transistor is in the **ACTIVE** state, find the transistor voltages and currents for

(a) $R_C = 4\text{K}\Omega$
(b) $R_C = 12\text{K}\Omega$

Determine the state of the transistor for each part.

Answer   **(a) ACTIVE**

$$V_E = 2\text{V},\ I_E = 1\text{mA},\ I_C = 1\text{mA},\ I_B = 0.01\text{mA},\ V_C = 6\text{V}$$

(b) SATURATED

The previous example and drill exercise show us that as the collector resistor increases, the collector voltage decreases. Because the emitter voltage in the **ACTIVE** transistor does not depend on the collector resistor, as collector voltage decreases, the transistor heads towards **SATURATION.**

The following example shows how the state of the transistor depends on the base voltage.

EXAMPLE 9–5
Variable Base Voltage Biasing

In the circuit shown in Figure 9–10, V_{BB} is set equal to these values:

(a) $V_{BB} = 0.5V$.   **(b)** $V_{BB} = 1.5V$.   **(c)** $V_{BB} = 3V$.

Determine the state of the transistor for each value of the base supply voltage.

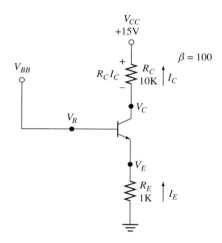

FIGURE 9–10

Solution

(a) For $V_{BB} = 0.5V$

Because the base voltage is less than 0.7V the transistor is **CUT-OFF.**

(b) For $V_{BB} = 1.5V$

The base voltage controls the emitter voltage, which controls the emitter current:

$$V_E = V_B - 0.7 = 1.5 - 0.7 = 0.8V$$

and

$$I_E = \frac{V_E}{R_E} = \frac{0.8}{1\,K} = 0.8mA$$

If the transistor is **ACTIVE,** the emitter current controls the collector current, which controls the base current:

$$I_C = I_E = 0.8mA$$

and

$$I_B = \frac{I_C}{\beta} = 0.008mA$$

The collector voltage is

$$V_C = V_{CC} - R_C I_C = 15 - 10K \times 0.8mA = 7V$$

Because $V_C > V_E$, the transistor is **ACTIVE** and our results are correct.

(c) For $V_{BB} = 3V$

$$V_E = V_B - 0.7 = 3 - 0.7 = 2.3V$$

and

$$I_E = \frac{V_E}{R_E} = \frac{2.3}{1K} = 2.3mA$$

Assuming that the transistor is **ACTIVE,** we get

$$I_C = I_E = 2.3\text{mA}$$

and

$$I_B = \frac{I_C}{\beta} = 0.02\text{mA}$$

The collector voltage is

$$V_C = V_{CC} - R_C I_C = 15 - 10\text{K} \times 2.3\text{mA} = -8\text{V}$$

Because $V_C < V_E$, the transistor is **SATURATED** and our results are *not* correct.

DRILL EXERCISE
Change the emitter resistor in Figure 9–10 (p. 269) to $R_E = 2\text{K}\Omega$, and redraw the circuit. Determine the state of the transistor and, if **ACTIVE,** the voltages and currents for

(a) $V_{BB} = 0.5\text{V}$ (b) $V_{BB} = 2.7\text{V}$ (c) $V_{BB} = 3.5\text{V}$

Answer   (a) **CUT-OFF**
             (b) **ACTIVE** $V_E = 2\text{V}$, $I_E = 1\text{mA}$, $I_C = 1\text{mA}$, $V_C = 5\text{V}$
             (c) **SATURATED**

For the transistor to act as a linear device, it must remain in the **ACTIVE** state. You can see from Example 9–5 (p. 269) that as the base voltage increases, the transistor will go from **CUT-OFF** to **ACTIVE** to **SATURATED.** The linear range of a transistor is given by the largest variation in the base voltage that keeps the transistor in the **ACTIVE** state.

REVIEW QUESTIONS

1. What transistor variables are most different between I_B control and V_B control?
2. What transistor variable is most affected by the presence of a collector resistor?
3. How do we find V_C when the circuit contains a collector resistor?
4. How does the value of R_C affect the state of the transistor?
5. How does the value of V_B affect the state of the transistor?

9–4 ■ THE VOLTAGE DIVIDER

To this point we have discussed two methods of biasing the transistor: I_B control and V_B control. Common to all of the biasing circuits we have showed you is that two batteries were available. One to supply V_{CC}, and another to supply V_{BB}.

Because batteries are expensive, take a lot of space, run down and need to be replaced, etc., we want to limit their number in any given circuit. For example, can we use one battery to supply the different voltages needed for V_{CC} and V_{BB}?

The answer lies in the voltage divider circuit shown in Figure 9–11A. One battery powers the circuit, and the voltage across R_2 provides the equivalent voltage of another battery. The voltage supplied by V_{CC} is divided between the two series resistors; V_{R2} is found from

$$V_{R2} = \frac{R_2}{R_1 + R_2} V_{CC}$$

FIGURE 9–11

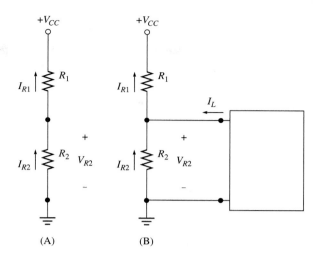

(A)   (B)

Remember, if two resistors are in series, they share the same current. That is

$$I_{R1} = I_{R2}$$

If R_1 and R_2 are not in series, then this circuit is not a voltage divider. An example of this is shown in Figure 9–11B. Here additional current enters the node between the two resistors. This means that the two resistor currents are not the same. Here

$$I_{R1} = I_{R2} + I_L$$

and the voltage divider equation does not hold.

Often, however, even though the two resistors do not form an exact voltage divider, as in Figure 9–11B, we can still approximate V_{R2} with the voltage divider equation. Whether this approximation is valid depends on the size of the load current. This is demonstrated in the next two examples.

EXAMPLE 9–6 **The Voltage** **Divider**	Find V_{R2} in Figure 9–12 if **(a)** $R_1 = R_2 = 5\text{K}\Omega$. **(b)** $R_1 = 100\text{K}\Omega$, $R_2 = 10\text{K}\Omega$.

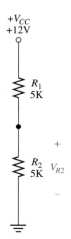

FIGURE 9–12

Solution

(a) For $R_1 = R_2 = 5K\Omega$

$$V_{R2} = \frac{R_2}{R_1 + R_2} \, V_{CC} = \frac{5K}{5K + 5K} 12 = \frac{1}{2} 12 = 6V$$

Note: Whenever the two resistors in a voltage divider are the same, the voltage across each resistor is one-half of the supply voltage.

(b) For $R_1 = 100K\Omega$, $R_2 = 10K\Omega$

$$V_{R2} = \frac{R_2}{R_1 + R_2} \, V_{CC} = \frac{10K}{100K + 10K} 12 = \frac{10}{110} 12 = 1.09V$$

Example 9–6 shows how we can use the voltage divider to provide two voltages for the cost of one battery. The circuit in Figure 9–12B, for example, can supply 12V from the battery and 1.09V from R_2. A potential problem arises with this technique when we connect R_2 to a load. This problem is explored in the next example.

EXAMPLE 9–7
Approximate Voltage Divider with Load

For the circuit shown in Figure 9–13 and the given load currents, find V_{R2}.

(a) $I_B = 0A$. **(b)** $I_B = 0.001mA$. **(c)** $I_B = 0.05mA$.

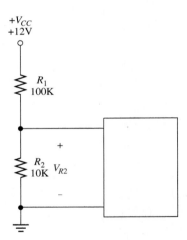

FIGURE 9–13

Solution

For completeness, we will derive the equation for V_{R2}. (You can also use the Thevenin analysis shown in Section 9–7, p. 284). What is important, for now, is not the derivation but the results. We first note that the currents have the following relation:

$$I_{R1} = I_{R2} + I_L$$

We now use Ohm's Law to get

$$\frac{V_{R1}}{R_1} = \frac{V_{R2}}{R_2} + I_L$$

Recognizing that

$$V_{R1} = V_{CC} - V_{R2}$$

we get

$$\frac{V_{CC} - V_{R2}}{R_1} = \frac{V_{R2}}{R_2} + I_L$$

Using the values in the circuit, we get

$$\frac{12 - V_{R2}}{100\text{K}} = \frac{V_{R2}}{10\text{K}} + I_L$$

Multiplying both sides by 100K results in

$$12 - V_{R2} = 10V_{R2} + 100\text{K}I_L$$

We rearrange to get

$$11V_{R2} = 12 - 100\text{K}I_L$$

Dividing by 11 gives us the final equation:

$$V_{R2} = 1.09 - 9.09\text{K}\,I_L$$

(a) For $I_L = 0$A, the second term on the right in the preceding equation disappears, and

$$V_{R2} = 1.09\text{V}$$

Note that if the load current is zero, R_1 and R_2 form an exact voltage divider (see Example 9–6B).

(b) For $I_L = 0.001$mA

$$V_{R2} = 1.09 - 9.09 \times \text{K } 0.001\text{mA} = 1.08\text{V}$$

A small load current will have a small effect on the voltage divider—compare 1.09 to 1.08V.

(c) For $I_L = 0.05$mA

$$V_{R2} = 1.09 - 9.09\text{K} \times 0.05\text{mA} = 0.64\text{V}$$

We see from parts (a) and (b) that if the load current is small, we can approximate the R_1, R_2 combination as a voltage divider. A large current, as in part (c), invalidates the voltage divider approach.

DRILL EXERCISE   For the circuit shown in Figure 9–13, find V_{R2} for

(a) $I_L = 0.005$mA   (b) $I_L = 0.01$mA

Answer   (a) $V_{R2} = 1.04$V   (b) $V_{R2} = 1$V

Example 9–7 and the drill exercise show that a small load current has a small effect on the output of the voltage divider. When the load current is zero, the two resistors form a pure voltage divider. We also saw from part (b) and the drill exercise that if the load current is smaller than 0.01mA (or 1/10 of the pure voltage divider I_{R2}), the voltage divider is barely affected. In other words, if $I_{R2} > 10I_L$, we can assume that R_1 and R_2 form a voltage divider. We can summarize as follows:

☐ Assume R_1 and R_2 form a voltage divider.

☐ $V_{R2} = \dfrac{R_2}{R_1 + R_2} V_{CC}$

☐ $I_{R2} = \dfrac{V_{R2}}{R_2}$

☐ If $I_{R2} > 10I_L$, answers are approximately correct.

☐ If $I_{R2} < 10I_L$, use the Thevenin approach (see Section 9–7).

One final note about load current and the voltage divider: the presence of a load current *always* reduces the voltage across R_2.

9–5 ■ SELF-BIASING THE TRANSISTOR

Figure 9–14A shows the most common type of V_B control biasing circuit, known as a **self-biasing circuit.** In this circuit we use a single supply to provide both the collector and base supply voltages. To help you understand how this circuit works, we have redrawn it in Figure 9–14B to show explicitly how V_{CC} supplies both the collector and base sides of the transistor.

FIGURE 9–14
The self-biased transistor

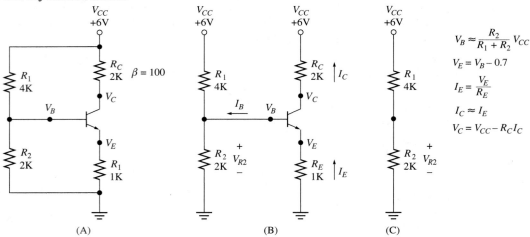

$$V_B \approx \frac{R_2}{R_1 + R_2} V_{CC}$$

$$V_E = V_B - 0.7$$

$$I_E = \frac{V_E}{R_E}$$

$$I_C \approx I_E$$

$$V_C = V_{CC} - R_C I_C$$

(A)   (B)   (C)

For a quick analysis of this circuit, we assume that V_{CC}, R_1, and R_2 form a voltage divider. From our discussion in Section 9–5, for this analysis to be reasonably accurate, the load current must be small enough to be ignored. The load current for this voltage divider is the transistor base current. Because base currents are very small, we can usually use this approach.

From Figure 9–14C we see that the base voltage is the voltage across R_2 and can be found from the voltage divider relation:

$$V_B = \frac{R_2}{R_1 + R_2} V_{CC}$$

so

$$V_B = \frac{R_2}{R_1 + R_2} V_{CC} = \frac{2K}{2K + 4K} 6 = \frac{2}{6} 6 = 2V$$

The self-biased circuit is still a V_B control circuit. Therefore, once we find V_B, we analyze the circuit the same way we did in Section 9–3 (p. 263). That is, the base voltage controls the emitter voltage:

$$V_E = V_B - 0.7 = 2 - 0.7 = 1.3V$$

and the emitter voltage controls the emitter current:

$$I_E = \frac{V_E}{R_E} = \frac{1.3}{1K} = 1.3mA$$

If the transistor is **ACTIVE,** the emitter current controls the collector current, which controls the base current:

$$I_C = I_E = 1.3\text{mA}$$

$$I_B = \frac{I_C}{\beta} = 0.013\text{mA}$$

Finally, the collector voltage is found from

$$V_C = V_{CC} - R_C I_C = 6 - 2\text{K} \times 1.3\text{mA} = 3.4\text{V}$$

Again we confirm that the transistor is **ACTIVE** because $V_C > V_E$. Note that if the transistor is **SATURATED,** all of our answers are wrong and there is no point validating the voltage divider approximation.

At this point, we should stop and examine our assumption that the base current is small enough to ignore. Is $I_{R2} > 10 I_B$? This is the requirement we obtained in Section 9–4 (p. 270). The load current here is the base current, so

$$I_{R2} = \frac{V_B}{R_2} = \frac{2}{2\text{K}} = 1\text{mA} > 10 I_B = 0.13\text{mA}$$

This result validates our use of the voltage divider.

The **DC Bias List** for this self-biased circuit is

$V_B = 2\text{V}$ from the voltage divider

$V_E = 1.3\text{V}$

$V_C = 3.4\text{V}$

$I_B = 0.013\text{mA}$

$I_C = 1.3\text{mA}$

$I_E = 1.3\text{mA}$

$V_{CE} = V_C - V_E = 3.4 - 1.3 = 2.1\text{V}$

$P = V_{CE} I_C = 2.1 \times 1.3\text{mA} = 2.73\text{mW}$

As with any biasing method, you must always check the state of the transistor. Our analysis always assumes that the transistor is in the **ACTIVE** state. If the transistor is **SATURATED,** the results we obtained would be useless.

EXAMPLE 9–8
Self-Biasing Circuit

For the circuit in Figure 9–15A (p. 276), determine the state of the transistor, all transistor currents and voltages, and the power dissipated by the transistor. Verify that R_2 and R_1 form an approximate voltage divider by comparing I_B and I_{R2}.

Solution

We assume that R_1 and R_2 form a voltage divider, and solve for the base voltage (Figure 9–15B)

$$V_B = \frac{R_2}{R_1 + R_2} V_{CC} = \frac{20\text{K}}{180\text{K} + 20\text{K}} 20 = \frac{20}{200} 20 = 2\text{V}$$

Now, V_B controls V_E:

$$V_E = V_B - 0.7 = 2 - 0.7 = 1.3\text{V}$$

and V_E controls I_E:

$$I_E = \frac{V_E}{R_E} = \frac{1.3}{2\text{K}} = 0.65\text{mA}$$

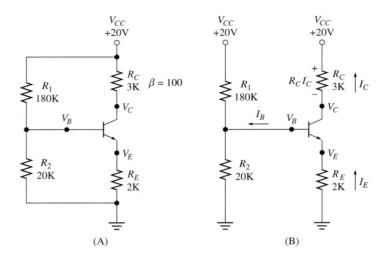

FIGURE 9–15

(A)   (B)

If the transistor is **ACTIVE,** the emitter current controls the collector current, which controls the base currents:

$$I_C = I_E = 0.65\text{mA}$$

and

$$I_B = \frac{I_C}{\beta} = 0.0065\text{mA}$$

We find the collector voltage to be

$$V_C = V_{CC} - R_C I_C = 20 - 3\text{K} \times 0.65\text{mA} = 18.1\text{V}$$

which confirms that the transistor is **ACTIVE.**
Comparing I_{R2} and I_B

$$I_{R2} = \frac{V_B}{R_2} = \frac{2}{20\text{K}} = 0.1\text{mA} > 10I_B = 0.065\text{mA}$$

Therefore, our voltage divider approximation is valid.
    The power dissipated by the transistor is found from

$$V_{CE} = V_C - V_E = 18.1 - 1.3 = 16.8\text{V}$$
$$P = V_{CE}I_C = 16.8 \times 0.65\text{mA} = 10.9\text{mW}$$

DRILL EXERCISE   For the circuit shown in Figure 9–15, change R_1 to 200KΩ, and redraw the circuit. Find the state of the transistor and all transistor voltages and currents. Compare I_{R2} to I_B to confirm the voltage divider approximation is valid.

Answer   **ACTIVE**

$V_B = 1.82\text{V}, V_E = 1.12\text{V}, V_C = 18.32\text{V}$

$I_E = 0.56\text{mA}, I_C = 0.56\text{mA}, I_B = 0.0056\text{mA}, I_{R2} = 0.09\text{mA}$

Voltage divider approximation is valid ($I_{R2} > 10I_B$).

    Up to this point we have validated the voltage divider approximation by comparing two currents, I_{R2} and I_B. Actually, it is easier to compare resistors rather than currents. In fact, the voltage divider approximation is valid if

$$R_2 < 10R_E$$

In Section 9–7 (p. 284), we will derive this relation. For now, we will give you an idea of why it is valid. The base current is related to the emitter current by

$$I_B = \frac{I_C}{\beta} = \frac{I_E}{\beta}$$

Because

$$I_{R2} = \frac{V_B}{R_2} \text{ and } I_E = \frac{V_E}{R_E} \rightarrow I_B = \frac{1}{\beta} \frac{V_E}{R_E}$$

the current requirement we have used to validate the voltage divider approximation becomes

$$I_{R2} > 10I_B \rightarrow \frac{V_B}{R_2} > 10 \frac{1}{\beta} \frac{V_E}{R_E}$$

We have seen that V_B and V_E have very similar values. If we assume they are approximately equal, we can eliminate these two voltages from the above, to get

$$\frac{1}{R_2} > \frac{10}{\beta} \frac{1}{R_E}$$

Rearranging this equation (if $\frac{1}{x} > \frac{1}{y}$, then $x < y$)

$$R_2 < \frac{\beta}{10} R_E$$

Because the typical value for β is 100 or greater, the requirement for a valid voltage divider approximation becomes

$$R_2 < 10R_E$$

DRILL EXERCISE

In Example 9–8 (p. 275), we used the voltage divider approximation to find the bias values for the transistor. Use the resistor requirement to determine if the approximation was valid.

Answer $R_2 = 20K = 10R_E$ The approximation is just valid.

In Section 9–4 (p. 270), you learned that if there is significant load current, the voltage across R_2 will be less than the voltage found from the voltage divider. Therefore, whenever we use the voltage divider approximation in self-biasing circuits, we are finding the maximum possible value for V_B. Its actual value will always be somewhat lower. If the actual base voltage is always smaller than the base voltage we calculate, we are always going to be further away from **SATURATION** than we calculate.

REVIEW QUESTIONS

1. Draw the self-biased V_B control transistor circuit.
2. What is V_B for this circuit?
3. What is the current requirement for the voltage divider approximation to be valid?
4. What is the resistor requirement for the voltage divider approximation to be valid?
5. Should you check for **SATURATION** in the self-biased circuit? Why?

9–6 ■ VOLTAGE AMPLIFICATION

Now that you know how to bias a transistor, let's see what happens if we set a bias point, then vary the base voltage up and down from that point. In the circuit of Figure 9–16A, we initially set $V_{BB} = 1.7$V as our base bias voltage. The base supply voltage sets the base voltage, and the base voltage controls the emitter voltage:

$$V_E = V_B - 0.7 = V_{BB} - 0.7 = 1.7 - 0.7 = 1\text{V}$$

Now, the emitter voltage controls the emitter current, so

$$I_E = \frac{V_E}{R_E} = \frac{1}{1\text{K}} = 1\text{mA}$$

Assuming the transistor is **ACTIVE,** the emitter current controls the collector current, which controls the base current:

$$I_C = I_E = 1\text{mA}$$

and

$$I_B = \frac{I_C}{\beta} = 0.01\text{mA}$$

The collector voltage is

$$V_C = V_{CC} - R_C I_C = 12 - 5\text{K} \times 1\text{mA} = 7\text{V}$$

Because $V_C > V_E$, our assumption that the transistor is **ACTIVE** is correct.

Now, in Figure 9–16B, we let V_{BB} increase from the bias level of 1.7V to a maximum of 2.4V in steps of 0.1V, then find the resulting transistor voltages V_E and V_C.

FIGURE 9–16

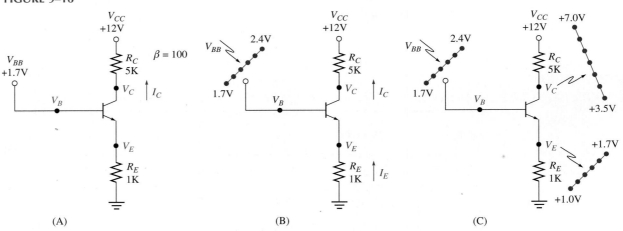

(A) (B) (C)

For each value of V_{BB}, we use the same procedure followed for the bias calculations. That is, V_{BB} controls V_E; V_E controls I_E controls I_C controls V_C. Remember to check for **SATURATION;** we are OK as long as $V_C > V_E$. We present the results in Figure 9–16C, and in tabular form in Table 9–1.

The values in Table 9–1 provide some interesting results. As base voltage increases, the emitter voltage also increases. But, as base voltage increases, collector voltage decreases. The inversion between base voltage and collector voltage is one of the basic operating principles of the transistor.

Examining the incremental changes that take place in the various voltages is also very instructive. For example, when V_B increases from 1.7 to 1.8V, V_E changes from 1.0 to 1.1V. The incremental changes in these voltages is

$$\Delta V_B = 1.8 - 1.7 = 0.1 \text{V}$$

and

$$\Delta V_E = 1.1 - 1.0 = 0.1 \text{V}$$

This shows us that the change in emitter voltage follows the change in base voltage; that is

$$\Delta V_E = \Delta V_B$$

We define **voltage gain** (A) as the ratio of the incremental change in output voltage to the incremental change in input voltage. That is, voltage gain is a measure of how much the output changes compared to how much the input changes. In this case the emitter voltage is the output, and the base voltage is the input, so we are interested in the voltage gain between the base and the emitter (A_E).

$$A_E = \frac{\Delta V_E}{\Delta V_B} = 1$$

Note that if we take any two values for V_B and the corresponding two values for V_E, we will always get $\Delta V_E = \Delta V_B$. This is what we mean by "linear operation." Anywhere within the linear operating region, the relation between the input and output voltage is the same.

TABLE 9–1

V_{BB} (V) =	V_B (V) = V_{BB}	V_E (V) = $V_{BB} - 0.7$	$I_C =$ I_E (mA) = V_E/R_E	V_C (V) = $V_{CC} - R_C I_C$
1.7	1.7	1.0	1	7.0
1.8	1.8	1.1	1.1	6.5
1.9	1.9	1.2	1.2	6.0
2.0	2.0	1.3	1.3	5.5
2.1	2.1	1.4	1.4	5.0
2.2	2.2	1.5	1.5	4.5
2.3	2.3	1.6	1.6	4.0
2.4	2.4	1.7	1.7	3.5

We now turn our attention to the collector voltage. As the base voltage increases from 1.7 to 2.4V, the collector voltage decreases from 7 to 3.5V (Figure 9–16C). If we look at incremental changes, we see, for example, that when V_B increases from 2.1 to 2.2V, V_C decreases from 5.0 to 4.5V. So

$$\Delta V_B = 2.2 - 2.1 = 0.1$$

and

$$\Delta V_C = 4.5 - 5 = -0.5$$

The negative sign for ΔV_C tells us that the collector voltage has decreased. The relation between the collector and base voltage changes is

$$\Delta V_C = \frac{-0.5}{0.1} V_B = -5 \Delta V_B$$

Voltage amplification between the base and the collector is, therefore

$$A_C = \frac{\Delta V_C}{\Delta V_B} = -5$$

That is, the transistor circuit has amplified the incremental voltage applied to the base. We remind you that amplification refers to the magnitudes of the input and output signals. The minus sign tells us that as the input increases, the output decreases. Although the transistor itself is a current amplifier, when we connect resistors to the transistor, the resistors convert transistor currents to circuit voltages.

EXAMPLE 9–9
Voltage Amplification

In the circuit of Figure 9–17, the base supply voltage is decreased from 1.7V to 1.0V in steps of 0.1V. Build a table similar to Table 9–1 (p. 279). Find the incremental amplification between the base and the emitter, and base and collector voltages.

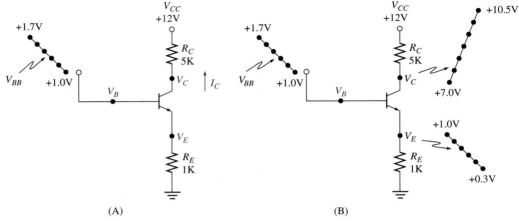

(A)   (B)

FIGURE 9–17

Solution The relations we need for this example are

$$V_B = V_{BB}$$

The base voltage controls the emitter voltage:

$$V_E = V_B - 0.7$$

and emitter voltage controls emitter current:

$$I_E = \frac{V_E}{R_E}$$

TABLE 9–2

V_{BB} (V) =	V_B (V) = V_{BB}	V_E (V) = $V_{BB} - 0.7$	$I_C =$ I_E (mA) = V_E/R_E	V_C (V) = $V_{CC} - R_C I_C$
1.7	1.7	1.0	1.0	7.0
1.6	1.6	0.9	0.9	7.5
1.5	1.5	0.8	0.8	8.0
1.4	1.4	0.7	0.7	8.5
1.3	1.3	0.6	0.6	9.0
1.2	1.2	0.5	0.5	9.5
1.1	1.1	0.4	0.4	10.0
1.0	1.0	0.3	0.3	10.5

Assuming the transistor is **ACTIVE**

$$I_C = I_E = \frac{V_E}{R_E} = \frac{V_E}{1\text{K}}$$

and

$$V_C = V_{CC} - I_C R_C$$

The results for the given base supply voltages are shown in Figure 9–17B and listed in Table 9–2.

In this example, you can see that emitter voltage decreases as base voltage decreases. Collector voltage, on the other hand, increases when base voltage decreases. Again, you see the inverse relation between V_C and V_B.

For more detail we compare the incremental change in V_B to incremental changes in V_E and V_C. We arbitrarily choose $V_B = 1.6$ and 1.7V to get

$$\Delta V_B = 1.6 - 1.7 = -0.1\text{V}$$
$$\Delta V_E = 0.9 - 1.0 = -0.1\text{V}$$

Once again, we see that the change in emitter voltage tracks, or follows, the change in the base voltage. That is

$$\Delta V_E = \Delta V_B$$

and

$$A_E = \frac{\Delta V_E}{\Delta V_B} = 1$$

At the collector, we get

$$\Delta V_C = 7.5 - 7.0 = 0.5\text{V}$$

The relation between the collector and base voltages is

$$\Delta V_C = \frac{3.5}{-0.7}\,\Delta V_B = -5\Delta V_B$$

and

$$A_C = \frac{\Delta V_C}{\Delta V_B} = -5$$

DRILL EXERCISE   For the circuit in Figure 9–17, let the base supply voltage drop from $V_{BB} = 1.5$V to $V_{BB} = 1.2$V.

(a) Find ΔV_E.   (b) Find A_E.   (c) Find ΔV_C.   (d) Find A_C.

Answer   (a) $\Delta V_E = -0.3$V,   (b) $A_E = 1$,
(c) $\Delta V_C = 1.5$V,   (d) $A_C = -5$

We see from Figures 9–16 (p. 278) and 9–17, and the drill exercise that whether base voltage increases or decreases, emitter voltage follows base voltage. That is $A_E = 1$. Note that emitter voltage does not depend on emitter or collector resistors.

Collector voltage is amplified with respect to base voltage, and collector voltage changes in the opposite direction to base voltage. As the next example shows, collector voltage gain depends on the value of the collector resistor. Although we don't show it in this chapter, collector voltage gain also depends on the value of the emitter resistor.

EXAMPLE 9–10

Effect of R_C on Voltage Amplification

In this example we examine how the collector resistor affects voltage amplification. To simplify the process we will analyze the circuit of Figure 9–18 for only 2 base voltages:

$$V_{BB} = 1.7V \text{ and } 1.8V.$$

For each base voltage, find ΔV_B, ΔV_C, and A_C when

(a) $R_C = 4K\Omega$ **(b)** $R_C = 6K\Omega$

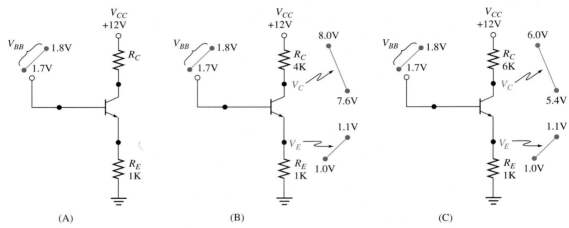

FIGURE 9–18

Solution   Common to both parts of this problem are

$$V_B = V_{BB}$$
$$V_E = V_B - 0.7$$
$$I_E = \frac{V_E}{R_E}$$
$$I_C = I_E$$

and

$$V_C = V_{CC} - R_C I_C$$

(a) For $R_C = 4K$
    When $V_{BB} = 1.7V$

$$V_B = 1.7V \text{ and } V_E = 1.0V$$
$$I_E = \frac{1.0}{1K} = 1mA \text{ and } I_C = I_E = 1mA$$
$$V_C = 12 - 4K \times 1mA = 8V$$

When $V_{BB} = 1.8\text{V}$

$$V_B = 1.8\text{V and } V_E = 1.1\text{V}$$

$$I_E = \frac{1.1}{1\text{K}} = 1.1\text{mA and } I_C = I_E = 1.1\text{mA}$$

$$V_C = 12 - 1.1\text{mA} \times 4\text{K} = 7.6\text{V}$$

The incremental changes in the voltages are

$$\Delta V_B = 1.8 - 1.7 = 0.1\text{V}$$

$$\Delta V_C = 7.6 - 8 = -0.4\text{V}$$

The voltage gain at the collector (Figure 9–18B) is

$$A_C = \frac{-0.4}{0.1} = -4$$

The amount of amplification between the base and the collector in this example is less than the amplification achieved when $R_C = 5\text{K}\Omega$.

(b) For $R_C = 6\text{K}\Omega$

When $V_{BB} = 1.7\text{V}$

$$V_B = 1.7\text{V and } V_E = 1.0\text{V}$$

$$I_E = \frac{1.0}{1\text{K}} = 1\text{mA and } I_C = I_E = 1\text{mA}$$

$$V_C = 12 - 6\text{K} \times 1.0\text{mA} = 6\text{V}$$

When $V_{BB} = 1.8\text{V}$

$$V_B = 1.8\text{V and } V_E = 1.1\text{V}$$

$$I_E = \frac{1.1}{1\text{K}} = 1.1\text{mA and } I_C = I_E = 1.1\text{mA}$$

$$V_C = 12 - 6\text{K} \times 1.1\text{mA} = 5.4\text{V}$$

The changes in the voltages are

$$\Delta V_B = 1.8 - 1.7 = 0.1\text{V}$$

and

$$\Delta V_C = 5.4 - 6 = -0.6\text{V}$$

The relation between the base and collector voltages (Figure 9–18C) is affected by R_C. That is

$$A_C = \frac{-0.6}{0.1} = -6$$

is more amplification between the base and the collector than the amplification achieved when $R_C = 5\text{K}\Omega \rightarrow A_C = -5$.

This example shows us that as R_C increases, the amplification between the base and the collector voltages increases. As R_C decreases, the amplification between the base and the collector voltages decreases. In fact, the voltage gain at the collector is given by

$$A_C \approx \frac{-R_C}{R_E}$$

REVIEW QUESTIONS

1. What is an incremental change in voltage?
2. How do we define voltage gain?
3. What is the voltage gain between the base and the emitter?
4. Why do we say the emitter voltage follows the base voltage?
5. When base voltage increases, what happens to collector voltage?
6. What are some typical values we obtained for the voltage gain between the base and the collector?
7. How does the collector resistor affect the collector voltage gain?
8. What is the voltage gain at the collector?

9-7 ■ EXACT BASE VOLTAGE CALCULATIONS

The Thevenin Approach

In previous sections, we discussed how to approximate V_{CC}, R_1, and R_2 as a voltage divider. Here, we will use the Thevenin equivalent of these elements to get exact answers.

Consider the self-biased circuit shown in Figure 9–19A. Earlier, we analyzed this circuit by assuming that V_{CC}, R_1, and R_2 form an approximate voltage divider. Now, we are going to find the Thevenin equivalent of this combination.

FIGURE 9–19

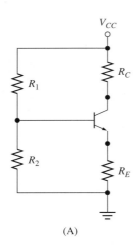

(A)

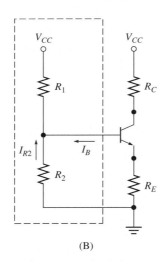

(B)

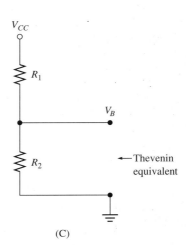

(C)

To clarify the Thevenin approach, we first redraw the circuit (in Figure 9–19B) to show explicitly the supply voltage and the two base resistors. It is this combination, shown in Figure 9–19C, that we will replace with a Thevenin equivalent. As you know, the Thevenin approach allows us to replace any circuit with a single voltage source in series with a single resistor (Figure 9–20). In general, it can be quite difficult to find a Thevenin equivalent. For the circuit of Figure 9–20, however, the equivalent is found from

$$V_{TH} = \frac{R_2}{R_1 + R_2} V_{CC}$$

That's right. The Thevenin voltage is simply the voltage divider voltage. The Thevenin resistor here turns out to be the parallel combination of the two resistors. That is

$$R_{TH} = \frac{R_1 R_2}{R_1 + R_2}$$

FIGURE 9–20

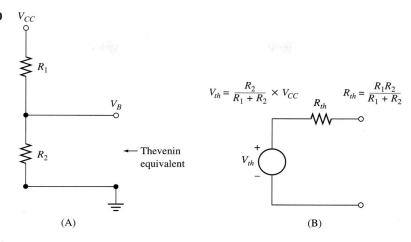

(A)   (B)

The two parts of Figure 9–21 show how we convert from the self-biased circuit to its Thevenin equivalent. In electronics, rather than use the notation of V_{TH} and R_{TH}, we use the notation of V_{BB} for the Thevenin voltage, and R_B for the Thevenin resistance (Figure 9–21C), where

$$V_{BB} = \frac{R_2}{R_1 + R_2} V_{CC}$$

and

$$R_B = \frac{R_1 R_2}{R_1 + R_2}$$

Now consider the circuit shown in Figure 9–22 (p. 286), which has both a base resistor and an emitter resistor. The base circuit is shown in Figure 9–22B. Although this analysis is slightly more complicated than those in previous sections and chapters, the principle is the same. From the base circuit, we can use Kirchhoff's voltage law to write

$$V_{BB} = R_B I_B + 0.7 + R_E I_E$$

FIGURE 9–21
Thevenin bias analysis

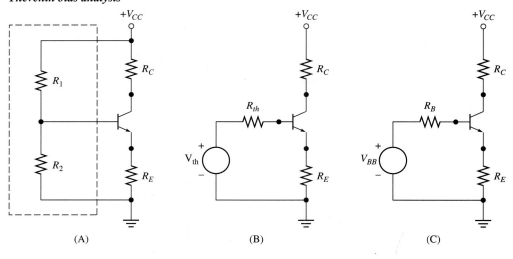

(A)   (B)   (C)

FIGURE 9–22

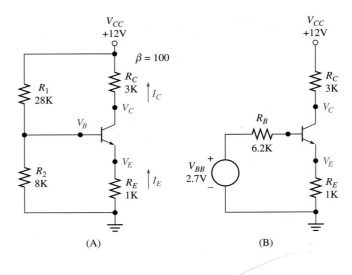

(A)   (B)

Now, if the transistor is **ACTIVE**

$$I_E = I_C \text{ and } I_C = \beta I_B$$

Therefore, we can write the base equation in terms of the base current:

$$V_{BB} = R_B I_B + 0.7 + R_E \beta I_B$$

Solving for the base current gives us

$$I_B = \frac{V_{BB} - 0.7}{R_B + \beta R_E}$$

and the collector current is

$$I_C = \beta \frac{V_{BB} - 0.7}{R_B + \beta R_E}$$

The collector voltage is still given by

$$V_C = V_{CC} - R_C I_C$$

Note that although we first found the base current, this is still a V_B control circuit, where

$$V_B = V_{BB} - R_B I_B$$

EXAMPLE 9–11
Thevenin
Equivalent I

For the circuit in Figure 9–22, find the collector current using the voltage divider approximation and the exact Thevenin approach.

Solution **Voltage Divider**
We find V_B from the voltage divider

$$V_B = \frac{8K}{28K + 8K} 12 = \frac{8}{36} 12 = 2.7V$$

Therefore, the collector current is

$$I_C = I_E = \frac{V_E}{R_E} = \frac{V_B - 0.7}{R_E} = \frac{2}{1K} = 2mA$$

Thevenin Approach

The Thevenin voltage and resistance are

$$V_{BB} = \frac{8K}{28K + 8K}\ 12 = \frac{8}{36}\ 12 = 2.7V$$

and

$$R_B = \frac{R_1 R_2}{R_1 + R_2} = \frac{28K \times 8K}{28K + 8K} = \frac{224K}{36} = 6.22K$$

The equivalent circuit is shown in Figure 9–22B. Rather than rederive the formula for I_C, we will simply repeat what we have already found:

$$I_C = \beta\frac{V_{BB} - 0.7}{R_B + \beta R_E} = 100\ \frac{2.7 - 0.7}{6.22K + 100 \times 1K} = \frac{200}{106.22K} = 1.88mA$$

We see that in this circuit the answer we obtained with voltage divider approximation is very close to the actual answer—2mA versus 1.88mA. Note that the actual collector current is smaller than that obtained with the voltage divider approximation. Actual base voltage is always less than the voltage calculated with the voltage divider.

EXAMPLE 9–12
Thevenin
Equivalent II

For the circuit in Figure 9–23, find the collector current using the voltage divider approximation and the exact Thevenin approach.

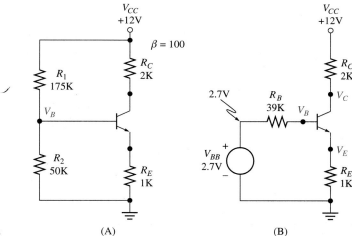

FIGURE 9–23   (A)   (B)

Solution **Voltage Divider**

We find V_B from the voltage divider:

$$V_B = \frac{50K}{175K + 50K}\ 12 = \frac{50}{225}\ 12 = 2.7V$$

Therefore, the collector current is

$$I_C = I_E = \frac{V_E}{R_E} = \frac{V_B - 0.7}{R_E} = \frac{2}{1K} = 2mA$$

Thevenin Approach

The Thevenin voltage and resistance are

$$V_{BB} = \frac{50K}{175K + 50K} \, 12 = \frac{50}{225} \, 12 = 2.7V$$

and

$$R_B = \frac{R_1 R_2}{R_1 + R_2} = \frac{175K \times 50K}{175K + 50K} = \frac{8750K}{225} \approx 39K$$

The equivalent circuit is shown in Figure 9–23B. The exact collector current is

$$I_C = \beta \frac{V_{BB} - 0.7}{R_B + \beta R_E} = 100 \, \frac{2.7 - 0.7}{39K + 100 \times 1K} = \frac{200}{139K} = 1.43mA$$

We see that in this circuit the answer we obtained with the voltage divider approximation was not close to the actual answer. In this case, we should use the exact approach.

DRILL EXERCISE

Redraw the circuit of Figure 9–23 with a new supply voltage of $V_{CC} = 9V$. Use the Thevenin approach to find the transistor voltages and currents. Is the transistor **ACTIVE?**

Answer   ACTIVE

$$V_{BB} = 2V, \; I_E = I_C = 0.94mA, \; I_B = 0.0094mA, \; V_B = 1.64V, \; V_C = 7.12V$$

Voltage Divider Versus the Exact Thevenin Approach

We finish off this section by deriving the requirement for using voltage divider approximation rather than the exact Thevenin approach. For the voltage divider approximation, base voltage is

$$V_B \approx \frac{R_2}{R_1 + R_2} \, V_{CC}$$

and collector current is

$$I_C = I_E = \frac{V_E}{R_E} = \frac{V_B - 0.7}{R_E}$$

so

$$I_C \approx \left(\frac{R_2}{R_1 + R_2} \, V_{CC} - 0.7 \right) \frac{1}{R_E}$$

In the exact Thevenin approach we present the equation in a form that we can compare more easily with the approximate equation:

$$I_C = \left(\frac{R_2}{R_1 + R_2} \, V_{CC} - 0.7 \right) \frac{\beta}{R_B + \beta R_E}$$

Now, if

$$R_B \ll \beta R_E$$

the last piece of the exact equation becomes

$$\frac{\beta}{R_B + \beta R_E} \rightarrow \frac{\beta}{\beta R_E} = \frac{1}{R_E}$$

which gives us the same result as the approximate equation. Therefore, if

$$R_B << \beta R_E$$

we can use the voltage divider approximation instead of the exact Thevenin approach. We usually accept a factor of 10 to indicate "much less than." Therefore, the requirement reduces to

$$R_B < \frac{\beta}{10} R_E$$

We can simplify this further when we realize that R_B is the parallel combination of R_1 and R_2. Because R_2 is typically much smaller than R_1, and because the smaller resistor in a parallel combination dominates, we can use R_2 instead of R_B. Also, because the typical value for β is 100, the requirement for the proper use of the voltage divider approximation reduces to

$$R_2 < 10R_E$$

DRILL EXERCISE: Use the resistor formula to determine whether the voltage divider approximations are valid for the circuits given

(a) Example 9.11 (p. 286)   **(b)** Example 9.12 (p. 287)

Answer   **(a)** Valid ($R_2 < 10R_E$)   **(b)** Invalid ($R_2 > 10R_E$)

9–8 ▪ DC TROUBLESHOOTING

In Chapter 8 we showed you how to use DC Troubleshooting to find problems in an electronic circuit. The technique is applicable to all electronic circuits in this textbook. As new elements are added to the circuit, we will consider additional modes of circuit failure.

Voltage Divider Problems

Bad Connection   Consider a voltage divider that is used to drive V_B control self-biased transistor circuits, such as the example in Figure 9–24A (p. 290). The estimated voltage across R_2 is

$$V_{R2} = \frac{3K}{3K + 9K} 12 = 3V$$

Figure 9–24B is an example in which the Measured Voltage is very different from the Estimated Voltage. The next step in DC Troubleshooting is to check the Measured Voltages at the resistors connected to the problem node; record these values on the connection diagram (Figure 9–24C). The connection diagram shows a discrepancy in the Measured Voltages at the terminals of resistors. There must, therefore, be a bad connection.

Wrong Resistor   Figure 9–25A (p. 290) shows a problem you may run across. The Estimated Voltage should be 3V, but the Measured Voltage shows a value of 0.3V. The Measured Voltage is not the supply voltage of 12V or the 0V of ground, as you would expect from a voltage divider with a bad connection between the resistors. What is happening here?

FIGURE 9–24

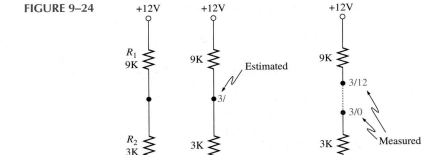

(A)   (B)   (C)

Figure 9–25B shows one problem that could cause the Measured Voltage that we see in Figure 9–25A. Instead of using a 9KΩ resistor, whoever constructed the circuit misread the last color band on the R_2 resistor and inserted a 90KΩ resistor. The Estimated Voltage for the actual resistors used in the circuit is 0.3V, just what is measured.

Whenever you find a discrepancy between the Estimated Voltage and the Measured Voltage in a circuit

☐ Check connections
☐ Check component values

FIGURE 9–25

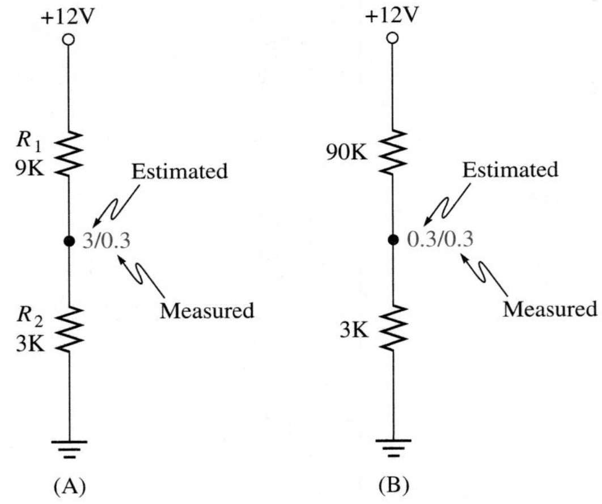

(A)   (B)

V_B Control Transistor Problems

The failures that can occur in a V_B control circuit fall into the same categories discussed in Chapter 8:

☐ Bad connections between resistors and transistor;
☐ Open circuit transistor failures at the collector, base, and the emitter; and
☐ Short circuit transistor failures between the collector and emitter, the emitter and base, and the collector and the base.

As shown in Chapter 7, we can use DC Troubleshooting to locate and diagnose all of the above problems. In these few pages, however, we cannot show you all possible combinations of the above failures. Examples incorporating a small sample of V_B control circuit failures should be enough to teach you the general principles of DC Troubleshooting.

Bad Connections at the Base First, consider the circuit shown in Figure 9–26A. The Estimated Voltage and the Measured Voltage show a discrepancy at the base. In Figure 9–26B the connection diagram shows that there is a bad connection between the base and R_1.

Because there is no voltage at the base, the transistor is **CUT-OFF** and all currents in the circuit are zero. This means that there are no voltage drops across the resistors in the circuit. The voltage at the bottom of R_1, therefore, is the supply voltage of 12V, and the voltage at the top of R_2 is 0V.

Now, let's examine the effect of a bad connection between R_2 and the base (Figure 9–27). The rather involved calculations on the circuit in Figure 9–27B show that this transistor is **SATURATED.** This is why the Measured Voltages at the collector and the emitter are the same. Not all circuits with this failure will **SATURATE,** however. The transistor state in a V_B control circuit with an open R_2 failure depends on the relative values of the remaining resistors.

Bad Connection at the Collector Figure 9–28 (p. 292) shows a circuit with mismatches in the Estimated Voltages and the Measured Voltages. Also shown is the connection diagram at the collector. The discrepancy here indicates that the connection between the collector and R_C is bad. The collector is open.

As we explained in Chapter 8, when the collector is open, the transistor will **SATURATE.** This will happen because an open circuit at the collector is equivalent to using a collector resistor of infinitely large resistance. As the value of R_C increases, the transistor

FIGURE 9–26

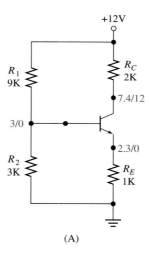

(A)

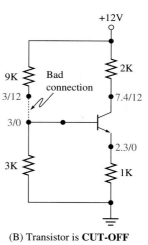

(B) Transistor is **CUT-OFF**

FIGURE 9–27

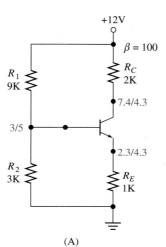

(A)

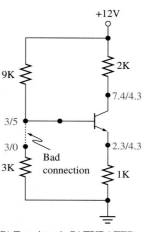

(B) Transistor is **SATURATED**

FIGURE 9–28

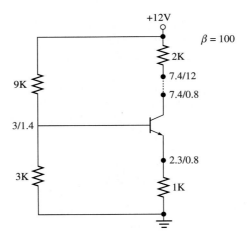

goes into **SATURATION.** This explains why the collector and emitter Measured Voltages are the same.

Bad Connection at the Emitter   The connection diagram in Figure 9–29 shows that this circuit has a bad connection at the emitter; that is, the emitter is not connected to ground. In this case, the transistor is in **CUT-OFF** and all *transistor* currents are zero. Because of this, R_1 and R_2 form an exact voltage divider. This is why the Measured Voltage at the base still matches the Estimated Voltage. Also, because all transistor currents are zero, there are no voltage drops across R_C and R_E, and we get the Measured Voltages shown.

FIGURE 9–29

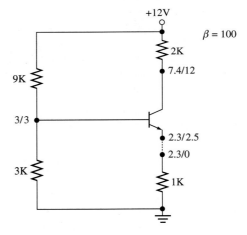

Also note that even with a bad emitter connection, the measured emitter voltage is still approximately a diode drop below the base voltage. If you connect a diode in the **ON** direction between a battery (greater than 0.7V) and a load resistor, the voltage at the load is always a diode drop below the battery voltage. This is true no matter how large the load resistor. Even as the resistor value heads to infinity, the load voltage remains a diode drop below the battery voltage.

Transistor Failures

Collector-to-Emitter Short   Figure 9–30 shows a circuit with DC Troubleshooting discrepancies at the collector and emitter. In this case we assume the connection diagram has shown that all connections are good. The Estimated Voltages indicate that the transistor should be in the **ACTIVE** state. The Measured Voltages at the collector and emitter are the same and, in fact, are those we would get if the transistor was **SATURATED.** However,

FIGURE 9–30

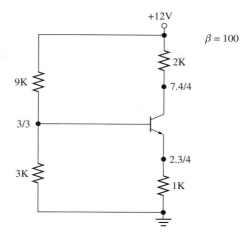

assuming that we are using the correct resistor values, this circuit cannot be in **SATURA-TION.** Therefore, we suspect an internal short circuit failure in the transistor.

A short circuit failure between the collector and the emitter is a common type of transistor failure. This is because the transistor dissipates most of its power between the collector and the emitter ($P = V_{CE}I_C$).

Note that the Measured Voltage at the emitter is greater than the voltage at the base. This indicates that the emitter diode must be **OFF.** The emitter diode will be **OFF** in all cases of collector-to-emitter internal short circuit.

Open Base Figure 9–31 shows a circuit with discrepancies at all transistor terminals. We assume that all connections are good. The fact that a voltage is measured at the emitter indicates that an emitter current exists. And because the measured emitter voltage is a diode drop below the base voltage, we conclude that the emitter diode is **ON.**

The Measured Voltage at the collector tells us that there is no collector current. Therefore, there is no voltage drop across the collector resistor. We conclude that the transistor in this circuit has suffered an open circuit at the collector. This is the second most common type of transistor failure.

FIGURE 9–31

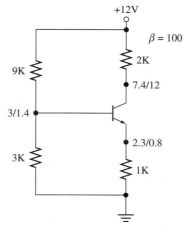

Figure 9–32 (p. 294) shows another circuit with discrepancies at the collector and the emitter. Again we assume that the connection diagram indicates that all connections are good. In this case the Measured Voltages at the collector and emitter must be the result of zero collector and emitter currents. The important clue here is that the measured emitter voltage is at 0V instead of a diode drop below the base voltage. This means that the base-emitter junction in the transistor must have opened up, perhaps due to excessive heat dissipation or mechanical failure.

FIGURE 9–32

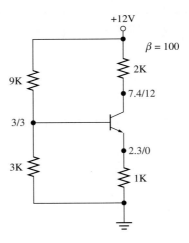

EXAMPLE 9–13
DC
Troubleshooting

The self-biased transistor circuits in Figure 9–33 show significant discrepancies between the Estimated Voltages and the Measured Voltages. All Measured Voltages were made on the transistor leads. Match the following transistor failures with the appropriate circuit.

(a) Open collector. **(b)** Open base. **(c)** R_1 failed open.

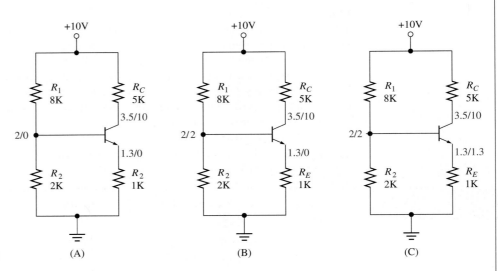

FIGURE 9–33

(A)   (B)   (C)

Solution

(a) Open collector: An open collector would result in zero collector current and the supply voltage appearing at the collector lead. Because this is true for all of the given circuits, we must consider the other transistor leads. A transistor with an open collector would still have an **ON** emitter diode; that is, the emitter voltage should be a diode drop below the base voltage. The only given transistor circuit that meets this result is in Figure 9–33C.

(b) Open base: A transistor with an open base is **CUT-OFF;** therefore, all transistor currents would be zero. In this case we expect the collector lead to be at V_{CC}, the emitter lead to be at ground potential, and the base lead to show the voltage divider value. The circuit that fits this description is in Figure 9–33B.

(c) R_1 failed open: If R_1 has failed open, there will be zero base voltage. Figure 9–33A matches this failure.

SUMMARY

■ The steps in determining the bias levels in V_B control circuits are

V_B $\qquad\qquad\qquad\qquad V_B = V_{CC}R_2/(R_1 + R_2)$

controls V_E

$$V_E = V_B - 0.7$$

controls I_E

$$I_E = V_E/R_E$$

controls I_C

if **ACTIVE**, $I_C = I_E$

controls I_B

if **ACTIVE**, $I_B = I_C/\beta$

■ Check to see if transistor is **ACTIVE** by finding V_C:

$$V_C = V_{CC} - R_C I_C$$

If $V_C > V_E$

The transistor is **ACTIVE** and all of the variables found above are correct.

If $V_C < V_E$

STOP! Transistor is **SATURATED.** Put down your pencil. The values you have calculated are wrong.

■ The voltage divider approximation in the self-biased circuit is valid if

$$R_2 < 10R_E$$

■ Emitter voltage changes follow base voltage changes. That is, the voltage gain between the emitter and base is $A_E = 1$

■ The voltage gain between the base and the collector is large and negative $(A_C = R_C/R_E)$. That is, small changes in the base voltage cause large changes in the collector voltage. Also, when base voltage increases, collector voltage decreases.

■ Problems in self-biased circuits are located with DC Troubleshooting. Remember also to check that resistors with the correct values are being used.

PROBLEMS

SECTION 9–2 Base Voltage Biasing (V_B Control)

1. For the circuit shown in Figure 9–34 (p. 296)
   (a) What voltage controls the DC bias voltages and currents of the transistor?
   (b) What is V_B?   (c) What is V_E?   (d) What is I_E?
   (e) What is I_C? What assumption are you making here?
   (f) What is I_B? What assumption are you making here?
   (g) What is the voltage across the emitter diode? Is this diode **ON** or **OFF?**
   (h) What is the voltage across the collector diode? Is this diode **ON** or **OFF?**
   (i) What is the state of the transistor? Why?
   (j) Write the complete DC Bias List for this circuit. That is, list the voltages and currents at each of the three transistor terminals.

2. For the transistor circuit shown in Figure 9–35 (p. 296)

$$V_{BB} = +15V, V_{CC} = +20V, R_E = 10K\Omega$$

   (a) List the values of the voltages and currents at each of the three terminals of the transistor. List your answers in the order that you find them.
   (b) Is the transistor **ACTIVE?** Why?
   (c) If the transistor is **ACTIVE,** give the complete DC Bias List for this circuit.

FIGURE 9–34

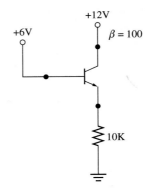

FIGURE 9–35

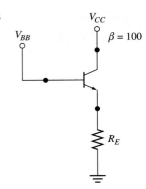

3. For the transistor circuit shown in Figure 9–35

$$V_{BB} = +12V, \; V_{CC} = +12V, \; R_E = 1K\Omega$$

(a) List the values of the voltages and currents at each of the three terminals of the transistor. List your answers in the order that you find them.

(b) Is the transistor **ACTIVE?** Why?

(c) If the transistor is **ACTIVE,** give the complete DC Bias List for this circuit.

4. For the transistor circuit shown in Figure 9–35

$$V_{BB} = 0V, \; V_{CC} = +12V, \; R_E = 1K\Omega$$

(a) List the values of the voltages and currents at each of the three terminals of the transistor. List your answers in the order that you find them.

(b) Is the transistor **ACTIVE?** Why?

(c) If the transistor is **ACTIVE,** give the complete DC Bias List for this circuit.

5. For the transistor circuit shown in Figure 9–35

$$V_{BB} = +20V, \; V_{CC} = +10V, \; R_E = 10K\Omega$$

(a) List the values of the voltages and currents at each of the three terminals of the transistor. List your answers in the order that you find them.

(b) Is the transistor **ACTIVE?** Why?

(c) If the transistor is **ACTIVE,** give the complete DC Bias List for this circuit.

SECTION 9–3   The Collector Resistor

6. For the transistor circuit shown in Figure 9–36

(a) What is V_B?   (b) What is V_E?   (c) What is I_E?

(d) What is I_C? What assumption are you making here?

(e) What is I_B? What assumption are you making here?

(f) What is V_C?   (g) What is V_{BE}? Is the emitter diode **ON** or **OFF?**

(h) What is V_{BC}? Is the collector diode **ON** or **OFF?**

(i) What is the state of the transistor? Why?

(j) Write the complete DC Bias List for this circuit. Be sure to account for current and voltage at all three transistor terminals.

7. For the circuit shown in Figure 9–37

$$V_{BB} = +4V, \; V_{CC} = +12V, \; R_C = 3K\Omega, \; R_E = 2K\Omega$$

(a) List the values of the voltages and currents at each of the three transistor terminals. List the variables in the order you find them.

(b) Is the transistor **ACTIVE?** Why?

(c) If the transistor is **ACTIVE,** write the complete DC Bias List for this circuit.

FIGURE 9–36

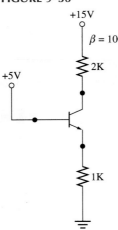

FIGURE 9–37

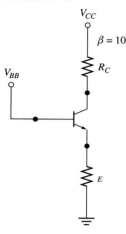

8. For the circuit shown in Figure 9–37

$$V_{BB} = +1V, V_{CC} = +9V, R_C = 5K\Omega, R_E = 1K\Omega$$

(a) List the values of the voltages and currents at each of the three transistor terminals. List the variables in the order you find them.

(b) Is the transistor **ACTIVE?** Why?

(c) If the transistor is **ACTIVE,** write the complete DC Bias List for this circuit.

9. For the circuit shown in Figure 9–37

$$V_{BB} = +0.5V, V_{CC} = +12V, R_C = 2K\Omega, R_E = 1K\Omega$$

(a) List the values of the voltages and currents at each of the three transistor terminals. List the variables in the order you find them.

(b) Is the transistor **ACTIVE?** Why?

(c) If the transistor is **ACTIVE,** write the complete DC Bias List for this circuit.

10. For the circuit shown in Figure 9–37

$$V_{BB} = +9V, V_{CC} = +9V, R_C = 2K\Omega, R_E = 1K\Omega$$

(a) List the values of the voltages and currents at each of the three transistor terminals. List the variables in the order you find them.

(b) Is the transistor **ACTIVE?** Why?

(c) If the transistor is **ACTIVE,** write the complete DC Bias List for this circuit.

11. For the circuit shown in Figure 9–37, $V_{CC} = +12V, R_C = 2K\Omega, R_E = 1K\Omega$. Find the base supply voltage (V_{BB}) that will put the transistor into the edge of **SATURATION.**

12. For the circuit shown in Figure 9–37, $V_{BB} = +6V, V_{CC} = +12V, R_E = 1K\Omega$. Find the value of the collector resistor (R_C) that will put the transistor into the edge of **SATURATION.**

13. For the transistor circuit shown in Figure 9–38

(a) Is this a situation of I_B control or V_B control? Why?

(b) List the values of the voltages and currents for the three terminals of the transistor. List variables in the order that they are found.

(c) What is the state of the transistor? Why?

(d) Write the complete DC Bias List.

14. For the transistor circuit shown in Figure 9–39

(a) Is this a situation of I_B control or V_B control? Why?

(b) List the values of the voltages and currents for the three terminals of the transistor. List variables in the order that they are found.

(c) What is the state of the transistor? Why?

(d) Write the complete DC Bias List.

FIGURE 9–38

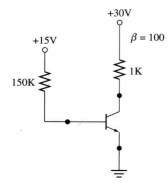

FIGURE 9–39

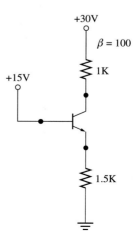

FIGURE 9–40

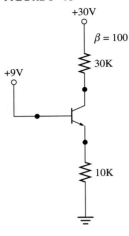

15. For the transistor circuit shown in Figure 9–40
   (a) Find the DC bias values, starting with the voltage at the base.
   (b) What is the state of the transistor?
   (c) Can you rely on the calculated DC bias values to be accurate? Why?
   (d) Why would you *not* make a DC Bias List for this circuit?

16. For the transistor circuit shown in Figure 9–40, let $R_C = 8K\Omega$.
   (a) Find the DC bias values.   **(b)** What is the state of the transistor?
   (c) Can you rely on the calculated DC bias values to be accurate? Why?
   (d) Write the DC Bias List for this circuit.

17. For the transistor circuit shown in Figure 9–40, change the base supply voltage to $V_{BB} = 3V$.
   (a) Find the DC bias values.   **(b)** What is the state of the transistor?
   (c) Can you rely on the calculated DC bias values to be accurate? Why?
   (d) Write the DC Bias List for this circuit.

SECTION 9–4 The Voltage Divider

18. For the resistor voltage divider circuit shown in Figure 9–41
   (a) Find I_{R2} when $I_L = 0mA$.
   (b) What is the current in R_1 and R_2 when $I_L = 0mA$? Is this the same as I_{R2}?
   (c) Let $I_L = 0.01mA$. Is this current large compared to the current in R_1 and R_2 when $I_L = 0mA$?
   (d) When $I_L = 0.01mA$, is V_{R2} changed very much from the value of V_{R2} when $I_L = 0mA$?
   (e) Let $I_L = 0.5mA$. Is this current considered to be large enough to change the value of V_{R2} calculated when $I_L = 0mA$?
   (f) When $I_L = 0.5mA$, is V_{R2} lower or higher than when $I_L = 0mA$? Why?

19. The load in Figure 9–41 is a resistor (R_L). How small can R_L get before the voltage divider approximation is invalid?

FIGURE 9–41

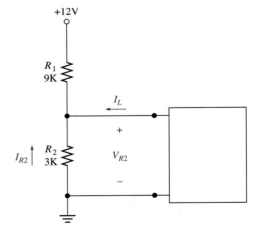

SECTION 9–5 Self-Biasing the Transistor

20. For the transistor circuit shown in Figure 9–42
   (a) What is the current in R_1 and R_2, assuming that I_B is negligible?
   (b) What is V_B, assuming I_B is negligible?
   (c) Find the DC bias values for the voltages and currents at all three transistor leads. Place a star (asterisk) by the steps where you assumed the transistor was **ACTIVE.**
   (d) Is the transistor operating in the **ACTIVE** state? Why?
   (e) Write the complete DC Bias List for this circuit.
   (f) Is I_B small compared to the current in I_{R2}?
   (g) Use the resistor requirement to determine if the base voltage divider approximation is valid.

21. For the transistor circuit shown in Figure 9–42, change the collector resistor to $R_C = 20K\Omega$.
    (a) Find the DC bias values for the voltages and currents at all three transistor leads. Place a star, or asterisk, by the steps where you assumed the transistor was **ACTIVE.**
    (b) What state is the transistor operating in?
    (c) Can you rely on the DC bias values that you calculated in part a?
    (d) Would you attempt to make a DC Bias List for this circuit?

22. For the transistor circuit shown in Figure 9–42, find R_C so that $V_C = 10V$.

23. For the transistor circuit shown in Figure 9–42, find R_C so that $V_C \approx V_E$.

24. Repeat Problem 20 for Figure 9–43.

25. For the transistor circuit shown in Figure 9–43, find R_C so that $V_C = 5V$.

26. For the transistor circuit shown in Figure 9–43, find R_C so that $V_C \approx V_E$.

FIGURE 9–42

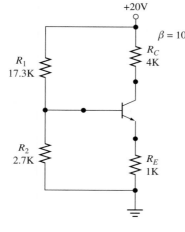

FIGURE 9–43

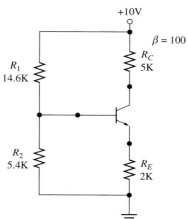

SECTION 9–6   Voltage Amplification

27. For the transistor circuit shown in Figure 9–42, assume that V_B has been increased by 0.2V by some external input.
    (a) How much does V_E change? Is this an increase or a decrease?
    (b) How much does I_E change?   (c) How much does I_C change?
    (d) How much does V_C change? Is this an increase or a decrease?
    (e) What is A_E, where $A_E = \dfrac{\Delta V_E}{\Delta V_B}$?   (f) What is A_C, where $A_C = \dfrac{\Delta V_C}{\Delta V_B}$?

28. Repeat Problem 27 for the circuit shown in Figure 9–43.

SECTION 9–7   Exact Base Voltage Calculations

29. For the circuit shown in Figure 9–42
    (a) Using the voltage divider approach, find the DC bias values, and write the DC Bias List for the circuit.
    (b) Draw the Thevenin equivalent circuit for the voltage divider formed by R_1 and R_2. Draw the equivalent circuit connected to the rest of the transistor circuit.
    (c) What is V_{TH}? What is V_{BB}?
    (d) What is R_{TH}? What is R_B?
    (e) Use I_B from part (a) to determine the voltage drop across R_B. Is this a significant voltage drop, or can it be neglected?
    (f) If the voltage drop across R_B is small, can you rely on the DC Bias List from part (a) to be accurate? Why?
    (g) Find the exact bias values, and compare with part (a).

30. Repeat Problem 29 for the circuit shown in Figure 9–43.

31. For the transistor circuit shown in Figure 9–44

    (a) Using the voltage divider approach, find the DC bias values. Write the DC Bias List for the circuit.

    (b) Draw the Thevenin equivalent circuit for the voltage divider formed by R_1 and R_2. Draw the equivalent circuit connected to the rest of the transistor circuit.

    (c) What is V_{TH}? What is V_{BB}?   **(d)** What is R_{TH}? What is R_B?

    (e) Use the I_B from part (a) to determine the voltage drop across R_B. Is this a significant voltage drop, or can it be neglected?

    (f) If the voltage drop across R_B is small, can you rely on the DC Bias List from part (a) to be accurate? Why?

    (g) Find the exact bias values, and compare to part (a).

FIGURE 9–44

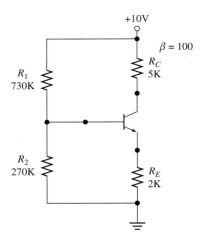

SECTION 9–8   DC Troubleshooting

32. Answer the following questions for the voltage divider shown in Figure 9–45.

    (a) What is the Estimated Voltage at the resistor connection in Figure 9–45A?

    (b) When you built this voltage divider, you accidentally used a resistor for R_2 that was 10 times too small. Which circuit shows the Measured Voltage that you would get in this case? Why?

    (c) When you built this voltage divider, you accidentally used a resistor for R_2 that was 10 times too large. Which circuit shows the Measured Voltage that you would get in this case? Why?

33. What does the connection diagram in Figure 9–46 tell you about the connection between R_1 and R_2? Explain your answer.

FIGURE 9–45   **FIGURE 9–46**

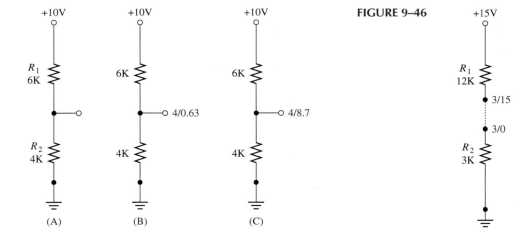

34. The connection diagram in Figure 9–47A shows that the voltage divider formed by R_1 and R_2 has a bad connection.

    (a) Draw the circuit diagram shown in Figure 9–47B. Write the Estimated Voltages at the three transistor leads.

    (b) On the same diagram write the Measured Voltages that you would get with the bad connection shown in Figure 9–47A. Remember, the Measured Voltages are recorded at the leads of the transistor.

FIGURE 9–47

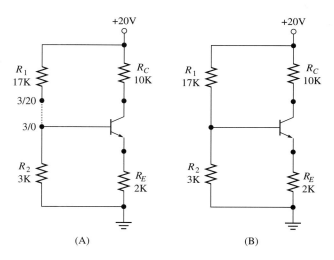

(A)   (B)

35. The connection diagram shown in Figure 9–48A shows that there is a bad connection between the base of the transistor and the voltage divider formed by R_1 and R_2.

    (a) Draw the circuit diagram shown in Figure 9–48B. Write the Estimated Voltages at the three transistor leads.

    (b) On the same drawing write the Measured Voltages that you would get with the bad connection shown in Figure 9–48A. Remember, the Measured Voltages are recorded at the leads of the transistor.

FIGURE 9–48

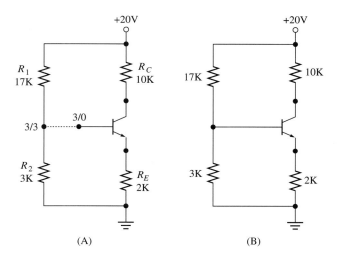

(A)   (B)

36. Figure 9–49 shows a transistor circuit with a number of possible DC Troubleshooting findings. Match the following possible problems to the appropriate circuit.

   (a) A properly working circuit with good matches between the Estimated Voltages and the Measured Voltages.

   (b) Bad connection at the emitter.   **(c)** Bad connection at the collector.

   (d) Bad connection at the base.

   (e) Open base connection inside the transistor package, a type of defect in transistors.

   (f) Short between the collector and the emitter, a type of defect in transistors.

FIGURE 9–49

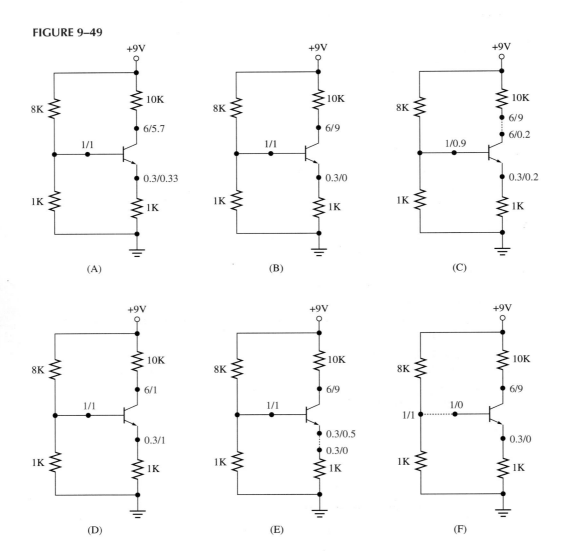

37. Each of the self-biased transistor circuits shown in Figure 9–50 (p. 303) shows a significant discrepancy between the Estimated Voltages and the Measured Voltages. All Measured Voltages were made on the transistor leads. Answer the following questions for these circuits.

   (a) Which circuit shows a short between the base and the emitter? Explain why the emitter, base, and collector have the Measured Voltages shown.

   (b) Which circuit shows an open collector inside the transistor? Explain why the emitter, base, and collector have the Measured Voltages shown.

   (c) Which circuit shows an open emitter inside the transistor? Explain why the emitter, base, and collector have the Measured Voltages shown.

FIGURE 9–50

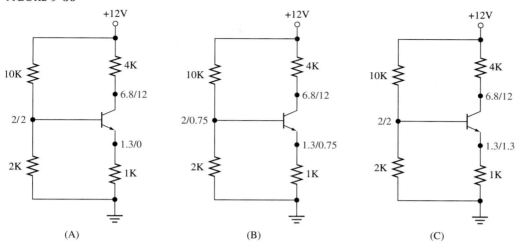

(A)   (B)   (C)

Computer Problems

38. Simulate the circuit of Figure 9–42 and repeat Problem 20.

39. Simulate the circuit of Figure 9–42 and repeat Problem 20 for
(a) $\beta = 50$ (b) $\beta = 200$

40. Simulate the circuit of Figure 9–43 and repeat Problem 25.

41. Simulate the circuit of Figure 9–43 and repeat Problem 25 for
(a) $\beta = 50$ (b) $\beta = 200$

42. Simulate the circuit of Problem 9–44 and repeat Problem 30.

43. Simulate the faulty circuit of Figure 9–47 and find the voltages in the circuit.

44. Simulate the faulty circuit of Figure 9–48 and find the voltages in the circuit.

45. Simulate the circuit of Figure 9–49. Find the circuit voltages under the following conditions:
(a) The base is open.   (b) The emitter is open.
(c) The collector is open.

46. Simulate the circuit of Figure 9–49. Find the circuit voltages under the following conditions:
(a) The base is shorted to the emitter.   (b) The base is shorted to the collector.
(c) The collector is shorted to the emitter.

CHAPTER

10

THE COMMON-EMITTER AMPLIFIER

To determine the maximum positive and negative swing of collector voltage.

To determine the AC behavior of the transistor.

To determine the gain of a common-emitter amplifier.

To determine the gain of a common-emitter amplifier with an emitter bypass capacitor.

To determine the gain of a common-emitter amplifier with a capacitor-coupled load.

To troubleshoot common-emitter amplifier circuits.

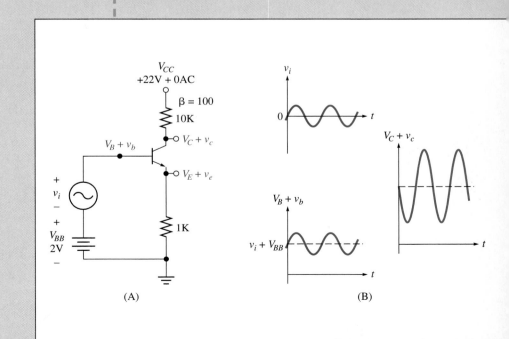

(A)

(B)

10–1 ■ INTRODUCTION

In the previous chapter we showed you how the transistor responded to incremental changes in base voltage. That is

$$\Delta V_E = A_E \, \Delta V_B \rightarrow A_E = \frac{\Delta V_E}{\Delta V_B}$$

where A_E had a value of 1; and

$$\Delta V_C = A_C \, \Delta V_B \rightarrow A_C = \frac{\Delta V_C}{\Delta V_B} = -\frac{R_C}{R_E}$$

where A_C had a typical value of -5. In this chapter, we will concentrate on collector voltage output, and delve deeper into the idea of incremental voltage changes.

We find it convenient to think of the general electrical signal as being composed of a DC component and an AC component. Consider, for example, the voltage wave form shown in Figure 10–1A. This wave form can be broken down into a DC component added to an AC component (Figure 10–1B). Note that the DC voltage is constant, whereas the AC component changes from -50 to 50mV.

FIGURE 10–1
Signal with AC and DC components

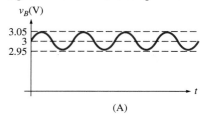

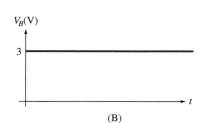

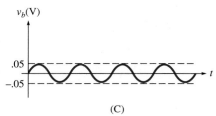

(A)	(B)	(C)

In electronics we use special notation to indicate whether we are working with the DC component, AC component, or the total signal. For example, for the wave shape shown in Figure 10–1A, we write

v_B	=	V_B	+	v_b
total		DC		AC

The uppercase voltage with the uppercase subscript (V_B) is a DC voltage, the lowercase voltage with the lowercase subscript (v_b) is an AC voltage, and the lowercase voltage with the uppercase subscript (v_B) is the total voltage.

Also note that all variables that are uppercase with uppercase subscripts are DC variables. Remember the difference between V_{CC}, V_{BB}, V_B, and V_{BE}. The double subscripts BB and CC indicate DC voltages that are supplied by a battery. Single subscripts indicate DC voltages that are measured circuit voltages.

Sinusoidal AC voltages are written in several ways. For example, the sinusoid in Figure 10–1B can be given

☐ in terms of its peak value of 50mV; that is $v_b = 50\text{mV}_\text{p}$;

☐ in terms of its peak-to-peak value of 100mV; that is $v_b = 100\text{mV}_\text{pp}$; or

☐ in terms of its root-mean-square (RMS) value; that is, $v_b = \dfrac{50}{\sqrt{2}}\text{mVRMS}$

In our discussions we will always use peak value to describe AC voltages and currents. Note that the peak value is always one-half of the total swing (peak-to-peak).

Look again at Figure 10–1A. What is the incremental change in the total voltage shown there? We can simply measure the peak-to-peak difference as

$$\Delta v_{B_{pp}} = 3.05 - 2.95 = 0.1 = 100mV$$

Note, however, that this same 100mV difference in maximum and minimum value is also exhibited in the AC component of the signal. The peak value of the incremental change in voltage is represented by the AC component v_b. That is

$$\Delta v_{B_p} = v_b$$

where

$$v_b = 50mV_p$$

in our notation.

In this chapter we will work with the AC components of the voltages and currents in transistor circuits.

EXAMPLE 10–1
DC and AC Signal Components

For the signal in Figure 10–2A, draw the DC and AC components.

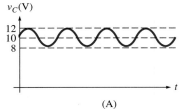

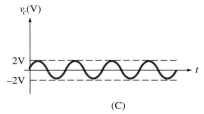

(A)   (B)   (C)

FIGURE 10–2

Solution The signal varies from 8 to 12V and is centered on 10V. The DC value is, therefore, 10V (Figure 10–2B). The AC component is a sinusoid that varies from −2 to 2V (Figure 10–2C), so

$$v = 2V_p$$

REVIEW QUESTIONS

1. What is A_E?

2. What is A_C?

3. Write the total collector voltage in terms of its DC and AC components.

4. What is the difference between the notations V_{CC} and V_C?

5. If v_C varies from 2-to-6V and is centered on 4V, what are the DC and AC components of this voltage?

10–2 ■ DC Bias Levels

Imagine that we want to build an amplifier so that patrons way up in the back row of a concert hall can hear their favorite rocker or opera singer. Our solution is to build a system in which the output of a microphone is fed into an amplifier that, in turn, produces a large enough voltage to drive a speaker (Figure 10–3). (Actually, this is a simplified presentation—usually we place a pre-amplifier between the microphone and the main amplifier.)

FIGURE 10–3

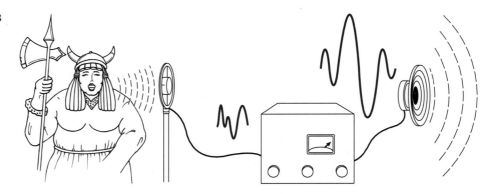

Our first attempt at building an amplifier is shown in Figure 10–4. The output of the microphone is a small-valued AC signal, which we use as the input base voltage. There is an immediate problem with this design. The input signal never reaches 0.7V. This means the transistor never turns on, and the collector output of the **CUT-OFF** transistor remains fixed at V_{CC}. This constant voltage will produce a whole lot of nothing at the speaker.

Even if we could produce a larger AC voltage with the microphone, we would still not succeed. In this case, the transistor would turn **ON** during the positive swing of the base voltage but would be **CUT-OFF** on the negative swing of the input signal. This would result in the half-wave rectified (clipped) voltage shown at the collector in Figure 10–4B. While the distortion caused by clipping may be acceptable for the heavy metal fan with pierced ears, nose, lip, etc., it would not please the opera buff.

Biasing

The solution to this problem is to bias the transistor in the **ACTIVE** region before we apply the AC signal. We can use the circuit in Figure 10–5 to provide the desired bias. Note that we have added a DC voltage to the AC microphone output. The microphone output is labeled v_i and is a small AC signal. The total base voltage is the series combination of v_i

FIGURE 10–4

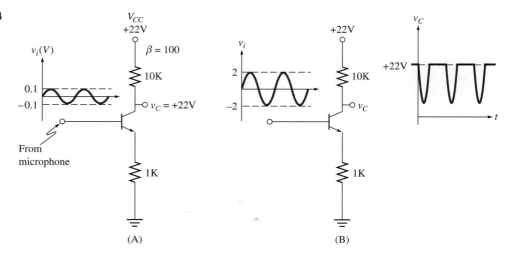

FIGURE 10–5

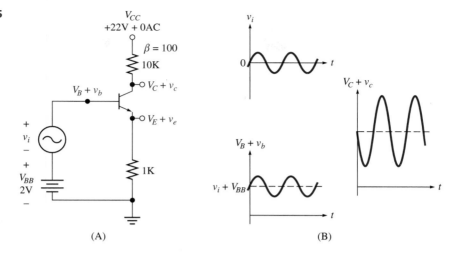

(A)   (B)

and the battery bias voltage of V_{BB}. The total base voltage now swings up and down from a fixed DC level.

We can also represent all other voltages in this circuit as a combination of a DC and an AC signal (Figure 10–5B); that is

☐ total base voltage = $V_B + v_b$

☐ total collector voltage = $V_C + v_c$

☐ total emitter voltage = $V_E + v_e$

We also find it instructive to represent the collector supply voltage as

☐ total collector supply voltage = $V_{CC} + 0V_p$

This reminds us that a DC supply voltage is the combination of a DC voltage and 0V AC voltage.

Rather than analyze the total circuit in Figure 10–5A, it is easier to analyze two separate circuits, the DC bias circuit and the AC amplifier circuit. This is equivalent to applying superposition to the transistor circuit. To find the bias levels, we first set all AC voltages to 0V. The important fact here is that if a voltage source is set to 0V, we replace it with a short circuit. Setting all AC voltages to 0V results in the DC bias circuit of Figure 10–6. This is now the same type of circuit that we analyzed in the last chapter. We find the DC voltage and current levels as follows:

$$V_B = V_{BB} = 2V$$

FIGURE 10–6

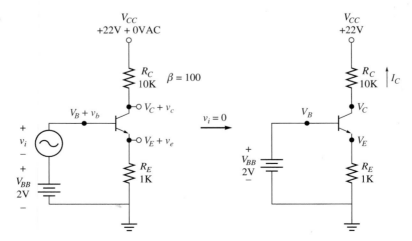

As in all V_B control bias circuits, the base voltage controls the emitter voltage:

$$V_E = V_B - 0.7 = 2 - 0.7 = 1.3\text{V}$$

The emitter voltage controls the emitter current:

$$I_E = \frac{V_E}{R_E} = \frac{1.3}{1\text{K}} = 1.3\text{mA}$$

And if the transistor is **ACTIVE,** the emitter current controls the collector current, which controls the base current:

$$I_C = I_E = 1.3\text{mA}$$

$$I_B = \frac{I_C}{\beta} = 0.013\text{mA}$$

Finally, we find the collector voltage from

$$V_C = V_{CC} - R_C I_C = 22 - 10\text{K} \times 1.3\text{mA} = 9\text{V}$$

Because $V_C > V_E$, the transistor is **ACTIVE** and we can proceed. These are the DC bias levels for this circuit. When the AC signal is applied to the base, the collector voltage will swing up and down from its bias level.

The DC bias levels in a transistor are also known as **quiescent levels,** and DC values are sometimes labeled I_{CQ} and V_{CQ}. The term *quiescent* means *still* or *quiet.* Because DC levels don't vary, DC is considered still, or quiescent. That is

$$I_C \equiv I_{CQ}$$
$$V_C \equiv V_{CQ}$$

EXAMPLE 10–2
Bias Level
Determination

Find the bias voltages and currents for the circuit shown in Figure 10–7A. Determine the state of the transistor.

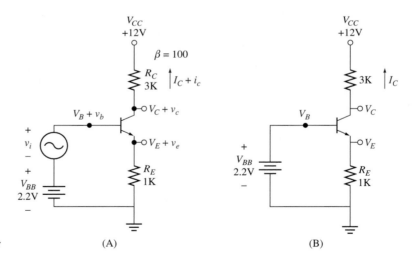

FIGURE 10–7 (A) (B)

Solution   We first set all AC voltages to 0V to create the DC bias circuit of Figure 10–7B. The analysis now proceeds:

$$V_B = V_{BB} = 2.2\text{V}$$

The base voltage controls the emitter voltage:

$$V_E = V_B - 0.7 = 2.2 - 0.7 = 1.5\text{V}$$

The emitter voltage controls the emitter current:

$$I_E = \frac{V_E}{R_E} = \frac{1.5}{1K} = 1.5\text{mA}$$

If **ACTIVE,** the emitter current controls the collector current, which controls the base current:

$$I_C = I_E = 1.5\text{mA}$$

$$I_B = \frac{I_C}{\beta} = 0.015\text{mA}$$

The collector bias voltage is

$$V_C = V_{CC} - R_C I_C = 12 - 3K \times 1.5\text{ma} = 7.5\text{V}$$

Because $V_C > V_E$, the transistor is **ACTIVE.**

DRILL EXERCISE

In the circuit in Figure 10–7, change the base supply voltage to 3.2V and redraw the circuit. Find the DC bias values for all transistor voltages and currents. What is the state of the transistor?

Answer $V_B = 3.2\text{V}$, $V_E = 2.5\text{V}$, $V_C = 4.5\text{V}$

$I_E = 2.5\text{mA}$, $I_C = 2.5\text{mA}$, $I_B = 0.025\text{mA}$

ACTIVE

DC Bias and Output Voltage Limits

Why does the electronic designer choose a specific bias level for her transistor circuit? You can find part of the answer by examining the relation between the DC bias level and the maximum change in the collector output voltage. Consider the circuit shown in Figure 10–8A. We find the DC bias levels from

$$V_B = V_{BB} = 1.7\text{V}$$
$$V_E = V_B - 0.7 = 1\text{V}$$
$$I_C = I_E = \frac{V_E}{R_E} = \frac{1}{1K} = 1\text{mA}$$

FIGURE 10–8

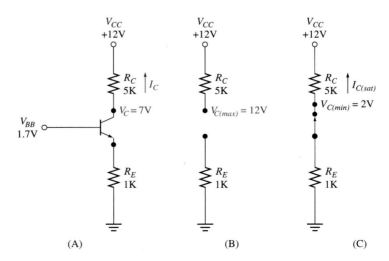

(A)   (B)   (C)

The bias level for the collector voltage is, therefore

$$V_C = V_{CC} - R_C I_C = 12 - 5K \times 1\text{mA} = 7\text{V}$$

Given that the collector voltage starts at 7V, how much higher and how much lower can V_C get? We know that as base voltage decreases, the transistor will go into **CUT-OFF**. As base voltage increases, the transistor will go into **SATURATION**. To remain a linear amplifier without distortion, the transistor must stay in the **ACTIVE** region as the base voltage varies. To determine the maximum and minimum values of V_C for linear operation, therefore, we must find the collector voltage as the transistor just goes into **CUT-OFF** and **SATURATION**. We will find, therefore, the collector voltages at the edge of **CUT-OFF** (Figure 10–8B, p. 311) and the edge of **SATURATION** (Figure 10–8C, p. 311).

□ Edge of **CUT-OFF:** When the transistor enters **CUT-OFF,** transistor current goes to 0A and the collector voltage rises to the supply voltage. That is, when $I_C = 0$

$$V_{Cmax} = V_{CC}$$

So

$$V_{Cmax} = V_{CC} = 12\text{V}$$

Note that the collector voltage is maximum when the transistor is **CUT-OFF.**

□ Edge of **SATURATION:** When the transistor first enters **SATURATION,** the collector voltage comes down to the emitter voltage; that is, the collector shorts to the emitter. Figure 10–8C (p. 311) shows that in **SATURATION** the collector circuit looks as though it is a voltage divider. That is, in **SATURATION**

$$V_{Cmin} = \frac{R_E}{R_E + R_C} V_{CC}$$

So

$$V_{Cmin} = \frac{R_E}{R_E + R_C} V_{CC} = \frac{1K}{1K + 5K} 12 = 2\text{V}$$

The collector voltage is smallest when the transistor just goes into **SATURATION.**

Note that modeling the collector circuit as a voltage divider is accurate only at the edge of **SATURATION.** At that point the base current is still very small, so we can ignore its contribution to the emitter current. Deeper into **SATURATION,** the base current can get large, so we must take it into account. (This will not be a problem for us in this chapter.) Remember that when used as an amplifier, the transistor is designed to stay in the **ACTIVE** region.

For this circuit, collector voltage can increase from 7V to 12V; that is, V_C can swing up 5V. Collector voltage can also decrease from 7V to 2V; that is, V_C can swing down 5V. In this circuit, collector voltage can swing symmetrically up and down by 5V. Figure 10–9 shows that for a moderate sinusoidal swing in base voltage, collector voltage also swings sinusoidally. For large swings in base voltage, however, the collector voltage clips at 12V on the top and 2V on the bottom.

We have just described a circuit in which the collector voltage output can swing symmetrically up and down by 5V. In many circuits, however, the collector voltage may be able to rise higher than it falls or vice-versa. In these circuits the maximum symmetrical swing is defined by the smallest amount that the collector voltage can change before clipping at the top or the bottom. This is demonstrated in the next two examples.

FIGURE 10–9

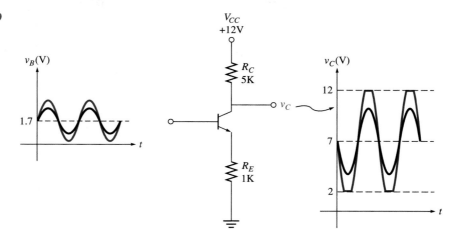

EXAMPLE 10–3
Maximum Voltage
Swing I

Find the bias levels and the maximum symmetrical collector voltage swing for the circuit shown in Figure 10–10

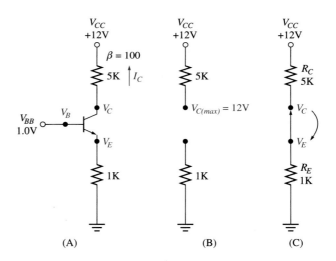

FIGURE 10–10   (A)   (B)   (C)

Solution   The bias levels are found from

$$V_B = V_{BB} = 1\text{V}$$
$$V_E = V_B - 0.7 = 1 - 0.7 = 0.3\text{V}$$
$$I_E = \frac{V_E}{R_E} = \frac{0.3}{1\text{K}} = 0.3\text{mA}$$

If **ACTIVE**

$$I_C = I_E = 0.3\text{mA}$$

and

$$I_B = \frac{I_C}{\beta} = 0.003\text{mA}$$

The collector bias voltage is

$$V_C = V_{CC} - R_C I_C = 12 - 5\text{K} \times 0.3\text{mA} = 10.5\text{V}$$

Because $V_C > V_E$, our assumption that the transistor is **ACTIVE** is correct. With this base voltage we are closer to **CUT-OFF** than in Figure 10–8 (p. 311).

We now find the maximum collector voltage from the **CUT-OFF** circuit of Figure 10–10B:

$$V_{Cmax} = V_{CC} = 12V$$

The collector voltage can rise from 10.5V to 12V, for a positive swing of 1.5V. The minimum collector voltage is found from the **SATURATED** circuit of Figure 10–10C.

$$V_{Cmin} = \frac{R_E}{R_E + R_C}V_{CC} = \frac{1K}{1K + 5K}12 = 2V$$

This tells us that the collector voltage can drop from 10.5V to 2V, for a negative swing of 8.5V. However, because the positive swing is only 1.5V, the maximum possible symmetrical swing is 1.5V.

Note that in this circuit, we have biased the transistor closer to **CUT-OFF.** That is, V_{BB} is smaller than in Figure 10–8. This means that the transistor in this circuit will reach **CUT-OFF** sooner than in earlier examples.

EXAMPLE 10–4
Maximum Voltage Swing II

Find the bias levels and the maximum symmetrical collector voltage swing for the circuit in Figure 10–11A.

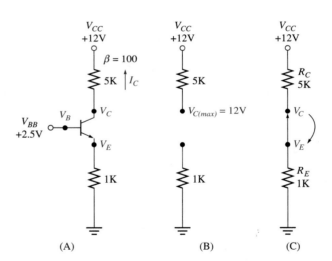

FIGURE 10–11   (A)   (B)   (C)

Solution   The bias levels are found from

$$V_B = V_{BB} = 2.5V$$
$$V_E = V_B - 0.7 = 2.5 - 0.7 = 1.8V$$

If **ACTIVE**

$$I_C = I_E = 1.8mA$$
$$I_B = \frac{I_C}{\beta} = 0.018mA$$

Finally, the collector bias voltage is

$$V_C = V_{CC} - R_C I_C = 12 - 5K \times 1.8mA = 3V$$

Because the $V_C > V_E$, the transistor is **ACTIVE**. With this base voltage, however, the collector voltage is now closer to the emitter voltage than in Figure 10–8. We are closer to **SATURATION** here.

We now find the maximum collector voltage from the **CUT-OFF** circuit of Figure 10–11B:

$$V_{Cmax} = V_{CC} = 12V$$

The collector voltage can rise from 3V to 12V, for a positive swing of 9V.

The minimum collector voltage is found from the **SATURATED** circuit of Figure 10–11C.

$$V_{Cmin} = \frac{R_E}{R_E + R_C}V_{CC} = \frac{1K}{5K + 1K}12 = 2V$$

The collector voltage can drop from 3V to 2V, for a negative swing of 1V. In this case, the negative swing is smaller. The maximum possible voltage swing is always the smaller of the positive and negative swings. In this example it equals 1V.

DRILL EXERCISE

For the circuit in Figure 10–11, change the collector resistor to $R_C = 3K\Omega$, and redraw the circuit.

(a) Find V_C, V_{Cmax}, V_{Cmin}.
(b) Find the maximum positive swing in the collector voltage.
(c) Find the maximum negative swing in the collector voltage.
(d) What is the maximum symmetrical swing?

Answer   (a) $V_C = 6.6V$, $V_{Cmax} = 12V$, $V_{Cmin} = 3V$
            (b) Maximum positive swing = 5.4V.
            (c) Maximum negative swing = 3.6V.
            (d) Maximum symmetrical swing = 3.6V.

REVIEW QUESTIONS

1. In what order do you find the bias levels of a V_B controlled transistor?
2. What is the correct setting for AC sources when you are doing DC bias calculations?
3. What happens to the collector voltage when the transistor goes into **SATURATION**?
4. What happens to the collector voltage when the transistor goes into **CUT-OFF?**
5. What determines the maximum symmetrical voltage swing at the collector?

10–3 ■ THE AC BEHAVIOR OF THE TRANSISTOR

As mentioned in the introduction, in Chapter 9 we found that

$$A = \frac{\Delta v_C}{\Delta v_B}$$

In this chapter we have dropped the subscript on the voltage gain *(A)* because we are only looking at the collector output. The voltage gains of Chapter 9 were found by letting the DC voltage at the base slowly rise or fall. We now want to find the gain as the base voltage varies sinusoidally.

In Chapter 7 we showed you how to model the **ACTIVE** transistor with a controlled current source $(I_C = \beta I_B)$ and a diode that represents the base-emitter PN junction (Figure 10–12A and B). Under DC conditions we can adequately model the diode with a battery representing the 0.7 turn-on voltage (Figure 10–12C).

Under **small-signal AC conditions,** the emitter diode does not behave as a 0.7V battery. As you shall see, small-signal AC refers to signals that cause only millivolt changes in the base-emitter voltage. To understand the small-signal behavior of the emitter diode, we must again look at the I-V characteristic of the PN junction.

FIGURE 10–12

(A)   (B)   (C)

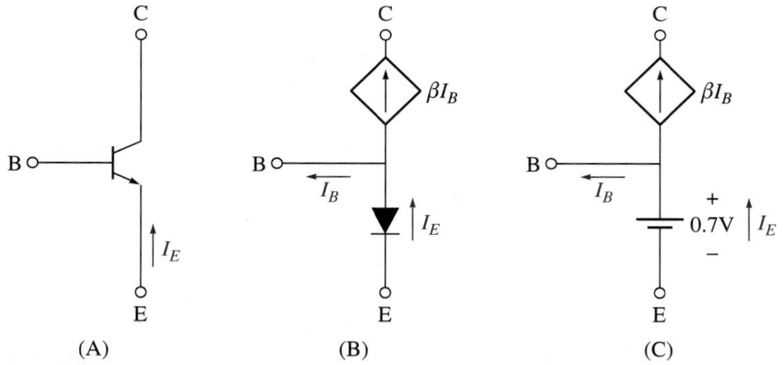

Because the base-emitter junction is similar to a diode, the relation between the emitter current and the base-emitter voltage looks familiar (Figure 10–13A). Compare this characteristic to the graph of current versus voltage for a resistor (Figure 10–13B). Because the resistor characteristic is a straight line, we can write the simple expression

$$I_R = \frac{1}{R}V_R$$

That is, the slope represents the inverse of the resistance. The inverse of resistance is termed **conductance** and is measured in Siemens (S).

FIGURE 10–13

(A)   (B)

The I_E - V_{BE} graph, on the other hand, is not a straight line; it is not linear. As a result, we cannot write a simple expression between I_E and V_{BE}. However, if we limit ourselves to small changes in I_E and V_{BE}, we can approximate small segments of the total I-V curve as straight lines. This is shown in Figure 10–14A for two values of I_E, 0.65mA and 3.2mA.

In Figure 10–14B we have isolated these two segments. One of these is centered on $I_E = 0.65$mA and the other on $I_E = 3.2$mA. The slope of each segment is found from

$$\text{slope} = \frac{\Delta i_E}{\Delta v_{BE}} = \frac{i_e}{v_{be}}$$

Because we are dealing with small variations, the voltage and current changes here are considered as small-signal AC quantities. The ratio of current to voltage can be represented as a conductance.

FIGURE 10–14

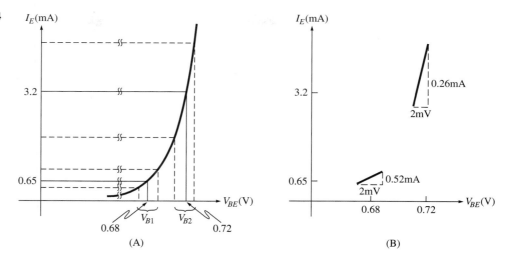

(A)   (B)

We can also get the same information about the line segments by using the ratio v_{be}/i_e. Because voltage over current represents resistance, we can model the small-signal AC behavior of the emitter diode with an AC emitter resistance (r_e) (Figure 10–15C):

$$r_e = \frac{v_{be}}{i_e}$$

For the straight line segments shown in Figure 10–14B, we find

$$I_E = 0.65\text{mA}: r_e = \frac{v_{be}}{i_e} = \frac{2\text{mV}}{0.052\text{mA}} = 38.5\Omega$$

and for

$$I_E = 3.2\text{mA}: r_e = \frac{v_{be}}{i_e} = \frac{2\text{mV}}{0.26\text{mA}} = 7.69\Omega$$

FIGURE 10–15

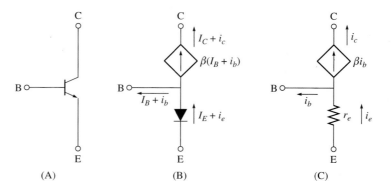

(A)   (B)   (C)

As you can see, the AC emitter resistance (r_e) does not have a fixed value but depends on the DC bias level of I_E. As the DC level of the emitter current increases, r_e decreases.

This is fine, you may be saying to yourself, but I really, really don't want to have to work with the PN junction curve in Figure 10–14. It's hard to find the curve for a given transistor, and the mathematics are messy. Fortunately, there is a straightforward formula for calculating the AC emitter resistance. At room temperature the AC emitter resistance is

$$r_e = \frac{26\text{mV}}{I_E}$$

where I_E is the DC bias level of the emitter current.

For the numbers used in Figure 10–14, we get

$$I_E = 0.65\text{mA}: r_e = \frac{26\text{mV}}{0.65\text{mA}} = 40\Omega$$

and for

$$I_E = 3.2\text{mA}: r_e = \frac{26\text{mV}}{3.2\text{mA}} = 8.13\Omega$$

which are very close to the values calculated previously.

As you have seen, we usually work with collector current rather than emitter current. This is no problem, because in the **ACTIVE** transistor $I_C \approx I_E$, and we find r_e from

$$r_e = \frac{26\text{mV}}{I_C}$$

EXAMPLE 10–5
Small-Signal AC Emitter Resistor

Find r_e for the following DC bias collector currents:

$$I_C = 10\text{mA, 1mA, 0.01mA, 1}\mu\text{A, and 1pA}$$

Solution   Because

$$r_e = \frac{26\text{mV}}{I_C}$$

we get

I_C	$r_e(\Omega)$
10mA	2.6
1mA	26
0.01mA	2.6K
1μA	26K
1pA	26M

The results of this example show that as the collector bias current decreases, the emitter resistance increases. For bias currents in the mA range, r_e is small. As the current decreases into the μA and pA range, r_e can get very large, reaching into the megaohm range.

If you take another look at the transistor AC model shown in Figure 10–15C, you will see that the collector current is still β times the base current. That is

$$i_c = \beta i_b$$

Technically, a transistor's DC current gain is different than its AC current gain. For example, the DC β may be 100, while the AC β can vary from 50 to 250. It is common practice, however, to assume that the DC and AC current gains are the same and use the same value of β for both DC and AC analysis. You must always keep in mind that any answer that depends on β will never be exact.

REVIEW QUESTIONS

1. How do we model the emitter diode under AC conditions?
2. What is the formula for the AC emitter resistor (use I_C)?
3. Under AC conditions, what is the relation between the collector current and the base current?

10–4 ■ THE COMMON-EMITTER AMPLIFIER

Figure 10–16 shows our basic transistor AC amplifier circuit. Here, base voltage is input voltage, and collector voltage is output voltage. This type of circuit is known as the **common-emitter amplifier.**

FIGURE 10–16

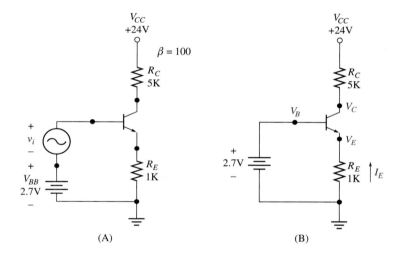

(A)   (B)

Even though we are primarily interested in the AC gain between the input and output voltages, we must always do a DC analysis first, to establish the bias level of the circuit. For the DC analysis, we set all AC sources to 0V, and we replace all 0-valued voltage sources with short circuits (Figure 10–16B). The DC bias level is found from

$$V_E = V_B - 0.7 = V_{BB} - 0.7 = 2.7 - 0.7 = 2V$$

$$I_E = \frac{V_E}{R_E} = \frac{2}{1K} = 2mA$$

If the transistor is **ACTIVE**

$$I_C = I_E = 2mA$$

$$I_B = \frac{I_C}{\beta} = 0.02mA$$

Finally, we find the collector voltage:

$$V_C = V_{CC} - R_C I_C = 24 - 5K \times 2mA = 14V$$

Because $V_C > V_E$, the transistor is **ACTIVE,** and we can proceed with the AC analysis. From the bias level of the collector current, we find the AC emitter resistance:

$$r_e = \frac{26mV}{I_C} = \frac{26mV}{2mA} = 13\Omega$$

At this point we could replace the transistor with its AC model (see Figure 10–15). However, you will probably find it easier to modify the original circuit to incorporate the AC properties of the transistor.

Figure 10–17A shows the original circuit with the total voltages, DC + AC at the base, collector, and emitter nodes. We also show that the AC component of V_{CC} is $0V_p$. To create the AC circuit, we need to make only two modifications to the original circuit. First, we set all DC voltages to 0V (Figure 10–17B). (Note that batteries are shorted, and we do not use the 0.7V emitter diode turn-on voltage in the AC model.)

FIGURE 10–17

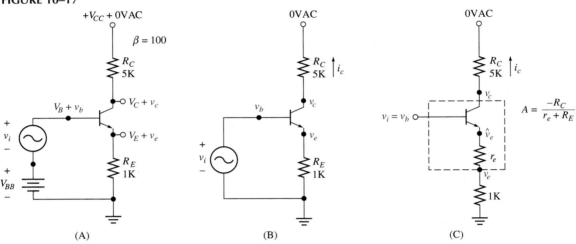

(A)   (B)   (C)

Second, and more important, we need to incorporate the AC emitter resistance into the circuit. The simplest way to do this is to add r_e to the emitter leg (Figure 10–17C). We see that this resistance is in series with the external emitter resistor (R_E). Therefore, the total resistance in the emitter is now

$$\text{total emitter resistance} = R_{Etotal} = r_e + R_E$$

Note that in this model

$$v_{\hat{e}} = v_b$$

We find the small-signal voltage gain as follows. The collector current equals the emitter current and is found from

$$i_c = i_e = \frac{v_{\hat{e}}}{r_e + R_E} = \frac{v_b}{r_e + R_E} = \frac{v_b}{R_{Etotal}}$$

Noting the polarities of v_c and i_c, the collector voltage is

$$v_c = -R_C i_c = \frac{-R_C}{R_{Etotal}} v_b$$

So the gain between the base and the collector is

$$A = \frac{v_c}{v_b} = -\frac{R_C}{R_{Etotal}}$$

and because $v_b = v_i$,

$$A = \frac{v_c}{v_i} = -\frac{R_C}{R_{Etotal}}$$

This is the same gain that we have already found; now, however, we use R_{Etotal}. For the values of r_e that we found from the DC analysis of Figure 10–16, we get

$$R_{Etotal} = r_e + R_E = 13 + 1K = 1.01K$$

So the gain in this circuit is

$$A = \frac{R_C}{R_{Etotal}} = -\frac{5K}{1.01K} = -4.95$$

You can see that if $r_e \ll R_E$, then r_e has very little effect on the gain.

EXAMPLE 10–6
The Common-Emitter Amplifier I

Find the AC voltage gain at the collector for the circuit in Figure 10–18.

Solution First, we must find the bias levels and determine r_e from the DC circuit of Figure 10–18B.

$$V_E = V_B - 0.7 = V_{BB} - 0.7 = 2.5 - 0.7 = 1.8V$$

$$I_E = \frac{V_E}{R_E} = \frac{1.8}{2K} = 0.9mA$$

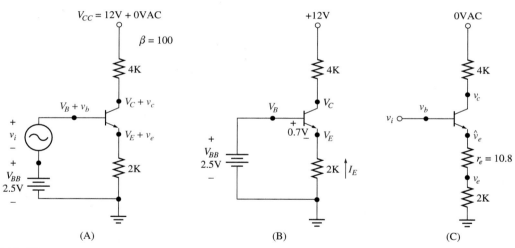

FIGURE 10–18

If the transistor is **ACTIVE**

$$I_C = I_E = 0.9mA$$

$$I_B = \frac{I_C}{\beta} = 0.009mA$$

We now find the collector voltage:

$$V_C = V_{CC} - R_C I_C = 12 - 4K \times 0.9mA = 0.4V$$

Because $V_C > V_E$, the transistor is **ACTIVE,** so we can proceed with the AC analysis. The AC emitter resistance is

$$r_e = \frac{26mV}{0.9mA} = 28.9\Omega = 0.0289K\Omega$$

The AC voltage gain (Figure 10–18C) is

$$A = -\frac{R_C}{R_{Etotal}} = -\frac{4K}{0.0289K + 2K} = -1.97$$

Again, because it is so small, the AC emitter resistance had little effect in this example.

EXAMPLE 10–7
The Common Emitter Amplifier II

Find the AC voltage gain at the collector for the circuit in Figure 10–19.

Solution First, we must find the bias levels and determine r_e from the DC circuit of Figure 10–19B.

$$V_E = V_B - 0.7 = V_{BB} - 0.7 = 1 - 0.7 = 0.3V$$

$$I_E = \frac{V_E}{R_E} = \frac{0.3}{2K} = 0.15mA$$

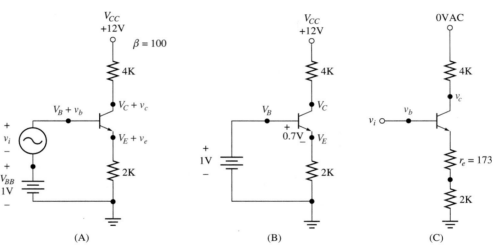

(A)   (B)   (C)

FIGURE 10–19

If the transistor is **ACTIVE**

$$I_C = I_E = 0.15mA$$

$$I_B = \frac{I_C}{\beta} = 0.0015mA$$

The collector current is found from

$$V_C = V_{CC} - R_C I_C = 12 - 4K \times 0.15mA = 11.4V$$

Because $V_C > V_E$, the transistor is **ACTIVE,** and we can proceed with the AC analysis. The AC emitter resistance is

$$r_e = \frac{26mV}{0.15mA} = 173\Omega = 0.173K\Omega$$

The AC voltage gain (Figure 10–19C) is

$$A = -\frac{R_C}{R_{Etotal}} = -\frac{4K}{0.173K + 2K} = -1.84$$

In this example, because of a small bias current, the larger AC emitter resistance reduced the gain.

DRILL EXERCISE   For the circuit in Figure 10–19, change the base supply voltage to $V_{BB} = 0.75V$, and re-draw the circuit. Find r_e and A.

Answer $r_e = 1.04K\Omega$, $A = -1.32$

REVIEW QUESTIONS

1. What is the total emitter resistance?
2. What is the voltage gain at the collector in terms of the total emitter resistance?
3. For collector bias current in the milliamp range, does the voltage gain depend very much on r_e? Why?

10–5 ■ THE EMITTER BYPASS CAPACITOR

Many discrete transistor amplifiers are constructed as shown in Figure 10–20. The capacitor that is connected in parallel with the emitter resistor is known as a **bypass capacitor** (C_E). To understand the reason for this name and to determine how the capacitor affects the gain, we first review the behavior of capacitors.

FIGURE 10–20

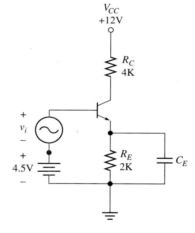

The Capacitor

In Chapter 3 we used the charge storage capability of the capacitor to build several important diode circuits. Here we are interested in how the capacitor behaves under DC and AC operating conditions.

The current-voltage relation for the capacitor is shown in Figure 10–21A. You know that the charge stored by a capacitor depends on the voltage across the capacitor. That is

$$Q = CV_{cap}$$

FIGURE 10–21

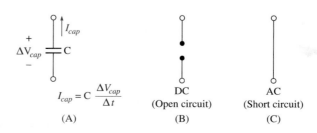

$$I_{cap} = C\,\frac{\Delta V_{cap}}{\Delta t}$$

(A)

DC (Open circuit) (B)

AC (Short circuit) (C)

where Q is the charge stored by the capacitor. Electrical current is a measure of how fast charges are moving and is defined as

$$I_{cap} = \frac{\Delta Q}{\Delta t}$$

Therefore, to find the capacitor current

$$I_{cap} = \frac{\Delta Q}{\Delta t} = \frac{\Delta(CV_{cap})}{\Delta t} = C\frac{\Delta V_{cap}}{\Delta t}$$

You can see that capacitor current depends on how fast capacitor voltage changes.

What is the capacitor current if the voltage across the capacitor is, for example, 5VDC? Because DC voltages don't change, we get

$$I_{cap} = C\frac{\Delta V_{cap}}{\Delta t} = C\frac{0}{\Delta t} = 0\text{A}$$

This tells us that because DC voltages don't change and because capacitor current depends on how fast capacitor voltages change, capacitor current is always 0A for DC voltages. At DC, therefore, all capacitors can be modeled as open circuits (Figure 10–21B, p. 323).

Capacitors are usually rated for the maximum DC voltage that they can handle. For this reason, it is a good idea always to determine the DC voltage that you expect to see across a capacitor.

The behavior of capacitors at AC is more complicated. Simply stated, under AC conditions the capacitor exhibits an AC "resistance," known as an **impedance** *(Z)*. The magnitude of this impedance is given by

$$|Z| = \frac{|V|}{|I|} = \frac{1}{2\pi f C}$$

where $|Z|$ is measured in Ohms, and f is the frequency (in Hz) of the AC voltage.

The formula for capacitor impedance shows that as the frequency of capacitor voltage increases, capacitor impedance decreases. We know that as resistance decreases, current increases (i.e., $I = V/R$). Similarly, as impedance decreases, the size of the current increases. This is consistent with our definition of capacitor current. As the frequency of capacitor voltage increases, capacitor voltage changes more and more rapidly. As the rate of change of capacitor voltage increases, capacitor current increases.

Let's return now to capacitor impedance to see what happens as we increase the frequency towards infinity:

$$\text{as } f \to \infty \quad |Z| = \frac{1}{2\pi f C} \to \frac{1}{2\pi \infty C} \to 0\Omega$$

That is, as frequency increases, impedance decreases towards 0Ω. Because a short circuit has 0Ω, we can model the capacitor at high frequencies as a short circuit (Figure 10–21C, p. 323). How high is high? In general, it depends on the particular capacitor circuit. For our purposes, however, we will assume that any AC voltage has a high enough frequency to short the capacitor. (In a later chapter we will treat the capacitor with more respect.) For now, all you really need to know about the capacitor is the following:

☐ DC: replace all capacitors with an open circuit.
☐ AC: replace all capacitors with a short circuit.

The Emitter Bypass Capacitor

We are now ready to return to the emitter bypass capacitor. Consider again the circuit of Figure 10–20, redrawn in Figure 10–22A. Now, for DC we make two modifications: turn off all AC sources, and open all capacitors—that is, remove them from the circuit

(Figure 10–22B). Note that capacitors do not affect DC bias levels, which we find from

$$V_E = V_B - 0.7 = V_{BB} - 0.7 = 4.5 - 0.7 = 3.8V$$

$$I_E = \frac{V_E}{R_E} = \frac{3.8}{2K} = 1.9mA$$

FIGURE 10–22

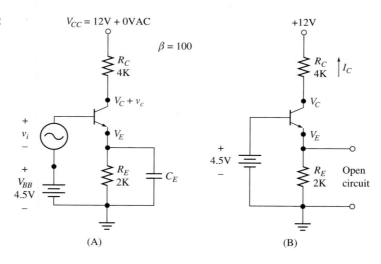

(A)   (B)

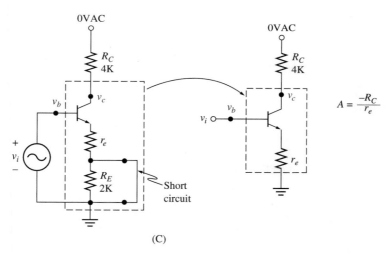

(C)

If the transistor is **ACTIVE**

$$I_C = I_E = 1.9mA$$

$$I_B = \frac{I_C}{\beta} = 0.019mA$$

The collector voltage is found from

$$V_C = V_{CC} - R_C I_C = 12 - 4K \times 1.9mA = 4.4V$$

Because $V_C > V_E$, the transistor is **ACTIVE.**

The DC voltage across the emitter bypass capacitor is found from

$$V_{C_E} = V_E = 3.8V$$

The AC emitter resistance is

$$r_e = \frac{26mV}{1.9mA} = 13.6K\Omega = 0.0137K\Omega$$

The emitter resistance is very small here, so we can ignore it, can't we? Not in this case. To see why, we now analyze the circuit under AC conditions. That is, we short all DC sources, insert the emitter resistance (r_e) into the emitter leg, and, in addition, short all capacitors (Figure 10–22C, p. 325).

Note that under AC conditions, the emitter resistor (R_E) is now shorted. All current approaching R_E now bypasses the resistor and instead goes through the shorted capacitor. Now you know why C_E is known as a bypass capacitor.

Figure 10–22C shows that the shorted capacitor effectively removes R_E from the circuit. The only resistor left in the emitter leg is r_e. This is why, no matter how small r_e is, it cannot be ignored if there is an emitter bypass capacitor.

With the bypass capacitor, the total emitter resistance is now

$$R_{Etotal} = r_e$$

So the voltage gain for the common-emitter amplifier with an emitter bypass capacitor is given by

Bypass capacitor $\qquad A = -\dfrac{R_C}{R_{Etotal}} = -\dfrac{R_C}{r_e}$

Because R_{Etotal} is now small, we get a very large voltage gain. For the circuit in Figure 10–22, the gain is

$$A = -\frac{R_C}{r_e} = -\frac{4\text{K}}{0.0137} = -292$$

Compare this gain with the gain of approximately -5 we achieved when we did not use the bypass capacitor.

EXAMPLE 10–8
The Emitter By-pass Capacitor

For the circuit in Figure 10–23, find the DC bias levels, the DC voltage across C_E, the AC emitter resistance (r_e), and the voltage gain. What would be the voltage gain if the bypass capacitor was removed from the circuit?

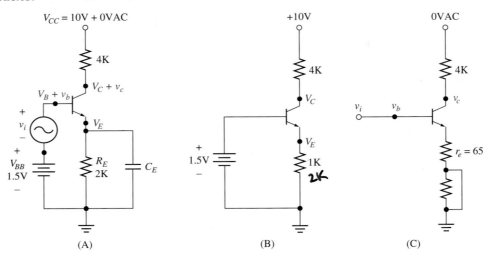

FIGURE 10–23

Solution For the DC analysis we turn off AC sources and open the bypass capacitor (Figure 10–23B).

$$V_E = V_B - 0.7 = V_{BB} - 0.7 = 1.5 - 0.7 = 0.8\text{V}$$

$$I_E = \frac{V_E}{R_E} = \frac{0.8}{2\text{K}} = 0.4\text{mA}$$

If the transistor is **ACTIVE**

$$I_C = I_E = 0.4\text{mA}$$

$$I_B = \frac{I_C}{\beta} = 0.004\text{mA}$$

The collector voltage is found from

$$V_C = V_{CC} - R_C I_C = 10 - 4\text{K} \times 0.4\text{mA} = 8.4\text{V}$$

Because $V_C > V_E$ the transistor is **ACTIVE.** The DC voltage across C_E is

$$V_{C_E} = V_E = 0.8\text{V}$$

The AC emitter resistance is

$$r_E = \frac{26\text{mV}}{I_C} = \frac{26\text{mV}}{0.4\text{mA}} = 65\Omega = 0.065\text{K}\Omega$$

For the AC analysis we turn off all DC sources, insert r_e in the emitter leg, and short the capacitor (Figure 10–23C). The gain is

$$A = -\frac{R_C}{R_{Etotal}} = -\frac{R_C}{r_e} = -\frac{4\text{K}}{0.065\text{K}} = -61.5$$

If the emitter bypass capacitor is removed, then

$$R_{Etotal} = r_e + R_E = 0.065\text{K} + 2\text{K} = 2.07\text{K}\Omega$$

and the gain would be

$$A = -\frac{4\text{K}}{2.07\text{K}} = -1.93$$

The presence of the emitter bypass capacitor greatly increases the gain of the transistor amplifier.

DRILL EXERCISE

For the circuit shown in Figure 10–23, change the base supply voltage to $V_{BB} = 1\text{V}$, and redraw the circuit. Find the AC emitter resistance (r_e) and the voltage gain (A).

Answer $r_e = 0.173\text{K}\Omega$ $A = -23.1$

REVIEW QUESTIONS

1. How do we model capacitors under DC conditions?
2. How do we model capacitors under AC conditions?
3. What is the DC voltage across C_E?
4. What is the total emitter resistance when R_E is bypassed with a capacitor?
5. What is the voltage gain when R_E is bypassed with a capacitor?
6. What happens to the voltage gain as the DC bias level of I_C increases? Why?
7. What happens to the voltage gain as the DC bias level of I_C decreases? Why?

10–6 ▪ INPUT CAPACITOR COUPLING—THE SELF-BIASED CIRCUIT

The common-emitter amplifier is usually built as a self-biased circuit (Figure 10–24, p. 328). In this circuit we place a coupling capacitor (C_i) between the AC voltage source and the base. Figure 10–24B shows why this capacitor is needed. Under DC conditions, we

FIGURE 10–24

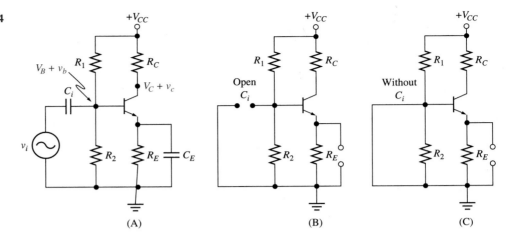

(A)   (B)   (C)

short all AC sources. If there is no coupling capacitor, shorting the input AC source will short out R_2. This effectively ties the base to ground, putting 0VDC at the base. The transistor will be **CUT-OFF** and will not work as a linear amplifier.

Figure 10–25 shows how the input coupling capacitor solves the DC bias problem. Under DC conditions we open all capacitors, which disconnects the AC source from the circuit (Figure 10–25B). The maximum DC voltage across the input coupling capacitor (C_i) is found when the AC signal generator is connected to the circuit and is set to 0V. The voltage across C_i is then the base voltage (V_B).

FIGURE 10–25

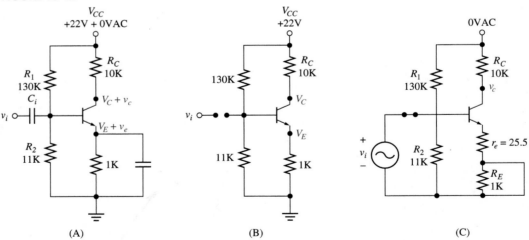

(A)   (B)   (C)

For the self-biased circuit we approximate the DC base voltage as

$$V_B = \frac{R_2}{R_1 + R_2} V_{CC} = \frac{11K}{11K + 130K}\, 22 = 1.72V$$

The rest of the DC analysis is

$$V_E = V_B - 0.7 = 1.7 - 0.72 = 1.02V$$

$$I_E = \frac{V_E}{R_E} = \frac{1}{1K} = 1mA$$

If the transistor is **ACTIVE**

$$I_C = I_E = 1.02\text{mA}$$

$$I_B = \frac{I_C}{\beta} = 0.0102\text{mA}$$

The collector voltage is

$$V_C = V_{CC} - R_C I_C = 22 - 10\text{K} \times 1.02\text{mA} = 11.8\text{V}$$

The voltage across C_i is

$$V_{C_i} = V_B = 1.7\text{V}$$

The AC emitter resistance is

$$r_e = \frac{26\text{mV}}{I_C} = \frac{26\text{mV}}{1.02\text{mA}} = 25.5\Omega = 0.0255\text{K}\Omega$$

For the AC analysis we turn off the DC supply (V_{CC}), insert r_e, and short both ca-pacitors. As before, the emitter resistor is bypassed. The only tricky part here is to recog-nize that when we short the input coupling capacitor (C_i), we directly connect the AC input voltage source to the base. That is

$$v_b = v_i$$

Note: For AC the base resistors R_1 and R_2 do not affect the base voltage, so they are not in-volved in the AC analysis.

The voltage gain is

$$A = -\frac{R_C}{R_{Etotal}} = -\frac{R_C}{r_e} = -\frac{10\text{K}}{0.0255\text{K}} = -392$$

EXAMPLE 10–9
The Input Coupling Capacitor

For the circuit in Figure 10–26, find the DC operating levels, the DC voltages across both capacitors, the AC emitter resistance, and the voltage gain.

Solution For the DC analysis we open all capacitors (Figure 10–26B) and use the voltage divider approximation to find

$$V_B = \frac{R_2}{R_1 + R_2} V_{CC} = \frac{30\text{K}}{30\text{K} + 240\text{K}} \, 9 = 1.0\text{V}$$

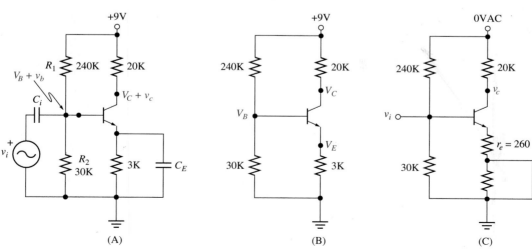

(A)   (B)   (C)

FIGURE 10–26

The rest of the DC analysis is

$$V_E = V_B - 0.7 = 1.0 - 0.7 = 0.3V$$

$$I_E = \frac{V_E}{R_E} = \frac{0.3}{3K} = 0.1mA$$

If the transistor is **ACTIVE**

$$I_C = I_E = 0.1mA$$

$$I_B = \frac{I_C}{\beta} = 0.001mA$$

The collector voltage is found from

$$V_C = V_{CC} - R_C I_C = 9 - 20K \times 0.1mA = 7V$$

Because $V_C > V_E$, the transistor is **ACTIVE.**
The DC voltages across C_i and C_E are

$$V_{C_i} = V_B = 1V$$
$$V_{C_E} = V_E = 0.3V$$

The AC emitter resistance is

$$r_e = \frac{26mV}{I_C} = \frac{26mV}{0.1mA} = 260\Omega = 0.26K\Omega$$

For the AC analysis, we turn off the DC sources, insert r_e, and short both capacitors (Figure 10–26C, p. 329). Noting that $v_b = v_i$, we find the gain to be

$$A = -\frac{R_C}{R_{Etotal}} = -\frac{R_C}{r_e} = -\frac{20K}{0.26K} = -76.9$$

DRILL EXERCISE

For the circuit in Figure 10–26, change the emitter resistor to $R_E = 6K\Omega$, and redraw the circuit. Find the AC emitter resistor (r_e) and the gain (A).

Answer $r_e = 520\Omega$, $A = -38.5$

REVIEW QUESTIONS

1. Why do we need to use a capacitor to couple the input to the base?
2. What do we do with all capacitors when we are finding the DC bias levels?
3. What do we do with all capacitors when we are finding the AC voltage gain?
4. What is the DC voltage across the input coupling capacitor (C_i)?

10–7 ■ THE LOAD RESISTOR

We have now built several transistor circuits that could be used to amplify the output of our opera diva's microphone. The amplifier's output voltage can now be used to drive a speaker so the patrons in the back row won't miss a warble. Figure 10–27 shows this system, in which the microphone provides v_i to the transistor amplifier.

Does the speaker we have connected to the transistor amplifier affect the gain of the amplifier? The answer is that yes it does. Any load we connect to the amplifier will affect the gain. To see this we will model the speaker—or any load—as a resistance (R_L). Figure

FIGURE 10–27

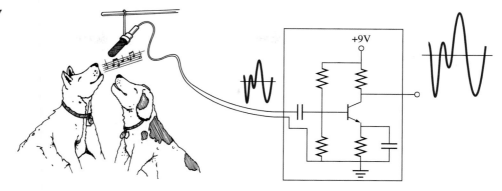

10–28 shows the resistor load coupled through a capacitor to the collector of the transistor. As a result, the load will not affect the circuit's DC bias levels. Figure 10–28B shows the DC circuit that we use for the following bias calculations. Note that we have opened all three capacitors and that v_i and R_L are not part of the DC circuit.

FIGURE 10–28

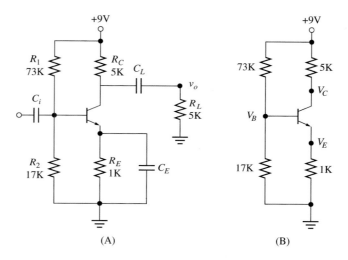

(A) (B)

$$V_B = \frac{R_2}{R_1 + R_2} V_{CC} = \frac{17K}{17K + 73K} \, 9 = 1.7V$$

The rest of the DC analysis is

$$V_E = V_B - 0.7 = 1.7 - 0.7 = 1.0V$$

$$I_E = \frac{V_E}{R_E} = \frac{1.0}{1K} = 1mA$$

If the transistor is **ACTIVE**

$$I_C = I_E = 1mA$$

$$I_B = \frac{I_C}{\beta} = 0.01mA$$

The collector voltage is found from

$$V_C = V_{CC} - R_C I_C = 12 - 5K \times 1mA = 7V$$

Because $V_C > V_E$, the transistor is **ACTIVE.**

The DC voltage across the output coupling capacitor is given by

$$V_{C_L} = V_C - V_o = V_C$$

because for DC the output voltage is 0V.

The AC emitter resistance is

$$r_e = \frac{26\text{mV}}{I_C} = \frac{26\text{mV}}{1.0\text{mA}} = 26\Omega = 0.026\text{K}\Omega$$

For the AC analysis we turn off the DC supply voltage, add the AC emitter resistance to the emitter leg, and short all capacitors (Figure 10–29A). You might notice something new in this figure. We have grounded the DC supply node. Remember that for the AC analysis we set DC voltages to 0V. Because the collector supply voltage is strictly DC, this voltage becomes 0V for AC analysis. As you know, ground points are at 0V; therefore, under AC conditions we have an effective ground at the collector supply node.

We have redrawn the AC collector circuit in Figure 10–29B. Because both R_C and R_L are connected to the collector on one side and ground on the other, these two resistors are in parallel. Total collector resistance is, therefore

$$R_{Ctotal} = R_C \parallel R_L = \frac{R_C R_L}{R_C + R_L}$$

FIGURE 10–29

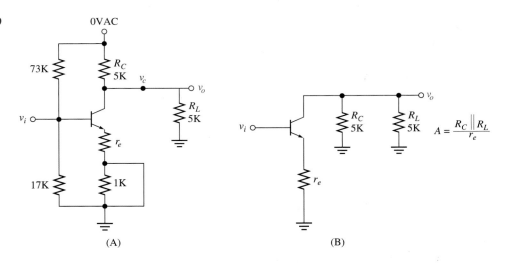

(A) (B)

The most general formula for the gain of the common-emitter amplifier is

$$A = -\frac{R_{Ctotal}}{R_{Etotal}}$$

For our circuit

$$R_{Ctotal} = R_C \parallel R_L = \frac{5\text{K} \times 5\text{K}}{5\text{K} + 5\text{K}} = 2.5\text{K}\Omega$$

and

$$R_{Etotal} = r_e = 0.026\text{K}\Omega$$

so the gain is

$$A = -\frac{R_{Ctotal}}{R_{Etotal}} = -\frac{2.5\text{K}}{0.026\text{K}} = -96.2$$

Note that the load resistor reduces the gain. This is the case for all common-emitter amplifiers.

EXAMPLE 10–10
The Load Resistor

For the circuit in Figure 10–30A, find the DC bias levels, the DC voltages across the capacitors, the AC emitter resistance, and the voltage gain.

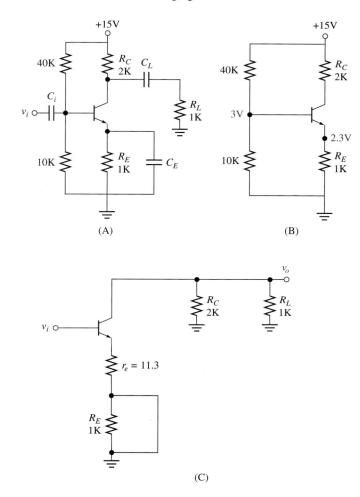

FIGURE 10–30   (C)

Solution   Figure 10–30B shows the DC circuit that we use for the following bias calculations. Note that we have opened all three capacitors and that v_i and R_L are not part of the DC circuit.

$$V_B = \frac{R_2}{R_1 + R_2} \, V_{CC} = \frac{10K}{10K + 40K} \, 15 = 3V$$

The rest of the DC analysis is

$$V_E = V_B - 0.7 = 3 - 0.7 = 2.3V$$

$$I_E = \frac{V_E}{R_E} = \frac{2.3}{1K} = 2.3mA$$

If the transistor is **ACTIVE**

$$I_C = I_E = 2.3mA$$

$$I_B = \frac{I_C}{\beta} = 0.023mA$$

The collector voltage is found from

$$V_C = V_{CC} - R_C I_C = 15 - 2K \times 2.3mA = 10.4V$$

Because $V_C > V_E$, the transistor is **ACTIVE.**
The DC voltages across the capacitors are

$$V_{C_i} = V_B = 3\text{V}$$
$$V_{C_E} = V_E = 2.3\text{V}$$
$$V_{C_L} = V_C = 10.4\text{V}$$

The AC emitter resistance is

$$r_e = \frac{26\text{mV}}{I_C} = \frac{26\text{mV}}{2.3\text{mA}} = 11.3\Omega = 0.0113\text{K}\Omega$$

For the AC analysis we turn off the DC supply voltage, add the AC emitter resistance to the emitter leg, and short all capacitors (Figure 10–30C). For this circuit

$$R_{Ctotal} = R_C \parallel R_L = 2\text{K} \parallel 1\text{K} = \frac{2\text{K} \times 1\text{K}}{2\text{K} + 1\text{K}} = 0.667\text{K}\Omega$$

and

$$R_{Etotal} = r_e = 0.0113\text{K}\Omega$$

The gain is

$$A = -\frac{R_{Ctotal}}{R_{Etotal}} = -\frac{0.667}{0.0113} = -59$$

DRILL EXERCISE

For the circuit in Figure 10–30 (p. 333), change the collector resistor to $R_C = 1\text{K}\Omega$, and redraw the circuit. Find the AC emitter resistor (r_e) and the gain (A).

Answer $r_e = 11.3\Omega$, $A = -42.2$

We have just discussed how the resistance of the load affects the overall gain of the common-emitter amplifier. There is, perhaps, a semantic problem with this discussion. After all, the amplifier you build or buy is a stand-alone unit. The load is not really part of the amplifier. The next chapter will explore how we can model the amplifier independently of the load and then separately analyze the effect of the load.

REVIEW QUESTIONS

1. Why do we use a capacitor to couple the load to the transistor?
2. What is the DC voltage across the output coupling capacitor?
3. What is the formula for R_{Ctotal}?
4. What is the formula for the voltage gain?
5. Does the load resistor reduce the voltage gain? Why?

10–8 ■ TROUBLESHOOTING

In Chapter 8 we introduced you to the technique of DC Troubleshooting. We then used the technique in Chapter 9 to locate problems in transistor bias circuits. In this chapter we are most interested in how the common-emitter amplifier processes AC signals. There are times, therefore, that we need to measure AC signals to locate a circuit problem. This process is called **signal tracing.** DC Troubleshooting, however, is still the primary method for locating problems in electronic circuits. If there is a problem with the DC bias in a circuit, the circuit will not properly amplify AC signals. If the DC bias levels in a circuit are correct, then the circuit will probably work. In some cases, even though the DC bias levels are correct, the circuit will not work for AC signals. These are the times when you must trace the AC signals through the circuit to find the problem. To do signal tracing, you must add the oscilloscope to your list of troubleshooting tools.

How can a circuit work for DC but not for AC? The answer is that these circuits contain capacitors, and capacitors behave differently under DC and AC conditions. Remember, capacitors act as open circuits for DC and short circuits for AC. Like any other electrical device, capacitors can fail. They can fail open in the following ways:

1. broken leads inside the capacitor;

2. bad connections to the capacitor, (e.g., cold solder joint), or

3. electrolytes can dry out.

When a capacitor fails, it usually fails open. However, occasionally capacitors can fail short if

1. external leads are touching each other;

2. an internal short occurs between plates; or

3. an electrolytic capacitor is connected backwards (the most common error).

Can DC Troubleshooting locate failed capacitors? Figure 10–31 shows a typical common-emitter amplifier circuit. Figure 10–31B shows the circuit with capacitors that

FIGURE 10–31

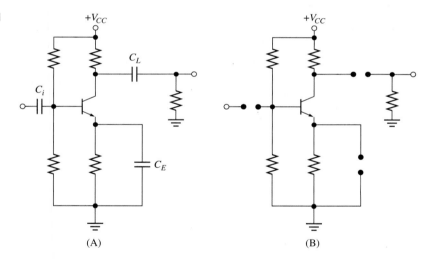

(A)   (B)

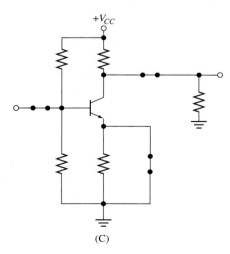

(C)

have failed open. What do we do when we find the DC bias levels in a circuit? We open the capacitors. Therefore, a capacitor that has failed open will not affect the DC levels in a circuit. DC Troubleshooting, therefore, will *not* detect a failed open capacitor. Then we must add AC signal tracing to our troubleshooting procedure.

Figure 10–31C (p. 335) shows the circuit with failed short capacitors. In this case, the unexpected shorts will definitely affect the DC behavior of the circuit so DC Troubleshooting will identify the problem.

You may be asking yourself that if the most common capacitor failure is to open, and if DC troubleshooting will not find this problem, why don't we just do AC signal tracing to troubleshoot amplifier circuits? The answer is that even in circuits with faulty DC behavior, some AC signals can get through the circuit. You may measure an AC signal and conclude that the circuit is working, when, in fact, DC Troubleshooting results would tell you that the circuit has a problem. We have often seen students waste an entire laboratory period making measurements on a circuit in which they see an AC output—an AC output that is meaningless.

There are several mechanisms that enable AC outputs to show up in circuits that actually are not working properly:

☐ AC signals can be conducted passively through capacitors and other components. Even a transistor that is not working can look like a lump of conducting material to AC signals.

☐ An improper connection to the DC power supply or to ground can lead to AC conduction through power supply or ground coupling.

☐ To find an output signal, students often turn up the gain of the oscilloscope until they see something. They then interpret what they see on the scope as an amplified signal, when, in fact, they are looking at a very small signal. It is the oscilloscope that has amplified the signal, not the faulty transistor circuit.

So far, in the linear amplifiers we have shown you in this chapter, we knew what AC signals we expect to see at each point in the circuit. Many electronic designs, however, incorporate non-linear circuits, such as the diode circuits discussed earlier in this textbook. In non-linear circuits, it is difficult to predict how an AC signal is supposed to look as it propagates through the circuit. Some common examples of non-linear circuits are

> AC-to-DC conversion
>
> peak detectors
>
> VU meters (used in stereo systems to indicate volume levels)
>
> multipliers, or mixers (used in radio circuits)
>
> sound-activated circuits ("Clap on—Clap off")
>
> infrared security devices
>
> amplifiers that are biased near **CUT-OFF** or **SATURATION**

In all cases, whenever you have to fix a circuit, your first step is to perform DC Troubleshooting. However, you must remember that bad DC measurements could be caused by shorted capacitors. If DC Troubleshooting does not show any discrepancies but the circuit still does not work, then you must trace the AC signals through the circuit.

EXAMPLE 10–11
Troubleshooting I

The transistor amplifier in Figure 10–32 was built and checked with DC Troubleshooting. The circuit was found to have good DC bias values. The AC voltage seen at the base, the collector, and at v_o was so small it could not be measured. What problem could account for these findings?

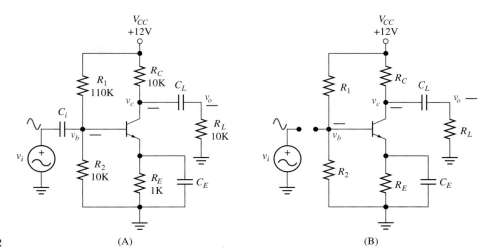

FIGURE 10–32   (A)   (B)

Solution Because DC Troubleshooting found no problems, we can conclude that one of the capacitors has failed. Because the AC signal did not even get to the base, we conclude that the signal source is not properly connected to the amplifier. The most likely cause of this problem is that the input capacitor (C_i) has failed open (Figure 10–32B).

EXAMPLE 10–12
Troubleshooting II

The transistor amplifier in Figure 10–33 was built and checked with DC Troubleshooting. The circuit was found to have good DC bias values. The AC voltage seen at the base was $10\,\text{mV}_\text{P}$. The AC voltage seen at the collector was $1.1\,\text{V}_\text{p}$. No AC voltage could be seen at v_o. What is the likely problem?

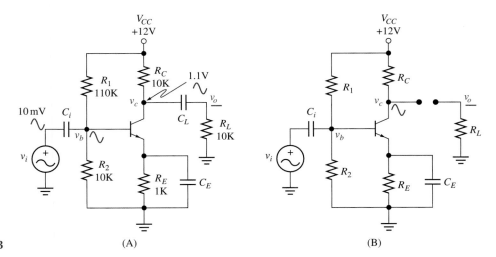

FIGURE 10–33   (A)   (B)

Solution Because DC troubleshooting found no problems, we can conclude that one of the capacitors has failed. The AC signal got to the base, was amplified by the transistor, and appeared at the collector. However, the AC signal did not appear at the load. We conclude that the load coupling capacitor has failed open (Figure 10–33B).

EXAMPLE 10–13
Troubleshooting III

The transistor amplifier in Figure 10–34 was built and checked with DC Troubleshooting. The circuit was found to have good DC bias values. The AC voltage seen at the base was 10mV$_p$. The AC voltage seen at the collector and at v_o was 100mV$_p$.

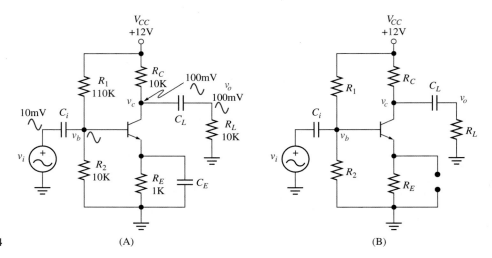

FIGURE 10–34   (A)   (B)

(a) What should the voltage gain be in this circuit?
(b) What should the AC voltage be at the output?
(c) What is the actual gain that you measured?
(d) What type of problem could account for the AC measurements?

Solution

(a) To find the expected voltage gain, we need to find r_e. We first do a DC analysis to find the DC collector current.

$$V_B = \frac{10K}{10K + 110K}\, 12 = 1V$$

$$V_E = V_B - 0.7 = 0.3V$$

$$I_E = \frac{V_E}{R_E} = \frac{0.3}{1K} = 0.3mA$$

If the transistor is **ACTIVE**

$$I_C = I_E = 0.3mA$$

The expected value for the collector voltage is

$$V_C = V_{CC} - R_C I_C = 12 - 10K \times 0.3mA = 9V$$

Because $V_C > V_E$, the transistor is **ACTIVE** and we can proceed. The AC emitter resistance is

$$r_e = \frac{26mV}{I_C} = \frac{26mV}{0.3mA} = 86.7\Omega = 0.0867K\Omega$$

The gain is found from

$$A = -\frac{R_{Ctotal}}{R_{Etotal}} = -\frac{R_C \parallel R_L}{r_e} = -\frac{5K}{0.0867K} = -57.7$$

(b) The input is $v_i = 10mV_p$. At the output we expect to see

$$v_o = Av_i = -57.7 \times 10mV_p = -0.577V_p$$

Note that the sign here is not important. We are interested in the peak value of the output. This estimated value for the AC output is much greater than the recorded AC output.

(c) The actual measured gain (ignoring the minus sign) is

$$A_{measured} = \frac{100\text{mV}}{10\text{mV}} = 10$$

The actual gain is a lot less than the expected gain.

(d) Because DC Troubleshooting did not indicate any discrepancies, we can conclude that the problem is with one of the capacitors. The clue is that the measured gain is much smaller than the expected gain. Remembering that amplifiers in which the emitter resistor is not bypassed have a low gain, we can conclude that the emitter bypass capacitor has failed open (Figure 10–34B).

EXAMPLE 10–14
Troubleshooting IV

Figure 10–35 shows a transistor amplifier, along with its DC Troubleshooting findings. Note that the signal generator is connected but turned down to 0V (for DC; therefore, C_i is connected to ground).

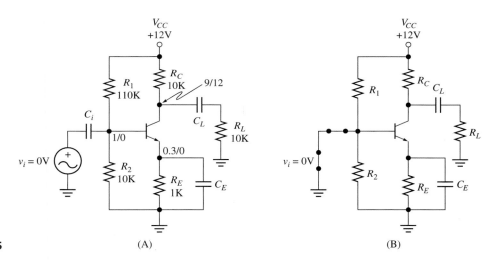

FIGURE 10–35

(A)   (B)

(a) Looking at the Estimated Voltages, what should be the state of the transistor? Looking at the Measured Voltages, what is the actual state of the transistor?

(b) You have already checked the transistor and connections and found that they are good. What is the likely problem here?

Solution

(a) The Estimated Voltages indicate that the transistor should be **ACTIVE** ($V_C > V_E$). But, because the Measured Voltage at the collector is equal to the supply voltage, we can conclude that the transistor is actually in **CUT-OFF.**

(b) Because the transistor has checked out and all connections are good, we suspect one of the capacitors is bad. If the capacitors were open, we would not expect any discrepancies between Measured Voltages and Estimated Voltages. What could account for the fact that the base voltage for DC is 0V? The answer is a short in the input coupling capacitor (C_i). Figure 10–35B shows how a short in the input capacitor provides a DC connection between ground and the base, causing the DC base voltage to be 0V.

Note: This failure will show up only when the signal source is connected to the circuit.

EXAMPLE 10–15
Troubleshooting V

Figure 10–36 shows a transistor amplifier and its DC Troubleshooting findings.

(a) Given the Estimated Voltages, what should be the state of the transistor? Looking at the Measured Voltages, what is the actual state of the transistor?

(b) If the transistor is good and all connections are good, what is the likely problem with this circuit?

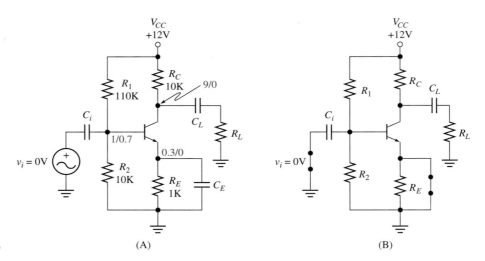

FIGURE 10–36

(A)   (B)

Solution

(a) The Estimated Voltages indicate that the transistor should be **ACTIVE.** However, the DC Troubleshooting findings confirm that the transistor is actually biased in **SATURATION.** You can see that the Measured Voltage at the collector equals the Measured Voltage at the emitter.

(b) Because the transistor and its connections are good, we suspect a problem with a capacitor. Because of the discrepancies between the Estimated and Measured Voltages, we can conclude that the problem is not a failed open capacitor. The Measured Voltage at the emitter indicates that the bypass capacitor (C_E) is shorted (Figure 10–36B). A shorted emitter bypass capacitor produces, in effect, a very small emitter resistor. This small emitter resistance will lead to large emitter and collector currents, which can put the transistor into **SATURATION.**

SUMMARY

■ Common-emitter collector voltage is an amplified version of voltage applied to the base.

■ The common-emitter amplifier is always biased in the **ACTIVE** region.

■ The transistor is **ACTIVE** as long as DC collector voltage is greater than the DC emitter voltage: $V_C > V_E$.

■ For AC signals the emitter diode is modeled as an AC emitter resistance:

$$r_e = \frac{26\text{mV}}{I_C}$$

■ The AC emitter resistance appears in series with the emitter resistor. The total emitter resistance is given by

$$R_{Etotal} = R_e + R_E \quad \text{without bypass capacitor}$$
$$R_{Etotal} = r_e \qquad\qquad \text{with bypass capacitor}$$

■ The total collector resistance is given by

$$R_{Ctotal} = R_C \qquad\qquad \text{without capacitor coupled load}$$
$$R_{Ctotal} = R_C \parallel R_L \quad \text{with capacitor coupled load}$$

■ The gain for the common-emitter amplifier is

$$A = -\frac{R_{Ctotal}}{R_{Etotal}}$$

■ AC signal tracing is used to find problems in circuits when DC Troubleshooting checks out.

PROBLEMS

SECTION 10–1   Introduction

1. Figure 10–37 shows a voltage v_S from a signal generator, similar to the one you use in the laboratory.

(a) What is the DC level, in volts, of this signal?

(b) What is the level of the AC sine wave in peak-to-peak volts?

(c) What is the level of the AC sine wave in peak volts?

(d) What is the level of the AC sine wave in RMS volts?

(e) What is the highest level reached by v_S?

(f) What is the lowest level reached by v_S?

(g) What is the average value of v_S?

FIGURE 10–37

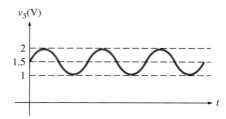

SECTION 10–2   DC Bias Levels

2. For the transistor circuit in Figure 10–38, set $v_i = 0V$, and find the DC bias levels for the transistor currents and voltages. Be sure to list the variables in the order given in the text.

3. For the transistor circuit in Figure 10–38, let $v_i = 4V_p$.

(a) What is the largest value of v_B?

(b) Find the following voltages and currents for the value of v_B found in part (a):

v_E

i_E

i_C

v_C

(c) When v_B is at its highest value, what state is the transistor driven into?

FIGURE 10–38

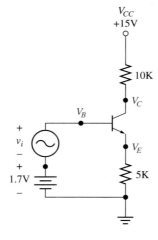

4. For the transistor circuit in Figure 10–38 (p. 341), let $v_i = 1.2V_p$.

   (a) What is the smallest value of v_B?

   (b) Find the following voltages and currents for the value of v_B found in part (a):

   v_E

   i_E

   i_C

   v_C

   (c) When v_B is at its lowest value, what state is the transistor driven into?

5. For the circuit in Figure 10–38 (p. 341), let $v_i = 0.5V_p$.

   (a) When v_i is at its highest value, find the following voltages and currents:

   v_B

   v_E

   i_E

   i_C

   v_C

   (b) What is the state of the transistor in part (a)? Why?

   (c) When v_B is at its smallest value, find the following voltages and currents:

   v_B

   v_E

   i_E

   i_C

   v_C

   (d) What is the state of the transistor in part (c)? Why?

6. For the circuit in Figure 10–38 (p. 341)

   (a) What is the quiescent value of V_C?

   (b) What is the maximum possible collector voltage?

   (c) What is the minimum possible collector voltage?

   (d) What is the maximum possible symmetrical voltage swing?

7. For the transistor in Figure 10–39, let $v_i = 0V$.

   (a) Find the following DC values:

   V_B

   V_E

   I_E

   I_C

   V_C

   (b) You now turn up the voltage at the signal generator. At what value of v_i will the transistor go into **SATURATION?** Why?

   (c) You continue to turn up the voltage at the signal generator. At what value of v_i will the transistor go into **CUT-OFF?** Why? Remember that v_i is an AC signal, it swings both positively and negatively.

FIGURE 10–39

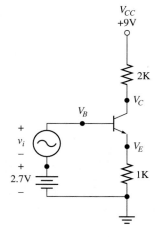

8. For the circuit in Figure 10–39

  (a) What is the quiescent value of V_C?

  (b) What is the maximum possible collector voltage?

  (c) What is the minimum possible collector voltage?

  (d) What is the maximum possible symmetrical voltage swing?

SECTION 10–3   The AC Behavior of the Transistor

9. Find the indicated slope and equivalent AC emitter resistance (r_e) for the I-V curve in Figure 10–40.

FIGURE 10–40

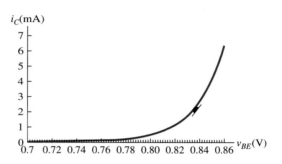

10. What is the AC emitter resistance (r_e) for the transistor circuit in Figure 10–38? Draw the AC circuit.

11. What is the AC emitter resistance (r_e) for the transistor circuit in Figure 10–39? Draw the AC circuit.

SECTION 10–4   The Common-Emitter Amplifier

12. For the transistor amplifier in Figure 10–38 (p. 341), what is the voltage gain (A) between the input voltage (v_i) and the collector voltage?

13. For the transistor amplifier in Figure 10–39, what is the voltage gain (A) between the input voltage (v_i) and the collector voltage?

14. What value of R_C in Figure 10–38 (p. 341) would produce a gain of -4?

15. What value of R_E in Figure 10–38 (p. 341) would produce a gain of -5?

16. What value of R_C in Figure 10–39 would produce a gain of -4?

17. What value of R_E in Figure 10–39 would produce a gain of -5?

FIGURE 10–41

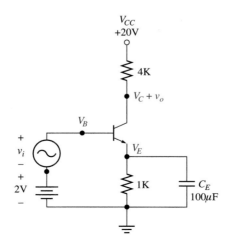

SECTION 10–5   The Emitter Bypass Capacitor

18. Redraw the circuit of Figure 10–41 for DC analysis.

19. Redraw the circuit of Figure 10–41 for AC analysis.

20. For the transistor amplifier in Figure 10–41

    (a) What is the voltage gain *(A)* with the bypass capacitor *(C_E)* in place?

    (b) What is the voltage gain *(A)* if C_E is removed?

    (c) What value did you use for I_E?   **(d)** What value did you use for r_e?

    (e) Is the transistor in the **ACTIVE** state? Why?

    (f) What is the DC voltage across the capacitor C_E?

21. Redraw the circuit of Figure 10–38 (p. 341), and add an emitter bypass capacitor. What is the gain?

22. Redraw the circuit of Figure 10–39 (p. 342), and add an emitter bypass capacitor. What is the gain?

SECTION 10–6   Input Coupling Capacitor—The Self-Biased Circuit

23. Draw the DC circuit for Figure 10–42.

24. For the transistor amplifier shown in Figure 10–42

    (a) Find V_B.

    (b) Find the DC voltage across the input capacitor C_i.

    (c) List the DC bias values for this transistor.

    (d) What is the DC voltage across the emitter bypass capacitor *(C_E)*?

    (e) What is r_e?

FIGURE 10–42

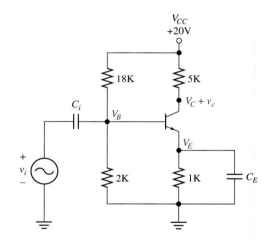

25. Draw the AC circuit for Figure 10–42.

26. For the transistor amplifier in Figure 10–42, let $v_i = 0.01V_p$.

    (a) What is the AC voltage seen at the base? Express your answer in peak volts.

    (b) What is the voltage gain *(A)* of this amplifier?

    (c) What is the AC peak voltage seen at the collector?

27. You want the circuit in Figure 10–42 to produce a gain of -500.

    (a) What r_e do you need?   **(b)** What I_C is needed to produce this r_e?

SECTION 10–7   The Load Resistor

28. Draw the DC circuit for Figure 10–43 (p. 345).

29. For the transistor amplifier in Figure 10–43

    (a) Find the DC bias values.   **(b)** What are the DC voltages across C_i, C_E, and C_L?

    (c) What is r_e?

30. Draw the AC circuit for Figure 10–43 (p. 345).

FIGURE 10–43

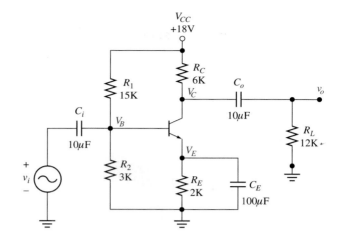

31. For the circuit in Figure 10–43

  (a) What is R_{Etotal}?   **(b)** What is R_{Ctotal}?   **(c)** What is the voltage gain (A)?

  (d) Let $v_i = 5mV_p$, and find the output in peak volts.

  (e) Find the output voltage in peak-to-peak volts.

  (f) What is the DC voltage across R_L?

32. In Figure 10–43 change the load resistor to $R_L = 3K\Omega$, and find the gain.

33. In Figure 10–43 change the load resistor to $R_L = 6K\Omega$, and find the gain.

34. You want the circuit in Figure 10–43 to have a gain of − 400.

  (a) What r_e do you need?   **(b)** What I_C is required to produce this r_e?

SECTION 10–8 Troubleshooting

35. The transistor amplifier in Figure 10–44 was built and was found to have good bias values. That is, DC Troubleshooting found no discrepancies between Measured Voltages and Estimated Voltages. However, the AC voltages seen at the base and at the collector were so small they could not be measured. What type of problem with the input coupling capacitor (C_i) could explain these findings? Why?

36. The transistor amplifier in Figure 10–44 was built and was found to have good bias values. That is, DC Troubleshooting found no discrepancies between Measured Voltages and Estimated Voltages. The AC voltage seen at the base was $10mV_p$. The AC voltage seen at the collector was $1.1V_p$. No AC voltage could be seen at the load. That is, $v_o = 0V$. What type of problem with the output coupling capacitor could explain this problem? Why?

37. The transistor amplifier in Figure 10–44 was built and was found to have good bias values. That is, DC Troubleshooting found no discrepancies between Measured Voltages and Estimated Voltages.

FIGURE 10–44

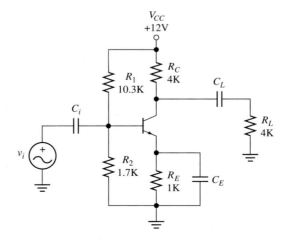

The AC voltage seen at the base was 10mV$_p$. The AC voltage seen at the collector was 40mV$_p$.

(a) What should be the voltage gain of this transistor?

(b) What should be the AC voltage at the output?

(c) What is the actual measured gain of the transistor?

(d) What type of problem with the emitter bypass capacitor (C_E) could cause the observed collector voltage? Why?

38. A transistor amplifier is shown in Figure 10–45, along with its DC Troubleshooting findings.

(a) What should be the state of the transistor?

(b) What is the state of the transistor?

(c) What type of problem with the input coupling capacitor (C_i) could explain these results? Why?

FIGURE 10–45

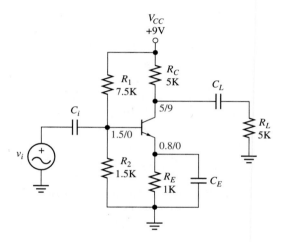

39. A transistor amplifier is shown in Figure 10–46, along with its DC Troubleshooting findings.

(a) What should be the state of the transistor?

(b) What is the state of the transistor?

(c) What type of problem with the emitter bypass capacitor (C_E) could explain these results? Why?

FIGURE 10–46

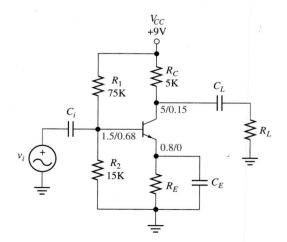

Computer Problems

40. Find the DC bias values for the transistor in Figure 10–38 (p. 341).

41. For the circuit in Figure 10–38, (p. 341) use transient analysis to find the collector voltages for the following *sinusoidal* signal generator inputs. Set the signal generator frequency to 1000Hz.

    (a) $v_i = 2V_p$ **(b)** $v_i = 0.5V_p$ **(c)** $v_i = 4V_p$

    Discuss your results.

42. For the circuit in Figure 10–41 (p. 343)

    (a) Find the bias levels.

    (b) Find the collector voltage output with the signal generator set to $1mV_p$ at the following frequencies:

        (i) 1Hz   **(ii)** 50Hz   **(iii)** 1000Hz

    (c) Find the gain at each frequency. Use either transient analysis or phasor analysis.

    (d) Discuss your results.

43. For the circuit in Figure 10–43 (p. 345)

    (a) Find the bias levels.

    (b) Find the collector voltage output with the signal generator set to $1mV_p$ at the following frequencies:

        (i) 1Hz   **(ii)** 50Hz   **(iii)** 1000 Hz

    (c) Find the gain at each frequency. Use either transient analysis or phasor analysis.

    (d) Discuss your results.

44. Find the DC values for the transistor voltages in the circuit of Figure 10–43 (p. 345) (set v_i to 0V) if

    (a) C_i has failed short.   **(b)** C_E has failed short.   **(c)** C_L has failed short.

45. Find the gain of the circuit in Figure 10–43 (p. 345) if

    (a) C_i has failed short.   **(b)** C_E has failed short.   **(c)** C_L has failed short.

    (d) C_i has failed open.   **(e)** C_E has failed open.   **(f)** C_L has failed open.

CHAPTER

11

THE BOX METHOD

CHAPTER **O**UTLINE

C HAPTER OBJECTIVES

To represent any amplifier with the Box Model.

To determine the input resistance, output resistance, and gain of the common-emitter amplifier.

To represent the common-emitter amplifier with a Box Model.

To determine the overall gain of the common-emitter amplifier in the presence of a load resistor and a signal source resistor.

To determine the effect of the emitter bypass capacitor on the input resistance and overall gain of the common-emitter amplifier.

To use Box Models to determine the overall gain, input resistance, and output resistance of multistage amplifiers.

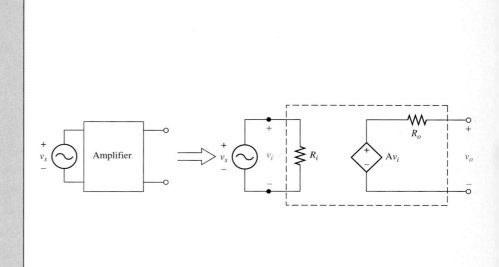

11–1 ▪ INTRODUCTION

We want to take another look at the system we built to amplify our opera singer's performance. Figure 11–1 shows the three basic elements of the audio system: the microphone that produces the input to the amplifier; the amplifier itself; and the speaker, which is driven by the amplifier output.

FIGURE 11–1

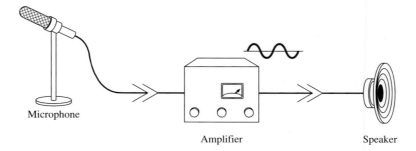

Microphone

Amplifier

Speaker

In Chapter 10 we analyzed this system as a single unit. What would happen if we changed the microphone and/or the speaker? We would have to perform an entirely new analysis to account for any changes. A better approach might be to analyze each component separately—the microphone, the amplifier, and the load—then find a way to combine the separate analyses.

The microphone and the speaker load are two terminal devices. As we discussed in Chapter 10, two terminal devices can be modeled with a Thevenin equivalent (Figure 11–2). The Thevenin equivalent contains a voltage source that represents signal sources inside the device; the Thevenin resistance represents the total resistance seen between the two terminals of the device.

FIGURE 11–2

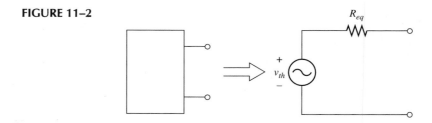

Figure 11–3A shows the Thevenin equivalent for the microphone. The voltage source represents the fact that a microphone produces an electrical voltage in response to its acoustic input. (This is the AC voltage source that we used as the amplifier input in Chapter 10.) Note, however, that we now include a source resistance. All electrical devices, even supposedly ideal sources such as batteries, have resistance. We will see in this chapter the effect of this previously-ignored source resistance on the overall gain of the system.

FIGURE 11–3

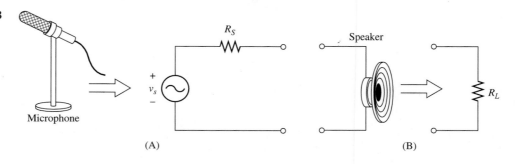

Microphone

(A)

(B)

Like most loads, a speaker is a passive device. That is, loads generally absorb power $(P_L = V_L I_L)$; they do not contain any independent sources of energy. The Thevenin equivalent of most loads, therefore, is simply a resistance (Figure 11–3B).

Because the input source and output load of the amplifier are two-terminal devices, they can be modeled with the Thevenin equivalent. The amplifier itself, on the other hand, is a four-terminal device. The amplifier has two input and two output terminals. Therefore, we must find a four-terminal equivalent to the Thevenin model.

11–2 ■ THE BOX MODEL

Any amplifier, from a single transistor to the fanciest system you can buy, has an input resistance, an output resistance, and a gain between the input and output voltages. These three amplifier parameters are used to create what we call the **Box Model,** shown in Figure 11–4. Here

$$R_i \equiv \text{amplifier input resistance}$$
$$R_o \equiv \text{amplifier output resistance}$$

and

$$A = \frac{v_o}{v_i} \equiv \text{amplifier voltage gain}$$

Note that the Box Model combines what resembles two Thevenin equivalents. The input side is represented with a resistance, and the output side is represented with a controlled voltage source and a resistance.

FIGURE 11–4
The Box Model

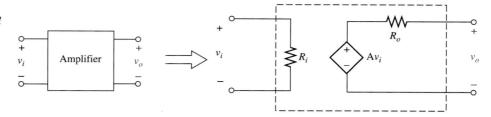

Let's examine the behavior of the Box Model when a voltage source is connected at the input and the output is left open-circuited (Figure 11–5, p. 352). Because the output is an open circuit, there is no output current, and, therefore, no voltage drop across R_o. Under the no-load condition then, the output voltage is

$$v_o = Av_i$$

and, because the voltage source is connected directly to the input terminal

$$v_i = v_s$$

so

$$v_o = Av_i = Av_s$$

and the voltage gain of the system is

$$A_{sys} = \frac{v_o}{v_s} = A$$

We see that the gain used in the Box Model is the gain of the amplifier when it has an ideal voltage source input and no load at the output.

FIGURE 11–5

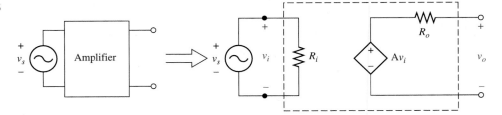

The Load Resistor

Let's see how a load resistor affects the gain of the overall system. Figure 11–6 shows the Box Model for an amplifier that has an input resistance of $10K\Omega$, an output resistance of $1K\Omega$, and a voltage gain of 100. The amplifier is driven by an ideal source and is connected to a $1K\Omega$ load.

We find the overall gain in two steps. We first note that because the source is connected directly to the amplifier input:

$$v_i = v_s$$

FIGURE 11–6

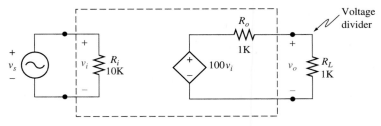

We now find the output voltage by recognizing that the right side of the circuit is a voltage divider. The output voltage is, therefore

$$v_o = \frac{1K}{1K + 1K} 100v_i = \frac{1}{2}100v_i = 50v_i = 50v_s$$

And the system gain is

$$A_{sys} = \frac{v_o}{v_s} = 50$$

Compare this gain to the no-load gain of $A = 100$. A load resistor always lowers the gain available from the amplifier. In a properly designed system, we can minimize the loss in gain.

EXAMPLE 11–1
The Load
Resistor I

For the Box Model in Figure 11–7

(a) Find the no-load gain (remove the load resistor).
(b) Find the overall gain when $R_L = 100\Omega$.
(c) Find the overall gain when $R_L = 4K\Omega$.

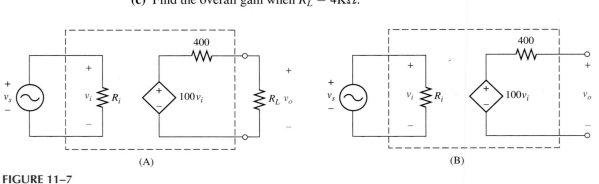

(A)   (B)

FIGURE 11–7

Solution

(a) No-load gain (Figure 11–7B).

Because the output current is 0A when there is no load, the voltage drop across R_o is 0V and

$$v_o = 100v_i = 100v_s$$

The no-load gain is

$$A_{no\ load} = \frac{v_o}{v_s} = 100$$

(b) For $R_L = 100\Omega$:

We find the output voltage from the voltage divider:

$$v_o = \frac{100}{100 + 400}\ 100v_i = \frac{1}{5}100v_i = 20v_i = 20v_s$$

The overall gain for the 100Ω load resistor is

$$A_{100\Omega} = \frac{v_o}{v_s} = 20$$

Note: If the load resistor is much smaller than the R_o of the amplifier, we lose most of our gain.

(c) For $R_L = 4\ K\Omega$:

We find the output voltage from the voltage divider:

$$v_o = \frac{4K}{4K + 400}\ 100v_i = \frac{4K}{4.4K}\ 100v_i = 90.9v_i = 90.9v_s$$

The overall gain for the 4KΩ load resistor is

$$A_{4K\Omega} = \frac{v_o}{v_s} = 91$$

Note: If the load resistor is much larger than R_o, the overall gain approximately equals the no-load gain.

This example shows that the overall gain of an amplifier system depends on the relation between the load resistor and the amplifier's output resistance. Figure 11–8 shows the general case of an amplifier that is driven by an ideal source and is connected to a load resistor. The output voltage is given by

$$v_o = \frac{R_L}{R_L + R_o}\ Av_i = \frac{R_L}{R_L + R_o}\ Av_s$$

and the system gain is

$$A_{sys} = \frac{v_o}{v_s} = \frac{R_L}{R_L + R_o}\ A$$

FIGURE 11–8

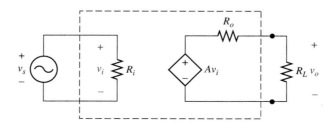

If the load resistor is much larger than the amplifier output resistance—that is, if

$$R_L \gg R_o$$

then the overall system gain is

$$A_{sys} = \frac{R_L}{R_L + R_o} A \approx \frac{R_L}{R_L} A = A$$

and the overall gain approximately equals the no-load gain.

If the load resistor is not much greater than the output resistance of the amplifier, then the overall gain is significantly lower than the no-load gain. The amplifier has been *loaded down*.

EXAMPLE 11–2
The Load
Resistor II

For the circuit in Figure 11–9, the load resistor is fixed at 100Ω. Find the system gain if we vary the amplifier output resistance as follows:

(a) $R_o = 1K\Omega$ **(b)** $R_o = 100\Omega$ **(c)** $R_o = 10\Omega$

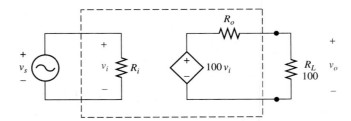

FIGURE 11–9

Solution   In all cases, we find the output from

$$v_o = \frac{R_L}{R_L + R_o} A v_i$$

and, because

$$v_i = v_s$$

$$v_o = \frac{R_L}{R_L + R_o} A v_s$$

the system gain is

$$A_{sys} = \frac{R_L}{R_L + R_o} A = \frac{100}{100 + R_o} 100$$

(a) For $R_o = 1K\Omega$:

$$A_{sys} = \frac{100}{100 + 1000} 100 = \underset{9.09}{\cancel{90.9}}$$

and we see that we lose most of the no-load gain of 100.

(b) For $R_o = 100\Omega$:

$$A_{sys} = \frac{100}{100 + 100} 100 = 50$$

If the amplifier's output resistance is equal to the load resistance, we lose one-half of the no-load gain.

(c) For $R_o = 10\Omega$:

$$A_{sys} = \frac{100}{100 + 10}\, 100 = 90.9$$

If the amplifier's output resistance is much less than the load resistance, we get to use most of the amplifier's no-load gain.

DRILL EXERCISE   For the circuit shown in Figure 11–9, find the system gain if

$$R_o = 5K\Omega \quad \text{and} \quad R_L = 10K\Omega$$

Answer $A_{sys} = 66.7$

Load Power

The typical amplifier, such as the one in Figure 11–1 (p. 350), is usually composed of two distinct amplifier sections. The first section is a voltage amplifier, which has the job of creating a large voltage from a small voltage. The second section is a power amplifier and is intended to maximize the *power* delivered to the external load (the speaker in our example). The input resistance of the power amplifier acts as the load for the voltage amplifier. Our previous discussion regarding the load resistor applies only when we are trying to maximize the voltage delivered to the load.

The Box Model that we use to represent a voltage amplifier can also be used to represent the power amplifier (we will discuss these amplifiers in Chapter 19). The difference here is that to maximize load voltage we want $R_L >> R_o$, but this relation will not maximize power delivered to the load. As you might recall from your circuits class, to maximize power delivered to a load, we want the load resistor to equal the Thevenin resistance (the output resistance) of the amplifier. That is, for

maximum power delivery: $R_L = R_o$

The maximum *average* power that is delivered to the load is then

maximum average load power: $P_{Lmax} = \dfrac{(AV_{imax})^2}{8R_L}$

For the rest of this chapter, we will always assume that we want to deliver maximum voltage to the load. We will revisit the power issue in Chapter 19.

The Source Resistor

We now turn our attention to the input source. Any real source of voltage contains an internal resistance, and, therefore, is properly modeled with a Thevenin equivalent circuit. If we assume our microphone has a $1K\Omega$ internal resistance, we get its Thevenin model shown in Figure 11–10A (p. 356).

In Figure 11–10B we connect the microphone to the amplifier input. For now we will analyze the circuit under no-load conditions. We recognize, once again, that we are dealing with a voltage divider, this time in the input circuit. We can find v_i from

$$v_i = \frac{R_i}{R_i + R_s}\, v_s = \frac{1K}{1K + 1K}\, v_s = \frac{1}{2}\, v_s$$

Now, because there is no load

$$v_o = 100 v_i = 100\,\frac{1}{2}v_s = 50 v_s$$

FIGURE 11–10

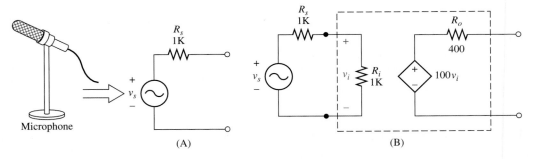

Microphone

(A)   (B)

And the system gain is

$$A_{sys} = \frac{v_o}{v_s} = 50$$

You can see that for this combination of source resistance and amplifier input resistance, we have lost on-half of the no-load gain. Figure 11–11 shows the general no-load case. The system gain is found from

$$v_i = \frac{R_i}{R_i + R_s} v_s$$

and

$$v_o = Av_i = \frac{R_i}{R_i + R_s} Av_s$$

so

$$A_{sys} = \frac{v_o}{v_s} = \frac{R_i}{R_i + R_s} A$$

FIGURE 11–11

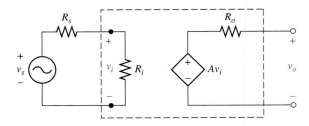

EXAMPLE 11–3
The Source Resistance I

For the circuit in Figure 11–12, find the system gain for the following amplifier input resistances:

(a) $R_i = 100\Omega$ **(b)** $R_i = 1K\Omega$

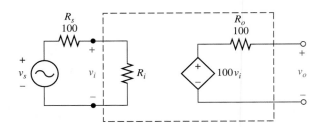

FIGURE 11–12

Solution Because there is no load

$$v_o = 100v_i$$

and, from the input voltage divider

$$v_i = \frac{R_i}{R_i + R_s} v_s = \frac{R_i}{R_i + 100} v_s$$

so

$$v_o = \frac{R_i}{R_I + 100} 100v_s$$

and the system gain is

$$A_{sys} = \frac{v_o}{v_s} = \frac{R_i}{R_i + 100} 100$$

(a) For $R_i = 100\Omega$, the system gain is

$$A_{sys} = \frac{100}{100 + 100} 100 = \frac{1}{2}100 = 50$$

and we have lost one-half of the no-load gain.

(b) For $R_i = 1$ KΩ, the system gain is

$$A_{sys} = \frac{1000}{1000 + 100} 100 = 0.909 \times 100 = 91$$

and the system gain approximately equals the no-load gain.

EXAMPLE 11–4
The Source
Resistance II

For the circuit in Figure 11–13, find the system gain for the following source resistances:

(a) $R_s = 10$KΩ **(b)** $R_s = 1$KΩ

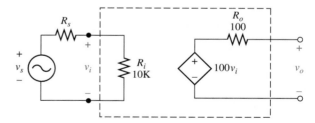

FIGURE 11–13

Solution Because there is no load

$$v_o = 100v_i$$

and, from the input voltage divider

$$v_i = \frac{R_i}{R_i + R_s} v_s = \frac{10K}{10K + R_s} v_s$$

so

$$v_o = \frac{10K}{10K + R_s} 100v_s$$

and the system gain is

$$A_{sys} = \frac{v_o}{v_s} = \frac{10K}{10K + R_s} \, 100$$

(a) For $R_s = 10K\Omega$, the system gain is

$$A_{sys} = \frac{10K}{10K + 10K} \, 100 = \frac{1}{2} 100 = 50$$

Again we have lost one-half the no-load gain.

(b) For $R_i = 1K\Omega$, the system gain is

$$A_{sys} = \frac{10K}{10K + 1K} \, 100 = 0.909 \times 100 = 90.9$$

and the system gain approximately equals the no-load gain.

DRILL EXERCISE   For the circuit in Figure 11–13 (p. 357), find the system gain if

$$R_i = 100K\Omega \quad \text{and} \quad R_s = 50K\Omega$$

Answer $A_{sys} = 66.7$

We can conclude from Examples 11–3 and 11–4 that in a well-designed system, the input resistance of the amplifier should be much greater than the resistance of the source.

☐ If you are given the input (the microphone, for example), you must choose an amplifier with a sufficiently high input resistance.

☐ If you are given the amplifier, you must then select a source (a microphone, for example) with an internal resistance that is much smaller than the amplifier input resistance.

The Complete System

In Figure 11–14, we show the complete system that includes the source resistance, the amplifier Box Model, and the output load resistance. While the circuit may look complicated, it is, in fact, composed of *two separate voltage dividers*. From the input side, we get

$$v_i = \frac{R_i}{R_i + R_s} \, v_s$$

From the output side, we get

$$v_o = \frac{R_L}{R_L + R_o} \, Av_i$$

FIGURE 11–14

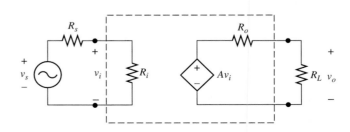

Putting these two voltage divider equations together

$$v_o = \frac{R_L}{R_L + R_o} A \frac{R_i}{R_i + R_s} v_s$$

Finally, we get the system gain

$$A_{sys} = \frac{v_o}{v_s} = \frac{R_i}{R_i + R_s} \frac{R_L}{R_L + R_o} A$$

In the well-designed voltage amplification system, we would have

$$R_i \gg R_s \quad \text{and} \quad R_L \gg R_o$$

and the system gain would be

$$A_{sys} \approx A$$

EXAMPLE 11–5
The Total System

The load and source resistances are given for the circuit in Figure 11–15. Find the system gain for the following three amplifiers; each amplifier has a no-load gain of 100. For each case, discuss whether the system is a well-designed voltage amplifier.

(a) $R_i = 100K\Omega$ and $R_o = 100\Omega$
(b) $R_i = 1K\Omega$ and $R_o = 10\Omega$
(c) $R_i = 100K\Omega$ and $R_o = 10\Omega$

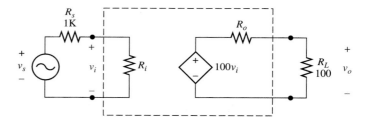

FIGURE 11–15

Solution   For all cases we find the system gain by analyzing the input and output voltage dividers separately, then combining the results. On the input side

$$v_i = \frac{R_i}{R_i + R_s} v_s = \frac{R_i}{R_i + 1K} v_s$$

On the output side

$$v_o = \frac{R_L}{R_L + R_o} A v_i = \frac{100}{100 + R_o} 100 v_i$$

Combining these results

$$v_o = \frac{R_i}{R_i + 1K} \frac{100}{100 + R_o} 100 v_i$$

The system gain is

$$A_{sys} = \frac{v_o}{V_{in}} = \frac{R_i}{R_i + 1K} \frac{100}{100 + R_o} 100$$

(a) For $R_i = 100\text{K}\Omega$ and $R_o = 100\Omega$, we get

$$A_{sys} = \frac{100\text{K}}{100\text{K} + 1\text{K}} \frac{100}{100 + 100} \, 100 = 0.99 \times \frac{1}{2} \times 100 = 49.5$$

In this case we have a good input design, but the output design is bad. Because R_o is the same size as the load resistor, we lose one-half of the no-load gain.

(b) For $R_i = 1\text{K}\Omega$ and $R_o = 10\Omega$

$$A_{sys} = \frac{1\text{K}}{1\text{K} + 1\text{K}} \frac{100}{100 + 10} \, 100 = \frac{1}{2} \times 0.909 \times 100 = 45.5$$

Here, the output is well-designed but the input resistance is too small, and once again we lose more than one-half of the no-load gain.

(c) For $R_i = 100\text{K}\Omega$ and $R_o = 10\Omega$

$$A_{sys} = \frac{100\text{K}}{100\text{K} + 1\text{K}} \frac{100}{100 + 10} \, 100 = 0.99 \times 0.909 \times 100 = 89.9$$

This is what we are looking for in an amplifier system design. The input resistance of the amplifier is high and the output resistance is low, resulting in a system gain that is close to the no-load gain.

DRILL EXERCISE   Find the system gain for the circuit in Figure 11–15 (p. 359) for

$$R_i = 10\text{K}\Omega \quad \text{and} \quad R_o = 1\text{K}\Omega$$

Answer $A_{sys} = 8.26$

REVIEW QUESTIONS

1. Draw the general Box Model.
2. What is the voltage gain of the Box Model?
3. When we connect a load resistor to the amplifier, what does the output circuit become?
4. What is the voltage gain of the amplifier with a load resistor?
5. For high voltage gain, what is the desired relation between R_o and R_L?
6. For maximum load power, what is the desired relation between R_o and R_L?
7. When we include a source resistor, what does the input circuit become?
8. When we include a source resistor, what is v_i?
9. For high gain, what is the desired relation between R_s and R_i?
10. What is the total voltage gain, including source and load resistors?

11–3 ■ THE COMMON-EMITTER AMPLIFIER

In Chapter 10 we showed you how to find the small-signal AC gain of the common-emitter amplifier. The Box Model described in this chapter has shown us the importance of the input and output resistances of an amplifier. We now want to determine the Box Model for the unloaded common-emitter amplifier (Figure 11–16). Note that the Box Model for the transistor only applies to the amplification of small-signal AC inputs. The Box Model is *not* used for DC analysis.

FIGURE 11–16

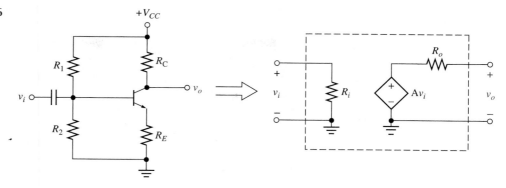

We begin with the self-biased common-emitter circuit shown in Figure 11–17A, in which the small-signal AC input is capacitively coupled to the base. Note that we have used dashes to sketch in the emitter bypass capacitor. We will first derive all AC relations in terms of the total emitter resistance (R_{Etotal}). We can then determine the effect of the presence of a bypass capacitor on the gain, input resistance, and output resistance.

To get the AC transistor circuit, we turn **OFF** the DC collector supply voltage, add the AC emitter resistance (r_e), and short the input capacitor (Figure 11–17B). Because the DC supply voltage (V_{CC}) has been turned **OFF,** we place an AC ground at the top of the circuit.

FIGURE 11–17

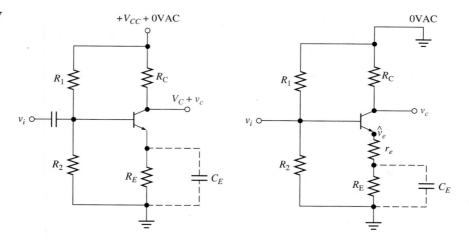

In Figure 11–18A (p. 362) we have redrawn the AC circuit to show that R_1 and R_2 are in parallel and to emphasize that the output voltage is the voltage across the collector resistor. Finally, because R_1 and R_2 are in parallel, they can be replaced with a single resistor (R_B) (Figure 11–18B). That is

$$R_B = R_1 \parallel R_2 = \frac{R_1 R_2}{R_1 + R_2}$$

No-Load Gain

We already know from Chapter 10 that the no-load gain of the emitter follower is given by

$$A = \frac{v_o}{v_i} = \frac{v_c}{v_b} = \frac{-R_C}{R_{Etotal}}$$

FIGURE 11–18

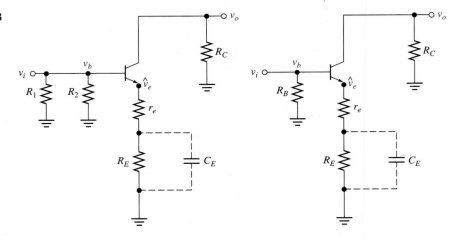

where the output voltage is the collector voltage and the input voltage is the base voltage.
If there is no bypass capacitor then $R_{Etotal} = r_e + R_E$, and

$$A = -\frac{R_C}{r_e + R_E} \qquad \text{without bypass capacitor}$$

where the gain is in the range of -5 to -10 for typical values of the various resistors.
If there is a bypass capacitor, $R_{Etotal} = r_e$ and

$$A = \frac{-R_C}{r_e} \qquad \text{with bypass capacitor}$$

where the gain is in the range of -50 to -500 for typical values. You can see that the presence of the emitter bypass capacitor greatly increases the gain of the common-emitter amplifier.

Output Resistance

To find the output resistance of an amplifier (the Thevenin resistance at the output), we turn **OFF** the input. A voltage source that is turned **OFF** is replaced with a short circuit (Figure 11–19A).

If there is no driving source at the base, then the base current must be 0A. If the base current is 0A, then the collector current must also be 0A. Therefore, when the AC input

FIGURE 11–19

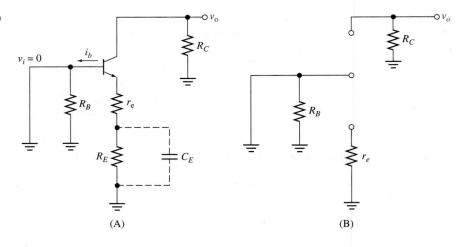

(A)   (B)

voltage is turned **OFF,** there are no transistor AC currents, and we replace the transistor with an open circuit (Figure 11–19B). From Figure 11–19B you can see that the output resistance is just the collector resistor; that is

$$R_o = R_C$$

Note that the output resistance does not depend on anything in the base or emitter of the transistor.

Input Resistance

The input resistance of an amplifier is defined as

$$R_i = \frac{v_i}{i_i}$$

where, for the common-emitter $v_i = v_b$ (Figure 11–20A). Note, however, that the base current is not the input current. Some of the input current goes through R_B.

FIGURE 11–20

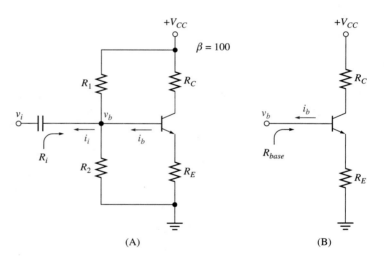

(A)   (B)

To simplify finding the total input resistance, we will first ignore R_B and find the input resistance at the base of the transistor (Figure 11–20B). That is, we want to find

$$R_{base} = \frac{v_b}{i_b}$$

Knowing the resistance at the base of a transistor is critical to understanding the behavior of transistor amplifiers. For this reason, we will show you two methods for finding R_{base}. The first is an informal method that should give you general insight into transistor behavior. The second method is a more formal derivation.

General Derivation of R_{base}

To find the resistance of any device, we can apply a known voltage and find the current. A large current indicates a small resistance; a small current indicates a large resistance. Consider the situation shown in Figures 11–21A and B (p. 364). In Figure 11–21A we apply a 1V change in voltage to a 1KΩ resistor. The resulting change in current is

$$\Delta I = \frac{\Delta V}{R} = \frac{1}{1K} = 1mA$$

FIGURE 11–21

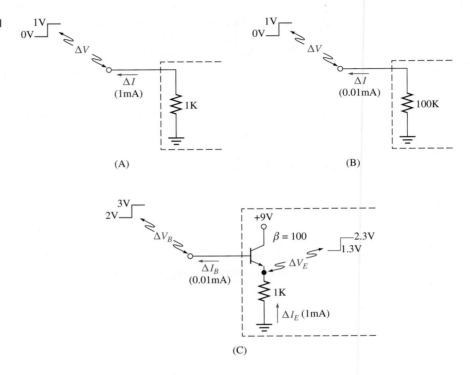

(A)

(B)

(C)

In Figure 11–21B we apply the same 1V change in voltage to a 100KΩ resistor. The change in current in this case is

$$\Delta I = \frac{\Delta V}{R} = \frac{1}{100\text{K}} = 0.01\text{mA}$$

Now, let us look at the transistor circuit shown in Figure 11–21C. If we apply a base voltage that increases from 2 to 3V, the emitter voltage will change from 1.3 to 2.3V. The change in the emitter current will be

$$\Delta I_E = \frac{\Delta V_E}{R_E} = \frac{2.3 - 1.3}{1\text{K}} = 1\text{mA}$$

If the transistor is **ACTIVE,** the change in the current at the base is

$$\Delta I_B = \frac{\Delta I_E}{\beta} = 0.01\text{mA}$$

At the base a 1V change in the base voltage results in a 0.01mA change in the base current. The apparent resistance at the base is, therefore

$$R_{base} = \frac{\Delta V_B}{\Delta I_B} = \frac{1\text{V}}{0.01\text{mA}} = 100\text{K}\Omega$$

The resistance you would measure at the base is actually 100 times greater than the resistance at the emitter! This is because for the same voltage the current at the base is 100 times smaller than the current at the emitter.

Formal Derivation of R_{base}

Figure 11–22A shows the AC common-emitter transistor circuit model. The key to the formal analysis is that the voltage labeled $\hat{v}_e$ in Figure 11–22B is equal to the base voltage, that is

$$\hat{v}_e = v_b$$

FIGURE 11–22

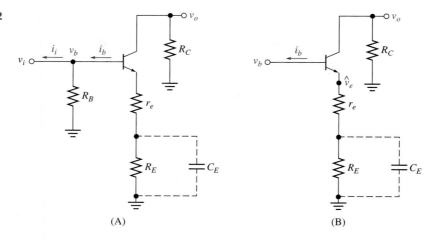

(A) (B)

Note that $\hat{v}_e$ is not a real-life voltage. Remember that r_e is how we model the small-signal behavior of the emitter diode and is internal to the transistor. Even so, we can use this voltage to solve for the base current. We first find the AC emitter current from

$$i_e = \frac{\hat{v}_e}{R_{Etotal}} = \frac{v_b}{R_{Etotal}}$$

so

$$v_b = i_e R_{Etotal}$$

Because $i_e = i_c = \beta i_b$

$$v_b = \beta i_b R_{Etotal}$$

and the AC resistance looking into the base of the transistor is

$$R_{base} = \frac{v_b}{i_b} = \frac{\beta i_b R_{Etotal}}{i_b}$$

Therefore

$$R_{base} = \beta R_{Etotal}$$

This equation describes one of the most important properties of the transistor. The AC resistance at the base of the transistor if β times greater than the resistance in the emitter. This is one advantage of the transistor: a small resistance in the emitter looks very large from the base. It's as though you were looking at the emitter through a telescope (see Figure 11–23).

FIGURE 11–23

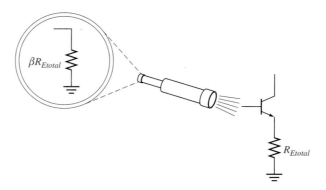

FIGURE 11–24

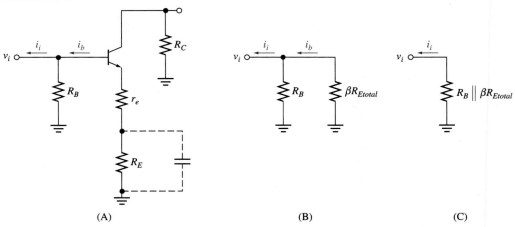

(A)   (B)   (C)

We can now replace the base side of the transistor with R_{base}, which is in parallel with the bias resistor equivalent (R_B) (Figure 11–24). The total input resistance is

$$R_i = R_B \parallel R_{base} = R_B \parallel \beta R_{Etotal}$$

The Box Model for the common-emitter amplifier is shown in Figure 11–25. The next two examples show the Box Model for a common-emitter amplifier with and without a bypass capacitor.

FIGURE 11–25
Common-emitter Box Model

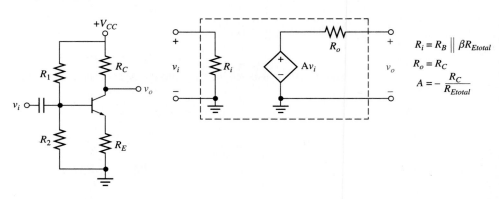

$R_i = R_B \parallel \beta R_{Etotal}$

$R_o = R_C$

$A = -\dfrac{R_C}{R_{Etotal}}$

EXAMPLE 11–6
Box Model of the Common-Emitter Amplifier

Find the small-signal input resistance, output resistance, and no-load gain for the common-emitter amplifier shown in Figure 11–26. Draw the Box Model for the amplifier.

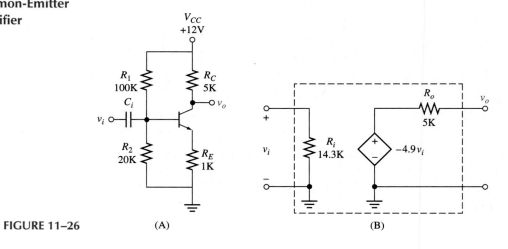

FIGURE 11–26   (A)   (B)

Solution We first find the DC bias levels as follows:

$$V_B = \frac{R_2}{R_2 + R_1} V_{CC} = \frac{20K}{20K + 100K} 12 = 2V$$

$$V_E = V_B - 0.7 = 2 - 0.7 = 1.3V$$

$$I_E = \frac{V_E}{R_E} = \frac{1.3}{1K} = 1.3mA$$

If the transistor is **ACTIVE**

$$I_C = I_E = 1.3mA$$

The collector voltage is

$$V_C = V_{CC} - R_C I_C = 12 - 5K \times 1.3mA = 5.5V$$

Because $V_C > V_E$, the transistor is **ACTIVE,** and the AC emitter resistance is

$$r_e = \frac{26mV}{I_C} = \frac{26mV}{1.3mA} = 20\Omega = 0.02K\Omega$$

The small-signal input resistance is given by

$$R_i = R_B \| \beta R_{Etotal}$$

where

$$R_B = R_1 \| R_2 = \frac{20K \times 100K}{20K + 100K} = 16.7K\Omega$$

and

$$R_{Etotal} = r_e + R_E = 1.02K\Omega$$

The input resistance is, therefore

$$R_i = 16.7K \| 100(1.02K) = 14.3K\Omega$$

The output resistance is the collector resistance, so

$$R_o = R_C = 5K\Omega$$

Finally, the no-load gain is

$$A = -\frac{R_C}{R_{Etotal}} = -\frac{5K}{1.02K} = -4.9$$

The Box Model is shown in Figure 11–26B.

DRILL EXERCISE

Redraw the common-emitter amplifier circuit of Figure 11–26 with $V_{CC} = 9V$. Find the AC emitter resistance, input resistance, output resistance, and the no-load gain.

Answer $r_e = 32.5\Omega$ $R_i = 14.4K\Omega$ $R_o = 5K\Omega$ $A = -4.8$

EXAMPLE 11–7
**Box Model
of the Bypassed
Common-Emitter
Amplifier**

In this example, we start with the circuit from Example 11–6 and add an emitter bypass capacitor (Figure 11–27, p. 368). Find the AC small-signal input resistance, output resistance, and no-load gain. Draw the Box Model for this amplifier.

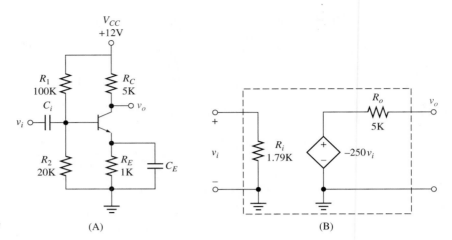

FIGURE 11–27   (A)   (B)

Solution   Under DC conditions the bypass capacitor is open, and the transistor bias circuit in this example is the same as the one in Example 11–6 (p. 366). We'll be a little lazy, therefore, and use the bias levels and AC emitter resistance we found there. That is

$$r_e = 20\Omega = 0.02K\Omega$$

For the AC analysis, we use the following:

$$R_B = R_1 \parallel R_2 = 16.7K\Omega$$

Because R_E is bypassed

$$R_{Etotal} = r_e = 0.02K\Omega$$

The Box Model parameters are

$$R_i = R_B \parallel \beta R_{Etotal} = 16.7K \parallel 100 \times 0.02K = 1.79K\Omega$$
$$R_o = R_C = 5K\Omega$$

and

$$A = -\frac{R_C}{R_{Etotal}} = -\frac{5K}{0.02K} = -250$$

The Box Model for the bypassed common-emitter amplifier is shown in Figure 11–27B.

DRILL EXERCISE   Redraw the circuit in Figure 11–27 with $V_{CC} = 9V$. Find the AC emitter resistance, input resistance, output resistance, and the gain.

Answer $r_e = 32.5\Omega$ $R_i = 2.7K\Omega$, $R_o = 5K\Omega$ $A = -154$

You can see from the previous two examples that the emitter bypass capacitor increases the amplifier gain but lowers the input resistance. We can summarize the Box Model parameters for the common-emitter amplifier as

No Bypass Capacitor	Bypass Capacitor
$A = -\dfrac{R_C}{r_e + R_E}$	$A = -\dfrac{R_C}{r_e}$
$R_i = R_B \parallel \beta(r_e + R_E)$	$R_i = R_B \parallel \beta r_e$
$R_o = R_C$	$R_o = R_C$

11–4 ■ COMMON-EMITTER AMPLIFIER WITH LOAD AND SOURCE RESISTORS

Load Resistor

In chapter 10, we analyzed the common-emitter when a load resistor was capacitively coupled to the collector (Figure 11–28A). We will now use the Box Model of the common-emitter amplifier developed in the last section to again analyze the effect of a load resistor on the system gain (Figure 11–28B).

FIGURE 11–28
Common-emitter amplifier with load resistor

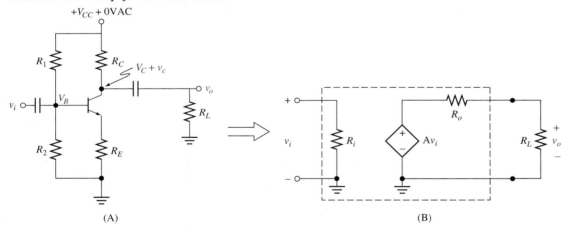

(A)   (B)

From the Box Model, we see that

$$v_o = \frac{R_L}{R_L + R_o} A v_i$$

and, because the input voltage is connected directly to the base

$$v_i = v_s$$

so

$$v_o = \frac{R_L}{R_L + R_o} A v_s$$

The system gain for the common-emitter amplifier with a load resistor is, therefore

$$A_{sys} = \frac{v_o}{v_s} = \frac{R_L}{R_L + R_o} A$$

Now, from the previous section, we know the Box Model parameters for the common-emitter amplifier:

$$R_{in} = R_B \parallel \beta R_{Etotal}$$
$$R_o = R_C$$

and

$$A = -\frac{R_C}{R_{Etotal}}$$

Using these values in the system gain equation, we get

$$A_{sys} = \left(\frac{R_L}{R_L + R_C}\right)\left(-\frac{R_C}{R_{Etotal}}\right) = -\frac{1}{R_{Etotal}}\frac{R_L R_C}{R_L + R_C}$$

You might recognize that the last fraction on the right side of the equation is the same formula we would get if R_C and R_L are in parallel. That is

$$\frac{R_L R_C}{R_L + R_C} = R_C \parallel R_L$$

so

$$A_{sys} = -\frac{R_C \parallel R_L}{R_{Etotal}}$$

The final answer that we obtained for the system gain is the same answer we found in Chapter 10.

EXAMPLE 11–8
Load Resistor

Find the Box Model for the common-emitter amplifier shown in Figure 11–29. Do not include the load resistor in these calculations. Connect the load resistor to the Box Model and find the system gain.

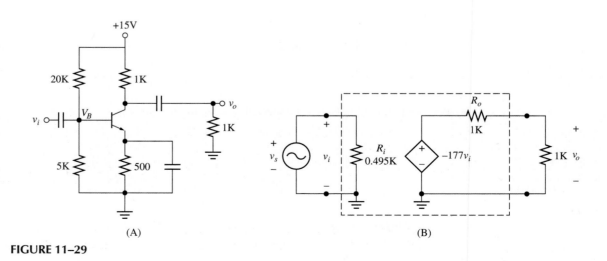

(A)   (B)

FIGURE 11–29

Solution The DC analysis (be sure to open all capacitors) is

$$V_B = \frac{5K}{5K + 20K} \, 15 = 3V$$

$$V_E = 3 - 0.7 = 2.3V$$

$$I_E = \frac{2.3}{0.5K} = 4.6mA$$

where $R_E = 500\Omega = 0.5K\Omega$.
If the transistor is **ACTIVE**

$$I_C = I_E = 4.6mA$$

and the collector voltage is

$$V_C = V_{CC} - R_C I_C = 15 - 1K \times 4.6mA = 10.4V$$

Because $V_C > V_E$, the transistor is **ACTIVE.** The AC emitter resistance is

$$r_e = \frac{26mV}{4.6mA} = 5.65\Omega = 0.00565K\Omega$$

Because the emitter resistor is bypassed

$$R_{Etotal} = r_e = 0.00565K\Omega$$

The base resistance is

$$R_B = R_1 \| R_2 = 5K \| 20K = 4K\Omega$$

The Box Model parameters are

$$R_i = R_B \| \beta R_{Etotal} = 4K \| 0.565K = 0.495K\Omega$$
$$R_o = R_C = 1K\Omega$$

and the gain is

$$A = -\frac{R_C}{R_{Etotal}} = -\frac{1K}{0.00565K} = -177$$

The Box Model is shown in Figure 11–29B. The system gain is found from

$$v_o = \frac{1K}{1K + 1K} \, (-177v_i) = -88.5v_i = -88.5v_s$$

so

$$A_{sys} = \frac{v_o}{v_s} = -88.5$$

DRILL EXERCISE

Redraw the circuit of Figure 11–29 with $V_{CC} = 9V$. Find the AC emitter resistance, input resistance, output resistance, and the system gain.

Answer $r_e = 11.8\Omega$ $R_i = 912\Omega$ $R_o = 1K\Omega$ $A_{sys} = -42.4$

Note: Because of the relatively large output resistance of the common-emitter, we have lost one-half of the no-load gain. This is typical with the common-emitter amplifier.

FIGURE 11–30

Common-emitter amplifier with source resistor

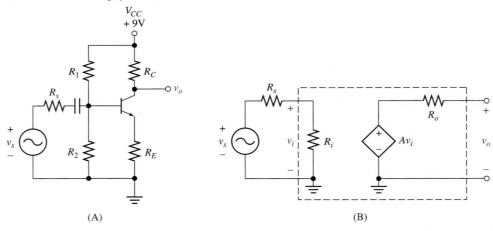

(A)   (B)

Source Resistor

Figure 11–30A shows a common-emitter amplifier that is driven by a signal source that has an internal resistance. Figure 11–30B shows the Box Model for the common-emitter amplifier and the source to which it is connected. We remind you that

$$R_{in} = R_B \parallel \beta R_{Etotal}$$

$$R_o = R_C$$

$$A = \frac{-R_C}{R_{Etotal}}$$

Here, from the input side we see that

$$v_i = \frac{R_i}{R_i + R_s} v_s$$

The output is, therefore

$$v_o = Av_i = A \frac{R_i}{R_i + R_s} v_s$$

Using the formula for A given above, we get

$$A_{sys} = \frac{v_o}{v_s} = A \frac{R_i}{R_i + R_s} = -\frac{R_C}{R_{Etotal}} \frac{R_i}{R_i + R_s}$$

EXAMPLE 11–9
Source Resistor

For the circuit in Figure 11–31, find the Box Model. Do not include the input source resistance. Draw the Box Model, connect the input source and its resistance, and find the system gain.

Solution   The DC analysis (be sure to open all capacitors) is

$$V_B = \frac{10K}{10K + 50K} 12 = 2V$$

$$V_E = 2 - 0.7 = 1.3V$$

$$I_E = \frac{1.3}{1K} = 1.3mA$$

EXAMPLE 11–10
The Complete Amplifier

Figure 11–33 shows an un-bypassed common emitter with a load resistor and a source resistor. Find the Box Model for the amplifier, excluding the load and source resistors. Use the Box Model to find the system gain.

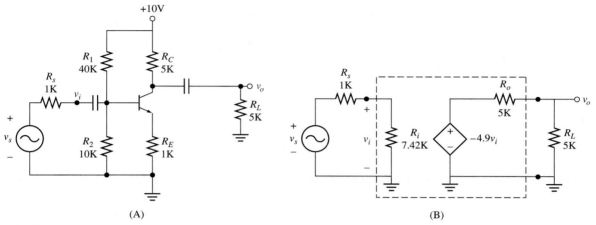

(A)   (B)

FIGURE 11–33

Solution   The DC analysis (be sure to open all capacitors) is

$$V_B = \frac{10K}{10K + 40K} \, 10 = 2V$$

$$V_E = 2 - 0.7 = 1.3V$$

$$I_E = \frac{1.3}{1K} = 1.3mA$$

If the transistor is **ACTIVE**

$$I_C = I_E = 1.3mA$$

and the collector voltage is

$$V_C = V_{CC} - R_C I_C = 10 - 5K \times 1.3mA = 3.5V$$

Because $V_C > V_E$, the transistor is **ACTIVE.** The AC emitter resistance is

$$r_e = \frac{26mV}{1.3mA} = 20\Omega = 0.02K\Omega$$

The Box Model parameters are

$$R_i = R_B \parallel \beta R_{Etotal} = 8K \parallel 100(1.02K) = 8K \parallel 102K = 7.42K\Omega$$

where $R_B = 10K \parallel 40K$ and $R_{Etotal} = r_e + R_E = 1.02K\Omega$

$$R_o = R_C = 5K\Omega$$

and the gain is

$$A = -\frac{R_C}{R_{Etotal}} = -\frac{5K}{1.02K} = -4.9$$

The Box Model connected to the source and load is shown in Figure 11–33B (p. 375). By analyzing the input and output voltage dividers, we can find the system gain.

$$v_i = \frac{7.42K}{7.42K + 1K} \, v_s = 0.881 v_s$$

$$v_o = \frac{5K}{5K + 5K} \, (-4.9v_i) = -2.45v_i$$

Substituting for v_i, we get

$$v_o = -2.45 \times 0.881 v_s = -2.16 v_s$$

and the system gain is

$$A_{sys} = \frac{v_o}{v_s} = -2.16$$

DRILL EXERCISE   Redraw the circuit shown in Figure 11–33 (p. 375) with $R_E = 2K\Omega$. Find the AC emitter resistance, input resistance, output resistance, and the system gain.

Answer $r_e = 40\Omega$ $R_i = 7.7K\Omega$ $R_o = 5K\Omega$ $A_{sys} = -1.1$

We can see two effects of not bypassing the emitter resistor in the previous example. The advantage is that the input resistance of the amplifier is high and we see most of the source voltage at the base of the transistor, i.e., $v_i = 0.881 v_s$. The disadvantage is that the gain is low because R_{Etotal} is large. There is, therefore, a trade-off between high input resistance and high gain for the single-stage common-emitter amplifier.

REVIEW QUESTIONS

1. What is the output resistance of the common-emitter amplifier?
2. What is the input resistance of the common-emitter amplifier?
3. What is the gain of the common-emitter amplifier?
4. What is the effect on the overall gain when a load resistor is connected to the common-emitter?
5. What is the effect on the overall gain when a source resistor is included?
6. What is the total gain of the common-emitter amplifier when both source and load resistors are included?

11–5 ■ MULTISTAGE AMPLIFIER

The real power of the Box Model can be shown in the analysis of a multistage amplifier. Figure 11–34A shows a two-stage amplifier in which the first common-emitter amplifier supplies the input to the second common-emitter amplifier. To simplify the analysis, each stage is identical, except for the fact that the emitter resistor is bypassed only in the second stage. Note that for DC analysis, the capacitor that couples the two stages is open. This means that we can do a separate DC analysis for each transistor stage (Figure 11–34B).

Because the two stages are identical for DC conditions, the DC analysis is the same for both. We find the DC bias levels for the first stage as follows:

FIGURE 11–34

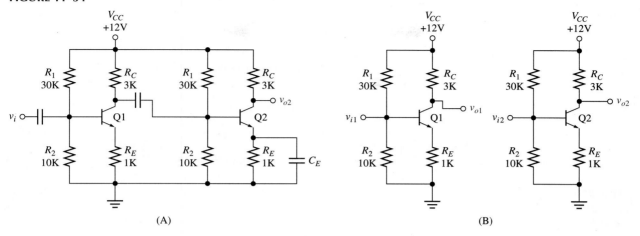

(A)

(B)

$$V_B = \frac{R_2}{R_1 + R_2} V_{CC} = \frac{10\text{K}}{10\text{K} + 30\text{K}} \, 12 = 3\text{V}$$

$$V_E = V_B - 0.7 = 3 - 0.7 = 2.3\text{V}$$

$$I_E = \frac{V_E}{R_E} = \frac{2.3}{1\text{K}} = 2.3\text{mA}$$

If the transistor is **ACTIVE**

$$I_C = I_E = 2.3\text{mA}$$

and the collector voltage is

$$V_C = V_{CC} - R_C I_C = 12 - 3\text{K} \times 2.3\text{mA} = 5.1\text{V}$$

Because $V_C > V_E$ the transistor is **ACTIVE.** The AC emitter resistance is

$$r_e = \frac{26\text{mV}}{I_C} = \frac{26\text{mV}}{2.3\text{mA}} = 11.3\Omega = 0.0113\text{K}\Omega$$

The DC bias levels and AC emitter resistance of the second stage are the same as those we found for the first stage.

Now the Box Model for each transistor is found from

$$R_i = R_B \parallel \beta R_{Etotal} = 7.5\text{K} \parallel 100 R_{Etotal}$$

$$R_o = R_C = 3\text{K}\Omega \text{ (output resistance for both stages)}$$

and

$$A = -\frac{R_C}{R_{Etotal}} = -\frac{3\text{K}}{R_{Etotal}}$$

For the first stage

$$R_{Etotal} = r_e + R_E = 1.011\text{K (no bypass capacitor)}$$

so

$$R_{i1} = 7.5\text{K} \parallel 100 \, (1.011\text{K}) = 6.98\text{K}\Omega$$

and

$$A_1 = -\frac{3\text{K}}{1.011\text{K}} = -3$$

For the second stage

$$R_{Etotal} = r_e = 0.0113\text{K}\Omega$$

so

$$R_{i2} = 7.5\text{K} \parallel 100(0.0113\text{K}) = 0.982\text{K}\Omega$$

$$A_2 = -\frac{3\text{K}}{0.011\text{K}} = -273$$

The Box Models for the first and second stage are shown in Figure 11–35A. Because the transistor stages have gains of -3 and -273, we might expect the overall gain to be $-3 \times -273 = 819$. We might—but we would be *wrong*.

FIGURE 11–35
Two-stage amplifier Box Models

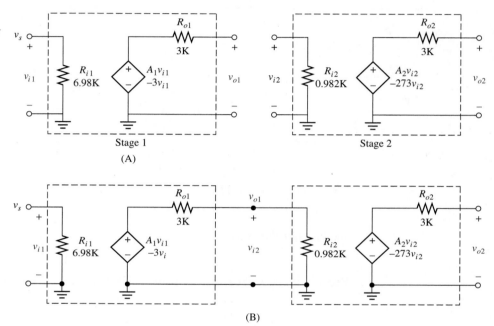

To make the proper analysis we must connect the two Box Models together (Figure 11–35B). Note that *the input resistance of stage two acts as the load for stage one*. We begin the analysis at the input of the first stage and work our way to the output of the second stage.

$$v_{i1} = v_s$$

Connecting the output of the first stage and the input of the second stage forms a voltage divider, so

$$v_{o1} = \frac{R_{i2}}{R_{i2} + R_{o1}} A_1 v_{i1} = \frac{0.982\text{K}}{0.982\text{K} + 3\text{K}} (-3v_s) = -0.74 v_s$$

Because the input to the second stage comes from the output of the first stage:

$$v_{i2} = v_{o1} = -0.74 v_s$$

The final output is, therefore

$$v_{o2} = A_2 v_{i2} = -273 (-0.74 v_s) = 202 v_s$$

The total two-stage gain is

$$A_{total} = \frac{v_{o2}}{v_s} = 202$$

As you see, the overall gain of the two-stage amplifier is less than the product of the individual stages.

Why? The second stage loads down the first stage and changes its effective gain.

Figure 11–35 shows that the input resistance of the two-stage amplifier is just the input resistance of the first stage. The output resistance of the two-stage amplifier is the output resistance of the second stage. We can now model the two-stage amplifier with a single Box Model, using the gain, input resistance, and output resistance, that we just found (Figure 11–36). Note that by using two stages, we have achieved both high input resistance and high gain.

FIGURE 11–36

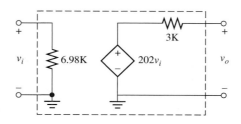

EXAMPLE 11–11
Two-Stage Amplifier

Find the overall Box Model for the two-stage common-emitter amplifier shown in Figure 11–37. Note that in this design both transistors use emitter bypass capacitors.

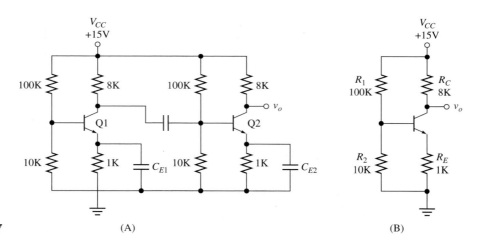

FIGURE 11–37 (A) (B)

Solution Because the two stages are coupled via capacitors, we can perform a separate DC analysis for each stage. (Because each stage is identical, we only have to analyze one stage.) The DC circuit for the first stage is shown in Figure 11–37B. The DC bias levels and AC emitter resistance are

$$V_B = \frac{R_2}{R_1 + R_2} V_{CC} = \frac{10K}{10K + 100K} 15 = 1.36V$$

$$V_E = V_B - 0.7 = 1.36 - 0.7 = 0.66V$$

$$I_E = \frac{V_E}{R_E} = \frac{0.66}{1K} = 0.66mA$$

If the transistor is **ACTIVE**

$$I_C = I_E = 0.66mA$$

and the collector voltage is

$$V_C = V_{CC} - R_C I_C = 15 - 8K \times 0.66\text{mA} = 9.72\text{V}$$

Because $V_C > V_E$, the transistor is **ACTIVE.** The AC emitter resistance is

$$r_e = \frac{26\text{mV}}{I_C} = \frac{26\text{mV}}{0.66\text{mA}} = 39.4\Omega = 0.0394\text{K}\Omega$$

For each stage (remember, they are identical), we get

$$R_i = R_B \parallel \beta R_{Etotal} = R_B \parallel \beta r_e = 9.09K \parallel 100(0.0394K) = 2.75K\Omega$$
$$R_o = 8K\Omega$$

$$A = -\frac{R_C}{r_e} = -\frac{8K}{0.0394K} = -203$$

The Box Models for the two stages are connected together (Figure 11–38A). The overall gain is found from

$$v_{i1} = v_s$$

$$v_{i2} = v_{o1} = \frac{R_{i2}}{R_{i2} + R_o} A v_i = \frac{2.75K}{2.75K + 8K} (-203v_s) = -51.9v_s$$

and

$$v_{o2} = A v_{i2} = -203 \, (-51.9v_s) = 10500v_s$$

The overall system gain is

$$A_{sys} = \frac{v_o}{v_s} = 10500$$

The Box Model for the two-stage amplifier is shown in Figure 11–38B.

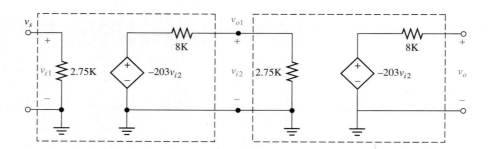

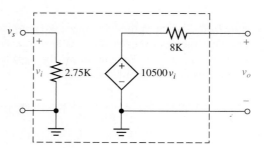

FIGURE 11–38

Note: We have achieved a very-high-gain amplifier at the cost of a small input resistance.

DRILL EXERCISE Redraw the circuit in Figure 11–37 (p. 379) with $V_{CC} = 9\text{V}$. Find the AC emitter resistance, input resistance, output resistance, and gain of each stage. Find the overall system gain.

Answer $r_e = 220\Omega$ $R_i = 6.43\text{K}\Omega$ $R_o = 8\text{K}\Omega$ $A = -36.4$ $A_{sys} = 590$

REVIEW QUESTIONS

1. Draw Box Models for two amplifiers.
2. Connect the Box Models together to represent a multistage amplifier.
3. What is the overall input resistance for the multistage amplifier?
4. What is the load resistance for the first amplifier?
5. What is the overall output resistance for the multistage amplifier?
6. What is the overall gain for the multistage amplifier?
7. Why isn't the overall gain of the multistage amplifier equal to the product of the gains of the individual stages?

SUMMARY

■ Any amplifier can be represented with a Box Model, which includes the input resistance (R_i), output resistance (R_o), and no-load gain of the amplifier (A).

■ A load resistor creates an output voltage divider and results in the system gain:

$$A_{sys} = \frac{R_L}{R_L + R_o} A$$

■ A source resistor creates an input voltage divider and results in the total system gain:

$$A_{sys} = \frac{R_i}{R_i + R_s} \frac{R_L}{R_L + R_L} A$$

■ The resistance seen at the base of a transistor is β times the total emitter resistance; that is

$$R_{base} = \beta R_{Etotal}$$

■ The common-emitter amplifier can be represented with a Box Model, in which

$$R_i = R_B \parallel \beta R_{Etotal}$$
$$R_o = R_C$$
$$A = -\frac{R_C}{R_{Etotal}}$$

■ If the emitter resistor is bypassed, common-emitter gain is high, but input resistance is low.

■ If the emitter resistor is not bypassed, common-emitter gain is low, but input resistance is high.

■ Multistage amplifiers can be represented by replacing each stage with its own Box Model and connecting these Box Models together. The input resistance of the following stage is the load resistance for the preceding stage.

PROBLEMS

SECTION 11–1 Introduction

1. You have a speaker that has a resistance of 14Ω. Draw the Thevenin equivalent for your speaker.
2. Your laboratory signal generator is set at 2V_{pp} amplitude and is set for a sine wave of 100Hz. The signal generator has an output impedance of 600Ω.
   (a) Draw the Thevenin equivalent of this voltage source.
   (b) What is v_s?   **(c)** What is R_s?

3. Your microphone has an output resistance of 100Ω. When it responds to a tone of 1KHz, the microphone produces a 10mV$_p$ sine wave.

   (a) Draw the Thevenin equivalent of the microphone.

   (b) What is v_s?   **(c)** What is R_s?

   (d) You observe the sine wave output of the microphone with an oscilloscope. What is the peak-to-peak voltage, and the time it takes for one complete cycle of the sine wave?

SECTION 11–2   The Box Model

4. Your stereo amplifier has two channels, both of which have an input resistance of 5KΩ and an output resistance of 1Ω. The volume control is set to amplify each channel's input signal by a factor of 80.

   (a) Draw the Box Model for one channel of your stereo amplifier.

   (b) What is R_i?   **(c)** What is R_o?   **(d)** What is the gain (A) of the amplifier?

   (e) If the output of each channel is 180° out of phase with the input signal (i.e., a positive input creates a negative output), is A positive or negative?

5. Consider the stereo amplifier from Problem 4. Each channel now drives a 4Ω speaker. How much of the amplified voltage actually reaches the speakers? Give your answer as both a fraction and a percentage of the amplified voltage.

6. Consider the one-channel audio amplifier shown in Figure 11–39. The amplifier is driving a 4Ω speaker through a pair of wires that are so thin (e.g. #28 gauge) that the wires have a total resistance of 5Ω.

   (a) If the amplified voltage is 10V$_p$, what is v_o?

   (b) Because of the voltage drop in the speaker wires, the load voltage does not equal the amplifier output voltage. What is the voltage that is delivered to the speaker (v_L)?

   (c) How much of the amplifier output voltage (v_o) is lost across the wires connecting the speaker and the amplifier?

FIGURE 11–39

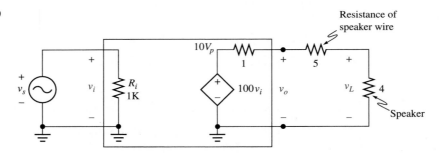

7. The input source in Figure 11–39 is a 10mV 1KHz voltage.

   (a) What is v_i?   **(b)** What is the output of the controlled voltage source?

   (c) What is v_o?   **(d)** What is v_L?

8. Consider the audio amplifier from Problem 6, shown in Figure 11–39. The speaker is connected to the amplifier by the same length of wire as before, but, this time, "Monster Cable" is used. Monster Cable is very thick and has a total resistance of only 0.05Ω.

   (a) Redraw the Box Model of the amplifier with the two resistors at the output, but give each resistor the values for this new situation.

   (b) For the same amount of amplifier voltage (i.e., 10V$_p$), how much voltage now gets to the speaker?

   (c) Compare v_o to v_L when you use Monster Cable.

   (d) What have you accomplished with the use of Monster Cable?

9. Repeat Problem 7 when the amplifier uses Monster Cable (defined in Problem 8).

10. A microphone and its pre-amplifier are shown in Figure 11–40.

   (a) What is the input to the microphone pre-amplifier (v_i) in peak millivolts?

   (b) What is the amplified voltage?

FIGURE 11–40

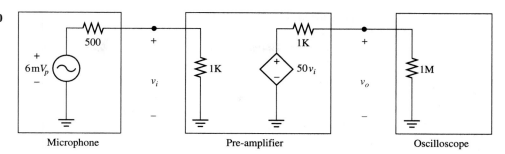

Microphone   Pre-amplifier   Oscilloscope

 (c) If the output of the pre-amplifier is connected to an oscilloscope, what will you find for v_o?

 (d) How much smaller is v_o than the amplifier voltage? Is this difference significant?

11. Figure 11–41 shows an amplifier that is connected between a transducer and a load.

 (a) What is v_s?   (b) What is v_i?   (c) What is the amplified voltage?

 (d) What is v_o?   (e) What is v_L?

 (f) What would v_o be if the load at the output is removed?

 (g) The nominal voltage gain for this amplifier is 10. What is the actual ratio of v_o/v_s?

 (h) Why is this ratio so much less than 10? Give at least two possible explanations.

FIGURE 11–41

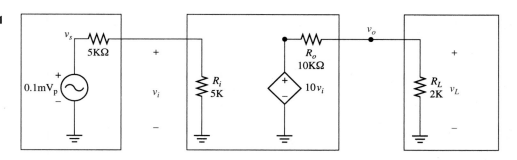

12. In Figure 11–41, what R_i would you need to deliver at least 95% of the input voltage to the amplifier input ($v_i = 0.95v_s$).

13. In Figure 11–41, what R_o would you need to deliver at least 98% of the amplifier output voltage to the load ($v_L = 0.98v_o$).

SECTION 11–3 The Common-Emitter Amplifier

14. For the common-emitter amplifier shown in Figure 11–42 (p. 384)

$$V_{CC} = +15V, R_1 = 12K\Omega, R_2 = 3K\Omega, R_C = 5K\Omega, R_E = 2K\Omega$$

 (a) What is I_E?   (b) Is the transistor **ACTIVE?**   (c) What is r_e?

 (d) What is R_{Etotal}?   (e) What is R_B?   (f) What is R_i?

 (g) What is R_o?   (h) What is the voltage gain, A?

 (i) Redraw the Box Model in the figure, and write in the proper values for R_i, A, and R_o.

15. Repeat Problem 14 if the parameters in Figure 11–42 are

$$V_{CC} = +9V, R_1 = 6.8K\Omega, R_2 = 2.2K\Omega, R_C = 2K\Omega, R_E = 1K\Omega$$

16. Repeat Problem 14 for the common-emitter amplifier in Figure 11–43 (p. 384), where

$$V_{CC} = +15V, R_1 = 12K\Omega, R_2 = 3K\Omega, R_C = 5K\Omega, R_E = 2K\Omega$$

17. Repeat Problem 14 if the parameters in Figure 11–43 (p. 384) are

$$V_{CC} = +9V, R_1 = 6.8K\Omega, R_2 = 2.2K\Omega, R_C = 2K\Omega, R_E = 1K\Omega$$

FIGURE 11–42

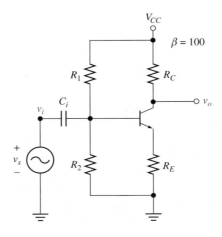

FIGURE 11–43

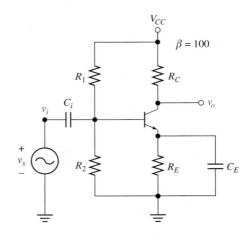

SECTION 11–4 Common-Emitter Amplifier with Load and Source Resistors

18. For the amplifier shown in Figure 11–44

$$V_{CC} = +15\text{V}, R_1 = 120\text{K}\Omega, R_2 = 30\text{K}\Omega, R_C = 5\text{K}\Omega, R_E = 2\text{K}\Omega$$

$$R_S = 1\text{K}\Omega, R_L = 10\text{K}\Omega$$

  (a) What is the resistance associated with the voltage source?

  (b) What is R_i for the amplifier?   **(c)** What is R_o for the amplifier?

  (d) What is the voltage gain (A) of the amplifier when R_L is removed?

  (e) Redraw the Box Model shown in the figure. Write in the proper values for: R_s, R_i, A, R_o, and R_L.

19. The amplifier shown in Figure 11–44 has the parameters given in Problem 18:

  (a) How much of the signal source voltage (v_s) gets to v_i, the input voltage of the amplifier?

  (b) If v_s is 10mV$_p$, what is v_i?

  (c) If v_s is 10mV$_p$, what is the no-load amplified voltage (Av_i)?

  (d) How much of Av_i (the no-load amplifier voltage) gets to v_o when there is no load?

  (e) How much of Av_i gets to v_o when the load resistor (R_L) is connected to the amplifier?

  (f) What is the total voltage gain for the entire system $(A_{sys} = v_o/v_s)$?

20. Repeat Problem 18 if the transistor current gain is $\beta = 250$.

21. Repeat Problem 19 if the transistor current gain is $\beta = 250$.

FIGURE 11–44

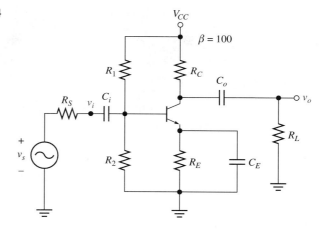

22. Redraw Figure 11–44 with the following parameters:

$$V_{CC} = +12V, R_1 = 103K\Omega, R_2 = 17K\Omega, R_C = 5K\Omega, R_E = 1K\Omega$$

    (a) What is R_i for the amplifier?   **(b)** What is R_o for the amplifier?

    (c) What is the voltage gain *(A)* of the amplifier when R_L is removed?

    (d) Redraw the Box Model shown in the figure. Write in the proper values for: R_i, *A,* and R_o.

23. For the circuit of Problem 22, what is the maximum source resistance that can be used if you want to deliver 98% of the source voltage to the amplifier $(v_i = 0.98v_S)$?

24. For the circuit of Problem 22, what is the minimum load resistor that can be used if you want to deliver 75% of the amplifier output to the load $(v_L = 0.75v_o)$?

25. Change the current gain of the transistor in Figure 11–44 to $\beta = 250$, and repeat Problem 22.

26. For the circuit of Problem 25, what is the maximum source resistance that can be used if you want to deliver 98% of the source voltage to the amplifier $(v_i = 0.98v_S)$?

Computer Problems

27. For the common-emitter transistor in Figure 11–42 and the parameters of Problem 14, use transient analysis, or small-signal (phasor) analysis, to find

    (a) The no-load gain.   **(b)** The resistance seen at the base (R_{base}).

    (c) The input resistance (R_i).

28. Find the output resistance of the transistor in Figure 11–42.

       Input resistance can be found by simply measuring the voltage and current at the input. This will not work for output resistance. We can find the output resistance of a transistor amplifier (excluding the load) by finding the Thevenin equivalent at the output as follows:

    (a) Apply a small-signal input (sinusoidal transient or phasor) of 1mV at 1000Hz. The low input voltage will keep the transistor in its **ACTIVE** region, and the frequency will ensure that coupling and bypass capacitors will short.

    (b) Measure the no-load voltage at the output. This is the Thevenin output voltage.

    (c) Connect a load resistor through a coupling capacitor (30μF) to the transistor output. (The coupling capacitor is needed to prevent changing the DC bias levels.) There are now two ways to proceed.

        (i) Run enough simulations, changing the load resistor for each run until the load voltage is one-half of the Thevenin voltage. The output resistance is equal to the load resistor at this point.

       (ii) Make a single simulation with a load resistor that you think is close in value to the output resistance. The load voltage is given by

$$v_o = \frac{R_L}{R_L + R_o} v_{TH}$$

    The only unknown in this equation is the output resistance R_o, which then can be found.

29. For the common-emitter transistor in Figure 11–43 (p.384) and the parameters of Problem 16 (p. 383), use transient analysis, or small-signal (phasor) analysis, to find

   (a) The no-load gain.   **(b)** The resistance seen at the base (R_{base}).

   (c) The input resistance (R_i).   **(d)** The output resistance (R_o).

30. For the common-emitter transistor in Figure 11–44 (p. 385) and the parameters of Problem 18 (p. 384), use transient analysis, or small-signal (phasor) analysis, to find

   (a) The no-load gain.   **(b)** The resistance seen at the base (R_{base}).

   (c) The input resistance (R_i).   **(d)** The output resistance (R_o).

   (e) Connect the source resistor, and find the total gain.

   (f) Remove the source resistor, connect the load resistor, and find the total gain.

   (g) Find the total gain with both source and load resistor connected.

CHAPTER

12

THE EMITTER FOLLOWER (THE COMMON–COLLECTOR AMPLIFIER)

C**HAPTER OBJECTIVES**

To use a buffer amplifier to isolate a load from its signal source.

To determine the desirable characteristics of the buffer amplifier.

To find the gain, input resistance, and output resistance of the emitter-follower buffer amplifier.

To use the emitter follower as a current buffer.

To use the emitter follower to improve Zener diode regulation.

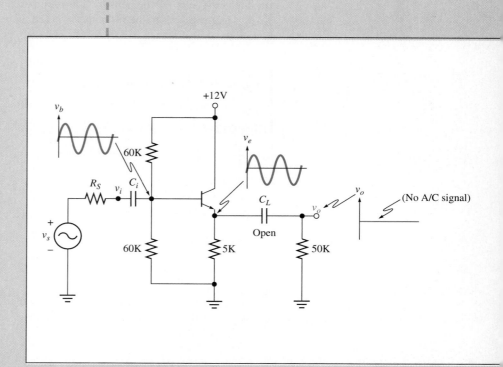

12–1 ■ INTRODUCTION—THE BUFFER AMPLIFIER

You have just landed a great job at the university medical center to help build and test one-of-a-kind amplifiers for special medical research projects. Your boss has built an amplifier with a voltage gain of 100. You and she go into the lab and connect the amplifier between the sensor attached to the volunteer subject (probably a poor graduate student) and the pen recorder (Figure 12–1).

You are ready to proceed with the experiment. You turn on the pen recorder; you turn on the amplifier; you turn on the graduate student. Nothing happens! The pen recorder does not move. Your boss does a quick check and announces that the subject is not dead and the electronics seem to be OK. She walks out muttering, "Laurie, you fix it."

FIGURE 12–1

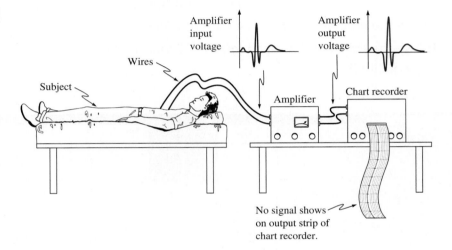

You're about to panic. "If my boss can't find the problem, how can I?" Then, you remember Chapter 11 of your introductory electronics textbook, and you know what to check for. There may be a mismatch in the input resistance of your boss's amplifier and the internal resistance of the sensor. Consider, for example, the possibility shown in Figure 12–2. The mismatch in the resistances on the input side results in

$$v_i = \frac{R_i}{R_i + R_{sensor}} \, v_{sensor} = \frac{100K}{100K + 1M} \, v_{sensor} = 0.0909 v_{sensor}$$

and most of the sensor signal is lost.

FIGURE 12–2

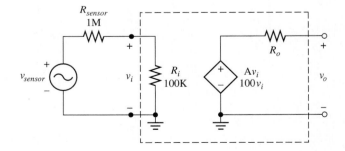

Figure 12–3 shows what can happen if there is also a mismatch between the output resistance of the amplifier and the resistance of the load (the pen recorder, in this case). Here the recorder voltage is

$$V_{recorder} = \frac{R_{recorder}}{R_{recorder} + R_o} Av_i = \frac{1K}{1K + 10K} 100 \times 0.0909 v_{sensor} = 0.826 v_{sensor}$$

The mismatch on the output side causes additional loss in the signal.

No wonder the pen recorder didn't move. The mismatches between input and output resistance destroy the effectiveness of the amplifier. Rather than telling your boss that her amplifier is useless, you can suggest connecting **buffer circuits** to the amplifier input or output, as needed. A buffer *isolates* a load from the signal source. That is, a buffer is used to prevent the loss of signal strength due to mismatches in the signal and load resistances.

FIGURE 12–3

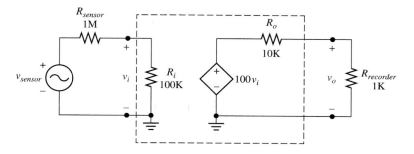

The ideal buffer can be modeled as shown in Figure 12–4A. The input resistance of the ideal buffer is infinity (open circuit), output resistance is 0Ω, and gain is 1. There is no real need for a buffer to have any gain. Figure 12–4B shows how a buffer can isolate a signal source from its load. Because the input resistance of the ideal buffer is infinity

$$v_{buff} = v_s$$

no matter how large the source resistance is. Because the output resistance of the ideal buffer is 0Ω, the load voltage is

$$v_o = v_{buff} = v_s$$

no matter how small the load resistance is.

FIGURE 12–4
The buffer amplifier

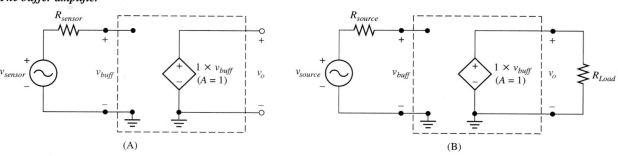

(A)   (B)

Of course, no electronic device is ideal. An actual buffer has very high input resistance, very low output resistance, and a gain close to 1.

Figure 12–5 (p. 392) shows the results of connecting a buffer between the sensor and the amplifier input. We now have two voltage dividers in the circuit. We analyze them one at a time. At the input to the first buffer

$$v_{buff} = \frac{100M}{100M + 1M} v_s = 0.99 v_s$$

FIGURE 12–5

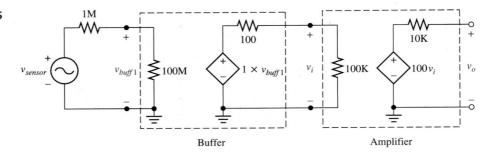

Buffer   Amplifier

At the input to the amplifier

$$v_i = \frac{100K}{100K + 100} v_{buff} = 1.0 \times 0.99v_s = 0.99v_s$$

Because of the buffer we have not lost very much of the input sensor signal.

We can also connect a second buffer between the output of the amplifier and the load (Figure 12–6). The amplifier output is the input to the second buffer. That is,

$$v_{buff2} = \frac{1M}{1M + 10K} Av_i = 0.99 \times 100 \times 0.99v_s = 98v_s$$

Finally, the voltage seen at the recorder is given by

$$v_o = \frac{1K}{1K + 10} v_{buff2} = 0.99 \times 98v_s = 97v_s$$

FIGURE 12–6

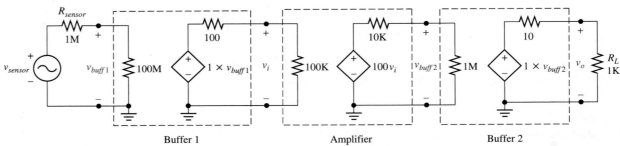

Buffer 1   Amplifier   Buffer 2

The output buffer has compensated for the relatively low resistance of the pen recorder. Through the use of buffers, we have provided the amplifier with high input and low output resistance.

EXAMPLE 12–1
The Buffer

Find the voltage gain for the voltage divider in Figure 12–7A. Then insert the buffer in Figure 12–7B, and find the voltage gain again.

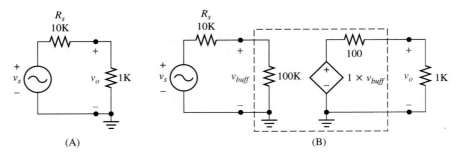

FIGURE 12–7   (A)   (B)

Solution   For the voltage divider of Figure 12–7A, we find

$$v_o = \frac{1K}{1K + 10K} v_{in} = 0.0909 v_{in}$$

The buffer in Figure 12–7B forms two voltage dividers. From the first, we find

$$v_{buff} = \frac{100K}{100K + 10K} v_{in} = 0.909 v_{in}$$

At the output side of the buffer, we get

$$v_o = \frac{1K}{1K + 0.1K} v_{buff} = 0.909 v_{buff}$$

The total output is found from

$$v_o = 0.909 v_{buff} = 0.909 \times 0.909 v_{in} = 0.826 v_{in}$$

Note that even with the buffer, we still lose approximately 17% of the input signal. This level is too great for adequate performance. The next example explores some criteria for buffer design.

EXAMPLE 12–2
Buffer Design

Consider the buffer inserted between a signal source and a load in Figure 12–8. Find the voltage loss between the signal source and the buffer input, the buffer output and the load, and the total loss between the signal source and the load, if

(a) $R_i = 10 R_{source}$ and $R_o = 0.10 R_{load}$
(b) $R_i = 100 R_{source}$ and $R_o = 0.01 R_{load}$

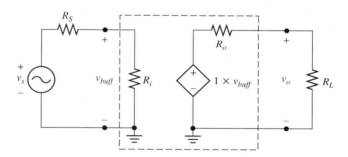

FIGURE 12–8

Solution   At the input to the buffer, we get

$$v_{buff} = \frac{R_i}{R_i + R_{source}} v_{in}$$

At the output of the buffer, we get

$$v_o = \frac{R_{load}}{R_{load} + R_o} v_{buff}$$

The overall voltage transfer from input to load is the product of the above; that is

$$v_o = \frac{R_{load}}{R_{load} + R_o} \frac{R_i}{R_i + R_{source}} v_{in}$$

(a) For $R_i = 10R_{source}$

$$v_{buff} = \frac{10R_{source}}{10R_{source} + R_{source}} \, v_{in} = \frac{10}{11} \, v_{in} = 0.909v_{in}$$

We have lost approximately 9% of the input signal. At the output of the buffer when

$$R_o = 0.10R_{load}$$

$$v_o = \frac{R_{load}}{R_{load} + 0.10R_{load}} \, v_{buff} = 0.909v_{buff}$$

and we have, again, lost another 9% at the buffer output. The total system transfer of voltage is given by

$$v_o = 0.909v_{buff} = 0.909 \times 0.909v_{in} = 0.826v_{in}$$

Therefore, the total loss is 17% of the input signal.

(b) For $R_i = 100R_{source}$

$$v_{buff} = \frac{100R_{source}}{100R_{source} + R_{source}} \, v_{in} = \frac{100}{101} \, v_{in} = 0.99v_{in}$$

We have now lost only 1% of the input signal. At the output of the buffer when $R_o = 0.01R_{load}$:

$$v_o = \frac{R_{load}}{R_{load} + 0.01R_{load}} \, v_{buff} = 0.99v_{buff}$$

and we have again lost only 1% at the buffer output. The total system transfer of voltage is given by

$$v_o = 0.99v_{buff} = 0.99 \times 0.99v_{in} = 0.98v_{in}$$

Therefore, the total loss is 2% of the input signal.

Note: To keep signal losses to a minimum, the well-designed buffer should have an input resistance that is 100 times greater than the source resistance and an output resistance that is 100 times smaller than the load resistance.

DRILL EXERCISE

You have a microphone with an internal resistance of 1KΩ. You want to connect it to an amplifier that has a 1KΩ input resistance. You are going to insert a buffer between the microphone and the amplifier. What do you want for the input and output resistances of the buffer?

Answer $R_i = 100\text{K}\Omega$ and $R_o = 10\Omega$

REVIEW QUESTIONS

1. When you connect two electronic devices together, why do you have to worry about their input and output resistances?

2. What is the desired relation between one device's output resistance and the other device's input resistance?

3. If the desired relation from Question 2 is not available, what can you insert between your two devices?

4. What properties does the ideal buffer have?

5. What makes a good buffer?

12–2 ■ THE EMITTER FOLLOWER (COMMON-COLLECTOR AMPLIFIER)

The transistor can be used to construct the buffer amplifier shown in Figure 12–9A. Note that in this circuit the load is coupled via a capacitor to the emitter. This circuit arrangement is known as the **common-collector amplifier.** Because capacitors are open for DC, the bias levels are found from the DC circuit of Figure 12–9B. You can see that this DC circuit is the self-biased transistor circuit that we analyzed earlier.

FIGURE 12–9
The common-collector (emitter-follower) amplifier

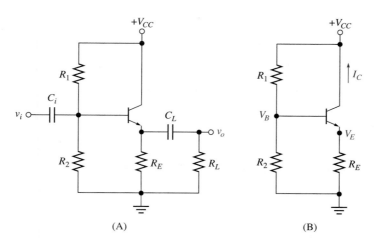

(A)   (B)

The gain of all properly-designed buffer amplifiers is close to the ideal value of 1; however, the input and output resistances can vary significantly from their ideal values. For this reason, we want to derive the Box Model for this transistor circuit. The Box Model provides the clearest representation of an amplifier's input and output resistances.

FIGURE 12–10
Emitter-follower gain

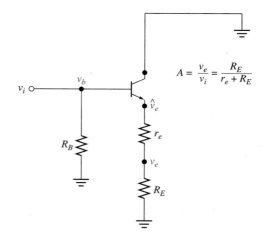

$$A = \frac{v_e}{v_i} = \frac{R_E}{r_e + R_E}$$

No-Load Gain

To determine the small-signal no-load gain of the common collector, we analyze the AC circuit in Figure 12–10A. Note that the biasing resistors, R_1 and R_2, are in parallel and have been replaced with

$$R_B = R_1 \parallel R_2$$

The no-load gain for the Box Model is defined as

$$A = \frac{v_o}{v_i}$$

For the common collector

$$v_i = v_b$$

and

$$v_o = v_e$$

so

$$A = \frac{v_e}{v_b}$$

Note that now the output voltage is taken from the emitter. From Figure 12–10 (p. 395) you can see that r_e and R_E form a voltage divider, so

$$v_e = \frac{R_E}{R_E + r_e} \hat{v}_e$$

As previously discussed, in the AC transistor circuit

$$\hat{v}_e = v_b$$

so

$$v_e = \frac{R_E}{R_E + r_e} v_b$$

The no-load gain for the common collector is

$$A = \frac{v_e}{v_b} = \frac{R_E}{R_E + r_e}$$

Note that for the common collector, the gain is positive. In the well-designed common collector

$$r_e << R_E$$

and the gain becomes

$$A = \frac{R_E}{R_E + r_e} \approx \frac{R_E}{R_E} = 1$$

That is, in the well-designed common collector, the gain is approximately 1 and the output voltage equals the input voltage. In other words, the output (v_e) follows the input (v_b). Because of this, buffers are often called *followers*. The common-collector circuit is more commonly called an **emitter follower.**

In Chapter 9 we showed that the emitter voltage is always a diode drop below the base voltage. That is, with the exception of the 0.7V difference, the emitter voltage looked the same as the base voltage. Even for DC inputs, therefore, the emitter voltage follows the base voltage. The name *emitter follower* comes from this fact.

EXAMPLE 12–3
Emitter-Follower Box Model Gain

The emitter follower in Figure 12–11 is designed to operate with a collector bias current of

$$I_C = 1\text{mA}$$

Find the system gain of the emitter follower for the following emitter resistors:

(a) $R_E = 2.6\text{K}\Omega$ **(b)** $R_E = 260\Omega$

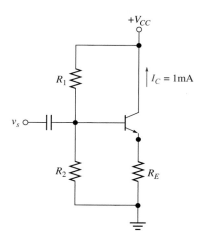

FIGURE 12–11

Solution   Because for each case the collector bias current is 1mA

$$r_e = \frac{26\text{mV}}{1\text{mA}} = 26\Omega = 0.026\text{K}\Omega$$

Therefore

(a) For $R_E = 2.6\text{K}\Omega$

$$A = \frac{R_E}{R_E + r_e} = \frac{2.6\text{K}}{2.6\text{K} + 0.026\text{K}} = 0.99$$

For an R_E that is 100 times greater than r_e, the gain is nearly 1.

(b) For $R_E = 260\Omega$

$$A_{sys} = \frac{R_E}{R_E + r_e} = \frac{260}{260 + 26} = 0.909$$

For an R_E that is 10 times greater than r_e, we have lost approximately 10% of the gain. In the well-designed emitter follower, R_E should be 100 times greater than r_e.

DRILL EXERCISE   Repeat Example 12–3 with a DC bias current of $I_C = 0.1\text{mA}$. What is the AC emitter resistance?

Answer $r_e = 260\Omega$ **(a)** $A = 0.909$ **(b)** $A = 0.5$

You should be aware that most textbooks do not give the no-load gain, as we do. Rather, they include the load resistor in their gain calculation. From Figure 12–9A (p. 395), you can see that for AC, the load coupling capacitor (C_L) shorts and puts R_L in parallel with R_E. The system gain between the base voltage and the output voltage, therefore, becomes

$$A_{sys} = \frac{v_o}{v_b} = \frac{R_E \parallel R_L}{r_e + R_E \parallel R_L}$$

Be aware, however, that the actual system gain of an emitter-follower circuit depends on the resistance of the signal source. The formula given above assumes that the signal-source resistance is 0Ω. Because we use a buffer when the signal source has significant internal resistance, this formula is not very useful. A better approach is to use the Box Model to account for all resistances.

Input Resistance

For reasons too complicated to discuss here, input and output resistances of the Box Model of the emitter follower depend on the source and load resistances. When we find the input resistance of the emitter follower, we must include the load resistor. When we find the output resistance of the emitter follower, we must include the source resistor. In other words, the Box Model of the emitter follower depends on the input and output circuit to which the emitter follower is connected.

To find the input resistance of the emitter follower, we analyze the AC circuit of Figure 12–12. Note that the load resistor must be included here; it is in parallel with the emitter resistor. In Figure 12–12B we have replaced the parallel connection with the single resistor $(\hat{R}_E)$, where

$$\hat{R}_E = R_E \parallel R_L = \frac{R_E R_L}{R_E + R_L}$$

FIGURE 12–12
Emitter-follower input resistance

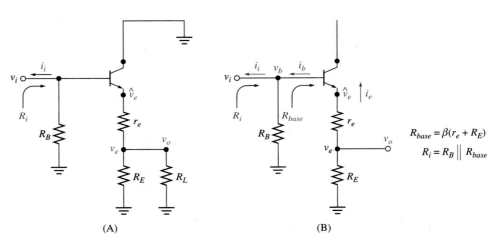

$$R_{base} = \beta(r_e + R_E)$$
$$R_i = R_B \parallel R_{base}$$

(A)   (B)

This circuit may look familiar. In fact, it is the same circuit we used to find the input resistance of the common-emitter amplifier of the previous chapter. The only difference is that we now have $\hat{R}_E$ instead of R_E. Rather than just repeat the input resistance formula, however, we will rederive it.

Consider, first, the resistance seen at the base (Figure 12–12B):

$$R_{base} = \frac{v_b}{i_b}$$

On the emitter side you can see that

$$\hat{v}_e = (r_e + \hat{R}_E)i_e$$

Because $i_e = \beta i_b$, and because $\hat{v}_e = v_b$

$$v_b = (r_e + \hat{R}_E)\beta i_b$$

and the input resistance at the base is

$$R_{base} = \frac{v_b}{i_b} = \beta(r_e + \hat{R}_E)$$

Once again we see that the resistance at the base is β times greater than the resistance in the emitter; that is

$$R_{base} = \beta R_{Etotal}$$

where, for the emitter follower

$$R_{Etotal} = r_e + \hat{R}_E = r_e + (R_E \| R_L)$$

Finally, the total input resistance is the parallel combination of R_B and R_{base}. That is,

$$\boxed{R_i = R_B \| R_{base} = R_B \| \beta R_{Etotal}}$$

You may remember that the parallel combination of two resistors is smaller than either of the individual resistors. In the emitter follower the presence of the base resistors R_1 and R_2 lowers the input resistance. In Problem 15 (p. 415), we show you how using both a positive and a negative supply voltage eliminates the need for the base resistors.

EXAMPLE 12–4
Emitter Follower
Input Resistance

Examine the effect of the base biasing resistors on the input resistance of the emitter follower. In Figure 12–13A find R_i if

$$I_C = 1\text{mA}$$

and

(a) $R_B = 10\text{K}\Omega$.   **(b)** $R_B = 50\text{K}\Omega$.   **(c)** $R_B = 500\text{K}\Omega$.   **(d)** $R_B = 5\text{M}\Omega$.

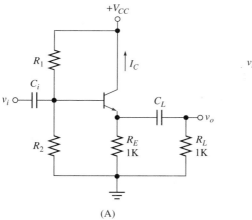 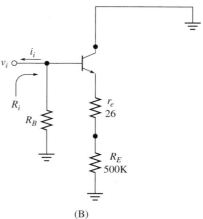

FIGURE 12–13   (A)   (B)

Solution   Because $I_C = 1\text{mA}$

$$r_e = \frac{26\text{mV}}{1\text{mA}} = 26\Omega$$

The AC circuit for the emitter follower is shown in Figure 12–13B. We first find the resistance seen at the base (R_{base}) from

$$\hat{R}_E = R_E \| R_L = \frac{1\text{K} \times 1\text{K}}{1\text{K} + 1\text{K}} = 500\Omega$$

$$R_{Etotal} = r_e + \hat{R}_E = 26 + 500 = 526\Omega = 0.526\text{K}\Omega$$

The resistance at the base is

$$R_{base} = \beta R_{Etotal} = 100 \times 0.526\text{K}\Omega = 52.6\text{K}\Omega$$

The total input resistance is given by

$$R_i = R_B \parallel R_{base}$$

(a) For $R_B = 10K\Omega$, the total input resistance is, therefore

$$R_i = 10K \parallel 52.6K = \frac{10K \times 52.6K}{10K + 52.6K} = 8.4K\Omega$$

Because the smallest resistor in a parallel combination dominates, input resistance is set by the smaller value of R_B, instead of the larger value of R_{base}.

(b) For $R_B = 50K\Omega$, the total input resistance is

$$R_i = 50K \parallel 52.6K = \frac{50K \times 52.6K}{50K + 52.6K} = 25.6K\Omega$$

Increasing R_B has increased the input resistance. Still, the input resistance is only one-half of R_{base}.

(c) For $R_B = 500K\Omega$, the total input resistance is

$$R_i = 500K \parallel 52.6K = \frac{500K \times 52.6K}{500K + 52.6K} = 47.6K\Omega$$

We have increased R_B to the point that the input resistance is approximately the same as R_{base}.

(d) For $R_B = 5M\Omega$, the total input resistance is

$$R_i = 5M \parallel 52.6K = \frac{5000K \times 52.6K}{5000K + 52.6K} = 52.1K\Omega$$

Here we see that further increases in R_B do not significantly change the input resistance.

DRILL EXERCISE   Repeat Example 12–4 (p. 399) if $I_C = 0.01mA$. Find the AC emitter resistance.

Answer $r_e = 2.6K\Omega$ **(a)** $R_i = 9.69K\Omega$ **(b)** $R_i = 43.1K\Omega$
                                   (c) $R_i = 191K\Omega$ **(d)** $R_i = 292K\Omega$

Output Resistance

As with emitter-follower input resistance, emitter-follower output resistance depends on the total circuit. Note, however, that we still do not include the load resistor in this calculation. Consider the circuit in Figure 12–14A and its AC equivalent in Figure 12–14B.

To find the output resistance, we must first turn off the input source (Figure 12–15A). You can see that the source resistance is now in parallel with the base biasing resistance. In Figure 12–15B we have combined these two resistances:

$$\hat{R}_B = R_{source} \parallel R_B$$

The major question here is how $\hat{R}_B$ affects the output resistance (Figure 12–16A). If you imagine standing inside the emitter and looking towards the base (Figure 12–16B), you will see the resistance:

$$R_{EB} = -\frac{\hat{v}_e}{i_e}$$

FIGURE 12–14

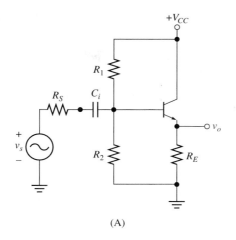

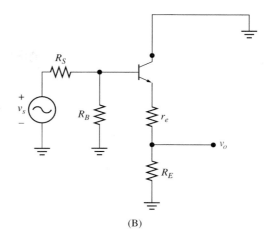

(A)   (B)

FIGURE 12–15
Emitter-follower output resistance

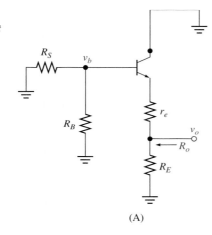

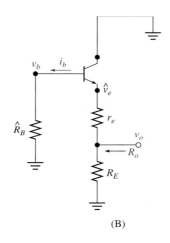

$$\hat{R}_B = R_S \parallel R_B$$

$$R_o = R_E \parallel \left(\frac{\hat{R}_B}{\beta} + r_e \right)$$

(A)   (B)

FIGURE 12–16

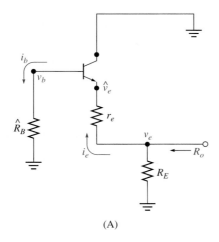

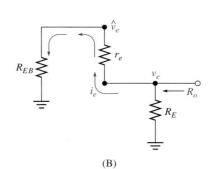

(A)   (B)

The minus sign is due to the current direction in R_{EB}. To find R_{EB} we first note that

$$\hat{v}_e = v_b$$

and

$$i_e = \beta i_b$$

From Figure 12–16A (p. 401), we see that

$$v_b = -\hat{R}_B i_b$$

Again, note the minus sign. Therefore

$$R_{EB} = -\frac{\hat{v}_e}{i_e} = -\frac{v_b}{\beta i_b} = -\frac{-\hat{R}_B i_b}{\beta i_b}$$

and we get

$$R_{EB} = \frac{\hat{R}_B}{\beta}$$

That is

☐ The resistance seen at the emitter is the total base resistance divided by β.

From Figure 12–16B (p. 401), we see that the R_{EB} and r_e are in series, and the combination is in parallel with R_E. Therefore, the total output resistance of the emitter follower is found from

$$R_o = R_E \parallel (r_e + R_{EB})$$

so

$$R_o = R_E \parallel \left(r_e + \frac{\hat{R}_B}{\beta} \right) \qquad \hat{R}_B = R_S \parallel R_B$$

EXAMPLE 12–5
Emitter-Follower
Output Resistance

Find the output resistance for the emitter follower in Figure 12–17.

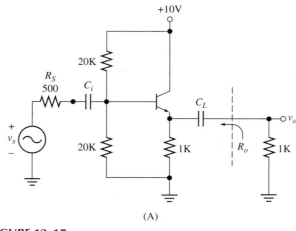

(A)

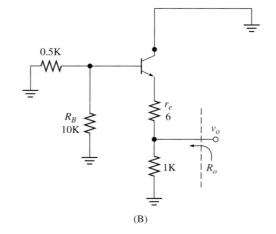

(B)

FIGURE 12–17

Solution From the DC analysis we find that

$$V_B = \frac{20\text{K}}{20\text{K} + 20\text{K}}\, 10 = 5\text{V}$$

$$V_E = V_B - 0.7 = 5 - 0.7 = 4.3\text{V}$$

$$I_C = \frac{4.3\text{V}}{1\text{K}} = 4.3\text{mA}$$

Note that choosing $R_1 = R_2$ in the base bias circuit gives us an emitter bias voltage that is approximately one-half of the collector supply voltage. This will allow a large symmetrical swing at the emitter.

The AC emitter resistance is

$$r_e = \frac{26\text{mV}}{4.3\text{mA}} = 6.05\Omega$$

The AC circuit is shown in Figure 12–17B, in which

$$R_B = 20\text{K} \parallel 20\text{K} = 10\text{K}\Omega$$

From the AC circuit, we see that

$$\hat{R}_B = R_S \parallel R_B = 0.5\text{K} \parallel 10\text{K} = 0.476\text{K}\Omega = 476\Omega$$

and

$$R_{EB} = \frac{\hat{R}_B}{\beta} = 4.76\Omega$$

Finally

$$R_o = R_E \parallel (r_e + R_{EB}) = 1000 \parallel (6.05 + 4.76) = 10.8\Omega$$

DRILL EXERCISE Redraw the circuit in Figure 12–17, changing the supply voltage to $V_{CC} = 2\text{V}$. Find the AC emitter resistance and the output resistance for this emitter follower.

Answer $r_e = 86.7\Omega$ $R_o = 83.8\Omega$

The Complete Box Model

Figure 12–18 (p. 404) shows the complete Box Model for the emitter follower. The Box Model parameters are summarized as follows:

$$A = \frac{R_E}{r_e + R_E}$$

$$R_i = R_B \parallel \beta(r_e + \hat{R}_E) \qquad \hat{R}_E = R_E \parallel R_L$$

$$R_o = R_E \parallel \left(r_e + \frac{R_{EB}}{\beta} \right) \qquad R_{EB} = R_S \parallel R_B$$

FIGURE 12–18
Emitter-follower box model

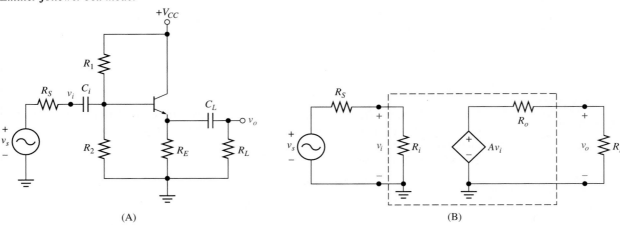

(A)　　　　　　　　　　　　　　　　　　(B)

EXAMPLE 12–6
The Complete Emitter Follower

To complete the analysis of the emitter follower, find the Box Model for the circuit shown in Figure 12–17A. Connect the source and the load to the Box Model, and find the system gain.

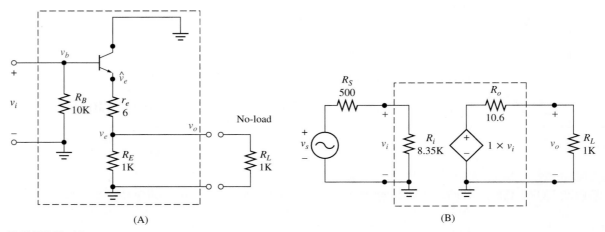

(A)　　　　　　　　　　　　　　　　　　(B)

FIGURE 12–19

Solution   Gain

To find the no-load gain of this emitter follower, we remove the signal source (with its internal resistance) and the load resistor, then analyze the AC circuit in Figure 12–19A. Because $\hat{v}_e = v_b$ and $r_e = 6\Omega$

$$v_e = \frac{R_E}{R_E + r_e} v_b = \frac{1\text{K}}{1\text{K} + 0.006\text{K}} v_b = v_b$$

Because $v_o = v_e$ and $v_i = v_b$, the Box Model gain for this emitter follower is

$$A = \frac{v_o}{v_i} = 1$$

As you can see, the no-load gain of the emitter follower is so close to 1 that we usually don't even bother solving for it.

Input Resistance

To find the input resistance, we must include the load resistor as follows:

$$\hat{R}_E = R_E \parallel R_L = 1K \parallel 1K = 0.5K\Omega$$

The resistance looking into the base is, therefore

$$R_{base} = \beta(r_e + \hat{R}_E) = 100 \times 0.506K\Omega = 50.6K\Omega$$

The total input resistance is found from

$$R_i = R_B \parallel R_{base} = 10K \parallel 50.6K = 8.35K\Omega$$

Output Resistance

The output resistance is found from

$$\hat{R}_B = R_S \parallel R_B = 0.5K \parallel 10K = 476\Omega$$

$$R_{EB} = \frac{\hat{R}_B}{\beta} = 4.76\Omega$$

Finally

$$R_o = R_E \parallel (r_e + R_{EB}) = 1000 \parallel 10.8 = 10.6\Omega$$

Figure 12–19B shows the complete Box Model, along with the signal generator and the load. The total system gain is given by the product of two voltage dividers.

$$v_i = \frac{8.35K}{8.35K + 0.5K} v_s = 0.944v_s$$

and

$$v_o = \frac{1K}{1K + .0106K} v_b = 0.99v_i$$

The overall output is found from

$$v_o = 0.99v_i = 0.99 \times 0.94v_s = 0.925v_s$$

and the total gain is

$$A_{sys} = \frac{v_o}{v_s} = 0.925$$

That is, with the emitter follower in place, 93% of the signal generator input gets to the output.

REVIEW QUESTIONS

1. Draw the common collector coupled via capacitors with a signal source and a load.
2. What is the gain between the emitter and base voltages of the common collector?
3. Why is the common collector more popularly known as the emitter follower?
4. What is the desired relation between r_e and R_E for a well-designed emitter follower?
5. What is the input resistance of the emitter follower?
6. What is the output resistance of the emitter follower?
7. What is unusual about the input and output resistances of the emitter follower?
8. What features of the emitter follower make it a good buffer?
9. Draw the Box Model of the emitter follower.
10. In the Box Model connect a signal source with its internal resistance to the input; connect a resistive load to the output. Find the system gain for this circuit.

12–3 ■ THE EMITTER FOLLOWER AS A CURRENT BUFFER

Anyone who has used a calculator or digital voltmeter is familiar with light-emitting diodes (LED). In these devices a microcomputer analyzes input data and decides whether a given LED should be turned **ON** or **OFF**.

Figure 12–20A shows a circuit which, at first glance, could accomplish the LED lighting task. When the computer decides to turn on the LED, the output line of the computer changes from 0V to 5V. The 5V computer output should cause the LED to light up. In fact, this circuit will not work. The LED needs approximately 2V across it to light up. The computer output is clearly large enough to supply the necessary 2V to the LED. The problem is with the current that the LED draws. You can see from the figure that the LED current is the current in the resistor, so

$$I_{LED} = \frac{5 - 2}{150} = 20\text{mA}$$

FIGURE 12–20
Emitter-follower current buffer

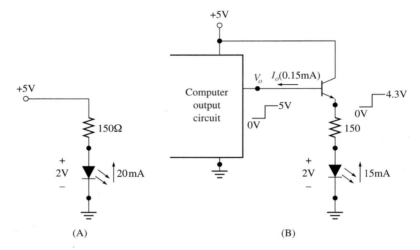

In general, microcomputers can only supply small currents. For example, we assume that this computer can output only 1mA. Because the LED requires approximately 20mA, it will not light up.

We need to insert a **current buffer** between the computer and the LED. A current buffer is a device that requires only a small input current to turn on but can provide a large output current. The emitter follower is often used as a current buffer. In Figure 12–20B we have inserted an emitter follower between the computer and the LED.

Now when the computer output is 0V, the emitter follower is in **CUT-OFF,** and there is no current to the LED. When the computer output increases to 5V, the emitter follower becomes **ACTIVE,** and the emitter voltage, which drives the LED, follows the computer output and rises to 4.3V (don't forget the emitter-diode voltage drop). The current that is drawn by the LED is now

$$I_{LED} = \frac{4.3 - 2}{150} = 15.3\text{mA}$$

We still need a large current to drive the LED. Now, however, the current that drives the LED is the emitter current of the transistor, and transistors can supply 15mA without breaking a sweat. The real question here is how much current does the computer have to supply to make the whole thing work. A glance at Figure 12–20B will show you that now the computer only has to supply the base current of the transistor. Because we need 15mA at the emitter, we only need 0.15mA at the base ($I_B = I_E/\beta$). The computer can easily supply the 0.15mA needed at the base of the transistor, so we have successfully used the emitter follower as a current buffer.

Where has the rest of the LED current gone? Remember that in the **ACTIVE** transistor the collector current is 99% of the emitter current. Therefore, 99% of the LED current now is diverted to the collector supply voltage. The computer only has to handle the 1% of the LED current that goes to the base.

EXAMPLE 12–7
The Current Buffer

Figure 12–21 shows another example of current buffering. This relay will close when the voltage across it is 3 to 5V. The resistance of the relay is 50Ω.

(a) Find the current range necessary to close the relay.
(b) Can the relay be driven by a computer that can output 5V at 5mA?
(c) If the answer to part B is no, insert a current buffer between the computer and the relay, and find the required computer current.

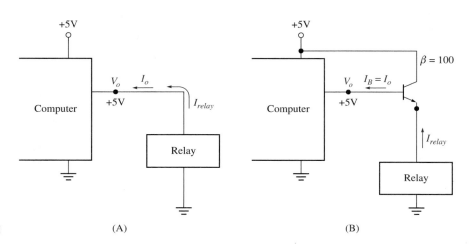

FIGURE 12–21　　　　　(A)　　　　　　　　　　　　　(B)

Solution

(a) The current required for the relay is

$$I_{relay} = \frac{V_{relay}}{50}$$

so, the current range is

$$I_{relay_{min}} = \frac{3}{50} = 60\text{mA}$$

to

$$I_{relay_{max}} = \frac{5}{50} = 100\text{mA}$$

(b) The computer cannot supply the required current.
(c) Figure 12–21B shows an emitter follower connected between the computer and the relay. The computer now has to supply only the base current of the transistor. This range is

$$I_B = \frac{I_{relay}}{\beta} = 0.6\text{mA to 1mA}$$

DRILL EXERCISE   Replace the relay in Figure 12–21 (p. 407) with one that has the same working range of voltage but has an internal resistance of 100Ω.

(a) Find the working current range for the relay.
(b) Insert an emitter follower (β = 100) between the relay and the driving computer, and find the required computer current.

Answer   **(a)** I_{relay} = 30mA to 50mA   **(b)** $I_{computer}$ = 0.3mA to 0.5mA

REVIEW QUESTIONS

1. When do we need a current buffer?
2. How does the emitter follower work as a current buffer?
3. Where does most of the load current go when using the emitter follower as a current buffer?

12–4 ■ Zener Voltage Regulation and the Current Buffer

In Chapter 5 we discussed how Zener diodes can be used to provide a regulated voltage to a load. The primary limitation to the use of Zener regulators is the amount of current available to drive the load. This situation is shown in Figure 12–22A. When the load draws no current, the 12V supply puts the Zener diode into breakdown, so the Zener diode is acting as an 8V battery. The current burning through the Zener in this case is equal to the resistor current and is given by

$$I_Z = I_R = \frac{12 - 8}{1K} = 4mA$$

FIGURE 12–22

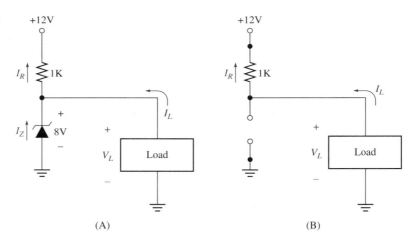

(A)   (B)

As long as the diode is in breakdown, the current in the resistor will be 4mA.

Now let's see what happens as the load starts to draw current. From KCL we know that the resistor current must now supply both the Zener diode and the load. As the load draws more current, less current is available to burn through the Zener diode. That is

$$I_Z = I_R - I_{Load} = 4mA - I_{Load}$$

How much current can the load draw before the Zener diode comes out of breakdown? When the load draws 4mA, all of the resistor current goes through the load and the Zener current goes to 0A:

$$I_Z = 4\text{mA} - I_{Load} = 4\text{mA} - 4\text{mA} = 0\text{A}$$

As soon as the current in the diode drops to 0A, the diode becomes an open circuit, and we lose regulation (Figure 12–22B). Therefore, we cannot use this Zener circuit to drive a load that requires more than 4mA. That is, the maximum current available from a Zener diode is the no-load current that burns through the diode.

Figure 12–23 shows how we can use a current-buffering emitter follower to increase the useful current range of the Zener circuit. From the Zener diode's point of view, the load it sees is the base of the transistor. That is, the load current for the Zener is now the base current of the transistor. We already know that this Zener circuit can supply up to 4mA to its load and still operate properly. Now, however, this 4mA-maximum is the maximum base current. This means that at the emitter of the transistor, there is approximately 400mA $(I_E = \beta I_B)$ available to drive the actual load. The emitter follower has allowed us to extend the useful range of the Zener circuit to 400mA of load current.

FIGURE 12–23

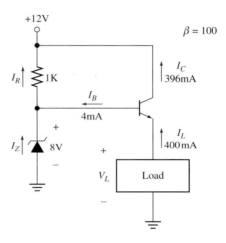

EXAMPLE 12–8
Zener Regulation

Consider the Zener regulation circuit in Figure 12–24.

(a) What is the maximum regulated load current available?
(b) Insert an emitter follower between the Zener diode and the load. Now what is the maximum load current available? (Assume $\beta = 100$.)
(c) What voltage now appears across the load?

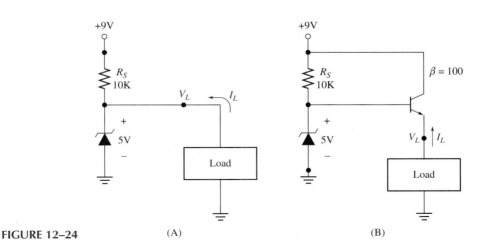

FIGURE 12–24 (A) (B)

Solution

(a) The maximum current available to the load is equal to the no-load current burning through the Zener diode.

$$I_{max} = \frac{9 - 5}{10K} = 0.4mA$$

(b) Figure 12–24B (p. 409) shows the emitter follower connected between the Zener circuit and the load. The no-load current burning through the Zener diode is now available to drive the base of the transistor. Because the load current is the emitter current of the transistor, the maximum load current is now

$$I_{max} = \beta I_B = 40mA$$

(c) Because the load voltage is the emitter voltage, the voltage that appears across the load is now

$$V_L = V_E = V_B - 0.7 = V_Z - 0.7 = 4.3V$$

DRILL EXERCISE

Replace the Zener diode in Example 12–8 (p. 409) with a 3V Zener diode and repeat the example.

Answer   **(a)** $I_{max} = 0.6mA$ **(b)** $I_{max} = 60mA$ **(c)** $V_L = 2.3V$

You may have noticed a problem with the current-buffered voltage regulator circuit of Figure 12–23. We choose a specific Zener diode to provide a specific voltage to a load. When we use the emitter follower, however, the voltage supplied to the load is now

$$V_{Load} = V_Z - 0.7$$

where 0.7V is the voltage drop across the emitter diode. For example, a 5V Zener diode will supply only 4.3V to a load that is coupled through an emitter follower. If we truly want 5V across the load we would have to use a 5.7V Zener.

A way to fix this problem is to insert a regular diode in series with the Zener diode (Figure 12–25). During operation the regular diode is **ON,** creating a diode voltage drop across itself. Therefore, the voltage at the base is now

$$V_B = V_Z + 0.7$$

FIGURE 12–25

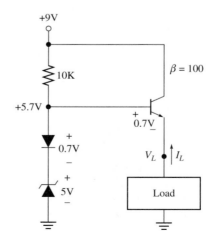

The load voltage, which is the emitter voltage of the transistor, is now

$$V_L = V_E = V_B - 0.7$$

Substituting for V_B, we get

$$V_L = V_Z + 0.7 - 0.7$$

And the load voltage is

$$V_L = V_Z$$

By inserting the regular diode in series with the Zener diode, we have ensured that the load voltage is the Zener diode voltage. While we have not stated so explicitly, the regular diode also helps to cancel the variations that occur in the base-emitter turn-on voltage. (Remember that our use of 0.7V for this voltage is an approximation.)

REVIEW QUESTIONS	
REVIEW QUESTIONS	**1.** Why do we use the Zener diode?
	2. In the basic Zener diode circuit, what is the maximum current available to a load?
	3. What will happen to the basic Zener diode circuit if the load draws more current than you found from Question 2?
	4. What purpose is served by inserting an emitter follower between the Zener circuit and the load?
	5. What voltage appears across the load in an emitter-follower buffered Zener regulator? Why?
	6. How can we correct for the loss in load voltage?

12–5 ■ TROUBLESHOOTING

Troubleshooting emitter-follower circuits is similar to troubleshooting the common-emitter circuits discussed in Chapter 10. The most important tool is DC Troubleshooting. If this technique does not reveal any problems, you would then connect the AC signal generator and trace AC signals through the circuit.

Because you are working with AC circuits, you need an oscilloscope to trace the AC signals. Of course, you can also use the oscilloscope to measure DC voltages. A switch on an oscilloscope allows you to use it in DC coupling or AC coupling modes. It is good practice to always keep the oscilloscope in the DC coupling mode. Switch to AC coupling only when you are interested in measuring just the AC portion of a signal.

In Chapter 10 we discussed the effect of an open or a short in the input coupling capacitor: An open in the input coupling capacitor would not show up in DC Troubleshooting but nonetheless no AC signal would get to the base. A short in the input coupling capacitor would short the bias resistors and change the DC behavior of the circuit. Here, the output coupling capacitor (C_L) now connects the emitter resistor to the load. Let's examine the effects of an open and a short in C_L.

Shorted C_L

Figure 12–26 (p. 412) shows the results of DC Troubleshooting a circuit with a shorted C_L. Note that the Estimated Voltages and Measured Voltages for the transistor show no discrepancies. However, the DC voltage across the load should be 0V, but it actually is at the same voltage as the emitter. This can only happen if the coupling capacitor is shorted.

Sometimes a capacitor will short out only partially. That is, it acts as a resistor instead of a short circuit. In this case the DC voltage across the load will not equal the emit-

FIGURE 12–26

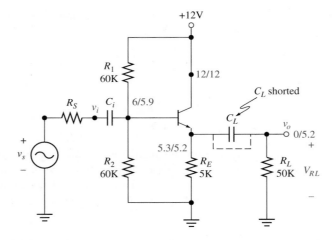

ter voltage. Any DC voltage at the output, however, indicates a bad capacitor. If the capacitor is good, the DC voltage at the load has to be 0V.

Open C_L

Figure 12–27 shows the results of tracing AC signals through the emitter-follower circuit. In this case we have assumed that DC Troubleshooting did not indicate any problems. Because there is an AC signal at the emitter but none at the load, we can conclude that the coupling capacitor (C_L) has failed open.

FIGURE 12–27

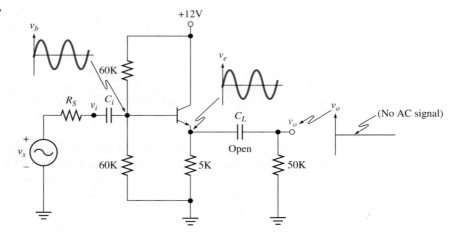

REVIEW QUESTIONS

1. What is the primary tool for troubleshooting?
2. Will DC Troubleshooting locate failed open capacitors? Why?
3. What will happen if the input coupling capacitor has failed short?
4. What will happen if the input coupling capacitor has failed open?
5. What will happen if the output coupling capacitor has failed short?
6. What will happen if the output coupling capacitor has failed open?

SUMMARY

■ A buffer amplifier is used to isolate the load from the signal source.

■ The ideal buffer has a no-load gain of 1, an infinite input resistance, and 0Ω output resistance.

■ A buffer can be built with a BJT by taking the output from the emitter.

- The BJT buffer is called the common collector.
- The common-collector amplifier is also called the emitter follower, because the emitter voltage follows the base voltage.
- The no-load gain of the emitter follower is given by

$$A = \frac{R_E}{r_e + R_E} \approx 1$$

- The input resistance of the emitter follower depends on the load, and is

$$R_i = R_B \parallel \beta(r_e + \hat{R}_E) \qquad \hat{R}_E = R_E \parallel R_L$$

- The output resistance of the emitter follower depends on the signal-source resistance, and is

$$R_o = R_E \parallel \left(r_e + \frac{\hat{R}_B}{\beta} \right) \qquad \hat{R}_B = R_S \parallel R_B$$

- The emitter follower can be used as a current buffer, where the emitter current supplies the load.
- The emitter-follower current buffer can be used to increase the load current available from a Zener diode voltage regulator.

PROBLEMS

SECTION 12–1 Introduction—The Buffer Amplifier

1. Consider the buffer amplifier in the circuit of Figure 12–28.
   - **(a)** What is the no-load voltage gain *(A)* of the buffer amplifier?
   - **(b)** What is the input resistance of the buffer?
   - **(c)** What is the output resistance of the buffer?
   - **(d)** How much of the signal source voltage *(v_s)* shows up at *v_i*? Give your answer as a fraction and as a percentage.
   - **(e)** What is the voltage across the controlled voltage source in the buffer?
   - **(f)** What would be the no-load output voltage of the buffer?
   - **(g)** With the load connected, how much of the voltage produced in the buffer shows up at *v_o*? Give your answer as a fraction and as a percentage.
   - **(h)** What is the overall ratio of *v_o/v_s*? This ratio is the system gain *(A_sys)*.

FIGURE 12–28

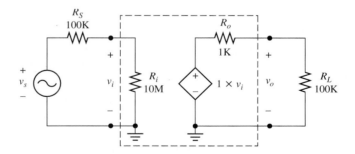

2. Repeat Problem 1 for the circuit in Figure 12–29.

FIGURE 12–29

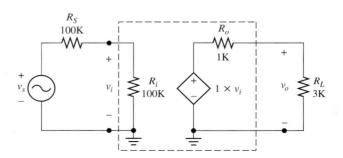

3. Consider the buffer amplifiers in Figures 12–28 and 12–29 (p. 413). Both buffer amplifiers have no-load gains of $A = 1$. Why does so much less of the signal voltage get to the output load from the second buffer circuit than from the first buffer circuit?

4. Consider the buffer amplifier in Figure 12–28 (p. 413). Let $v_s = 100\text{mV}_p$.
   (a) What is v_i of the buffer?   **(b)** What is v_o?

5. Consider the buffer amplifier in Figure 12–29 (p. 413). Let $v_s = 100\text{mV}_p$.
   (a) What is v_i of the buffer?   **(b)** What is v_o?

SECTION 12–2   The Emitter Follower (The Common-Collector Amplifier)

6. You have a signal source with an internal resistance of $R_S = 200\Omega$. What R_i do you need so that v_i will be at least 95% of v_s?

7. You have a load resistor of $2\text{K}\Omega$. What R_o should your amplifier have if you want to deliver 98% of the buffer-amplifier output voltage to the load?

8. Redraw Figure 12–6 (p. 392) with $R_{sensor} = 10\text{M}\Omega$ and $R_L = 50\Omega$.
   (a) What is v_{buff1}?   **(b)** What is v_i?   **(c)** What is v_{buff2}?   **(d)** What is v_o?

9. Consider the buffer built with the emitter-follower amplifier shown in Figure 12–30, where

$$R_1 = 100\text{K}\Omega, \; R_2 = 100\text{K}\Omega, \; R_E = 10\text{K}\Omega, \; R_S = 1\text{K}\Omega, \; R_L = 10\text{K}\Omega$$

   (a) What is I_E?   **(b)** What is r_e?   **(c)** What is R_{base}?
   (d) What is the parallel combination of R_1 and R_2?   **(e)** What is R_i?
   (f) What is the total resistance at the base of the transistor?
   (g) Looking from the emitter, what resistance do you see?
   (h) What is R_o? Be sure to include R_E in your answer.
   (i) What is the no-load gain *(A)* for the buffer amplifier?
   (j) Draw the box model for the emitter-follower buffer amplifier. Be sure to show your values for R_i, R_o, and A.
   (k) What is the overall gain (v_o/v_s)?

FIGURE 12–30

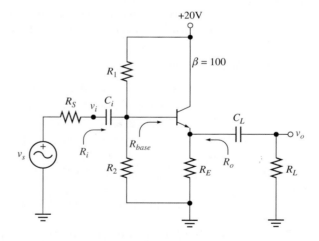

10. In Figure 12–30

$$R_1 = 8\text{K}\Omega, \; R_2 = 12\text{K}\Omega, \; R_E = 1\text{K}\Omega, \; R_S = 500\Omega, \; R_L = 1\text{K}\Omega$$

   Repeat Problem 9.

11. In Figure 12–30, $R_S = 200\Omega$, what R_i do you need to deliver 90% of the source voltage to the base of the emitter follower?

12. In Figure 12–30, $R_L = 2\text{K}\Omega$, what R_o do you need to deliver 95% of the emitter-follower output to the load?

13. An emitter follower has a base resistance of $R_{base} = 50\text{K}\Omega$. What $R_B \; (R_B = R_1 \| R_2)$ do you need to achieve an input resistance of $R_i = 40\text{K}\Omega$?

14. An emitter follower has the following parameters:

$$R_1 = 200K\Omega, R_2 = 200K\Omega, r_e = 26\Omega, R_E = 1K\Omega, R_L = 1K\Omega$$

What is the minimum β required to achieve an input resistance of $R_i = 65K\Omega$?

FIGURE 12–31

15. Figure 12–31 shows an emitter follower constructed with two supply voltages. This circuit does not require base resistors to provide the biasing.

(a) When $v_i = 0V$, what is V_E (assume the transistor is **ACTIVE**)?

(b) What is V_{R_E} (do not forget the $-9V$ supply voltage)?

(c) What is I_E?   (d) What is r_e?   (e) What is R_i?

SECTION 12–3   The Emitter Follower as a Current Buffer

16. Figure 12–32A shows an instrument panel indicator light bulb. The bulb has a current of 50mA at a voltage of 4.3V.

(a) What is the resistance of the bulb?

(b) Figure 12–32B shows a logic circuit that drives the indicator bulb through a transistor. When the output of the logic circuit (V_o) is 5V, find the following:
V_E
I_E
I_C

(c) Is the transistor operating in the **ACTIVE** region? Why?

(d) What current must the logic circuit supply when the indicator bulb is lit?

FIGURE 12–32

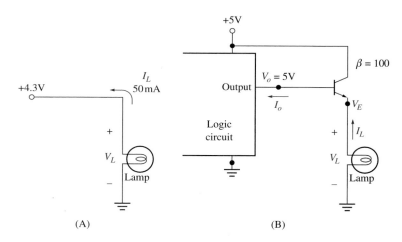

(A)   (B)

17. The output voltage of the logic circuit in Figure 12–32B falls to 0V. Find the following:

(a) V_E, V_B, V_C of the transistor.

(b) V_L and I_L for the indicator bulb.

(c) What is the output current (I_o) of the logic circuit?

(d) In what state is the transistor operating? Why?

FIGURE 12–33

+9V

1K

V_Z

R_L

SECTION 12–4   Zener Voltage Regulation and the Current Buffer

18. In Figure 12–33, $V_Z = 4V$

(a) What is the Zener diode no-load current?

(b) What is the maximum current available to the load?

(c) What is the minimum R_L that can be used in this circuit?

19. Redraw Figure 12–33 with an emitter follower between the Zener diode and the load. If $\beta = 100$

(a) What is the no-load Zener diode current?

(b) What is the maximum available base current?

(c) What is the maximum available load current?

(d) What is the load voltage?

(e) What is the minimum load resistor (R_L) that can be used in this circuit?

20. Redraw Figure 12–33 (p. 415), with an emitter follower between the Zener diode and the load resistor. If $R_L = 10\Omega$, $V_Z = 6V$, and the transistor is **ACTIVE**:

(a) What is the load voltage?

(b) What is the load current?

(c) If $\beta = 200$, what is the base current?

(d) What is the current in the Zener diode?

(e) How small can the current gain (β) of the transistor become before the Zener diode comes out of breakdown?

SECTION 12–5   Troubleshooting

21. In the laboratory at school you have built the emitter-follower amplifier in Figure 12–34. As you have been trained, before you connect the signal generator, you perform DC Troubleshooting. The resulting Estimated Voltages and Measured Voltages are recorded in the figure.

FIGURE 12–34

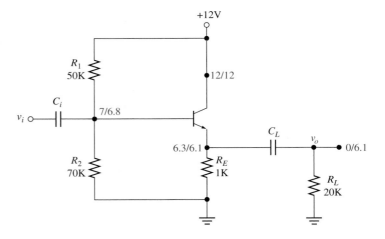

The voltages at the leads of the transistor are in good agreement, but V_{R_L} is not 0V, as it should be. What type of problem with capacitor C_L can cause this problem? Why?

22. DC Troubleshooting of the emitter-follower amplifier in Figure 12–35 indicates good agreement between the Estimated Voltages and the Measured Voltages at the transistor leads. As expected, you also see 0V across R_L.

You now connect the signal generator to the input (v_i in the figure). On your oscilloscope you can see the input signal at the base and at the emitter. There is no signal across the load, however.

(a) Is there a problem with the input capacitor (C_i)? Why?

(b) Is there a problem with the output capacitor (C_L)? Why?

(c) What type of problem exists with C_L?

FIGURE 12–35

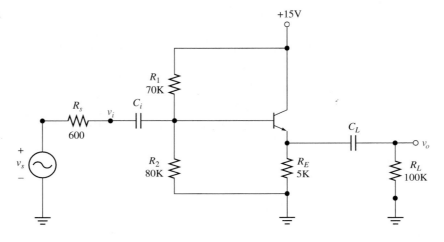

23. Consider the emitter-follower amplifier in Figure 12–36, along with its DC Troubleshooting results. Here you would not connect the signal generator to the circuit until you located the problems shown by DC Troubleshooting. From the list below, which problems could explain the results shown in the figure? Explain your answers.

(a) Resistor R_1 is not properly connected to R_2 and the base of the transistor.

(b) The collector of the transistor is not connected to the supply voltage (V_{CC}).

(c) The transistor has an open base inside the transistor.

(d) Capacitor C_L has failed open.

(e) The load resistor (R_L) is not properly connected to ground.

(f) Capacitor C_L has failed short.

FIGURE 12–36

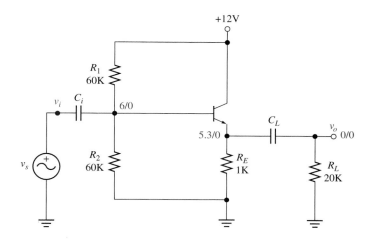

Computer Problems

24. Use the appropriate controlled-source model to simulate the circuit of Figure 12–28 (p. 413). Find v_o if the input source is a 1V 1KHz sine wave.

25. Use the appropriate controlled-source model to simulate the circuit of Figure 12–5 (p. 392). Find v_{buff1}, v_i, and v_o if the input source is a 1V 1KHz sine wave.

26. Use the computer to find the DC bias values for the emitter follower in Figure 12–30 (p. 414) and the parameters given in Problem 9 (p. 414).

27. For the emitter follower in Figure 12–30 (p. 414) and the parameters given in Problem 9 (p. 414).

(a) Remove the signal-source resistor and the load resistor, and find the no-load gain.

(b) Reconnect the load, and find the input resistance.

(c) Reconnect the signal-source resistor, remove the load resistor, and find the output resistance seen at the emitter. (See the discussion of output resistance in Problem 28 of Chapter 11.)

(d) Connect all resistors, and find the system gain.

(e) Compare your answers to Problem 9 (p. 414).

28. For the emitter follower in Figure 12–30 (p. 414) and the parameters given in Problem 9 (p. 414), short C_i, and find the voltages at the transistor terminals (do not forget to set $v_s = 0$V).

29. For the emitter follower in Figure 12–30 (p. 414) and the parameters given in Problem 9 (p. 414), short C_L, and find the voltages at the transistor terminals (do not forget to set $v_s = 0$V).

30. Simulate the circuit of Figure 12–31 (p. 415), and find the DC bias values for the transistor.

31. Use the computer to find the answers in Problem 20 (p. 416). Start with $\beta = 200$, and make several simulation runs to determine the smallest allowable β in this circuit.

CHAPTER

13

IMPROVED BJT AC MODELS

C HAPTER OBJECTIVES

To draw the family of curves that describe i_E versus v_{BE}, i_B versus v_{BE}, and i_c versus v_{CE}.

To derive the simple AC BJT model from simplified transistor curves.

To derive the hybrid-pi model from the transistor curves.

To use h-parameters to model the transistor.

To compare the hybrid-pi model to the simple model and to the h-parameter model.

To read manufacturer's transistor data sheets.

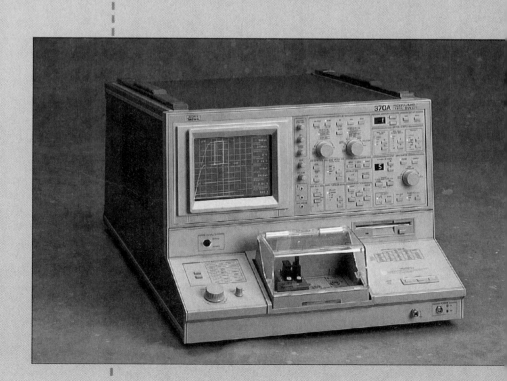

13–1 ▪ INTRODUCTION

In this chapter we will show you models for the NPN transistor that provide more information than the simple model presented in Chapter 7. We include this chapter to provide those students who want a deeper understanding of transistor behavior and modeling.

The good news is that the DC model we have used for the **ACTIVE** transistor $(V_{BE} = 0.7V, I_E = I_C = \beta I_B)$ is accurate enough for DC bias calculations. Nothing is gained by using a more detailed model. Most, if not all, practitioners in electronics technology use this model to find the bias levels in a transistor circuit.

On the other hand, you can use several different models to determine the AC behavior of the **ACTIVE** transistor circuit. The two most common models are:

☐ The hybrid-pi model

☐ The h-parameter model

We will show you these two new models and compare them to each other, and to the simple AC model from Chapter 7. Bear in mind that whatever model is used in AC analysis, the final results for input resistance, output resistance, and gain of a transistor circuit must be the same.

13–2 ▪ BJT TRANSISTOR CHARACTERISTIC CURVES

We can derive a model for an electronic device in one of two ways: (1) by understanding the physics of the device, or (2) by using standard electrical components that produce the transistor voltage-current relations. Here we take the latter approach.

Consider the transistor test circuit shown in Figure 13–1A. Note that this circuit is an I_B control circuit. We will first hold V_{CC} constant, while we vary V_{BB} and measure the resulting base-emitter voltage (v_{BE}), emitter current (i_E), and base current (i_B). We then plot both i_E versus v_{BE} (Figure 13–1B) and plot i_B versus v_{BE} (Figure 13–1C). These two plots describe the dependence of the transistor currents on the base-emitter voltage.

FIGURE 13–1

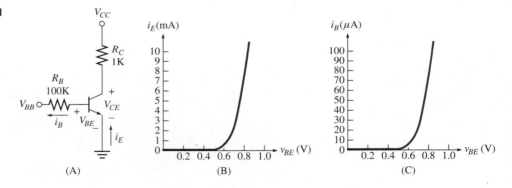

(A)　　　　　(B)　　　　　(C)

Figure 13–1B should look familiar to you. It is the same plot we used in Chapter 7 to derive AC emitter resistance (r_e). This plot represents the I-V characteristic of the emitter diode. The plot in Figure 13–1C seems to look the same as the plot in Figure 13–1B, but look carefully. In Figure 13–1C we are measuring the base current, not the emitter current. Therefore, the current values in Figure 13–1C are *one-hundredth* of the current values in Figure 13–1B $(i_B \approx i_E/\beta)$.

It should be noted that if we vary the collector bias voltage (V_{CC}), these plots will change slightly. Figure 13–2 shows the result of using three different values for V_{CC}, which in turn creates three different values of V_{CE}. As V_{CE} increases, we get less base current for the same v_{BE}, which results in the family of plots shown. Note that the effect of V_{CE} on the base current is very small.

FIGURE 13–2
i_B-v_BE curve

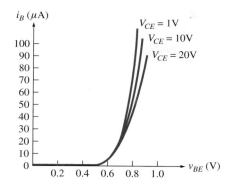

The plots in Figures 13–1 and 13–2 show the V-I relations for a transistor's base-emitter circuit. We now want to see what happens in the transistor's collector-emitter circuit. In this case we want to measure and plot the collector current (i_C) as a function of the collector-emitter voltage (V_{CE}). This plot is more complicated because the collector current depends both on the collector-emitter voltage and on the base current.

We begin by holding the base bias voltage V_{BB} constant and by varying the collector supply voltage. A fixed V_{BB} will produce a fixed base current. In Figure 13–1A V_{BB} is fixed at 1.7V. This will produce a base current of

$$I_B = \frac{V_{BB} - 0.7}{100K} = 0.01\text{mA}$$

FIGURE 13–3
i_C-v_CE curve

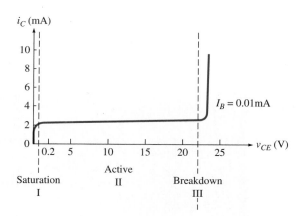

We now increase V_{CC} from 0V to 50V and plot the resulting i_C versus v_{CE} (Figure 13–3). Before we discuss this curve, note a few facts from Figure 13–1A:

1. Because the emitter is grounded, $v_{CE} = v_C$ and $v_{BE} = v_B$.

2. The voltage across the collector diode is $v_{BC} = v_B - v_C = v_{BE} - v_{CE}$.

3. The collector diode is **ON** when the voltage across it exceeds a certain threshold. The turn-on voltage for the collector diode is approximately 0.5V. Therefore, the collector diode is **ON** when $v_{BC} \geq 0.5$V (remember, the transistor is **SATURATED** when the collector diode is **ON**).

4. The collector diode is **OFF** when the voltage across it is less than the turn-on voltage. For the collector diode this occurs when $v_{BC} < 0.5$V (remember, the transistor is **ACTIVE** when the collector diode is **OFF**).

5. The collector diode goes into breakdown when v_{BC} exceeds the collector diode breakdown voltage.

Now look at Region I of Figure 13–3 (p. 421). Here, $0 < v_{CE} < 0.2$V. And, because

$$v_B = v_{BE} = 0.7\text{V}$$

we get

$$v_{BC} = v_{BE} - v_{CE} = 0.7 - v_{CE}$$

In this region the collector diode voltage will vary from

$$v_{BC} = 0.7 - 0.2 = 0.5\text{V} \quad \text{to} \quad v_{BC} = 0.7 - 0 = 0.7\text{V}$$

So the collector diode will be **ON** and the transistor will be **SATURATED.**

In Region II of Figure 13–3 (p. 421), the collector-emitter voltage increases from 0.2V to approximately 50V. The collector diode voltage will, therefore, vary from

$$v_{BC} = 0.7 - 0.2 = 0.5\text{V} \quad \text{to} \quad v_{BC} = 0.7 - 50 = -43.7\text{V}$$

Here the collector diode will be **OFF** and the transistor will be **ACTIVE.** Note that in this region the current increases slightly as the collector voltage increases.

FIGURE 13–4

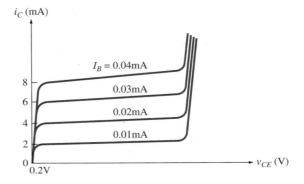

FIGURE 13–5
Curve tracer

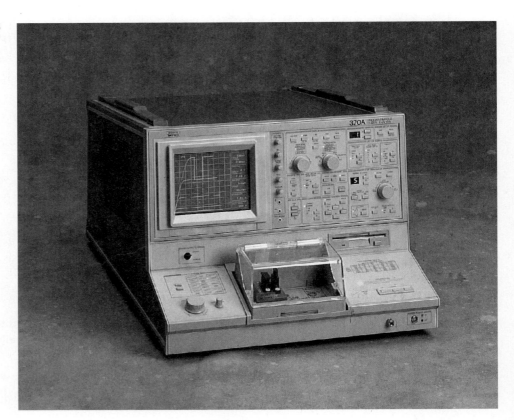

In Region III the collector-emitter reverse-bias voltage increases to the point that the collector diode goes into breakdown and we get a sudden rush of collector current.

We now run a series of tests. For each test we increase the base current (by increasing V_{BB}) and measure and plot collector current as a function of collector-emitter voltage. You already know that as we increase base current, we increase collector current. This leads to the final collector I-V curve shown in Figure 13–4. Note that this plot is actually a family of curves, in which the base bias current (I_B) determines which curve you are on. The basic shape of each curve is the same as that shown in Figure 13–3. If you have a **curve tracer** available in the laboratory or at work, you can use it to produce this set of characteristic curves for any transistor (Figure 13–5).

REVIEW QUESTIONS

1. Why is the circuit of Figure 13–1A an I_B control circuit?
2. Draw an i_B vs. v_{BE} curve.
3. Draw an i_E vs. v_{BE} curve.
4. How are these two plots related?
5. Draw the family of curves that describe i_C vs. v_{CE}.
6. Describe the three regions of these curves.
7. How do you determine which curve in the family of curves you would use for a given circuit?

13–3 ■ THE SIMPLE BJT MODEL

In the simplest model for the BJT (the one we have been using), we simplify the characteristic curves in several ways.

1. We assume that the small variations in the family of curves of i_E versus v_{BE} (Figure 13–2, p. 420) can be ignored. Because the effect of v_{CE} on i_B is so small, we can adequately model the base circuit with a single curve.

2. We represent the exponential curve as a series of straight lines (Figure 13–6A).

FIGURE 13–6
Simple BJT model

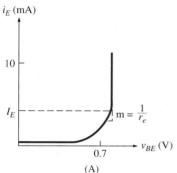
(A)

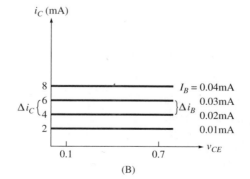
(B)

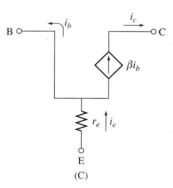
(C)

The simple BJT model is created by representing the linearized curves with circuit elements. These straight lines can be represented with a resistor, where the resistance is given by

$$r_e = \frac{\Delta v_{BE}}{\Delta i_E}$$

We already know that this resistance is given by

$$r_e = \frac{26\text{mV}}{I_E}$$

where I_E is the DC bias level of the emitter current. (Note that the value of r_e is temperature-dependent; as the transistor heats up, the AC emitter resistance increases.)

3. Finally, we idealize the i_C versus v_{CE} family of curves as horizontal lines (Figure 13–6B). Note that we do not include the breakdown region.

The idealized family of horizontal collector curves tells us what we already know: The ratio of collector current to base current is constant. That is

$$\frac{\Delta i_C}{\Delta i_B} = \beta$$

And, because we represent changes in current as AC quantities

$$i_c = \beta i_b$$

We can model this relation with a current-controlled current source. The simple BJT model, the model we have been using, is shown in Figure 13–6C (p. 423).

EXAMPLE 13–1
The Simple BJT
AC Model

For the circuit in Figure 13–7A, replace the transistor with the simple BJT model. Assume that $I_E = 5\text{mA}$, and find the voltage gain *(A)*.

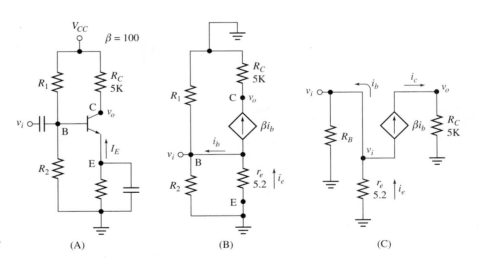

FIGURE 13–7   (A)   (B)   (C)

Solution We first find the AC emitter resistance

$$r_e = \frac{26\text{mV}}{I_E} = \frac{26\text{mV}}{5\text{mA}} = 5.2\Omega$$

We now replace the transistor with the simple model, and, because we are doing an AC analysis, short V_{CC} to ground and short the capacitors (Figure 13–7B). Finally, we rearrange the circuit (Figure 13–7C), where $R_B = R_1 \| R_2$.

From Figure 13–7C we find the emitter current by recognizing that the input voltage appears at the top of r_e:

$$i_e = \frac{v_i}{r_e}$$

Because we know that

$$i_c \approx i_e$$

we get

$$i_c \approx i_e = \frac{v_i}{r_e}$$

The output voltage is given by

$$v_o = -i_c R_C = -\frac{v_i}{r_e} R_C$$

The gain is

$$A = \frac{v_o}{v_i} = -\frac{R_C}{r_e}$$

So, for our circuit

$$A = -\frac{R_C}{r_e} = -\frac{5K}{5.2} = -962$$

DRILL EXERCISE   If the bias current in Figure 13–7A is $I_E = 10\mu A$, find the AC emitter resistance, replace the transistor with its model, and find the gain.

Answer $r_e = 2.6K\Omega, A = -1.92$

REVIEW QUESTIONS

1. For the simple BJT model, how do we treat the family of i_E vs. v_{BE} plots?
2. What circuit element do we use to represent the linearized i_E vs. v_{BE} curve?
3. Does r_e depend on DC bias values? What is the formula for r_e?
4. What information do we get from the idealized family of i_C vs. v_{CE} plots? How do we model this information?
5. Draw the simple BJT model.

13–4 ■ THE HYBRID-PI BJT MODEL

The Hybrid-Pi Model

When we use the **hybrid-pi model,** we look at base current rather than emitter current. In its most complete form, the hybrid-pi model is quite complicated. It includes such considerations as the resistances associated with transistor leads, the bulk resistance of the silicon, and capacitors to model the frequency behavior of the transistor. So when we discuss the transistor's frequency behavior, we will use a modified hybrid-pi model. The most common electronic analysis computer programs use the hybrid-pi model or a variation of it.

FIGURE 13–8

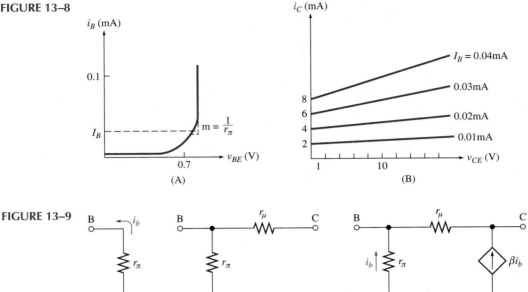

(A)

(B)

FIGURE 13–9

(A)   (B)   (C)

For now, we just want to use enough of the model to account for the curves in Figures 13–2 (p. 421) and 13–4 (p. 422). We linearize the transistor curves as shown in Figure 13–8 and build the hybrid-pi model (Figure 13–9) with the following steps:

1. We first ignore the variations in i_B due to v_{CE} in Figure 13–8A and define an AC *base* resistor as

$$r_\pi = \frac{\Delta v_{BE}}{\Delta i_B} = \frac{26\text{mv}}{I_B}$$

where I_B is the DC bias level for the base current. We can also (and more usually do) define the AC base resistance in terms of the collector current DC bias level:

$$r_\pi = \beta \frac{26\text{mV}}{I_C}$$

As shown in Figure 13–9A, the AC base resistor is connected between the base and the emitter. Note that

$$r_\pi = \beta r_e$$

2. By connecting a resistor (r_μ) between the collector and the base (Figure 13–9B), we model the fact that changing v_{CE} changes the base current.

3. The relation between the collector current and the base current is still given by

$$i_c = \beta i_b$$

and is modeled with a current-controlled current source (Figure 13–9C).

4. Finally, we represent the slopes in the collector current curves with an output resistor (r_o). Note from Figure 13–8B that the slopes are not constant. This means that r_o depends on the DC bias level of the collector current and is given approximately by

$$r_o \approx \frac{V_A}{I_C}$$

where I_C is the DC bias level of the collector current and V_A is a device parameter known as the **Early Voltage.** The Early Voltage (named for an individual, not because it gets out of bed before 7:00 AM) is typically greater than 50V.

EXAMPLE 13–2
The Early Voltage and r_o

(a) A transistor has a DC bias current of I_C = 2mA and an Early Voltage of V_A = 200V. What is the output resistance of this transistor?

(b) A transistor has an Early Voltage of V_A = 200V. What DC bias current do you need to achieve an output resistance of r_o = 1MΩ?

Solution

(a) For a DC current of I_C = 2mA and an Early Voltage of V_A = 100V

$$r_o = \frac{V_A}{I_C} = \frac{200}{2\text{mA}} = 100\text{K}\Omega$$

(b) To find I_C that would produce r_o = 1MΩ, we rearrange the output resistance formula to get

$$I_C = \frac{V_A}{r_o}$$

So

$$I_C = \frac{V_A}{r_o} = \frac{200}{1\text{M}} = 0.2\text{mA}$$

The hybrid-pi model is shown in Figure 13–10A. Even this simplified hybrid-pi model is still quite complicated, so we would never use it when we analyze transistor circuits by hand. Typical values for the resistors in this model are

$r_\pi \approx 1\text{K}\Omega$
$r_o \approx 100\text{K}\Omega$
$r_\mu \approx 10\text{M}\Omega$

You can see that r_o is large and r_μ is very large. We usually ignore these resistors and work with the basic hybrid-pi model shown in Figure 13–10B.

FIGURE 13–10
BJT hybrid-pi model

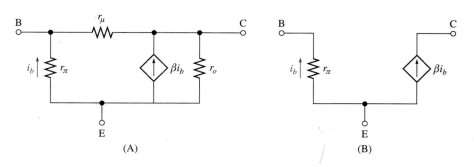

(A)   (B)

EXAMPLE 13–3
The Hybrid-Pi Model

For the circuit in Figure 13–7A (p. 424), replace the transistor with the hybrid-pi model. Assume that r_o and r_μ are so large that they can be ignored. Also assume that I_C = 5mA, and find the voltage gain *(A)*. Compare this result to the gain you found in Example 13–1 (p. 424).

Solution The AC base resistor (r_π) is found from

$$r_\pi = \beta \frac{26mV}{I_C} = 100 \frac{26mV}{5mA} = 520\Omega$$

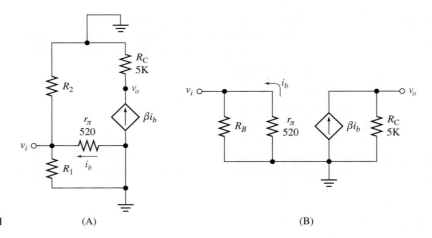

FIGURE 13–11   (A)   (B)

Figure 13–11A shows the result of replacing the transistor with the basic hybrid-pi model. Note that for AC we short the collector supply and short the capacitors. We have rearranged the AC circuit in Figure 13–11B so that $R_B = R_2 \| R_1$. Because the input voltage is applied directly to r_π

$$i_b = \frac{v_i}{r_\pi}$$

The output voltage is given by

$$v_o = -\beta i_b R_C = -\beta \frac{v_i}{r_\pi} R_C$$

The gain is given by

$$A = -\beta \frac{R_C}{r_\pi}$$

For our circuit

$$A = -\beta \frac{5K}{520} = -962$$

which is the same gain we found in Example 13–1.

EXAMPLE 13–4
The Hybrid-Pi
Model with r_o

Repeat the previous example, but now include r_o where

(a) $r_o = 100K\Omega$ **(b)** $r_o = 10K\Omega$

Discuss the effect of r_o on the gain of the transistor amplifier.

Solution We show the final AC circuit model in Figure 13–12. You can see from the figure that r_o and R_C are in parallel. This results in

$$A = -\beta \frac{R_C \| r_o}{r_\pi}$$

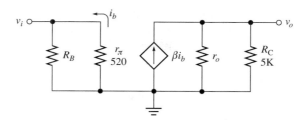

FIGURE 13–12

(a) For $r_o = 100\text{K}\Omega$

$$A = -\beta\frac{R_C\|r_o}{r_\pi} = -100 \times \frac{(100\text{K}\|5\text{K})}{520} = -100 \times \frac{4.76\text{K}}{520} = -915$$

Including the large r_o in the calculations gives us a smaller gain, -915 instead of -962. Because r_o is large, the variation in the gain calculation is less than 5%.

(b) For $r_o = 10\text{K}\Omega$

$$A = -\beta\frac{R_C\|r_o}{r_\pi} = -100 \times \frac{(10\text{K}\|5\text{K})}{520} = -100 \times \frac{3.33\text{K}}{520} = -640$$

Including the smaller r_o in the calculations gives us a much smaller gain, -640 instead of -962. Because r_o is not very large, the variation in the gain calculation is large, greater than 33%.

The conclusion we can draw from this example is that r_o can be ignored only if it is much greater than R_C.

DRILL EXERCISE   If the bias current in Figure 13–7A (p. 424) is $I_C = 10\mu\text{A}$, find the AC base resistance, replace the transistor with its hybrid-pi model, and find the gain if

(a) $r_o = 1\text{M}\Omega$.   **(b)** $r_o = 100\text{K}\Omega$.   **(c)** $r_o = 10\text{K}\Omega$.

Answer $r_\pi = 260\text{K}\Omega$ **(a)** $A = -1.9$ **(b)** $A = -1.8$ **(c)** $A = -1.3$

Alternate Hybrid-Pi Model

Figure 13–13 shows that the current-controlled current source in the hybrid-pi model $(i_C = \beta i_b)$ can be changed to a voltage-controlled current source $(i_C = g_m v_{be})$. To see this, we use Ohm's Law to find i_b in the original model:

$$i_b = \frac{v_{be}}{r_\pi}$$

We now substitute for i_b at the controlled source:

$$\beta i_b = \beta\frac{v_{be}}{r_\pi} = g_m v_{be}$$

FIGURE 13–13
Alternate hybrid-pi model

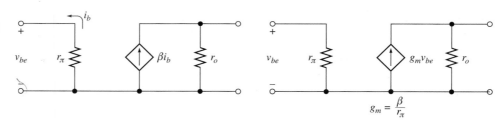

$$g_m = \frac{\beta}{r_\pi}$$

Because $g_m v_{be}$ is a current, g_m has units of conductance. Also, because g_m relates to output current and to input voltage, g_m is known as a transconductance and is given by

$$g_m = \frac{\beta}{r_\pi}$$

This alternate model for the hybrid-pi is the model we use when we analyze the frequency behavior of the transistor.

REVIEW QUESTIONS

1. Which current do we use to derive the AC base resistance (r_π)? What is the formula for r_π?

2. How do we model the fact that v_{CE} will affect the i_B vs. v_{BE} plots?

3. How do we model the relation between i_c and i_b?

4. How do we model slopes in the i_C vs. v_{CE} plots? What is the formula for r_o?

5. Draw the hybrid-pi model.

6. Assume that r_o and r_μ can be ignored, and draw the simplified hybrid-pi model.

7. What is the alternate hybrid-pi model, and how do you find it?

13–5 ■ THE H-PARAMETER BJT MODEL

Many electronic devices and circuits have an input side and an output side, as represented in Figure 13–14A. This circuit is known generically as a **two port circuit:** side 1 forms one "porthole" into the circuit and side 2 provides the other. Assuming the circuit is linear, we can derive a model without looking inside the box. We apply sources to both sides and measures the voltages and currents on each side (Figure 13–14B). The sources can be any combination of voltage and current sources.

FIGURE 13–14

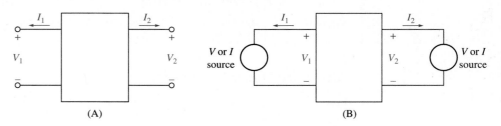

(A)   (B)

For example, consider the circuit in Figure 13–15, in which side 1 is driven with a current source and side 2 is driven with a voltage source. Since I_1 and V_2 are known, we measure V_1 and I_2.

Using the known and measured values, we can find the four **h-parameters:** $h_{11}, h_{12}, h_{21}, h_{22}$ (how they are found is not important here). We then use the h-parameters to write

$$V_1 = h_{11}I_1 + h_{12}V_2$$
$$I_2 = h_{21}I_1 + h_{22}V_2$$

FIGURE 13–15

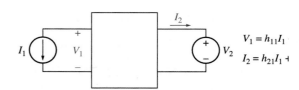

$$V_1 = h_{11}I_1$$
$$I_2 = h_{21}I_1 \dashv$$

Note that the h-parameters have different units: h_{11} has units of resistance (Ω); h_{12} is unitless; h_{21} is unitless; h_{22} has units of conductance (S).

EXAMPLE 13–5
h-Parameter Model

Measurements of a circuit give us the following h-parameters:

$$h_{11} = 10\text{K}\Omega, \ h_{12} = 0.5, \ h_{21} = 100, \ h_{22} = 2\text{mS}$$

Find V_1 and I_2 if $I_1 = 1\text{mA}$ and $V_2 = 2\text{V}$.

Solution   We find V_1 from

$$V_1 = h_{11}I_1 + h_{12}V_2 = 10\text{K}\Omega \times 1\text{mA} + 0.5 \times 2 = 11\text{V}$$

We find I_2 from

$$I_2 = h_{21}I_1 + h_{22}V_2 = 10 \times 1\text{mA} + 2\text{mS} \times 2 = 14\text{mA}$$

We will show you a circuit model for this h-parameter equation after we look at the transistor. While computer programs usually work with the hybrid-pi model, manufacturers usually use h-parameters to describe their transistors.

Figure 13–16A shows conceptually the circuit used to derive the h-parameters for a transistor. Note that the bias circuitry is not shown. Remember that a transistor is not a linear device. The relations between AC voltages and currents depend on DC bias levels. For this reason, manufacturers provide h-parameter data for several different bias levels.

FIGURE 13–16
BJT h-parameter model

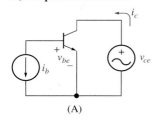 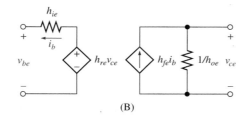 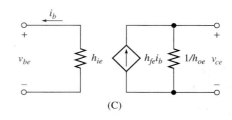

(A)   (B)   (C)

Figure 13–16A shows that we can visualize a transistor as a two-port by letting the emitter lead be common to both sides, hence the designation **common-emitter configuration.** We write the h-parameter equations as

$$v_{be} = h_{ie}i_b + h_{re}v_{ce}$$
$$i_c = h_{fe}i_b + h_{oe}v_{ce}$$

where the subscript e indicates the use of the common-emitter configuration; i indicates that the input current creates an input voltage; r indicates that the output voltage affects the input voltage (reverse direction); f indicates that the input current creates an output current (forward direction); and o indicates the relation between the output current and voltage. Note that h_{re} models the small effect we saw in Figure 13–2, where the collector voltage changes the base current. Typical values for the h-parameters are

$$h_{ie} \approx 1\text{K}\Omega$$
$$h_{re} \approx 10^{-5}$$
$$h_{fe} \approx 100$$
$$h_{oe} \approx 0.01\text{mS}$$

The parameter h_{ie} is measured in units of ohms, and h_{oe} is measured in units of Siemens; therefore, these two parameters are modeled with resistors. The other two parameters are modeled with controlled sources. The h-parameter transistor model is shown in Figure 13–16B (p. 431). However, because h_{re} is so small, we usually remove the controlled voltage source in the input side, producing the simplified h-parameter circuit in Figure 13–15C (p. 430).

EXAMPLE 13–6
The BJT h-Parameter Model

The h-parameters for the transistor in the circuit of Figure 13–17A are

$$h_{ie} = 2K\Omega$$
$$h_{fe} = 200$$
$$h_{oe} = 0.01mS$$

(h_{re} can be ignored.) Replace the transistor with its h-parameter model, and find the voltage gain.

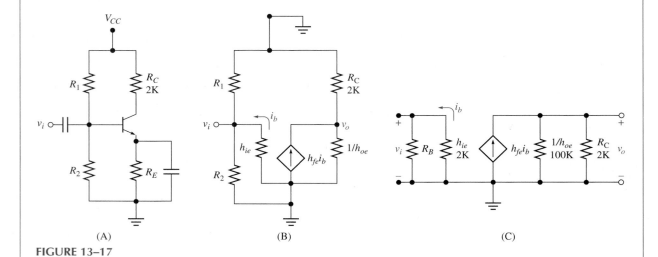

(A)   (B)   (C)

FIGURE 13–17

Solution In Figure 13–17B we have set the supply voltage to 0V, shorted the capacitors, and replaced the transistor with its h-parameter model. We rearrange the circuit as shown in Figure 13–16C to see that

$$i_b = \frac{v_i}{h_{ie}}$$

We also see that $1/h_{oe}$ and R_c are in parallel, so

$$v_o = -h_{fe}i_b \times (1/h_{oe})\|R_c$$

Substituting for i_b

$$v_o = -h_{fe}\frac{(1/h_{oe})\|R_c}{h_{ie}}v_i$$

For the values given

$$1/h_{oe} = 100K\Omega \quad \text{so} \quad (1/h_{oe})\|R_c = 100K \,\| \, 2K \approx 2K\Omega$$

and the output voltage is

$$v_o = -200 \times \frac{2\,K}{2\,K}v_i = -200v_i$$

Therefore, the gain is

$$A = -\frac{v_o}{v_i} = -200$$

REVIEW QUESTIONS

1. Draw a general two-port, and give the h-parameter equation describing the I-V relations.
2. How can you represent the transistor as a two-port? Write the transistor h-parameter equations.
3. Do transistor h-parameters depend on the DC bias level? Why?
4. Which h-parameter(s) can usually be ignored? Why?
5. Draw the h-parameter model for a transistor. Give typical values.

13–6 ■ MODEL COMPARISONS

Simple Model <—> Hybrid-Pi Model

The only way to compare these two models is by simplifying the hybrid-pi to its most basic elements, r_π and βi_b (Figure 13–18).

FIGURE 13–18

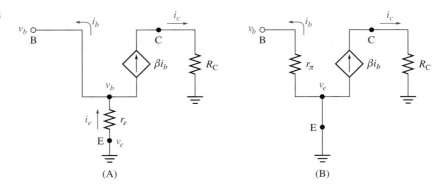

(A)   (B)

In Figure 13–17 we have grounded the emitters and connected a load resistor to the collector of each model. For both models we can see that the collector current is

$$i_c = \beta i_b$$

The key question here is this: Do both models give us the same base current (does $v_e = 0\text{V}$ in both models)? To find the answer, note that for the simple model the voltage at the top of r_e is v_b. Therefore, from Figure 3–18A

$$i_e = \frac{v_b}{r_e}$$

and

$$i_e = i_b + \beta i_b = (\beta + 1)i_b$$

Substituting for i_e

$$(\beta + 1)i_b = \frac{v_b}{r_e}$$

Solving for i_b from the simple model

$$i_b = \frac{v_b}{(\beta + 1)r_e}$$

From the basic hybrid-pi model in Figure 13–18B ($v_e = 0V$)

$$i_b = \frac{v_b}{r_\pi}$$

For the two models to be the same

$$r_\pi = (\beta + 1)r_e \approx \beta r_e$$

This result is consistent with the fact that a resistor in the emitter leg "looks" β times larger from the base.

Hybrid-Pi Model <—> h-Parameter Model

Figure 13–19 shows the hybrid-pi model and the h-parameter model. Because we can only make a comparison if we can ignore r_μ and h_{re}, these elements have been eliminated from their respective models. We can now make the following approximate comparisons by direct inspection of the two models:

$$r_\pi \approx h_{ie}$$
$$\beta \approx h_{fe}$$
$$r_o \approx \frac{1}{h_{oe}}$$

FIGURE 13–19

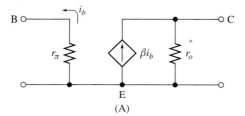 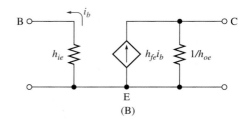

If you use a computer program, most of which are based on the hybrid-pi model, you specify the β and the Early Voltage (V_A) for your transistor. The computer calculates r_π and r_o after the DC bias level of the collector current is found:

$$r_\pi = \beta \frac{26mV}{I_C}$$

and

$$r_o = \frac{V_A}{I_C}$$

REVIEW QUESTIONS

1. Draw the most basic hybrid-pi model.
2. What is the relation between r_π from the hybrid-pi model and r_e from the simple model?
3. Draw the hybrid-pi model without r_μ.
4. Draw the h-parameter model without h_{re}.
5. What are the relations between the remaining model elements?
6. How are r_π and r_o found?
7. What is the alternate hybrid-pi model, and how do you find g_m?

13–7 ■ BJT Data Sheets

Interpreting Data Sheets

Figure 13–20 (p. 436) shows the data sheets for a Fairchild family of general-purpose transistors, the 2N3903 and 2N3904. We presented a shortened version of these data sheets in Chapter 7. Following is a brief description of the Electrical Characteristics.

h_{ie}: Small-signal input resistance ($r_\pi = h_{ie}$, $r_e = h_{ie}/h_{fe}$).

h_{re}: Models the effect that collector voltage has on base current. Because h_{re} is so small (varying from 0.1 to 8×10^{-4}), we always ignore it in a hand analysis.

h_{fe}: Small-signal AC current gain ($h_{fe} = \beta_{AC}$; we have usually assumed that $\beta_{AC} = 100$).

h_{oe}: Small-signal output conductance ($r_o = 1/h_{oe}$): Note for the values given for h_{oe}, the output resistance varies from 250KΩ to 1MΩ. We can usually ignore such large output resistances when we analyze by hand.

$V_{(BR)CEO}$, $V_{(BR)CBO}$, $V_{(BR)EBO}$: The reverse-bias voltages that put the various transistor PN-junctions into breakdown ($V_{(BR)CEO}$ is the breakdown voltage found in Figure 13–3, p. 421).

I_{CEX}, I_{BL}: Collector and base currents that actually exist when the transistor is in **CUT-OFF.** We have usually assumed these currents to be 0A.

h_{FE}: The h-parameter DC current gain ($h_{FE} = \beta_{DC}$; we have usually assumed that $\beta_{DC} = 100$).

$V_{CE(sat)}$: The collector-emitter voltage when the transistor is **SATURATED** (we have assumed that $V_{CE(sat)} = 0$V).

$V_{BE(sat)}$: The emitter diode turn-on voltage (we have assumed that $V_{BE(sat)} = 0.7$V).

C_{ob} and C_{ib}: Internal capacitances that we have ignored. We will use this data when we discuss the frequency response of the transistor.

Determining the Early Voltage

If you want to include the output resistance of a transistor in a computer simulation, you must determine the Early Voltage (V_A) of the device. To find V_A, we first note that

$$r_o = \frac{V_A}{I_C} \rightarrow V_A = r_o I_C$$

Because

$$r_o = \frac{1}{h_{oe}}$$

FIGURE 13–20

MAXIMUM RATINGS

Rating	Symbol	Value	Unit
Collector-Emitter Voltage	V_{CEO}	40	Vdc
Collector-Base Voltage	V_{CBO}	60	Vdc
Emitter-Base Voltage	V_{EBO}	6.0	Vdc
Collector Current—Continuous	I_C	200	mAdc
Total Device Dissipation @ T_A = 25°C Derate above 25°C	P_D 5.0	625 mW/°C	mW
*Total Device Dissipation @ T_C = 25°C Derate above 25°C	P_D 12	1.5 mW/°C	Watts
Operating and Storage Junction Temperature Range	T_J, T_{stg}	−55 to +150	°C

*THERMAL CHARACTERISTICS

Characteristic	Symbol	Max	Unit
Thermal Resistance, Junction to Ambient	$R_{\theta JA}$	200	°C/W
Thermal Resistance, Junction to Case	$R_{\theta JC}$	83.3	°C/W

*Indicates Data in addition to JEDEC Requirements.

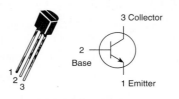

2N3903
2N3904*

CASE 29-04, STYLE 1
TO-92 (TO-226AA)

3 Collector
2 Base
1 Emitter
1 2 3

GENERAL PURPOSE TRANSISTORS

NPN SILICON

*This is a Motorola designated preferred device.

ELECTRICAL CHARACTERISTICS (T_A = 25°C unless otherwise noted.)

Characteristic		Symbol	Min	Max	Unit
OFF CHARACTERISTICS					
Collector-Emitter Breakdown Voltage(1) (I_C = 1.0 mAdc, I_B = 0)		$V_{(BR)CEO}$	40	—	Vdc
Collector-Base Breakdown Voltage (I_C = 1.0 μAdc, I_E = 0)		$V_{(BR)CBO}$	60	—	Vdc
Emitter-Base Breakdown Voltage (I_E = 1.0 μAdc, I_C = 0)		$V_{(BR)EBO}$	6.0	—	Vdc
Base Cutoff Current (V_{CE} = 30 Vdc, V_{EB} = 3.0 Vdc)		I_{BL}	—	50	nAdc
Collector Cutoff Current (V_{CE} = 30 Vdc, V_{EB} = 3.0 Vdc)		I_{CEX}	—	50	nAdc
ON CHARACTERISTICS					
DC Current Gain(1)		h_{FE}			—
(I_C = 0.1 mAdc, V_{CE} = 1.0 Vdc)	2N3903 2N3904		20 40	— —	
(I_C = 1.0 mAdc, V_{CE} = 1.0 Vdc)	2N3903 2N3904		35 70	— —	
(I_C = 10 mAdc, V_{CE} = 1.0 Vdc)	2N3903 2N3904		50 100	150 300	
(I_C = 50 mAdc, V_{CE} = 1.0 Vdc)	2N3903 2N3904		30 60	— —	
(I_C = 100 mAdc, V_{CE} = 1.0 Vdc)	2N3903 2N3904		15 30	— —	
Collector-Emitter Saturation Voltage(1) (I_C = 10 mAdc, I_B = 1.0 mAdc) (I_C = 50 mAdc, I_B = 5.0 mAdc)		$V_{CE(sat)}$	— —	0.2 0.3	Vdc
Base-Emitter Saturation Voltage(1) (I_C = 10 mAdc, I_B = 1.0 mAdc) (I_C = 50 mAdc, I_B = 5.0 mAdc)		$V_{BE(sat)}$	0.65 —	0.85 0.95	Vdc
SMALL-SIGNAL CHARACTERISTICS					
Current-Gain — Bandwidth Product (I_C = 10 mAdc, V_{CE} = 20 Vdc, f = 100 MHz)	2N3903 2N3904	f_T	250 300	— —	MHz
Output Capacitance (V_{CB} = 5.0 Vdc, I_E = 0, f = 1.0 MHz)		C_{obo}	—	4.0	pF
Input Capacitance (V_{EB} = 0.5 Vdc, I_C = 0, f = 1.0 MHz)		C_{ibo}	—	8.0	pF
Input Impedance (I_C = 1.0 mAdc, V_{CE} = 10 Vdc, f = 1.0 kHz)	2N3903 2N3904	h_{ie}	1.0 1.0	8.0 10	k ohms
Voltage Feedback Ratio (I_C = 1.0 mAdc, V_{CE} = 10 Vdc, f = 1.0 kHz)	2N3903 2N3904	h_{re}	0.1 0.5	5.0 8.0	X 10^{-4}
Small-Signal Current Gain (I_C = 1.0 mAdc, V_{CE} = 10 Vdc, f = 1.0 kHz)	2N3903 2N3904	h_{fe}	50 100	200 400	—
Output Admittance (I_C = 1.0 mAdc, V_{CE} = 10 Vdc, f = 1.0 kHz)		h_{oe}	1.0	40	μmhos
Noise Figure (I_C = 1.0 μAdc, V_{CE} = 5.0 Vdc, RS = 1.0 k ohms, f = 1.0 kHz)	2N3903 2N3904	NF	— —	6.0 5.0	dB

we get

$$V_A = \frac{I_C}{h_{oe}}$$

The data sheet for the 2N3904 shows that h_{oe} varies from 1-to-40μS at a DC collector current of 1mA. Assuming a middle value for h_{oe} of 20μS, we find

$$V_A = \frac{I_C}{h_{oe}} = \frac{1\text{mA}}{20\mu\text{S}} = 50\text{V}$$

SUMMARY

- i_E versus v_{BE} is a PN-junction exponential curve.
- i_B versus v_{BE} is also a PN-junction exponential curve, where current levels are $1/\beta$ of the i_E versus v_{BE} curve.
- Increasing V_{CE} decreases i_B, resulting in a family of i_B versus v_{BE} curves.
- i_C versus v_{CE} curves can be divided into three regions:

SATURATION ($v_{CE} < 0.2$V)	
ACTIVE	(0.2V $< v_{CE} < BV_{CEO}$)
BREAKDOWN ($v_{CE} = BV_{CEO}$)	

- i_E versus v_{CE} is a family of curves that depend on I_B.
- The simple AC BJT model represents several simplifications of the transistor characteristic curves: The effect of V_{CE} on the i_E versus v_{BE} curve is ignored; the exponential curve is linearized (modeled with r_e); the i_C versus v_{CE} curves are assumed to be horizontal and are represented by $i_C = \beta i_B$.

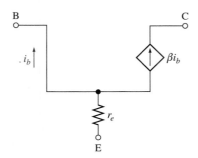

- The hybrid-pi BJT model uses the i_B versus v_{BE} family of curves: r_π represents the straight line approximation of these curves; r_μ represents the effect of V_{CE} on these curves. The i_C versus v_{CE} curves are represented with a controlled current source ($i_c = \beta i_b$) and an output resistance (r_o) that represents the slope of the output characteristics.

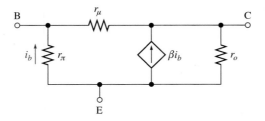

■ $r_\pi \approx \beta r_e$ and $r_o \approx \dfrac{V_A}{I_C}$

■ The h-parameter BJT model is derived from experimental data:

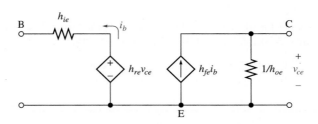

■ The relation between the h-parameter model elements and the hybrid-pi model elements (ignoring h_{re} and r_μ) are

$$r_\pi \approx h_{ie}$$

$$\beta \approx h_{fe}$$

$$r_o \approx \dfrac{1}{h_{oe}}$$

$$V_A \approx \dfrac{I_C}{h_{oe}}$$

PROBLEMS

SECTION 13–3 The Simple BJT Model

1. An idealized family of transistor curves is shown in Figure 13–21.
   (a) Use the i_E vs. v_{BE} curve to find r_e if $I_E = 1$mA.
   (b) Find β from the i_C vs. v_{CE} plot.
   (c) Draw the simple BJT model using the values obtained for r_e and β.

FIGURE 13–21

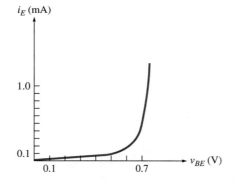
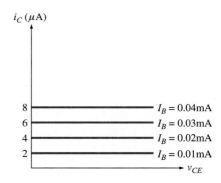

2. Use the i_E vs. v_{BE} curve of Figure 13–21 to find r_e if $I_E = 0.2$mA
3. For the transistor circuit in Figure 13–22

$$R_1 = 110\text{K}\Omega, \ R_2 = 10\text{K}\Omega, \ R_C = 5\text{K}\Omega, \ R_E = 1\text{K}\Omega, \ \beta = 100$$

   (a) Find the DC bias levels.   (b) Find the AC emitter resistance (r_e).
   (c) Replace the transistor with the simple BJT model and draw the AC circuit.
   (d) Find the following in terms of v_i:
   i_e
   i_c
   v_o
   (e) What is the gain of this amplifier?

4. Repeat Problem 3 if the circuit in Figure 13–22 has the following parameters:

$$R_1 = 9.8\text{K}\Omega, R_2 = 2.2\text{K}\Omega, R_C = 2\text{K}\Omega, R_E = 500\Omega, \beta = 200$$

5. The circuit of Figure 13–22 has the following parameters:

$$R_1 = 83\text{K}\Omega, R_2 = 37\text{K}\Omega, R_C = 5\text{K}\Omega, R_E = 1\text{K}\Omega, \beta = 100$$

    (a) Find the DC bias levels in this circuit.   **(b)** What is the state of the transistor?

    (c) Can we replace the transistor with the simple BJT model? Why?

6. For the transistor circuit in Figure 13–23

$$R_1 = 88\text{K}\Omega, R_2 = 12\text{K}\Omega, R_C = 10\text{K}\Omega, R_E = 1\text{K}\Omega, \beta = 200$$

    (a) Find the DC bias levels.   **(b)** Find the AC emitter resistance (r_e).

    (c) Replace the transistor with the simple BJT model and draw the AC circuit.

    (d) Find the following in terms of v_i:

        v_e

        i_e

        i_c

        v_o

    (e) What is the gain of this amplifier?

FIGURE 13–22

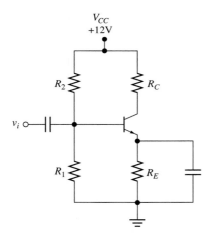

FIGURE 13–23

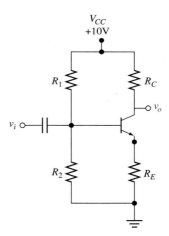

7. Repeat Problem 6 if the circuit of Figure 13–23 has the following parameters:

$$R_1 = 92.5\text{K}\Omega, R_2 = 7.5\text{K}\Omega, R_C = 5\text{K}\Omega, R_E = 1\text{K}\Omega, \beta = 200$$

SECTION 13–4   The Hybrid-Pi BJT Model

8. A BJT has a DC current of $I_C = 0.1\text{mA}$. If $\beta = 200$ and $V_A = 100\text{V}$, find r_π and r_o.

9. If $\beta = 100$, what DC collector current (I_C) do you need to produce $r_\pi = 2\text{K}\Omega$?

10. If the Early Voltage is $V_A = 150\text{V}$, what DC collector current (I_C) do you need to produce $r_o = 1\text{M}\Omega$?

11. A linearized family of transistor curves is shown in Figure 13–24 (p. 440).

    (a) Use the i_B vs. v_{BE} curve to find r_π if $I_B = 0.01\text{mA}$.

    (b) Approximate β from the i_C vs. v_{CE} plot.

    (c) Use the i_C vs. v_{CE} curve at $I_B = 0.01\text{mA}$ to find r_o.

    (d) Draw the hybrid-pi model using these values.

FIGURE 13–24

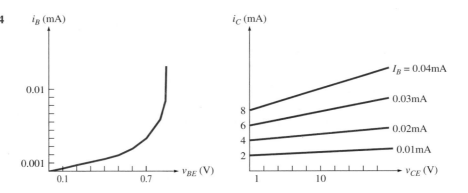

12. For the transistor circuit shown in Figure 13–22 (p. 439) and the parameters of Problem 3 (p. 438)

  (a) Find the DC bias levels.   (b) Find the AC base resistance (r_π).

  (c) Replace the transistor with the hybrid-pi model ($r_o = 200\text{K}\Omega$), and draw the AC circuit.

  (d) Find the following in terms of v_i:

  i_b

  i_c

  v_o

  (e) What is the gain of this amplifier?

13. Repeat Problem 12 for the circuit of Figure 13–22 (p. 439) and the parameters of Problem 4 (p. 439).

14. For the transistor circuit shown in Figure 13–22 (p. 439) and the parameters of Problem 3 (p. 438), if $V_A = 50\text{V}$, find r_o and the gain of the amplifier.

15. For the transistor circuit shown in Figure 13–23 (p. 439) and the parameters of Problem 6 (p. 439)

  (a) Find the DC bias levels.   (b) Find the AC emitter resistance (r_e).

  (c) Ignoring r_o, replace the transistor with the hybrid-pi model ($\beta = 200$), and draw the AC circuit.

16. The hybrid-pi model of a transistor amplifier with an unbypassed emitter resistor is shown in Figure 13–25. Assume r_o can be ignored.

  (a) Find i_e in terms of i_b.   (b) Find v_{R_E} and v_{be} in terms of i_b.

  (c) Noting that $v_i = v_{be} + v_{R_E}$, find v_i in terms of i_b.

  (d) Invert this relation to find i_b in terms of v_i.

  (e) Find the input resistance (R_i).   (f) Find v_o in terms of v_i.

  (g) Find the gain of this amplifier.

FIGURE 13–25

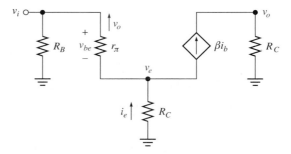

SECTION 13–5   The h-Parameter BJT Model

17. An h-parameter model has the following data:

$$h_{11} = 2\text{K}\Omega, \; h_{12} = 1, \; h_{21} = 200, \; h_{22} = 0.1\text{mS}$$

  (a) Write the h-parameter equations.   (b) Find v_1 and i_2 if $i_1 = 2\text{mA}$ and $v_2 = 1\text{V}$.

18. A 2N3904 transistor is to be biased at $I_C = 1\text{mA}$. Using the maximum values given in the data sheets of Figure 13–20 (p. 436), draw the h-parameter model (include all parameters) for this transistor.

19. A 2N3903 transistor is to be biased at $I_C = 1\text{mA}$. Using the maximum values given in the data sheets of Figure 13–20 (p. 436), draw the h-parameter model (include all parameters) for this transistor.

20. For the transistor circuit in Figure 13–22 (p. 439) and the parameters given in Problem 3 (p. 438)

    (a) Find the DC bias levels.   **(b)** What is h_{ie}?

    (c) Replace the transistor with its h-parameter model, assuming $h_{fe} = 200$ and $h_{oe} = 0.02 \times 10^{-5}\text{S}$.

    (d) Find the following in terms of v_i:

        i_b

        i_c

        v_o

    (e) What is the gain of this amplifier?

21. Repeat Problem 20 with the parameters of Problem 4 (p. 439).

SECTION 13–6   Model Comparisons

22. A simple BJT model has the following parameters: $r_e = 500\Omega$, $\beta = 200$.

    (a) Draw the simple BJT model.

    (b) Replace the simple BJT model with the equivalent hybrid-pi model. What parameters have you ignored?

    (c) Replace the simple BJT model with the equivalent h-parameter model. What parameters have you ignored?

23. Use the minimum data given for the 2N3904 biased at 1mA given in Figure 13–20 (p. 436) to answer the following:

    (a) Draw the h-parameter model for this device.   **(b)** Derive the hybrid-pi for this device.

    (c) Draw the simple BJT model for this device.

24. Use the minimum data given for the 2N3903 biased at 1mA given in Figure 13–20 (p. 436) to answer the following:

    (a) Draw the h-parameter model for this device.   **(b)** Derive the hybrid-pi for this device.

    (c) Draw the simple BJT model for this device.

Computer Problems

25. For your available computer program

    (a) List the BJT device models that are contained in your program's library.

    (b) What BJT model is used in your program?

    (c) Give the forward current gain and Early Voltage for five of the BJTs in your library.

26. Simulate the circuit of Figure 13–22 (p. 439) with the data given in Problem 3. Find the gain and input resistance of this amplifier.

27. Replace the transistor with its simple BJT model (use the controlled sources available in your program) and repeat Problem 26.

28. Replace the transistor with its hybrid-pi model (use the controlled sources available in your program) and repeat Problem 26.

29. Replace the transistor with its h-parameter model (use the controlled sources available in your program) and repeat Problem 26.

30. Simulate the circuit of Figure 13–23 (p. 439) with the data given in Problem 6 (p. 439). Find the gain and input resistance of this amplifier.

31. Replace the transistor with its simple BJT model (use the controlled sources available in your program) and repeat Problem 30.

32. Replace the transistor with its hybrid-pi model (use the controlled sources available in your program) and repeat Problem 30.

33. Replace the transistor with its h-parameter model (use the controlled sources available in your program) and repeat Problem 30.

Chapter

14

The PNP Transistor

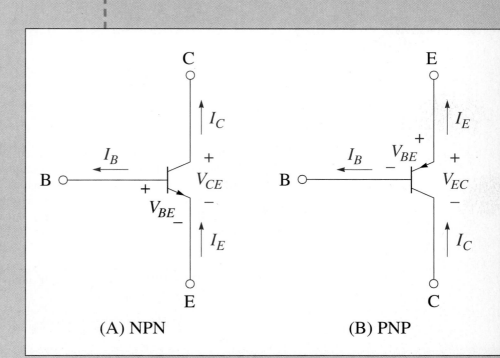

(A) NPN   (B) PNP

14–1 ■ INTRODUCTION

In Chapter 7 we showed you how you could sandwich a P-type semiconductor between two N-type semiconductors to create the NPN BJT (Figure 14–1A). You can also create a BJT by sandwiching an N-type semiconductor between two P-type semiconductors to create a PNP BJT. The layers of the PNP transistor and its circuit schematic are shown in Figure 14–1B. Note that for the PNP shown in Figure 14–1B, we have drawn the emitter of the transistor at the top. We do this because this is how the PNP is connected in most circuits. The arrow on the emitter lead of the PNP shows the direction of the emitter diode.

FIGURE 14–1
The PNP transistor

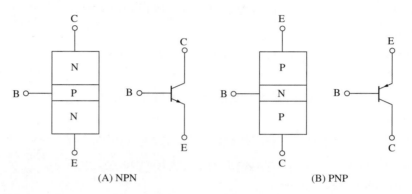

(A) NPN (B) PNP

The key difference between the NPN and PNP transistors is the direction of the emitter and collector diodes (Figure 14–2). The emitter diode of the NPN turns **ON** when the base voltage is a diode-voltage drop greater than the emitter voltage (Figure 14–2A). But in the PNP the emitter diode turns **ON** when the emitter voltage is a diode-voltage drop greater than the base voltage. The collector diode **ON** direction for the PNP is also the reverse of the NPN. That is, the collector diode for the PNP turns **ON** when the collector voltage is greater than the base voltage.

FIGURE 14–2

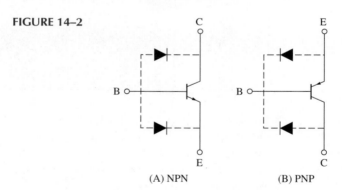

(A) NPN (B) PNP

Figure 14–3 shows a comparison of transistor currents and voltages in the NPN and PNP. Currents in the PNP flow in the opposite direction to currents in the NPN. Note that in the **ON** direction current still flows through the PNP emitter diode. Figure 14–3B also shows that the voltage across the emitter diode is now V_{EB}.

The PNP transistor operates in the same three states as the NPN transistor which are determined in the same way—by the emitter and collector diodes. For the NPN and the PNP the states are

CUT-OFF:	Emitter diode and collector diode are **OFF.**
SATURATED:	Emitter diode and collector diode are **ON.**
ACTIVE:	Emitter diode is **ON**—collector diode is **OFF.**

FIGURE 14–3
PNP voltages and currents

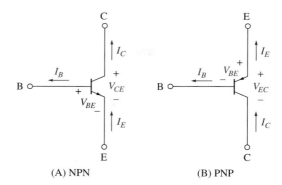

(A) NPN   (B) PNP

Note that the terms describing the states of the PNP are identical to the words describing the states of the NPN.

For linear amplifiers the PNP transistor is always operated in the **ACTIVE** region. When the PNP is used as a switch, the transistor is operated between the **CUT-OFF** and **SATURATED** states.

Except for the fact that the directions of the current and voltage are reversed, the PNP transistor behaves the same as the NPN. When the PNP is **CUT-OFF,** all transistor currents are 0A, and the transistor acts as an open circuit. When the PNP is **ACTIVE,** the collector current is β times the base current, and the emitter and collector currents are approximately equal. When the PNP is **SATURATED,** the collector voltage equals the emitter voltage. In summary, for the PNP transistor (Figure 14–3)

$$
\begin{aligned}
&\textbf{CUT-OFF:} && V_{EB} < 0.7\text{V} \\
& && I_B = I_C = I_E = 0\text{A} \\
&\textbf{SATURATED:} && V_{EB} = 0.7\text{V},\ V_{CB} = 0.5\text{V} \\
& && V_{EC} \approx 0\text{V} \rightarrow V_E \approx V_C \\
&\textbf{ACTIVE:} && V_{EB} = 0.7\text{V}\ (V_{BE} = -0.7\text{V}) \\
& && I_C = \beta I_B \\
& && I_E \approx I_C \\
& && V_E > V_C
\end{aligned}
$$

A data sheet for a typical PNP transistor is shown in Figure 14–27 (p. 472).

Most textbooks show the PNP with a negative collector supply voltage. Biased this way the PNP and NPN bias calculations are identical. The only difference is that we reverse the direction of the transistor currents and the polarities of the transistor voltages (Figure 14–4, p. 446). Note that the PNP in Figure 14–4B is drawn with the emitter down.

The **DC Bias Lists** for these two circuits are

NPN	*PNP*
$V_B = +2$V	$V_B = -2$V
$V_E = +1.3$V	$V_E = -1.3$V
$I_C \approx I_E = 1.3$mA	$I_C \approx I_E = 1.3$mA
$V_C = 5.5$V	$V_C = -5.5$V

When the PNP circuit is biased with a negative V_{CC}, it produces the same current and voltage magnitudes as the equivalent NPN circuit. The only differences are that we have changed the defined current directions, and the transistor terminal voltages are negative instead of positive.

FIGURE 14-4

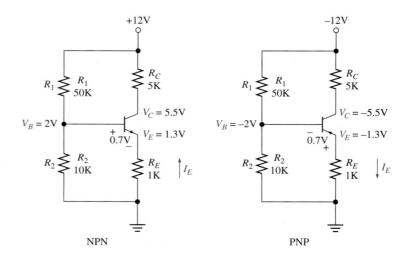

NPN

PNP

There is one problem with this analysis. The vast majority of electronic circuits are powered with positive voltage supplies. Even when a negative supply is used, it is usually used in conjunction with a positive supply. So, when the PNP circuit has a positive supply voltage, we must modify the bias analysis procedure. In the next two sections we will show you the V_B-control DC bias circuits for the PNP transistor and the common-emitter and emitter-follower circuits. Following these sections we will show you the PNP I_B-control circuit and its use in switching circuits. We will then show you some circuits that combine both NPN and PNP transistors, and discuss the advantages of having two "flavors" of bipolar junction transistors.

REVIEW QUESTIONS

1. Draw the three-layer sandwiches for the NPN and PNP transistors.
2. What are the directions of the emitter and collector diodes for the PNP transistor?
3. Draw the circuit schematic for the PNP transistor, with the emitter up. Label all currents and voltages.
4. In what three states does the PNP operate?
5. How do these states depend on the emitter and collector diodes?
6. What are the transistor currents when the transistor is **CUT-OFF?**
7. How are the transistor currents related when the transistor is **ACTIVE?**
8. How are the voltages related when the transistor is **SATURATED?**
9. How do the bias currents and voltages in a negative-supply PNP circuit compare to the bias currents and voltages in the equivalent NPN circuit?

14-2 ■ V_B Control of PNP Transistors

Before we begin analyzing PNP transistor circuits, we again want to emphasize the essential sameness of the PNP and the NPN. The steps of causality for V_B control are the same in the PNP as in the NPN. Even the words we use to describe V_B control are the same. The base voltage controls the emitter voltage. The emitter voltage controls the emitter current. If the transistor is **ACTIVE,** the emitter current controls the collector current; the base current is then found from the collector current.

Remember that V_B control is used to bias linear amplifiers. The transistors in V_B control circuits, therefore, should always be **ACTIVE.** If your analysis shows that the transistor is **SATURATED,** there is something wrong with the circuit. There is no need to proceed further.

The procedure for determining the DC bias levels of a V_B-control PNP transistor circuit is

☐ Find V_B

      V_B controls V_E

      V_E controls I_E

      I_E controls I_C if **ACTIVE** $I_C \approx I_E$

      I_C controls I_B if **ACTIVE** $I_B = I_C/\beta$

☐ Check to see if transistor is **ACTIVE** by finding V_C

      If $V_C < V_E$

      The transistor is **ACTIVE** and all variables found above are correct.

      If $V_C > V_E$

      Stop! The transistor is **SATURATED.** Put down your pencil. Values you have calculated are wrong.

The clearest difference in the PNP procedure is that for the PNP to be **ACTIVE,** emitter voltage must be greater than collector voltage. This is the reverse of what we need for the NPN. What is not so clear is the fact that you have to be careful when using Ohm's Law to find the emitter current.

Figure 14–5 shows a very simple PNP circuit with V_B control. Note that the emitter resistor is connected to the positive voltage supply. For practical reasons, we still call this supply voltage V_{CC}. To find the DC bias levels, we first find the base voltage:

$$V_B = V_{BB} = 5V$$

FIGURE 14–5

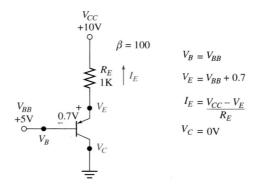

Assuming the emitter diode is **ON,** the emitter voltage is a diode drop *above* the base voltage:

$$\boxed{V_E = V_B + 0.7}$$

so

$$V_E = V_B + 0.7 = 5.7V$$

We want to emphasize here that in the **ACTIVE** PNP, emitter voltage is greater than base voltage.

Emitter voltage controls emitter current, which can be found from Ohm's Law. **BE CAREFUL HERE!** For this PNP connection the emitter resistor is *not grounded.* Rather, the emitter resistor is connected to the positive voltage supply. Therefore

$$\boxed{I_E = \frac{V_{CC} - V_E}{R_E}}$$

so

$$I_E = \frac{10 - 5.7}{1K} = 4.3mA$$

We now return to the original circuit in Figure 14–5 (p. 447). If the transistor is **ACTIVE**

$$I_C \approx I_E = 4.3mA$$

and

$$I_B = \frac{I_C}{\beta} = 0.043mA$$

To check the state of the transistor, we find the collector voltage. Because the collector in this circuit is grounded

$$V_C = 0V$$

Because $V_E > V_C$, the transistor is **ACTIVE** and our values are correct.

Let's now examine the more realistic self-biased PNP V_B-control circuit shown in Figure 14–6. We first find the base voltage by noting that V_B is the voltage across R_2, and using the voltage divider

$$V_B = \frac{R_2}{R_1 + R_2} V_{CC}$$

so

$$V_B = \frac{R_2}{R_1 + R_2} V_{CC} = \frac{50K}{50K + 10K} 12 = 10V$$

FIGURE 14–6
PNP self-bias circuit

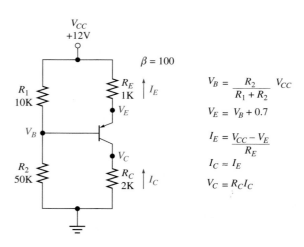

The base voltage controls the emitter voltage:

$$V_E = V_B + 0.7$$

so

$$V_E = V_B + 0.7 = 10 + 0.7 = 10.7V$$

The emitter voltage controls the emitter current:

$$I_E = \frac{V_{CC} - V_E}{R_E}$$

so

$$I_E = \frac{V_{CC} - V_E}{R_E} = \frac{12 - 10.7}{1\text{K}} = 1.3\text{mA}$$

If the transistor is **ACTIVE**

$$I_C \approx I_E = 1.3\text{mA}$$

$$I_B = \frac{I_C}{\beta} = 0.013\text{mA}$$

To check the state of the transistor, we find the collector voltage (note that R_C is grounded):

$$V_C = R_C I_C$$

so

$$V_C = R_C I_C = 2\text{K} \times 1.3\text{mA} = 2.6\text{V}$$

Because $V_E > V_C$, the transistor is **ACTIVE,** and the DC bias values we have found are correct.

EXAMPLE 14–1
V_B **Control I**

Find the DC bias values for the PNP circuit in Figure 14–7. Determine the state of the transistor.

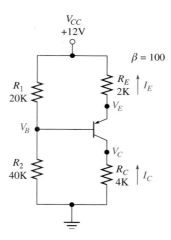

FIGURE 14–7

Solution

$$V_B = \frac{R_2}{R_1 + R_2} V_{CC} = \frac{40\text{K}}{40\text{K} + 20\text{K}} \, 12 = 8\text{V}$$

The base voltage controls the emitter voltage:

$$V_E = V_B + 0.7 = 8 + 0.7 = 8.7\text{V}$$

The emitter voltage controls the emitter current:

$$I_E = \frac{V_{CC} - V_E}{R_E} = \frac{12 - 8.7}{2\text{K}} = 1.65\text{mA}$$

If the transistor is **ACTIVE**

$$I_C \approx I_E = 1.65\text{mA}$$

$$I_B = \frac{I_C}{\beta} = 0.0165\text{mA}$$

To check the state of the transistor, we find the collector voltage:

$$V_C = R_C I_C = 4K \times 1.65mA = 6.6V$$

Because $V_E > V_C$, the transistor is **ACTIVE,** and the DC bias values we have found are correct.

DRILL EXERCISE

Redraw the circuit in Figure 14–7 (p. 449), changing the supply voltage to $V_{CC} = 9V$. Find the bias levels of the circuit, and determine the state of the transistor.

Answer $V_B = 6V$, $V_E = 6.7V$, $I_E = 1.15mA$
$I_C = 1.15mA$, $I_B = 0.0115mA$, $V_C = 4.6V$

ACTIVE

EXAMPLE 14–2

V_B **Control II**

Assume the transistor is **ACTIVE,** and find the DC bias values for the PNP circuit in Figure 14–8. Determine the actual state of the transistor.

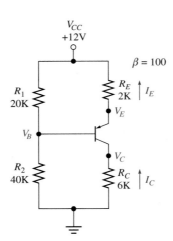

FIGURE 14–8

Solution
$$V_B = \frac{R_2}{R_1 + R_2} V_{CC} = \frac{40K}{40K + 20K} 12 = 8V$$

The base voltage controls the emitter voltage:

$$V_E = V_B + 0.7 = 8 + 0.7 = 8.7V$$

The emitter voltage controls the emitter current:

$$I_E = \frac{V_{CC} - V_E}{R_E} = \frac{12 - 8.7}{2K} = 1.65mA$$

If the transistor is **ACTIVE**

$$I_C \approx I_E = 1.65mA$$

$$I_B = \frac{I_C}{\beta} = 0.0165mA$$

To check the state of the transistor, we find the collector voltage:

$$V_C = R_C I_C = 6K \times 1.65\text{mA} = 9.9\text{V}$$

Because $V_E < V_C$ (8.7V < 9.9V), the transistor is **SATURATED,** and the DC bias values we have found are *not* correct.

DRILL EXERCISE   Redraw the circuit in Figure 14–8, changing the supply voltage to $V_{CC} = 9$V. Assume the transistor is **ACTIVE,** and find the bias levels of the circuit. Are these values correct?

Answer   If **ACTIVE**

$$V_B = 6\text{V}, \quad V_E = 6.7\text{V}, \quad I_E = 1.15\text{mA}$$
$$I_C = 1.15\text{mA}, \quad I_B = 0.012\text{mA}, \quad V_C = 6.9\text{V}$$

Because $V_E < V_C$, the transistor is **SATURATED,** so these values are not correct.

We summarize the DC bias calculations for the PNP V_B-control circuit as

$$V_B = \frac{R_2}{R_1 + R_2} V_{CC}$$
$$V_E = V_B + 0.7$$
$$I_E = \frac{V_{CC} - V_E}{R_E}$$

If **ACTIVE** $I_C \approx I_E$

If **ACTIVE** $I_B = \dfrac{I_C}{\beta}$

$$V_C = R_C I_C$$

If $V_E > V_C$, transistor is **ACTIVE,** and the results are correct.

If $V_E < V_C$, transistor is **SATURATED,** and the results are not correct.

REVIEW QUESTIONS

1. What is the procedure for finding the bias levels in a V_B control PNP circuit?
2. How do you check the state of the PNP?
3. Given the base voltage, what is the emitter voltage?
4. Given the emitter voltage, what is the emitter current?
5. If there is a collector resistor, what is the collector voltage in the **ACTIVE** transistor? How should this voltage compare to the emitter voltage?
6. How do you know if a PNP transistor is **SATURATED?**
7. For the self-biased PNP circuit, how do you find the base voltage?

14–3 ■ DOUBLED-SIDED SUPPLY BIASING

Figure 14–9 (p. 452) shows a biasing circuit for a PNP transistor that uses both a positive and a negative supply voltage. We make the reasonable assumption that the base current is small enough that we can neglect the voltage drop across R_B. Therefore, in this V_B control circuit

$$V_B \approx 0$$

FIGURE 14–9
PNP double-sided supply voltages

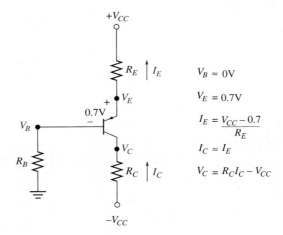

$V_B \approx 0V$

$V_E = 0.7V$

$I_E = \dfrac{V_{CC} - 0.7}{R_E}$

$I_C \approx I_E$

$V_C = R_C I_C - V_{CC}$

And, because the emitter voltage is a diode drop above the base voltage

$$V_E = V_B + 0.7 \approx 0.7V$$

As with the single-sided supply circuit

$$I_E = \frac{V_{CC} - V_E}{R_E} = \frac{V_{CC} - 0.7}{R_E}$$

If the transistor is **ACTIVE**

$$I_C \approx I_E$$

The major change in this circuit is that V_C is no longer the voltage across R_C. Because the bottom end of R_C is connected to the negative supply voltage:

$$I_C = \frac{V_C - (-V_{CC})}{R_C} = \frac{V_C + V_{CC}}{R_C}$$

Rearranging, we find V_C as

$$V_C = R_C I_C - V_{CC}$$

EXAMPLE 14–3
Double-Sided Supply Voltages

Find the DC bias levels, and determine the state of the transistor for the circuit in Figure 14–10.

Solution   Assuming the base current is negligible

$$V_B = 0V$$

The base voltage controls the emitter voltage:

$$V_E = V_B + 0.7 = 0.7V$$

The emitter voltage controls the emitter current:

$$I_E = \frac{V_{CC} - 0.7}{R_E} = \frac{10 - 0.7}{4K} = 2.33mA$$

If the transistor is **ACTIVE**

$$I_C \approx I_E = 2.33mA$$

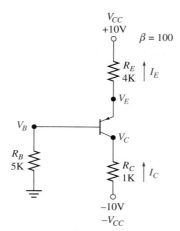

FIGURE 14-10

The collector voltage is

$$V_C = R_C I_C - V_{CC} = 1K \times 2.33mA - 10 = -7.67V$$

Because $V_E > V_C$, the transistor is **ACTIVE.**

DRILL EXERCISE   Redraw the circuit in Figure 14–10, changing the positive and negative supply voltages to $V_{CC} = 12V$. Find the DC bias levels of your circuit, and determine the state of the transistor.

Answer $V_B = 0V$, $V_E = 0.7V$, $I_E = 2.83mA$, $I_C = 2.83mA$, $V_C = -9.17V$
ACTIVE

14-4 ■ THE PNP COMMON-EMITTER AMPLIFIER

The AC behavior of the PNP transistor is identical to the AC behavior of the NPN transistor. That is, for the **ACTIVE** transistor, the emitter diode is modeled with an AC emitter resistance (r_e), in which

$$r_e = \frac{26mV}{I_E}$$

Consider the common-emitter amplifier in Figure 14–11A (p. 454). Note that the collector is now on the bottom and is coupled via a capacitor to the load. We first find the DC bias levels, and check the state of the transistor.

DC Bias Analysis

To find the DC bias levels, we open all capacitors (Figure 14–11B, p. 454) and proceed as follows:

$$V_B = \frac{R_2}{R_1 + R_2} V_{CC} = \frac{60K}{60K + 10K} 12 = 10.3V$$

The base voltage controls the emitter voltage:

$$V_E = V_B + 0.7 = 10.3 + 0.7 = 11V$$

FIGURE 14–11

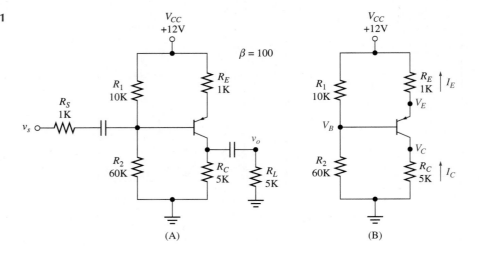

(A)   (B)

The emitter voltage controls the emitter current:

$$I_E = \frac{V_{CC} - V_E}{R_E} = \frac{12 - 11}{1K} = 1mA$$

If the transistor is **ACTIVE,** the collector and base currents are

$$I_C \approx I_E = 1mA$$

and

$$I_B = \frac{I_C}{\beta} = 0.01mA$$

The collector voltage is

$$V_C = R_C I_C = 5K \times 1mA = 5V$$

Because $V_E > V_C$, the transistor is **ACTIVE** and we proceed to the AC analysis.

AC Analysis—The Box Model

The AC behavior of the PNP transistor is identical to the AC behavior of the NPN. For instance, a PNP common-emitter amplifier produces the same Box Model as an NPN common-emitter amplifier, and a PNP emitter follower produces the same Box Model as an NPN emitter follower.

The AC emitter resistance of a PNP transistor is also found by

$$r_e = \frac{26mV}{I_E}$$

so

$$r_e = \frac{26mV}{I_E} = \frac{26mV}{1K} = 26\Omega$$

The AC circuit for our common emitter is shown in Figure 14–12A. Note that the base biasing resistors are now in parallel and have been replaced with $R_B = R_1\|R_2 = 8.57K\Omega$. We want to find the Box Model for this PNP amplifier, so we exclude the input source and load from the analysis. The Box Model parameters for the PNP common-emitter amplifier (see Figure 14–12B) are identical to the parameters for the NPN transistor.

FIGURE 14–12
PNP common-emitter
Box Model

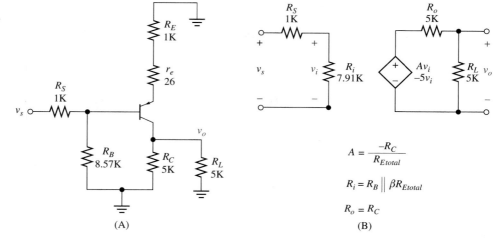

(A)

(B)

$$A = -\frac{R_C}{R_{Etotal}}$$

$$R_i = R_B \parallel \beta R_{Etotal}$$

$$R_o = R_C$$

Gain

The Box Model gain is given by

$$A = -\frac{R_C}{R_{Etotal}}$$

where R_{Etotal} is the total resistance at the emitter and (when there is no emitter bypass capacitor) is given by

$$R_{Etotal} = r_e + R_E$$

If there is an emitter-resistor bypass capacitor, then

$$R_{Etotal} = r_e$$

The negative sign in the gain indicates that the output goes down when the input goes up, and the output goes up when the input goes down. That is, this amplifier inverts the input signal.

Input Resistance

The Box Model input resistance is given by

$$R_i = R_B \parallel \beta R_{Etotal}$$

Output Resistance

The Box Model output resistance is

$$R_o = R_C$$

For our PNP common-emitter amplifier, the Box Model parameters are found from

$$R_{Etotal} = r_e + R_E = 0.026K + 1K = 1.03K\Omega$$

$$R_B = R_1 \parallel R_2 = 60K \parallel 10K = 8.57K\Omega$$

so

$$A = -\frac{R_C}{R_{Etotal}} = -\frac{5\text{K}}{1.03\text{K}} = -4.85$$

$$R_i = R_B \parallel \beta R_{Etotal} = 8.57\text{K} \parallel 103\text{K} = 7.91\text{K}\Omega$$

$$R_o = R_C = 5\text{K}\Omega$$

Again, the negative sign in the gain indicates that the input signal is inverted. That is, when the input signal increases, the output signal decreases, and vice versa.

System Gain

The Box Model, including the source and load resistors, is shown in Figure 14–12B (p. 455). The system gain is found by analyzing the two voltage dividers to get

$$v_o = \frac{R_i}{R_i + R_s}\frac{R_L}{R_L + R_o}Av_s$$

So the total system gain is found from

$$A_{sys} = \frac{R_i}{R_i + R_s}\frac{R_L}{R_L + R_o}A$$

For our circuit

$$A_{sys} = \frac{R_i}{R_i + R_s}\frac{R_L}{R_L + R_o}A = \frac{8\text{K}}{8\text{K} + 1\text{K}}\frac{5\text{K}}{5\text{K} + 5\text{K}}(-5) = 2.22$$

EXAMPLE 14–4
The PNP
Common-Emitter
Amplifier

In Figure 14–13 an emitter bypass capacitor is now added to the common-emitter amplifier we just analyzed. Find the Box Model parameters and the system gain for this amplifier.

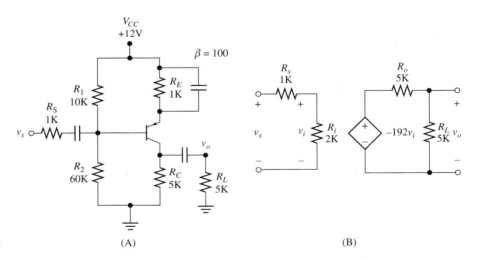

FIGURE 14–13   (A)   (B)

Solution   Because the circuits in Figure 14–12 and 14–13 are the same for DC calculations, we will use the results we found previously:

$$I_E = 1\text{mA}$$

and the AC emitter resistance is

$$r_e = \frac{26\text{mV}}{I_E} = \frac{26\text{mV}}{1\text{mA}} = 26\Omega = 0.026\text{K}\Omega$$

Because the emitter resistor is now bypassed

$$R_{Etotal} = r_e = 0.026\text{K}\Omega$$

and

$$A = -\frac{R_C}{R_{Etotal}} = -\frac{5\text{K}}{0.026\text{K}} = -192$$

The input resistance is

$$R_i = R_B \parallel \beta R_{Etotal} = 8.57\text{K} \parallel 2.6\text{K} = 1.99\text{K}\Omega$$

and the output resistance is

$$R_o = R_C = 5\text{K}\Omega$$

Note that the bypass capacitor has increased the gain but lowered the input resistance.
The Box Model, with the source and load resistors, is shown in Figure 14–13B. The system gain is found from

$$A_{sys} = \frac{R_i}{R_i + R_s} \frac{R_L}{R_L + R_o} A$$

$$= \frac{2\text{K}}{2\text{K} + 1\text{K}} \frac{5\text{K}}{5\text{K} + 5\text{K}} \ (192)$$

$$= -64$$

DRILL EXERCISE

Redraw the circuit in Figure 14–13, changing the supply voltage to $V_{CC} = 9$V. Find the DC bias levels, the AC emitter resistance, the Box Model parameters, and the system gain.

Answer $V_B = 7.71$V, $V_E = 8.41$V, $I_E = 0.586$mA
$I_C = 0.586$mA, $I_B = 0.00586$mA, $V_C = 2.93$V
$r_e = 44.4\Omega$, $A = -113$, $R_i = 2.92$KΩ, $R_o = 5$KΩ, $A_{sys} = -42.1$

REVIEW QUESTIONS

1. What is the AC emitter resistance of the PNP?
2. What is the total resistance looking into the base of the PNP?
3. Write the formula for the Box Model gain for the PNP common emitter.
4. Write the formula for the Box Model input resistance of the PNP common emitter.
5. Write the formula for the Box Model output resistance for the PNP common emitter.
6. How does an emitter resistor bypass capacitor affect the Box Model parameters?
7. Including the source and load resistors, write the formula for the system gain.

14–5 ■ THE PNP EMITTER FOLLOWER (COMMON-COLLECTOR)

Figure 14–14 shows the emitter follower constructed with a PNP transistor. As we shall see, the PNP emitter follower exhibits the same properties as the NPN emitter follower: voltage gain close to 1, high input resistance, and low output resistance.

FIGURE 14–14
DNP emitter follower

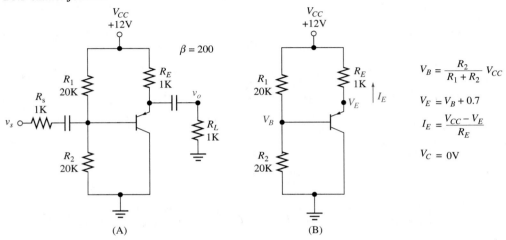

(A)　　　　　(B)

DC Bias Analysis

To find the DC bias levels of this circuit, we analyze the circuit in Figure 14–14B. We first find the base voltage:

$$V_B = \frac{R_2}{R_1 + R_2} V_{CC} = \frac{20K}{20K + 20K} 12 = 6V$$

The base voltage controls the emitter voltage:

$$V_E = V_B + 0.7 = 6.7V$$

The emitter voltage controls the emitter current:

$$I_E = \frac{V_{CC} - V_E}{V_E} = \frac{12 - 6.7}{1K} = 5.3mA$$

If the transistor is **ACTIVE**

$$I_C \approx I_E = 5.3mA$$

and

$$I_B = \frac{I_C}{\beta} = \frac{5.3mA}{200} = 0.0265mA$$

Because the collector is grounded

$$V_C = 0V$$

So, the transistor is **ACTIVE** and we proceed to the AC analysis.

AC Analysis—the Box Model

Figure 14–15A shows the AC circuit for this emitter follower. Note that

$$R_B = R_2 \| R_1 = 10K\Omega$$

FIGURE 14–15
PNP emitter follower Box Model

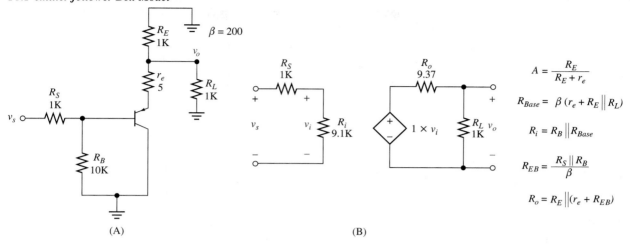

(A)   (B)

The AC emitter resistance is given by

$$r_e = \frac{26\text{mV}}{I_E} = \frac{26\text{mV}}{5.3\text{mA}} = 4.91\Omega$$

Gain

The Box Model gain for the PNP emitter follower is the same as for the NPN emitter follower. That is

$$A = \frac{R_E}{R_E + r_e}$$

For this circuit, we get

$$A = \frac{R_E}{R_E + r_e} = \frac{1\text{K}}{1\text{K} + 0.00491\text{K}} \approx 1$$

Again, note that the emitter-follower Box Model gain is positive and very close to 1.

Input Resistance

The Box Model input and output resistances for the emitter follower depend on the load and source resistors. That is, for the emitter follower the resistance seen at the base is β times the total resistance (including the load resistor) at the emitter:

$$R_{base} = \beta(r_e + R_E \| R_L)$$

For this circuit

$$R_{base} = 200(0.00491\text{K} + 0.5\text{K}) = 101\text{K}\Omega$$

The Box Model input resistance is given by

$$R_i = R_B \| R_{base}$$

For this circuit

$$R_i = 10\text{K} \| 101\text{K} = 9.1\text{K}\Omega$$

Output Resistance

The resistance seen at the emitter (R_{EB}) is equal to the total resistance at the base (including the source resistor) divided by β. That is

$$R_{EB} = \frac{R_s \parallel R_B}{\beta}$$

For this circuit

$$R_{EB} = \frac{1\text{K} \parallel 10\text{K}}{200} = 4.55\Omega$$

To find the output resistance, R_{EB} is now added to r_e, and the result is placed in parallel with R_E:

$$R_o = (R_{EB} + r_e) \parallel R_E$$

For this circuit

$$R_o = (4.55 + 4.91) \parallel 1\text{K} = 9.37\Omega$$

System Gain

The emitter-follower Box Model, along with the source and load resistors, is shown in Figure 14–15B (p. 459). The system gain is found from

$$A_{sys} = \frac{R_i}{R_i + R_s} \frac{R_L}{R_L + R_o} A$$

so

$$A_{sys} = \frac{R_i}{R_i + R_s} \frac{R_L}{R_i + R_o} A = \frac{9.1\text{K}}{9.1\text{K} + 1\text{K}} \frac{1\text{K}}{1\text{K} + 0.00937\text{K}} 1 = 0.891$$

and we see that the system gain for the emitter follower is positive and close to 1.

EXAMPLE 14–5
The PNP Emitter Follower

For the circuit in Figure 14–16, find the DC bias levels, the Box Model parameters, and the system gain.

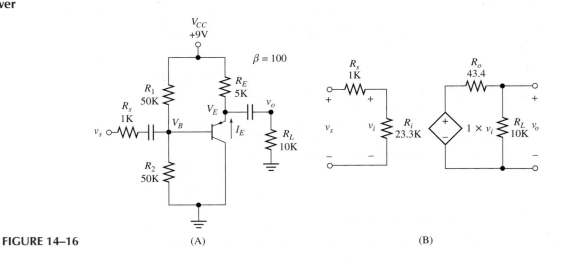

FIGURE 14–16   (A)   (B)

Solution   DC Analysis

The base voltage is

$$V_B = \frac{R_2}{R_1 + R_2} \, V_{CC} = \frac{50K}{50K + 50K} \, 9 = 4.5V$$

The base voltage controls the emitter voltage:

$$V_E = V_B + 0.7 = 5.2V$$

The emitter voltage controls the emitter current:

$$I_E = \frac{V_{CC} - V_E}{R_E} = \frac{9 - 5.2}{5K} = 0.76mA$$

If the transistor is **ACTIVE**

$$I_C \approx I_E = 0.76mA$$

and

$$I_B = \frac{I_C}{\beta} = 0.0076mA$$

Because the emitter is grounded, $V_C = 0V$ and the transistor is **ACTIVE.** We can proceed with the AC analysis. The AC emitter resistance is

$$r_e = \frac{26mV}{I_E} = \frac{26mV}{0.76mA} = 34.2\Omega$$

The Box Model gain is found from

$$A = \frac{R_E}{R_E + r_e} = \frac{5K}{5.034K} = 0.99$$

To find the Box Model input resistance, we first find the resistance at the base (do not forget to include the load resistor):

$$R_{base} = \beta(r_e + R_E \parallel R_L) = 100(0.034K + 5K \parallel 10K) = 337K\Omega$$

and

$$R_i = R_B \parallel R_{base} = 25K \parallel 337K = 23.3K\Omega$$

To find the Box Model output resistance, we first find the total resistance at the base (including the source resistor) and divide by β:

$$R_{EB} = \frac{R_s \parallel R_B}{\beta} = \frac{1K \parallel 25K}{100} = 9.62\Omega$$

We add this result to r_e, and put that combination in parallel with R_E to find the output resistance:

$$R_o = (R_{EB} + r_e) \parallel R_E = (9.62 + 34.2) \parallel 5K = 43.4\Omega$$

For this PNP emitter follower the Box Model, along with the source and load resistors, is shown in Figure 14–16B. The system gain is found from

$$A_{sys} = \frac{R_i}{R_i + R_s} \, \frac{R_L}{R_L + R_o} \, A = \frac{23K}{23K + 1K} \, \frac{10K}{10K + 0.0434K} \, 0.99 = 0.945$$

DRILL EXERCISE

Redraw the PNP emitter follower of Figure 14–16A (p. 460), changing the supply voltage to $V_{CC} = 12V$. Find the DC bias levels, the Box Model parameters, and the system gain.

Answer $V_B = 6V$, $V_E = 6.7V$, $I_E = 1.06mA$
$I_C = 1.06mA$, $I_B = 0.0106mA$, $V_C = 0V$
$r_e = 24.5\Omega$, $A = 1$, $R_i = 23.3K\Omega$, $R_o = 34.7\Omega$, $A_{sys} = 0.956$

REVIEW QUESTIONS

1. What is the approximate gain of the emitter follower?
2. Is the input resistance of the emitter follower low or high?
3. Is the output resistance of the emitter follower low or high?
4. What is the formula for the Box Model gain?
5. What is the formula for the Box Model input resistance? What does this resistance depend on?
6. What is the formula for the Box Model output resistance? What does this resistance depend on?
7. Draw the Box Model, including source and load resistors, for the PNP emitter follower.
8. Derive a formula for the system gain of the emitter follower. Use the two voltage dividers in the Box Model.

14–6 ■ I_B CONTROL

Most switching applications use I_B control of the transistor. The key to I_B control is that there is no emitter resistor. The input voltage applied to R_B creates a base current, which in turn controls the behavior of the transistor.

Figure 14–17 shows a practical I_B control circuit. As the input is varied, we can put the transistor into any one of its three states. (We remind you that the state of a transistor depends on the condition of the emitter and the collector diodes.) In this circuit,

$$V_E = V_{CC} = 12V$$

If the emitter diode is **ON**

$$V_B = V_E - 0.7 = 12 - 0.7 = 11.3V$$

FIGURE 14–17

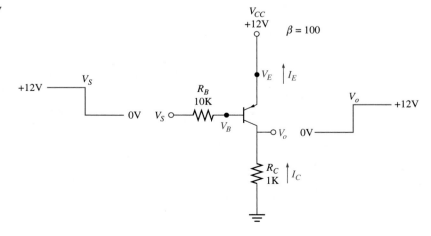

We know that if the emitter diode is **ON,** the base current must go from the source to the base. This will only happen if the source voltage is smaller than the base voltage. That is, the emitter diode will be **ON** only if

$$V_S < V_B$$

In this circuit, if

$$V_S < 11.3\text{V}$$

the emitter diode will be **ON.** If

$$V_S > 11.3\text{V}$$

the emitter diode will be **OFF** and the transistor will be in **CUT-OFF.**

In I_B control circuits, usually used to construct electronic switches, the transistor acts between the **CUT-OFF** and **SATURATED** states. Consider the input voltage swing in Figure 14–17. For

$$V_S = 12\text{V} > 11.3\text{V}$$

The emitter diode is **OFF** and the transistor is **CUT-OFF;** all transistor currents are 0A. The output voltage is, therefore

$$V_o = R_C I_C = 0\text{V}$$

When

$$V_S = 0\text{V} < 11.3\text{V}$$

the emitter diode is **ON,** and we find the base current from

$$I_B = \frac{V_B - V_S}{R_B} = \frac{11.3 - 0}{10\text{K}} = 1.13\text{mA}$$

If the transistor is **ACTIVE**

$$I_C = \beta I_B = 113\text{mA}$$

With the collector current this large

$$V_C = R_C I_C = 1\text{K} \times 113\text{mA} = 113\text{V}!$$

This cannot be. The transistor must be **SATURATED,** so

$$V_E = V_C = 12\text{V}$$

When the transistor **SATURATES,** it acts as though a switch has closed between the collector and the emitter. As the base current rises from 0mA (on its way to 113mA), the collector diode turns **ON,** the transistor **SATURATES** and

$$I_C \neq \beta I_B$$

In **SATURATION** $V_C = V_E$, and the collector current is found from

$$I_{C(sat)} = \frac{V_C}{R_C} = \frac{12}{1\text{K}} = 12\text{mA}$$

This transistor switch creates an **inverter.** That is, when the input is high (12V), the output is low (0V). When the input is low (0V), the output is high (12V).

EXAMPLE 14–6
Comparison of the NPN and PNP Inverters

For the NPN and PNP circuits in Figure 14–18 (p. 464), find the output voltage when the input is high (12V) and low (0V). Discuss the difference in these two inverters.

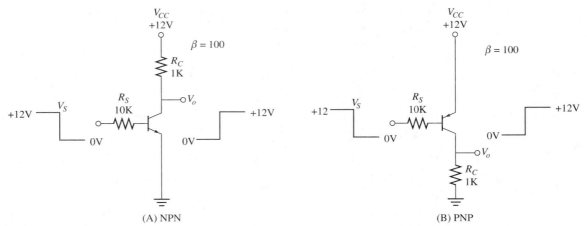

FIGURE 14–18

Solution   **NPN Circuit (Figure 14–18A)**

When the input voltage is high—that is when

$$V_S = 12V$$

the emitter diode of the transistor turns **ON** and

$$I_B = \frac{V_S - 0.7}{R_B} = \frac{12 - 0.7}{10K} = 1.13mA$$

This current is enough to put the NPN into **SATURATION,** so

$$V_o = V_C = V_E = 0V$$

When the input voltage is low—that is when

$$V_S = 0V$$

the NPN transistor is **CUT-OFF,** so all currents are 0A. This leads to

$$V_o = V_C = V_{CC} = 12V$$

When the input is high, the output is low. When the input is low, the output is high. This NPN circuit is an inverter.

   PNP Circuit (Figure 14–18B)

   When the input voltage is high—that is when

$$V_S = 12V$$

the PNP transistor is **CUT-OFF,** so all currents are 0A. This leads to

$$V_o = V_C = 0V$$

When the input is low—that is when

$$V_S = 0V$$

the emitter diode of the PNP is turned **ON** and

$$V_B = V_{CC} - 0.7 = 11.3V$$

The base current is

$$I_B = \frac{V_B - V_S}{10K} = \frac{11.3}{10K} = 1.13mA$$

Because this current is large enough to **SATURATE** the PNP transistor

$$V_C = V_E = 12V$$

When the input is high, the output is low. When the input is low, the output is high. This PNP circuit is also an inverter.

Comparison

Both the NPN and the PNP circuits are inverters. The difference is that the load resistor in the NPN has current in it and dissipates power when the input is high. The load resistor in the PNP circuit, on the other hand, has current in it and dissipates power when the input is low. The availability of the two circuits gives the designer the option to choose under which input condition the load resistor will dissipate power.

REVIEW QUESTIONS

1. How can you tell that a transistor bias circuit is I_B control?
2. When does the emitter diode turn **ON?**
3. When the emitter diode is **ON,** what is V_B?
4. What is I_B?
5. In what states does an I_B-control transistor usually operate?
6. When the transistor is **CUT-OFF,** what is V_C?
7. When the transistor is **SATURATED,** what is V_C?
8. What are the input-output results in the inverter circuit?
9. What is the difference between an NPN and a PNP inverter?

14–7 ■ COMBINED NPN-PNP CIRCUITS

Although it is true that all of the basic circuits we have analyzed can be built either with NPN transistors or PNP transistors, most single-transistor circuits use the NPN transistor. When the PNP transistor is used, it is usually used in combination with the NPN transistor.

V_B Control

Consider, for example, the V_B control circuits in Figure 14–19 (p. 466). Remember that V_B control circuits are used in linear applications. The circuit in Figure 14–19A is a simple NPN emitter follower that has high input resistance, low output resistance, and close to unity gain. One disadvantage of the circuit is that the output voltage is 0.7V below the input voltage. That is

$$V_o = V_E = V_S - 0.7$$

Now, consider the NPN-PNP circuit shown in Figure 14–19B (p. 466). Note that the emitter-follower NPN drives the PNP. Because the output of the PNP is taken from the emitter, the PNP circuit is also an emitter follower. This combination has high input resistance, low output resistance, and a gain close to 1.

Let's trace the input voltage to the output of the PNP. At the emitter of the NPN

$$V_{E1} = V_S - 0.7$$

The emitter voltage of the NPN is applied to the base of the PNP; that is

$$V_{B2} = V_{E1} = V_S - 0.7$$

FIGURE 14–19

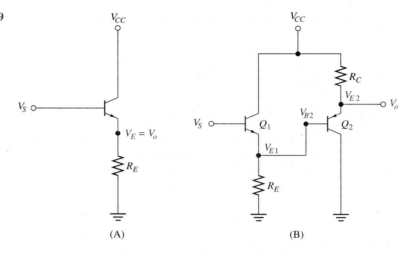

(A) (B)

The output voltage is

$$V_o = V_{B2} + 0.7 = V_S - 0.7 + 0.7 = V_S$$

The output voltage of the combined circuit is now the same as the input voltage. The voltage loss encountered in the single NPN circuit has been compensated by the addition of the PNP. Correcting for emitter-diode voltage drops is one of the reasons we combine NPN and PNP transistors.

I_B Control

Now consider the I_B control NPN-PNP circuit in Figure 14–20. As with other I_B control circuits, this one represents a switching application of the two transistors. Note that the collector current of the NPN is the base current of the PNP. When the input is low—that is when

$$V_S = 0V$$

FIGURE 14–20

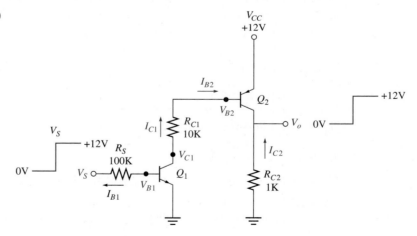

the NPN transistor is **CUT-OFF,** and all of its currents are 0A. This means that

$$I_{B2} = I_{C1} = 0A$$

and the PNP is also **CUT-OFF.** Because there is no PNP collector current

$$V_o = 0V$$

When the input voltage goes high—that is when

$$V_S = 12\text{V}$$

the NPN **SATURATES,** the collector current of the NPN increases greatly and **SATU-RATES** the PNP. When the PNP **SATURATES,** we get

$$V_o = 12\text{V}$$

Therefore, when the input goes low in this NPN-PNP circuit, the output goes low. When the input goes high, the output goes high. This circuit can be used as a non-inverting switch.

Current Gain

This circuit also provides a greater current gain than is available with a single stage. When the input is high

$$I_{B1} = \frac{V_S - 0.7}{R_S} = \frac{11.3}{100\text{K}} = 0.113\text{mA}$$

For this base current the NPN **SATURATES** ($V_{C1} = 0\text{V}$), and the NPN collector current is found from

$$I_{C1} = \frac{V_{B2} - V_{C1}}{10\text{K}} = \frac{V_{B2}}{10\text{K}}$$

Because

$$V_{B2} = 12 - 0.7 = 11.3\text{V}$$

the NPN collector current is

$$I_{C1} = \frac{V_{B2}}{10\text{K}} = 1.13\text{mA}$$

The current gain between the base and collector of the NPN is

$$\frac{I_{C1}}{I_{B1}} = \frac{1.13\text{mA}}{0.113\text{mA}} = 10$$

Note that in a **SATURATED** transistor, the current gain is less than β. Now the NPN collector current is the base current of the PNP—that is

$$I_{B2} = I_{C1} = 1.13\text{mA}$$

which **SATURATES** the PNP so

$$V_{C2} = V_{E2} = 12\text{V}$$

The collector current of the **SATURATED** PNP is found from

$$I_{C2} = \frac{V_{C2}}{1\text{K}} = \frac{V_o}{1\text{K}} = \frac{12}{1\text{K}} = 12\text{mA}$$

The overall current gain is

$$\frac{I_{C2}}{I_{B1}} = \frac{12\text{mA}}{0.113\text{mA}} = 106$$

The current gain for the single-stage NPN is 10, compared to a current gain of 106 for the NPN-PNP combination.

14–8 ■ TROUBLESHOOTING

Troubleshooting PNP circuits is exactly the same as troubleshooting NPN circuits. The very first thing you do is DC Troubleshooting. Connect the AC signal source and the load *only* if the Estimated Voltages and Measured Voltages at the transistor leads are close (Figure 14–21).

FIGURE 14–21

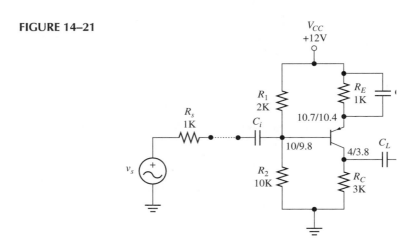

DC Troubleshooting will help you locate problems caused by bad connections, failed transistors, and, usually, shorted capacitors. If the Estimated Voltages and Measured Voltages are close but the AC output is not what you expected, one of the capacitors has probably failed open. To locate this problem, you would trace the AC signals through the circuit.

Bad Connections

Consider, for example, the DC Troubleshooting results shown in Figure 14–22. In Figure 14–22A the Measured Voltage at the collector lead shows a large discrepancy. The connection diagram at the collector shows 0V at R_C. This indicates a bad connection between the collector resistor and the transistor.

FIGURE 14–22

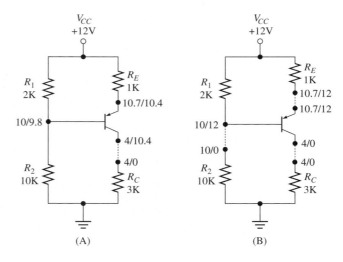

(A)   (B)

Figure 14–22B shows the transistor circuit with discrepancies in Estimated Voltages and Measured Voltages at all transistor leads. The connection diagrams at these leads indicate a bad connection between R_2 and the base.

Transistor Open Failures

Figure 14–23 shows some examples of transistor failures. In Figure 14–23A the Measured Voltage at the external base lead is correct. The Measured Voltages at the collector and emitter indicate that there is no collector or emitter current. Therefore, the transistor is **CUT-OFF.** This will happen if the transistor has an open inside the base.

Figure 14–23B shows the results of DC Troubleshooting a circuit in which the transistor has failed with an open circuit inside the emitter. You can see that a failed open base or a failed open emitter produces the same DC Troubleshooting results.

Figure 14–23C shows the results of DC Troubleshooting a circuit in which the transistor has failed with an open circuit inside the collector.

FIGURE 14–23

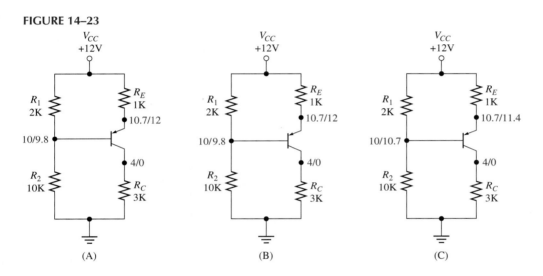

Transistor Short Failures

Internal shorts between two transistor leads will result in the same Measured Voltages at those leads. The results of internal shorts between the collector and emitter are shown in Figure 14–24A, between the base and collector in Figure 14–24B, and between the base and the emitter in Figure 14–24C.

FIGURE 14–24

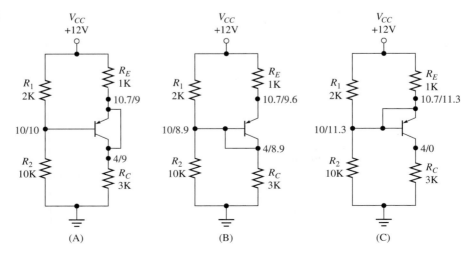

AC Signal Tracing

Figure 14–25 shows a circuit with good DC Troubleshooting results but poor AC behavior. If there is no AC signal at the base, the input coupling capacitor has failed open. If there is an AC signal at the collector but no AC signal at the load, the output coupling capacitor has failed open. If there is an AC signal at the output but the gain is much lower than expected, the bypass capacitor has probably failed open.

FIGURE 14–25

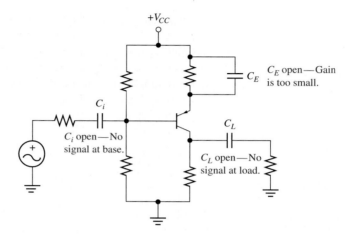

EXAMPLE 14–7
Troubleshooting

For the circuit in Figure 14–26, assume all connections have been checked and are good. Determine the likely cause of the discrepancies between the Estimated Voltages and the Measured Voltages.

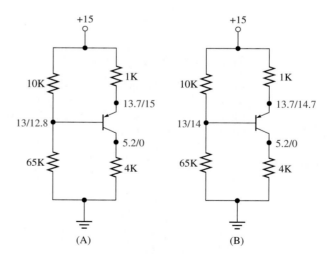

FIGURE 14–26   (A)   (B)

Solution

(a) The circuit in Figure 14–26A shows a discrepancy at both the emitter and collector leads. Because the Measured Voltage at the emitter is the same as the supply voltage, there can be no current in R_E. We can conclude that the emitter is open inside the transistor. The open emitter would also lead to no current in R_C, and we would get the collector Measured Voltage shown.

(b) In the circuit in Figure 14–26B, we also measure 0V across R_C. This means that there is no collector current in this circuit. The measured emitter voltage is less than V_{CC}, indicating that there is an emitter current. Also note that the measured emitter voltage is a diode drop above the base voltage, indicating that the emitter diode is **ON**. These facts lead to the conclusion that the transistor has an internal open at the collector.

REVIEW QUESTIONS

1. How do you record Estimated Voltages and Measured Voltages?
2. If the Estimated Voltage and Measured Voltage show a discrepancy at a transistor lead, what do you do next?
3. What do you expect to see in DC Troubleshooting if there is an internal open inside the base, the emitter, the collector?
4. What do you expect to see in DC Troubleshooting if there is a base-emitter short, a base-collector short, a collector-emitter short?
5. When can you connect the AC signal source?
6. Describe the effects on a circuit of failed open capacitors.

SUMMARY

■ The state of the PNP transistor depends on the emitter and collector diodes:

CUT-OFF	Emitter diode and collector diode are **OFF.**
SATURATED	Emitter diode and collector diode are **ON.**
ACTIVE	Emitter diode is **ON**—collector diode is **OFF.**

■ For V_B control of the PNP transistor with a positive supply

$$V_B = \frac{R_2}{R_1 + R_2} V_{CC}$$

$$V_E = V_B + 0.7$$

$$I_E = \frac{V_{CC} - V_E}{R_E}$$

If **ACTIVE**, $I_C \approx I_E$ and $I_B = \frac{I_C}{\beta}$

$$V_C = R_C I_C$$

ACTIVE if $V_E > V_C$

■ Common-Emitter Box Model parameters:

$$A = -\frac{R_C}{R_{Etotal}}$$

$$R_o = R_C$$

$$R_i = R_B \,\|\, \beta R_{Etotal}$$

■ Emitter-Follower Box Model parameters:

$$A \approx 1$$

$$R_o = \left(\frac{R_s \,\|\, R_B}{\beta} + r_e \right) \,\|\, R_E$$

$$R_i = R_B \,\|\, \beta (r_e + R_E \,\|\, R_L)$$

■ For I_B control of the PNP transistor switch

CUT-OFF:	V_S is high	$V_C = 0V$
SATURATION:	V_S is low	$V_C = V_{CC}$

FIGURE 14–27

MAXIMUM RATINGS

Rating	Symbol	Value	Unit
Collector-Emitter Voltage	V_{CEO}	40	Vdc
Collector-Base Voltage	V_{CBO}	40	Vdc
Emitter-Base Voltage	V_{EBO}	5.0	Vdc
Collector Current — Continuous	I_C	200	mAdc
Total Device Dissipation @ T_A = 25°C Derate above 25°C	P_D	625 5.0	mW mW/°C
Total Power Dissipation @ T_A = 60°C	P_D	250	mW
Total Device Dissipation @ T_C = 25°C Derate above 25°C	P_D	1.5 12	Watts mW/°C
Operating and Storage Junction Temperature Range	T_J, T_{stg}	−55 to +150	°C

*THERMAL CHARACTERISTICS

Characteristic	Symbol	Max	Unit
Thermal Resistance, Junction to Ambient	$R_{\theta JA}$	200	°C/W
Thermal Resistance, Junction to Case	$R_{\theta JC}$	83.3	°C/W

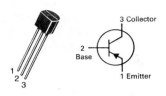

2N3905
2N3906★

GENERAL PURPOSE
TRANSISTORS

PNP SILICON

★**This is a Motorola
designated preferred device.**

ELECTRICAL CHARACTERISTICS (T_A = 25°C unless otherwise noted.)

Characteristic		Symbol	Min	Max	Unit
OFF CHARACTERISTICS					
Collector-Emitter Breakdown Voltage (1) (I_C = 1.0 mAdc, I_B = 0)		$V_{(BR)CEO}$	40	—	Vdc
Collector-Base Breakdown Voltage (I_C = 10 μAdc, I_E = 0)		$V_{(BR)CBO}$	40	—	Vdc
Emitter-Base Breakdown Voltage (I_E = 10 μAdc, I_C = 0)		$V_{(BR)EBO}$	5.0	—	Vdc
Base Cutoff Current (V_{CE} = 30 Vdc, V_{EB} = 3.0 Vdc)		I_{BL}	—	50	nAdc
Collector Cutoff Current (V_{CE} = 30 Vdc, V_{EB} = 3.0 Vdc)		I_{CEX}	—	50	nAdc
ON CHARACTERISTICS(1)					
DC Current Gain (I_C = 0.1 mAdc, V_{CE} = 1.0 Vdc)	2N3905 2N3906	h_{FE}	30 60	— —	—
(I_C = 1.0 mAdc, V_{CE} = 1.0 Vdc)	2N3905 2N3906		40 80	— —	
(I_C = 10 mAdc, V_{CE} = 1.0 Vdc)	2N3905 2N3906		50 100	150 300	
(I_C = 50 mAdc, V_{CE} = 1.0 Vdc)	2N3905 2N3506		30 60	— —	
(I_C = 100 mAdc, V_{CE} = 1.0 Vdc)	2N3905 2N3906		15 30	— —	
Collector-Emitter Saturation Voltage (I_C = 10 mAdc, I_B = 1.0 mAdc) (I_C = 50 mAdc, I_B = 5.0 mAdc)		$V_{CE(sat)}$	— —	0.25 0.4	Vdc
Base-Emitter Saturation Voltage (I_C = 10 mAdc, I_B = 1.0 mAdc) (I_C = 50 mAdc, I_B = 5.0 mAdc)		$V_{BE(sat)}$	0.65 —	0.85 0.95	Vdc
SMALL-SIGNAL CHARACTERISTICS					
Current-Gain — Bandwidth Product (I_C = 10 mAdc, V_{CE} = 20 Vdc, f = 100 MHz)	2N3905 2N3906	f_T	200 250	— —	MHz
Output Capacitance (V_{CB} = 5.0 Vdc, I_E = 0, f = 1.0 MHz)		C_{obo}	—	4.5	pF
Input Capacitance (V_{EB} = 0.5 Vdc, I_C = 0, f = 1.0 MHz)		C_{ibo}	—	10.0	pF
Input Impedance (I_C = 1.0 mAdc, V_{CE} = 10 Vdc, f = 1.0 kHz)	2N3905 2N3906	h_{ie}	0.5 2.0	8.0 12	k ohms
Voltage Feedback Ratio (I_C = 1.0 mAdc, V_{CE} = 10 Vdc, f = 1.0 kHz)	2N3905 2N3906	h_{re}	0.1 0.1	5.0 10	X 10^{-4}
Small-Signal Current Gain (I_C = 1.0 mAdc, V_{CE} = 10 Vdc, f = 1.0 kHz)	2N3905 2N3906	h_{fe}	50 100	200 400	—
Output Admittance (I_C = 1.0 mAdc, V_{CE} = 10 Vdc, f = 1.0 kHz)	2N3905 2N3906	h_{oe}	1.0 3.0	40 60	μmhos
Noise Figure (I_C = 100 μAdc, V_{CE} = 5.0 Vdc, R_S = 1.0 k ohm, f = 1.0 kHz)	2N3905 2N3906	NF	— —	5.0 4.0	dB

PROBLEMS

SECTION 14–1 Introduction

1. The data sheet for two PNP transistors is shown in Figure 14–27. For the 2N3905 with a collector current of $I_C = 1\text{mA}$
   - (a) What is the DC current gain?
   - (b) What is the AC current gain?
   - (c) What is $V_{C(sat)}$?
   - (d) What is $V_{EB(on)}$?
   - (e) What is the output resistance $(r_o = \dfrac{1}{h_{oe}})$?

2. Repeat Problem 1 for the 2N3906.

3. Compare Figure 14–27 with Figure 13–20 (Chapter 13). Which NPN is matched to the 2N3905?

SECTION 14–2 V_B Control of PNP Transistors

4. For the PNP transistor circuit in Figure 14–28, use the steps of V_B control to find the following DC bias values if $V_{BB} = +5\text{V}$:
   - (a) V_B.
   - (b) V_E.
   - (c) The voltage across R_E.
   - (d) I_E.
   - (e) I_C.
   - (f) I_B.
   - (g) V_C.
   - (h) What is the voltage across the emitter diode (i.e., what is V_{EB})?
   - (i) What is the voltage on the collector diode (i.e., what is V_{BC})?
   - (j) In what state is the transistor operating? Why?

5. Repeat Problem 4 if $V_{BB} = +10\text{V}$.

6. For the PNP transistor circuit in Figure 14–29, use the steps of V_B control to find the following DC bias values if $V_{CC} = 10\text{V}$, $V_{BB} = 5\text{V}$, $R_E = 2\text{K}\Omega$, $R_C = 1\text{K}\Omega$:
   - (a) V_B.
   - (b) V_E.
   - (c) The voltage across R_E.
   - (d) I_E.
   - (e) I_C.
   - (f) I_B.
   - (g) The voltage across R_C.
   - (h) V_C.
   - (i) What is the voltage across the emitter diode (i.e., what is V_{EB})?
   - (j) What is the voltage on the collector diode (i.e., what is V_{BC})?
   - (k) In what state is the transistor operating? Why?

7. Repeat Problem 6 if $V_{CC} = 12\text{V}$, $V_{BB} = 5.3\text{V}$, $R_E = 5\text{K}\Omega$, $R_C = 1\text{K}\Omega$.

8. Repeat Problem 6 if $V_{CC} = 12\text{V}$, $V_{BB} = 1.3\text{V}$, $R_E = 5\text{K}\Omega$, $R_C = 1\text{K}\Omega$.

9. For the PNP transistor circuit in Figure 14–30, use the steps of V_B control to find the following DC bias values:
   - (a) V_B.
   - (b) V_E.
   - (c) The voltage across R_E.
   - (d) I_E.
   - (e) I_C.
   - (f) I_B.
   - (g) V_C.
   - (h) Is the emitter diode **ON** or **OFF**? What is V_{EB}?
   - (i) Is the collector diode **ON** or **OFF**? What is V_{BC}?
   - (j) In what state is the transistor operating? Why?

10. Choose a new R_E in Figure 14–30 so that $I_C = 5\text{mA}$.

FIGURE 14–28

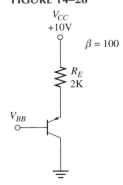

FIGURE 14–29

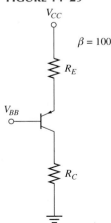

FIGURE 14–30

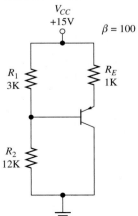

11. For the PNP transistor circuit in Figure 14–31, use the steps of V_B control to find the following DC bias values if

$$V_{CC} = 15V, R_1 = 3K\Omega, R_2 = 12K\Omega, R_E = 1K\Omega, R_C = 2K\Omega:$$

(a) V_B.   **(b)** V_E.   **(c)** The voltage across R_E.   **(d)** I_E.

(e) I_C.   **(f)** I_B.   **(g)** The voltage across R_C.   **(h)** V_C.

(i) Is the emitter diode **ON** or **OFF?** What is V_{EB}?

(j) Is the collector diode **ON** or **OFF?** What is V_{BC}?

(k) In what state is the transistor operating? Why?

12. Repeat Problem 11 if

$$V_{CC} = 9V, R_1 = 1.7K\Omega, R_2 = 7.3K\Omega, R_E = 1K\Omega, R_C = 5K\Omega.$$

13. The PNP transistor shown in Figure 14–32 is **SATURATED.**

(a) List the results of the V_B control procedure to show that this transistor is **SATURATED.**

(b) Using the results of part (a), explain how you know that this transistor is **SATURATED.**

(c) If this transistor is **SATURATED,** are the results of the V_B control procedure accurate?

FIGURE 14–31

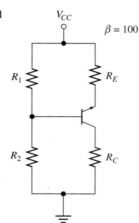

FIGURE 14–32

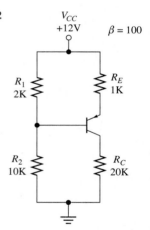

SECTION 14–3   Double-Sided Supply Biasing

14. In Figure 14–33, $V_{CC} = 12V, R_E = 1K\Omega, R_C = 4K\Omega, R_B = 10K\Omega, \beta = 200$. Find

(a) V_B.   **(b)** V_E.   **(c)** The voltage across R_E.   **(d)** I_E.

(e) I_C.   **(f)** I_B.   **(g)** The voltage across R_C.   **(h)** V_C.

15. What value of R_C would put the transistor in Figure 14–33 into **SATURATION?**

SECTION 14–4   The PNP Common-Emitter Amplifier

16. For the PNP common-emitter amplifier in Figure 14–34,

$$V_{CC} = 20V, R_1 = 20K\Omega, R_2 = 180K\Omega, R_E = 1K\Omega, R_C = 10K\Omega, R_L = 10K\Omega:$$

(a) Use the steps of V_B control to find the DC bias values. List these values in the form of the DC Bias List.

(b) Is the transistor **ACTIVE?** Why?   **(c)** What is r_e?   **(d)** What is R_{Etotal}?

(e) What is the voltage gain (A)?   **(f)** What is the input resistance (R_i)?

(g) What is the output resistance (R_o)?

(h) Draw the Box Model for this amplifier. Be sure to label the Box Model parameters with the values you found for R_i, A, R_o.

(i) Find the total system gain (A_{sys}).

17. The source in Figure 14–34 has an internal resistance of $R_S = 10K\Omega$. Draw the complete Box Model, including the source resistance, and find the total system gain (A_{sys}).

FIGURE 14–33

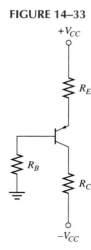

FIGURE 14–34

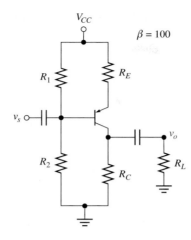

$\beta = 100$

18. Repeat Problem 16 if

$$V_{CC} = 9V,\ R_1 = 22K\Omega,\ R_2 = 68K\Omega,\ R_E = 3K\Omega,\ R_C = 6K\Omega,\ R_L = 6K\Omega.$$

19. For the PNP common-emitter amplifier shown in Figure 14–35
   (a) Use the steps of V_B control to find the DC bias values. List these values in the form of the DC Bias List.
   (b) Is the transistor **ACTIVE?** Why?　　**(c)** What is r_e?　　**(d)** What is R_{Etotal}?
   (e) What is the voltage gain *(A)*?　　　　**(f)** What is the input resistance (R_i)?
   (g) What is the output resistance (R_o)?
   (h) Draw the Box Model for this amplifier. Be sure to label the Box Model parameters with the values you found for R_i, *A*, R_o.
20. Repeat Problem 16 if the emitter resistor is bypassed.

FIGURE 14–35

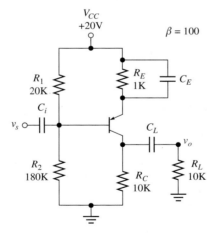

SECTION 14–5　The PNP Emitter Follower (Common-Collector)

21. For the PNP emitter follower shown in Figure 14–36 (p. 476)
   (a) Use the steps of V_B control to find the DC bias values. List these values in the form of the DC Bias List.
   (b) What is r_e?　　　**(c)** What is the Box Model voltage gain *(A)*?
   (d) What is R_{base}?　　**(e)** What is the input resistance (R_i)?
   (g) What is R_{EB}?　　**(h)** What is the output resistance (R_o)?
   (i) Draw the complete Box Model for this amplifier. Be sure to label the Box Model parameters with the values you found for R_i, *A*, R_o. Also show v_s, R_s, R_L in their proper locations.
   (j) Find v_i as a fraction of v_s.　　**(k)** Find v_o as a fraction v_s.
   (l) What is the total system gain (A_{sys})?

22. For the circuit of Figure 14–36, what is the maximum allowable R_S if you want at least 95% of the source voltage to appear at the base?

23. Redraw Figure 14–36 with $V_{CC} = 12V$, $R_1 = R_2 = 50K\Omega$, $R_E = R_L = 5K\Omega$ and repeat Problem 21.

FIGURE 14–36

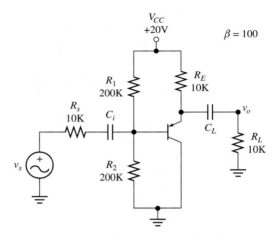

SECTION 14–6 I_B **Control**

24. Figure 14–37 shows a PNP transistor used as a switch or an inverter.

(a) Draw the base input circuit. Be sure it shows a resistor, a diode, a connection to the power supply, and the input voltage (V_S).

(b) Let $V_S = 9V$. Find the values of

I_B

V_B

I_C

V_C

(c) Let $V_S = 0V$. Find the values of

I_B

V_B

I_C

V_C

25. In Figure 14–37

(a) What is $I_{C(sat)}$ at the edge of **SATURATION?**   (b) What is I_B?

(c) When $V_S = 0V$, what R_B is needed to just put the transistor into **SATURATION?**

26. Choose a new R_C in Figure 14–37 that will just put the transistor into **SATURATION** when $v_s = 0V$.

FIGURE 14–37

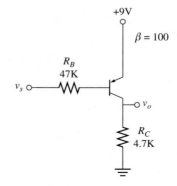

SECTION 14–8   Troubleshooting

27. Figure 14–38 shows the results of the DC Troubleshooting of two PNP circuits. Assuming all connections are good, match the failure listed below with the proper set of Estimated and Measured Voltages. Explain your answers.

(a) Open collector.

(b) Open base.

FIGURE 14–38

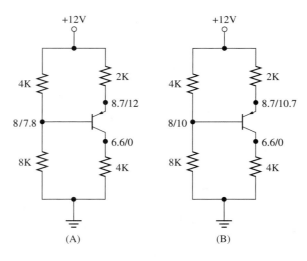

(A)   (B)

28. Figure 14–39 shows the results of the DC Troubleshooting of two PNP circuits. Assuming all connections are good, match the failure listed below with the proper set of Estimated and Measured Voltages. Explain your answers.

(a) Base to collector short.

(b) Collector to emitter short.

(c) Base to emitter short.

FIGURE 14–39

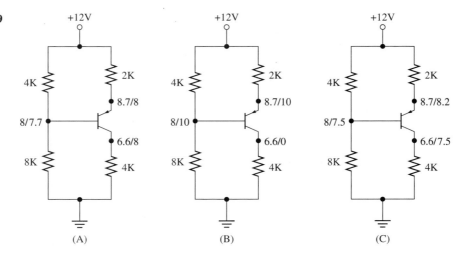

(A)   (B)   (C)

29. For the PNP circuits shown in Figure 14–40 (p. 478), DC Troubleshooting has found no discrepancies. You make AC measurements and get the results shown at the input, base, and output. For each circuit, explain what the problem could be.

FIGURE 14–40

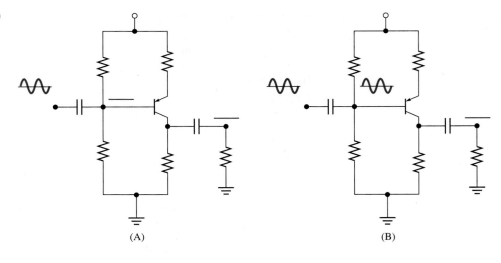

(A) (B)

FIGURE 14–41

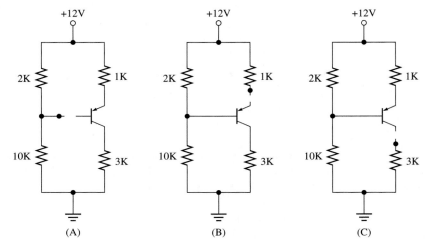

(A) (B) (C)

FIGURE 14–42

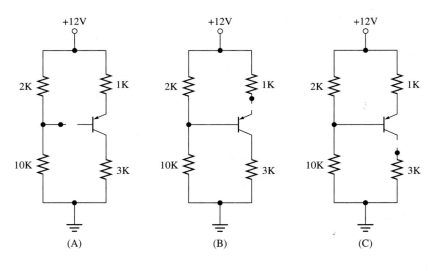

(A) (B) (C)

Computer Problems

30. Simulate the PNP bias circuit in Figure 14–31 (p. 474) with the parameters of Problem 11 (p. 474). Find all bias currents and voltages.

31. Simulate the PNP bias circuit in Figure 14–34 (p. 475) with the parameters of Problem 16 (p. 474). Find all bias currents and voltages. In what state is the transistor operating?

32. Simulate the PNP common-emitter circuit of Figure 14–35 (p. 475). Use a 1mV 1KHz sinewave for the input. Find the gain of the amplifier.

33. Repeat Problem 31 for the PNP emitter follower of Figure 14–36 (p. 476). Why is the gain here much larger?

34. Simulate the circuit of Figure 14–19 (p. 466). Find the DC bias voltages and currents, and determine the states for both transistors when

    (a) $V_S = 0V$ **(b)** $V_S = 12V$

35. The circuits in Figure 14–41 mimic internal opens inside the base, emitter, and collector of a PNP transistor. Simulate each circuit and confirm the results given for these open failures.

36. The circuits in Figure 14–42 mimic internal shorts between the collector and emitter, the base and collector, and the base and emitter. Simulate each circuit and confirm the results given for these short failures.

Chapter

15

THE COMMON–BASE AMPLIFIER

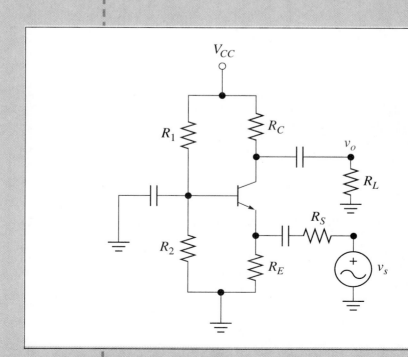

C HAPTER OBJECTIVES

To construct a common-base amplifier.

To find the DC bias levels of the common-base amplifier.

To find the Box Model parameters of the common-base amplifier.

To determine the advantages and disadvantages of the common-base amplifier.

To determine the total system gain of the common-base amplifier.

To compare the various amplifier configurations.

To troubleshoot the common-base amplifier.

15–1 ▪ INTRODUCTION

Oh no!, you may be saying to yourself. I thought that I could build amplifiers with the common-emitter and emitter-follower circuits. Why is there another chapter on amplifiers?

Indeed, using the principles of the common emitter and emitter follower, you can construct most of today's amplifiers. You have already seen some of these combinations in previous chapters, and we will show you even more in the next chapter. However, the common-emitter amplifier has one problem that we haven't yet discussed: it does not work well at very high frequencies.

The common-base amplifier solves this problem. This system can be used to amplify high-frequency signals, such as cable TV and other video signals. In Chapter 28 we will discuss in more detail the frequency-dependent behavior of the transistor. There we will show you why the common base works better than the common emitter at high frequency. For now we just want to show you the DC and AC behaviors of the common-base amplifier.

Figure 15–1 compares common-emitter and common-base amplifiers. Figure 15–1A shows a common-emitter amplifier, in which the input is coupled via a capacitor to the base and a load is coupled via a capacitor to the collector. Figure 15–1B shows a transistor circuit in which the input is coupled via a capacitor to the emitter. The load is still connected to the collector. The circuit in Figure 15–1B is the **common-base amplifier.**

FIGURE 15–1
Common-emitter and common-base amplifiers

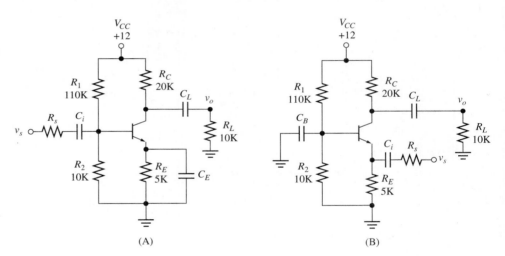

(A)   (B)

15–2 ▪ DC BIASING OF THE COMMON-BASE AMPLIFIER

The NPN Common-Base Amplifier

The common-base amplifier in Figure 15–1B is constructed with an NPN transistor. For DC analysis, we open all capacitors (Figure 15–2). As you can see, when all of the capacitors are opened, the common-base bias circuit is identical to the bias circuit of a common emitter. Both are examples of V_B bias control. The analysis proceeds as before.

FIGURE 15-2

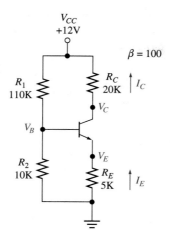

We first find V_B from the voltage divider:

$$V_B = \frac{R_2}{R_1 + R_2} V_{CC}$$

so

$$V_B = \frac{R_2}{R_1 + R_2} V_{CC} = \frac{10K}{110K + 10K} 12 = 1V$$

V_B controls V_E:

$$V_E = V_B - 0.7$$

so

$$V_E = V_B - 0.7 = 1 - 0.7 = 0.3V$$

V_E controls I_E:

$$I_E = \frac{V_E}{R_E}$$

so

$$I_E = \frac{V_E}{R_E} = \frac{0.3}{5K} = 0.06mA$$

If the transistor is **ACTIVE**, I_E controls I_C and I_B:

$$I_C \approx I_E = 0.06mA$$

and

$$I_B = \frac{I_C}{\beta} = 0.6\mu A$$

To check the state of the transistor, we find V_C:

$$V_C = V_{CC} - R_C I_C$$

so

$$V_C = V_{CC} - R_C I_C = 12 - 20K \times 0.06mA = 10.8V$$

Because $V_C > V_E$, the DC analysis is correct, so we can proceed with the AC analysis.

The PNP Common-Base Amplifier

The common-base amplifier in Figure 15–3A is built with a PNP transistor. For the DC analysis of the PNP circuit, we open all capacitors and work with the circuit in Figure 15–3B. For DC, the PNP common-base circuit is identical to the PNP common-emitter circuit that we described in the previous chapter. The DC analysis follows.

FIGURE 15–3

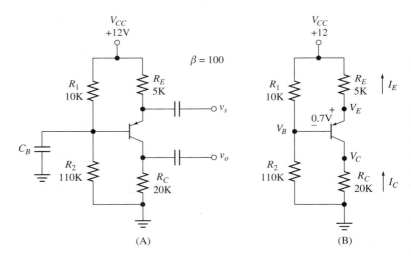

(A) (B)

We first find V_B from the voltage divider:

$$V_B = \frac{R_2}{R_1 + R_2} V_{CC}$$

so

$$V_B = \frac{R_2}{R_1 + R_2} V_{CC} = \frac{110K}{10K + 110K} \, 12 = 11V$$

V_B controls V_E:

$$V_E = V_B + 0.7$$

so

$$V_E = V_B + 0.7 = 11 + 0.7 = 11.7V$$

Note: For the PNP, the emitter voltage is 0.7V above the base voltage.

V_E controls I_E:

Remember, for the PNP, V_E is not the voltage across R_E. Instead, V_E is the voltage at one end of R_E, while V_{CC} is the voltage at the other end. Therefore,

$$V_{RE} = V_{CC} - V_E$$

and

$$I_E = \frac{V_{RE}}{R_E} = \frac{V_{CC} - V_E}{R_E}$$

so

$$I_E = \frac{V_{CC} - V_E}{R_E} = \frac{12 - 11.7}{5K} = 0.06mA$$

If the transistor is **ACTIVE,** I_E controls I_C and I_B:

$$I_C \approx I_E = 0.06\text{mA}$$

and

$$I_B = \frac{I_C}{\beta} = 0.6\mu\text{A}$$

To check the state of the transistor, we find V_C:

$$V_C = R_C I_C$$

so

$$V_C = R_C I_C = 20\text{K} \times 0.06\text{mA} = 1.2\text{V}$$

Because $V_E > V_C$ (for the PNP), the DC analysis is correct and we can proceed with the AC analysis.

EXAMPLE 15–1
DC Bias Levels in the Common-Base Amplifier

Find the DC bias levels of the NPN common-base amplifier in Figure 15–4. Be sure to check the state of the transistor.

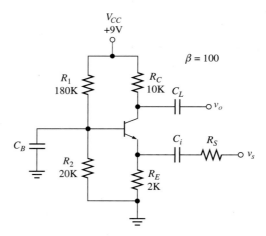

FIGURE 15–4

Solution   With all capacitors open, we first find V_B:

$$V_B = \frac{R_2}{R_1 + R_2} V_{CC} = \frac{20\text{K}}{180\text{K} + 20\text{K}} 9 = 0.9\text{V}$$

V_B controls V_E:

$$V_E = V_B - 0.7 = 0.9 - 0.7 = 0.2\text{V}$$

V_E controls I_E:

$$I_E = \frac{V_E}{R_E} = \frac{0.2}{2\text{K}} = 0.1\text{mA}$$

If the transistor is **ACTIVE,** I_E controls I_C and I_B:

$$I_C \approx I_E = 0.1\text{mA}$$

and

$$I_B = \frac{I_C}{\beta} \doteq 1\mu\text{A}$$

To check the state of the transistor, we find V_C:

$$V_C = V_{CC} - R_C I_C = 9 - 10K \times 0.1mA = 8V$$

Because $V_C > V_E$, the transistor is **ACTIVE** and the DC analysis is correct.

DRILL EXERCISE

Redraw the common-base circuit of Figure 15–4 (p. 485), and change the supply voltage to $V_{CC} = 18V$. Find the DC bias levels, and determine the state of the transistor.

Answer $V_B = 1.8V$, $V_E = 1.1V$, $I_C \approx I_E = 0.55mA$, $I_B = 5.5\mu A$, $V_C = 12.5V$

ACTIVE

REVIEW QUESTIONS

1. How does the DC behavior of the common-base amplifier differ from the DC behavior of the common-emitter amplifier?

2. Drawn an NPN common-base amplifier.

3. Draw a PNP common-base amplifier.

15–3 ■ AC BEHAVIOR OF THE COMMON-BASE AMPLIFIER—BOX MODEL

As we pointed out in Chapter 14, the AC behavior of the NPN and PNP transistors is identical. We will just discuss, therefore, the AC behavior of the NPN common-base amplifier. Figure 15–5A shows the NPN common-base amplifier from Figure 15–2, with the source and load removed. For the common base, as for the common emitter, the source and load do not affect the Box Model. Because we have already performed the DC analysis and found that

$$I_E = 0.06mA$$

we know that the AC emitter resistance is

$$r_e = \frac{26mV}{I_E}$$

FIGURE 15–5

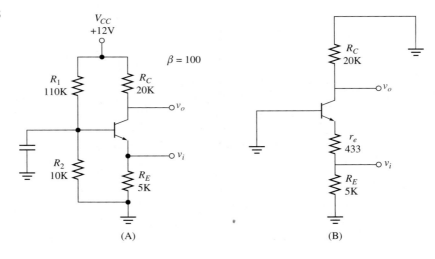

(A) (B)

so

$$r_e = \frac{26\text{mV}}{I_E} = \frac{26\text{mV}}{0.06\text{mA}} = 433\Omega$$

The AC circuit for the common-base amplifier is shown in Figure 15–5B. Note that for AC analysis of the common-base amplifier, the base is connected to ground and the input signal is applied to the emitter. Because the base is grounded, the base biasing resistors are shorted and can be ignored. As with the common-emitter amplifier, the output is taken from the collector.

Gain

Because the input signal is applied to the emitter, the input voltage will create an AC emitter current. To find this current we note that the base is grounded (Figure 15–5B), so the input voltage is applied directly across the AC emitter resistance:

$$i_e = -\frac{v_i}{r_e}$$

The minus sign occurs because the emitter current is heading away from the input voltage.
Now, because the collector current equals the emitter current

$$i_c = i_e = -\frac{v_i}{r_e}$$

As with the common-emitter, the output voltage is found from

$$v_o = -R_C i_c = -R_C\left(-\frac{v_i}{r_e}\right)$$

So

$$v_o = \frac{R_C}{r_e} v_i$$

Note that the two minus signs cancel each other. The voltage gain for the common-base amplifier is, therefore

$$A = \frac{v_o}{v_i} = \frac{R_C}{r_e}$$

For the circuit we are analyzing

$$A = \frac{R_C}{r_e} = \frac{20\text{K}}{0.433\text{K}} = 46.2$$

The magnitude of the gain of the common base turns out to be the same as for the common emitter. Both amplifier types provide a high gain; however, the common-base gain is positive. That is, when input voltage increases in a common base, output voltage also increases. The common-base circuit is a non-inverting amplifier. On the other hand, the gain of the common emitter is negative. That is, when the input voltage to the common emitter increases, the output voltage decreases.

The difference in sign in the gains is not the major difference between the common-base and common-emitter amplifiers. We can't show you why right now, but the gain of the common emitter drops significantly as input frequency increases. The gain of the common base stays high, even as the input frequency gets very large.

Output Resistance

Because the output of the common base is taken from the collector, as is the output of the common emitter, the output resistance of the common base is the same as the output resistance of the common emitter. That is

$$R_o = R_C$$

For our circuit

$$R_o = R_C = 20\text{K}\Omega$$

Input Resistance

To find the input resistance, we return to the AC circuit, reproduced in Figure 15–6A. Because the base is grounded, the AC emitter resistance is effectively connected to ground (Figure 15–6B). You can see from this figure that r_e and R_E are in parallel. The input resistance of the common-base amplifier is, therefore

$$R_i = r_e \| R_E$$

FIGURE 15–6

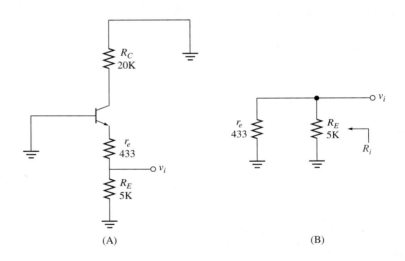

(A)   (B)

For our circuit

$$R_i = r_e \| R_E = 433\|5\text{K} = 398\Omega$$

Here we see a very important difference between the common base and the common emitter. The input resistance of the common emitter, being proportional to βR_E, is very high. The input resistance of the common base is dependent on r_e, which is very low. The low input resistance of the common-base amplifier limits its usefulness, as we will show in the next section.

The Box Model for the common-base configuration is shown in Figure 15–7. The Box Model parameters of the common-base amplifier are

$$A = \frac{R_c}{r_e} \qquad \text{Gain is high—no inversion.}$$
$$R_o = R_C \qquad \text{Output resistance is high.}$$
$$R_i = r_e \| R_E \qquad \text{Input resistance is very low.}$$

FIGURE 15–7
Common-base Box Model

$$A = \frac{R_C}{r_e}$$

$$R_o = R_C$$

$$R_i = r_e \parallel R_E$$

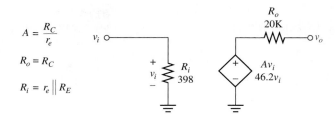

EXAMPLE 15–2
Box Model of the Common-Base Amplifier

Find the Box Model for the common-base amplifier of Example 15–8A. (This figure is identical to Figure 15–1.)

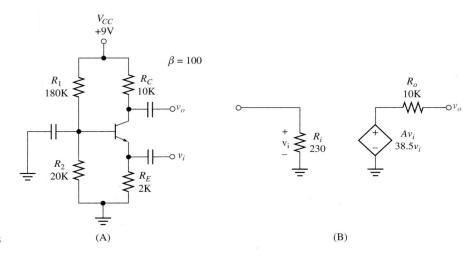

FIGURE 15–8
(A)   (B)

Solution In Example 15–1 we found that

$$I_E = 0.1\text{mA}$$

The AC emitter resistance is, therefore

$$r_e = \frac{26\text{mV}}{I_E} = 260\Omega$$

The Box Model gain is

$$A = \frac{R_C}{r_e} = \frac{10\text{K}}{0.26\text{K}} = 38.5$$

The Box Model output resistance is

$$R_o = R_C = 10\text{K}\Omega$$

The Box Model input resistance is

$$R_i = r_e \parallel R_E = 260 \parallel 2\text{K} = 230\Omega$$

The Box Model is shown in Figure 15–8B.

DRILL EXERCISE

Redraw the common-base circuit of Figure 15–8A (p. 489), and change the supply voltage to 18V. The collector current is $I_C = 0.55$mA; find the AC emitter resistance and the Box Model parameters.

Answer $r_e = 47.3\Omega$, $A = 211$, $R_o = 10K\Omega$, $R_i = 46.2\Omega$

REVIEW QUESTIONS

1. What is the voltage gain of the common-base amplifier?
2. What is the difference between the voltage gain of common-base and common-emitter amplifiers?
3. What is the significance of a negative voltage gain?
4. What is the significance of a positive voltage gain?
5. What is the output resistance of the common-base amplifier?
6. What is the input resistance of the common-base amplifier?
7. Draw the Box Model of the common-base amplifier.
8. What is the disadvantage of the common-base amplifier?

15–4 ▪ THE COMPLETE COMMON-BASE AMPLIFIER

Figure 15–9A shows the common-base amplifier originally shown in Figure 15–5 except that now there are a signal source resistance at the input and a load resistor coupled to the output. We will analyze the AC behavior of this circuit by first replacing the common-base circuit with the Box Model we derived in the last section (for Figure 15–7, p. 489). We then connect the signal source and load to complete the analysis (Figure 15–9B). At the input, we get

$$v_i = \frac{R_i}{R_i + R_s} v_s$$

At the output we get

$$v_o = \frac{R_L}{R_L + R_o} A v_i$$

FIGURE 15–9

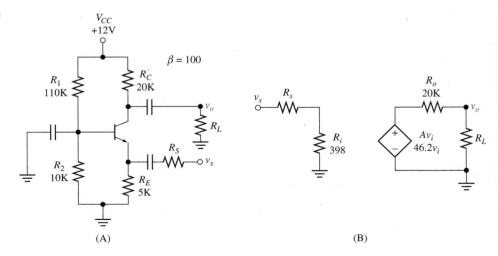

(A)   (B)

The system gain is found by combining these equations:

$$A_{sys} = \frac{R_i}{R_i + R_s} \frac{R_L}{R_L + R_o} A$$

You already know from your study of the common emitter that for good amplification the input resistance of the amplifier (R_i) should be greater than the resistance of the signal source (R_s). If we try to use the common-base configuration as a microphone pre-amplifier, for example, we will have a very poor amplifier because the internal resistance of a microphone is in the KΩ range. For example, a microphone internal resistance of $R_s = 1\text{K}\Omega$ would result in

$$v_i = \frac{R_i}{R_i + R_s} v_s = \frac{398}{398 + 1\text{K}} v_s = 0.285 v_s$$

That is, less than 29% of the input signal will be amplified. Fortunately, the high-frequency signal sources used in typical common-base applications have low internal resistances. Video cable, for example, has a resistance of only $R_S = 75\Omega$. In this case the input voltage divider would produce

$$v_i = \frac{R_i}{R_i + R_s} v_s = \frac{398}{398 + 75} v_s = 0.841 v_s$$

Approximately 84% of the input signal is amplified.

EXAMPLE 15–3
Complete AC Analysis of the Common-Base Amplifier

For the common-base circuit of Figure 15–9A, find the system gain if

$$R_s = 50\Omega \quad \text{and} \quad R_L = 1\text{M}\Omega$$

Solution   We can use the Box Model shown in Figure 15–9B. On the input side

$$v_i = \frac{R_i}{R_s} = \frac{398}{398 + 50} v_s = 0.888 v_s$$

At the output

$$v_o = \frac{R_L}{R_L + R_o} A v_i = \frac{1\text{M}}{1\text{M} + 20\text{K}} 46.2 v_i = 45.3 v_i$$

Therefore

$$v_o = 45.3 v_i = 45.3 \times 0.888 v_s = 40.2 v_s$$

The system gain is

$$A_{sys} = \frac{v_o}{v_s} = 40.2$$

DRILL EXERCISE

For the common-base circuit in Figure 15–9, find the system gain if $R_s = 300\Omega$ and $R_L = 100\text{K}\Omega$

Answer $A_{sys} = 22$

15–5 ■ AMPLIFIER COMPARISONS

You now know the three basic amplifier configurations: the common emitter (CE); the common collector (CC), also known as the emitter follower (EF); and the common base (CB). The following table summarizes the properties of these amplifiers:

	CE	CC (or EF)	CB
Input	Base	Base	Emitter
Output	Collector	Emitter	Collector
Inversion	Yes	No	No
Gain	High	1	High
Input resistance	High	High	Low
Output resistance	High	Low	High
Frequency range	Medium	Medium	High

15–6 ■ TROUBLESHOOTING

The DC bias circuit of the common-base amplifier is identical to the DC bias circuit of the common-emitter amplifier (compare Figures 15–10A and 15–10B). Therefore, you conduct DC Troubleshooting for a common-base amplifier exactly as DC Troubleshooting for a common-emitter amplifier. The only thing new here is the location of the input coupling capacitor. If C_s has failed open, then the AC signal applied at the input will not appear at the emitter of the transistor (Figure 15–11).

The following examples will give you some practice in troubleshooting the common-base amplifier.

FIGURE 15–10

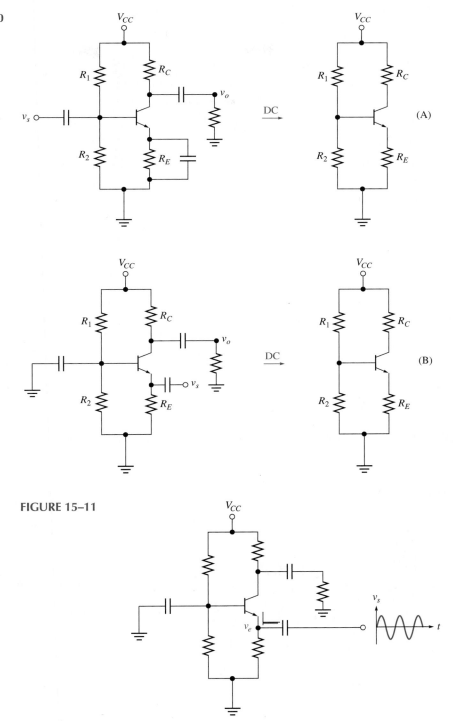

FIGURE 15–11

EXAMPLE 15–4

DC Troubleshooting the Common-Base

The results of DC Troubleshooting of two common-base circuits are shown in Figure 15–12 (p. 494). Assuming that all connections and resistors are good, find the likely cause of the discrepancies.

Solution

(a) Because the voltage at the base is at ground potential (0V), it is likely that the base capacitor (C_B) is shorted. With the base shorted to ground, the transistor will be

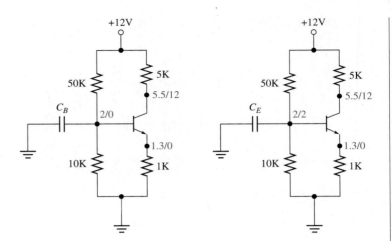

FIGURE 15–12

CUT-OFF, and all transistor currents will be 0A. This would explain seeing the supply voltage at the collector and seeing 0V at the emitter.

(b) In this case the Measured Voltage at the base is correct. The collector and emitter voltage still indicate that there is no collector or emitter current. This will occur if either the base or emitter lead of the transistor is broken inside the transistor.

EXAMPLE 15–5
AC Signal Tracing in the Common-Base Amplifier

For the circuits in Figure 15–13, DC Troubleshooting found no discrepancies in Estimated Voltages and Measured Voltages. Explain the likely cause for the AC behavior shown.

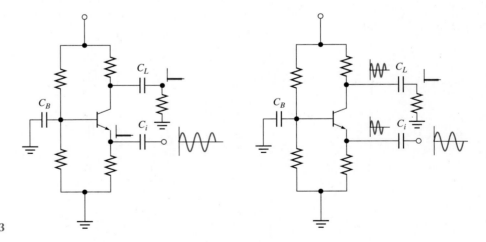

FIGURE 15–13

(a) Here the signal applied from the source is not seen at the emitter. The only failure that could cause this signal loss is an open in the signal-source capacitor (C_s).

(b) In this case the input signal is seen at the emitter, the amplified signal is seen at the collector, but no AC signal is seen at the load. This must mean that the load capacitor (C_L) has failed open.

REVIEW QUESTIONS

1. Why is DC Troubleshooting the common-base circuit the same as DC Troubleshooting the common-emitter circuit?

2. What will happen if the base capacitor in the common-base amplifier fails short?

3. What will you observe if the source capacitor in the common-base amplifier fails open?

4. What will you observe if the load capacitor in the common-base amplifier fails open?

SUMMARY

- The common-emitter amplifier is limited to moderate frequency ranges.
- The common-base amplifier, in which the input signal is applied to the emitter, operates over a higher frequency range.
- The bias circuit of the common-base amplifier is identical to the bias circuit of the common emitter.
- The Box Model parameters for the common-base circuit are

$$A = \frac{R_C}{r_e}$$

$$R_i = r_e \| R_E$$

$$R_o = R_C$$

- The common-base amplifier has a non-inverting high gain but a very low input resistance.
- The system gain of the common-base amplifier is

$$A_{sys} = \frac{v_o}{v_s} = \frac{R_i}{R_i + R_s} \frac{R_L}{R_L + R_o} A$$

- DC Troubleshooting of common-base amplifier circuits is the same as DC Troubleshooting of common-emitter amplifiers.

PROBLEMS

SECTION 15–2   DC Biasing of the Common-Base Amplifier

1. For the NPN common-base amplifier circuit in Figure 15–14

$$V_{CC} = 12V, R_1 = 50K\Omega, R_2 = 10K\Omega, R_C = 5K\Omega, R_E = 1K\Omega$$

Use the steps of V_B control to find the bias voltage and currents. List your answers in the appropriate order.

FIGURE 15–14

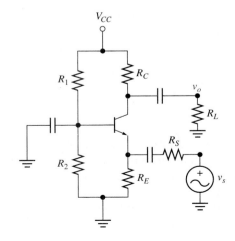

2. For the NPN common-base amplifier circuit in Figure 15–14 (p. 495)

$$V_{CC} = 9V, R_1 = 30K\Omega, R_2 = 6K\Omega, R_C = 5K\Omega, R_E = 1K\Omega$$

Use the steps of V_B control to find the bias voltage and currents. List your answers in the appropriate order.

3. For the PNP common-base amplifier circuit in Figure 15–15

$$V_{CC} = 12V, R_1 = 10K\Omega, R_2 = 50K\Omega, R_C = 5K\Omega, R_E = 1K\Omega$$

Use the steps of V_B control to find the bias voltage and currents. List your answers in the appropriate order.

FIGURE 15–15

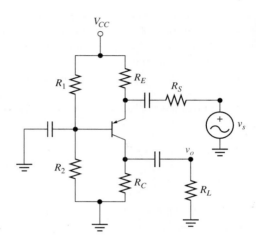

4. For the PNP common-base amplifier circuit in Figure 15–15

$$V_{CC} = 22V, R_1 = 1.7K\Omega, R_2 = 20.3K\Omega, R_C = 10K\Omega, R_E = 1K\Omega$$

Use the steps of V_B control to find the bias voltage and currents. List your answers in the appropriate order.

5. What is the difference between the relation of V_E to V_B in the NPN and PNP transistors?

6. How do you find I_E in the NPN and the PNP transistor circuits?

SECTION 15–3   AC Behavior of Common-Base Amplifier—Box Model

7. For the NPN common-base amplifier shown in Figure 15–14 (p. 495) and the parameters of Problem 1

  (a) What is r_e?

  (b) What is the Box Model gain (A)? What can you say about the sign of the gain?

  (c) What is the Box Model input resistance (R_i)? Is it low or high?

  (d) What is the Box Model output resistance (R_o)?

  (e) Draw the Box Model.

8. Repeat Problem 7 if Figure 15–14 has the parameters of Problem 2.

9. Repeat Problem 7 for Figure 15–15 with the parameters of Problem 3.

10. Repeat Problem 7 for Figure 15–15 with the parameters of Problem 4.

SECTION 15–4   The Complete Common-Base Amplifier

11. A Box Model for a common-base amplifier has the following parameters:

$$R_i = 200\Omega, R_o = 5K\Omega, A = 20.$$

  (a) Draw the complete model if $R_S = 10\Omega$ and $R_L = 5K\Omega$.

  (b) What is the total system gain?

12. For the Box Model of Problem 11, what is the maximum R_s that can be used if you want to deliver at least 98% of the signal voltage to the emitter of the common base?

13. Connect a load and a signal source to the Box Model of Problem 7, and find the system gain (A_{sys}) if $R_L = 5K\Omega$ and

    (a) $R_s = 10\Omega$ **(b)** $R_s = 1K\Omega$

14. Connect the load and the signal source to the Box Model of Problem 8, and find the system gain (A_{sys}) if $R_L = 5K\Omega$ and

    (a) $R_s = 10\Omega$ **(b)** $R_s = 1K\Omega$

15. Connect the load and the signal source to the Box Model of Problem 9, and find the system gain (A_{sys}) if $R_L = 5K\Omega$ and

    (a) $R_s = 10\Omega$ **(b)** $R_s = 1K\Omega$

16. Connect the load and the signal source to the Box Model of Problem 10 and find the system gain (A_{sys}) if $R_L = 10K\Omega$ and

    (a) $R_s = 10\Omega$ **(b)** $R_s = 1K\Omega$

SECTION 15–5 Amplifier Comparisons

17. What amplifier configuration would you use if you needed

    (a) High gain and high input resistance.   **(b)** High gain and high frequency range.

    (c) A buffer amplifier with a gain of 1.

18. A common-emitter amplifier and a common-base amplifier both have a total system gain of 5. If the input to each amplifier is a 1mV 1KHz voltage, draw the input voltage and the two amplifier output voltages on the same plot.

SECTION 15–6 Troubleshooting

19. Match the DC Troubleshooting results in Figure 15–16 with the following failures. Explain your selection.

    (a) Base-to-collector short.   **(b)** Base-to-emitter short.

    (c) Open base inside the transistor.

FIGURE 15–16

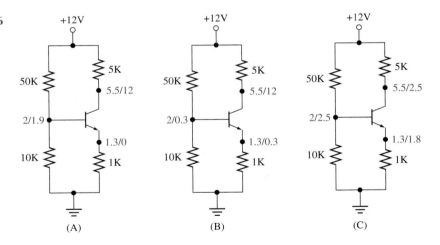

    (A)   (B)   (C)

20. Match the DC Troubleshooting results in Figure 15–17 (p. 488) with the following failures. Explain your selection.

    (a) Open collector inside the transistor.   **(b)** Collector-to-emitter short.

    (c) Open emitter inside the transistor.

21. Match the AC signal tracing results in Figure 15–18 (p. 488) with the following failures. Explain your answer.

    (a) C_i has failed open.   **(b)** C_L has failed open.

FIGURE 15–17

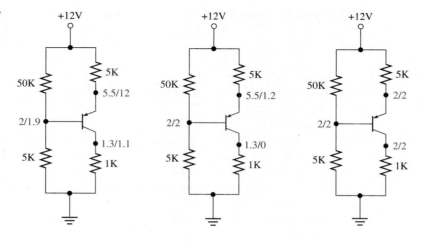

FIGURE 15–18

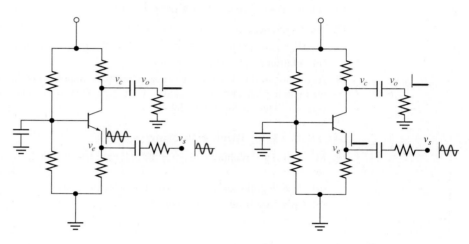

FIGURE 15–19

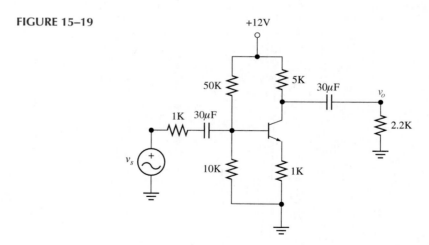

Computer Problems

22. For the common-base circuit in Figure 15–3 (p. 484):

   (a) Calculate the DC bias levels of this circuit.

   (b) Calculate the box model parameters.

   (c) Use the computer to find the DC bias levels, and compare to your calculations.

   (d) Use the computer to find the Box Model parameters, and compare to your calculations. To find the gain and the input resistance (you cannot easily find output resistance), set the signal source to a 1mV 1KHz sinewave. Input resistance can be found as input voltage magnitude divided by input current magnitude.

23. Simulate the common-emitter amplifier in Figure 15–19. Set all external capacitors to 30μF. Read your computer program instruction manual to learn how to set the internal base and emitter junction capacitors to 3pF (that's 3×10^{-12}F). Use AC Analysis to find the output over a frequency range of 1KHz to 100MHz.

24. Repeat Problem 23 for the common-base amplifier of Figure 15–20.

25. Compare the results of Problems 23 and 24.

FIGURE 15–20

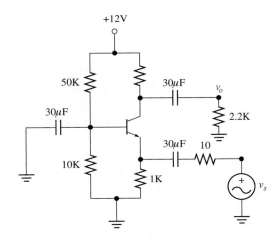

CHAPTER

16

SPECIALIZED TRANSISTOR CIRCUITS

CHAPTER OUTLINE

To determine the DC and AC behavior of the Darlington connection.

To determine the output of a Darlington amplifier.

To determine the DC bias levels of the cascode amplifier.

To determine the AC behavior of the cascode amplifier.

To determine the DC bias levels of the collector-feedback circuit.

To determine the AC behavior of the collector-feedback circuit.

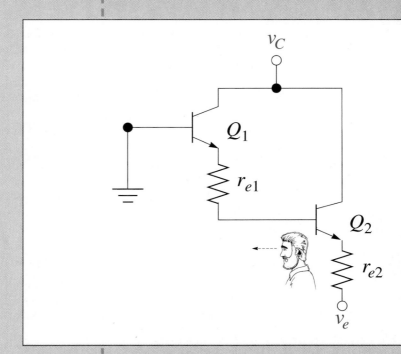

16–1 ■ INTRODUCTION

In this chapter we discuss three transistor circuits that are occasionally seen in amplifiers. The first circuit, known as the *Darlington connection,* is used to create a "super transistor" that has a very high current gain. The second circuit, the *cascode amplifier,* has properties that combine the advantages of the common emitter with the advantages of the common base. The last circuit, a *collector-feedback circuit,* is an alternative way to bias a transistor. We will present a DC analysis of this circuit and spend some time discussing its AC properties.

16–2 ■ THE DARLINGTON CONNECTION

Figure 16–1A shows a two-transistor configuration in which the collectors are tied together. The emitter of the first transistor (Q_1) provides the input to the base of the second transistor (Q_2). We want to show you how this two-transistor combination can be analyzed as a single "super transistor"—**a Darlington connection.** For this reason, we have labeled the three external terminal voltages as V_C, V_B, and V_E and the three external terminal currents I_C, I_B, and I_E.

FIGURE 16–1
Darlington connection

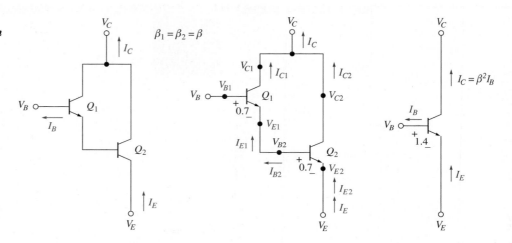

In Figure 16–1B we have labeled the internal voltages and currents with the subscript appropriate to the transistor. From this figure you can see that the emitter current of Q_1 is the base current of Q_2. That is

$$I_{C1} \approx I_{E1} = I_{B2}$$

Therefore, assuming both transistors are **ACTIVE** and have the same β

$$I_{C2} = \beta I_{C1}$$

And because

$$I_{C1} = \beta I_B$$

we see that

$$I_{C2} = \beta I_{C1} = \beta^2 I_B$$

The external collector current (I_C) is the sum of the two internal collector currents. Because I_{C1} is the base current of Q_2, it is much smaller than I_{C2}. Therefore, the external collector current is approximately equal to the collector current of Q_2. That is

$$I_C = I_{C1} + I_{C2} \approx i_{r2}$$

Therefore

$$I_C = \beta^2 I_B$$

We have produced a device that has a current gain of β^2!

We now look at the voltage required to turn on the two transistors. As you can see from Figure 16–1B, the base-emitter turn-on voltage of Q_1 is in series with the base-emitter turn-on voltage of Q_2. Therefore, the total voltage required to turn on both transistors is

$$V_{BE} = 0.7 + 0.7 = 1.4V$$

The Darlington configuration produces an equivalent transistor with a current gain of β^2 and a turn-on voltage of 1.4V (Figure 16–1C).

Equivalent AC Emitter Resistance

Finding the equivalent AC emitter resistance of the Darlington connection is a bit tricky. The key is to remember that the AC emitter resistance of any transistor is the small-signal resistance seen at the emitter when the base is grounded. For example, when you look into the emitter of Q_1, with the base of Q_1 grounded, you see r_{e1} (Figure 16–2A).

FIGURE 16–2

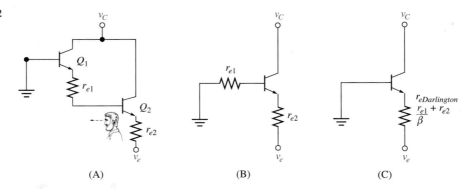

(A)   (B)   (C)

Therefore, when you look into the emitter of Q_2, you get the circuit shown in Figure 16–2B. The equivalent AC emitter resistance that we want is actually the output resistance at the emitter of Q_2. From your study of the emitter follower, you know that the resistance in the base is divided by β and is then added in series with r_{e2}. The overall AC emitter resistance of the Darlington connection is given by

$$r_{e_{Darlington}} = \frac{r_{e1}}{\beta} + r_{e2}$$

You might be saying to yourself that because $\dfrac{r_{e1}}{\beta}$ is much smaller than r_{e2}, we can ignore the contribution from r_{e1}. This is a reasonable assumption, but it is wrong. In the next section we will show you why. The AC single-transistor equivalent of the Darlington connection is shown in Figure 16–2C.

REVIEW QUESTIONS

REVIEW QUESTIONS

1. Draw the Darlington connection, and label all internal and external voltages and currents.
2. What is the total current gain of the Darlington connection?
3. What is the base-emitter turn-on voltage of the Darlington connection?
4. Draw a single-transistor equivalent for the Darlington connection, and label all important parameters.
5. What is the equivalent AC emitter resistance of the Darlington connection?

16–3 ■ DARLINGTON COMMON-EMITTER AMPLIFIER

Figure 16–3A shows the Darlington connection used in a common-emitter amplifier. We will first find the DC bias levels of the amplifier and then determine its Box Model parameters.

FIGURE 16–3
Darlington common-emitter amplifier

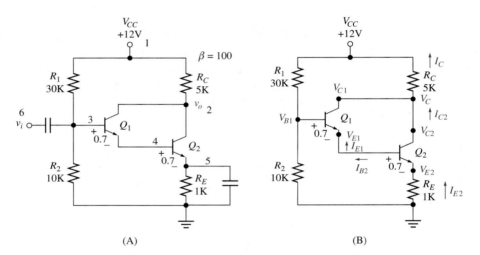

(A)　　　　　　　　(B)

DC Bias Levels

We first open the bypass capacitor (Figure 16–3B), then apply the V_B control procedure. The external base voltage equals the base voltage of Q_1:

$$V_B = V_{B1} = \frac{R_2}{R_1 + R_2} V_{CC} = \frac{10K}{10K + 30K} 12 = 3V$$

The base voltage of Q_1 controls the emitter voltage of Q_1:

$$V_{E1} = V_{B1} - 0.7 = 3 - 0.7 = 2.3V$$

Now because

$$\boxed{V_{B2} = V_{E1}}$$

we get

$$V_{B2} = V_{E1} = 2.3V$$

Because V_{E2} is the same as the external emitter voltage

$$V_E = V_{E2} = V_{B2} - 0.7 = V_{E1} - 0.7 = 2.3 - 0.7 = 1.6V$$

Note that the external emitter voltage is 1.4V below the external base voltage. That is, the effective emitter-diode voltage of the Darlington connection is 1.4V. Therefore, V_B must exceed this voltage to turn on both transistors. The external emitter voltage controls the external emitter current, which is also the emitter current of Q_2:

$$I_E = I_{E2} = \frac{V_E}{R_E} = \frac{1.6}{1\text{K}} = 1.6\text{mA}$$

If *both* transistors are **ACTIVE**, then

$$I_C \approx I_{C2} = I_{E2} = 1.6\text{mA}$$

and, because I_{E1} is the base current of Q_2

$$I_{E1} = \frac{I_{E2}}{\beta} = 0.016\text{mA}$$

To check the state of the transistors, we must compare the collector voltage to the emitter voltage for each transistor. The collector voltages for the transistors are the same, so

$$V_{C1} = V_{C2} = V_C = V_{CC} - R_C I_C = 12 - 5\text{K} \times 1.6\text{mA} = 4\text{V}$$

Because

$$V_{C2} > V_{E2} \qquad \text{and} \qquad V_{C1} > V_{E1}$$

Q_1 and Q_2 are **ACTIVE.**

AC Analysis—the Box Model

To find the Box Model of the Darlington common emitter, we draw the AC circuit, replacing the two transistors with the equivalent single-transistor AC model (Figure 16–4A). Once we find the equivalent AC emitter resistance, we can use the standard common-emitter formulas.

FIGURE 16–4

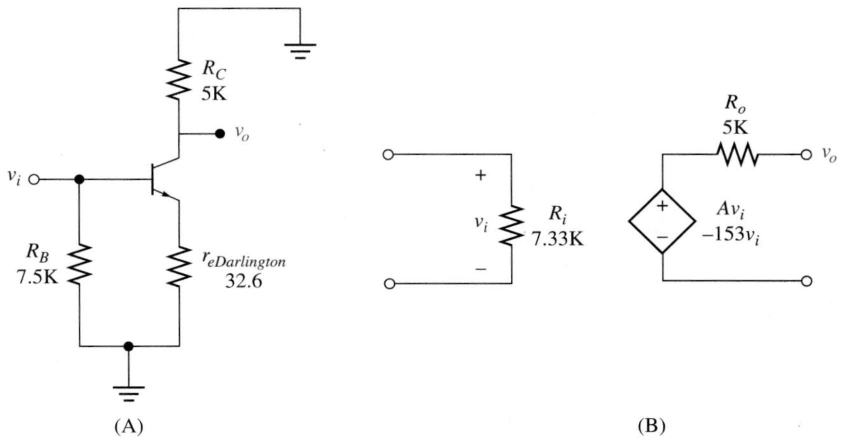

(A)   (B)

$$r_{e2} = \frac{26\text{mV}}{I_{E2}} = \frac{26\text{mV}}{I_E} = \frac{26\text{mV}}{1.6\text{mA}} = 16.3\Omega$$

$$r_{e1} = \frac{26\text{mV}}{I_{E1}} = \frac{26\text{mV}}{0.016\text{mA}} = 1630\Omega$$

Now

$$r_{e_{Darlington}} = \frac{r_{e1}}{\beta} + r_{e2} = 16.3 + 16.3 = 32.6\Omega$$

Note that $\dfrac{r_{e1}}{\beta}$ actually equals r_{e2}. This is because I_{E1} is β times smaller than I_{E2}. The formula for the equivalent AC emitter resistance can be simplified to

$$r_{e_{Darlington}} = 2r_{e2}$$

Because this circuit is still a common-emitter amplifier, the Box Model gain is

$$A = -\frac{R_C}{R_{Etotal}} = -\frac{R_C}{r_{e_{Darlington}}}$$

so

$$A = -\frac{R_C}{r_{e_{Darlington}}} = -\frac{5\text{K}}{0.0326\text{K}} = -153$$

The output resistance of this Darlington amplifier is still

$$R_o = R_C$$

so

$$R_o = R_C = 10\text{K}\Omega$$

The input resistance is given by (remember, the current gain is β^2)

$$R_i = R_B \| (\beta^2 R_{Etotal})$$

For the values in our circuit

$$R_i = R_B \| (\beta^2 R_{Etotal}) = R_B \| (\beta^2 r_{e_{Darlington}}) = 7.5\text{K} \| 326\text{K} = 7.33\text{K}\Omega$$

Because the Darlington amplifier has a current gain of β^2, it has a very high voltage gain and a high input resistance. The Box Model is shown in Figure 16–4B (p. 505).

EXAMPLE 16–1
The Darlington Common-Emitter Amplifier

For the Darlington amplifier in Figure 16–5, find the DC bias levels of both transistors, the states of the transistors, and the AC emitter resistances of both transistors. Draw the overall Box Model for the circuit.

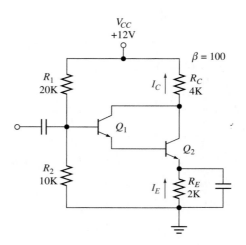

FIGURE 16–5

Solution **DC Bias Levels**

The base voltage of Q_1 is

$$V_{B1} = \frac{R_2}{R_1 + R_2} V_{CC} = \frac{10K}{10K + 20K} 12 = 4V$$

The base voltage of Q_1 controls the emitter voltage of Q_1:

$$V_{E1} = V_{B1} - 0.7 = 4 - 0.7 = 3.3V$$

The emitter voltage of Q_1 is the base voltage of Q_2:

$$V_{B2} = V_{E1} = 3.3V$$

The base voltage of Q_2 controls the emitter voltage of Q_2:

$$V_{E2} = V_{B2} - 0.7 = 3.3 - 0.7 = 2.6V$$

The emitter voltage of Q_2 controls the emitter current of Q_2:

$$I_{E2} = \frac{V_{E2}}{R_E} = \frac{2.6}{2K} = 1.3mA$$

If both transistors are **ACTIVE**

$$I_C = I_{C2} = I_{E2} = 1.3mA$$

Because I_{E1} is equal to I_{B2}

$$I_{E1} = I_{B2} = \frac{I_{E2}}{\beta} = 0.013mA$$

To check the state of the transistor, we note that the two collector voltages are equal and are given by

$$V_{C1} = V_{C2} = V_C = V_{CC} - R_C I_C = 12 - 4K \times 1.3mA = 6.8V$$

Both transistors are **ACTIVE** because

$$V_{C1} > V_{E1} \qquad \text{and} \qquad V_{C2} > V_{E2}$$

AC Analysis

The AC emitter resistances for the two transistors are

$$r_{e1} = \frac{26mV}{0.013mA} = 2000\Omega$$

and

$$r_{e2} = \frac{26mV}{1.3mA} = 20\Omega$$

The AC emitter resistance of the Darlington connection is

$$r_{e_{Darlington}} = \frac{r_{e1}}{\beta} + r_{e2} = 2r_{e2} = 40\Omega$$

The Box Model gain is

$$A = -\frac{R_C}{R_{Etotal}} = -\frac{R_C}{r_{e_{Darlington}}} = -\frac{4K}{40} = -100$$

The Box Model output resistance is

$$R_o = R_C = 4K\Omega$$

The Box Model input resistance is

$$R_i = R_B \| \beta^2 r_{e_{Darlington}} = (20K \| 10K) \| 400K = 6.56K\Omega$$

DRILL EXERCISE

Redraw the circuit in Figure 16–5 (p. 506), changing the supply voltage to $V_{CC} = 9V$. Find the bias voltage and currents for each transistor, the AC emitter resistance of the Darlington connection, and the Box Model parameters of the Darlington connection.

Answer $V_{B1} = 3V, V_{E1} = 2.3V, V_{E2} = 1.6V$
$I_{E2} = 0.8mA, I_{C2} = 0.8mA, I_{E1} = 0.008mA$
$V_{C1} = V_{C2} = 5.8V, r_{e_{Darlington}} = 65\Omega$
$A = -61.5, R_o = 4K\Omega, R_i = 6.6K\Omega$

REVIEW QUESTIONS

1. Draw the Darlington common-emitter amplifier circuit.
2. Describe the steps in determining the bias levels of the two transistors in the Darlington connection.
3. How do you check the state of the two transistors?
4. Give two formulas for the AC emitter resistance of the Darlington connection.
5. What is the Box Model gain of the Darlington common emitter?
6. What is the Box Model output resistance of the Darlington common emitter?
7. What is the Box Model input resistance of the Darlington common emitter?

16–4 ■ THE CASCODE AMPLIFIER

In Chapter 15 we introduced the common-base amplifier and compared it to the common-emitter amplifier. Both amplifiers have high gain. The common emitter has high input resistance but poor high-frequency performance. The common base, on the other hand, has good high-frequency response but low input resistance.

If we combine the common base with the common emitter we can have our cake and eat it too. That is, if we apply the input signal to a common emitter, then use the common emitter to provide the input to a common base, we get a device that has good frequency behavior and high input resistance. This combination is known as a **cascode amplifier.**

Figure 16–6 shows a typical cascode amplifier. The input signal is connected to the base of Q_1, and the collector voltage of Q_1 provides the input signal to the emitter of Q_2. The collector voltage of Q_2 is the output voltage of the amplifier.

DC Analysis

Both transistors in the cascode amplifier are biased with V_B control. While we could have used a separate voltage divider to bias each transistor, the arrangement shown in Figure 16–6 reduces the number of required base resistors. Let's look at each transistor one at a time to find the bias levels in the circuit.

The key to the DC analysis is that we assume both base currents can be ignored. For example, to find the DC bias levels in Q_1, we can work with the partial circuit shown in Figure 16–7A. Note that the capacitor at the base of Q_2 is open. Because we ignore the base current into Q_2, R_1 and R_2 are treated as though they are in series and can be added together.

FIGURE 16–6
Cascode amplifier

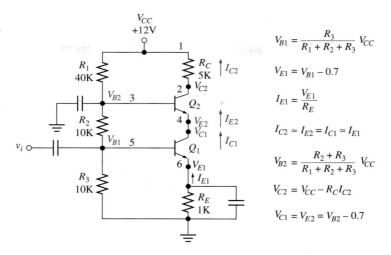

$$V_{B1} = \frac{R_3}{R_1 + R_2 + R_3} V_{CC}$$

$$V_{E1} = V_{B1} - 0.7$$

$$I_{E1} = \frac{V_{E1}}{R_E}$$

$$I_{C2} \approx I_{E2} = I_{C1} \approx I_{E1}$$

$$V_{B2} = \frac{R_2 + R_3}{R_1 + R_2 + R_3} V_{CC}$$

$$V_{C2} = V_{CC} - R_C I_{C2}$$

$$V_{C1} = V_{E2} = V_{B2} - 0.7$$

FIGURE 16–7

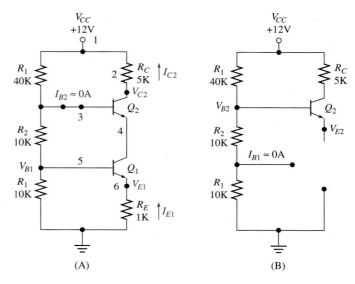

(A)   (B)

The base voltage for Q_1 is the voltage across R_3, so

$$V_{B1} = \frac{R_3}{R_1 + R_2 + R_3} V_{CC}$$

For the values given

$$V_{B1} = \frac{R_3}{R_1 + R_2 + R_3} V_{CC} = \frac{10K}{10K + 10K + 40K} 12 = 2V$$

V_{B1} controls V_{E1}:

$$V_{E1} = V_{B1} - 0.7$$

and the emitter current is found from

$$I_{E1} = \frac{V_{E1}}{R_E}$$

And if *both* transistors are **ACTIVE**

$$I_{C1} \approx I_{E1} \quad \text{and} \quad I_{E2} = I_{C1}$$

So

$$I_{C2} \approx I_{E2} = I_{C1} \approx I_{E1}$$

For the values given

$$I_{C2} \approx I_{E2} = I_{C1} \approx I_{E1} = \frac{V_{B1} - 0.7}{R_E} = \frac{2 - 0.7}{1K} = 1.3mA$$

We find the collector voltage of Q_2 from

$$V_{C2} = V_{CC} - R_C I_{C2}$$

So

$$V_{C2} = V_{CC} - R_C I_{C2} = 12 - 5K \times 1.3mA = 5.5V$$

To complete the DC analysis, we need to find the base voltage of Q_2. To find V_{B2} we analyze the partial circuit shown in Figure 16–7B (p. 509). Here you can see that R_2 and R_3 are in series. The base voltage of Q_2 is the voltage across this series combination ($R_2 + R_3$). The voltage divider gives us, therefore

$$V_{B2} = \frac{R_2 + R_3}{R_1 + R_2 + R_3} V_{CC}$$

For the values given

$$V_{B2} = \frac{R_2 + R_3}{R_1 + R_2 + R_3} V_{CC} = \frac{10K + 10K}{10K + 10K + 40K} 12 = 4V$$

Now, V_{B2} controls V_{E2}

$$V_{E2} = V_{B2} - 0.7$$

So

$$V_{E2} = 4 - 0.7 = 3.3V$$

Finally, we recognize that

$$V_{C1} = V_{E2}$$

So

$$V_{C1} = V_{E2} = 3.3V$$

To check the state of the transistors, we compare collector voltages to emitter voltages. Because

$$V_{C2} > V_{E2} \quad \text{and} \quad V_{C1} > V_{E1}$$

both transistors are **ACTIVE.**

EXAMPLE 16–2
DC Analysis of the
Cascode Amplifier

Find the DC bias levels for both transistors of the cascode amplifier in Figure 16–8. Determine the state of both transistors.

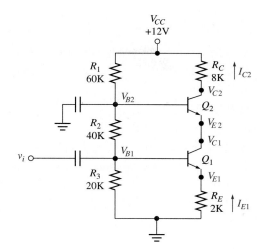

FIGURE 16–8

Solution We begin by finding the base voltage of Q_1:

$$V_{B1} = \frac{R_3}{R_1 + R_2 + R_3} V_{CC} = \frac{20K}{20K + 40K + 60K} 12 = 2V$$

$$V_{E1} = V_{B1} - 0.7 = 2 - 0.7 = 1.3V$$

$$I_{C2} = I_{E1} = \frac{V_{E1}}{R_E} = \frac{1.3}{2K} = 0.65mA$$

$$V_{C2} = V_{CC} - R_C I_{C2} = 12 - 8K \times 0.65mA = 6.8V$$

We now find the base voltage of Q_2:

$$V_{B2} = \frac{R_2 + R_3}{R_1 + R_2 + R_3} V_{CC} = \frac{20K + 40K}{20K + 40K + 60K} 12 = 6V$$

Finally

$$V_{C1} = V_{E2} = V_{B2} - 0.7 = 6 - 0.7 = 5.3V$$

Because

$$V_{C2} > V_{E2} \qquad \text{and} \qquad V_{C1} > V_{E1}$$

both transistors are **ACTIVE.**

DRILL EXERCISE Redraw the circuit in Figure 16–8, changing the supply voltage to $V_{CC} = 9V$. Find the DC bias levels in the circuit, and determine the state of the transistors.

Answer $V_{B1} = 1.5V$, $V_{E1} = 0.8V$, $I_{C2} = I_{E1} = 0.4mA$
$V_{C2} = 5.8V$, $V_{B2} = 4.5V$, $V_{C1} = V_{E2} = 3.8V$

Both transistors are **ACTIVE.**

16-5 ■ THE CASCODE AMPLIFIER—AC ANALYSIS

Figure 16–9 shows the AC circuit of the cascode amplifier from Figure 16–6. You can see that the input signal is applied to the base of the common emitter. The output of the common emitter (v_{C1}) becomes the input signal to the common base. Because of the capacitor at its base (Figure 16–6), note that for AC the base of Q_2 is tied to ground. This means that R_1 is not part of the AC circuit and

$$R_B = R_2 \| R_3$$

FIGURE 16–9

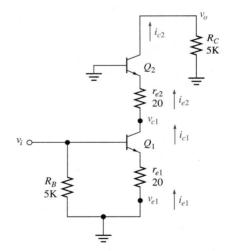

For the values in Figure 16–6

$$R_B = R_2 \| R_3 = 10K \| 10K = 5K\Omega$$

From Figure 16–9 you can see that the same emitter current flows in both transistors. This means that the AC emitter resistances of the two transistors are the same. That is

$$r_{e2} = r_{e1}$$

For the bias currents found from Figure 16–6 (p. 509)

$$r_{e2} = r_{e1} = \frac{26\text{mV}}{I_{E1}} = \frac{26\text{mV}}{1.3\text{mA}} = 20\Omega$$

The simplest way to analyze the cascode amplifier is to first consider the common base as the load for the common emitter. That is, the input resistance of the common base (r_{e2} in this case) is the load resistor for the common emitter (Figure 16–10A). For the common emitter, we get

$$v_{c1} = -\frac{r_{e2}}{R_{Etotal}} v_i$$

And, because R_E has been bypassed

$$v_{c1} = -\frac{r_{e2}}{r_{e1}} v_i = -v_i$$

FIGURE 16–10

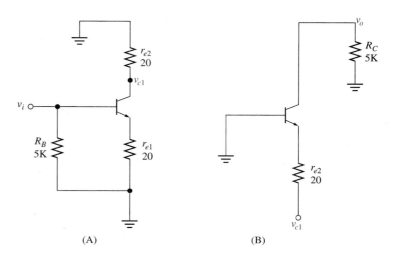

(A)   (B)

Because the AC emitter resistances of Q_1 and Q_2 are the same, the gain for the common emitter is

$$\frac{v_{c1}}{v_i} = -\frac{r_{e2}}{r_{e1}} = -1$$

The input resistance of the cascode amplifier is the input resistance of the common emitter:

$$R_i = R_B \| \beta R_{Etotal}$$

In this circuit, with a bypassed R_E

$$R_i = R_B \| \beta r_{e1}$$

For the values given

$$R_i = 5K \| 2K = 1.43K\Omega$$

The input resistance of the cascode amplifier is high, and it can be made higher by not bypassing R_E. We have corrected the low input resistance problem associated with the common base.

Now the collector voltage of the common emitter becomes the input signal to the common base (Figure 16–10B). The output voltage and output resistance of the common base are the output voltage and output resistance of the cascode amplifier.

$$R_o = R_C$$

For our circuit

$$R_o = R_C = 5\text{K}\Omega$$

and

$$v_o = \frac{R_C}{r_{e2}} v_{c1} = \frac{R_C}{r_{e2}} \left(-\frac{r_{e2}}{R_{Etotal}} v_i\right) = -\frac{R_C}{R_{Etotal}} v_i$$

So the Box Model gain for this cascode combination is

$$A = -\frac{R_C}{R_{Etotal}}$$

Because the emitter resistor is bypassed

$$A = -\frac{R_C}{r_{e1}}$$

For our circuit

$$A = -\frac{R_C}{r_{e1}} = -\frac{5\text{K}}{20} = -250$$

You can see that this cascode amplifier has high input resistance, high output resistance, high gain, and inverts the input signal. Because we use the common base, we also get the high-frequency response that we need.

EXAMPLE 16–3
Cascode Amplifier—AC Analysis

For the cascode amplifier in Figure 16–8 (p. 511), find the AC emitter resistances of each transistor and the Box Model parameters for the amplifier. In Example 16–2 the emitter current for this circuit was found to be $I_{E2} = I_{E1} = 0.65\text{mA}$.

Solution Because the same emitter current flows in both transistors, the AC emitter resistances are equal and are given by

$$r_{e1} = r_{e2} = \frac{26\text{mV}}{I_E} = \frac{26\text{mV}}{0.65\text{mA}} = 40\Omega$$

Because R_1 is bypassed

$$R_B = R_2 \| R_3 = 40\text{K} \| 20\text{K} = 13.3\text{K}\Omega$$

Note that in this circuit R_E is not bypassed, so

$$R_{E1total} = R_E + r_{e1} = 2\text{K} + 40 \approx 2\text{K}\Omega$$

The input resistance, therefore, is

$$R_i = R_B \| \beta R_{E1total} = 13.3\text{K} \| 200\text{K}\Omega = 12.5\text{K}\Omega$$

Because the emitter resistor is not bypassed, we achieve a higher input resistance. Output resistance is

$$R_o = R_C = 8\text{K}\Omega$$

Gain is

$$A = -\frac{R_C}{R_{E1total}} = -\frac{8\text{K}}{2\text{K}} = -4$$

While we have increased the input resistance by not bypassing the emitter resistor, we have lowered the gain.

REVIEW QUESTIONS

1. In the cascode configuration what is the comparison between the two collector currents? The two AC emitter resistances?

2. If the emitter resistor is bypassed, what are the Box Model parameters for the cascode amplifier? Is there inversion?

3. If the emitter resistor is not bypassed, what are the Box Model parameters of the cascode amplifier? Is there inversion?

4. Describe what happens to the Box Model parameters when the emitter resistor is not bypassed.

16–6 ■ COLLECTOR-FEEDBACK BIASING

Figure 16–11A shows a biasing scheme in which the base is connected to the collector through the resistor R_B. In this circuit the collector voltage is fed back to the base. While this is still an example of V_B control, it is easier to derive the DC bias values by first finding I_C.

FIGURE 16–11
Collector-feedback biasing

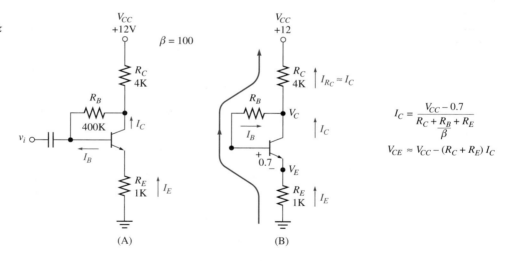

$$I_C = \frac{V_{CC} - 0.7}{R_C + R_B + R_E}$$
$$\frac{}{\beta}$$

$$V_{CE} \approx V_{CC} - (R_C + R_E)\,I_C$$

Figure 16–11B shows the currents in the three resistors in the circuit. By following a path through R_E, V_{BE}, R_B, and R_C, we can write the voltage equation:

$$R_E I_E + 0.7 + R_B I_B + R_C I_{R_C} = V_{CC}$$

From KCL, we know that

$$I_{R_C} = I_B + I_C \approx I_C$$

Now, because

$$I_E \approx I_C \quad \text{and} \quad I_B = \frac{I_C}{\beta}$$

we can substitute these relations into the voltage equation to get

$$R_E I_C + 0.7 + R_B \frac{I_C}{\beta} + R_C I_C = V_{CC}$$

Factoring and rearranging

$$\left(R_E + \frac{R_B}{\beta} + R_C \right) I_C = V_{CC} - 0.7$$

Therefore, the collector current is

$$I_C = \frac{V_{CC} - 0.7}{R_E + \dfrac{R_B}{\beta} + R_C}$$

For the values in our circuit

$$I_C = \frac{V_{CC} - 0.7}{R_E + \dfrac{R_B}{\beta} + R_C} = \frac{12 - 0.7}{1K + \dfrac{400K}{100} + 4K} = 1.26\text{mA}$$

EXAMPLE 16–4
Collector-Feedback Biasing

For the collector-feedback biasing circuit of Figure 16–12, find the DC bias values.

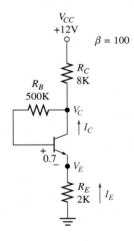

FIGURE 16–12

Solution   The collector current is

$$I_C = \frac{V_{CC} - 0.7}{R_E + \dfrac{R_B}{\beta} + R_C} = \frac{12 - 0.7}{8K + 5K + 2K} = 0.753\text{mA}$$

The emitter voltage is

$$V_E = R_E I_C = 2K \times 0.753\text{mA} = 1.51\text{V}$$

The collector voltage is

$$V_C = V_{CC} - R_C I_C = 12 - 8K \times 0.753\text{mA} = 5.98\text{V}$$

DRILL EXERCISE   Redraw the circuit of Figure 16–12, changing the supply voltage to 9V. Find the DC bias levels in the circuit.

Answer $I_C = 0.553mA$, $V_E = 1.12V$, $V_C = 4.58V$

REVIEW QUESTIONS

1. Draw the collector-feedback bias circuit.
2. What is the voltage drop across each resistor?
3. What is the formula for I_C?

16–7 ■ AC ANALYSIS OF COLLECTOR-FEEDBACK BIASING—THE MILLER EFFECT

In the collector-feedback biased circuit, the input signal is applied to the base, and the output is taken from the collector (Figure 16–13). Therefore, the collector-feedback biased circuit is still a common-emitter amplifier. Because current in the feedback resistor is small, we can ignore R_B to find the approximate Box Model output parameters. The DC analysis of this circuit shows that

$$I_E \approx I_c = 1.26mA$$

FIGURE 16–13

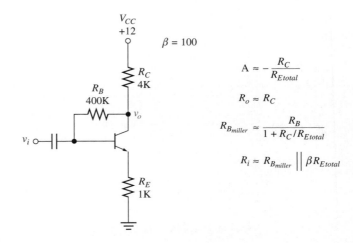

Therefore, the AC emitter resistance is

$$r_e = \frac{26mV}{I_E} = \frac{26mV}{1.26mA} = 20.6\Omega$$

Ignoring R_B, the Box Model gain of this amplifier is given by

$$A \approx -\frac{R_C}{R_{Etotal}}$$

And, because R_E is not bypassed

$$R_{Etotal} = R_E + r_e = 1K + 20.6 \approx 1K\Omega$$

and

$$A \approx -\frac{R_C}{R_{Etotal}} = -\frac{4K}{1K} = -4$$

Because we ignore R_B, the output resistance is approximately

$$R_o \approx R_C$$

For the values given

$$R_o = R_C = 4K\Omega$$

Input Resistance—The Miller Effect

We cannot ignore R_B when we find the input resistance. In fact, because of the collector feedback, the input resistance in this circuit is a lot smaller than you might think. To get a general grasp of the effect of feedback on input resistance, we first look at the current in the base resistor of the regular common emitter.

For this circuit the base resistor is tied to ground (Figure 16–14A). Let's assume that the input voltage has a 1mV amplitude, and find the current in the base resistor for several values of R_B. Because

$$i_{R_B} = \frac{v_i}{R_B} = \frac{1mV}{R_B}$$

FIGURE 16–14
The Miller effect

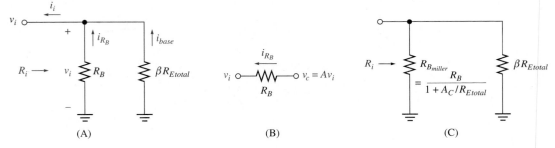

| (A) | (B) | (C) |

we get the following:

R_B	i_{R_B}
10KΩ	0.1μA
2KΩ	0.5μA
100Ω	10.0μA

Now let's look at R_B in the collector-feedback bias circuit, where, rather than being connected to ground, R_B is connected between v_i and v_c (Figure 16–14B). Here the current in R_B is given by

$$i_{R_B} = \frac{v_i - v_c}{R_B}$$

We know that

$$v_c = A v_i$$

Therefore,

$$i_{R_B} = \frac{v_i - A v_i}{R_B}$$

Factoring out v_i

$$i_{R_B} = (1 - A)\frac{v_i}{R_B}$$

As an example, let's assume that the input voltage amplitude is 1mV and $R_B = 10K\Omega$. From our table above, the current in R_B for the regular common emitter (Figure 16–14A) would be

$$\text{Regular common emitter: } i_{R_B} = \frac{v_i}{R_B} = \frac{1mV}{10K} = 0.1\mu A$$

Now let's assume the gain of the amplifier is −4. The current in the base resistor for the collector-feedback circuit (Figure 16–4B) is

$$\text{Collector-feedback circuit: } i_{R_B} = (1 - A)\frac{v_i}{R_B} = (1 - (-4))\frac{1mV}{10K} = 0.5\mu A$$

This means that because of feedback from the collector, the current produced in R_B by the input voltage is 5 times greater than if R_B is tied to ground.

Because the same voltage produces a greater current, the effective resistance due to R_B has decreased significantly. In fact, the current produced in our collector-feedback example is the same current produced in the regular common emitter when the base resistor was reduced to 2KΩ.

The reduction of the input resistance due to feedback is known as the **Miller effect.** For the collector-feedback circuit, we define the Miller resistance as

$$R_{B_{Miller}} = \frac{v_i}{i_{R_B}} = \frac{R_B}{(1 - A)}$$

And, because

$$A \approx \frac{-R_C}{R_{Etotal}}$$

$$R_{B_{Miller}} \approx \frac{R_B}{1 + R_C/R_{Etotal}}$$

For the circuit of Figure 16–13 (p. 517), the Miller resistance is

$$R_{B_{Miller}} \approx \frac{R_B}{(1 + R_C/R_{Etotal})} = \frac{400\ K}{(1 + 4K/1K)} = 80K\Omega$$

The total input resistance (Figure 16–14C) is now

$$R_i = R_{B_{Miller}} \| \beta R_{Etotal}$$

For the values in our circuit, the input resistance is

$$R_i = R_{B_{Miller}} \| \beta R_{Etotal} = 80\text{K} \| 100\text{K} = 44.4\text{K}\Omega$$

The Miller effect acts to decrease the input resistance of the collector-feedback biased circuit.

EXAMPLE 16–5
Collector
Feedback—
AC Analysis

For the collector-feedback circuit of Figure 16–15, find the AC emitter resistance, $R_{B_{Miller}}$, and the Box Model parameters. Use the DC bias levels you found in Example 16–4 (p. 518).

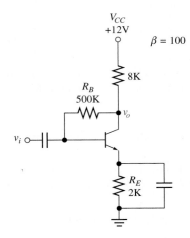

FIGURE 16–15

Solution   In Example 16–4 we found that $I_E = 0.753$mA. Therefore

$$r_e = \frac{26\text{mV}}{I_E} = \frac{26\text{mV}}{0.753\text{mA}} = 34.5\Omega$$

Because the emitter resistor is bypassed here

$$R_{Etotal} = r_e = 34.5\Omega$$

And

$$A = -\frac{R_C}{R_{Etotal}} = -\frac{8\text{K}}{34.5} = -232$$

The Miller effect produces

$$R_{B_{Miller}} = \frac{R_B}{1 + R_C/R_{Etotal}} = \frac{500\text{K}}{(1 + 8\text{K}/34.5)} = 2.15\text{K}\Omega$$

You can see that for large gains, Miller resistance gets very small. The total input resistance is

$$R_i = R_{B_{Miller}} \| \beta R_{Etotal} = 2.15\text{K} \| 3.45\text{K} = 1.32\text{K}\Omega$$

Finally the output resistance is

$$R_o \approx R_C = 8\text{K}\Omega$$

DRILL EXERCISE   Redraw the collector-feedback circuit of Figure 16–15, changing the supply voltage to 9V. Find the AC emitter resistance, $R_{B_{Miller}}$, and the Box Model Parameters.

Answer $r_e = 47\Omega$, $R_{B_{Miller}} = 2.92\text{K}\Omega$, $A = -170$, $R_i = 1.8\text{K}\Omega$, $R_o = 8\text{K}\Omega$

1. What type of amplifier is the collector-feedback circuit?
2. What is the gain of this amplifier?
3. Does R_B affect the gain? Why?
4. Why does the amplifier gain affect the input resistance? What is this effect called?
5. What is $R_{B_{Miller}}$?
6. What is the input resistance of the collector-feedback circuit?
7. How does the Miller effect change the input resistance?

SUMMARY

▪ The Darlington connection is built by tying the collectors of two transistors together. The emitter of one transistor supplies the input to the base of the other.

▪ The Darlington connection behaves as a super transistor in which

$$I_C = \beta^2 I_B$$
$$V_{BE} = 1.4V$$

▪ The Darlington common-emitter amplifier has the following Box Model parameters

$$r_{e_{Darlington}} = 2r_{e2}$$

$$A = -\frac{R_C}{R_{Etotal}}$$

$$R_i = R_B \| \beta^2 R_{Etotal}$$

$$R_o = R_C$$

▪ In the cascode amplifier the output of a common-emitter amplifier is the input to a common-base amplifier.

▪ The cascode amplifier Box Model parameters are the same as the common-emitter parameters, except the cascode amplifier has good high-frequency response.

▪ The collector current in a collector-feedback circuit is given by

$$I_C \approx \frac{V_{CC} - 0.7}{R_C + \dfrac{R_B}{\beta} + R_E}$$

▪ The Box Model gain and output resistance of the collector-feedback common-emitter amplifier are approximately the same as the gain and output resistance of the regular common-emitter amplifier.

▪ The Miller effect reduces the apparent base resistance of the collector-feedback circuit. That is

$$R_{B_{Miller}} = \frac{R_B}{1 - A} \approx \frac{R_B}{1 + R_C/R_{Etotal}}$$

▪ The input resistance is

$$R_i = R_{B_{Miller}} \| \beta R_{Etotal}$$

PROBLEMS

SECTION 16–2   The Darlington Connection

1. Redraw the circuit of Figure 16–16, and label the individual currents for both Q_1 and Q_2.

   (a) What is the relation between I_{C1} and I_B?

   (b) Without making any approximations, what is the relation between I_{E1} and I_B?

   (c) What is the relation between I_{B2} and I_B?   (d) What is the relation between I_{C2} and I_B?

   (e) What is the relation between I_C and I_B?   (f) What is the relation between I_E and I_{E2}?

   (g) Without making any approximations, what is the relation between I_E and I_B?

   (h) Use the results of part (e) to determine the over-all current gain of the Darlington connection.

FIGURE 16–16

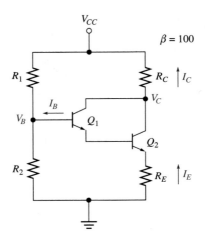

SECTION 16–3   Darlington Common-Emitter Amplifier

2. For the Darlington circuit shown in Figure 16–16

$$V_{CC} = +12V, R_C = 5K\Omega, R_E = 1K\Omega, R_1 = 300K\Omega, R_2 = 100K\Omega$$

   (a) Find the bias voltages and currents in the following order:

V_{B1}	I_{B2}
V_{E1}	I_{E1}
V_{B2}	I_{C1}
V_{E2}	I_{B1}
I_{E2}	V_{C2}
I_{C2}	V_{C1}

   (b) Find these external terminal voltages and currents:

V_B	I_C
V_E	I_B
I_E	V_C

3. Redraw the circuit of Figure 16–16, replacing the two-transistor combination with the equivalent "super transistor." Use the device values given in Problem 2.

   (a) Find these external terminal voltages and currents:

V_B	I_C
V_E	I_B
I_E	V_C

   (b) Compare these results to part (b) of the previous problem.

4. Repeat Problem 2 if

$$V_{CC} = +20V, R_C = 9K\Omega, R_E = 1K\Omega, R_1 = 176K\Omega, R_2 = 24K\Omega$$

5. For the Darlington common-emitter amplifier in Figure 16–17

$$V_{CC} = +12V, R_C = 5K\Omega, R_E = 1K\Omega, R_1 = 300K\Omega, R_2 = 100K\Omega$$

   (a) Find $r_{e_{Darlington.}}$ **(b)** Find the Box Model parameters: A, R_i, R_o.

6. Repeat Problem 5 if the emitter resistor (R_E) is bypassed.

7. For the Darlington common-emitter amplifier in Figure 16–17

$$V_{CC} = +20V, R_C = 9K\Omega, R_E = 1K\Omega, R_1 = 176K\Omega, R_2 = 24K\Omega$$

   (a) Find $r_{e_{Darlington.}}$ **(b)** Find the Box Model parameters: A, R_i, R_o.

8. Repeat Problem 7 if the emitter resistor (R_E) is bypassed.

FIGURE 16–17

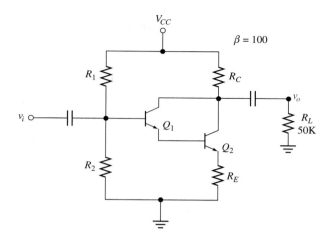

9. Figure 16–18 shows a Darlington emitter follower. If

$$V_{CC} = +12V, R_C = 5K\Omega, R_E = 1K\Omega, R_1 = 300K\Omega, R_2 = 100K\Omega$$

(a) Redraw the circuit, replacing the Darlington combination with an equivalent "super transistor."

(b) Find the Box Model parameters: A, R_i, R_o. (Remember that the input and output resistance depend on the source and load resistors.)

10. Figure 16–18 shows a Darlington emitter follower. If

$$V_{CC} = +20V, R_C = 9K\Omega, R_E = 1K\Omega, R_1 = 176K\Omega, R_2 = 24K\Omega$$

(a) Redraw the circuit, replacing the Darlington combination with an equivalent "super transistor."

(b) Find the Box Model parameters: A, R_i, R_o. (Remember that the input and output resistance depend on the source and load resistors.)

11. Figure 16–19 shows a modification of the Darlington connection. A resistor (R_{E2}) has been added to the circuit.

(a) Find the bias-current levels in this circuit in the same order requested in Problem 1. Note that now

$$I_{E1} = I_{B2} + I_{RE1}$$

(b) Find the current gain

$$\frac{I_C}{I_B}$$

(c) Compare this current gain to the current gain of the regular Darlington (β^2).

FIGURE 16–18

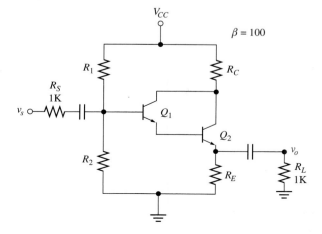

FIGURE 16–19

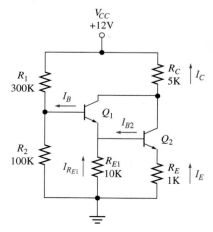

SECTION 16–4 The Cascode Amplifier

12. For the cascode circuit shown in Figure 16–20:

(a) Find the bias voltages and currents in the following order:

V_{B1}	I_{C2}
V_{E1}	I_{B2}
I_{E1}	V_{C2}
I_{C1}	V_{B2}
I_{B1}	V_{E2}
I_{E2}	V_{C1}

(b) In what states are the transistors operating? Why?

13. For the circuit in Figure 16–20, if $R_1 = 60K\Omega$ and $R_2 = 50K\Omega$, as shown

(a) Choose a new R_3 to set the collector current to $I_C = 0.5$mA.

(b) What are V_{C1} and V_{C2} now?

14. For the circuit in Figure 16–20, if $R_2 = 50K\Omega$ and $R_3 = 10K\Omega$, as shown

(a) Choose a new R_1 to set the collector current to $I_C = 0.5$mA.

(b) What are V_{C1} and V_{C2} now?

FIGURE 16–20

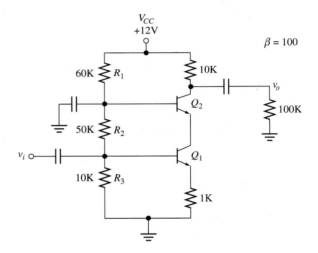

SECTION 16–5 The Cascode Amplifier—AC Analysis

15. For the cascode amplifier in Figure 16–20:

(a) What are r_{e1} and r_{e2}?

(b) Find the Box Model parameters, and draw the Box Model.

(c) What is the maximum source resistance that should be used with this amplifier? Why?

16. Bypass the emitter resistor in the cascode amplifier of Figure 16–20, and repeat Problem 15.

SECTION 16–6 Collector-Feedback Biasing

17. For the collector-feedback bias circuit of Figure 16–21, $\beta = 100$. Find the collector current.

18. For the collector-feedback bias circuit of Figure 16–21, $\beta = 400$. Find the collector current.

19. For the collector-feedback bias circuit of Figure 16–21, $\beta = 100$.

(a) Find the Box Model parameters, and draw the Box Model.

(b) What is the maximum source resistance that should be used with this amplifier? Why?

20. Bypass the emitter resistor of the collector-feedback bias circuit in Figure 16–21 and repeat Problem 19.

21. For the collector-feedback bias circuit of Figure 16–21, $\beta = 400$.

(a) Find the Box Model parameters, and draw the Box Model.

(b) What is the maximum source resistance that should be used with this amplifier? Why?

FIGURE 16–21

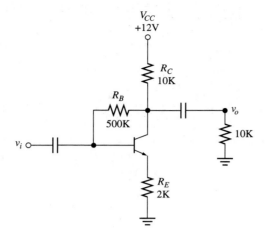

22. Bypass the emitter resistor of the collector-feedback bias circuit shown in Figure 16–21 and repeat Problem 21.

Computer Problems

23. Simulate the Darlington circuit of Figure 16–16 (p. 522), using the device parameters of Problem 2 (p. 522). Find the bias levels and compare your results to Problem 2.

24. For the Darlington common-emitter amplifier of Figure 16–17 (p. 523) and the device parameters given in Problem 2 (p. 522):

   (a) Use the computer to find the Box Model gain and input resistance.

   (b) Use the computer to find the output for a 1mV 1KHz sinewave input.

25. For the Darlington emitter follower of Figure 16–18 (p. 523), and the device parameters of Problem 10 (p. 523):

   (a) Use the computer to find the bias levels.

   (b) Use the computer to find the Box Model gain and input resistance.

   (c) Use the computer to find the output for a 1mV 1KHz sinewave input.

26. Use the computer to find the bias levels in the modified Darlington amplifier of Figure 16–19 (p. 523).

27. Simulate the cascode amplifier of Figure 16–20.

   (a) Find the DC bias levels.   **(b)** Find the Box Model gain and input resistance.

28. Simulate the cascode amplifier shown in Figure 16–20. Set $\beta = 100$. Set all external capacitors to 30µF. Read your computer program instruction manual to learn how to set the internal base and emitter junction capacitors to 3pF (that's 3×10^{-12}F). Use AC Analysis to find the output over a frequency range of 1KHz to 100 MHz.

29. Simulate the collector-feedback bias circuit of Figure 16–21.

   (a) Find all bias voltages and currents, including the current in R_B.

   (b) Find the Box Model gain and input resistance.

30. Repeat Problem 29 if $\beta = 400$.

CHAPTER

17

THE DIFFERENTIAL AMPLIFIER

To use a differential amplifier to measure differences in two input voltages.

To derive the Box Model parameters for differential-mode inputs.

To find the gain for common-mode inputs.

To determine the common-mode rejection ratio for a differential amplifier.

To determine the Box Model parameters for general inputs.

To determine the uses of integrated circuit differential amplifiers.

To troubleshoot differential amplifier circuits.

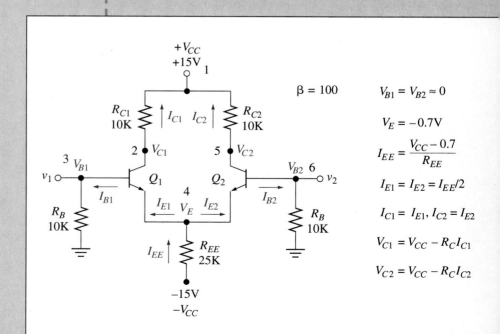

$$\beta = 100 \qquad V_{B1} = V_{B2} \approx 0$$

$$V_E = -0.7\text{V}$$

$$I_{EE} = \frac{V_{CC} - 0.7}{R_{EE}}$$

$$I_{E1} = I_{E2} = I_{EE}/2$$

$$I_{C1} = I_{E1}, I_{C2} = I_{E2}$$

$$V_{C1} = V_{CC} - R_C I_{C1}$$

$$V_{C2} = V_{CC} - R_C I_{C2}$$

17–1 ■ INTRODUCTION

So far, all of the amplifiers we have looked at (and there have been many) have one thing in common. They are all designed to amplify a signal from a voltage source that is connected to ground (Figure 17–1). Consider the filter circuit in Figure 17–2. Because the driving voltage is very small, we will need an amplifier to measure the capacitor and inductor voltages. To measure the capacitor voltage, we can connect a common-emitter amplifier as shown in the figure. This will work because the capacitor voltage is measured with respect to ground.

FIGURE 17–1

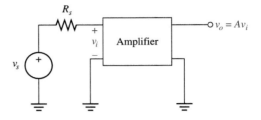

FIGURE 17–2

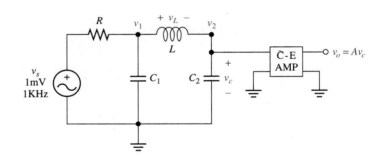

Now how can you connect the common-emitter amplifier to measure the inductor voltage? Take a few minutes. Give up? What about measuring v_1 first, then measuring v_2, and finding v_L from

$$v_L = v_1 - v_2$$

This might work in theory but is not very practical. It is a whole lot easier to use an amplifier like the one in Figure 17–3, an amplifier that has two inputs and produces an output that amplifies the difference in the two inputs. That is

$$v_o = A(v_1 - v_2)$$

Figure 17–3B shows how this type of amplifier can be used to directly measure the inductor voltage. It should come as no surprise that this amplifier is known as a *difference amplifier.* If the difference in the input voltages is very small, we refer to this type of amplifier as a **differential amplifier.**

The differential amplifier is also used to amplify signals that are contaminated with noise, especially biomedical signals. For example, we want to amplify EEG (electroencephalogram) brain wave signals. If we use a common-emitter amplifier (Figure 17–4A), we will get a very noisy output. This is because the actual voltage measured at the EEG electrode is composed of the brain wave signal plus electrical noise (Figure 17–4B). That is

$$v_{EEG} = v_{brainwave} + v_{noise}$$

Noise is a fact of life in all real systems.

FIGURE 17–3

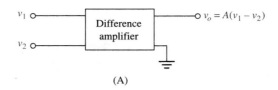

(A)

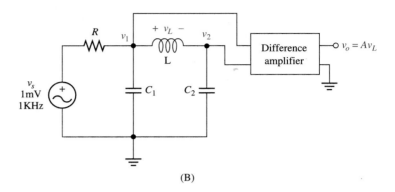

(B)

FIGURE 17–4

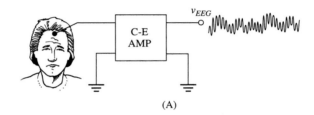

(A)

Brainwave   Noise

(B)

Now let's use a differential amplifier (Figure 17–5). One electrode is placed where we want to measure the brain wave signal. The other electrode is placed elsewhere on the body. Electrode 1 measures the same noisy EEG signal as in Figure 17–4B:

$$v_1 = v_{EEG} = v_{brainwave} + v_{noise}$$

Electrode 2 is placed so it only measures noise. Because noise is random, it is the same at both electrodes, so

$$v_2 = v_{noise}$$

FIGURE 17–5

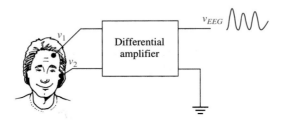

The differential amplifier magnifies the difference in the two input signals:

$$v_o = A(v_1 - v_2) = A(v_{brainwave} + v_{noise} - v_{noise}) = Av_{brainwave}$$

The result is the noise is eliminated, and we get the nice clean output signal shown.

The differential amplifier is a very useful device in itself. It also serves as the input stage to the most useful of all analog devices, the operational amplifier. In this chapter we will describe some simple differential amplifier circuits and try to give you a basic feel for how the circuitry works. Exact analysis of these circuits can be very involved and are best handled with computer analysis, as presented in the problems at the end of the chapter.

17–2 ■ BIASING THE DIFFERENTIAL AMPLIFIER

Figure 17–6 shows the basic configuration for an NPN differential amplifier. Note that for this configuration, we need both positive and negative supply voltages ($\pm V_{CC}$). (Some authors label the negative supply as $-V_{EE}$). Also note the symmetry in this circuit. Because the emitters are tied together

$$V_{E1} = V_{E2} = V_E$$

FIGURE 17–6
The differential amplifier

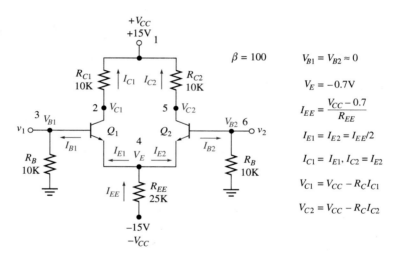

We assume that the two transistors are identical, and from the symmetry of the configuration we can conclude that

$$V_{C1} = V_{C2}$$
$$I_{E1} = I_{E2}$$
$$I_{C1} = I_{C2}$$
$$I_{B1} = I_{B2}$$

This symmetry makes DC bias calculations straightforward.

We assume that because the base current is very small, we can ignore the voltage drops across the base resistors. That is, we assume

$$V_{B1} = V_{B2} \approx 0V$$

For DC calculations we can still assume that the emitter diode turn-on voltage for each transistor is 0.7V. Therefore

$$V_E = V_{B1} - 0.7 = V_{B2} - 0.7 = 0 - 0.7 = -0.7V$$

We now know the voltage across R_{EE}. The current in R_{EE} is given by

$$I_{EE} = \frac{-0.7 - (-V_{CC})}{R_{EE}}$$

so

$$I_{EE} = \frac{V_{CC} - 0.7}{R_{EE}}$$

For the values in our circuit

$$I_{EE} = \frac{V_{CC} - 0.7}{R_{EE}} = \frac{15 - 0.7}{25K} = 0.572\text{mA}$$

Because of the symmetry in the circuit, I_{EE} must split equally between Q_1 and Q_2. Therefore

$$I_{E1} = I_{E2} = \frac{I_{EE}}{2}$$

For the values given

$$I_{E1} = I_{E2} = \frac{I_{EE}}{2} = \frac{0.572\text{mA}}{2} = 0.286\text{mA}$$

If both transistors are **ACTIVE**

$$I_{C1} \approx I_{E1} = 0.286\text{mA}$$

and

$$I_{C2} \approx I_{E2} = 0.286\text{mA}$$

The base currents are ($\beta = 100$)

$$I_{B1} = \frac{I_{C1}}{\beta} = 2.86\mu\text{A} \qquad \text{and} \qquad I_{B2} = \frac{I_{C2}}{\beta} = 2.86\mu\text{A}$$

This confirms our assumption that the base current is small enough to neglect. To check the state of each transistor, we find the collector voltages from

$$V_{C1} = V_{CC} - R_C I_{C1} \qquad \text{and} \qquad V_{C2} = V_{CC} - R_C I_{C2}$$

For our circuit

$$V_{C1} = V_{C2} = 15 - 10K \times 0.286\text{mA} = 12.1\text{V}$$

Because $V_C > V_E$ for each transistor, we conclude that both transistors are **ACTIVE.**

EXAMPLE 17–1
NPN Differential Amplifier Biasing

Find the bias voltages and currents for the differential amplifier circuit in Figure 17–7 (p. 532). Note that Q_1 does not have a collector resistor in this circuit.

Solution   The fact that there is no collector resistor on Q_1 will only affect the calculation for V_{C1}. Because of the symmetry in the base-emitter circuits of Q_1 and Q_2, we still have

$$I_{E1} = I_{E2} = \frac{I_{EE}}{2}$$

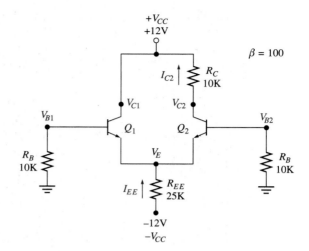

FIGURE 17–7

Ignoring the base current, the emitter voltage for both transistors is

$$V_E = -0.7V$$

and

$$I_{EE} = \frac{V_{CC} - 0.7}{R_{EE}} = \frac{12 - 0.7}{25K} = 0.452mA$$

So

$$I_{E1} = I_{E2} = \frac{I_{EE}}{2} = 0.226mA$$

Assuming both transistors are **ACTIVE**

$$I_{C1} \approx I_{E1} = 0.226mA$$

and

$$I_{C2} \approx I_{E2} = 0.226mA$$

The base currents are

$$I_{B1} = I_{B2} = 2.26\mu A$$

To check the states of the transistors, we find the collector voltages. In this circuit

$$V_{C1} = V_{CC} = 12V$$

and

$$V_{C2} = V_{CC} - R_C I_{C2} = 12 - 10K \times 0.226mA = 9.74V$$

Because for both transistors $V_C > V_E$, both transistors are **ACTIVE.**

DRILL EXERCISE   Redraw the circuit of Figure 17–7, and change the supply voltage to ±9V. Find the bias currents and voltages. What are the states of the two transistors?

Answer $V_E = -0.7V$, $I_{EE} = 0.332mA$, $I_{E1} = I_{E2} = 0.166mA$
$I_{C1} = I_{C2} = 0.166mA$, $I_{B1} = I_{B2} = 1.66\mu A$
$V_{C1} = 9V$, $V_{C2} = 7.34V$

Q_1 and Q_2 **ACTIVE**

EXAMPLE 17–2
Current Source Biasing

Most differential amplifiers are actually biased with a current source (Figure 17–8). In the next chapter we will show you how to build current sources. For now, assume that the current source supplies the DC current shown. Find all bias currents and voltages.

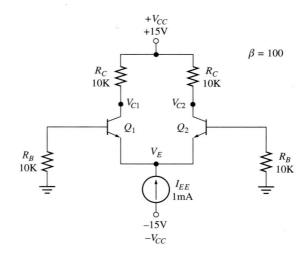

FIGURE 17–8

Solution   In this circuit I_{EE} is given by the value of the current source. That is

$$I_{EE} = 1\text{mA}$$

The emitter voltage is still

$$V_E = -0.7\text{V}$$

The other bias levels are derived from the symmetry of the circuit:

$$I_{E1} = I_{E2} = \frac{I_{EE}}{2} = 0.5\text{mA}$$

If the transistors are **ACTIVE**

$$I_{C1} \approx I_{E1} = 0.5\text{mA} \qquad \text{and} \qquad I_{C2} \approx I_{E2} = 0.5\text{mA}$$
$$I_{B1} = I_{B2} = 5\mu\text{A}$$

The collector voltages are

$$V_{C1} = V_{C2} = 15 - 10\text{K} \times 0.5\text{mA} = 10\text{V}$$

Both transistors are **ACTIVE.**

EXAMPLE 17–3
PNP Differential Amplifier Biasing

Figure 17–9 (p. 534) shows a differential amplifier constructed with PNP transistors. Find the bias currents and voltages.

Solution   For the PNP transistor the emitter voltage is a diode drop above the base voltage.

$$V_E = 0.7\text{V}$$

The current in R_{EE} is

$$I_{EE} = \frac{V_{CC} - 0.7}{R_{EE}} = \frac{11.3}{25\text{K}} = 0.452\text{mA}$$

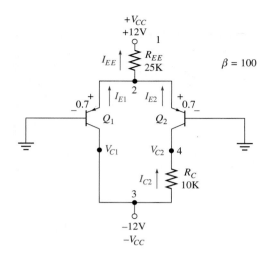

FIGURE 17–9

The emitter currents are

$$I_{E1} = I_{E2} = \frac{I_{EE}}{2} = \frac{0.452\text{mA}}{2} = 0.226\text{mA}$$

If the transistors are **ACTIVE**

$$I_{C1} \approx I_{E1} = 0.226\text{mA} \quad \text{and} \quad I_{C2} \approx I_{E2} = 0.226\text{mA}$$
$$I_{B1} = I_{B2} = 2.26\mu\text{A}$$

To check the state of the transistors, we find the collector voltages. Because there is no resistor at the collector of Q_1

$$V_{C1} = -V_{CC} = -12\text{V}$$

The collector voltage of Q_2 is found from

$$V_{C2} = -V_{CC} + R_C I_{C2} = -12 + 10\text{K} \times 0.226\text{mA} = -9.74\text{V}$$

Because the collector voltage is less than the emitter voltage for both PNP transistors, the transistors are **ACTIVE.**

DRILL EXERCISE   Redraw the circuit of Figure 17–9, and change the supply voltage to ±9V. Find the bias currents and voltages. What are the states of the two transistors?

Answer $V_E = 0.7\text{V}$, $I_{EE} = 0.332\text{mA}$, $I_{E1} = I_{E2} = 0.166\text{mA}$
$I_{C1} = I_{C2} = 0.166\text{mA}$, $I_{B1} = I_{B2} = 1.66\mu\text{A}$
$V_{C1} = -9\text{V}$, $V_{C2} = -7.34\text{V}$

Q_1 and Q_2 **ACTIVE**

REVIEW QUESTIONS

1. Draw the basic NPN differential amplifier circuit.
2. What does the symmetry of the circuit imply?
3. If you ignore the base currents, what is V_E?
4. Why can the base currents be ignored?
5. How do you find I_{EE}?
6. How do you find the emitter currents I_{E1} and I_{E2}?
7. Draw the basic PNP differential amplifier circuit. What are I_{EE}, I_{E1}, I_{E2}, V_{C1}, V_{C2}?

17–3 ■ THE DIFFERENTIAL MODE

Figure 17–10A shows the block diagram of the ideal differential amplifier. The output of this device is an amplified version of the difference in the input signals. A more realistic differential amplifier is shown in Figure 17–10B.

FIGURE 17–10

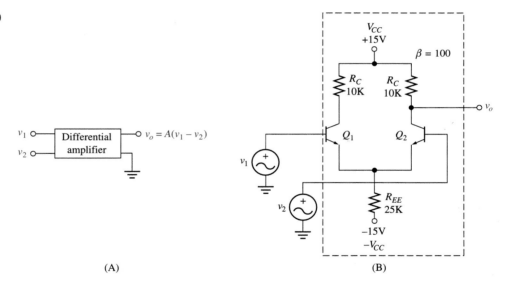

(A)

(B)

In the previous section we showed you how to find the bias levels for this circuit. Note that the absence of the base resistors does not affect the bias calculations. When the input voltage sources are turned off, they connect the bases to ground, and the emitter voltage is still $-0.7V$.

We now want to show you how to find the gain of the amplifier. For any two arbitrary inputs, it is very difficult to find the output voltage. For this reason, we usually look at two very special cases first:

$$\text{differential mode:} \quad v_2 = -v_1$$

and

$$\text{common mode:} \quad v_2 = v_1$$

We will discuss the differential mode in this section and the common mode in the next section. In the following section we discuss the general situation, in which the two inputs are not related.

FIGURE 17–11
Differential mode

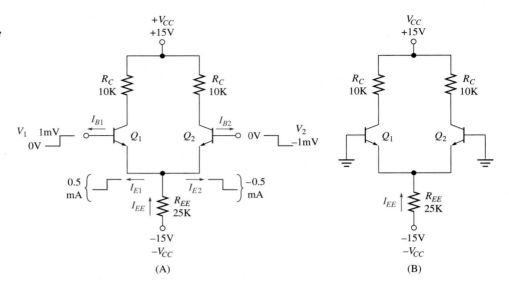

(A)   (B)

First, let's look at the circuit shown in Figure 17–11A. When the input voltages are turned off

$$V_1 = V_2 = 0\text{V}$$

and we get the bias circuit in Figure 17–11B. This is essentially the same bias circuit that we saw in Figure 17–6. The relevant bias values are

$$V_E = -0.7\text{V}$$
$$I_{EE} = 0.572\text{mA}$$

and

$$I_{E1} = I_{E2} = 0.286\text{mA}$$

We now change the inputs to

$$V_1 = 1\text{mV} \qquad \text{and} \qquad V_2 = -V_1 = -1\text{mV}$$

Let's look at the corresponding changes in the emitter currents. At this point we don't know how much they will change (unless you use a computer); however, we can make the following generalizations:

☐ Because V_1 increases, I_{E1} will increase.

☐ Because V_2 decreases, I_{E2} will decrease.

☐ Because of the symmetry of the circuit and the inputs, the size of the decrease in I_{E2} will be the same size as the increase in I_{E1}.

That is, the new emitter currents will be

$$I_{E1} = 0.286\text{mA} + \Delta I \qquad \text{and} \qquad I_{E2} = 0.286\text{mA} - \Delta I$$

ΔI is the unknown change in the emitter currents.

What is happening to the current in R_{EE} while all of this is going on? We know that

$$I_{EE} = I_{E1} + I_{E2}$$

so

$$I_{EE} = 0.286\text{mA} + \Delta I + 0.286\text{mA} - \Delta I = 0.572\text{mA}$$

That's right. While all of these changes occur in the input voltages and emitter currents, the current in R_{EE} doesn't change at all. In a sense, the changing emitter currents go from V_2 to V_1, bypassing R_{EE} (Figure 17–12A).

The following fact provides the key to the analysis of the differential mode. Because current in R_{EE} does not change under differential-mode conditions, emitter voltage must remain constant. That is, the change in emitter voltage is

$$\Delta V_E = 0V$$

FIGURE 17–12
Virtual ground

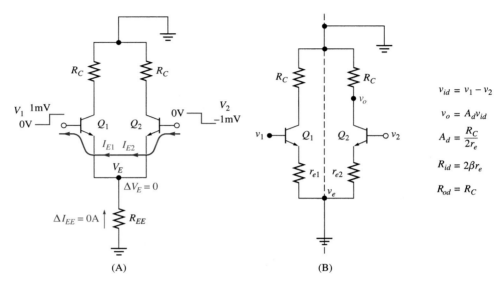

(A)   (B)

Even though the input signals here are not AC, AC analysis is a useful tool for determining amplifier response to any small change in input signals. As before, we represent changes in voltages with lowercase AC quantities. In this example

$$v_1 = 1mV$$
$$v_2 = -v_1 = -1mV$$

and, most importantly

$$v_e = 0V$$

We can now draw the AC circuit for the differential-mode amplifier. Because emitter voltage is held at 0V for AC analysis, we can consider the emitter to be grounded. This is known as a **virtual ground,** because the emitter is grounded *only* in the AC model, not in the actual circuit. Figure 17–12B shows the AC model, including the AC emitter resistances.

Because the AC differential-mode amplifier is virtually grounded at both the top and bottom of the circuit, we really have two separate common-emitter amplifiers, represented by the dotted line in Figure 17–12B. The formal name for this (not something you really need to know) is *1/2-circuit analysis.* At this point we can simply analyze one half of the circuit by calculating the answers based on our knowledge of common-emitter amplifiers. That is

$$v_{c1} = -\frac{R_C}{r_e}v_1$$

and

$$v_{c2} = -\frac{R_C}{r_e}v_2$$

To carry out our numerical example, we remind you that the DC bias emitter currents were

$$I_{E1} = I_{E2} = 0.29\text{mA}$$

So

$$r_e = r_{e1} = r_{e2} = \frac{26\text{mV}}{0.286\text{mA}} = 90.9\Omega$$

Therefore

$$v_{c1} = -\frac{R_C}{r_e} v_1 = -\frac{10\text{K}}{90.9} v_1 = -110v_1 = -110\text{mV}$$

and

$$v_{c2} = -\frac{R_C}{r_e} v_1 = -\frac{10\text{K}}{90.9} v_2 = -110v_2 = +110\text{mV}$$

We aren't quite finished yet, because we are not dealing with two common-emitter amplifiers. We are dealing with a single differential amplifier. Only the special circumstance of the differential-mode input ($v_2 = -v_1$) enables us to use the 1/2-circuit analysis. The actual input to the differential amplifier is considered to be

$$v_{id} = v_1 - v_2$$

If we look at the numbers

$$v_{id} = v_1 - v_2 = 1\text{mV} - (-1\text{mV}) = 2\text{mV}$$

Most differential amplifiers have only a single output. In this circuit we take the output from the collector of Q_2. Again, looking at the numbers we get the differential-mode gain:

$$A_d = \frac{v_o}{v_{id}} = \frac{v_{c2}}{v_{id}} = \frac{110\text{mV}}{2\text{mV}} = 55$$

You can see that the differential-mode gain is one half of the common-emitter gain.

What all this finally tells us is that the differential-mode gain is given by

$$A_d \equiv \frac{v_o}{v_{id}} = \frac{1}{2} \frac{R_C}{r_e}$$

Note: Because we take the output from Q_2, as is usually done, the differential-mode gain is positive.

EXAMPLE 17–4
Differential-Mode Gain

For the circuit in Figure 17–13, find the differential-mode gain.

Solution Because we are taking the output from Q_2, there is no need for a collector resistor on Q_1. To find the relevant bias currents, we set the input sources to 0V, connecting the two bases to ground. Therefore

$$V_E = -0.7\text{V}$$

and

$$I_{EE} = \frac{V_{CC} - 0.7}{R_{EE}} = \frac{11.3}{200\text{K}} = 0.0565\text{mA}$$

So

$$I_{E1} = I_{E2} = \frac{I_{EE}}{2} = 0.0283\text{mA}$$

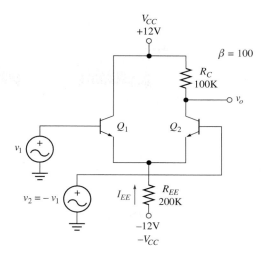

FIGURE 17–13

The AC emitter resistance, therefore, is

$$r_e = \frac{26\text{mV}}{I_{E1}} = \frac{26\text{mA}}{0.0283\text{mA}} = 919\Omega$$

The differential-mode gain is

$$A_d = \frac{v_o}{v_{id}} = \frac{1}{2}\frac{R_C}{r_e} = \frac{1}{2}\frac{100\text{K}}{919} = 54.4$$

where $v_{id} = v_1 - v_2$.

DRILL EXERCISE Redraw the circuit in Figure 17–13, changing the supply voltages to ±9V. Find the emitter currents, the AC emitter resistance, and the differential-mode gain.

Answer $I_{EE} = 0.0415\text{mA}$, $I_{E1} = I_{E2} = 0.0208\text{mA}$, $r_e = 1.25\text{K}\Omega$, $A_d = 40$

The Differential-Mode Box Model

We have just found the Box Model gain for differential-mode inputs:

$$A_d = \frac{1}{2}\frac{R_C}{r_e}$$

Output resistance is still

$$R_o = R_C$$

The input resistance of the differential-mode amplifier is the resistance that exists *between* the two input sources. If we track the AC current path between the two input sources (see Figures 17–12, p. 537, and 17–14, p. 540), we see that the source current effectively goes from v_2, into the base of Q_2, through Q_2's AC emitter resistance, through the AC emitter resistance of Q_1, out of Q_1's base, and finally to v_1. In this pathway the total resistance

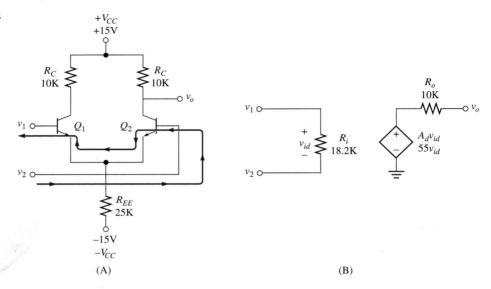

FIGURE 17–14
Box Model

(A)

(B)

is the sum of the two AC emitter resistances. Because resistance seen at the base is β times the resistance seen in the emitter

$$R_i = 2\beta r_e$$

For the values given in the circuit, the Box Model parameters are

$$A_d = \frac{1}{2}\frac{R_C}{r_e} = \frac{1}{2}\frac{10\text{K}}{90.9} = 55$$

$$R_o = R_C = 10\text{K}\Omega$$

$$R_i = 2\beta r_e = 2 \times 100 \times 90.9 = 18.2\text{K}\Omega$$

The Box Model for the differential-mode amplifier is shown in Figure 17–14B. Note that in this model there are two inputs and one output.

Limits on Input Voltage Difference

The circuits we have looked at in this chapter will not work for just any inputs. The differential input voltage must be limited to the millivolt range. To see this, let's reexamine the numerical example we used previously. Figure 17–15 repeats Figure 17–10, except that here we have not assigned a value to the changes in V_1 and V_2.

Our concern here is what happens to the emitter diode voltages as the input voltages change. We have always assumed that emitter diode voltages are 0.7V. This is not always the case. In fact, the emitter diode voltages can range from approximately 0.4V to 1V. Keep this in mind as you read on.

Let's assume that when the input signals are set to 0V, each emitter diode drop is 0.7V and that

$$V_E = -0.7\text{V}$$

Now we let V_1 increase to 1mV and let V_2 decrease to -1mV. What will happen if we still assume both emitter diodes have a 0.7V—or 700mV—drop? From side one we get

$$V_E = V_1 - 700\text{mV} = 1\text{mV} - 700\text{mV} = -699\text{mV}$$

From side two we get

$$\dot{V}_E = V_2 - 700\text{mV} = -1\text{mV} - 700\text{mV} = -701\text{mV}$$

FIGURE 17–15

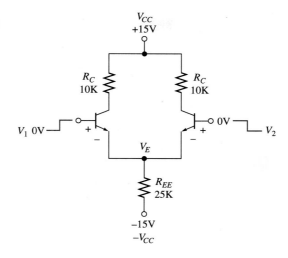

This cannot be! Emitter voltage cannot be two values at once. What actually happens is that the emitter diode voltages change to accommodate the changing input signals. For example, Q_1 might have an emitter-diode turn-on voltage of 0.701V while the emitter-diode turn-on voltage of Q_2 is 0.699V.

Now let the inputs change to $V_1 = +1V$ and $V_2 = -1V$. If we assume that the emitter diode drop for Q_1 is 0.7V, then

$$V_E = 1 - 0.7 = 0.3V$$

If we assume that the emitter diode drop for Q_2 is also 0.7V, then

$$V_E = -1 - 0.7 = -1.7V$$

There is no way for the emitter diodes to accommodate this large a difference in the input voltages; Q_1 will **SATURATE,** and Q_2 will be **CUT-OFF.** In general, the input differential voltage ($V_1 - V_2$) must be less than a few hundred millivolts for these differential amplifiers to work.

REVIEW QUESTIONS

1. How are the two input signals in the differential mode related?
2. What is the change in current in R_{EE} in the differential mode?
3. Why do we say there is a virtual ground at the emitters in the differential mode?
4. How is differential-mode input defined?
5. What is the useful range of $v_{id} = v_1 - v_2$?
6. Draw the differential-mode Box Model. Give the parameter formulas.

17–4 ■ COMMON-MODE GAIN

At the beginning of the chapter, we showed you how the differential amplifier can be used to eliminate the noise in a signal. This happens, we said, because the differential amplifier amplifies only the *difference* in two inputs; signals that are common to both inputs are eliminated.

This is not completely true. Some of the signal that is common to both inputs does get through to the output. This output is the *common-mode output* of the amplifier. In a well-designed differential amplifier, the common-mode gain (A_c) is very small compared to the differential-mode gain. The ratio of differential-mode gain to the common-mode gain is an important specification of a differential amplifier. This ratio is known as the **common-mode rejection ratio** *(CMRR)* and is given by

$$CMRR = \left| \frac{A_d}{A_c} \right|$$

Because the common-mode rejection ratio is a very large number, we usually use the logarithm to express this ratio. (This is similar to expressing earthquake strength with the Richter scale. On the Richter scale a temblor rated at 5 is ten times stronger than a temblor rated at 4.) The logarithmic unit used to express the common-mode rejection ratio is the **decibel.** The decibel measure for the common-mode rejection ratio is given by

$$CMRR_{dB} = 20 \log \left| \frac{A_d}{A_c} \right|$$

Note that the common-mode rejection ratio uses absolute values for the gains. For this measure it does not matter whether or not the amplifier inverts the input signal.

The following table shows the relation between the two measurements:

CMRR	*CMRR_dB*
10	20dB
10^3	60dB
10^5	100dB
10^7	140dB

To give you a better feel for the common-mode rejection ratio, we will derive the common-mode gain for our differential amplifier (Figure 17–16). In this case, we will let both inputs increase by 1V. Note that the differential-mode input here is

$$V_{id} = V_1 - V_2 = 1 - 1 = 0V$$

Because the 1V input is common to both sides, we call this input the **common-mode input:**

$$V_{ic} = V_1 = V_2 = 1V$$

FIGURE 17–16
Common mode

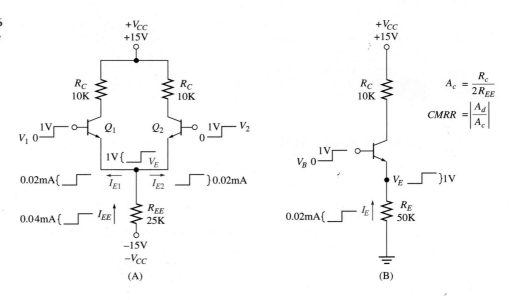

(A)

(B)

Remember, to find a gain we look at the changes in transistor currents and voltages as the input voltage changes.

Consider the regular common-emitter amplifier shown in Figure 17–16B. A 1V change in the base voltage creates a 1V change at the emitter. This 1V change at the emitter creates a change in the emitter current of

$$i_e = \frac{v_e}{R_E} = \frac{1}{50\text{K}} = 0.02\text{mA}$$

As we know, these changes lead to the common-emitter gain of

$$A_{CE} = -\frac{R_C}{R_E} = -\frac{10\text{K}}{50\text{K}} = -0.2$$

where we can ignore the AC emitter resistance because R_E is so large.

With the common-emitter as a background, let's return to the differential amplifier in Figure 17–16A. We want to find the changes that occur in the emitter currents when both inputs increase by 1V. We expect the emitter voltage will also increase by 1V (i.e., the emitter voltage will change from -0.7V to 0.3V). This will lead to a change in the current in R_{EE} of

$$i_{R_{EE}} = \frac{v_e}{R_{EE}} = \frac{1}{25\text{K}} = 0.04\text{mA}$$

Because of the circuit symmetry, the two emitter currents must be equal and must equal one half of the current in R_{EE}. That is

$$i_{e1} = i_{e2} = \frac{i_{R_{EE}}}{2} = \frac{0.04\text{mA}}{2} = 0.02\text{mA}$$

If you compare the emitter currents in the differential amplifier to the emitter current in the common emitter (Figure 17–14B, p. 540), you can see that an R_{EE} of 25KΩ creates the same emitter current as a common emitter R_E of 50KΩ. That is, the effective emitter resistance of the differential amplifier is

$$R_{E\text{effective}} = 2R_{EE}$$

So the common-mode gain is given by

$$A_c = -\frac{R_C}{2R_{EE}}$$

For the numbers in this example

$$A_c = -\frac{R_C}{2R_{EE}} = -\frac{10\text{K}}{50\text{K}} = -0.2$$

You can see that the common-mode gain is very small, but not exactly 0.

Earlier we found the differential-mode gain for this circuit was

$$A_d = \frac{1}{2}\frac{R_C}{r_e} = \frac{1}{2}\frac{10\text{K}}{90.9} = 55$$

The common-mode rejection ratio for this circuit is, therefore

$$CMRR = \left|\frac{A_d}{A_c}\right| = \frac{55}{0.2} = 275$$

Note that for the *CMRR* we do not include the sign of A_c. In decibels, the common-mode rejection ratio is

$$CMRR_{dB} = 20 \log \left| \frac{A_d}{A_c} \right| = 20 \log \frac{55}{0.2} = 48.8 \text{dB}$$

EXAMPLE 17–5
Common-Mode Gain

The differential amplifier of Example 17–4 is repeated in Figure 17–17. Find the common-mode gain and the common-mode rejection ratio.

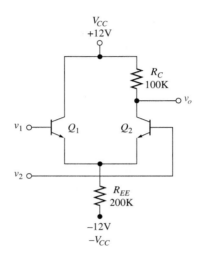

FIGURE 17–17

Solution The common-mode gain is given by

$$A_c = -\frac{R_C}{2R_{EE}} = -\frac{100\text{K}}{400\text{K}} = -0.25$$

In Example 17–4 we found that the differential-mode gain was

$$A_d = 54.4$$

so the common-mode rejection ratio is

$$CMRR = \frac{54.4}{0.25} = 218$$

or

$$CMRR_{dB} = 20 \log CMRR = 46.8 \text{dB}$$

DRILL EXERCISE

A differential amplifier has a differential-mode gain of $A_d = 200$ and a common-mode rejection ratio of $CMRR = 500$. What is the common-mode gain? What is the common-mode rejection ratio in decibels?

Answer $A_c = -0.4$, $CMRR_{dB} = 54\text{dB}$

17–5 ■ THE GENERAL DIFFERENTIAL AMPLIFIER

So far we have only considered two special cases of inputs: differential mode and common mode. In both of these cases, the two inputs had a special correlation with each other. We now consider the general case, in which the two inputs are *not* related (Figure 17–18A).

FIGURE 17–18
General differential amplifier

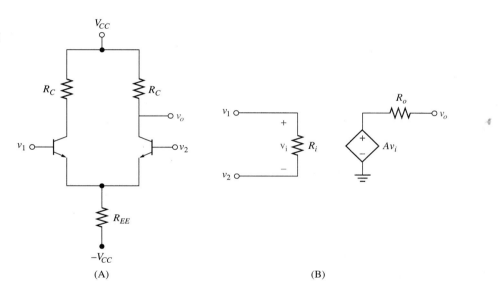

(A)   (B)

In the general case the actual output is a combination of the differential-mode and common-mode outputs. Because the common-mode gain is so small, we always assume that the general case output is the same as the differential-mode output. Therefore, the Box Model of the differential amplifier for the general case (Figure 17–18B) is the same as the Box Model for the differential-mode case. That is

$$v_i = v_1 - v_2$$

$$A = \frac{1}{2}\frac{R_C}{r_e} \qquad v_o = A(v_1 - v_2)$$

$$R_o = R_C$$

$$R_i = 2\beta r_e$$

(Note that we have dropped the subscript "d" from the symbol for the gain.)

To see how this works, let's consider the circuit of Figure 17–19A (p. 546), where the base of Q_2 is grounded and a signal source is connected to Q_1. The first question you may be asking is how can we get an output from Q_2 when its base is grounded. The answer is given in the AC model of Figure 17–19B (p. 546).

The key to analyzing this circuit is the fact that the emitter connection is *not* a virtual ground. When the input increases, the emitter voltage (v_e) also increases. This emitter volt-

FIGURE 17–19

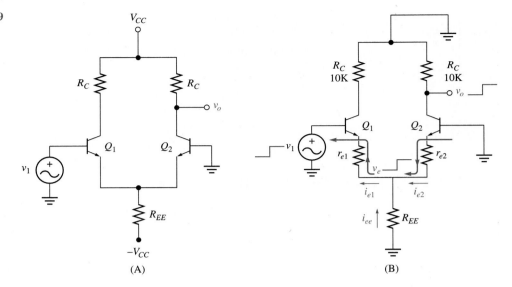

(A)   (B)

age will draw a current through R_{EE}, i_{ee}, and through the base of Q_2 (i_{e2}). The emitter current in Q_2 produces a collector current in Q_2, which, in turn, creates an output voltage at the collector of Q_2.

We can now demonstrate why the general case behaves approximately the same as the differential-mode case. Summing the currents at the emitter connection, we see that

$$i_{e1} = i_{ee} + i_{e2}$$

Because R_{EE} is much greater than r_{e2}, i_{ee} will be very small. If we neglect this current, we get

$$i_{e1} \approx i_{e2}$$

This is approximately the same result that exists between the emitter currents in the differential-mode case, justifying our statement that the general case is similar to the differential-mode case.

EXAMPLE 17–6
General
Differential
Amplifier
Response

For the differential amplifier circuit of Figure 17–20

(a) Find the DC bias levels.
(b) Find the Box Model parameters (ignore the source and load resistances).
(c) Find the common-mode rejection ratio.
(d) Draw the Box Model, connect the source and load, and find the total system gain.

Solution
(a) With the bases grounded

$$V_E = -0.7V$$

$$I_{EE} = \frac{V_{CC} - 0.7}{R_{EE}} = \frac{11.3}{500K} = 0.0226\text{mA}$$

$$I_{E1} = I_{E2} = \frac{I_{EE}}{2} = 0.0113\text{mA}$$

If the transistors are **ACTIVE**

$$I_{C1} \approx I_{E1} = 0.0113\text{mA} \quad \text{and} \quad I_{C2} \approx I_{E2} = 0.0113\text{mA}$$

To check the states of the transistors, we find the collector voltages:

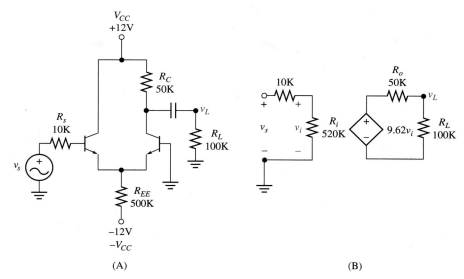

FIGURE 17–20

(A) (B)

$$V_{C1} = V_{CC} = 12V$$
$$V_{C2} = V_{CC} - R_C I_{C2} = 12 - 50K \times 0.0131mA = 11.4V$$

Because the collector voltages are greater than the emitter voltage, both transistors are **ACTIVE.**

(b) The AC emitter resistance for each transistor is

$$r_e = \frac{26mV}{0.01mA} = 2.6K\Omega$$

The Box Model parameters are

$$A = \frac{1}{2}\frac{R_C}{r_e} = \frac{1}{2}\frac{50K}{2.6K} = 9.62$$
$$R_o = R_C = 50K\Omega$$

and

$$R_i = 2\beta r_e = 520K\Omega$$

(c) To find the common-mode rejection ratio, we first find the common-mode gain:

$$A_c = -\frac{R_C}{2R_{EE}} = -\frac{50K}{1000K} = -0.05$$

The common-mode rejection ratio is

$$CMRR = \left|\frac{A}{A_c}\right| = \frac{9.62}{0.05} = 192 \quad \text{or} \quad CMRR_{dB} = 20 \log CMRR = 46dB$$

(d) The Box Model, along with the input source and load, is shown in Figure 17–20B. Note that one input is grounded. The system gain is found from

$$v_i = \frac{R_i}{R_i + R_s} v_s = \frac{520K}{520K + 10K} v_s = 0.981v_s$$

and

$$v_o = \frac{R_L}{R_L + R_o} Av_i = \frac{100K}{100K + 50K} \times 9.62 \times 0.981v_s = 6.29v_s$$

The system gain is

$$A_{sys} = \frac{v_o}{v_i} = 6.29$$

DRILL EXERCISE

Redraw the circuit in Figure 17–20 (p. 547), changing the supply voltages to ±24V. Find the DC bias levels, the Box Model parameters, the common-mode rejection ratio, and the total system gain.

Answer $I_{e1} = I_{E2} = 0.0233\text{mA}$, $V_{C1} = 24\text{V}$, $V_{C2} = 22.8\text{V}$
$r_e = 1.12\text{K}\Omega$, $A = 22.3$, $R_o = 50\text{K}\Omega$, $R_i = 226\text{K}\Omega$
$CMRR = 446 = 53\text{dB}$, $A_{sys} = 14.2$

REVIEW QUESTIONS

1. For the differential mode, how are the input voltages related?
2. For the common mode, how are the input voltages related?
3. In the general case the output voltage is a combination of what two special-case responses?
4. How can we approximate the general case Box Model? Why?
5. When, and only when, is the emitter connection considered a virtual ground?
6. Even if the base of Q_2 is grounded, there is still an output at the collector of Q_2. Why?
7. Draw the Box Model for the general differential amplifier response.

17–6 ■ INTEGRATED CIRCUIT DIFFERENTIAL AMPLIFIERS

The **operational amplifier,** or Op-Amp, is an integrated circuit that is essentially a very high-gain differential amplifier. Figure 17–21 shows the schematic of an Op-Amp, which has two inputs, one output, and positive and negative voltage supplies. The output voltage is given by

$$v_o = A(v_2 - v_1) = Av_i$$

where v_i is the input differential voltage. The most common Op-Amp, the 741, has a differential gain of approximately

$$A = 50,000!$$

FIGURE 17–21
Operational amplifier

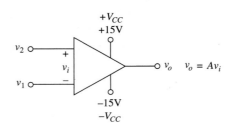

A gain this high is really not very useful. Consider that the maximum output voltage of the Op-Amp is the supply voltage. For our circuit, the maximum output voltage is

$$v_{omax} = 15V$$

Because $v_{omax} = Av_{imax}$, the maximum allowable input voltage is

$$v_{imax} = \frac{v_{omax}}{A} = \frac{15}{50,000} = 0.3\text{mV}$$

Because of the limitation on the input differential voltage, the Op-Amp is rarely used by itself. In Chapters 23–25 we discuss the Op-Amp and its applications in some detail.

The **instrumentation amplifier** is another integrated circuit differential amplifier. Unlike the very-high-gain Op-Amp, instrumentation amplifiers come in fixed or variable gains that typically range from 1 to 10,000.

Figure 17–22 (p. 550) shows the Burr Brown 1NA101 instrumentation amplifiers. Let's examine the simplest of the devices shown, the TO-100 package. The differential input is connected between pin 5 and pin 10. The output is taken from pin 8. The plus and minus supply voltages (note the use of $\pm V_{CC}$) are connected at pins 9 and 6, respectively. The ground connection is pin 7.

The gain of this instrumentation amplifier is set by connecting an external resistor (R_G) between pins 1 and 4. The differential gain is given by

$$A = 1 + \frac{40\text{K}}{R_G}$$

and can be set anywhere between 1 and 1000.

When we set both input voltages to 0V, we expect to see 0V at the output. In fact, in any practical device there will always be a small output voltage, even when there is no input voltage. This voltage is known as the **output offset voltage.** We can set the output voltage back to 0V by applying a small input voltage, known as the **input offset voltage.**

Many integrated circuits provide external terminals specifically designed for nulling the output offset voltage—that is, setting the output voltage to 0V when there is no input voltage. On the TO-100 1NA101, the offset adjustment terminals are pins 2 and 3.

Figure 17–23 (p. 551) shows data for several instrumentation amplifiers. An explanation of some of the data follows:

Column 3 **Gain Range:** This is the differential gain for the device. Some devices have gains that can be continuously set from 1 to 10,000, or 1 to 1000; others have gains that can be set to one of several fixed levels.

Column 4 **Gain Error:** This is a measure of the device's precision when the gain has been set to 100. A 0.5% error means that the actual gain might vary from 99.5 to 100.5.

Column 5 **Gain Drift:** This is a measure of how much the gain changes as the temperature changes.

Column 6 **Non-Linearity:** If the output of an amplifier is $v_o = Av_i$, then a plot of v_o vs. v_i will be a straight line. Any deviation of the actual plot of v_o vs. v_i from a straight line is a non-linearity.

Column 7 **CMR:** This is the common-mode rejection ratio in dB ($CMRR_{dB}$), found when the differential gain is set to 10.

Column 9 **Dynamic Response:** This is the useful frequency range of the amplifier when the gain is set to 100. In general, as you increase the gain, you decrease the dynamic range.

FIGURE 17–22

INA101

High Accuracy
INSTRUMENTATION AMPLIFIER

FEATURES

- LOW DRIFT: 0.25μV/°C max
- LOW OFFSET VOLTAGE: 25μV max
- LOW NONLINEARITY: 0.002%
- LOW NOISE: 13nV/√Hz
- HIGH CMR: 106dB AT 60Hz
- HIGH INPUT IMPEDANCE: $10^{10}\Omega$
- 14-PIN PLASTIC AND CERAMIC DIP SOL-16, TO-100 PACKAGES

APPLICATIONS

- STRAIN GAGES
- THERMOCOUPLES
- RTDs
- REMOTE TRANSDUCERS
- LOW-LEVEL SIGNALS
- MEDICAL INSTRUMENTATION

DESCRIPTION

The INA101 is a high accuracy instrumentation amplifier designed for low-level signal amplification and general purpose data acquisition. Three precision op amps and laser-trimmed metal film resistors are integrated on a single monolithic integrated circuit.

The INA101 is packaged in TO-100 metal, 14-pin plastic and ceramic DIP, and SOL-16 surface-mount packages. Commercial, industrial and military temperature range models are available.

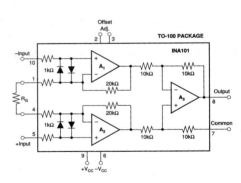

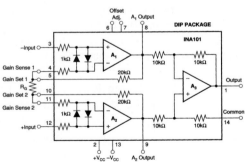

International Airport Industrial Park • Mailing Address: PO Box 11400 • Tucson, AZ 85734 • Street Address: 6730 S. Tucson Blvd. • Tucson, AZ 85706
Tel: (602) 746-1111 • Twx: 910-952-1111 • Cable: BBRCORP • Telex: 066-6491 • FAX: (602) 889-1510 • Immediate Product Info: (800) 548-6132

Source: Reprinted with permission of Burr-Brown Corp.

FIGURE 17–23

INSTRUMENTATION AMPLIFIERS

Boldface = NEW

Description	Model	Gain Range	Gain Error G=100 25°C, max(%)	Gain Drift G=100 (ppm/°C)	Non-Linearity G=100 max(%)	Input Parameters CMR[3] min(dB)	Input Parameters Offset Voltage vs Temp max(µV/°C)	Dynamic Response G=100 −3dB BW (kHz)	Temp Range[1]	Pkg	Page No.
Very High Accuracy	INA114	1–10,000[2]	0.5	25	±0.002	96	±0.25 + 5/G	10	Ind	DIP, SOIC	4.74
	INA115	1–10,000[2]	0.5	25	±0.002	96	±0.25 + 5/G	10	Ind	SOIC	4.87
	INA131	100	0.024	10	±0.002	110	±0.25	70	Ind	DIP, SOIC	4.135
	INA120	1, 10, 100, 1000	0.5	30	±0.01	96	±0.25 ± 10/G	20	Ind	DIP	4.125
	INA104	1–1000[2]	0.15	22[3]	±0.003	96	±0.25 ± 10/G	25	Ind	DIP	A
	INA101	1–1000[2]	0.1	22[3]	±0.003	96	±0.25 ± 10/G	25	Ind, Mil	DIP TO-100, DIP	4.4
Electrometer Input I_s = 10fA typ	INA116	1–10,000[2]	0.5	25	±0.005	90	±5 ± 20/G	70	Ind	DIP/SOIC	4.98
Dual	INA2128	1–10,000[2]	0.5	10	±0.002	97	±0.75 ± 20G	200	XInd	DIP/SOIC	4.145
	INA2141	10, 100	0.15	10	±0.002	97	1[7]	200	XInd	DIP/SOIC	4.147
Low Quies-cent Power	INA118	1–10,000	0.5	25	±0.002	97	±0.5 + 20/G	100	Ind	DIP, SOIC	4.114
	INA102	1, 10, 100, 1000	0.15	15	±0.02	90	±2 ± 5/G	3	Com, Ind	DIP, SOIC	4.10
Low Noise, Low Distortion	INA103	1–1000[2]	0.1	25	±0.004	100	±0.5 + 10/G typ	800	Ind	DIP	4.22
		1–1000[2]	0.25	25	±0.010	90	±0.5 + 20/G typ	800	Com	DIP, SOIC	
Fast Settling FET Input	INA110	1, 10, 100, 200, 500	0.1	20	±0.01	96	±2 + 50/G	470	Ind	DIP, SOIC	4.52
	INA111	1–10,000[2]	0.5	25	±0.01	96	±5 + 100/G	450	Ind	DIP, SOIC	4.63
Unity-Gain Difference Amp	INA105	1 V/V, fixed	0.01[3]	5	±0.001[3]	86[4]	10	1000[3]	Ind	TO-99, DIP SOIC	4.34
	3627	1 V/V, fixed	0.01[3]	5	±0.001[3]	100	20	800[3]	Ind	TO-99	A
Gain of 10 Diff. Amp	INA106	10 V/V, fixed	0.025	10	±0.001	100[4]	0.20	500	Ind	DIP, SOIC	4.46
High Com. Mode Volt. Diff. Amp (200VDC CMV)	INA117	1 V/V, fixed	0.02	10	±0.001	86[4]	20	200[3]	Ind	TO-99, DIP SOIC	4.100
4–20mA Loop Receiver	RCV420	4–20mA in 0–5V Out	0.05	25	±0.002	86	25[5]	150	Com, Ind	DIP	4.190

NOTES: (1) Com = 0°C to +70°C, Ind = −25°C to +85°C, Mil = −55°C to +125°C. (2) Set with external resistor. (3) Unity-gain. (4) No source imbalance. (5) DC to 60Hz, Gain = 10, (or specified gain of device). (6) RTO. (7) G = 100.
"A" indicates a product that is not included in the 1995 Data Books—contact factory for data sheet.

Source: Reprinted with permission of Burr-Brown Corp.

EXAMPLE 17–7
Instrumentation Amplifiers

Use the data sheet in Figure 17–23 to answer the following questions:

(a) What is the common-mode gain for the 1NA101?
(b) Which devices can be used to produce a differential gain of 500?
(c) Which device has a fixed gain of 10 and can be used up to 500KHz?

Solution

(a) When the gain is set to 10, the 1NA101 has a common-mode rejection ratio of

$$CMRR_{dB} = 96dB$$

To find the common-mode gain, we must first convert this dB measure to find *CMRR*. Because

$$CMRR_{dB} = 20 \log CMRR$$

We divide by 20 and take the anti-log to get

$$CMRR = 10^{CMRR_{dB}/20}$$

So for this device

$$CMRR = 10^{96/20} = 10^{4.8} = 63,100$$

Now, we know that

$$CMRR = \frac{A_d}{A_c}$$

so

$$A_c = -\frac{A_d}{CMRR} = -\frac{10}{63,100} = -1.58 \times 10^{-4}$$

(b) The following devices have variable gains that can be set to 500: 1NA114, 1NA115, 1NA104, 1NA101, 1NA116, 1NA2128, 1NA118, 1NA103, 1NA111. The 1NA110 has a fixed gain that can be set to 500.

(c) The 1NA106 has a fixed gain of 10 and a dynamic range of 500KHz.

REVIEW QUESTIONS

1. What is an operational amplifier?
2. What is the differential gain of a typical operational amplifier?
3. Why is a very high gain a problem?
4. What is an instrumentation amplifier?
5. On a data sheet, what is the Gain Error?
6. On a data sheet, what is the Dynamic Response?
7. Given $CMRR_{dB}$, how do you find $CMRR$?
8. Given the differential gain and $CMRR$, how do you find the common-mode gain?

17–7 ■ TROUBLESHOOTING

We will use the circuit of Figure 17–24 to discuss DC Troubleshooting of differential amplifiers. The Estimated Voltages are shown in Figure 17–24A. We will assume that all connections are good and that any discrepancy between Estimated Voltages and Measured Voltages are due to transistor failure. Because of the symmetry of these circuits, we will only discuss failures of Q_1.

Open Failures

Open Base or Open Emitter   Figure 17–24B shows the Measured Voltage values we would get if either the base or emitter failed open. In either case, Q_1 will be **CUT-OFF,** and there will be no currents in this transistor. As expected, we see the supply voltage at the collector of Q_1. Note that the Measured Voltages at the base and the emitter do not show any discrepancy. The reason for this is that the base is grounded and, therefore, held at 0V. Because the emitter of Q_1 is connected to the emitter of Q_2, the emitter voltage looks OK.

The failure of Q_1 also affects the Measured Voltage at the collector of Q_2. This is because all of I_{EE} now goes to Q_2, increasing I_{C2}, which increases the voltage drop across the collector resistor. This results in a decreased Measured Voltage at the collector of Q_2.

FIGURE 17–24

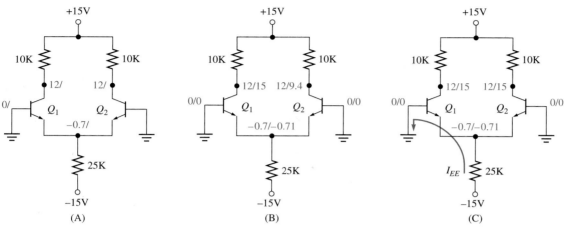

(A) (B) (C)

Open Collector Figure 17–24C shows the Measured Voltages if the collector of Q_1 has failed open. With no collector current in Q_1, V_{C1} will equal the supply voltage. Even though the collector has failed open, the emitter diode of Q_1 is still **ON.** This diode provides a low-resistance path to ground for I_{EE}. Therefore, no current goes to Q_2, so Q_2 is in **CUT-OFF.** This explains why we see the supply voltage at the collector of Q_2.

Short Failures

Base-Emitter Short Figure 17–25A shows the results due to a base-emitter short. Because the emitter is shorted to the base, which is connected to ground, we see 0V at the emitter. With 0V across their emitter diodes, both Q_1 and Q_2 are in **CUT-OFF,** so the supply voltage appears at both collectors.

FIGURE 17–25

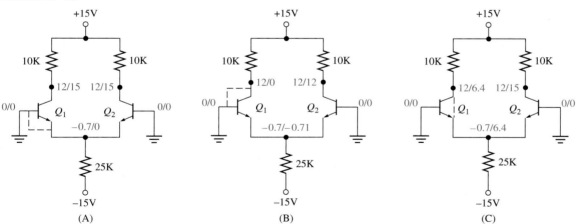

(A) (B) (C)

Base-Collector Short Figure 17–25B shows the results due to a short between the collector and base. Because the collector is now tied to ground, we measure 0V at the collector of Q_1. In this case, the Measured Voltages at Q_2 show no discrepancy.

Collector-Emitter Short Figure 17–25C shows the results of a collector-to-emitter short. This short creates the same conditions as a transistor in **SATURATION.** That is, we see that the Measured Voltages at Q_1's collector and emitter are the same. The large positive voltage at the emitter turns Q_2's emitter diode **OFF.** This puts Q_2 into **CUT-OFF,** resulting in the collector voltage seen.

EXAMPLE 17–8

Troubleshooting
Differential
Amplifier Circuits

Figure 17–26 shows several examples of DC Troubleshooting of differential amplifier circuits. Determine the likely cause of failure in each case. Assume that all connections are good and that any discrepancy is due to a failure in Q_1.

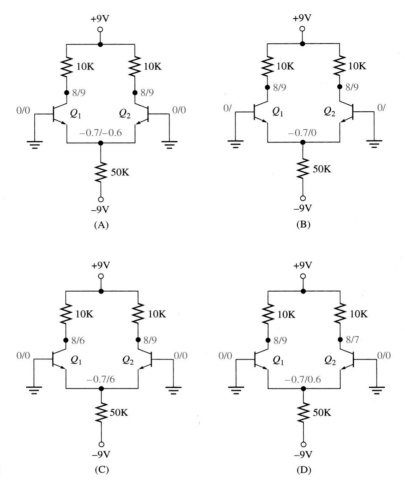

FIGURE 17–26

Solution

(a) Because both collectors are at the supply voltage, we can conclude that Q_1 has a failed open collector.

(b) Because the Measured Voltage at the emitter connection is 0V, the emitter of Q_1 must be shorted to the base of Q_1.

(c) Because the Measured Voltages at the collector and emitter of Q_1 are the same, we conclude that there is a collector-emitter short in Q_1.

(d) The Measured Voltage at the collector of Q_1 shows that there is no collector current. Because the Measured Voltage at the collector of Q_2 is lower than expected, I_{C2} must be greater than expected. This would happen if there is an open in the base or emitter of Q_1.

Troubleshooting Integrated Circuit Amplifiers

Troubleshooting an integrated circuit amplifier is very straightforward.

☐ Check the connections between the supply pins on the chip and the bias supply voltage. You can do this with an ohmmeter (be sure all voltage sources are disconnected), or you can check the voltages at the supply pins with a voltmeter.

☐ Check the connection between the ground pin on the chip and the circuit ground. Once you disconnect all of the voltage sources, check the ground connection with an ohmmeter.

☐ Check the connection between the input pins on the chip and the signal sources. With the signal source and supply voltages disconnected, check the signal connections with an ohmmeter. With all source and supply voltages turned on, use an oscilloscope to determine if the input signal is reaching the chip pins.

☐ Check the connections of all other external circuitry.

☐ If all connections are good, the integrated circuit has probably failed.

REVIEW QUESTIONS

1. What are the consequences of a failed open base?
2. What are the consequences of a failed open emitter?
3. What are the consequences of a failed open collector?
4. What are the consequences of a base-emitter short?
5. What are the consequences of a base-collector short?
6. What are the consequences of a collector-emitter short?
7. What is the procedure for troubleshooting an integrated circuit amplifier?

SUMMARY

■ The differential amplifier amplifies the difference in two input voltages:

$$v_o = A(v_1 - v_2)$$

■ The differential-mode ($v_1 = -v_2$) Box Model parameters are

$$A_d = \frac{1}{2}\frac{R_C}{r_e}$$

$$R_o = R_C$$

$$R_i = 2\beta r_e$$

■ The common-mode ($v_1 = v_2$) gain is

$$A_c = \frac{-R_C}{2R_{EE}}$$

The common-mode rejection ratio is given by

$$CMRR = \left|\frac{A_d}{A_c}\right| \qquad \text{or} \qquad CMRR_{dB} = 20 \log CMRR$$

■ The general Box Model parameters are approximately equal to the differential-mode Box Model parameters.

$$A \approx \frac{1}{2}\frac{R_C}{r_e}$$

$$R_o \approx R_C$$

$$R_i \approx 2\beta r_e$$

■ An operational amplifier is a very high-gain integrated circuit differential amplifier.

■ An instrumentation amplifier is an integrated circuit differential amplifier with a fixed (or variable within a limited range) gain.

PROBLEMS

SECTION 17–1 Introduction

1. You need to measure the indicated voltages in the low-voltage circuit shown in Figure 17–27.
   (a) Which measurements can be made with a single-input amplifier?
   (b) Which measurements must be made with a differential input amplifier?

FIGURE 17–27

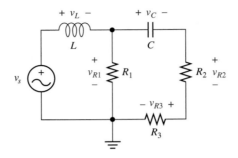

SECTION 17–2 Biasing the Differential Amplifier

2. For the NPN differential amplifier circuit in Figure 17–28

$$V_{CC} = 15\text{V}, R_C = 15\text{K}\Omega, R_B = 10\text{K}\Omega, R_{EE} = 50\text{K}\Omega$$

   (a) Find the bias voltage and currents in the following order:
   V_E
   I_{EE}
   I_{E1} and I_{E2}
   I_{C1} and I_{C2}
   I_{B1} and I_{B2}
   V_{C1} and V_{C2}

   (b) What are the states of the transistors? Explain your answer.

   (c) What are the voltage drops on the base resistors? Were you justified in neglecting this voltage drop? Why?

3. Repeat Problem 2 if the circuit values in Figure 17–28 are

$$V_{CC} = 9\text{V}, R_C = 10\text{K}\Omega, R_B = 2\text{K}\Omega, R_{EE} = 100\text{K}\Omega$$

4. If the supply voltages in Figure 17–28 are $V_{CC} = 15\text{V}$
   (a) Find R_{EE} to set the individual emitter currents to $I_{E_1} = I_{E_2} = 0.1\text{mA}$.
   (b) Find R_C to set the collector voltages to $V_{C_1} = V_{C_2} = 0\text{V}$.

FIGURE 17–28

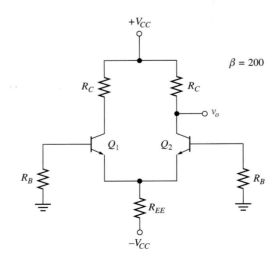

5. For the PNP differential amplifier circuit shown in Figure 17–29

$$V_{CC} = 15V, R_C = 15K\Omega, R_B = 10K\Omega, R_{EE} = 50K\Omega$$

   (a) Find the bias voltage and currents in the following order:

      V_E

      I_{EE}

      I_{E1} and I_{E2}

      I_{C1} and I_{C2}

      I_{B1} and I_{B2}

      V_{C1} and V_{C2}

   (b) What are the states of the transistors? Explain your answer.

   (c) What are the voltage drops on the base resistors? Were you justified in neglecting this voltage drop? Why?

6. Repeat Problem 5 if the circuit values in Figure 17–29 are

$$V_{CC} = 9V, R_C = 10K\Omega, R_B = 2K\Omega, R_{EE} = 100K\Omega$$

FIGURE 17–29

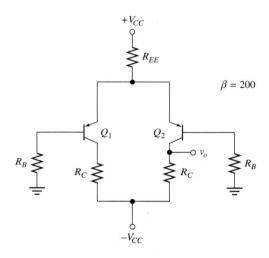

7. Figure 17–30 shows a differential amplifier that is biased with a current source. We want the collector bias voltage to be

$$V_{C2} = 0V$$

Select the proper value for the current source to achieve this voltage.

FIGURE 17–30

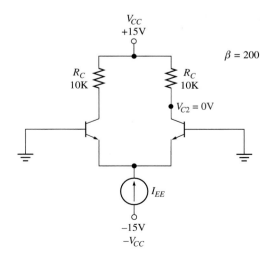

SECTION 17–3 The Differential Mode

8. Find the differential-mode Box Model parameters for the differential amplifier of Figure 17–28 (p. 556), where $V_{CC} = 15V$, $R_C = 15K\Omega$, $R_B = 10K\Omega$, $R_{EE} = 50K\Omega$.

9. Find the differential-mode Box Model parameters for the differential amplifier of Figure 17–28 (p. 556), where $V_{CC} = 9V$, $R_C = 10K\Omega$, $R_B = 2K\Omega$, $R_{EE} = 100K\Omega$.

10. Find the differential-mode Box Model parameters for the differential amplifier of Figure 17–29 (p. 557), where $V_{CC} = 15V$, $R_C = 15K\Omega$, $R_B = 10K\Omega$, $R_{EE} = 50K\Omega$.

11. Find the differential-mode Box Model parameters for the differential amplifier of Figure 17–29 (p. 557), where $V_{CC} = 9V$, $R_C = 10K\Omega$, $R_B = 2K\Omega$, $R_{EE} = 100K\Omega$.

SECTION 17–4 The Common-Mode Gain

12. Find the common-mode gain and common-mode rejection ratio for the differential amplifier of Figure 17–28 (p. 556), where

$$V_{CC} = 15V, R_C = 15K\Omega, R_B = 10K\Omega, R_{EE} = 50K\Omega$$

13. Find the common-mode gain and common-mode rejection ratio for the differential amplifier of Figure 17–28 (p. 556), where

$$V_{CC} = 9V, R_C = 10K\Omega, R_B = 2K\Omega, R_{EE} = 100K\Omega$$

14. Find the output voltage of the differential amplifier of Figure 17–28 (p. 556) if

$$V_{CC} = 15V, R_C = 15K\Omega, R_B = 10K\Omega, R_{EE} = 50K\Omega, \text{ and}$$

   (a) v_1 is a 5mV 1KHz sine wave, and $v_2 = -v_1$.
   (b) v_1 is a 5mV 1KHz sine wave, and $v_2 = v_1$.

15. Find the common-mode gain and common-mode rejection ratio for the differential amplifier of Figure 17–29 (p. 557), where

$$V_{CC} = 15V, R_C = 15K\Omega, R_B = 10K\Omega, R_{EE} = 50K\Omega$$

16. Find the common-mode gain and common-mode rejection ratio for the differential amplifier of Figure 17–29 (p. 557), where

$$V_{CC} = 9V, R_C = 10K\Omega, R_B = 2K\Omega, R_{EE} = 100K\Omega$$

17. Find the output voltage of the differential amplifier of Figure 17–29 (p. 557) if

$$V_{CC} = 15V, R_C = 15K\Omega, R_B = 10K\Omega, R_{EE} = 50K\Omega, \text{ and}$$

   (a) v_1 is a 10mV 1KHz sine wave and $v_2 = -v_1$.
   (b) v_1 is a 10mV 1KHz sine wave and $v_2 = v_1$.

SECTION 17–5 The General Differential Amplifier

18. Given the differential amplifier shown in Figure 17–31
   (a) Find the Box Model parameters, and draw the Box Model.
   (b) Find the common-mode gain and common-mode rejection ratio.
   (c) Connect a source with a 50KΩ source resistance and a 100KΩ load resistor to the Box Model. Find the system gain.

19. In Figure 17–31 we want to deliver 99% of the input signal to the differential amplifier (the base of Q_1). What is the maximum allowable source resistance (R_S)?

20. For the differential amplifier of Figure 17–31
   (a) Choose a new R_{EE} so that the bias values of the collector voltages are

$$V_{C_1} = V_{C_2} = 0V$$

   (b) Repeat Problem 18 with this new value of R_{EE}.

21. For any set of v_1 and v_2, we can define a differential-mode input and a common-mode input as follows:

$$v_{id} = v_1 - v_2$$

and

$$v_{ic} = \frac{v_1 + v_2}{2}$$

The total output voltage is given by

$$v_o = A_d v_{id} + A_c v_{ic}$$

An amplifier that has a differential-mode gain of $A_d = 100$ and a common-mode rejection ratio of $CMRR = 200$. The input voltages are 1KHz sine waves with amplitudes of $V_{1p} = 5\text{mV}$ and $V_{2p} = 2\text{mV}$.

(a) What is v_{id}?   **(b)** What is v_{ic}?   **(c)** What is A_C?

(d) What is the total output voltage?

FIGURE 17–31

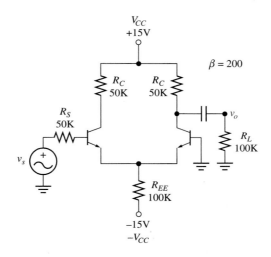

SECTION 17–6   Integrated Circuit Differential Amplifiers

22. An integrated circuit differential amplifier has supply voltages of $\pm15\text{V}$. What is the maximum allowable input differential voltage if the gain of the amplifier is

(a) $A = 10$ **(b)** $A = 1000$ **(c)** $A = 10,000$ **(d)** $A = 200,000$

23. Use Figure 17–23 (p. 551) to find amplifier models that will satisfy the following requirements:

(a) Gain = 100, Dynamic Range = 70KHz.

(b) Gain = 100, Dynamic Range = 800KHz.   **(c)** Non-Linearity less than 0.0025%.

SECTION 17–7   Troubleshooting

24. Determine the likely cause of the DC Troubleshooting results shown in Figure 17–32 (p. 560). Assume that all connections are good and that Q_2 is good.

25. Determine the likely cause of the DC Troubleshooting results shown in Figure 17–33 (p. 560). Assume that all connections are good and that Q_2 is good.

Computer Problems

26. Simulate the circuit of Figure 17–28 (p. 556), and find all bias voltages and currents. Let

$$V_{CC} = 15\text{V}, R_C = 15\text{K}\Omega, R_B = 10\text{K}\Omega, R_{EE} = 50\text{K}\Omega$$

27. Simulate the circuit of Figure 17–29 (p. 557), and find all bias voltages and currents. Let

$$V_{CC} = 15\text{V}, R_C = 15\text{K}\Omega, R_B = 10\text{K}\Omega, R_{EE} = 50\text{K}\Omega$$

FIGURE 17–32

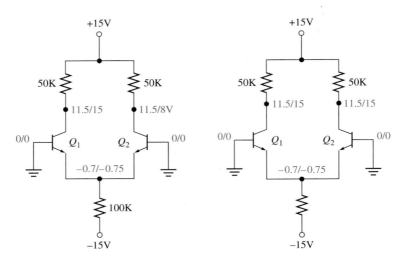

FIGURE 17–33

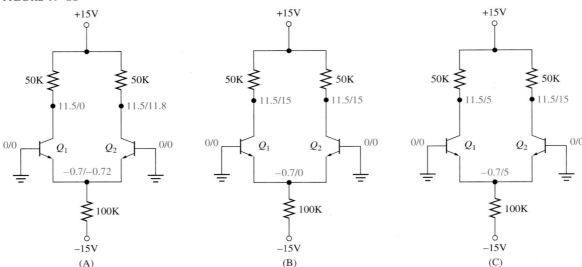

(A) (B) (C)

28. Simulate the circuit of Figure 17–28 (p. 556), and find the output voltage for the following DC inputs:

(a) $v_1 = 1mV$ and $v_2 = -1mV$ (b) $v_1 = 5mV$ and $v_2 = -5mV$

(c) $v_1 = 10mV$ and $v_2 = -10mV$ (d) $v_1 = 15mV$ and $v_2 = -15mV$

(e) $v_1 = 20mV$ and $v_2 = -20mV$

(f) For each set of inputs, what is the input differential voltage?

(g) Does the output stop changing at some point? What does this tell you about the limit on the input differential voltage?

29. If your computer program has a transfer function analysis function, use it on the circuit of Figure 17–28 (p. 556) to find the gain, input resistance, and output resistance. Let

$$V_{CC} = 15V, R_C = 15K\Omega, R_B = 10K\Omega, R_{EE} = 50K\Omega$$

Compare your simulation results with your calculated results.

30. Simulate the circuit of Figure 17–31 (p. 559) (let $C = 30\mu F$). Plot the output for the following inputs:

(a) $v_s = 1mV$ 1KHz sine wave. (b) $v_s = 50mV$ 1KHz sine wave.

(c) $v_s = 100mV$ 1KHz sine wave. (d) $v_s = 500mV$ 1KHz sine wave.

FIGURE 17–34

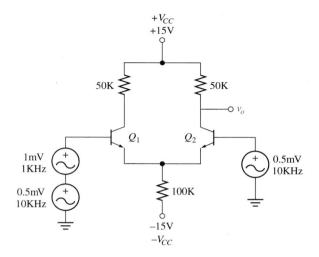

31. Simulate the circuit of Figure 17–34. The two inputs on side 1 simulate a noisy signal. The source on side 2 simulates the noise only. (This is the EEG situation we discussed in the Introduction.)

(a) Plot the voltage at the base of Q_1.   (b) Plot the output voltage.

(c) Discuss the results.

32. In this problem use the computer to confirm the DC Troubleshooting results for a failed open transistor.

(a) Simulate the circuit of Figure 17–24A (p. 553) to establish the Estimated Voltages at the bases, emitter, and collectors.

(b) Open the base of the transistor Q_1, and find the DC bias voltages at the bases, emitter, and collectors.

(c) Open the emitter of the transistor Q_1, and find the DC bias voltages at the bases, emitter, and collectors. Compare the results to part (b).

(d) Open the collector of the transistor Q_1, and find the DC bias voltages at the bases, emitter, and collectors.

33. In this problem, confirm the results of a failed short transistor.

(a) Simulate the circuit of Figure 17–24A (p. 553) to establish the Estimated Voltages.

(b) Connect a short between the base and the emitter of Q_1, and find the DC bias voltages at the bases, emitter, and collectors.

(c) Connect a short between the base and the collector of Q_1, and find the DC bias voltages at the bases, emitter, and collectors.

(d) Connect a short between the emitter and collector of Q_1, and find the DC bias voltages at the bases, emitter, and collectors.

CHAPTER

18

CURRENT SOURCES

CHAPTER OBJECTIVES

To construct a constant current source with an NPN
V_B-control bias circuit.

To construct a constant current source with a PNP
V_B-control bias circuit.

To temperature-compensate the current source.

To construct a current mirror.

To bias a differential amplifier with a current source.

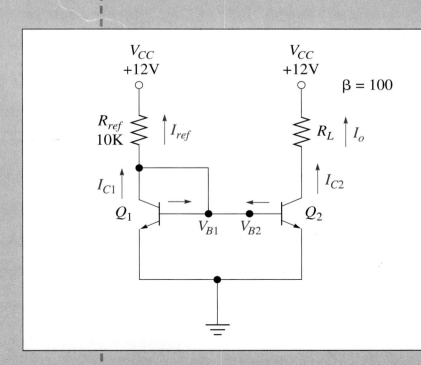

18–1 ■ INTRODUCTION

We're going to take a break from amplifiers for a while. In this chapter we want to discuss the construction and use of DC current sources.

When you think of sources of electrical energy, you probably think of the battery. The battery supplies an approximately constant voltage that is independent of the load to which it is connected. That is, a 9V transistor battery will supply 9V to any transistor circuit. Up to this point, we have powered and biased all of our circuits with battery supply voltages.

We now want to show you how the transistor can be used to create a source of constant, or DC, current, independent of the load. The ideal constant current source is shown schematically in Figure 18–1. The voltage across the load resistor is given by

$$V_L = R_L I_o = R_L \times 5\text{mA}$$

FIGURE 18–1
Constant current source

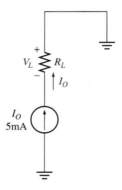

Because the ideal current source always supplies the same amount of current, as the load resistor increases, the load voltage increases, as shown in the table below:

R_L	V_L
100Ω	0.5V
1KΩ	5V
10KΩ	50V
100KΩ	500V
1MΩ	5000V

Whoa!, you may be saying to yourself. Surely not 5000V. Well, yes, if the current source is ideal. Of course, no source is ideal.

Probably you have all (much to your regret) experienced that sinking feeling when the battery in your car runs down and your engine refuses to turn over. A real battery can supply only a finite amount of current. Exceed this amount, and your 12V battery becomes a 10V, 5V, 4V, 3V . . . battery.

Likewise, there is a finite limit to the amount of voltage a real current source can produce. Try to exceed this voltage, and your constant current source is no longer a constant current source. For this reason we will determine the maximum voltage that our current sources can provide.

18–2 ■ THE BASIC NPN TRANSISTOR CURRENT SOURCE

V_B Control

The basic NPN transistor current source is shown in Figure 18–2. This circuit should look very familiar by now. It is our old friend, the V_B control circuit. Because the load is connected to the collector, the load current is the collector current; that is

$$I_o = I_C$$

FIGURE 18–2
NPN constant current source

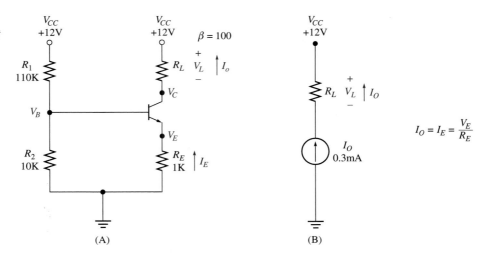

To find the load current, then, we use the usual NPN V_B control procedure:

$$V_B = \frac{R_2}{R_1 + R_2} V_{CC} = \frac{10K}{110K + 10K} 12 = 1V$$

$$V_E = V_B - 0.7 = 0.3V$$

$$I_E = \frac{V_E}{R_E} = \frac{0.3}{1K} = 0.3mA$$

It is very important that this circuit remain in the **ACTIVE** state. If the transistor **SATURATES,** it will no longer act as a constant current source. If the transistor is **ACTIVE**

$$I_C \approx I_E = 0.3mA$$

and the load current is, therefore

$$I_o = I_C = 0.3mA$$

This transistor circuit acts as the constant current source shown schematically in Figure 18–2B. Note that even though we use the transistor to provide a constant current, we still need the V_{CC} battery to make everything work.

We repeat that this transistor circuit will only supply a constant current if the transistor remains in the **ACTIVE** state. Remember, for the NPN transistor to be **ACTIVE**

$$V_C > V_E$$

So, for this current source the requirement is that $V_C > 0.3V$. To find V_C, we first find the load voltage:

$$V_L = R_L I_o = R_L \times 0.3mA$$

Then we find V_C from

$$V_C = V_{CC} - V_L = 12 - R_L \times 0.3mA$$

The load and collector voltages for several load resistors are tabulated below. We also show the state of the transistor for each load.

R_L	V_L	V_C	State
10KΩ	3V	9V	ACTIVE
20KΩ	6V	6V	ACTIVE
30KΩ	9V	3V	ACTIVE
40KΩ	**12V**	**0V**	**SATURATED**

Because the transistor saturates when $R_L = 40$KΩ, for this load the transistor current is *not* constant and is *not* 0.3mA. This circuit will only act as a constant current source for loads up to approximately 40KΩ. Another way to say this is that our 0.3mA current source can only supply approximately 9V to the load.

EXAMPLE 18–1
The NPN Current Source

For the circuit shown in Figure 18–3

(a) Find the load current.
(b) Find the load voltage, and determine the state of the transistor for the following loads. Tabulate your results.

$$R_L = 1\text{K}\Omega,\ 2\text{K}\Omega,\ 4\text{K}\Omega,\ 5\text{K}\Omega$$

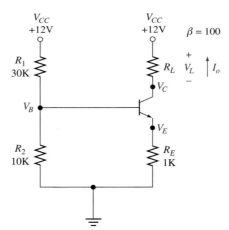

FIGURE 18–3

Solution

(a) To find the load current (that is, the collector current), we assume the transistor is **ACTIVE**:

$$V_B = \frac{R_2}{R_1 + R_2} V_{CC} = \frac{10\text{K}}{30\text{K} + 10\text{K}}\, 12 = 3\text{V}$$

$$V_E = V_B - 0.7 = 3 - 0.7 = 2.3\text{V}$$

$$I_E = \frac{V_E}{R_E} = \frac{2.3}{1\text{K}} = 2.3\text{mA}$$

If the transistor is **ACTIVE**

$$I_o = I_C \approx I_E = 2.3\text{mA}$$

(b) The load voltage is given by

$$V_L = R_L I_o = R_L \times 2.3\text{mA}$$

and the collector voltage is

$$V_C = V_{CC} - V_L = 12 - R_L \times 2.3\text{mA}$$

For the transistor to be **ACTIVE**

$$V_C > 2.3\text{V}$$

The results are tabulated below:

R_L	V_L	V_C	State
1KΩ	2.3V	9.7V	**ACTIVE**
2KΩ	4.6V	7.4V	**ACTIVE**
4KΩ	9.2V	2.8V	**ACTIVE**
5KΩ	**11.5V**	**0.5V**	**SATURATED**

This circuit will not act as a constant current source for the 5KΩ load.

DRILL EXERCISE

Redraw the constant current circuit of Figure 18–3, changing the supply voltage to $V_{CC} = 6$V. Find the load current. Find the state of the transistor and the load voltage, if **ACTIVE,** for $R_L = 3$KΩ, 5KΩ, 7KΩ.

Answer $I_o = 0.8$mA

$$V_L = 2.4\text{V, } \textbf{ACTIVE; } V_L = 4\text{V, } \textbf{ACTIVE; } \textbf{SATURATED}$$

Maximum Load Voltage and Resistance

While it is unlikely that you will ever be required to find the maximum available voltage from a constant current source, it is a worthwhile learning effort. Looking back at the constant current source in Figure 18–2, we see that

$$V_C = V_{CC} - V_L$$

We know that a transistor **SATURATES** when

$$V_C = V_E$$

Comparing these two results, at the edge of **SATURATION** we get

$$V_{CC} - V_L = V_E$$

Because we get the maximum permissible load voltage at the edge of **SATURATION,** the maximum load voltage is found from

$$V_{L\text{max}} = V_{CC} - V_E$$

For the circuit of Figure 18–3, the maximum available load voltage is, therefore,

$$V_{L\text{max}} = V_{CC} - V_E = 12 - 0.3 = 11.7\text{V}$$

Maximum allowable load resistance is found from Ohm's Law:

$$R_{Lmax} = \frac{V_{Lmax}}{I_o}$$

For our circuit

$$R_{Lmax} = \frac{V_{Lmax}}{I_o} = \frac{11.7}{0.3\text{mA}} = 39\text{K}\Omega$$

EXAMPLE 18–2
Maximum Load
Voltage and
Current

For the circuit of Figure 18–3, find the maximum available load voltage and the maximum allowable load resistance.

Solution   From Example 18–1 we found that

$$V_E = 2.3\text{V}$$

and

$$I_o = 2.3\text{mA}$$

The maximum available load voltage is, therefore

$$V_{Lmax} = V_{CC} - V_E = 12 - 2.3 = 9.7\text{V}$$

The maximum allowable load resistor is

$$R_{Lmax} = \frac{V_{Lmax}}{I_o} = \frac{9.7}{2.3\text{mA}} = 4.2\text{K}\Omega$$

DRILL EXAMPLE

Redraw the circuit of Figure 18–3 (p. 566), changing the supply voltage to $V_{CC} = 6\text{V}$. Find the maximum available load voltage and the maximum allowable load resistor.

Answer $V_{Lmax} = 5.2\text{V}, R_{Lmax} = 6.5\text{K}\Omega$

**REVIEW
QUESTIONS**

1. What type of bias circuit is the transistor current source?
2. What transistor current equals the load current?
3. What state must the transistor be in to behave as a current source?
4. What is the maximum available load voltage?
5. What is the maximum allowable load resistance?

18–3 ▪ THE BASIC PNP TRANSISTOR CURRENT SOURCE

Figure 18–4 shows another advantage of having two transistor "flavors" available. The load in the NPN current source (Figure 18–2) must be connected between the collector supply voltage and the transistor. The load in the PNP current source, however, is connected to ground (Figure 18–4). The desired location of the load will determine whether an NPN or a PNP current source will be used.

FIGURE 18–4
PNP constant current source

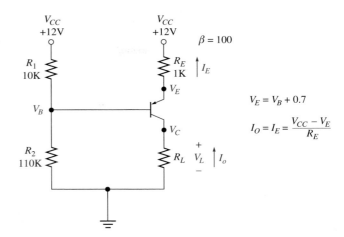

$$V_E = V_B + 0.7$$

$$I_O = I_E = \frac{V_{CC} - V_E}{R_E}$$

V_B Control

The load current is determined by following the V_B control procedure for the PNP transistor. Again, for this circuit to act as a current source, the transistor must be **ACTIVE.** We begin by finding the base voltage:

$$V_B = \frac{R_2}{R_1 + R_2} V_{CC} = \frac{110\text{K}}{10 + 110\text{K}} \, 12 = 11\text{V}$$

The emitter voltage is now 0.7V *above* the base voltage:

$$V_E = V_B + 0.7 = 11 + 0.7 = 11.7\text{V}$$

The emitter current is found from

$$I_E = \frac{V_{CC} - V_E}{R_E} = \frac{12 - 11.7}{1\text{K}} = 0.3\text{mA}$$

If the transistor is **ACTIVE,**

$$I_o = I_C \approx I_E = 0.3\text{mA}$$

The value of the load resistor will determine the load voltage and the state of the transistor. Remember, for the PNP the transistor is **ACTIVE** if

$$V_E > V_C$$

For the PNP current source, the load voltage is the same as the collector voltage, so

$$V_C = V_L = R_L I_o = 1\text{K} \times 0.3\text{mA} = 0.3\text{V}$$

The transistor is, therefore, **ACTIVE.**

Maximum Load Voltage and Resistance

Once again, maximum load voltage is available at the edge of **SATURATION.** At this point collector voltage equals emitter voltage. Because load voltage is equal to collector voltage, the maximum available load voltage for the PNP current source is

$$V_{L\text{max}} = V_E$$

which for this circuit gives us

$$V_{L\text{max}} = V_E = 11.7\text{V}$$

The maximum allowable load resistor for the PNP current source is found from

$$R_{Lmax} = \frac{V_{Lmax}}{I_o}$$

which for our circuit is

$$R_{Lmax} = \frac{V_{Lmax}}{I_o} = \frac{11.7}{0.3\text{mA}} = 39\text{K}\Omega$$

EXAMPLE 18–3
The PNP Current Source

For the current source of Figure 18–5

(a) Find the load current and load voltage.
(b) Find the maximum available load voltage and the maximum allowable load resistance.

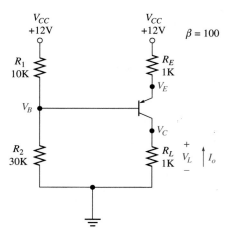

FIGURE 18–5

Solution

(a) We follow the V_B control procedure for the PNP transistor:

$$V_B = \frac{R_2}{R_1 + R_2} V_{CC} = \frac{30\text{K}}{10\text{K} + 30\text{K}} 12 = 9\text{V}$$

$$V_E = V_B + 0.7 = 9 + 0.7 = 9.7\text{V}$$

$$I_E = \frac{V_{CC} - V_E}{R_E} = \frac{12 - 9.7}{1\text{K}} = 2.3\text{mA}$$

If the transistor is **ACTIVE**

$$I_o = I_C \approx I_E = 2.3\text{mA}$$

The load voltage is

$$V_L = R_L I_o = 1\text{K} \times 2.3\text{mA} = 2.3\text{V}$$

(b) The maximum available voltage is given by

$$V_{Lmax} = V_E = 9.7\text{V}$$

The maximum allowable load resistance is given by

$$R_{Lmax} = \frac{V_{Lmax}}{I_o} = \frac{9.7}{2.3\text{mA}} = 4.2\text{K}\Omega$$

DRILL EXERCISE

Redraw the circuit shown in Figure 18–5, increasing the supply voltage to $V_{CC} = 24V$. Find the load current and the maximum available load voltage and the maximum allowable load resistance.

Answer $I_o = 5.3mA$, $V_{Lmax} = 18.7V$, $R_{Lmax} = 3.5K\Omega$

REVIEW QUESTIONS

1. What type of bias circuit is the PNP transistor current source?
2. What is the difference between an NPN and a PNP current source?
3. What transistor current equals the load current?
4. What state must the transistor be in to behave as a current source?
5. What is the maximum available load voltage?
6. What is the maximum allowable load resistance?

18–4 ■ NEGATIVE VOLTAGE SUPPLY CURRENT SOURCES

NPN

Figure 18–6 shows the implementation of an NPN current source with a negative voltage supply. We include this circuit (and the PNP circuit that follows) because it is typically seen in integrated circuit designs. Note that in this circuit the load is connected between the current source and ground.

FIGURE 18–6
NPN current source with negative supply

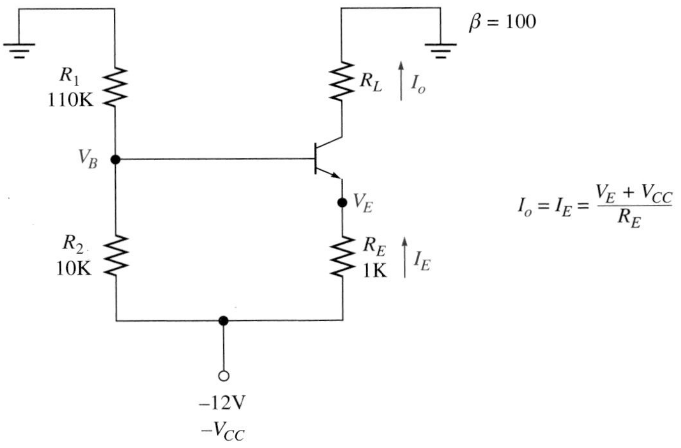

$$I_o = I_E = \frac{V_E + V_{CC}}{R_E}$$

The analysis procedure is still the V_B control procedure. Some care must be exercised, however. Do not just repeat formulas.

The base voltage now is the voltage across R_1. That is

$$V_B = \frac{R_1}{R_1 + R_2}(-V_{CC}) = \frac{110K}{110K + 10K}(-12) = -11V$$

As with all NPN transistors

$$V_E = V_B - 0.7$$

so

$$V_E = V_B - 0.7 = -11 - 0.7 = -11.7V$$

The emitter current is now found from

$$I_E = \frac{V_E - (-V_{CC})}{R_E} = \frac{V_E + V_{CC}}{R_E}$$

so

$$I_E = \frac{V_E + V_{CC}}{R_E} = \frac{-11.7 + 12}{1K} = 0.3\text{mA}$$

Assuming the transistor is **ACTIVE,** the load current is

$$I_o = I_C \approx I_E = 0.3\text{mA}$$

EXAMPLE 18–4

Negative Voltage Supply NPN Current Source

Find the load current for the circuit in Figure 18–7.

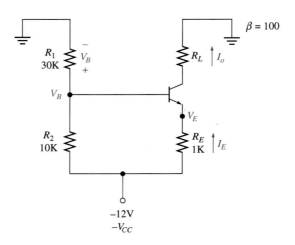

FIGURE 18–7

Solution The base voltage is the voltage across R_1. That is

$$V_B = \frac{R_1}{R_1 + R_2}(-V_{CC}) = \frac{30K}{30K + 10K}(-12) = -9\text{V}$$

$$V_E = V_B - 0.7 = -9 - 0.7 = -9.7\text{V}$$

The emitter current is now found from

$$I_E = \frac{V_E + V_{CC}}{R_E} = \frac{-9.7 + 12}{1K} = \frac{2.3}{1K} = 2.3\text{mA}$$

Assuming the transistor is **ACTIVE,** the load current is

$$I_o = I_C \approx I_E = 2.3\text{mA}$$

DRILL EXERCISE

Redraw the circuit of Figure 18–7, changing the supply voltage to −15V. Find the load current.

Answer $I_o = 3.05\text{mA}$

PNP

The circuit in Figure 18–8 shows a PNP current source that is biased with a negative supply. Here the load is connected between the current source and the supply voltage. Once again, V_B is the voltage across R_1:

$$V_B = \frac{R_1}{R_1 + R_2}(-V_{CC})$$

so

$$V_B = \frac{R_1}{R_1 + R_2}(-V_{CC}) = \frac{10K}{10K + 110K}(-12) = -1V$$

FIGURE 18–8
PNP current source with
negative supply

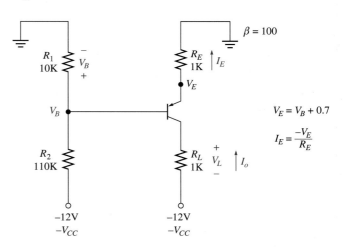

For the PNP, the emitter voltage is 0.7V above the base voltage:

$$V_E = V_B + 0.7$$

so

$$V_E = V_B + 0.7 = -1 + 0.7 = -0.3V.$$

Here, the emitter current is

$$I_E = \frac{0 - V_E}{R_E} = \frac{-V_E}{R_E}$$

so

$$I_E = \frac{-V_E}{R_E} = \frac{0.3}{1K} = 0.3mA$$

If the transistor is **ACTIVE,** the load current is

$$I_o = I_C \approx I_E = 0.3mA$$

EXAMPLE 18–5
**Negative Voltage
Supply PNP
Current Source**

Find the load current for the circuit in Figure 18–9 (p. 574).

Solution   The base voltage is found from

$$V_B = \frac{R_1}{R_1 + R_2}(-V_{CC}) = \frac{10K}{10K + 30K}(-12) = -3V$$

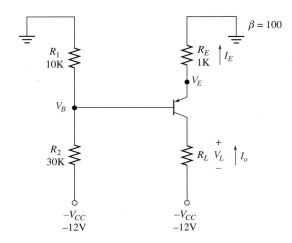

FIGURE 18–9

For the PNP, the emitter voltage is 0.7V above the base voltage:

$$V_E = V_B + 0.7 = -3 + 0.7 = -2.3V$$

The emitter current is

$$I_E = \frac{-V_E}{R_E} = \frac{2.3}{1K} = 2.3mA$$

If the transistor is **ACTIVE,** the load current is

$$I_o = I_C \approx I_E = 2.3mA$$

DRILL EXERCISE Redraw the circuit shown in Figure 18–9, changing the supply source to −24V. Find the load current.

Answer $I_o = 5.3mA$

We have now shown you how to implement the basic current source with either a positive or a negative supply voltage. In each case, by using an NPN or a PNP transistor you can place the load between the current source and ground or between the current source and the supply voltage.

REVIEW QUESTIONS

1. Draw an NPN current source with a negative supply voltage.
2. Where is the load connected in this circuit?
3. What is the base voltage in this circuit?
4. What is the emitter voltage in this circuit?
5. What is the emitter current in this circuit?
6. Draw a PNP current source with a negative supply voltage.
7. Where is the load connected in this circuit?
8. What is the base voltage in this circuit?
9. What is the emitter voltage in this circuit?
10. What is the emitter current in this circuit?

18–5 ■ TEMPERATURE COMPENSATION

Consider the NPN current source shown again in Figure 18–10A. The base voltage is

$$V_B = \frac{R_2}{R_1 + R_2} V_{CC} = \frac{10K}{110K + 10K} 12 = 1V$$

So far, so good. This base voltage can be controlled with a high degree of accuracy. The problem is that emitter voltage depends on emitter diode turn-on voltage. In general terms

$$V_E = V_B - V_{BE(ON)} = 1 - V_{BE(ON)}$$

FIGURE 18–10
Temperature compensation

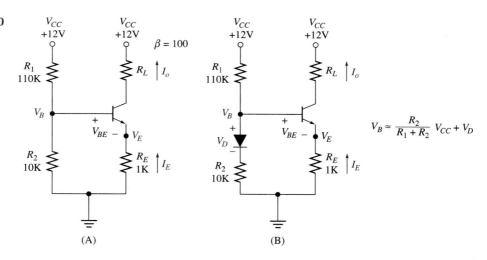

(A)

(B)

$$V_B \approx \frac{R_2}{R_1 + R_2} V_{CC} + V_D$$

We have usually assumed that the emitter diode turn-on voltage is $V_{BE(ON)} = 0.7V$. In reality, as discussed in Chapter 17, this value is an approximation. More importantly, emitter diode turn-on voltage is very sensitive to changes in ambient temperature and can range from 0.4V to 1V.

Because load current equals emitter current, and because emitter current depends on emitter voltage, changes in $V_{BE(ON)}$ will cause changes in the load current. Below, we tabulate the emitter current that results in Figure 18–10A as $V_{BE(ON)}$ varies:

$V_{BE(ON)}$	V_E	I_E
0.4V	0.6V	0.6mA
0.6V	0.4V	0.4mA
0.8V	0.2V	0.2mA
0.9V	0.1V	0.1mA

You can see that for reasonable variations in the emitter diode turn-on voltage, the emitter current, and hence the load current, has a 6:1 variation. This is completely unacceptable in a constant current source. Your radio might work at the beach on a summer day but wouldn't work in the winter on a ski slope. This situation would force you to buy a different radio for each indoor or outdoor temperature. Very expensive.

Figure 18–10B shows a fairly simple fix to the problem of temperature variation. You can see that we have inserted a diode into the base circuit. Although the fix is a simple circuit addition, the analysis of this circuit is somewhat involved. We'll lead you through the procedure in Problem 13 at the end of the chapter. For now, we just state that here the base voltage is given by

$$V_B = \frac{R_2}{R_1 + R_2} V_{CC} + \frac{R_1}{R_1 + R_2} V_D$$

where V_D is the diode turn-on voltage. The first part of V_B is familiar to you. The second term is the contribution of the diode.

We can greatly simplify this expression if we assume that $R_1 \gg R_2$. In this case

$$V_B \approx \frac{R_2}{R_1 + R_2} V_{CC} + V_D$$

For our circuit, the base voltage is now

$$V_B \approx \frac{R_2}{R_1 + R_2} V_{CC} + V_D = \frac{10K}{110K + 10K} 12 + V_D = 1 + V_D$$

The emitter voltage then is

$$V_E = 1 + V_D - V_{BE(ON)}$$

You can see that the diode voltage and the base-emitter voltage tend to cancel each other out. The emitter current is

$$I_E = \frac{V_E}{R_E} = \frac{1 + V_D - V_{BE(ON)}}{1K}$$

If $V_D = V_{BE(ON)}$, then

$$I_E = \frac{1}{1K} = 1mA$$

The emitter diode drop has been eliminated from the determination of I_E! Now this voltage can vary all it wants to as temperature changes without affecting the emitter current.

Even if V_D and $V_{BE(ON)}$ are not identical, variations in $V_{BE(ON)}$ will track variations in V_D if the diode and transistor are constructed in the same manner. As shown in the following example, these variations will cancel each other out.

EXAMPLE 18–6

Temperature Compensation

For the current source shown in Figure 18–11

(a) Find the emitter current if

$$V_D = 0.8V \text{ and } V_{BE(ON)} = 0.5V$$

(b) Assume that when the temperature increases, both V_D and $V_{BE(ON)}$ increase by 0.1V. Find the emitter current, and compare to part (a).

Solution

(a) Because R_1 is much greater than R_2, we use the approximation to find V_B:

$$V_B = \frac{R_2}{R_1 + R_2} V_{CC} + V_D = \frac{20K}{280K + 20K} 15 + 0.8 = 1 + 0.8 = 1.8V$$

The emitter voltage is

$$V_E = V_B - V_{BE(ON)} = 1.8 - 0.5 = 1.3V$$

The emitter current is

$$I_E = \frac{V_E}{R_E} = \frac{1.3}{1K} = 1.3mA$$

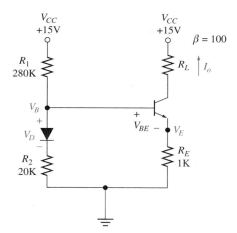

FIGURE 18–11

(b) Now, the semiconductor voltages have changed to $V_D = 0.9$V and $V_{BE(ON)} = 0.6$V. The base voltage is now

$$V_B = 1 + V_D = 1 + 0.9 = 1.9\text{V}$$

The emitter voltage is

$$V_E = V_B - V_{BE(ON)} = 1.9 - 0.6 = 1.3\text{V}$$

You can see that the emitter voltage has not changed. The emitter current, therefore, also does not change:

$$\frac{I_E}{R_E} = \frac{1.3}{1\text{K}} = 1.3\text{mA}$$

For the temperature compensation circuit to work, both the diode and transistor must vary in the same way as temperature changes. There is no way to guarantee this if you just select a diode and transistor arbitrarily. However, we can provide excellent temperature compensation by using two matched transistors. Figure 18–12A shows how we construct a diode by tying the base of a transistor to its collector. We then use this diode-connected transistor in the temperature compensation circuit (Figure 18–12B).

FIGURE 18–12
Diode-connected
transistor

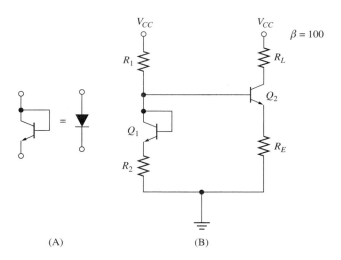

(A) (B)

18–6 ■ THE CURRENT MIRROR

In Figure 18–12B (p. 577) we showed you how to construct a temperature-compensated current source by using a transistor to mimic a diode. It turns out that in that circuit we can eliminate two of the resistors and still have an excellent current source. This is commonly done in integrated circuits to save on resistors. Resistors take up a lot of real estate on a chip, so integrated-circuit designers use the minimum they can get away with.

The circuit in Figure 18–13A is known as a **current mirror.** It is a constant current source that both reduces the number of required resistors and produces an output current that is unaffected by emitter diode turn-on voltages.

FIGURE 18–13
Current mirror

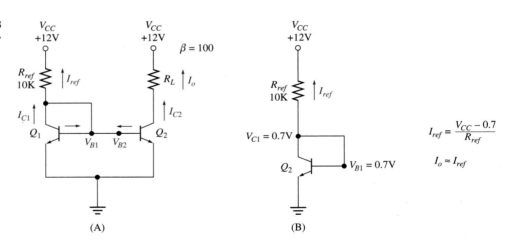

(A)   (B)

You can see that in the current mirror the collector and base of transistor Q_1 are tied together. In Figure 18–13B, we have isolated Q_1. If we assume that the transistor is **ACTIVE** and use our standard assumption that the emitter diode turn-on voltage is 0.7V, then

$$V_{B1} = V_{E1} + 0.7V = 0.7V$$

Because the base and collector of Q_1 are tied together

$$V_{C1} = V_{B1} = V_{BE(ON)} = 0.7V$$

This means that the current in the resistor is given by

$$I_{ref} = \frac{V_{CC} - V_{C1}}{R_{ref}}$$

so

$$I_{ref} = \frac{V_{CC} - V_{C1}}{R_{ref}} = \frac{12 - 0.7}{10K} = 1.13mA$$

What happens to this current if we let the emitter diode turn-on voltage vary? The answer can be seen in the following table:

$V_{BE(ON)}$	I_{ref}
0.4V	1.16mA
0.6V	1.14mA
0.8V	1.12mA
0.9V	1.11mA

Because the small emitter-diode voltage is subtracted directly from the large supply voltage, large variations in $V_{BE(ON)}$ cause negligible variations in the resistor current.

Now, where does most of I_R go? You can see from Figure 18–13A that this current supplies the collector of Q_1 and the bases of Q_1 and Q_2. If we make our standard assumption that base currents can be ignored, we can say that most of I_R comes from the collector of Q_1. That is,

$$I_{C1} \approx I_{ref}$$

For our circuit, then

$$I_{C1} \approx I_{ref} = 1.13\text{mA}$$

Now comes the tricky part. How do we find the current in Q_2, which is also the load current? Nothing you have learned about transistor biasing will help you here. Before we show you the somewhat complicated derivation, we'll give you the answer (kind of like *Jeopardy*).

$$I_o = I_{C2} = I_{C1} \approx I_{ref}$$

You can get a general feel for this relation by considering the base circuits of Q_1 and Q_2. Because the transistors are identical, and because the base circuits of the two transistors are the same, it doesn't take a great leap of faith to assume

$$I_{B1} = I_{B2}$$

Again, because the transistors are identical, if the base currents are the same, then

$$I_{C1} = I_{C2}$$

The exact derivation comes from the defining equation for the collector current of a transistor:

$$I_C = I_R(e^{V_{BE}/V_T})$$

where I_R and V_T are device parameters. If two transistors are matched, then these parameters are the same for both transistors. Now, returning to Figure 18–13A, you can see that

$$V_{B1} = V_{B2} \quad \text{and} \quad V_{E1} = V_{E2}$$

so that

$$V_{BE1} = V_{BE2}$$

Finally, because

$$I_{C1} = I_R(e^{V_{BE1}/V_T}) \quad \text{and} \quad I_{C2} = I_R(e^{V_{BE2}/V_T})$$

and all parameters are the same for both currents

$$I_{C1} = I_{C2}$$

So, for the current mirror in Figure 18–13A (p. 578)

$$I_o = I_{C2} = I_{C1} = I_{ref} = 1.13\text{mA}$$

EXAMPLE 18–7
Negative Supply NPN Current Mirror

Find I_o for the current mirror of Figure 18–14.

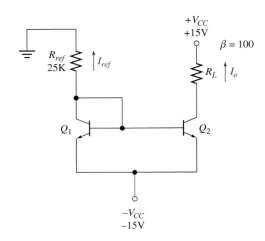

FIGURE 18–14

Solution   In this circuit

$$V_{B1} = V_{E1} + 0.7 = -15 + 0.7 = -14.3\text{V}$$

The collector voltage of Q_1 equals the base voltage. So

$$V_{C1} = V_{B1} = -14.3\text{V}$$

The current in the reference resistor is

$$I_{ref} = \frac{0 - V_{C1}}{25\text{K}} = \frac{14.3}{25\text{K}} = 0.57\text{mA}$$

Finally, the output current approximately equals the reference current:

$$I_o = I_{ref} = 0.57\text{mA}$$

DRILL EXERCISE

Redraw the circuit of Figure 18–14, changing the supply voltage to -9V. Find the output current.

Answer $I_o = 0.33\text{mA}$

EXAMPLE 18–8
PNP Current Mirror

The current mirror in Figure 18–15 is constructed with PNP transistors. Find R_{ref} so that the output current is 1μA.

Solution   We solve this problem assuming that R_{ref} is known. In this circuit

$$V_B = 15 - 0.7 = 14.3\text{V}$$

Because the collector of Q_1 is tied to the base of Q_1

$$V_{C1} = V_{B1} = 14.3\text{V}$$

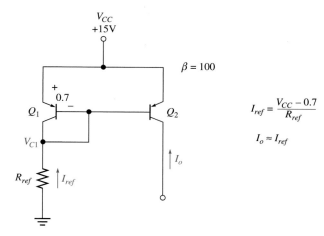

FIGURE 18–15

The current in R_{ref} is

$$I_{ref} = \frac{14.3}{R_{ref}}$$

Because we want an output current of 1μA

$$I_{ref} = I_o = \frac{14.3}{R_{ref}} = 1\mu A$$

Solving for R_{ref} we get

$$R_{ref} = \frac{14.3}{1\mu A} = 14.3 M\Omega$$

DRILL PROBLEM Redraw the circuit of Figure 18–15, changing the supply voltage to 9V. Find the resistor so that the output current is 1mA.

Answer $R_{ref} = 8.3 K\Omega$

REVIEW QUESTIONS

1. Draw an NPN current mirror.
2. Assume Q_1 is the transistor with the base and collector tied together. What is the relation between V_{E1} and V_{B1}?
3. What is the relation between V_{C1} and V_{B1}?
4. What is I_{ref}?
5. What is the relation between the output current and I_{ref}?
6. Give a general justification for this relation.

18–7 ■ CURRENT SOURCE BIASING OF DIFFERENTIAL AMPLIFIERS

We finish this chapter with some examples of differential amplifier circuits that are biased with a current source. We start with the circuit shown in Figure 18–16A (p. 582).

FIGURE 18–16
Differential amplifier biasing

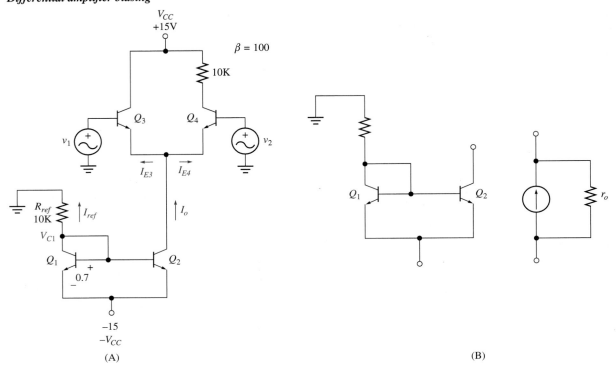

(A)

(B)

DC Bias

In this circuit Q_1 and Q_2 form a current mirror that will provide the bias current to the differential amplifier constructed with Q_3 and Q_4. The reference current is found from

$$V_{C1} = -15 + 0.7 = -14.3\text{V}$$

So

$$I_{ref} = \frac{V_{C1}}{R_{ref}} = \frac{14.3}{10\text{K}} = 1.4\text{mA}$$

Therefore, the differential amplifier bias current is

$$I_o = I_{ref} = 1.4\text{mA}$$

From Chapter 17, we know that the bias current splits equally between Q_3 and Q_4:

$$I_{E1} = I_{E2} = \frac{I_o}{2} = \frac{1.4\text{mA}}{2} = 0.7\text{mA}$$

Box Model

The Box Model parameters are found from the differential-mode analysis. The AC emitter resistance is

$$r_e = \frac{26\text{mV}}{I_E} = \frac{26\text{mV}}{0.7\text{mA}} = 37\Omega$$

The differential-mode gain is

$$A_d = \frac{R_C}{r_e} = \frac{10\text{K}}{37} = 270$$

The differential-mode input resistance is

$$R_i = 2\beta r_e = 7.4\text{K}\Omega$$

The output resistance is

$$R_o = R_C = 10\text{K}\Omega$$

Common-Mode Rejection Ratio

To find the common-mode rejection ratio, we need to find the common-mode gain:

$$A_c = \frac{R_C}{2R_{EE}} = ?$$

We don't know the resistance between the emitter connection and ground *(REE)*. A more detailed look at the output transistor of the current mirror will tell us what we need to know.

Current Source Output Resistance

We have assumed that the current mirror provides us with an ideal current source. In fact, this is not true. All real current sources have output resistance associated with them (Figure 18–16B). A transistor's output resistance is given in its specification sheet; typically, the output resistance is greater than 100KΩ. This gives us, at a minimum

$$R_{EE} = r_o = 100\text{K}\Omega$$

so

$$A_c = -\frac{R_C}{2R_{EE}} = -\frac{10\text{K}}{200\text{K}} = -0.05$$

This gives us a common-mode rejection rate of

$$CMRR = \left| \frac{A}{A_c} \right| = \frac{270}{0.05} = 5400$$

or

$$CMRR_{dB} = 20 \log CMRR = 75\text{dB}$$

EXAMPLE 18–9
Differential Amplifier

For the circuit in Figure 18–16A, we want a differential mode gain of

$$A_d = 100$$

Find R_{ref} to achieve this gain.

Solution We will work this problem backwards. That is, given the desired gain, we first find the required AC emitter resistance:

$$A_d = \frac{R_C}{r_e}$$

so

$$r_e = \frac{R_C}{A_d} = \frac{10\text{K}}{100} = 100\Omega$$

Now, from

$$r_e = \frac{26\text{mV}}{I_E}$$

we can find the required emitter current:

$$I_E = \frac{26\text{mV}}{r_e} = \frac{26\text{mV}}{100} = 0.26\text{mA}$$

The required bias current comes from

$$I_E = \frac{I_{EE}}{2}$$

so

$$I_o = I_{EE} = 2I_E = 0.52\text{mA}$$

Finally, we can find R_{ref} from

$$I_o = I_{ref} = \frac{14.3}{R_{ref}}$$

so

$$R_{ref} = \frac{14.3}{I_{ref}} = \frac{14.3}{0.52\text{mA}} = 27.5\text{K}\Omega$$

DRILL EXERCISE Find R_{ref} in Figure 18–16 (p. 582) so that the differential amplifier has a gain of 500.

Answer $R_{ref} = 5.5\text{K}\Omega$

SUMMARY

■ An ideal DC current source provides a constant current output that is independent of the load.

■ A constant current source can be built with the basic V_B-control NPN or PNP bias circuit.

■ The maximum available output voltage and maximum allowable load resistance for the NPN current source are

$$V_{L\max} = V_{CC} - V_E \quad \text{and} \quad R_{L\max} = \frac{V_{L\max}}{R_L}$$

■ The maximum available output voltage and maximum allowable load resistance for the PNP current source are

$$V_{L\max} = V_E \quad \text{and} \quad R_{L\max} = \frac{V_{L\max}}{R_L}$$

■ A diode inserted in the base circuit can be used to compensate for temperature variations.

■ The current mirror is a current source built with a single resistor and two transistors, one of them connected as a diode.

■ The output resistance of the current mirror is the output resistance of the transistor, typically greater than 100 KΩ.

PROBLEMS

SECTION 18–1 Introduction

1. Figure 18–17 shows a realistic current source with an output resistance of 100KΩ.

   (a) Find and tabulate the load current and load voltage for the following load resistors:

   $$R_L = 1\text{K}\Omega, 10\text{K}\Omega, 100\text{K}\Omega, 1\text{M}\Omega$$

   (b) What can you conclude about a realistic current source?

2. Find the maximum possible load resistor in Figure 18–17 so that the load current is at least 99% of the source current.

FIGURE 18–17

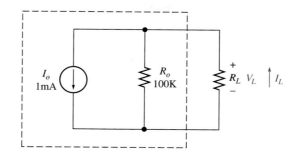

SECTION 18–2 The Basic NPN Transistor Current Source

3. For the current source in Figure 18–18
   (a) Find the output current.   **(b)** Find the maximum available load voltage.
   (c) Find the maximum allowable load resistance.
4. For the current source in Figure 18–18, choose a new R_E to set the output current to $I_o = 10\mu A$.
5. For the current source in Figure 18–18, choose a new R_1 to set the output current to $I_o = 10\mu A$.
6. For the current source in Figure 18–18, choose a new R_2 to set the output current to $I_o = 10\mu A$.

FIGURE 18–18

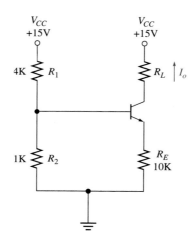

SECTION 18–3 The Basic PNP Transistor Current Source

7. For the current source shown in Figure 18–19 (p. 586)
   (a) Find the output current.   **(b)** Find the maximum available load voltage.
   (c) Find the maximum allowable load resistance.
8. For the current source in Figure 18–19, choose a new R_E to set the output current to $I_o = 10\mu A$.
9. For the current source in Figure 18–19, choose a new R_1 to set the output current to $I_o = 10\mu A$.
10. For the current source in Figure 18–19, choose a new R_2 to set the output current to $I_o = 10\mu A$.

SECTION 18–4 Negative Voltage Supply Current Sources

11. For the NPN current source shown in Figure 18–20 (p. 586)
   (a) Find the output current.   **(b)** Find the maximum available load voltage.
   (c) Find the maximum allowable load resistance.
12. For the PNP current source shown in Figure 18–21 (p. 586)
   (a) Find the output current.   **(b)** Find the maximum available load voltage.
   (c) Find the maximum allowable load resistance.

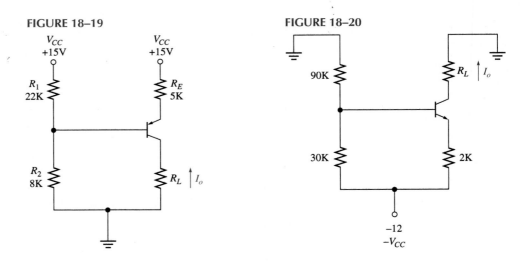

FIGURE 18–19   FIGURE 18–20

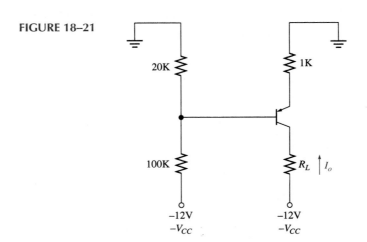

FIGURE 18–21

SECTION 18–5   Temperature Compensation

13. In this problem you will derive the exact formula for the diode-compensated base voltage. The base voltage in Figure 18–22 can be found using superposition. That is

$$V_B = V_{B1} + V_{B2}$$

   (a) From Figure 18–22B, find V_{B1}.

   (b) From Figure 18–22C, find V_{B2}. Note that V_{B2} is the voltage across R_1.

   (c) Add these two answers to find V_B.

14. For the diode-compensated current source of Figure 18–23

$$V_D = 0.5V \qquad \text{and} \qquad V_{BE(ON)} = 0.8V$$

   (a) Find I_o using the approximation.

   (b) Find I_o using the exact formula. Is the approximation valid?

   (c) If both V_D and $V_{BE(ON)}$ increase by 0.2V, find I_o. Compare this result to part (a) or part (b), as appropriate.

15. For the diode-compensated current source of Figure 18–24

$$V_D = 0.5V \qquad \text{and} \qquad V_{BE(ON)} = 0.8V$$

FIGURE 18–22

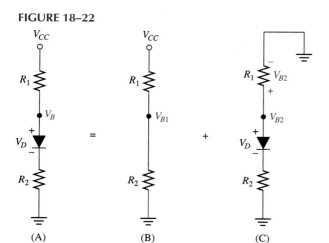

(A)   =   (B)   +   (C)

FIGURE 18–23

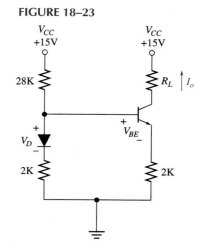

FIGURE 18–24

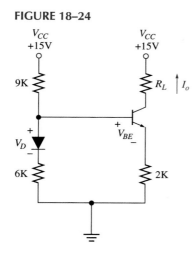

FIGURE 18–25

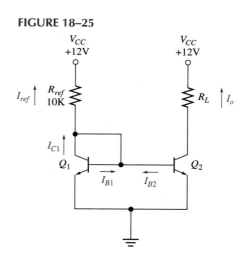

(a) Find I_o using the approximation.

(b) Find I_o using the exact formula. Is the approximation valid?

(c) If both V_D and $V_{BE(ON)}$ increase by 0.2V, find I_o. Compare this result to part (a) or part (b), as appropriate.

SECTION 18–6   The Current Mirror

16. Find the output current for the current mirror of Figure 18–25.

17. For the current mirror of Figure 18–25

    (a) Find the base currents of Q_1 and Q_2 (assume that $\beta = 100$).

    (b) Find the current in the short circuit that connects the base to the collector of Q_1.

    (c) Subtract this current from I_{ref}. This current is I_{C1}.

    (d) What percentage of I_{ref} is I_{C1}?

18. You want an output current of 5μA. Select R_{ref} in Figure 18–25 to achieve this current.

19. Find the output current of the current mirror in Figure 18–26 (p. 588).

20. Find the output current of the PNP current mirror in Figure 18–27 (p. 588).

21. Find the output current of the PNP current mirror in Figure 18–28 (p. 588).

22. For the rather elaborate current source in Figure 18–29 (p. 588)

    (a) Find I_{C2}.   (b) Find I_o.   (c) Find V_L.

FIGURE 18–26

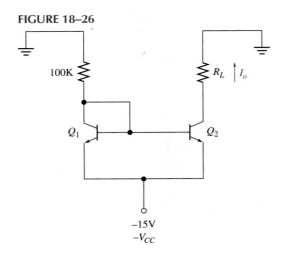

FIGURE 18–27

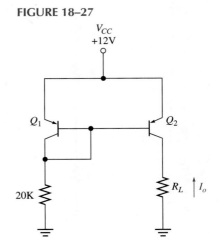

FIGURE 18–28

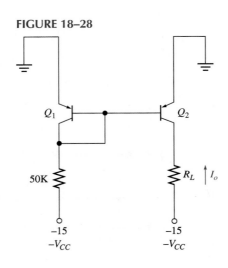

FIGURE 18–29

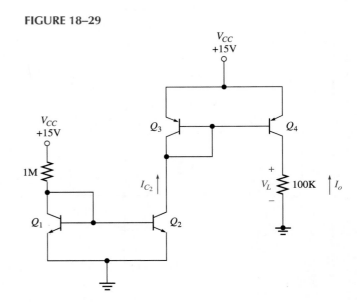

SECTION 18–7 **Current Source Biasing of Differential Amplifiers**

23. For the current-source biased differential amplifier in Figure 18–30
  (a) Find I_o.
  (b) Find the emitter currents of Q_3 and Q_4.
  (c) Find the Box Model parameters for the differential amplifier.
  (d) Assume that Q_2 has an output resistance of 500KΩ, and find the common-mode gain.
  (e) Find the common-mode rejection ratio.

Computer Problems

24. Simulate the current source of Figure 18–2A (p. 565).
  (a) For $R_L = 1$KΩ, find I_o and V_L.
  (b) Change the current gain to $\beta = 50$, and repeat part (a).
  (c) Change the current gain to $\beta = 200$, and repeat part (a).
  (d) As β varies does the output current change much?
25. In Figure 18–2A (p. 565), change the base resistors to $R_1 = 1.1$MΩ and $R_2 = 100$KΩ. Repeat Problem 24.

FIGURE 18–30

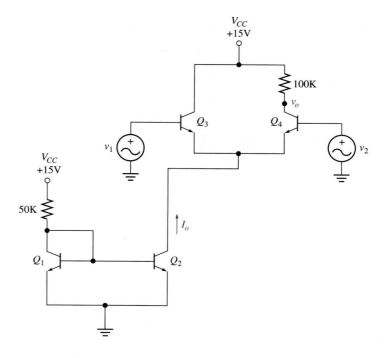

26. Simulate the temperature-compensated circuit of Figure 18–11 (p. 577). Use a diode-connected transistor, as shown in Figure 18-12.

   (a) Find the output current.

   (b) If your program allows, vary the temperature and find the output current from 0°C to 60°C.

27. If we want to build a current source that produces a current in the μA range, the reference resistor of the current mirror will need to be in the MΩ range. While this might be reasonable using discrete components, it is much too large a resistor to fabricate on an integrated circuit chip. The current source shown in Figure 18–31 is known as a Widlar current source. It produces very small output currents for reasonably sized resistors. This circuit is too complicated to analyze by hand. Rather, use the computer to find the output current for the following emitter resistors:

$$R_E = 1K\Omega, \ 10K\Omega, \ 100K\Omega$$

FIGURE 18–31

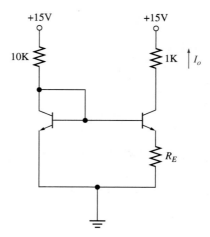

CHAPTER

19

POWER AMPLIFIERS

C HAPTER OBJECTIVES

To determine the desirable properties of an output stage.

To determine the properties of the Class A and Class B amplifier.

To determine the maximum symmetrical output voltage swing of common-emitter amplifiers (Class A amplifiers).

To determine the maximum symmetrical output voltage swing of an emitter-follower amplifier (a Class A amplifier).

To determine the supply power, load power, power efficiency, and transistor power dissipation of the Class A output stage.

To determine the supply power, load power, power efficiency, and transistor power dissipation of the transformer-coupled Class A output stage.

To determine the supply power, load power, power efficiency, and transistor power dissipation of the Class B amplifier.

To bias the Class B amplifier.

To troubleshoot output stages.

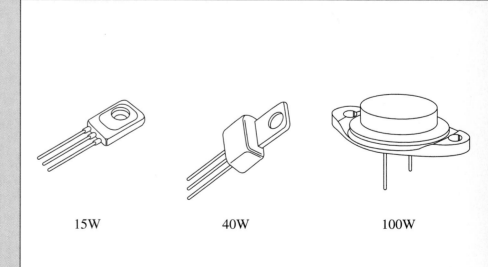

15W   40W   100W

19–1 ■ INTRODUCTION

More power! That's the tag line in one of today's popular TV comedies. It also expresses very well what we in the electronics industry constantly strive for, more and more efficient use of electrical power. While it might take only a few watts of power to drive the headphones on portable radios, it can take 100 watts or more to power the huge speakers that can make your hair stand on end and your ears bleed (Figure 19–1).

FIGURE 19–1
Power

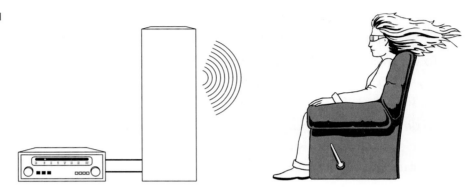

The typical amplifier is constructed in several stages (Figure 19–2). The pre-amplifier is designed to increase the very small signal generated by a turntable, microphone, or other devices to a useful value. The most important property of the pre-amplifier is its input resistance.

FIGURE 19–2
Typical amplifier

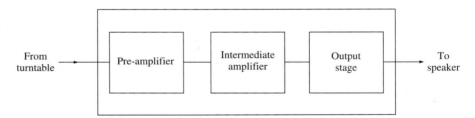

The intermediate stages of amplification further increase the signal strength and shape the frequency response. The important properties for these amplifiers are the amount of gain produced and the frequency behavior (to be discussed in a later chapter).

Finally, the output stage must deliver enough power to drive the load. We examine this last stage in this chapter. The important properties that we look for in the power stage are

☐ *Linearity of the amplifier.* That is, if we apply a sine wave to the input, do we get a sine wave at the load? This is very important in hi-fidelity applications such as classical music and opera, less important to the slam dancers among us.

☐ *Maximum power that can be delivered to the load.* You can't drive the aforementioned 100-watt speakers with a little personal radio. The power delivered to a load is proportional to the maximum voltage that can be applied to the load, so we must first find the maximum voltage.

☐ *Efficiency of power delivery.* We don't like to waste power. We would like to see that most of the power that the amplifier uses is delivered to the load. We shall see that amplifiers that are very linear are not very efficient.

FIGURE 19–3
Power transistors

 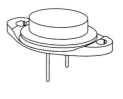

15W   40W   100W

☐ *Power dissipation in the transistors that make up the amplifier.* Don't send a boy to do a man's job. The transistors used in signal amplification applications will likely melt when used in power applications. Figure 19–3 shows some low-power, medium-power, and high-power transistors. In general, transistor size grows as power handling capability grows. Power transistors tend to have lower current gains ($30 < \beta < 120$) than transistors built for signal amplification ($100 < \beta < 400$).

☐ *Power drain on the supply sources.* If your electronic device is plugged into the wall, the power drain may not be that important, involving only slightly higher electricity bills. However, if your device is powered by batteries, you have a great interest in minimizing the power drain. Popping the batteries out of your Game Boy may be a minor inconvenience, but popping the battery out of a pacemaker implanted in someone's chest cavity is quite another matter.

An amplifier is classified based on where its bias levels (or Q-points) are set.

☐ **Class A:** biased so that the transistor remains in the **ACTIVE** region, allowing for maximum symmetrical swing. The common emitter, emitter follower, and common base are all Class A amplifiers. Class A amplifiers are very linear but not very power efficient.

☐ **Class B:** biased at or near **CUT-OFF.** The transistor will turn on only during a portion of the input cycle. The Class B amplifier is very power efficient but is not very linear. Class B amplifiers require two transistors to provide full output swing.

☐ **Class AB:** biased slightly into the **ACTIVE** region. The transistor will still be on for only a portion of the input cycle, but it remains on for a longer portion of the cycle than a Class B amplifier. Class AB amplifiers also require two transistors to provide full output swing. These amplifiers are more linear than the Class B and more efficient than the Class A amplifiers.

☐ **Class C:** biased deeply into **CUT-OFF.** The transistor will only turn on for a very small portion of the input cycle. The Class C amplifier is used to build tuned amplifiers and oscillators.

In this chapter we will discuss Class A and Class B amplifiers in detail. Class C amplifiers are discussed in Chapter 29.

REVIEW QUESTIONS

1. What is the primary responsibility of the output stage of an amplifier?
2. What are some important properties that we look for in an output stage?
3. How is the Class A amplifier biased?
4. How is the Class B amplifier biased?

19–2 ■ THE COMMON-EMITTER AMPLIFIER— MAXIMUM VOLTAGE SWING

The maximum power that can be delivered to a load depends on the maximum voltage that can be applied to the load. Therefore, we begin our discussion of power amplifiers by determining the maximum voltage swing that the common emitter, *a Class A amplifier,* can provide to a load. We already covered some of this material in Chapter 9; however, repetition helps us to understand any topic more completely.

Figure 19–4 shows a simple, and unrealistic, common-emitter power amplifier. In power amplifiers we are not interested in amplification but rather in the maximum and minimum voltage that can be applied to the load. In this circuit the load is the collector resistor, and the load voltage is the collector voltage.

FIGURE 19–4
Maximum collector voltage swing

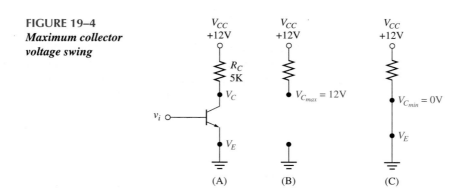

(A)   (B)   (C)

Symmetrical Swing

In all transistor circuits the maximum and minimum voltage outputs occur when the transistor just goes into **CUT-OFF** or **SATURATION.** When the transistor is in **CUT-OFF,** all currents are 0A, and we get the maximum collector voltage (Figure 19–4B):

$$\textbf{CUT-OFF} \qquad V_{c_{max}} = V_{CC} = 12V$$

The minimum collector voltage occurs when the transistor enters **SATURATION,** at which time the collector voltage comes down to the emitter voltage (Figure 19–4C):

$$\textbf{SATURATION} \qquad V_{c_{min}} = V_E = 0V$$

Therefore, the output collector voltage can swing from 0V to V_{CC} (Figure 19–5A). Note that we are using a lowercase subscript in the formula. We do this to indicate that we are dealing with the total collector voltage. In this chapter we will not separate the DC component of a signal (V_C) from the AC component of the signal (v_c).

FIGURE 19–5

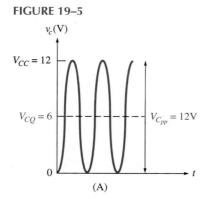

(A)

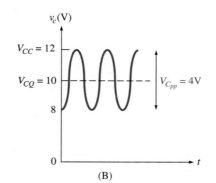

(B)

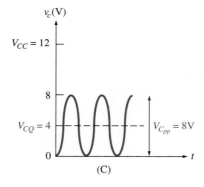

(C)

Can the output voltage always swing between these limits? Not if we are interested in symmetrical swing. Symmetrical swing is *required* for linearity. That is, if we have a pure sinusoidal input signal, the output must also be sinusoidal. If we exceed the symmetrical output limits, the output signal will be clipped at the top or bottom, causing distortion.

As long as we stay within the limits of the output symmetrical swing, the Class A amplifier is very linear. In practice, the Class A amplifier does produce distortion if the output swing is too large. This is due to the nonlinearities in the transistor I-V curves and the variation of β with collector current magnitude.

Remember, the collector voltage sits on its DC bias level. That is, with no input signal, the collector voltage equals

Quiescent (No Input) $\qquad V_c = V_{CQ}$

where we use the subscript "Q" to emphasize that we are looking at the DC bias (or quiescent) level. For maximum swing, the DC bias, or Q-point, should be half-way between the minimum and maximum collector voltages. Then, and only then, can the collector voltage swing up to V_{CC} and down to 0V. That is

For Maximum Swing $\qquad V_{CQ} = \dfrac{V_{CC}}{2}$

So, for this circuit

$$V_{CQ} = \frac{V_{CC}}{2} = 6V$$

We define the symmetrical swing as the peak-to-peak change in the voltage:

$$V_{cpp} = V_{c_{max}} - V_{c_{min}}$$

For this circuit the maximum possible symmetrical swing is

$$V_{cpp_{max}} = V_{c_{max}} - V_{c_{min}} = V_{CC} = 12V$$

The next example shows what happens if the Q-point is not centered.

EXAMPLE 19–1
Maximum
Symmetrical Swing

For Figure 19–4, find the maximum possible *symmetrical* swing in V_c for

(a) $V_{CQ} = 6V$ (b) $V_{CQ} = 10V$ (c) $V_{CQ} = 4V$

Solution

(a) When $V_{CQ} = 6V$, V_c can swing from 6V up to 12V, for a positive swing of 6V. V_c can swing from 6V down to 0V, for a negative swing of 6V. Therefore, V_c can swing symmetrically from 0 to 12V (Figure 19–5A):

$$V_{cpp} = 12 - 0 = 12V$$

Note that here

$$V_{CQ} = \frac{V_{CC}}{2}$$

(b) When $V_{CQ} = 10V$, V_c can swing from 10V up to 12V, for a positive swing of 2V. V_c can swing from 10V down to 0V, for a negative swing of 10V. To be symmetrical, the total swing must be limited to the smaller of the positive and negative swings. Therefore, V_c can swing symmetrically 2V up and down from 10V, for a total swing of 8 to 12V (Figure 19–5B). The swing is

$$V_{cpp} = 12 - 8 = 4V$$

(c) When $V_{CQ} = 4V$, V_c can swing from 4V up to 12V, for a positive swing of 8V. V_c can swing from 4V down to 0V, for a negative swing of 4V. To be symmetrical, we must limit the total swing to the smaller of the positive and negative swings. Therefore, V_c can swing symmetrically 4V up and down from 4V, for a total swing of 0 to 8V (Figure 19–5C, p. 594). The swing is

$$V_{cpp} = 8 - 0 = 8V$$

DRILL EXERCISE   The supply voltage in Figure 19–4 (p. 594) has been changed to 9V. Find the output voltage range and the maximum possible symmetrical swing if

$$V_{CQ} = 4.5V, 7V, 3V$$

Answer $0\,V \rightarrow 9V,\ V_{cpp} = 9V;\ 5V \rightarrow 9V,\ V_{cpp} = 4V;\ 0V \rightarrow 6V,\ V_{cpp} = 6V$

We now look at the more realistic circuit in Figure 19–6A. At the moment we are not concerned with the DC bias levels. We just want to see what happens at the output as the input signal varies sinusoidally.

FIGURE 19–6

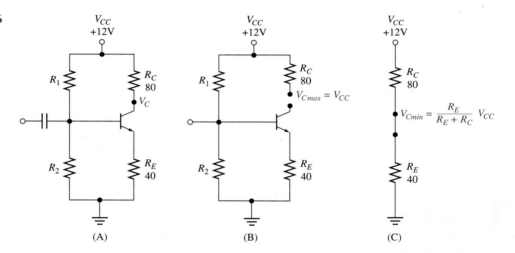

(A)   (B)   (C)

As the input signal goes negative, the transistor is driven towards **CUT-OFF.** At this time, all transistor currents are 0A (Figure 19–6B), and the output voltage goes to

CUT-OFF $V_{c\text{max}} = V_{CC}$

For this circuit

$$V_{c_{\text{max}}} = 12V$$

As the input signal goes positive, the transistor is driven into **SATURATION.** At the edge of **SATURATION,** the collector voltage comes down to the emitter voltage (Figure 19–6C). The collector circuit now looks like a voltage divider, and

SATURATION $V_{c_{\text{min}}} = \dfrac{R_E}{R_E + R_C} V_{CC}$

For the values in this circuit

$$V_{c_{\text{min}}} = \frac{R_E}{R_E + R_C} V_{CC} = \frac{40}{40 + 80} 12 = 4V$$

The maximum possible symmetrical swing for this circuit occurs when the Q-point is centered between the maximum and minimum values. To find the center between two values, you add the values and divide by two; that is

For Maximum Symmetrical Swing	$V_{CQ} = \dfrac{V_{c_{max}} + V_{c_{min}}}{2}$

so for maximum swing in this circuit, the Q-point should be

$$V_{CQ} = \frac{V_{c_{max}} + V_{c_{min}}}{2} = \frac{12 + 4}{2} = 8V$$

For this Q-point we will get the maximum possible symmetrical swing, which is given by

$$V_{cpp_{max}} = V_{c_{max}} - V_{c_{min}} = 12 - 4 = 8V$$

EXAMPLE 19–2

Maximum Swing—Common-Emitter Amplifier

For Figure 19–6 find and sketch the maximum possible collector voltage swing if the base bias resistors are chosen so that

(a) $V_{CQ} = 8V$ **(b)** $V_{CQ} = 10V$ **(c)** $V_{CQ} = 6V$

Solution   For the values given in the circuit, the maximum possible output voltage occurs at **CUT-OFF** and is

$$V_{c_{max}} = V_{CC} = 12V$$

The minimum voltage occurs at **SATURATION** and is

$$V_{c_{min}} = \frac{R_E}{R_E + R_C} V_{CC} = \frac{40}{40 + 80} 12 = 4V$$

(a) For a Q-point of 8V, the largest positive swing in collector voltage is 4V. The largest negative swing in collector voltage is 4V. In this case the positive and negative swings are the same, so the maximum symmetrical swing is

$$V_{cpp} = 8V$$

Note that the Q-point here is centered between the maximum and minimum values.

(b) For a Q-point of 10V, the largest positive swing in collector voltage is 2V. The largest negative swing in collector voltage is 6V. In this case the positive swing is smaller, so the maximum symmetrical swing is

$$V_{cpp} = 4V$$

Note that the Q-point here is above the center.

(c) For a Q-point of 6V, the largest positive swing in collector voltage is 12V. The largest negative swing in collector voltage is 2V. In this case the negative swing is smaller, so the maximum symmetrical swing is

$$V_{cpp} = 4V$$

Note that the Q-point here is below the center.

DRILL EXERCISE

Redraw the circuit in Figure 19–6, changing the supply voltage to $V_{CC} = 18V$. Find the maximum and minimum collector voltages. Find the maximum possible symmetrical swing for the following Q-points: $V_{CQ} = 12V, 15V, 8V$

Answer $V_{c_{max}} = 18V$, $V_{c_{min}} = 6V$, $V_{cpp} = 12V, 6V, 4V$

Capacitor Coupling and the Bypass Capacitor

Figure 19–7A shows the full-blown common-emitter amplifier with capacitor coupling to the load and an emitter bypass capacitor. Because we now want to look at the total voltages, not just the DC or AC components, we cannot treat the capacitors as opens or shorts.

FIGURE 19–7
Capacitor voltages

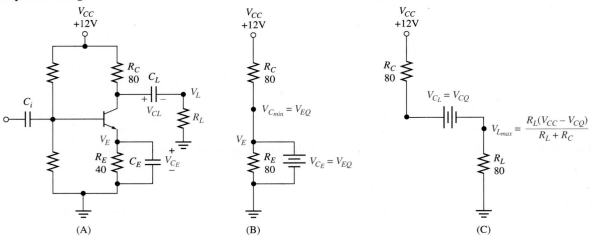

(A) (B) (C)

The key to working with the capacitors in this circuit is to remember that capacitors store voltage and, once charged, act as batteries. Before the input is applied, the capacitors are charged by the bias circuitry. The bypass capacitor charges to

$$V_{C_E} = V_{EQ}$$

and the coupling capacitor charges to

$$V_{C_L} = V_{CQ} - V_{LQ}$$

Because the output voltage is 0V for DC (that is $V_{LQ} = 0$V), the coupling capacitor charges to

$$V_{C_L} = V_{CQ}$$

Once the capacitors are fully charged, they act as batteries.

When the transistor is driven into **SATURATION,** the collector voltage comes down to the emitter voltage (Figure 19–7B). Note that we represent the emitter bypass capacitor with a battery. In **SATURATION,** therefore

$$V_{c_{min}} = V_{EQ}$$

We are primarily interested in the voltage swing of the output voltage. You can see from Figure 19–7A that, in general

$$V_L = V_c - V_{C_L} = V_c - V_{CQ}$$

where the coupling capacitor acts as a battery. The minimum load voltage is, therefore

SATURATION $V_{Lmin} = V_{cmin} - V_{CQ} = V_{EQ} - V_{CQ}$

When the transistor is driven into **CUT-OFF,** all transistor currents go to 0A, and we get the collector circuit shown in Figure 19–7C. This is an all-series circuit with two sources. To find the load voltage, we can subtract the capacitor voltage from the supply voltage and then use a voltage divider:

$$\textbf{CUT-OFF} \qquad V_{L_{\max}} = \frac{R_L}{R_L + R_C}(V_{CC} - V_{CQ})$$

The maximum possible symmetrical swing is shown in Figure 19–8A and is given by

$$V_{Lpp_{\max}} = V_{L_{\max}} - V_{L_{\min}}$$

FIGURE 19–8

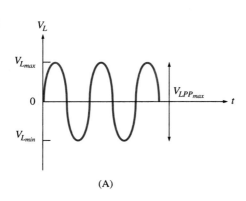

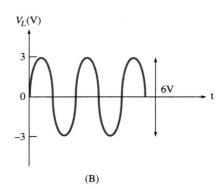

(A)   (B)

EXAMPLE 19–3
Common Emitter with Capacitors I

For the circuit in Figure 19–7A, assume the base bias resistors are chosen so that

$$V_{EQ} = 3V \qquad \text{and} \qquad V_{CQ} = 6V$$

Find $V_{L_{\min}}$, $V_{L_{\max}}$, and the symmetrical swing.

Solution The minimum load voltage is found from

$$V_{L_{\min}} = V_{EQ} - V_{CQ} = 3 - 6 = -3V$$

The maximum load voltage is found from

$$V_{L_{\max}} = \frac{R_L}{R_L + R_C}(V_{CC} - V_{CQ}) = \frac{80}{80 + 80}(12 - 6) = 3V$$

For these bias values the maximum positive and negative swings are the same (Figure 19–8B). This gives us the largest possible symmetrical swing we can obtain from this circuit:

$$V_{Lpp_{\max}} = 6V$$

EXAMPLE 19–4
Common Emitter with Capacitors II

For the circuit in Figure 19–9 (p. 600), find V_{EQ}, V_{CQ}, $V_{L_{\min}}$, $V_{L_{\max}}$, and the symmetrical swing.

Solution We'll let you do the work to find that for these bias resistors

$$V_{EQ} = 2.3V \qquad \text{and} \qquad V_{CQ} = 7.4V$$

The minimum load voltage is, therefore

$$V_{L_{\min}} = V_{EQ} - V_{CQ} = 2.3 - 7.4 = -5.1V$$

The maximum load voltage is

$$V_{L_{\max}} = \frac{R_L}{R_L + R_C}(V_{CC} - V_{CQ}) = \frac{80}{80 + 80}(12 - 7.4) = 2.3V$$

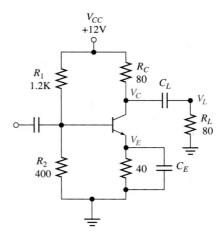

FIGURE 19–9

Because the positive swing is smaller than the negative swing, the positive swing sets the symmetrical swing:

$$V_{Lpp} = 4.6\text{V}$$

DRILL EXAMPLE

Redraw the circuit in Figure 19–9 and set the bias resistors to $R_1 = 800\Omega$ and $R_2 = 400\Omega$. Find V_{EQ}, V_{CQ}, $V_{L_{min}}$, $V_{L_{max}}$, and the symmetrical swing.

Answer $V_{EQ} = 3.3\text{V}$, $V_{CQ} = 5.4\text{V}$, $V_{L_{min}} = -2.1\text{V}$, $V_{L_{max}} = 3.3\text{V}$, $V_{Lpp} = 4.2\text{V}$

We have shown you how to find the maximum, minimum, and symmetrical swing for the load voltage in three different common-emitter circuits, shown in Figures 19–4 (p. 594), 19–6 (p. 596), and 19–7 (p. 598). The largest load voltage swing is attained in the simplest circuit (Figure 19–4). That is, the largest possible output swing that can be obtained in a common-emitter amplifier is equal to the DC collector bias voltage:

Maximum Common-Emitter Symmetrical Swing $V_{Lpp\,max} = V_C$

REVIEW QUESTIONS

1. What do we mean when we say an amplifier is linear?
2. What is the maximum output voltage, minimum output voltage, and maximum symmetrical swing for the common-emitter circuit in Figure 19–4 (p. 594)?
3. What is the maximum output voltage, minimum output voltage, and maximum symmetrical swing for the common-emitter circuit in Figure 19–6 (p. 596)?
4. How do we treat the capacitors in Figure 19–7 (p. 598)?
5. To what does the emitter bypass capacitor charge? The load coupling capacitor?
6. What is the maximum output voltage, minimum output voltage, and maximum symmetrical swing for the common-emitter circuit in Figure 19–7 (p. 598)?
7. What is the largest symmetrical swing available from a common-emitter amplifier?

19–3 ■ THE EMITTER FOLLOWER—MAXIMUM VOLTAGE SWING

Because amplification is often not required in the output stage of an amplifier, we frequently use an emitter follower to provide power to the load. Consider the emitter-follower circuit (also a Class A amplifier) in Figure 19–10.

FIGURE 19–10
Emitter follower

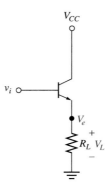

The load in this circuit is the emitter resistor, and the output voltage is the emitter voltage. The maximum output voltage occurs when the transistor goes into **SATURATION:**

$$\textbf{SATURATION} \qquad V_{L\,\text{max}} = V_{CC}$$

The minimum output voltage occurs when the transistor goes into **CUT-OFF:**

$$\textbf{CUT-OFF} \qquad V_{L\,\text{min}} = 0\text{V}$$

As with the common emitter, the maximum possible symmetrical swing is attained when the transistor is biased in the center of the swing; that is

$$\textbf{For Maximum Swing} \qquad V_{EQ} = \frac{V_{CC}}{2}$$

and the maximum possible symmetrical swing is

$$\textbf{Maximum Swing} \qquad V_{Lpp\,\text{max}} = V_{CC}$$

EXAMPLE 19–5
Emitter Follower—
Maximum Swing

For the emitter follower in Figure 19–11 (p. 602), find V_{EQ}, $V_{L\,\text{min}}$, $V_{L\,\text{max}}$, and the symmetrical swing for the following bias resistors.

(a) $R_1 = 80\Omega$, $R_2 = 100\Omega$.   **(b)** $R_1 = 100\Omega$, $R_2 = 100\Omega$.
(c) $R_1 = 50\Omega$, $R_2 = 100\Omega$.

Solution   In all cases, the minimum and maximum load voltages are

$$V_{L\,\text{min}} = 0\text{V}$$

and

$$V_{L\,\text{max}} = V_{CC} = 12\text{V}$$

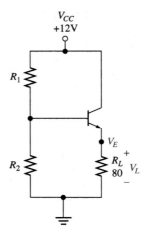

FIGURE 19–11

(a) For $R_1 = 80\Omega$, $R_2 = 100\Omega$

$$V_B = \frac{R_2}{R_1 + R_2}V_{CC} = \frac{100}{80 + 100}12 = 6.7V$$

so

$$V_{EQ} = V_B - 0.7 = 6V$$

In this case both the positive and negative swings are 6V, so the symmetrical swing is

$$V_{Lpp} = 12V$$

Note that here the bias resistors were selected to bias the transistor in the center of the largest possible swing.

(b) For $R_1 = 100\Omega$, $R_2 = 100\Omega$

$$V_B = \frac{R_2}{R_1 + R_2}V_{CC} = \frac{100}{100 + 100}12 = 6V$$

so

$$V_{EQ} = V_B - 0.7 = 5.3V$$

In this case the positive swing is 6.7V and the negative swing is 5.3V, so the symmetrical swing is determined by the negative swing:

$$V_{Lpp} = 10.6V$$

(c) For $R_1 = 50\Omega$, $R_2 = 100\Omega$

$$V_B = \frac{R_2}{R_1 + R_2}V_{CC} = \frac{100}{50 + 100}12 = 8V$$

so

$$V_{EQ} = V_B - 0.7 = 7.3V$$

In this case, the positive swing is 4.7V and the negative swing is 7.3V, so the symmetrical swing is set by the positive swing:

$$V_{Lpp} = 9.4V$$

DRILL EXERCISE   Redraw the circuit of Figure 19–11, and set the bias resistors to $R_1 = 60\Omega$, $R_2 = 30\Omega$. Find the relevant Q-points and the symmetrical swing in the output voltage.

Answer $V_{BQ} = 4V$, $V_{EQ} = 3.3V$, $V_{Lpp} = 6.6V$

REVIEW QUESTIONS

1. Why is the emitter follower a Class A amplifier?
2. Where is the load resistor placed in the emitter follower?
3. What is the maximum possible load voltage?
4. What is the minimum possible load voltage?
5. What is the maximum attainable voltage swing?
6. Where must the Q-point be located to obtain the maximum voltage swing?

19–4 ■ CLASS A POWER CONSIDERATIONS

We begin our discussion of power distribution in Class A amplifiers by looking at the simple common-emitter amplifier in Figure 19–12. As we have previously found, this circuit will provide us with the maximum possible output voltage swing of

$$V_{Lpp_{max}} = V_{Cpp} = V_{CC}$$

FIGURE 19–12

$$P_{CC} = V_{CC}I_{CQ}$$

$$P_D = V_{CEQ}I_{CQ}$$

$$P_L = \frac{V_{LP}^2}{2R_L}$$

$$\% \ efficiency = 100\ \frac{P_L}{P_{CC}}$$

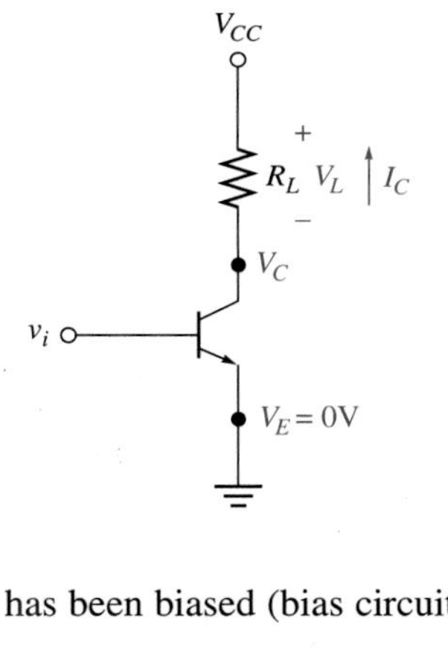

$$P_{CC_{max}} = \frac{V_{CC}^2}{2R_L}$$

$$P_{D_{max}} = \frac{V_{CC}^2}{4R_L}$$

$$P_{L_{max}} = \frac{V_{CC}^2}{8R_L}$$

$$\% \ efficiency = 25\%$$

We assume that the circuit has been biased (bias circuitry not shown) in the center of this swing. That is

For Maximum Swing $V_{CQ} = \dfrac{V_{CC}}{2}$

We will need to find the DC bias collector current (I_{CQ}) for some of the following power calculations. Because

$$I_C = \frac{V_{CC} - V_C}{R_L}$$

we get the following:

For Maximum Swing $I_{CQ} = \dfrac{V_{CC} - V_{CQ}}{R_L} = \dfrac{V_{CC}}{2R_L}$

We want to find the following information:

☐ The average power supplied by the source (P_{CC}).

☐ The average power dissipated by the transistor (P_D).

☐ The average AC power delivered to the load (P_L).

☐ The power efficiency $\left(\dfrac{P_L}{P_{CC}}\right)$.

First, a quick review of average power. Electrical power is defined as

$$p(t) = v(t)i(t).$$

Average power is what drives loads, drains batteries, and heats (possibly destroying) components. If both the voltage and the current are constant, then average power is given by

$$P = V_{DC}I_{DC}$$

If both the voltage and current are sinusoidal, average power is found from their root-mean-square values:

$$P = V_{rms}I_{rms} = \frac{V_p I_p}{2}$$

where V_p and I_p are the peak values of the voltage and current, and

$$V_{rms} = \frac{V_p}{\sqrt{2}} \qquad \text{and} \qquad I_{rms} = \frac{I_p}{\sqrt{2}}$$

The peak value of a sinusoid is one-half of the peak-to-peak value.

We can now return to the common-emitter amplifier in Figure 19–12. The power supply current in this circuit is the collector current of the transistor (we ignore the current drawn by the unshown bias resistors). Therefore

$$P_{CC} = V_{CC}I_{C_{DC}}$$

The DC value of the collector current is the bias Q-point current, so

$$P_{CC} = V_{CC}I_{CQ}$$

If the circuit is biased for maximum swing

$$P_{CC_{\text{max}}} = \frac{V^2_{CC}}{2R_L} = V_{CC} = \frac{V_{CC}}{2R_L}$$

This is the amount of power that a battery must be able to provide. Actually, the battery also has to supply the bias resistor currents, but these are usually very small. Note that for the Class A amplifier there is a significant drain on the battery, even when there is no signal present. This is why Class A power amplifiers are not used with battery supplies.

To avoid melting the output transistor, we need to use a large enough device or provide for cooling through a heat sink or fan. The important specification here is the average power dissipated in the transistor. For the Class A amplifier, maximum power dissipation occurs when no signal is present. That is, maximum power dissipation occurs at the Q-point:

$$P_D = V_{CEQ}I_{CQ}$$

Because the emitter in this circuit is grounded ($V_{CEQ} = V_{CQ}$) in this circuit

$$P_D = V_{CQ}I_{CQ}$$

Maximum power dissipation occurs when the circuit is biased for maximum swing:

$$P_{D_{max}} = \frac{V_{CC}}{2}\frac{V_{CC}}{2R_L} = \frac{V^2_{CC}}{4R_L}$$

The average AC power dissipated in a load depends on the voltage across the load. That is

$$P_L = V_{Lrms}I_{Lrms} = \frac{V^2_{Lrms}}{R_L} = \frac{V^2_L}{2R_L}$$

For our circuit the maximum peak-to-peak voltage is V_{CC}, so

$$V_{Lp_{max}} = \frac{V_{CC}}{2}$$

and the maximum average AC power delivered to the load is

$$P_{L_{max}} = \frac{V^2_{Lp_{max}}}{2R_L} = \frac{(V_{CC}/2)^2}{2R_L} = \frac{V^2_{CC}}{8R_L}$$

Finally, to compare how much of the power supplied by the battery actually gets to the load, we define power efficiency as the percentage

$$\% \ efficiency = 100\frac{P_L}{P_{CC}}$$

Note that for efficiency calculations, we always exclude the power supplied to the bias resistors. Here the maximum efficiency is

$$\% \ efficiency = 100\frac{P_{L_{max}}}{P_{CC_{max}}} = \frac{\dfrac{V^2_{CC}}{8R_L}}{\dfrac{V^2_{CC}}{2R_L}} = 100 \times \frac{2}{8} = 25$$

So, for the Class A amplifier

$$\% \ efficiency = 25\%$$

That's right. Under the best of circumstances, the Class A amplifier only delivers one-fourth of the source power to the load. Remember, as we add an emitter resistor and a capacitively coupled load, the maximum load voltage swing decreases, reducing the efficiency even further.

EXAMPLE 19–6
Power
Calculations for
the Class A
Amplifier

Figure 19–13 (p. 606) repeats the circuit of Figure 19–6. The bias resistors have been chosen to give us maximum possible output swing. Find the following:

V_{EQ}

I_{CQ}

V_{CQ}

V_{CEQ}

P_D

P_{CC} (excluding the bias resistors)

V_{Cp}

P_L

% efficiency

P_{CC} (including the bias resistors)

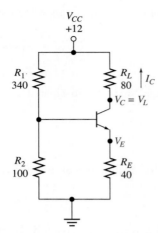

FIGURE 19–13

Solution   The base voltage is

$$V_B = \frac{R_2}{R_1 + R_2} V_{CC} = \frac{100}{340 + 100} 12 = 2.7\text{V}$$

The bias emitter voltage is, therefore

$$V_{EQ} = V_B - 0.7 = 2\text{V}$$

The bias collector current is

$$I_{CQ} = I_{EQ} = \frac{V_{EQ}}{R_E} = \frac{2}{40} = 50\text{mA}$$

The collector voltage is

$$V_{CQ} = V_{CC} - R_L I_{CQ} = 12 - 80 \times 50\text{mA} = 8\text{V}$$

The collector-emitter voltage is

$$V_{CEQ} = V_{CQ} - V_{EQ} = 8 - 2 = 6\text{V}$$

We can now find the power dissipated in the transistor:

$$P_D = V_{CEQ} I_{CQ} = 6 \times 50\text{mA} = 300\text{mW}$$

The power supplied by the battery (excluding the bias resistors) is

$$P_{CC} = V_{CC} I_{CQ} = 12 \times 50\text{mA} = 600\text{mW}$$

Note that one-half of this power is dissipated (lost) in the transistor.
The maximum load (collector) voltage occurs at **CUT-OFF** and is given by

$$V_{L_{max}} = V_{CC} = 12\text{V}$$

The minimum load (collector) voltage occurs at **SATURATION** and is given by

$$V_{L_{min}} = \frac{R_E}{R_E + R_L} V_{CC} = \frac{40}{40 + 80} 12 = 4\text{V}$$

The load voltage can swing from 8V up to 12V, for a positive swing of 4V. The load voltage can swing from 8V down to 4V, for a negative swing of 4V. As we stated earlier, the bias resistors were selected to center the Q-point. The maximum peak-to-peak swing is, therefore, 8V. The peak value for the load voltage is then

$$V_{Lp} = \frac{V_{Lpp}}{2} = 4\text{V}$$

The maximum average AC power delivered to the load is

$$P_L = \frac{V^2_{Lp}}{2R_L} = \frac{16}{2 \times 80} = 100\text{mW}$$

The power efficiency in this circuit is

$$\% \text{ efficiency} = 100\frac{P_L}{P_{CC}} = 100\frac{100\text{mW}}{600\text{mW}} = 16.7\%$$

You can see that the addition of an emitter resistor has reduced the efficiency of this Class A amplifier.

Finally, we want to find the power supplied to the bias resistors. If we ignore the base current, R_1 and R_2 can be treated as though they are in series and can be added. The power delivered to these resistors is

$$P_{(R_1 + R_2)} = \frac{V^2_{CC}}{R_1 + R_2} = \frac{144}{440} = 330\text{mW}$$

The total power that must be supplied by the battery is

$$P_{CC} + P_{(R_1 + R_2)} = 600 + 330 = 930\text{mW}$$

DRILL EXERCISE

Redraw the circuit in Figure 19–13, changing the supply voltage to 24V. Find P_{CC}, P_D, V_{Lpp}, P_L, % efficiency.

Answer

$P_{CC} = 2.83\text{W}$, $P_D = 1.16\text{mW}$, $V_{Lpp} = 13\text{V}$, $P_L = 264\text{mW}$, % efficiency = 9.33%

REVIEW QUESTIONS

1. What power calculations do we want to make for power amplifiers?
2. For DC voltages and currents, what is the power?
3. For AC voltages and currents, what is the average power dissipated?
4. What is P_{CC}?
5. What is P_D?
6. What is P_L?
7. What is the maximum power efficiency obtainable in the Class A amplifier?
8. Besides poor efficiency, what is another disadvantage of the Class A amplifier?

19–5 ■ TRANSFORMER COUPLING

Figure 19–14 (p. 608) shows a common-emitter amplifier that is coupled to the load through a transformer. As we will shortly show you, transformer coupling can increase the power efficiency of the Class A amplifier to 50%. The disadvantage to this circuit is that good hi-fidelity transformers are very large and very heavy. In this day of miniaturization, we rarely see this type of power amplifier. However, in biomedical applications transformers (coupled coils) are often used to transmit power to an implanted neural stimulator.

FIGURE 19–14
Transformer-coupled load

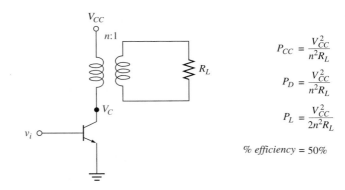

$$P_{CC} = \frac{V_{CC}^2}{n^2 R_L}$$

$$P_D = \frac{V_{CC}^2}{n^2 R_L}$$

$$P_L = \frac{V_{CC}^2}{2n^2 R_L}$$

% efficiency = 50%

Analysis of the Transformer-Coupled Amplifier

Analysis of the transformer-coupled amplifier is not straightforward. This is because for DC the transformer acts as a short circuit, so there is no DC resistance in the collector circuit. We will, therefore, analyze this circuit in a somewhat roundabout way.

We begin by determining the maximum swing in V_c. When the transistor is driven into **SATURATION,** the collector shorts to ground, so

$$\textbf{SATURATION} \qquad V_{c_{min}} = 0\text{V}$$

The tricky part is what happens when the transistor goes into **CUT-OFF.** Because the transformer primary is an inductor, and because inductors store energy, when the transistor opens, the collector voltage actually goes to

$$\textbf{CUT-OFF} \qquad V_{c_{max}} = 2V_{CC}$$

Yes. The collector voltage actually goes to twice the supply voltage! To understand how this happens, let's look at Figure 19–15, where we have isolated the primary circuit of the transformer.

As the transistor heads towards **SATURATION,** the collector current increases until the collector shorts to ground (Figure 19–15A). The increasing collector current creates an increasing magnetic field, which in turn induces an increasing voltage (V_1) across the coil. As pointed out in Chapter 4, the induced voltage is related to the rate of change of the current. That is

$$V_1 = L_1 \frac{\Delta I_c}{\Delta t}$$

We don't need to use this equation. We just want to emphasize that as the rate of change of the collector current increases in time, the magnetic field in the coil also increases, and the

FIGURE 19–15

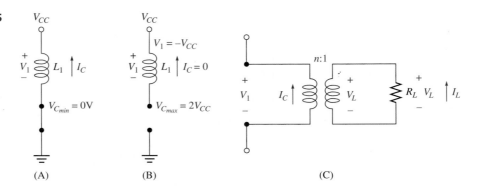

(A)   (B)   (C)

voltage across the primary coil increases as well. You can see from the figure that the coil voltage eventually reaches

$$V_1 = V_{CC}$$

Now the input reverses direction and starts to drive the transistor towards **CUT-OFF.** At this time, the collector current decreases towards 0A. The rate of change of the collector current is now negative, causing the magnetic field in the coil to decrease.

Because the magnetic field is decreasing, the polarity of the voltage induced in the coil is reversed. In addition, because the rate of decrease is the same as the rate of increase, the magnitude of the voltage induced must equal the magnitude of the voltage induced on the charging cycle. That is, now the voltage across the coil goes to

$$V_1 = -V_{CC}$$

From Figure 19–15B, at **CUT-OFF** the collector voltage is

$$\textbf{CUT-OFF} \qquad V_{c_{max}} = V_{CC} - V_1 = V_{CC} - (-V_{CC}) = 2V_{CC}$$

The maximum possible swing in the collector voltage is, therefore,

$$V_{cpp} = 2V_{CC}$$

To achieve this maximum swing, we want to bias the transistor so that

$$V_{CQ} = \frac{1}{2}V_{cpp} = V_{CC}$$

Now let's jump to the secondary of the transformer (Figure 19–15C). We first note that the primary voltage (V_1) changes from $+V_{CC}$ to $-V_{CC}$, so V_1 also has a swing of

$$V_{1pp} = 2V_{CC}$$

In transformers, as you know, secondary voltage is related to primary voltage by

$$V_L = \frac{1}{n}V_1$$

where n is the *turns ratio* of the transformer. So

$$V_{Lpp} = \frac{1}{n}V_{1pp} = \frac{2V_{CC}}{n}$$

The load current is given by

$$I_L = \frac{V_L}{R_L}$$

so

$$I_{Lpp} = \frac{V_{Lpp}}{R_L} = \frac{2V_{CC}}{nR_L}$$

We now return to the primary, where

$$I_c = \frac{1}{n}I_L$$

That is, if primary voltage is greater than secondary voltage, primary current is smaller than secondary current. This keeps the power on both sides of the transformer the same. The swing in the collector current is, therefore

$$I_{cpp} = \frac{1}{n}I_{Lpp} = \frac{2V_{CC}}{n^2 R_L}$$

It appears then that the effective resistance, or **reflected resistance,** in the collector is

$$R_{Creflected} = \frac{V_{Cpp}}{I_{Cpp}} = n^2 R_L$$

That is, the transformer can change the apparent resistance of a load. This is known as **impedance matching.**

We have now come full circle and can determine the desired collector bias current:

$$I_{CQ} = \frac{I_{cpp}}{2} = \frac{V_{CC}}{n^2 R_L}$$

Power Calculations

Now that we know the maximum peak-to-peak load voltage and the Q-point voltage and current, we can determine the power relationships for this circuit.

The power dissipated by the transistor is

$$P_D = V_{CEQ}I_{CQ} = V_{CQ}I_{CQ} = V_{CC}\frac{V_{CC}}{n^2 R_L} = \frac{V^2_{CC}}{n^2 R_L}$$

The supply power is

$$P_{CC} = V_{CC}I_{CQ} = \frac{V^2_{CC}}{n^2 R_L}$$

Interestingly, when there is no input signal in this circuit, all of the power from the supply goes to the transistor. This is because the transformer is a short circuit for DC. When the input signal is turned on, power is diverted to the load. The maximum load power is

$$P_L = \frac{V^2_{Lp}}{2R_L} = \frac{V^2_{CC}}{2n^2 R_L}$$

where $V_{Lp} = V_{Lpp}/2 = V_{CC}/n$.

The efficiency for the transformer-coupled Class A amplifier is

$$\% \ efficiency = 100\frac{P_L}{P_{CC}} = \frac{V_{CC}^2/2n^2 R_L}{V_{CC}^2/n^2 R_L} = 100\frac{1}{2} = 50\%$$

The transformer-coupled Class A amplifier has a maximum power efficiency of 50%. Again, the disadvantage of this circuit is the large size of transformers.

EXAMPLE 19–7
Transformer-Coupled Class A Amplifier

For the circuit of Figure 19–14

$$V_{CC} = +12\text{V}, R_L = 8\Omega, n = 6$$

Assume the transistor is biased for maximum swing, and find V_{cpp}, V_{Lpp}, I_{Cpp}, P_D, P_{CC}, P_L, % efficiency.

Solution Because of the energy storage capability of the transformer

$$V_{cpp} = 2V_{CC} = 24\text{V}$$

The collector voltage Q-point is

$$V_{CQ} = \frac{V_{cpp}}{2} = 12\text{V}$$

The primary voltage of the transformer has the same swing as the collector voltage. The load voltage swing is, therefore

$$V_{Lpp} = \frac{2V_{CC}}{n} = \frac{24}{6} = 4\text{V}$$

The collector current swing depends on the load current swing:

$$I_{Lpp} = \frac{V_{Lpp}}{R_L} = \frac{4}{8} = 0.5\text{A}$$

so

$$I_{cpp} = \frac{1}{n}I_{Lpp} = \frac{0.5}{6} = 0.0833\text{A}$$

Note that because of the transformer, the transistor current is less than the load current. The collector current Q-point is

$$I_{CQ} = \frac{I_{cpp}}{2} = 0.0417\text{A}$$

The transistor power is

$$P_D = V_{CEQ}I_{CQ} = V_{CQ}I_{CQ} = 12 \times 0.0417 = 0.5\text{W}$$

For the following power calculations, we will use basic power definitions rather than just plugging into the formulas we derived.

The supply power is

$$P_{CC} = V_{CC}I_{CQ} = 12 \times 0.0417 = 0.5\text{W}$$

The load power is

$$P_L = \frac{V^2_{Lp}}{2R_L} = \frac{2^2}{2 \times 8} = 0.25\text{W}$$

Finally, the efficiency is

$$\% \text{ efficiency} = 100\frac{P_L}{P_{CC}} = 100\frac{0.25}{0.5} = 50\%$$

DRILL EXERCISE Change the turns ratio in Figure 19–14 (p. 608) to $n = 1$, and repeat the preceding example.

Answer $V_{cpp} = 24\text{V}, V_{Lpp} = 24\text{V}, I_{cpp} = 3\text{A}$
$P_D = 18\text{W}, P_{CC} = 18\text{W}, P_L = 9\text{W}, \textit{efficiency} = 50\%$

REVIEW QUESTIONS

1. On what does the voltage induced in an inductor depend?

2. Explain why the peak-to-peak swing in the collector voltage of the transformer-coupled Class A amplifier is twice the supply voltage.

3. What is the relation between the secondary and primary voltages of a transformer?

4. What is the maximum output voltage swing?

5. What is the relation between the primary and secondary currents of a transformer? What is the maximum collector current swing?

6. What is the relation between the resistance seen at the primary of a transformer and the load connected to the secondary?

7. What are V_{CQ} and I_{CQ} for maximum swing?

8. What are P_D, P_{CC}, P_L?

9. What is the efficiency of the transformer-coupled Class A amplifier?

10. What is the disadvantage of this amplifier?

19–6 ■ THE CLASS B AMPLIFIER

Operation

The basic *Class B amplifier* is shown in Figure 19–16A. The use of an NPN and a matched PNP transistor makes for a very simple circuit. Note the use of two power supplies in this circuit. Also note that the load resistor is connected to the emitters of the transistors. We do not look for any gain from the Class B; rather, we are interested in increasing the efficiency of power use.

When the input is turned off (Figure 19–16B), the bases of the two transistors are grounded. So

$$V_{BQ} = 0V$$

Because of the symmetry in the collector circuit, the emitter voltage is halfway between the positive and negative supply voltages. That is

$$V_{EQ} = \frac{V_{cc} + (-V_{cc})}{Z} \, 0V$$

Therefore, the voltage across the emitter diode of each transistor is 0V, and both transistors are **CUT-OFF**.

Now we turn on the input signal. As the input signal goes positive, the NPN transistor (Q_1) will turn on. Because the base voltage is positive, the PNP transistor (Q_2) will be

FIGURE 19–16
Class B amplifier (push-pull)

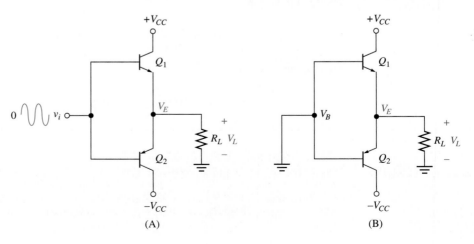

(A) (B)

CUT-OFF. This will give us the circuit shown in Figure 19–17A. Because the output is taken from the emitter, the output voltage will follow the input voltage (we ignore, for now, the 0.7V emitter diode-drop of Q_1). The output voltage (Figure 19–18B) resembles the output of a diode rectifier.

FIGURE 19–17

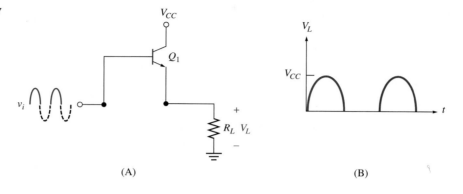

(A)   (B)

When the input signal goes negative, the PNP transistor (Q_2) turns on. Because the base voltage is now negative, the NPN transistor (Q_1) is **CUT-OFF,** and we get the circuit of Figure 19–18A. The output voltage (Figure 19–18B) gives us the missing negative half of the input cycle.

FIGURE 19–18

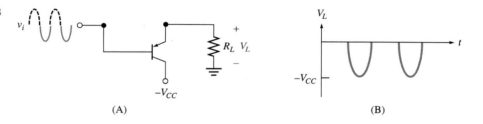

(A)   (B)

Ignoring the emitter diode voltage drops, the full output is shown in Figure 19–19. In the Class B amplifier, only one transistor is on at a time. One transistor supplies the positive output; the other transistor supplies the negative output. For this reason, the Class B is often called a **push-pull amplifier.**

FIGURE 19–19
Push-pull power formulas.

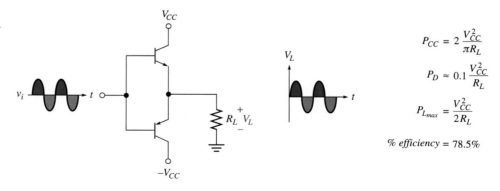

$$P_{CC} = 2\,\frac{V_{CC}^2}{\pi R_L}$$

$$P_D \approx 0.1\,\frac{V_{CC}^2}{R_L}$$

$$P_{L_{max}} = \frac{V_{CC}^2}{2R_L}$$

% efficiency = 78.5%

Originally Class B amplifiers were constructed from two similar vacuum tubes, then from two NPN transistors. Because this particular circuit uses one NPN and one PNP transistor, it is referred to as a **complementary symmetry push-pull amplifier.**

Power Considerations

All power calculations for the Class B amplifier depend on the magnitudes of the output voltage and the collector current. In the discussion below we will give you the general power relationships and also calculate the maximum power that can be supplied or dissipated.

Load Power   Because the load voltage is still, at least approximately, a sinusoid, the load power is given by

$$P_L = \frac{V^2_{Lp}}{2R_L}$$

To find the maximum possible load power, we examine Figure 19–19 (p. 613) to see that the output voltage can swing from $-V_{CC}$ to $+V_{CC}$, giving us a maximum swing of

$$V_{Lpp_{max}} = 2V_{CC}$$

The maximum load power is then

$$P_{L_{max}} = \frac{V^2_{Lp_{max}}}{2R_L} = \frac{V^2_{CC}}{2R_L}$$

Supply Power   The Class B amplifier is biased in **CUT-OFF.** This means that when there is no input signal, both transistors are off and there is no current. That is, for each transistor

$$I_{CQ} = 0\text{A}$$

When there is no signal, we say that the amplifier is in **standby mode.** Because there is no current flow in standby, the power drain from each source is

Standby Power $P_{CC} = 0\text{W}$

This is the major advantage to the Class B amplifier. When there is no signal, there is no drain on the batteries. Battery life is, therefore, greatly extended. This is why all battery-driven amplifiers use the Class B output stage.

There is an average power drain from any battery. Otherwise batteries would last forever, and battery companies would be out of business. The average power drain from a battery is given by

$$P_{CC} = V_{CC}I_{DC}$$

In the Class A amplifier, the DC value of the transistor current is the same as the Q-point current. This is not true in the Class B amplifier.

Let's take a look at the collector current in Q_1, which looks the same as a half-wave rectified signal (Figure 19–20). We saw in Chapter 2 that the average, or DC, value of a half-wave rectified signal is given by

$$I_{C_{DC}} = \frac{I_{Cp}}{\pi}$$

And because

$$I_{Cp} = \frac{V_{Lp}}{R_L}$$

FIGURE 19–20

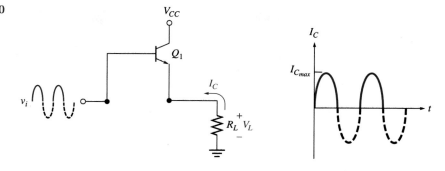

we get

$$I_{C_{DC}} = \frac{V_{Lp}}{\pi R_L}$$

So the power supplied by each battery is

$$P_{CC} = V_{CC} \frac{V_{Lp}}{\pi R_L}$$

The maximum power needed from the battery occurs when the output voltage is greatest. Because

$$V_{Lp_{max}} = V_{CC}$$

we get the maximum average power drain from *each* battery:

$$P_{CC_{max}} = V_{CC} \frac{V_{CC}}{\pi R_L} = \frac{V^2_{CC}}{\pi R_L}$$

Because each battery supplies the same amount of power, the total power supplied by the batteries is

$$P_{CCtotal_{max}} = 2 \frac{V^2_{CC}}{\pi R_L}$$

The maximum power efficiency for the Class B amplifier is, therefore

$$\% \; efficiency = 100 \; \frac{P_{L_{max}}}{P_{CCtotal_{max}}} = 100 \; \frac{V^2_{CC}/2R_L}{2V^2_{CC}/\pi R_L}$$

So

$$\% \; efficiency = 100 \; \frac{\pi}{4} = 78.5\%$$

Besides having no standby power drain, the Class B amplifier is extremely efficient in delivering power to the load.

Transistor Power   It is easier to calculate the power dissipated by the transistor if you consider that the power supplied by the sources is equal to the power dissipated by the two transistors and the load.

$$P_{CCtotal} = P_{Q1} + P_{Q2} + P_L$$

Because of the symmetry of the circuit, the power dissipated by the transistors is the same $(P_{Q1} = P_{Q2})$ and is given by

$$P_D = \frac{P_{CCtotal} - P_L}{2}$$

The maximum power dissipated in the transistors is difficult to derive. At quiescence the collector current is 0A, so there is no power dissipation. The output voltage reaches a maximum when the transistor saturates. At this time $V_{CE} = 0V$, so again there is no power dissipation. Maximum power dissipation in the transistors occurs when the output is approximately two-thirds of the maximum swing. The voltage at which maximum power dissipation occurs is

$$V_{Lp} = \frac{2}{\pi}V_{CC}$$

The maximum power dissipated by each transistor turns out to be

$$P_D \approx 0.2P_{L_{max}} = 0.1\frac{V^2_{CC}}{R_L}$$

EXAMPLE 19–8
Class B Maximum Power Calculations

For the circuit shown in Figure 19–19 (p. 613)

$$V_{CC} = +12V \text{ and } R_L = 80\Omega$$

Find the maximum load voltage swing, the total maximum average supply power, the maximum average load power, and the maximum average power dissipated in each transistor.

Solution   The maximum possible output swing is

$$V_{Lpp} = 2V_{CC} = 24V$$

Maximum power calculations are made using the maximum peak load voltage:

$$V_{Lp_{max}} = 12V$$

The average load power is

$$P_{L_{max}} = \frac{V^2_{Lp_{max}}}{2R_L} = \frac{12^2}{2 \times 80} = 0.9W$$

The total supply power is

$$P_{CCtotal_{max}} = 2\frac{V^2_{CC}}{\pi R_L} = 2\frac{12^2}{\pi \times 80} = 1.15W$$

The maximum efficiency is

$$\% \text{ efficiency} = 100\frac{P_{L_{max}}}{P_{CCtotal_{max}}} = 100\frac{0.9}{1.15} = 78.3\%$$

The power dissipated by each transistor is

$$P_D = \frac{P_{CCtotal} - P_L}{2} = \frac{1.15 - 0.9}{2} = 0.125W$$

EXAMPLE 19–9

Class B Power Calculations

For the circuit shown in Figure 19–19 (p. 613)

$$V_{CC} = +12\text{V and } R_L = 80\Omega$$

Find the load power, supply power, efficiency, and power dissipated in each transistor for the following output voltage swings (note that these are peak values). Discuss the power dissipated in the transistors, and compare to the previous example.

(a) $V_{Lp} = 2\text{V}$ **(b)** $V_{Lp} = 8\text{V}$

Solution

(a) For $V_{Lp} = 2\text{V}$

The actual load power is

$$P_L = \frac{V^2_{Lp}}{2R_L} = \frac{2^2}{2 \times 80} = 25\text{mW}$$

To find the actual supply power, we need to find the average (DC) collector current:

$$I_{C_{DC}} = \frac{I_{Cp}}{\pi} = \frac{V_{Lp}}{\pi R_L} = \frac{2}{\pi \times 80} = 7.96\text{mA}$$

The total supply power is

$$P_{CCtotal} = 2V_{CC}I_{C_{DC}} = 2 \times 12 \times 7.96\text{mA} = 191\text{mW}$$

The power efficiency is

$$\% \text{ efficiency} = 100\frac{P_L}{P_{CCtotal}} = 100\frac{25}{191} = 13.1\%$$

The supply power, load power, and efficiency are all lower when the output voltage is not driven to its maximum possible value. The power dissipated in each transistor is

$$P_D = \frac{P_{CCtotal} - P_L}{2} = \frac{191 - 25}{2} = 83\text{mW}$$

(b) For $V_{Lp} = 8\text{V}$

The actual load power is

$$P_L = \frac{V^2_L}{2R_L} = \frac{8^2}{2 \times 80} = 400\text{mW}$$

To find the actual supply power, we need to find the average (DC) collector current:

$$I_{C_{DC}} = \frac{I_{Cp}}{\pi} = \frac{V_{Lp}}{\pi R_L} = \frac{8}{\pi \times 80} = 32\text{mA}$$

The total supply power is

$$P_{CCtotal} = 2V_{CC}I_{C_{DC}} = 2 \times 12 \times 32 = 768\text{mW}$$

The power efficiency is

$$\% \text{ efficiency} = 100\frac{P_L}{P_{CCtotal}} = 52\%$$

As the output peak increases the supply power, load power, and efficiency all increase. However, they are still lower than when the output voltage is driven to its maximum possible value.

The power dissipated in each transistor is

$$P_D = \frac{P_{CCtotal} - P_L}{2} = \frac{768 - 400}{2} = 184\text{mW}$$

Comparing the results of this example and the previous example, we get

	V_{Lp}	P_D
Maximum swing	12V	125mW
	2V	83mW
2/3 maximum swing	8V	184mW

This confirms our statement that the transistors dissipate maximum power when the output swing is two-thirds of maximum.

DRILL EXERCISE Redraw the circuit in Figure 19–19 (p. 613), changing the supply voltages to $+24$V. For a load resistor of 80Ω and a 12V sinusoidal input, find the load power, supply power, efficiency, and transistor power.

Answer $P_L = 0.9$mW, $P_{CCtotal} = 2.29$W, % *efficiency* $= 39.3\%$, $P_D = 0.695$W

Distortion in the Class B Amplifier

So far, we have ignored the emitter diode voltage drop in Q_1 and Q_2. Let's now take a closer look at the load voltage for two different inputs (Figure 19–21). The key point here is that Q_1 will not turn on until the input voltage reaches 0.7V. At this time the load voltage is equal to

$$v_L = v_i - 0.7$$

FIGURE 19–21

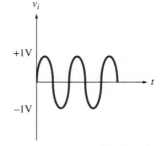

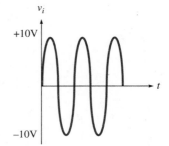

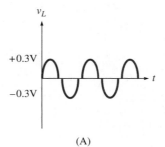

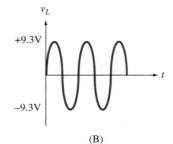

(A)   (B)

Likewise, Q_2 will not turn on until the input reaches -0.7V. At this time the load voltage is equal to

$$V_L = v_i + 0.7$$

For a 1V input, only a small part of the input signal is seen at the output (Figure 19–21A). The larger input of 10V produces an output that is less distorted and looks much more like the input (Figure 19–21B). This distortion is known as **crossover distortion** and results from the need to turn one transistor off and the other one on.

The distortion in the Class B amplifier, then, depends on the size of the input signal. That is, low magnitude (soft) inputs are more distorted than large magnitude (loud) inputs. That is why one stereo amplifier manufacturer has designed an output stage that sends soft musical passages to a Class A output stage and loud musical passes to a Class B output stage.

REVIEW QUESTIONS

1. Draw the basic NPN–PNP Class B output stage.
2. What is the name given to this type of output stage?
3. What are the quiescent states of the transistors in the Class B output stage?
4. When the input goes positive, which transistor becomes **ACTIVE?**
5. When the input goes negative, which transistor becomes **ACTIVE?**
6. What is the maximum load power?
7. When there is no input signal, what is the power drain on the supply batteries?
8. What is the maximum power drain on the supply batteries? When does this occur?
9. What is the maximum efficiency of the Class B output stage?
10. What is the maximum power dissipation in each transistor?
11. What causes crossover distortion in the output of the Class B amplifier?

19–7 ■ PUSH-PULL AMPLIFIER BIASING

In Figure 19–16B (p. 612) we showed you that the basic Class B amplifier is biased so that both transistors are **CUT-OFF.** The advantage here is that there is no current drain on the batteries when there is no input signal. That is, the standby power drain is 0W. The disadvantage with this biasing scheme is the crossover distortion introduced before the transistors turn on.

If we are willing to put up with a slight standby current drain, we can bias the transistors so that they are both just on. This will eliminate most, but not all, of the crossover distortion. Some authors still call these Class B amplifiers, while others label them *Class AB amplifiers,* because the transistors are biased slightly into the **ACTIVE** region. We will simply call all of the following output stages Class B amplifiers.

There are almost as many ways to bias a Class B amplifier as there are electronic designers. We will cover the most common of these techniques. The first circuit we look at is shown in Figure 19–22A (p. 620). The dashed line in the figure emphasizes the symmetry of the bias circuit for the NPN and the bias circuit of the PNP. Because of the symmetry, we can find V_{B1Q} from Figure 19–22B (p. 620):

$$V_{B1Q} = \frac{R_1}{R_1 + R_2}V_{CC} = \frac{0.7\text{K}}{11.3\text{K} + 0.7\text{K}}\,12 = 0.7\text{V}$$

And the emitter bias voltage is

$$V_{EQ} = V_{B1Q} - 0.7 = 0\text{V}$$

FIGURE 19–22
Biasing at the edge of
CUT-OFF.

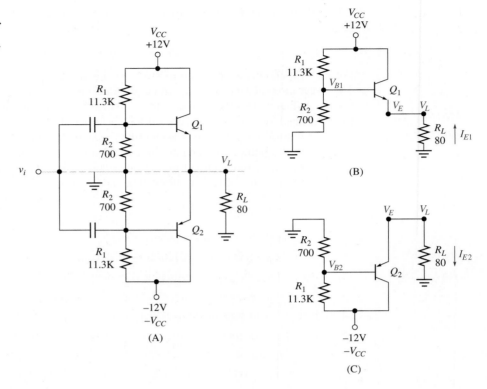

(A)

(B)

(C)

Therefore

$$I_{EQ} = \frac{V_{EQ}}{R_L} = 0A$$

and Q_1 is biased at the edge of **CUT-OFF.**

The equivalent bias circuit for the PNP transistor (Q_2) is shown in Figure 19–22C. We'll derive the bias values in the next example. However, there isn't any real need to perform a separate analysis here. This circuit is identical, except for the negative voltages in the NPN circuit. The bias values for Q_2 are, therefore

$$V_{B2Q} = -0.7V$$
$$V_{EQ} = 0V$$

and

$$I_{EQ} = 0A$$

EXAMPLE 19–10
Class B Biasing

Derive the bias values given for the partial Class B circuit shown in Figure 19–22C.

Solution   The base voltage here is the voltage across R_2, so

$$V_{B2Q} = \frac{R_2}{R_1 + R_2}(-V_{CC}) = \frac{0.7K}{11.3K + 0.7K}(-12) = -0.7V$$

Because the emitter voltage is a diode drop above the base voltage for a PNP transistor

$$V_{EQ} = V_{B2Q} + 0.7 = -0.7 + 0.7 = 0V$$

$$I_{EQ} = \frac{V_{EQ}}{R_L} = 0A$$

DRILL EXERCISE   Redraw the circuit of Figure 19–22A using the following values:

$$V_{CC} = 24V, R_2 = 0.7K\Omega, R_1 = 23.3K\Omega.$$

Find the bias voltages and currents.

Answer $V_{B1Q} = 0.7V, V_{B2Q} = -0.7V, V_{EQ} = 0V, I_{EQ} = 0A$

Diode Biasing

A very popular biasing method, particularly in integrated circuits, is to use diodes in the biasing scheme (Figure 19–23A). Because of the symmetry in this circuit (see the dashed line), the voltage between the diodes must be halfway between $+12V$ and $-12V$, so

$$V_1 = 0V$$

The base voltage at Q_1 is a diode drop above V_1, so

$$V_{B1Q} = V_1 + 0.7 = 0.7V$$

And the emitter voltage, as expected, is

$$V_{E1Q} = V_{B1Q} - 0.7 = 0V$$

The base voltage at Q_2 is a diode drop below V_1, so

$$V_{BQ2} = V_1 - 0.7 = -0.7V$$

And, again the emitter voltage

$$V_{E2Q} = V_{B2Q} + 0.7 = 0V$$

As long as the diodes have matching properties with the transistors, this circuit will automatically bias itself at the edge of **CUT-OFF.** As we discussed in Chapter 18, placing diodes in the base circuit helps to compensate for changes in temperature. And, as we discussed in the previous chapter, the best way to ensure that the diodes match the transistors is to use transistors connected as diodes. Figure 19–23B shows one straightforward way to accomplish this connection.

FIGURE 19–23
Biasing with diodes

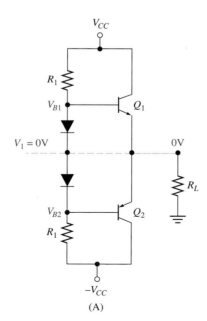

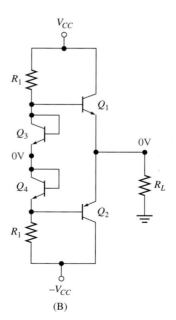

(A)   (B)

Single-Supply Biasing

Figure 19–24A shows how to build a Class B output stage when only a single positive supply is available. Again, because of the symmetry in the collector circuit, we expect the emitter voltage to be approximately halfway between V_{CC} and 0V, so

$$V_{EQ} = \frac{V_{CC}}{2} = 6V$$

FIGURE 19–24
Single-supply biasing

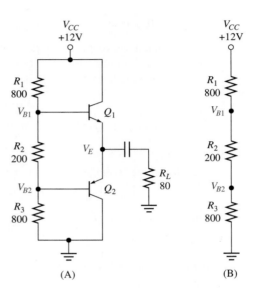

(A)   (B)

Because the emitter voltage is not 0V, we capacitor-couple the load to the output stage.

Assuming that the base currents are negligible, we can find the base bias voltages from the bias circuit shown in Figure 19–24B. The base voltage on Q_1 is

$$V_{B1Q} = \frac{R_2 + R_3}{R_1 + R_2 + R_3} V_{CC} = \frac{800 + 200}{800 + 200 + 800} 12 = 6.7V$$

The base voltage at Q_2 is

$$V_{B2Q} = \frac{R_3}{R_1 + R_2 + R_3} V_{CC} = \frac{800}{800 + 200 + 800} 12 = 5.3V$$

We can find the emitter voltage from either

$$V_{EQ} = V_{B1Q} - 0.7 = 6.7 - 0.7V = 6V$$

or

$$V_{EQ} = V_{B2Q} + 0.7 = 5.3 + 0.7 = 6V$$

EXAMPLE 19–11
Single-Supply
Diode Biasing

Find the bias levels in the circuit of Figure 19–25.

Solution We will use symmetry to find the answer here. The point between the two diodes should be at a voltage that is halfway between the supply and ground, so

$$V_1 = 6V$$

The base voltage for Q_1 is a diode drop above V_1:

$$V_{B1Q} = V_1 + 0.7 = 6 + 0.7 = 6.7V$$

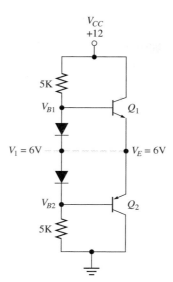

FIGURE 19–25

The base voltage for Q_2 is a diode drop below V_1:

$$V_{B2Q} = V_{B1} - 0.7 = 6 - 0.7 = 5.3V$$

The emitter voltage is found either from

$$V_E = V_{B1Q} - 0.7 = 6.7 - 0.7 = 6V$$

or

$$V_E = V_{B2Q} + 0.7 = 5.3 + 0.7 = 6V$$

The Bridge Amplifier and Phase Splitter

All of the power amplifiers in this chapter have driven loads that are connected to ground. It is possible to increase the power delivered to a load by connecting the load between two properly phased matched amplifiers (Figure 19–26). This arrangement is known as a **bridge amplifier.**

In the bridge amplifier, v_i is the input to the amplifier on the left and its inverse, $-v_i$, is the input to the amplifier on the right. Therefore, when the output on the left increases, the output on the right decreases.

The voltage across the load is given by

$$v_L = v_{o1} - v_{o2}$$

Let's assume that each amplifier can produce a maximum output voltage of 15V. When v_{o1} goes to $+15V$, v_{o2} goes to $-15V$. The maximum load voltage is now

$$v_{L_{max}} = 15 - (-15) = 30V$$

FIGURE 19–26
Bridge amplifier

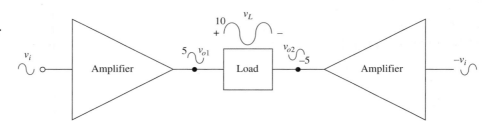

The bridge amplifier produces twice the maximum load voltage as the other amplifiers we have discussed.

Figure 19–27A shows how we can use a single-transistor amplifier to produce output signals that are the inverse of each other. Because v_1 is taken from the emitter, we know that

$$v_1 \approx v_i$$

FIGURE 19–27

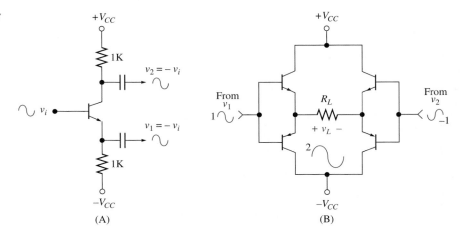

(A)   (B)

Because the collector and emitter resistors are the same, the collector output (ignoring r_e) is approximately

$$v_2 \approx -\frac{R_C}{R_E} v_i = -\frac{1\text{K}}{1\text{K}} v_i = -v_i$$

A circuit that can produce an output of $+v_i$ and $-v_i$ is known as a **phase splitter.**

Figure 19–27B shows a bridge amplifier built with two matching push-pull complementary amplifiers. The push-pull amplifiers can be driven by the phase splitter shown.

REVIEW QUESTIONS

1. With double-supply biasing, what is the quiescent emitter voltage?
2. What two advantages are there to using diodes in the biasing circuitry?
3. Draw a transistor connected to act as a diode.
4. With a signal supply source, what is the quiescent emitter voltage?
5. Why is the load capacitively coupled here?
6. Draw a single-supply resistor-biased Class B amplifier. What are the formulas for V_{B1Q} and V_{B2Q}?
7. Draw a single-supply diode-biased Class B amplifier. What is the voltage between the diodes? Why?
8. What is a bridge amplifier?
9. What is a phase splitter?

19–8 ■ TROUBLESHOOTING

The Class A amplifiers are common-emitter or emitter-follower amplifiers. Because you already know how to troubleshoot these amplifiers, we will limit this discussion to the Class B amplifier.

Short Circuit Failures

Figure 19–28 shows the results of DC Troubleshooting several Class B amplifier circuits. With the double supply we expect to see 0V at the output, 0.7V at the base of Q_1, and -0.7V at the base of Q_2. We now look at the Measured Voltages for three different short circuit failures of Q_1. In the following we assume that Q_2, all resistors, and all connections are good.

FIGURE 19–28
DC Troubleshooting Class B amplifiers: short circuit failures

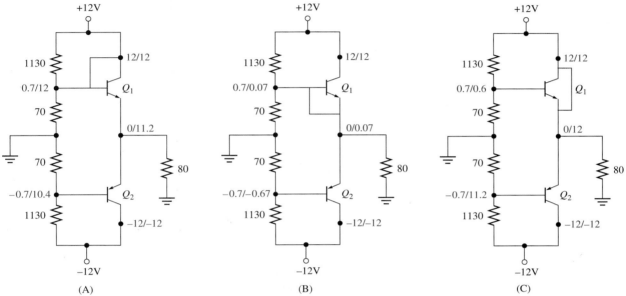

Q_1 Base-Collector Short Figure 19–28A shows the Measured Voltages that result if the base of Q_1 is internally shorted to the collector. The base voltage is the same as the supply voltage. The emitter voltage is a diode drop below the base voltage because the emitter diode of Q_1 is **ON.** Also note that Q_2 is now **ACTIVE,** and the base voltage of Q_2 is a diode drop below the emitter voltage.

Q_1 Base-Emitter Short Figure 19–28B shows the Measured Voltages that result if the base of Q_1 is shorted to the emitter. At first glance it seems that the Measured Voltage at the emitter is close to the Estimated Voltage. The key result here, however, is that the Measured Voltage at the emitter is the same as the Measured Voltage at the base of Q_1, indicating a short between these points. The load resistor is now directly connected to the base biasing resistors, changing the voltage divider relationship at this base. This is why the base voltage of Q_1 is only 0.07V.

Note that the emitter diode of Q_2 is still **ON,** so the base voltage of Q_2 is a diode drop below the emitter voltage.

Q_1 Collector-Emitter Short Figure 19–28C shows the Measured Voltages that result if the collector of Q_1 is shorted to the emitter. Because of this short, we see the supply voltage at the load. Because of this large emitter voltage, the PNP transistor Q_2 is **ACTIVE.** Because Q_2 is **ACTIVE,** the base voltage of Q_2 is a diode drop below the emitter voltage.

EXAMPLE 19–12
Q_2 Short Failures

Figure 19–29 shows the results of DC Troubleshooting a Class B amplifier. For each result, determine the likely cause of the failure. Assume all resistors and connections are good. Assume the DC Troubleshooting results are due to failures of Q_2.

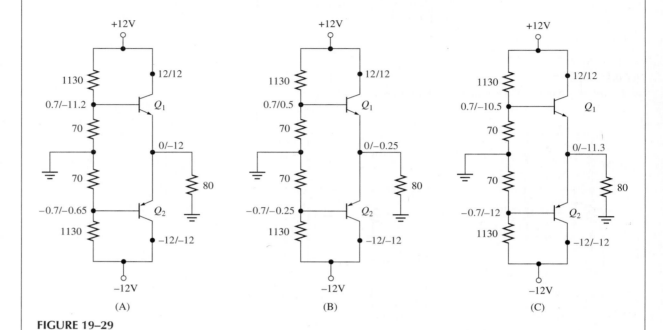

FIGURE 19–29

Solution Figure 19–29A. Because the load voltage is the same as the negative supply voltage, we can conclude that the collector of Q_2 has shorted to its emitter.

Figure 19–29B. Because the Measured Voltages at the base and emitter of Q_2 are the same, we conclude that there is a base-emitter short here.

Figure 19–29C. Because the Measured Voltages at the collector and base are the same, the base and collector must be shorted together.

We have also examined diode biasing and single-supply biasing of Class B amplifiers. The DC Troubleshooting results of failed short transistors in these circuits will show some differences from the results we have presented here. The basics of recognizing shorts by comparing Measured Voltages at the two points of interest still holds. We will examine some of these circuits in the problems at the end of the chapter.

Open Circuit Failures

Because the output stage of an amplifier handles the most power, the transistors in the output stage are the most likely to fail. This is especially true if an amplifier's output is accidentally short-circuited. When this happens, one of the transistors in the output stage is likely to fail open. (Be careful in the lab. Students short-circuiting the output is all too common. This can really upset your lab instructor.)

Because the transistors in a Class B amplifier are biased in **CUT-OFF,** DC Troubleshooting will not indicate failures that cause one or both of the transistors to open up. For this reason, an open transistor in the output stage is most easily recognized with AC signal tracing.

Figure 19–30A shows the output of a Class B amplifier in which Q_1 as failed open. You see a half-wave rectified output. Because the negative half-cycle is retained, Q_2 must be operating properly.

Figure 19–30B shows the output of a class B amplifier in which Q_2 has failed open. Again, you see a half-wave rectified output. Because the positive half-cycle is still there, Q_1 must be operating properly.

FIGURE 19–30
Open circuit failures

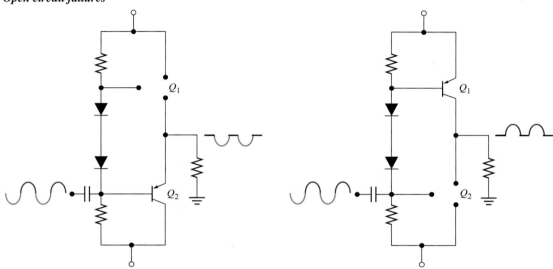

REVIEW QUESTIONS

1. What do you expect to see in the Measured Voltages if the base of a transistor is shorted to the collector of the transistor?

2. What do you expect to see in the Measured Voltages if the base of a transistor is shorted to the emitter of the transistor?

3. What do you expect to see in the Measured Voltages if the collector of a transistor is shorted to the emitter of the transistor?

4. Why can't you use DC Troubleshooting to locate a failed open transistor in a Class B amplifier? What technique must you use to locate failed open transistors in a Class B amplifier?

6. Draw the output signal you would see if Q_1 fails open.

7. Draw the output signal you would see if Q_2 fails open.

SUMMARY

■ The Class A amplifier is biased so that it always remains in the **ACTIVE** state.

■ The Class B amplifier is biased at or near **CUT-OFF.**

■ The maximum load voltage swing attainable in a Class A amplifier is

$$V_{Lpp} = V_C$$

■ Maximum power calculations for the Class A amplifier are

$$P_{CC_{max}} = \frac{V^2_{CC}}{2R_L}$$

$$P_{L_{max}} = \frac{V^2_{CC}}{8R_L}$$

$$\% \text{ efficiency} = 25\%$$

$$P_D = \frac{V^2_{CC}}{4R_L}$$

■ Transformers have the following properties, where V_1 is the primary voltage, V_2 is the secondary voltage, and n is the turns ratio:

$$V_2 = \frac{V_1}{n}$$

$$I_1 = \frac{I_2}{n}$$

$$R_{primary} = n^2 R_{secondary}$$

■ The collector voltage in the transformer-coupled Class A output stage has a maximum peak-to-peak voltage of

$$V_{C_{pp_{max}}} = 2V_{CC}$$

■ The transformer-coupled Class A output stage has a maximum efficiency of

$$\% \text{ efficiency} = 50\%$$

■ The complementary-symmetry Class B amplifier is constructed with an NPN and a PNP transistor biased in or near **CUT-OFF.** Each transistor produces one-half of the output cycle.

■ The maximum power calculations for the Class B output stage are

$$P_{CC_{max}} = \frac{V^2_{CC}}{2\pi R_L}$$

$$P_{L_{max}} = \frac{V^2_{CC}}{2R_L}$$

$$\% \text{ efficiency} = 78.5\%$$

$$P_D \approx 0.1 \frac{V^2_{CC}}{R_L}$$

■ DC Troubleshooting the Class B amplifier will locate only short circuit failures.

■ Open circuit failures of an output transistor in the Class B amplifier can be located with AC signal tracing.

PROBLEMS

SECTION 19–1   Introduction

1. Describe the properties we look for in a power amplifier.
2. Describe the behavior and uses of
   (a) Class A amplifiers.   (b) Class B amplifiers.
   (c) Class AB amplifiers.   (d) Class C amplifiers.

SECTION 19–2   The Common-Emitter Amplifier—Maximum Voltage Swing

3. For the Class A common-emitter amplifier in Figure 19–31, find
   (a) $V_{C_{max}}$ at **CUT-OFF.**   (b) $V_{C_{min}}$ at **SATURATION.**
   (c) V_{CQ} for maximum symmetrical output voltage swing.
   (d) The maximum possible output voltage symmetrical swing.
4. For the circuit of Figure 19–31, find
   (a) The voltage gain.
   (b) The maximum allowable *input* voltage swing $\left(\text{remember } A = \frac{v_o}{v_i}\right)$.

FIGURE 19–31

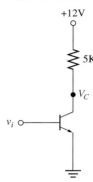

5. For the Class A common-emitter amplifier in Figure 19–32, find
    (a) $V_{C_{max}}$ at **CUT-OFF.**   **(b)** $V_{C_{min}}$ at **SATURATION.**
    (c) V_{CQ} for maximum symmetrical output voltage swing.
    (d) The maximum possible output voltage symmetrical swing.

6. For the circuit of Figure 19–32, $R_1 = 51$ KΩ and $R_2 = 9$ KΩ. Find
    (a) I_{CQ} and V_{CQ}.   **(b)** The maximum positive swing in the output voltage.
    (c) The maximum negative swing in the output voltage.
    (d) The symmetrical output voltage swing.

7. For the circuit of Figure 19–32, $R_1 = 53.5$ KΩ and $R_2 = 605$ KΩ. Find
    (a) I_{CQ} and V_{CQ}.   **(b)** The maximum positive swing in the output voltage.
    (c) The maximum negative swing in the output voltage.
    (d) The symmetrical output voltage swing.

8. For the circuit of Figure 19–32, and the parameters of Problem 6, find
    (a) The voltage gain.
    (b) The maximum allowable *input* voltage swing $\left(\text{remember } A = \dfrac{v_o}{v_i}\right)$.

9. For the Class A common-emitter amplifier in Figure 19–33, find
    (a) $V_{C_{max}}$ at **CUT-OFF.**   **(b)** $V_{C_{min}}$ at **SATURATION.**
    (c) V_{CQ} for maximum symmetrical output voltage swing.
    (d) The maximum possible output voltage symmetrical swing.

10. For the circuit of Figure 19–33, find
    (a) I_{CQ} and V_{CQ}.   **(b)** The quiescent voltage on the capacitor.
    (c) The maximum positive swing in the output voltage.
    (d) The maximum negative swing in the output voltage.
    (e) The symmetrical output voltage swing.

11. For the circuit of Figure 19–33, find
    (a) The voltage gain.
    (b) The maximum allowable *input* voltage swing (remember $A = \dfrac{v_o}{v_i}$).

FIGURE 19–32

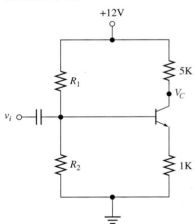

FIGURE 19–33

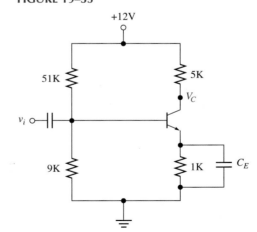

12. For the Class A common-emitter amplifier shown in Figure 19–34 (p. 630), find
    (a) $V_{C_{max}}$ at **CUT-OFF.**   **(b)** $V_{C_{min}}$ at **SATURATION.**
    (c) V_{CQ} for maximum symmetrical collector voltage swing.
    (d) The maximum possible output voltage symmetrical swing.

13. For the circuit of Figure 19–34, find

(a) I_{CQ} and V_{CQ}. (b) The quiescent voltages on the capacitors.

(c) The maximum positive swing in the collector voltage.

(d) The maximum negative swing in the collector voltage.

(e) The symmetrical collector voltage swing. (f) The symmetrical load voltage swing.

FIGURE 19–34

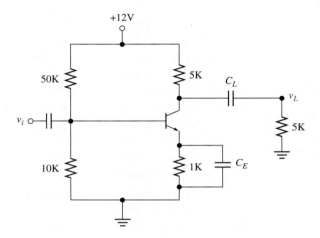

SECTION 19–3 The Emitter Follower—Maximum Voltage Swing

14. For the Class A emitter follower in Figure 19–35, find

(a) $V_{E_{max}}$ at **CUT-OFF.** (b) $V_{E_{min}}$ at **SATURATION.**

(c) The maximum possible output voltage symmetrical swing.

FIGURE 19–35

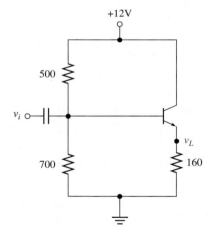

15. For the emitter follower of Figure 19–35, find

(a) I_{CQ} and V_{CQ}. (b) The maximum positive swing in the output voltage.

(c) The maximum negative swing in the output voltage.

(d) The symmetrical output voltage swing.

16. Redraw the emitter follower of Figure 19–35, and capacitively couple a 100 Ω load to the emitter.

(a) What is the quiescent voltage on the load capacitor?

(b) What is the symmetrical output voltage swing (the output voltage is the voltage across the 100Ω load resistor)?

SECTION 19–4 Class A Power Considerations

17. For the circuit of Figure 19–31 (p. 629), find the maximum values for

   (a) P_{CC}.   **(b)** P_L.   **(c)** % efficiency.   **(d)** P_D.

18. If the Q-point in Figure 19–31 (p. 629) is set to $V_{CQ} = 5, 7.5,$ and 12V (bias circuitry not shown), find

   (a) P_{CC}.   **(b)** P_L.   **(c)** % efficiency.   **(d)** P_D.

19. For the circuit of Figure 19–32 (p. 629) and the parameters of Problem 6 (p. 629), find

   (a) P_{CC}.   **(b)** P_L.   **(c)** % efficiency.   **(d)** P_D.

20. For the circuit of Figure 19–32 (p. 629) and the parameters of Problem 6 (p. 629), find

   (a) The power dissipated in the bias resistors.   **(b)** The total power to be supplied by V_{CC}.

21. For the circuit of Figure 19–32 (p. 629), find the Q-point, the symmetrical output voltage swing, P_{CC}, P_L, efficiency, P_D, if $R_1 = 51\text{K}\Omega$ and R_2 is set at

   (a) $R_2 = 12\text{K}\Omega$.   **(b)** $R_2 = 6\text{K}\Omega$.

22. For the circuit of Figure 19–33 (p. 629), find

   (a) P_{CC}.   **(b)** P_L.   **(c)** % efficiency.   **(d)** P_D.

23. For the circuit of Figure 19–33 (p. 629), find

   (a) The power dissipated in the bias resistors.   **(b)** The total power to be supplied by V_{CC}.

24. For the circuit of Figure 19–34, find

   (a) P_{CC}.   **(b)** P_L.   **(c)** % efficiency.   **(d)** P_D.

25. For the circuit of Figure 19–34, find

   (a) The power dissipated in the bias resistors.   **(b)** The total power to be supplied by V_{CC}.

26. The emitter-follower output stage of Figure 19–35 is part of a laboratory amplifier that you have built for Dr. Fumblefingers. The good doctor, unfortunately, is technically challenged and proceeds to connect the output of the emitter follower to ground.

   (a) Find I_{CQ} and V_{CQ} with the emitter output shorted to ground (that is $V_E = 0$V). Note that you cannot use the base resistor voltage divider here. Use the exact formula given in Chapter 9, Section 7 (R_E here is effectively 0Ω).

   (b) What is P_D now?

   (c) In your original circuit you used a transistor that was rated at 5W. What will happen to the transistor when the output is shorted to ground?

SECTION 19–5 Transformer Coupling

27. For the transformer circuit of Figure 19–36, find

   (a) The secondary voltage (V_2).   **(b)** The secondary current (I_2).

   (c) The primary current (I_1).   **(d)** The resistance seen at the input (V_1/I_1).

28. For the transformer-coupled Class A amplifier of Figure 19–37, assume the circuit is biased for maximum swing, and find

   (a) $V_{L_{max}}$, $V_{L_{min}}$ and the symmetrical load voltage swing.

   (b) P_{CC}, P_L, % efficiency, P_D.

FIGURE 19–36

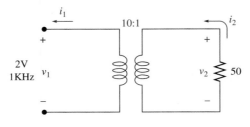

FIGURE 19–37

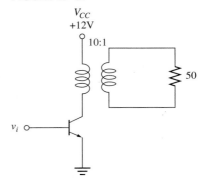

29. For the transformer-coupled circuit of Figure 19–38, find

(a) V_{EQ}, I_{CQ}, V_{CQ}. (b) V_c and V_1 at **SATURATION.**

(c) The maximum swing in V_1 and v_L. (d) P_{CC}, P_L, % efficiency, P_D.

FIGURE 19–38

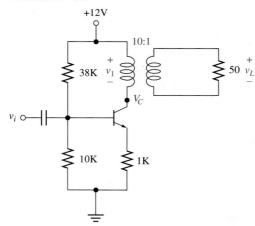

FIGURE 19–39

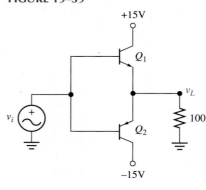

SECTION 19–6   The Class B Amplifier

30. For the Class B amplifier of Figure 19–39

(a) When is Q_1 **ACTIVE?** When is Q_2 **ACTIVE?**

(b) Find P_{CC}, $P_{CCtotal}$, P_L, % efficiency, P_D.

31. For the amplifier of Figure 19–39, plot the input voltage vs. time; directly below plot the output voltage vs. time for input sinusoids with the following peak values (include the 0.7V emitter diode voltages):

(a) 10V. (b) 1V. (c) 15V.

SECTION 19–7   Push-Pull Amplifier Biasing

32. For the amplifier of Figure 19–40, find

(a) V_{B1Q}, V_{B2Q}, V_{EQ}, I_{C1Q}, I_{C2Q}. (b) P_{CC}, $P_{CCtotal}$, P_L, % efficiency, P_D.

(c) The currents in the bias resistors. (d) The total average power supplied by each battery.

33. For the amplifier of Figure 19–41, find

(a) V_{B1Q}, V_{B2Q}, V_{EQ}, I_{C1Q}, I_{C2Q}. (b) P_{CC}, $P_{CCtotal}$, P_L, % efficiency, P_D.

(c) When is Q_1 **ACTIVE?** When is Q_2 **ACTIVE?**

FIGURE 19–40

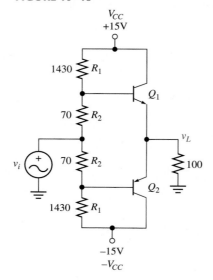

FIGURE 19–41

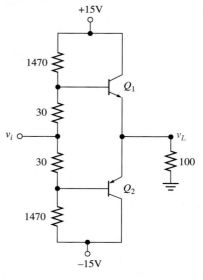

34. For the amplifier of Figure 19–42, find

    (a) V_{B1Q}, V_{B2Q}, V_{EQ}, I_{C1Q}, I_{C2Q}.   **(b)** P_{CC}, $P_{CCtotal}$, P_L, *efficiency*, P_D.

35. For the single-supply amplifier of Figure 19–43, find

    (a) V_{B1Q}, V_{B2Q}, V_{EQ}, I_{C1Q}, I_{C2Q}.   **(b)** P_{CC}, $P_{CCtotal}$, P_L, *efficiency*, P_D.

    (c) The current in the bias resistors.   **(d)** The total average power supplied by the battery.

FIGURE 19–42

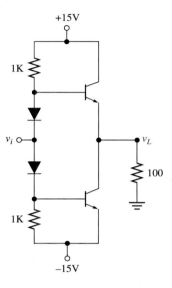

Darlington Output Stage

36. One of the problems that we see in output stages is the large base current that is required to drive the output amplifier.

    (a) The power transistor of the Class A amplifier shown in Figure 19–44A (p. 634) has a β of 25. Assume the amplifier is biased for maximum swing, and find the peak load current and the peak base current.

    (b) In Figure 19–44B (p. 634) we have replaced the single output transistor with a Darlington pair. If $\beta_1 = 100$ and $\beta_2 = 25$, find the peak load current and the peak base current required to drive Q_1.

37. Draw a push-pull amplifier that uses Darlington pairs.

FIGURE 19–43

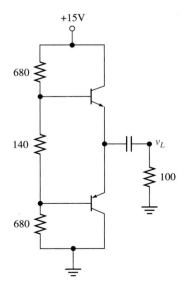

FIGURE 19–44

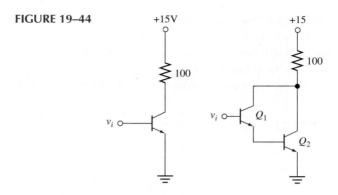

SECTION 19–8   Troubleshooting

38. The Class A amplifier circuits in Figure 19–45 show several possible DC Troubleshooting results. If all connections, resistors, and capacitors are good, what transistor failure could explain each result?

FIGURE 19–45

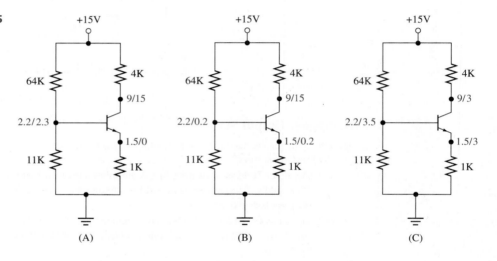

(A)   (B)   (C)

39. DC Troubleshooting of the Class A amplifier circuits in Figure 19–46 shows no discrepancies. The results of AC signal tracing turns up the problems indicated. What is the likely cause of each AC signal tracing result?

FIGURE 19–46

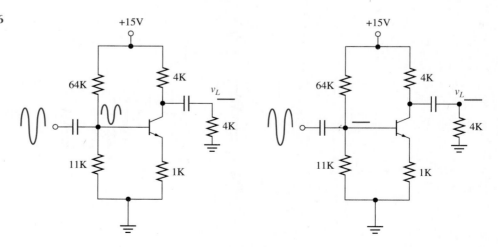

40. The transformer-coupled Class A amplifier shown in Figure 19–47 shows no DC Troubleshooting discrepancies.

  (a) What transistor failure could explain the fact that no signal shows up at the load? Why wouldn't this problem show up during DC Troubleshooting?

  (b) What other problem in this circuit could explain why no signal is seen at the load?

FIGURE 19–47

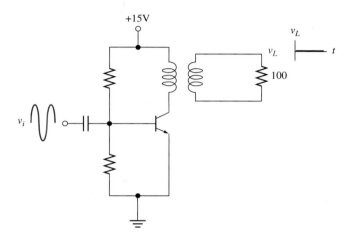

41. What transistor failures could explain the DC Troubleshooting results of the Class B amplifiers shown in Figure 19–48?

42. Figure 19–49 (p. 636) shows DC Troubleshooting results for several diode-biased Class B amplifiers. What transistor failures could explain these results?

Computer Problems

43. Simulate the circuit of Figure 19–32 (p. 629) with the parameters of Problem 6 (p. 629).

  (a) Find the Q-point voltages and currents.

  (b) Apply a 1KHz sinusoidal input through a 30μF capacitor and plot the output voltage for input magnitudes of

  (i) 0.5V.   **(ii)** 1V.   **(iii)** 2V.   **(iv)** 3V.

  (c) Discuss the allowable swing for the input voltage.

FIGURE 19–48

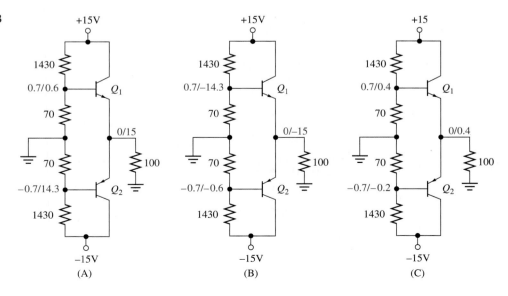

FIGURE 19–49

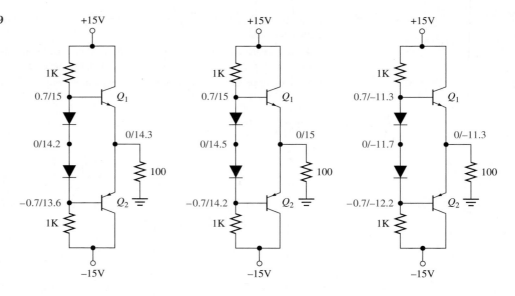

44. Simulate the circuit of Figure 19–33 (p. 629) with $C_E = 30\ \mu F$.

    (a) Find the Q-point voltages and currents.

    (b) Apply a 1KHz sinusoidal input through a 30μF capacitor, and plot the output voltage for input magnitudes of

        (i) 1mV.   **(ii)** 10mV.   **(iii)** 20mV.   **(iv)** 30mV.

    (c) Discuss the allowable swing for the input voltage, and compare to the previous problem.

45. Simulate the circuit of Figure 19–34 (p. 630) with $C_E = 30\ \mu F$, $C_L = 3\ \mu F$.

    (a) Find the Q-point voltages and currents.

    (b) Apply a 1 KHz sinusoidal input through a 30μF capacitor and plot the output voltage for input magnitudes of

        (i) 1mV.   **(ii)** 10mV.   **(iii)** 30mV.   **(iv)** 60mV.

    (c) Discuss the allowable swing for the input voltage, and compare to the two previous problems.

46. Simulate the circuit of Figure 19–40 (p. 632). Note that you will not be able to label two resistors as R_1 or R_2.

    (a) Find the Q-point currents and voltages.

    (b) Halve the value for the two resistors labeled R_1 in Figure 19–40, and find the Q-point currents and voltages.

    (c) Double the value for the two resistors labeled R_1 in Figure 19–40, and find the Q-point currents and voltages.

47. Simulate the circuit of Figure 19–42 (p. 633).

    (a) Find the Q-point voltages and currents.

    (b) Apply a 1KHz sinusoidal input through a 30μF capacitor, and plot the output voltage for input magnitudes of

        (i) 1V.   **(ii)** 5V.   **(iii)** 10V.   **(iv)** 15V.

    (c) Discuss the allowable swing for the input voltage.

48. Replace the diodes in Figure 19–42 (p. 633) with diode-connected transistors, and repeat Problem 47.

PART 3

FIELD EFFECT TRANSISTORS (FET)

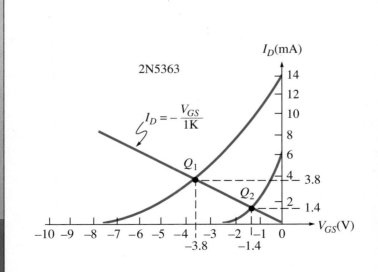

CHAPTER

20

THE FIELD EFFECT TRANSISTOR

CHAPTER OUTLINE

CHAPTER OBJECTIVES

To determine the DC behavior of the Junction Field Effect Transistor (JFET).

To determine the states of the JFET.

To use the JFET I-V curve to bias an N-channel JFET with two sources.

To use the JFET I-V curve to bias an N-channel JFET with a single source.

To use the JFET I-V curve to bias an N-channel JFET with a gate voltage divider.

To bias a P-channel JFET.

To DC Troubleshoot JFET circuits.

JFET AMPLIFIERS

POWER MOSFETS

SMALL-SIGNAL MOSFETS

CMOS DEVICES

20–1 ▪ INTRODUCTION

When the authors were very young, vacuum tubes ruled the electronic waves. Then, during the 1950s the bipolar junction transistor began to replace the vacuum tube, both in the marketplace and in the college curriculum. Also during this time engineers, never ones to leave well enough alone, created an entirely different type of transistor, the Field Effect Transistor (FET).

The FET is a three-terminal device in which a voltage applied at one terminal, the gate (G), controls the current flow between the other two terminals, the source (S) and the drain (D) (Figure 20–1A). Comparing the labels used for the FET and the BJT (Figure 20–1B), we get the following equivalences:

FET	BJT
Drain (D)	Collector (C)
Gate (G)	Base (B)
Source (S)	Emitter (E)

The primary functional difference between the FET and the BJT is that no current (actually, a very, very small current) enters the gate of the FET. A typical BJT base current might be a few microamps, while FET gate currents are typically a thousand times smaller.

FIGURE 20–1
FET-BJT comparison

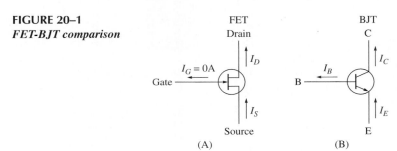

FETs come in several varieties: JFET, D-MOSFET, E-MOSFET, CMOS, and VFET (Figure 20–2A). Each type comes in two "flavors" (N- and P-channel), equivalent to the NPN and PNP BJTs. Figure 20–2B shows the circuit schematic for the N-channel JFET, D-MOSFET, and E-MOSFET. At this point you might be feeling a bit like one of the soldiers in Tennyson's Light Brigade—transistors in front of you, transistors to the left of you, transistors to the right of you.

FIGURE 20–2
Field effect transistors

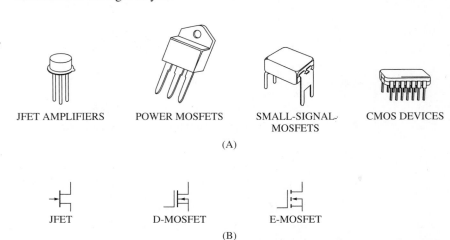

Well, there is some good news and some bad news. The good news is that all FETs exhibit essentially the same small-signal AC behavior as the BJT. As we will show you in Chapter 22, you can determine the response of an FET amplifier by making some notation changes to the formulas we derived for the equivalent BJT amplifier. The bad news is that the DC behavior of an FET is quite different from the DC behavior of the BJT. In fact, there are some differences in DC behavior, even among the various FET types.

In this chapter we will discuss the DC behavior and the DC Troubleshooting of the JFET. In Chapter 21 we will show you the DC behavior and DC Troubleshooting of the MOSFETs. Finally, in Chapter 22 we will discuss the AC behavior of FETs and compare FET applications to BJT applications.

20–2 ■ THE JUNCTION FIELD EFFECT TRANSISTOR (JFET)

Figure 20–3A shows how we form a JFET transistor with a different arrangement of PN junctions. In this construction we start with an N-type piece of silicon that forms a conductive channel between the **drain (D)** and the **source (S)**. We then diffuse a collar of P-type material into the N-channel. The lead connected to this P-type collar is the **gate (G)** of the device. This particular device is known as an N-channel Junction Field Effect Transistor, or **N-channel JFET.**

FIGURE 20–3
The N-channel

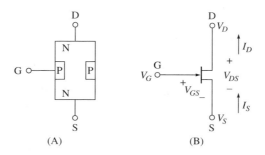

(A)   (B)

You can see that there is one PN junction between the gate and the drain and another PN junction between the gate and the source. As we stated in the introduction, the unique property of the FET is that there is no gate current. This means that both PN junctions of the FET must be **OFF.** This is crucial to the proper operation of all FETs.

For the FET, the PN junctions are never forward biased.

The circuit symbol for the N-channel JFET is shown in Figure 20–3B. Note that

$$V_{DS} = V_D - V_S$$
$$V_{GS} = V_G - V_S$$

Also note that there is no gate current, so

$$I_D = I_S$$

And we usually don't bother labeling I_S.

To determine the behavior of this transistor, we set up the testing circuit shown in Figure 20–4A (p. 642). Here, we connect the gate and the source to ground, so

$$V_{GS} = V_G - V_S = 0V$$

Because the voltage across the gate-source PN junction is less than 0.7V, this junction is not forward biased.

FIGURE 20–4

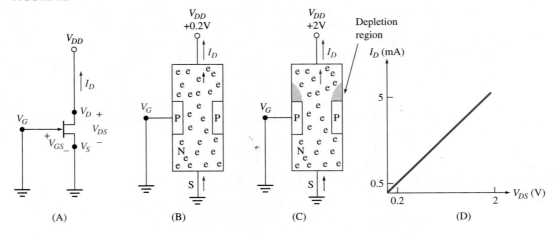

We now apply a DC voltage to the drain (V_{DD}) and measure the drain current as we vary V_{DD}. Here, the drain-to-source voltage is

$$V_{DS} = V_D - V_S = V_{DD} - 0 = V_{DD}$$

Note that a positive voltage at the drain reverse-biases the gate-drain PN junction.

As shown in Figure 20–4B, when we apply a small V_{DS}, there is a conducting channel between the drain and the source. For small values of V_{DS}, the FET behaves as a simple resistor between the drain and the source. Therefore, as V_{DS} is increased from 0V, the drain current (I_D) increases in an approximately linear manner.

As we continue to increase V_{DS}, we increase the reverse-bias voltage on the gate-drain junction. Any reverse-biased PN junction is surrounded by a depletion region, which acts as an insulator. As the reverse-bias voltage is increased, the depletion region surrounding the junction increases (Figure 20–4C).

Figure 20–4C shows how the depletion region around the gate-drain PN junction extends into the conducting channel. The depletion region, therefore, narrows the channel at the drain end of the FET. When V_{DS} is small, the depletion region is small and does not appreciably narrow the channel. For this range of values of V_{DS}, the resistance of the FET remains approximately constant. This leads to the linear I_D-V_{DS} curve shown in Figure 20–4D.

As V_{DS} increases to a value known as the **pinch-off voltage** ($V_{DS_{po}}$), the depletion region increases to the point where the conducting channel at the drain end is completely pinched off (Figure 20–5A). At this point, any further increase in V_{DS} will not produce any additional current.

FIGURE 20–5
JFET I_D-V_{DS}
characteristic

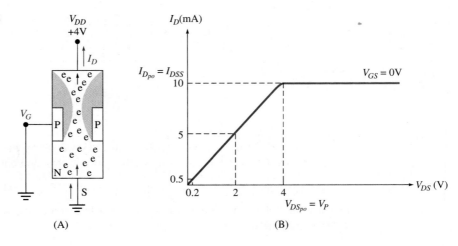

It is important to note that even though the channel at the drain end of the device is pinched-off, there is still current flow to the drain. The positive drain voltage is able to pull the electrons in the channel through this small pinched-off region. This is why the current levels off rather than dropping to 0A. Figure 20–5B shows that the I_D-V_{GS} curve becomes approximately constant for $V_{DS} > V_{DS_{po}}$. This maximum current (I_{DSS}) is known as the *saturation current* of the JFET. (Do not confuse this term with the **SATURATED** state of the BJT.)

This experiment shows another big difference between the BJT and the JFET. You can see that even though the gate of the JFET is held at 0V, there is still a significant current. In fact, this current is the maximum current that can flow in this device. The maximum pinch-off voltage and maximum saturation current in a JFET usually occur when there is no applied gate voltage and are given special names:

$$V_{GS} = 0V \quad \rightarrow \quad I_{D_{max}} = I_{DSS}$$
$$\text{and } V_{DS_{pomax}} = V_p$$

The values I_{DSS} and V_p are given in JFET data sheets. Note that when the term "pinch-off voltage" is used, it usually refers to the maximum pinch-off voltage (V_p).

Remember, if we do not apply a base voltage to the BJT, there is no collector current. The JFET is normally **ON,** whereas the BJT is normally **OFF.**

We now connect another battery (V_{GG}) to the gate (Figure 20–6A). The gate voltage is set by V_{GG}, so

$$V_{GS} = V_G - V_S = V_{GG} - 0 = V_{GG}$$

We can produce a family of curves by choosing several values for V_{GG} and then repeating the experiment of Figure 20–4. As we make the gate voltage more *negative* with respect to the source, we increase the reverse bias on the gate-source PN junction. Again, as a PN junction is reverse biased, a depletion region is created around the junction. This depletion region extends more extensively into the channel than the depletion region caused by the drain voltage (Figure 20–6B).

FIGURE 20–6

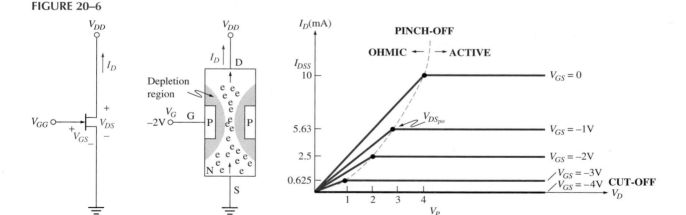

(A) (B) (C)

Here, before we even apply a drain voltage, the gate-source depletion region has significantly narrowed the conducting channel. This effectively raises the channel's resistance and reduces the current for a given V_{DS}. An increase either in V_{DS} or in the magnitude of V_{GS} produces a depletion region in the conducting channel. Therefore, when we apply a negative V_{GS}, it takes a smaller V_{DS} to reach pinch-off. This results in a lower pinch-off voltage and current.

Figure 20–6C (p. 643) shows the I_D-V_{DS} family of curves for a typical JFET. The pinch-off current and voltage for the given values of V_{GS} are

V_{GS}	$I_{D_{po}}$	$V_{DS_{po}}$
0V	10mA (I_{DSS})	4V (V_p)
−1V	5.63mA	3V
−2V	2.5mA	2V
−3V	0.625mA	1V
−4V ($V_{GS(off)}$)	0A	not defined

Eventually we reach a value of V_{GS} that completely blocks all current flow. This value of V_{GS} is known as $V_{GS(off)}$ and is a device parameter given in JFET data sheets. You might also note that $V_p = -V_{GS(off)}$.

In Figure 20–6C (p. 643) we have connected the pinch-off voltages with a dotted line. We label the region to the left of the dotted line as the **OHMIC** region of the JFET. The standard—and somewhat confusing—label for the region of the I_D-V_{DS} curve to the right of $V_{DS_{po}}$ is the *pinch-off region*. This region is actually equivalent to the **ACTIVE** region of the BJT I_C-V_{CE} curve and is where we want to operate if we want to build JFET amplifiers. For this reason, we will label the region to the right of $V_{DS_{po}}$ the **ACTIVE** region of the JFET.

JFET Operating States

You already know that the BJT can operate in one of three states: **ACTIVE, CUT-OFF,** and **SATURATION.** The JFET also has three states of operation, but there are some differences between the two devices.

ACTIVE Region In the BJT the active region extends all the way down to $V_{CE} \approx 0V$. This is true for any applied base voltage and any collector current. On the other hand, in the JFET the active region only extends down to $V_{DS_{po}}$. That is, the JFET is **ACTIVE** when

$$V_{DS} \geq V_{DS_{po}}$$

The pinch-off value can range from approximately 0V up to approximately 10V. This uncertainty is one disadvantage of the JFET. Fortunately, there is a simple way to determine the pinch-off voltage:

$$V_{DS_{po}} = V_p + V_{GS}$$

where, for the N-channel JFET, V_p and $V_{DS_{po}}$ are positive and V_{GS} is negative.

EXAMPLE 20–1

Determining the Pinch-Off Voltage

An N-channel JFET has a maximum pinch-off voltage of $V_p = 4$. Find $V_{DS_{po}}$ for

(a) $V_{GS} = 0V$ **(b)** $V_{GS} = -1V$ **(c)** $V_{GS} = -2V$ **(d)** $V_{GS} = -3V$

Solution   The pinch-off voltages for the N-channel JFET are positive and are given by

$$V_{DS_{po}} = V_p + V_{GS}$$

(a) For $V_{GS} = 0V$

$$V_{DS_{po}} = V_p + V_{GS} = 4 + 0 = 4V = V_p$$

Again, the maximum pinch-off voltage occurs when there is no gate voltage.

(b) For $V_{GS} = -1\text{V}$

$$V_{DS_{po}} = V_p - V_{GS} = 4 + (-1) = 3\text{V}$$

(c) For $V_{GS} = -2\text{V}$

$$V_{DS_{po}} = V_p + V_{GS} = 4 + (-2) = 2\text{V}$$

(d) For $V_{GS} = -3\text{V}$

$$V_{DS_{po}} = V_p + V_{GS} = 4 + (-3) = 1\text{V}$$

These values of pinch-off voltage are the same as those shown in Figure 20–6C (p. 643).

In the BJT **ACTIVE** region, the collector current can be controlled by either the base current or the base-emitter voltage. In the JFET **ACTIVE** region, the drain current is controlled only by the gate-source voltage. We'll examine this relation in the next section.

CUT-OFF Region Both the BJT and the JFET act as open circuits when they are in **CUT-OFF.** In practice, the JFET has smaller leakage currents (currents that flow in **CUT-OFF**) than the BJT. The JFET is in **CUT-OFF** when

$$|V_{GS}| \geq |V_{GS(off)}|$$

BJT SATURATION Region—JFET OHMIC Region

When the BJT is **SATURATED,** the resistance between the collector and the emitter is so small that we assume that there is a short between these two terminals ($V_C \approx V_E$ in Figure 20–7A).

FIGURE 20–7
***BJT SATURATION—
JFET OHMIC region
models***

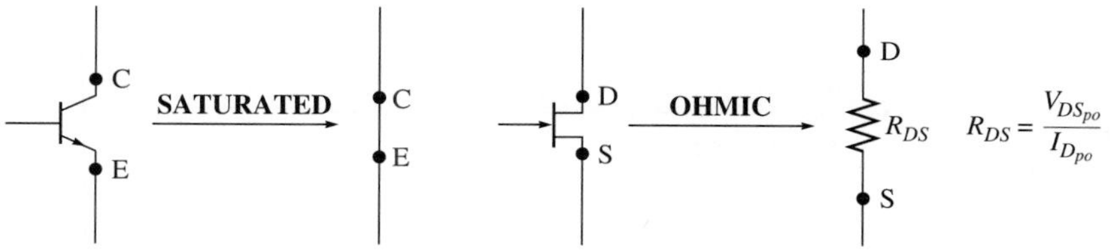

(A)   (B)

In the **OHMIC** region of the JFET, there is still a measurable resistance (R_{DS}) between the drain and the source (Figure 20–7B). This is the resistance of the conducting channel between these two terminals. Because we plot current versus voltage, the approximately straight lines in the **OHMIC** region of the I-V curve shown in Figure 20–6C gives us the conductance between the drain and source. The resistance is the inverse slope of these lines and is approximately

$$R_{DS} \approx \frac{V_{DS_{po}}}{I_{D_{po}}}$$

where $I_{D_{po}}$ is the current at the knee of the I_D-V_{DS} curve. This relation is approximate because the I_D-V_{DS} curve in the **OHMIC** region is not a perfectly straight line.

EXAMPLE 20–2
The OHMIC Region

Find the drain-to-source resistance (R_{DS}) for the family of curves shown in Figure 20–6C (p. 643) when

(a) $V_{GS} = 0\text{V}$ (b) $V_{GS} = -1\text{V}$ (c) $V_{GS} = -2\text{V}$ (d) $V_{GS} = -3\text{V}$

Solution   We use the relation

$$R_{DS} \approx \frac{V_{DS_{po}}}{I_{D_{po}}}$$

and the values given in Figure 20–6C to find the drain-to-source resistance.

(a) For $V_{GS} = 0\text{V}$, $V_{DS_{po}} = V_p = 4\text{V}$, and $I_D = I_{DSS} = 10\text{mA}$. Therefore

$$R_{DS} \approx \frac{V_{DS_{po}}}{I_{D_{po}}} = \frac{4}{10\text{mA}} = 400\Omega$$

(b) For $V_{GS} = -1\text{V}$, we see that $V_{DS_{po}} = 3\text{V}$ and $I_{D_{po}} = 5.63\text{mA}$. Therefore

$$R_{DS} \approx \frac{V_{DS_{po}}}{I_{D_{po}}} = \frac{3}{5.63\text{mA}} = 533\Omega$$

(c) For $V_{GS} = -2\text{V}$, we see that $V_{DS_{po}} = 2\text{V}$ and $I_D = 2.5\text{mA}$. Therefore

$$R_{DS} \approx \frac{V_{DS_{po}}}{I_{D_{po}}} = \frac{2}{2.5\text{mA}} = 800\Omega$$

(d) For $V_{GS} = -3\text{V}$, we see that $V_{DS_{po}} = 1\text{V}$ and $I_D = 0.625\text{mA}$. Therefore

$$R_{DS} \approx \frac{V_{DS_{po}}}{I_{D_{po}}} = \frac{1}{0.625\text{mA}} = 1.6\text{K}\Omega$$

You can see that the minimum R_{DS} occurs when $V_{GS} = 0\text{V}$. As the gate voltage is made more negative, R_{DS} increases into the KΩ range. These are significant values that must be considered when the JFET is not in the **ACTIVE** region.

A summary comparison of the BJT and JFET is given below:

BJT	JFET
Normally **OFF**	Normally **ON**
ACTIVE: $\begin{cases} V_{BE} = 0.7\text{V} \\ V_{CE} > 0\text{V} \end{cases}$	**ACTIVE:** $\begin{cases} \lvert V_{GS} \rvert < \lvert V_{GS(off)} \rvert \\ V_{DS_{po}} = V_p + V_{GS} \\ V_{DS} > V_{DS_{po}} \end{cases}$
CUT-OFF: $\begin{cases} V_{BE} < 0.7\text{V} \\ I_B = I_C = 0\text{A} \end{cases}$	**CUT-OFF:** $\begin{cases} \lvert V_{GS} \rvert \geq \lvert V_{GS(off)} \rvert \\ I_D = 0\text{A} \end{cases}$
SATURATION: $V_{CE} = 0\text{V}$	
	OHMIC: $\begin{cases} V_{DS} \leq V_{DS_{po}} \\ R_{DS} \approx \dfrac{V_{DS_{po}}}{I_{D_{po}}} \end{cases}$

1. Draw a simple picture that shows the construction of the N-channel JFET. Label the three terminal leads.

2. Draw the circuit symbol for the N-channel JFET. Label the terminals and show I_D, V_{GS}, and V_{DS}.

3. What is the gate current of the JFET?

4. What is true about all PN junctions in an FET?

5. Using your FET picture from Question 1, draw the depletion region that occurs when you apply a positive drain-source voltage. Describe pinch-off.

6. Using your FET picture from Question 1, draw the depletion region that occurs when a negative gate-source voltage is applied.

7. Sketch a typical I_D-V_{DS} curve if $I_{DSS} = 20\text{mA}$ and $V_p = -3\text{V}$.

8. What is the maximum drain current? When does it occur?

9. What is the minimum drain current? When does it occur?

10. What are the three states of operation for the JFET?

11. When is the JFET **ACTIVE?**

12. When is the JFET **CUT-OFF?**

13. When is the JFET in the **OHMIC** region?

14. How is the JFET modeled in the **OHMIC** region?

2–3 ■ ACTIVE REGION DRAIN CURRENT VS. GATE VOLTAGE (I_D-V_{GS})

The I_D-V_{GS} Curve

The DC behavior of the JFET in the **ACTIVE** region is more easily determined by looking at the relation between I_D and V_{GS}. We fix the drain voltage at some convenient value that puts us in the **ACTIVE** region of the JFET. We now vary V_{GS} and measure and plot I_D (Figure 20–8). Note that V_{GS} is always negative (keeping the gate-source PN junction reverse biased). The drain current is maximum ($I_D = I_{DSS}$) when $V_{GS} = 0\text{V}$ and goes to zero when $V_{GS} = V_{GS(off)}$. Typical JFET values for $V_{GS(off)}$ vary from -3V to -10V.

FIGURE 20–8
N-channel JFET I_D-V_{GS} characteristic

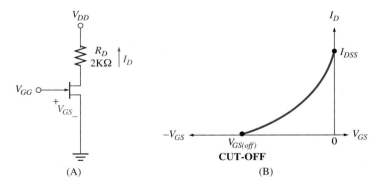

(A)   (B)

Figure 20–9 (p. 648) shows the relation between the I_D-V_{GS} and the I_D-V_{DS} curves for a JFET that has the parameters $V_{GS(off)} = -4\text{V}$ and $I_{DSS} = 10\text{mA}$. Once the gate-source voltage (V_{GS}) has been determined, the drain current is found from the I_D-V_{GS} curve. This

FIGURE 20–9

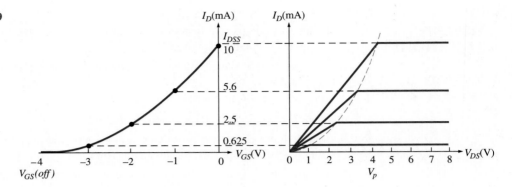

value provides us with the constant drain current ($I_{D_{po}}$) on the I_D-V_{DS} curve. Also, once we know V_{GS}, we can find the pinch-off voltage $V_{DS_{po}}$ from

$$V_{DS_{po}} = V_p + V_{GS}$$

Again, the dotted line in I_D-V_{DS} curve connects the pinch-off voltages and determines the boundary between the **OHMIC** and **ACTIVE** regions. The shape of the dotted line is the same as the I_D-V_{DS} curve.

EXAMPLE 2–3
Constructing the
I_D-V_{DS} **Curve**

Figure 20–10A shows the I_D-V_{GS} curve for a given JFET. Construct the I_D-V_{DS} curve for $V_{GS} = -9V, -6V, -3V, 0V$.

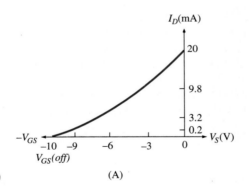

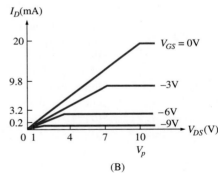

FIGURE 20–10   (A)   (B)

Solution   We can see from the I_D-V_{GS} curve that

$$I_{DSS} = 20\text{mA} \qquad \text{and} \qquad V_{GS(off)} = -10\text{V}$$

Because the maximum pinch-off voltage is equal in magnitude to the gate-source **CUT-OFF** voltage

$$V_p = -V_{GS(off)} = 10\text{V}$$

The values of I_D are ready directly from the I_D-V_{GS} curve. The pinch-off voltages are found from

$$V_{DS_{po}} = V_p + V_{GS} = 10 + V_{GS}$$

For the values of the gate-source voltage given, we get

V_{GS}	I_D	$V_{DS_{po}}$
−9V	0.2mA	1V
−6V	3.2mA	4V
−3V	9.8mA	7V
0V	20mA	10V

To construct the I_D-V_{DS} curve, first draw an approximately straight line from the origin to the point corresponding to a given I_D-$V_{DS_{po}}$ point. Complete the curve by extending to the right an approximately horizontal line from this point.

The resulting family of I_D-V_{DS} curves is shown in Figure 20–10B.

The CUT-OFF, PINCH-OFF Confusion

There is a lot of confusion in textbooks and in manufacturer's data sheets between the terms **CUT-OFF** and **PINCH-OFF.** The way we use these terms clarifies the very definite difference between **CUT-OFF** and **PINCH-OFF.**

☐ **PINCH-OFF** (V_p): The *drain-to-source voltage* (V_{DS}) that closes off the conducting channel at the drain end of the JFET when $V_{GS} = 0$V. At pinch-off, maximum JFET current flows. For drain-to-source voltages greater than V_p, JFET current is approximately constant, and the JFET is in the **ACTIVE** mode. For the N-channel JFET, V_p is positive.

☐ **CUT-OFF** ($V_{GS(off)}$): The *gate-to-source voltage* (V_{GS}) that completely closes off the conducting channel and reduces JFET current to 0A. For gate-to-source voltages with magnitudes greater than $V_{GS(off)}$, the JFET will be in **CUT-OFF.** For the N-channel JFET, $V_{GS(off)}$ is negative.

The confusion arises because the pinch-off voltage and the cut-off voltage for a given device have the same magnitude:

$$|V_p| = |V_{GS(off)}|$$

In fact, some manufacturer's data sheets give the value of V_p, while others give the value of $V_{GS(off)}$. One manufacturer even uses the symbol $V_{GS(off)}$ and calls this symbol "the pinch-off voltage."

Many textbooks perpetuate this confusion by labeling the point at which the I_D-V_{GS} curve (Figure 20–8, p. 647) goes to zero as pinch-off and label the gate-to-source voltage at this point as V_p.

REVIEW QUESTIONS

1. Draw a typical N-channel JFET I_D-V_{GS} curve.

2. For this curve to be accurate, in what state is the JFET operating?

3. When does the maximum current occur on this plot? What is the label used for the maximum current?

4. When is the current 0A? What is the voltage that sets the current to 0A?

5. Sketch the I_D-V_{DS} family of curves that correspond to your I_D-V_{GS} curve. Label the pinch-off voltage V_p.

6. Explain the difference between V_p and $V_{GS(off)}$.

20–4 ■ BASIC N-CHANNEL JFET BIASING

Because there is no gate current in the JFET, all JFETs are biased with gate voltage control circuits. Calculating the bias levels in a JFET circuit requires more work than finding the bias levels in a BJT circuit. You can see why when you compare the I-V characteristics for the JFET and the BJT (Figure 20–11). Because the I-V curve of the BJT is nearly vertical when the BJT is **ACTIVE,** we can model the base-emitter junction as a battery. This means that once we know the base voltage, we immediately find the emitter voltage ($V_E = V_B - 0.7$).

FIGURE 20–11
BJT I-V curve

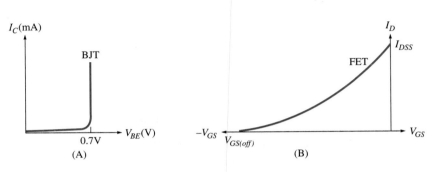

Unfortunately for us, the I-V curve for the JFET is a continually changing function that cannot be modeled with a simple circuit element. To find the bias levels in a JFET circuit, we work with the curve. Consider the basic N-channel JFET circuit shown in Figure 20–12, along with the I-V curve for this particular JFET. In this circuit we see that

$$V_G = V_{GG} = -2V$$

FIGURE 20–12

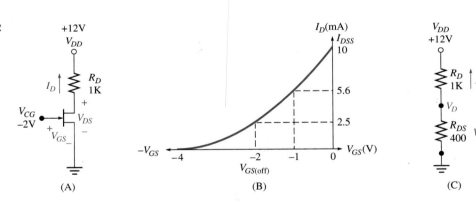

and

$$V_S = 0V$$

Now, because

$$V_{GS} = V_G - V_S$$

we find

$$V_{GS} = 0 - 2 = -2V$$

We can now use the I-V curve to find I_D. Figure 20–12B shows that when $V_{GS} = -2V$

$$I_D = 2.5mA$$

When we analyzed BJT circuits, we always found V_{CE} and checked the state of the transistor. To check the state of the JFET, we first find $V_{DS_{po}}$:

$$V_{DS_{po}} = V_p + V_{GS}$$

and because

$$V_p = -V_{GS(off)} = 4V$$

we find

$$V_{DS_{po}} = V_p + V_{GS} = 4 + (-2) = 2V$$

We now find V_{DS} from the circuit

$$V_{DS} = V_{DD} - R_D I_D$$

For our circuit

$$V_{DS} = V_{DD} - R_D I_D = 12 - 1K \times 2.5 \text{ mA} = 9.5V$$

Because $V_{DS} > V_{DS_{po}}$, the transistor is **ACTIVE.**

EXAMPLE 20–4
N-Channel JFET Biasing

Find the bias levels and the state of the JFET for the circuit in Figure 20–12 if

(a) $V_{GS} = -4V$ **(b)** $V_{GS} = -1V$ **(c)** $V_{GS} = 0V$

Solution

(a) For $V_{GS} = -4V$, we see from the graph in Figure 20–12B that

$$V_{GS} = V_{GS(off)} \qquad \text{so} \qquad I_D = 0A$$

The JFET is in **CUT-OFF.** Because $I_D = 0A$, we find V_{DS} as

$$V_{DS} = V_{DD} - R_D I_D = V_{DD} = 12V$$

(b) For $V_{GS} = -1V$, we see from the graph in Figure 20–12B that

$$I_D = 5.6mA$$

The pinch-off voltage is

$$V_{DS_{po}} = V_p + V_{GS} = 4 + (-1) = 3V$$

We find V_{DS} from

$$V_{DS} = V_{DD} - R_D I_D = 12 - 1K \times 5.6mA = 6.4V$$

Because $V_{DS} > V_{DS_{po}}$, the JFET is **ACTIVE.**

(c) For $V_{GS} = 0V$, we see from the graph in Figure 20–12B that

$$I_D = I_{DSS} = 10mA \qquad \text{and} \qquad V_{DS_{po}} = V_p = 4V$$

We find V_{DS} from

$$V_{DS} = V_{DD} - R_D I_D = 12 - 1K \times 10mA = 2V$$

Stop! Because we found that $V_{DS} < V_{DS_{po}}$, the JFET is in the **OHMIC** region. The values calculated above are wrong. In the **OHMIC** region, the JFET acts as a resistor with a value given by

$$R_{DS} \approx \frac{V_{DS_{po}}}{I_{D_{po}}} = \frac{4}{10mA} = 400\Omega$$

In Figure 20–12C (p. 650) we have replaced the JFET with R_{DS}. The actual **OHMIC** region drain current and drain-source voltage are given by

$$\text{OHMIC region: } I_D = \frac{V_{DD}}{R_{DS} + R_D} \quad \text{and} \quad V_{DS} = \frac{R_{DS}}{R_{DS} + R_D} V_{DD}$$

So

$$I_D = \frac{V_D}{R_{DS} + R_D} = \frac{12}{0.4K + 1K} = 8.57\text{mA}$$

$$V_{DS} = \frac{R_{DS}}{R_{DS} + R_D} V_{DD} = \frac{0.4K}{0.4K + 1K} 12 = 3.42\text{V}$$

DRILL EXERCISE

Redraw the circuit in Figure 20–12 (p. 650), letting $R_D = 1.5\text{K}\Omega$. Repeat Example 20–4 (p. 651).

Answer   (a) **CUT-OFF** $V_{GS} = -4\text{V}$, $I_D = 0\text{A}$, $V_{DS} = 12\text{V}$
(b) **ACTIVE** $V_{GS} = -1\text{V}$, $I_D = 5.6\text{mA}$, $V_{DS} = 3.6\text{V}$
(c) **OHMIC** $R_{DS} = 400\Omega$, $I_D = 6.32\text{mA}$, $V_{DS} = 2.53\text{V}$

EXAMPLE 20–5
The OHMIC Region

A gate-source voltage of $V_{GS} = -1\text{V}$ is applied to the JFET in Figure 20–12A (p. 650). We want the JFET to operate in the **ACTIVE** region. What is the largest value of R_D that can be used in this circuit?

Solution   To keep the JFET in the **ACTIVE** region, we need to make sure that

$$V_{DS} > V_{DS_{po}}$$

For this circuit with $V_{GS} = -1\text{V}$

$$V_{DS_{po}} = 4 + (-1) = 3\text{V}$$

We saw in Example 20–4B (p. 651) that when $V_{GS} = -1\text{V}$, $I_{D_{po}} = 5.6\text{mA}$. The drain-source resistance is found from

$$R_{DS} = \frac{V_{DS_{po}}}{I_{D_{po}}} = \frac{3}{5.6\text{mA}} = 536\Omega$$

We can find the value of R_D that will put the JFET at the boundary between the **OHMIC** and the **ACTIVE** regions ($V_{DS} = V_{DS_{po}}$) by using Figure 20–12C (p. 650):

$$V_{DS} = V_{DS_{po}} \rightarrow V_{DS_{po}} = \frac{R_{DS}}{R_{DS} + R_D} V_{DD} = \frac{0.536K}{0.536K + R_D} 12 = 3$$

Solving for the drain resistor

$$6.43\text{K} = 1.61\text{K} + 3R_D$$

so

$$R_D = \frac{6.43\text{K} - 1.61\text{K}}{3} = 1.61\text{K}\Omega$$

If R_D is larger than 1.61 KΩ, the drain-source voltage will drop below $V_{DS_{po}}$ and the JFET will move out of the **ACTIVE** region and into the **OHMIC** region.

REVIEW QUESTIONS

1. Why are FET circuits biased with gate voltage control circuits?
2. Once you know V_{GS}, how do you find I_D?
3. In what three states can the JFET operate?
4. How do you know that a JFET circuit is not in the **OHMIC** region?
5. How do you model the JFET when it is in the **OHMIC** region?
6. How do you find R_{DS}?

20–5 ■ SINGLE-SOURCE JFET BIASING (GRAPHICAL)

Determining the Q-Point

Figure 20–13 shows one way to bias a JFET with a single source. This is when bias calculations become a bit more difficult. Let's first assume that you have the I-V curve available, in which case we can find the Q-point (the DC bias levels) with a scale and a straightedge.

FIGURE 20–13
Graphical Q-point analysis

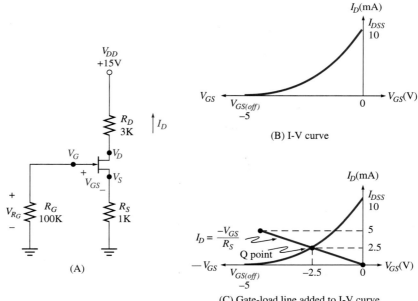

(A)

(B) I-V curve

(C) Gate-load line added to I-V curve

We begin by finding the gate voltage. Because there is no gate current, there is no voltage drop across the gate resistor:

$$V_{R_G} = R_G I_G = 0V$$

so

$$V_G = V_{R_G} = 0V$$

The voltage at the source is found from Ohm's Law:

$$V_S = R_S I_D$$

The gate-to-source voltage is, therefore

$$V_{GS} = V_G - V_S = -R_S I_D$$

Rearrange to solve for the drain current:

$$I_D = -\frac{V_{GS}}{R_S}$$

This is the equation of a straight line, which we will label the **gate load-line.**

We now have two relations for I_D: the I-V curve given in Figure 20–13B (p. 653), and the gate load-line relation just found. The actual I_D and V_{GS} must satisfy *both* of these relations. To find these values, we can plot the gate load-line on the same graph as the I-V curve.

The simplest way to plot $I_D = -V_{GS}/R_S$ is to locate two points on the graph, then connect a straight line between them. We suggest using

$$V_{GS} = 0\text{V} \rightarrow I_D = 0\text{A}$$

and

$$V_{GS} = V_{GS(off)} \rightarrow I_D = \frac{-V_{GS(off)}}{R_S}$$

Note that because $V_{GS(off)}$ of the N-channel JFET is negative, I_D will be positive. For our circuit these values are

$$V_{GS} = 0\text{V} \rightarrow I_D = 0\text{A}$$

and

$$V_{GS} = V_{GS(off)} \rightarrow I_D = -\frac{V_{GS(off)}}{R_S} = -\frac{(-5)}{1\text{K}} = 5\text{mA}$$

Figure 20–13C (p. 653) shows these two points and the load-line connecting them. Because there is only one correct set of values for V_{GS} and I_D, and because both the I-V curve and the load-line must produce these same values, the Q-point *must* be the intersection of these two plots, as shown. We can now read the bias values from the graph:

$$I_D = 2.5\text{mA}$$

and

$$V_{GS} = -2.5\text{V}$$

We always finish bias calculations by finding V_{DS} to check the state of the JFET. Remember, to determine the state of the JFET, we first must find $V_{DS_{po}}$:

$$V_{DS_{po}} = V_p + V_{GS} = 5 + (-2.5) = 2.5\text{V}$$

We find the drain-source voltage from

$$V_{DS} = V_{DD} - (R_D + R_S)I_D$$

Here

$$V_{DS} = V_{DD} - (R_D + R_S)I_D = 15 - (3\text{K} + 1\text{K}) \times 2.5\text{mA} = 5\text{V}$$

Because $V_{DS} > V_{DS_{po}}$, the JFET is **ACTIVE.**

EXAMPLE 20–6
Graphical Bias Calculations

Find the Q-point and bias levels for the JFET circuit in Figure 20–14.

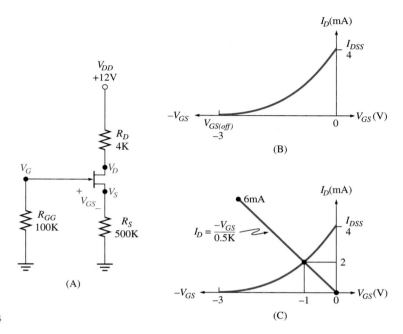

FIGURE 20–14

Solution V_S is found from Ohm's Law to be

$$V_S = R_S I_D = 0.5K I_D$$

Therefore

$$V_{GS} = V_G - V_S = -0.5K I_D$$

so

$$I_D = -\frac{V_{GS}}{0.5K}$$

The two points of interest are

$$V_{GS} = 0V \rightarrow I_D = 0A$$

and

$$V_{GS} = V_{GS(off)} \rightarrow I_D = \frac{-(-3)}{0.5K} = 6mA$$

These two points and their connecting load-line are shown on together with the I-V curve in Figure 20–14C. The Q-point intersection gives us the bias values:

$$I_D = 2mA \qquad \text{and} \qquad V_{GS} = -1V$$

To check the state of the transistor

$$V_{DS_{po}} = V_p + V_{GS} = 3 + (-1) = 2V$$

and

$$V_{DS} = V_{DD} - (R_D + R_S)I_D = 12 - (4K + 0.5K) \times 2mA = 3V$$

Because $V_{DS} > V_{DS_{po}}$, the transistor is **ACTIVE.**

But I Can't Find the JFET I-V Curve!

Don't worry, you can construct your own JFET I-V curve. Luckily for us, the JFET I-V curve can be approximated as a parabola. Figure 20–15A shows a simple parabola; note in particular that the parabola is flat at the origin. Figure 20–15B shows a parabola that has been shifted horizontally to the left. Finally, Figure 20–15C shows the JFET I-V curve superimposed on the parabola. You can see that the parabola is an excellent model for the JFET I-V curve. So, to construct a JFET I-V curve, sketch a parabola that touches the horizontal axis at $V_{GS(off)}$ and intersects the vertical axis at I_{DSS}. You may find it helpful to use one additional point on the graph:

$$I_D = \frac{1}{4}I_{DSS} \text{ when } V_{GS} = \frac{1}{2}V_{GS(off)}$$

FIGURE 20–15
Parabolic curve and the JFET I_D-V_{GS} curve

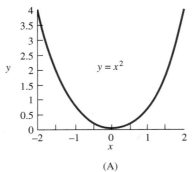
(A)

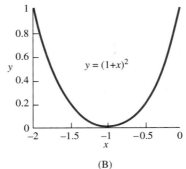
(B)

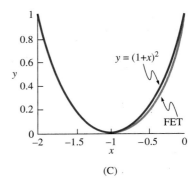
(C)

FIGURE 20–16

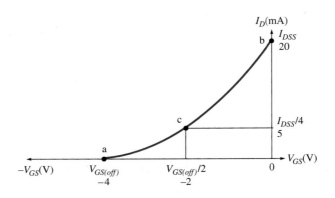

EXAMPLE 20–7
Constructing the JFET I-V Curve

Sketch the JFET I-V curve if the device parameters are $V_{GS(off)} = -4V$ and $I_{DSS} = 20mA$.

Solution   In the graph of Figure 20–16, we have marked the following points:

> Point a: $V_{GS} = V_{GS(off)} = -4V$, $I_D = 0A$
> Point b: $V_{GS} = 0V$, $I_D = I_{DSS} = 20mA$
> Point c: $V_{GS} = \frac{1}{2}V_{GS(off)} = -2V$, $I_D = \frac{1}{4}I_{DSS} = 5A$

and have sketched in the JFET I-V curve.

REVIEW QUESTIONS

1. For the single-source biasing circuit, what is V_G?
2. What is V_S? V_{GS}?
3. Write I_D in terms of V_{GS} and R_S.
4. Sketch the JFET I-V curve.
5. On the same plot, draw the gate load-line.
6. Indicate the Q-point on your graph. Also indicate the bias values for V_{GS} and I_D.
7. What geometric form does the JFET I-V curve resemble?
8. What is I_D when $V_{GS} = \frac{1}{2}V_{GS(off)}$?

20–6 ■ SINGLE-SOURCE JFET BIASING (ANALYTICAL)

The mathematically inclined can find the Q-point of a JFET circuit without using the I-V curve. The equation of the JFET I-V curve is given by

$$I_D = I_{DSS}\left(1 - \frac{V_{GS}}{V_{GS(off)}}\right)^2$$

When we apply this formula to the JFET I-V curve of Figure 20–16, the JFET equation gives us

$$I_D = I_{DSS}\left(1 - \frac{V_{GS}}{V_{GS(off)}}\right)^2 = 10mA\left(1 - \frac{V_{GS}}{(-5)}\right)^2 = 10mA\left(1 + \frac{V_{GS}}{5}\right)^2$$

Don't forget that $V_{GS(off)}$ is negative. From Ohm's Law, we know that

$$I_D = -\frac{V_{GS}}{R_S} = -\frac{V_{GS}}{1K}$$

Substituting the Ohm's Law relation into the JFET equation gives us

$$\frac{-V_{GS}}{1K} = 10mA\left(1 + \frac{V_{GS}}{5}\right)^2$$

We find it easiest to divide through by 10mA (10mA × 1K = 10V) and then square the right side to get

$$\frac{-V_{GS}}{10} = 1 + \frac{2}{5}V_{GS} + \frac{V_{GS}^2}{25}$$

Then we combine the V_{GS} terms and rearrange

$$0.04 V^2_{GS} + 0.5 V_{GS} + 1 = 0$$

Quadratic equations have two solutions; for this equation, we get

$$V_{GS} = -2.5\text{V}$$

and

$$V_{GS} = -10\text{V}$$

Which answer is correct? If you remember that V_{GS} has to lie between 0V and $V_{GS(off)} = -5\text{V}$, we reject the -10V answer to get

$$V_{GS} = -2.5\text{V} \qquad \text{and} \qquad I_D = -\frac{V_{GS}}{R_S} = \frac{-(-2.5)}{1\text{K}} = 2.5\text{mA}$$

The simplest way to solve a quadratic equation is to use a calculator that has a polynomial solver in it. If you don't have such a calculator, you can use the standard quadratic formula:

$$a V^2_{GS} + b V_{GS} + c = 0$$

where for this JFET circuit

$$a = \frac{1}{V^2_{GS(off)}}$$

$$b = \frac{-2}{V_{GS(off)}} + \frac{1}{R_S I_{DSS}}$$

$$c = 1$$

The solution is given by (we give only the result that lies between 0V and $V_{GS(off)}$)

$$V_{GS} = \frac{-b + \sqrt{b^2 - 4ac}}{2a}$$

EXAMPLE 20–8
Analytical Bias Calculation

For the circuit in Figure 20–17, $V_{GS(off)} = -3\text{V}$ and $I_{DSS} = 10\text{mA}$. Find the bias levels for this circuit.

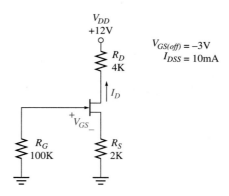

FIGURE 20–17

Solution We first determine a, b, and c:

$$a = \frac{1}{V^2_{GS(off)}} = \frac{1}{(-3)^2} = 0.111$$

$$b = \frac{-2}{V_{GS(off)}} + \frac{1}{R_S I_{DSS}} = \frac{-2}{-3} + \frac{1}{2K \times 10mA} = 0.667 + 0.05 = 0.717$$

$$c = 1$$

and V_{GS} is found from

$$V_{GS} = \frac{-b + \sqrt{b^2 - 4ac}}{2a} = -2.04V$$

The drain current is found from

$$I_D = \frac{-V_{GS}}{R_S} = -\frac{(-2.04)}{2K} = 1.02mA$$

DRILL EXERCISE Repeat Example 20–8 if $V_{GS(off)} = -5V$ and $I_{DSS} = 20mA$.

Answer $V_{GS} = -3.52V$, $I_D = 1.76mA$

REVIEW QUESTIONS

1. Write the JFET I_D-V_{GS} equation.
2. Write I_{DS} in terms of V_{GS} and R_S.
3. Use the two equations to eliminate I_D.
4. Combine all V_{GS} terms and write the final formula.
5. If the standard form for a quadratic equation is $ax^2 + bx + c = 0$, identify a, b, and c from your final JFET formula.
6. How many solutions are there to a quadratic equation? How do you know which solution for V_{GS} is correct?
7. Write the formula that gives us the correct V_{GS} in terms of a, b, and c.

20–7 ■ VOLTAGE DIVIDER BIASING

Occasionally you will se a JFET biased with a voltage divider at the gate (Figure 20–18A, p. 660). In this type of biasing, the gate voltage is not zero. Here

$$V_G = \frac{R_2}{R_1 + R_2} V_{DD}$$

Because V_S is still given by

$$V_S = R_S I_D$$

the gate-to-source voltage is now

$$V_{GS} = V_G - V_S = V_G - R_S I_D$$

FIGURE 20–18

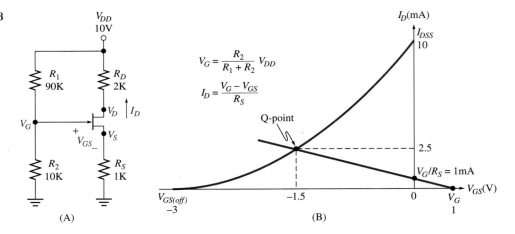

(A)

(B)

Solving for I_D

$$I_D = \frac{V_G}{R_S} - \frac{V_{GS}}{R_S}$$

which is, again, a straight line equation. Now, however, there is the constant term V_G/R_S in the equation. The two points we can use to plot this new gate load-line are

$$V_{GS} = V_G \rightarrow I_D = 0\text{A}$$

and

$$V_{GS} = 0\text{V} \rightarrow I_D = \frac{V_G}{R_S}$$

For our circuit

$$V_G = \frac{R_2}{R_1 + R_2} V_{DD} = \frac{10\text{K}}{10\text{K} + 90\text{K}} 10 = 1\text{V}$$

The two points we need to plot the load-line are

$$V_{GS} = V_G = 1\text{V} \rightarrow I_D = 0\text{A}$$

and

$$V_{GS} = 0\text{V} \rightarrow I_D = \frac{V_G}{R_S} = \frac{1}{1\text{K}} = 1\text{mA}$$

The gate load-line equation is plotted on top of the JFET I-V curve in Figure 20–18B. Note that now the line starts from the positive V_{GS} axis. The Q-point bias values are

$$V_{GS} = -1.5\text{V} \quad \text{and} \quad I_D = 2.5\text{mA}$$

To check the state of the transistor, we find

$$V_{DS_{po}} = V_p + V_{GS} = 3 + (-1.5) = 1.5\text{V}$$

and

$$V_{DS} = V_{DD} - (R_D + R_S)I_D = 10 - (2\text{K} + 1\text{K}) \times 2.5\text{mA} = 2.5\text{V}$$

Because $V_{DS} > V_{DS_{po}}$, the JFET is **ACTIVE.**

EXAMPLE 20–9
Voltage Divider Biasing

Find the DC bias levels for the circuit in Figure 20–19.

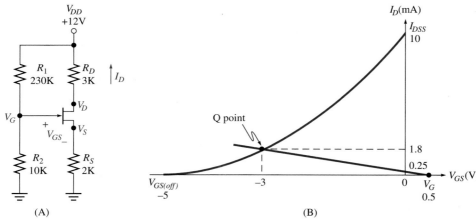

(A)   (B)

FIGURE 20–19

Solution   We first find V_G:

$$V_G = \frac{R_2}{R_1 + R_2} V_{DD} = \frac{10\text{K}}{230\text{K} + 10\text{K}} \, 12 = 0.5\text{V}$$

The two points we need to plot the load-line are

$$V_{GS} = V_G = 0.5\text{V} \rightarrow I_D = 0\text{A}$$

and

$$V_{GS} = 0\text{V} \rightarrow I_D = \frac{V_G}{R_S} = \frac{0.5}{2\text{K}} = 0.25\text{mA}$$

The load-line and the JFET curve are plotted together in Figure 20–19B. The Q-point values are

$$V_{GS} = -3\text{V} \qquad \text{and} \qquad I_D = 1.8\text{mA}$$

To check the state of the transistor

$$V_{DS_{po}} = V_p + V_{GS} = 5 + (-3) = 2\text{V}$$

and

$$V_{DS} = V_{DD} - (R_D + R_S)I_D = 12 - (3\text{K} + 2\text{K}) \times 1.8\text{mA} = 3\text{V}$$

Because $V_{DS} > V_{DS_{po}}$ the transistor is **ACTIVE.**

Analytical Calculation of Bias Levels

The additional term in the gate load-line equation still results in the following quadratic equation:

$$aV^2_{GS} + bV_{GS} + c = 0$$

where for this JFET circuit

$$a = \frac{1}{V^2_{GS(off)}}$$

$$b = \frac{-2}{V_{GS(off)}} + \frac{1}{R_S I_{DSS}}$$

$$c = 1 - \frac{V_G}{R_S I_{DSS}}$$

The solution is given by (we give only the result that lies between 0 and $V_{GS(off)}$)

$$V_{GS} = \frac{-b + \sqrt{b^2 - 4ac}}{2a}$$

EXAMPLE 20–10
Voltage Divider Biasing—Analytical Calculation

Calculate the bias levels for the circuit in Figure 20–19 (p. 661).

Solution We first find V_G:

$$V_G = \frac{R_2}{R_1 + R_2} V_{DD} = \frac{10K}{230K + 10K} 12 = 0.5V$$

We now find a, b, and c:

$$a = \frac{1}{V^2_{GS(off)}} = \frac{1}{-5^2} = 0.04$$

$$b = \frac{-2}{V_{GS(off)}} + \frac{1}{R_S I_{DSS}} = \frac{-2}{-5} + \frac{1}{2K \times 10mA} = 0.45$$

$$c = 1 - \frac{V_G}{R_S I_{DSS}} = 1 - \frac{0.5}{2K \times 10mA} = 0.975$$

The solution is given by (we give only the result that lies between 0 and $V_{GS(off)}$)

$$V_{GS} = \frac{-b + \sqrt{b^2 - 4ac}}{2a} = -2.93V$$

The drain current is found from

$$I_D = \frac{V_G - V_{GS}}{R_S} = \frac{0.5 - (-2.93)}{2K} = 1.72mA$$

DRILL EXERCISE

Redraw the circuit of Figure 20–19 (p. 661). If $V_{GS} = -3$ and $I_{DS} = 5mA$, calculate a, b, c, and the bias levels.

Answer $a = 0.111$, $b = 0.766$, $c = 0.95$, $V_{GS} = -1.62V$, $I_D = 1.06mA$

The Universal JFET I-V Curve

For your convenience, we give you an I-V curve that applies to *all* N-channel JFETs (Figure 20–20). To use it, make a copy of this graph, and label $V_{GS(off)}$ and I_{DSS} with the appropriate values for your device.

FIGURE 20–20

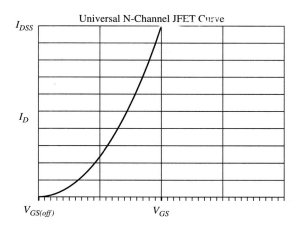

REVIEW QUESTIONS

1. In voltage divider biasing, what is V_G?
2. What is V_S? I_D?
3. What two points can you use to draw the gate load-line?
4. Plot the new gate load-line equation on top of an FET I-V curve. Indicate the Q-point.
5. Write the formulas for a, b, and c in the quadratic equation for V_{GS}.
6. Write the formula for V_{GS}.

20–8 ■ THE P-CHANNEL JFET

Just as the BJT comes in two "flavors," the JFET comes as an N-channel device and a P-channel device. Figure 20–21A shows the construction of the P-channel JFET and its circuit symbol (note the direction of the gate arrow). The P-channel device is used with *negative* supply voltages at the drain. This means that I_D for the P-channel goes from the drain to the source. To **CUT-OFF** the P-channel, we apply a *positive* voltage between the gate and the source. Figure 20–21B shows the gate-source I-V curve for the P-channel JFET.

FIGURE 20–21
P-channel JFET

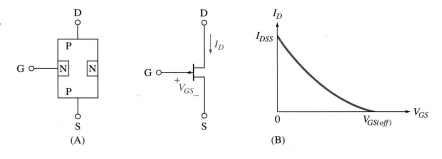

(A)   (B)

Finding Bias Values

We find the bias values for the P-channel JFET the same way that we determine the Q-point for the N-channel JFET. For the P-channel in Figure 20–22 (p. 664)

$$I_D = -\frac{V_S}{R_S}$$

FIGURE 20–22

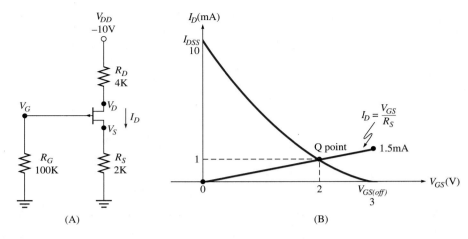

(A)   (B)

and

$$V_{DS} = V_{DD} + R_D I_{DS}$$

The drain-source pinch-off voltage is found from

$$V_{DS_{po}} = V_p + V_{GS}$$

where $V_{DS_{po}}$ and V_p are negative and V_{GS} is positive. For the JFET to be **ACTIVE**

$$|V_{DS}| > |V_{DS_{po}}|$$

Note the use of absolute values in this relation.

EXAMPLE 20–11
P-Channel Biasing,
($V_G = 0V$)

Find the bias levels for the P-channel circuit in Figure 20–22.

Solution   We first find V_{GS}:

$$V_{GS} = V_G - V_S = -V_S$$

Therefore

$$V_S = -V_{GS}$$

The drain current is given by

$$I_D = \frac{-V_S}{R_S} = \frac{V_{GS}}{R_S}$$

Note that the slope of this equation is positive.
For our circuit

$$I_D = \frac{V_{GS}}{2K}$$

The two points we need to draw the load-line are

$$V_{GS} = 0V \rightarrow I_D = 0A$$

and

$$V_{GS} = V_{GS(off)} = 3V \rightarrow I_D = \frac{V_{GS(off)}}{R_S} = \frac{3}{2K} = 1.5mA$$

Note that $V_{GS(off)}$ is now positive. The load-line is plotted along with the I-V curve in Figure 20–22B. The Q-point values are

$$V_{GS} = 2\text{V} \qquad \text{and} \qquad I_D = 1\text{mA}$$

To check the state of the JFET, we find

$$V_{DS_{po}} = V_p + V_{GS} = -3 + 2 = -1\text{V}$$

and

$$V_{DS} = V_{DD} + (R_D + R_S)I_D = -10 + (4\text{K} + 2\text{K}) \times 1\text{mA} = -4\text{V}$$

Because $|V_{DS}| > |V_{DS_{po}}|$, the JFET is **ACTIVE.**

DRILL EXERCISE   Redraw the circuit of Figure 20–22, changing the source resistor to $R_S = 1\text{K}\Omega$. Find the bias values.

Answer $V_{GS} = 1.7\text{V}, I_D = 1.7\text{mA}$

EXAMPLE 20–12   Find the bias levels for the circuit in Figure 20–23.

**Voltage Divider
P-Channel Biasing**

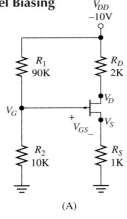

(A)

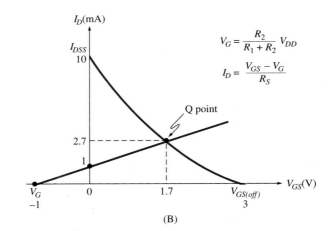

(B)

FIGURE 20–23

Solution   We first find the gate voltage:

$$V_G = \frac{R_2}{R_1 + R_2} V_{DD}$$

For our circuit

$$V_G = \frac{R_2}{R_1 + R_2} V_{DD} = \frac{10\text{K}}{90\text{K} + 10\text{K}}(-10) = -1\text{V}$$

The source voltage is still given by

$$V_S = V_G - V_{GS}$$

Now

$$I_D = \frac{-V_S}{R_S}$$

so

$$I_D = \frac{-V_G}{R_S} + \frac{V_{GS}}{R_S}$$

The two points we need to draw the gate load-line are

$$V_{GS} = 0V \rightarrow I_D = \frac{-V_G}{R_S}$$

and

$$V_{GS} = V_G \rightarrow I_D = 0A$$

Note that V_G is negative. For our circuit

$$V_{GS} = 0V \rightarrow I_D = \frac{-V_G}{R_S} = \frac{-(-1)}{1K} = 1mA$$

and

$$V_{GS} = -V_G = -(-1) = 1V \rightarrow I_D = 0A$$

The load-line and the I-V curve are shown in Figure 20–23B (p. 665). The Q-point values are

$$V_{GS} = 1.7V \qquad \text{and} \qquad I_D = 2.7mA$$

To check the state of the JFET, we find

$$V_{DS_{po}} = V_p + V_{GS} = -3 + 1.7 = -1.3V$$

and

$$V_{DS} = V_{DD} + (R_D + R_S)I_D = -10 + 3K \times 2.7 = -1.9V$$

Because $|V_{DS}| > |V_{DS_{po}}|$, the JFET is **ACTIVE.**

DRILL EXERCISE   Redraw the circuit of Figure 20–23 (p. 665), changing the supply voltage to $V_{DD} = -20V$. Find the bias values.

Answer $V_{GS} = 1.4V$, $I_D = 3.4mA$

For your convenience, we give you the universal P-channel I-V curve in Figure 20–24. To use it, make a copy, and write in the the values for $V_{GS(off)}$ and I_{DSS}.

FIGURE 20–24

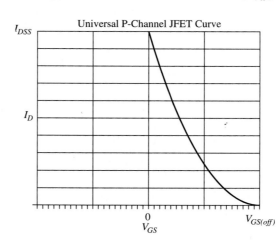

Universal P-Channel JFET Curve

Analytical Calculations

We still solve the quadratic equation:

$$a\,V_{GS}^2 + b\,V_{GS} + c = 0$$

Now, however, for the P-channel JFET, the solution is given by

$$V_{GS} = \frac{-b - \sqrt{b^2 - 4ac}}{2a}$$

where for the P-channel JFET

$$a = \frac{1}{V_{GS(off)}^2}$$

$$b = \frac{-2}{V_{GS(off)}} - \frac{1}{R_S I_{DSS}}$$

and

$$c = 1 + \frac{V_G}{R_S I_{DSS}}$$

EXAMPLE 20–13

Analytical Bias Calculation
($V_G = 0V$)

Calculate the bias values for the circuit of Figure 20–22 (p. 664).

Solution   We first find a, b, and c:

$$a = \frac{1}{V_{GS(off)}^2} = \frac{1}{3^2} = 0.111$$

$$b = \frac{-2}{V_{GS(off)}} - \frac{1}{R_S I_{DSS}} = \frac{-2}{3} - \frac{1}{2K \times 10mA} = -0.716$$

Because $V_G = 0V$

$$c = 1 + \frac{V_G}{R_S I_{DSS}} = 1$$

V_{GS} is found from

$$V_{GS} = \frac{-b - \sqrt{b^2 - 4ac}}{2a} = 2.04V$$

and the drain current is

$$I_D = -\frac{V_G - V_{GS}}{R_S} = -\frac{-2.04}{2K} = 1.02mA$$

DRILL EXERCISE

Redraw the circuit of Figure 20–22 (p. 664), changing the source resistor to $R_S = 500\Omega$ and the drain resistor to $R_D = 2K\Omega$. Calculate a, b, c, and the bias levels.

Answer $a = 0.111$, $b = -0.866$, $c = 1$, $V_{GS} = 1.41V$, $I_D = 2.82mA$

EXAMPLE 20–14
Voltage Divider Biasing— Analytical

Calculate the bias values for the circuit of Figure 20–23 (p. 665).

Solution We first find V_G:

$$V_G = \frac{R_2}{R_1 + R_2} V_{DD} = \frac{10K}{90K + 10K}(-10) = -1V$$

We now find a, b, and c:

$$a = \frac{1}{V^2_{GS(off)}} = \frac{1}{3^2} = 0.11$$

$$b = \frac{-2}{V_{GS(off)}} - \frac{1}{R_S I_{DSS}} = \frac{-2}{3} - \frac{1}{2K \times 10mA} = -0.716$$

and

$$c = 1 + \frac{V_G}{R_S I_{DSS}} = 1 + \frac{-1}{1K \times 10mA} = 0.9$$

V_{GS} is found from

$$V_{GS} = \frac{-b - \sqrt{b^2 - 4ac}}{2a} = 1.71V$$

and the drain current is

$$I_D = -\frac{V_G - V_{GS}}{R_S} = -\frac{-1 - 1.71}{1K} = 2.71mA$$

DRILL EXERCISE

Redraw the circuit of Figure 20–23 (p. 665), changing the supply voltage to $V_{DD} = -20V$. Find a, b, c, and the bias values.

Answer $a = 0.111$, $b = -0.716$, $c = 0.8$, $V_{GS} = 1.44V$, $I_D = 3.41mA$

REVIEW QUESTIONS

1. Draw a simple picture that shows the construction of the P-channel JFET. Label the three terminal leads.
2. Draw the circuit symbol for the P-channel JFET. Label the terminals and show I_D, V_{GS}, and V_{DS}. Sketch the P-channel JFET I-V curve.
3. What are the polarities of V_{DD}, V_{GS}, V_{DS}?
4. What is Ohm's Law for the P-channel I_D (assume V_G is *not* 0)?
5. What is the sign of the slope of the gate load-line Ohm's Law relation?
6. What two points are used to draw this load-line?
7. Draw a plot with the I-V curve and the load-line. Indicate the Q-point and bias values.
8. What are a, b, and c for the P-channel JFET?
9. What is the formula for V_{GS}?

20–9 ▪ JFET DATA SHEETS AND COMPUTER MODELS

Figure 20–25 shows the manufacturer's specification sheet for several Motorola N-channel and P-channel JFETs. For now, just examine the last three columns of the specifications.

FIGURE 20–25

Low-Frequency/Low-Noise

JFETs

STYLE 3:
PIN 1. DRAIN
    2. SOURCE
    3. GATE
    4. CASE LEAD

STYLE 5:
PIN 1. SOURCE
    2. GATE 1
    3. DRAIN
    4. CASE

N-CHANNEL JFETs TO-206AF (TO-72) Case 20-03, Style 3

Device	$\|Y_{fs}\|$ mmho Min	$\|Y_{os}\|$ μmho Max	C_{iss} pF Max	C_{rss} pF Max	$V_{(BR)GSS}$ V Min	$V_{GS(off)}$ V Min	$V_{GS(off)}$ V Max	I_{DSS} mA Min	I_{DSS} mA Max
2N4117#	0.07	3	3	1.5	40	0.6	1.8	0.03	0.09
2N4117A#	0.07	3	3	1.5	40	0.6	1.8	0.03	0.09
2N4118#	0.08	5	3	1.5	40	1	3	0.08	0.24
2N4118A#	0.08	5	3	1.5	40	1	3	0.08	0.24
2N4119#	0.1	10	3	1.5	40	2	6	0.2	0.6
2N4119A#	0.1	10	3	1.5	40	2	6	0.2	0.6
2N4220	1	10	6	2	30	—	4	0.5	3
2N4220A	1	10	6	2	30	—	4	0.5	3
2N5358	1	10	6	2	40	0.5	3	0.5	1
2N5359	1.2	10	6	2	40	0.8	4	0.6	1.6
2N5360	1.4	20	6	2	40	0.8	4	0.5	2.5
2N5361	1.5	20	6	2	40	1	6	2.5	5
2N3821***#	1.5	10	6	3	50	—	4	0.5	2.5
2N4221	2	20	6	2	30	—	6	2	6
2N4221A	2	20	6	2	30	—	6	2	6
2N5362	2	40	6	2	40	2	7	4	8
2N4222	2.5	40	6	2	30	—	8	5	15
2N4222A	2.5	40	6	2	30	—	8	5	15
2N5363	2.5	40	6	2	40	2.5	8	7	14
2N5864	2.7	60	6	2	40	2.5	8	9	18
2N3822***#	3	20	6	3	50	—	6	2	10
2N5556##	1.5	20	6	3	30	0.2	4	0.5	2.5

P-CHANNEL JFETs TO-206AF (TO-72) Case 20-03, Style 5

Device	$\|Y_{fs}\|$ mmho Min	$\|Y_{os}\|$ μmho Max	C_{iss} pF Max	C_{rss} pF Max	$V_{(BR)GSS}$ V Min	$V_{GS(off)}$ V Min	$V_{GS(off)}$ V Max	I_{DSS} mA Min	I_{DSS} mA Max
2N5265	0.9	75	7	2	60	—	3	0.5	1
2N3909	1	100	32	16	20	—	8	0.3	15
2N5266	1	75	7	2	60	—	3	0.8	1.6
2N3330***	1.5	40	20	—	20	—	6	2	6
2N5267	1.5	75	7	2	60	—	6	1.5	3
2N5268	2	75	7	2	60	—	6	2.5	5
2N3909A	2.2	100	9	3	20	—	8	1	15
2N5269	2.2	75	7	2	60	—	8	4	8
2N5270	2.5	75	7	2	60	—	8	7	14

***JAN, JTX, JTXV   #Case Style 1   ##Case Style 5

JFET Specifications

☐ $V_{(BR)GSS}$: the breakdown voltage for the gate-to-source PN junction. Remember, the normal operation of an FET requires that no PN junction is forward biased. As you know, if the reverse bias on a PN junction gets large enough, the junction will go into breakdown, and the device will stop functioning properly.

☐ $V_{GS(off)}$: cut-off voltage. Note that $V_{GS(off)}$ is given as a positive voltage for both the N-channel and the P-channel JFETs. You must remember to use a negative voltage for the N-channel $V_{GS(off)}$. In some cases a minimum and maximum value is given. In other cases only a maximum value is given. The **CUT-OFF** voltage ranges from a low of 0.2V to a high of 8V.

☐ I_{DSS}: the maximum JFET current, found by setting the gate voltage to 0V. This current varies from a low of 0.2mA to a high of 15mA.

When you use the values given to construct the JFET I-V curve or to plug into the analytical solution, be sure to use the minimum $V_{GS(off)}$ with the minimum I_{DSS} or the maximum $V_{GS(off)}$ with the maximum I_{DSS}.

Very often we will use both the minimum and maximum values of the device parameters to establish a range for the Q-point. Consider Figure 20–26, which shows the two gate I-V curves that correspond to the minimum and maximum $V_{GS(off)}$ and I_{DSS} for the 2N5363 (Figure 20–25, p. 669). The figure also shows a typical gate load-line. In this case the gate voltage can vary from -3.8 to -1.4V. The drain current can vary from 3.8 to 1.4mA.

FIGURE 20–26

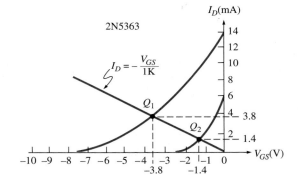

JFET Computer Model

Figure 20–27 shows the parameters used to model the JFET in the computer program Electronic Work Bench.

Threshold Voltage (VTO)   The threshold voltage is the term this computer program uses for the gate-source cut-off voltage. That is

$$V_{GS(off)} \equiv VTO$$

Transconductance Coefficient (β)   What term is missing? You don't see a term for I_{DSS}. Many computer programs use a different formulation for I_D:

$$I_D = \beta(V_{GS(off)} - V_{GS})^2$$

The term β here has nothing to do with β used in BJT models. For the FET computer model, β is the *transconductance coefficient* and has the units of A/V^2. In Chapter 22 we will introduce the term *transconductance* (g_m), which is a small-signal AC parameter. Even though the terms are similar, there is no direct relation between β and g_m.

FIGURE 20–27

N-Channel JEFT Model 'J2N5458'		
Threshold voltage (VTO):	–3.5	V
Transconductance coefficient (β):	0.00045	A/V
Channel-length modulation (λm):	0.0103	1/V
Drain ohmic resistance (rd):	4.94	Ω
Source ohmic resistance (rs):	4.45	Ω
Gate-junction saturation current (Is):	5e-15	A
Zero-bias gate-drain junction capacitance (Cgd):	3e-12	F
Zero-bias gate-source junction capacitance (Cgs):	1.5e-12	F
Gate-junction potential (ϕB):	1	V

Accept | Cancel

We have used the formulation

$$I_D = I_{DSS}\left(1 - \frac{V_{GS}}{V_{GS(off)}}\right)^2$$

Comparing the two formulas for the drain current, we can derive the following relations:

$$\beta = \frac{I_{DSS}}{V^2_{GS(off)}} \qquad \text{or} \qquad I_{DSS} = \beta V^2_{GS(off)}$$

EXAMPLE 20–15
Data Sheet Parameters and Computer Model Parameters

Refer to Figures 20–25 (p. 669) and 20–27 to answer the following:

(a) Find VTO and β for the 2N4220 (use the maximum values in Figure 20–25).
(b) Find $V_{GS(off)}$ and I_{DSS} for the J2N5458.

Solution

(a) We see from Figure 20–25 (p. 669) that the 2N4220 is an N-channel JFET with the following parameters:

$$\left|V_{GS(off)}\right| = 4V \qquad \text{and} \qquad I_{DSS} = 3mA$$

Because this device is an N-channel JFET

$$VTO = V_{GS(off)} = -4V$$

The transconductance coefficient is found from

$$\beta = \frac{I_{DSS}}{V^2_{GS(off)}} = \frac{3mA}{(-4)^2} = 0.000188 A/V^2$$

(b) From Figure 20–27, we see that for the J2N5458

$$VTO = -3.5V \qquad \text{and} \qquad \beta = 0.00045 A/V^2$$

Therefore

$$V_{GS(off)} = VTO = -3.5V$$

The drain current is found from

$$I_{DSS} = \beta V^2_{GS(off)} = 0.00045 \times (-3.5)^2 = 5.51 mA$$

1. What are some typical values for the gate-source PN junction breakdown voltage?
2. What are typical values for the cut-off voltage ($V_{GS(off)}$)?
3. What are typical values for the maximum drain current (I_{DSS})?
4. Computer models use the transconductance coefficient (β). How can you find β from the information given in a manufacturer's data sheet?
5. Given β, how can you find I_{DSS}?

20–10 ■ TROUBLESHOOTING THE JFET

Out-of-Circuit Tests

Both N-channel and P-channel JFETs can be tested with an analog ohmmeter or digital multimeter. Figure 20–28A shows what readings to expect when you test a JFET with an analog ohmmeter. If the JFET is good you will see

☐ A low resistance reading between the drain and the source (normally **ON** channel between D and S).

☐ A low reading in one direction and a high reading in the other direction from the gate to the source (PN junction).

☐ A low reading in one direction and a high reading in the other direction from the gate to the drain (PN junction).

FIGURE 20–28

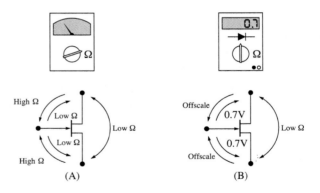

(A)   (B)

Figure 20–28B shows what to expect when you test a good JFET with a digital multimeter that has a diode tester.

☐ A low resistance reading between the drain and the source (normally **ON** channel between D and S).

☐ A reading of approximately 0.7V between the gate and the source in one direction and an offscale reading in the other direction (PN junction).

☐ A reading of approximately 0.7V between the gate and the drain in one direction and an offscale reading in the other direction (PN junction).

DC Troubleshooting of JFET Circuits

We apply DC Troubleshooting to JFET circuits in a slightly different manner than we apply it to BJT circuits. Figure 20–29 shows the results of DC Troubleshooting a JFET circuit where the JFET has an internal open in the gate (A), the drain (B), and the source (C). Note that the Estimated Voltages at the drain and the source are not shown. In BJT circuits we can find the Estimated Voltages by assuming that $V_{BE} = 0.7$V. In JFET circuits, we need to know the FET parameters ($V_{GS(off)}$ and I_{DSS}) to find the Estimated Voltages at the drain and the source. When you troubleshoot JFET circuits, it is unlikely that you will know what particular JFET you are using. Even so, we can make the following general statements.

FIGURE 20–29
JFET open failures

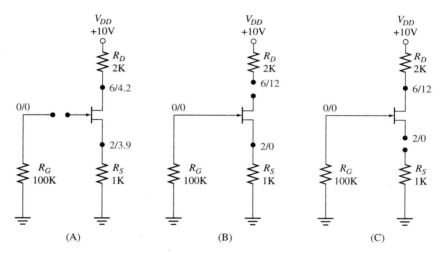

Open Gate   Because there is no current into the gate in a good JFET, we will not see any discrepancies in the Measured Voltages at the gate of a bad JFET. With the gate open, the JFET will act as a simple resistor between the drain and the source. The complicated physics that levels off the drain current (Figure 20–5B, p. 642) will not be operative. This small resistance of the JFET channel will drop only a small voltage, so the source and drain voltage will be very close to each other.

Open Drain or Open Source   As you can see from Figures 20–29B and C, an internal open at the drain or source will prevent the existence of any drain current ($I_D = 0$A). In this case, we expect to see the supply voltage at the drain and 0V at the source.

FIGURE 20–30
JFET short failures

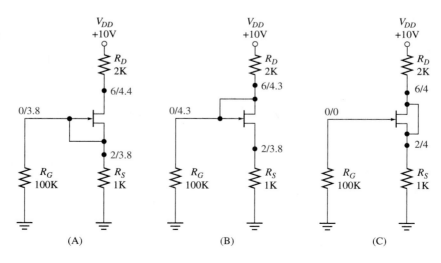

Shorts Figure 20–30 (p. 673) shows the results of DC Troubleshooting a JFET circuit in which the JFET has an internal short between the gate and source (A), gate and drain (B), and drain and source (C). The causes of these results are a bit obscure, but the interpretation is fairly straightforward. An internal short between any two terminals will result in the same Measured Voltage at the two terminals.

EXAMPLE 20–16
DC
Troubleshooting

Figure 20–31 shows the results of DC Troubleshooting several JFET circuits. Determine a likely cause for the observed measurements. Assume the resistors and all connections are good.

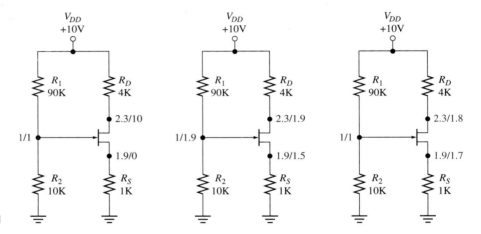

FIGURE 20–31

Solution

(a) The measurements indicate that there is no drain current. Therefore, either the drain or the source is internally open.

(b) Because the drain and gate measurements are the same, we can conclude that there is an internal short between the gate and the drain.

(c) Because the drain voltage is close to the source voltage, the likely cause here is an internal open at the gate.

REVIEW QUESTIONS

1. Describe the Measured Voltages you would find if the JFET has an internal open at the gate, source, or drain.

2. Describe the Measured Values you would find if the JFET has an internal short between the gate and source, the gate and drain, the drain and source.

SUMMARY

■ The JFET is normally **ON**. When the gate voltage is 0V, the drain current is maximum and is I_{DSS}.

■ A negative gate voltage is applied to the N-channel JFET to cut off the channel. Drain current is 0A when $V_{GS} = V_{GS(off)}$.

■ A positive gate voltage is applied to the P-channel JFET to cut off the channel.

■ The JFET I-V curve can be approximated as a parabola.

■ When V_{GS} is known, I_D is read directly from the JFET I-V curve.

■ When the gate is grounded, the bias values are found by drawing the following gate load-lines on the same graph as the JFET I-V curve:

$$I_D = \frac{-V_{GS}}{R_S} \qquad \text{N-Channel JFET}$$

$$I_D = \frac{V_{GS}}{R_S} \qquad \text{P-Channel JFET}$$

■ When a voltage divider at the gate is used, the load-line equations are

$$I_D = \frac{V_G}{R_S} - \frac{V_{GS}}{R_S} \qquad \text{N-channel JFET}$$

$$I_D = \frac{-V_G}{R_S} + \frac{V_{GS}}{R_S} \qquad \text{P-channel JFET}$$

where $V_G = \dfrac{R_2}{R_1 + R_2} V_{DD}$

■ Bias values can be calculated from the quadratic equation

$$aV^2_{GS} + bV_{GS} + c = 0$$

■ For the N-channel JFET

$$a = \frac{1}{V^2_{GS(off)}}$$

$$b = \frac{-2}{V_{GS(off)}} + \frac{1}{R_S I_{DSS}}$$

$$c = 1 - \frac{V_G}{R_S I_{DSS}}$$

■ For the P-channel JFET

$$a = \frac{1}{V^2_{GS(off)}}$$

$$b = \frac{-2}{V_{GS(off)}} - \frac{1}{R_S I_{DSS}}$$

$$c = 1 + \frac{V_G}{R_S I_{DSS}}$$

■ The solution for V_{GS} must lie between 0 and $V_{GS(off)}$.
■ Typical values for I_{DSS} range from 0.2mA to 15mA.
■ Typical values for $V_{GS(off)}$ range from 0.2V to 10V.
■ DC Troubleshooting procedures for the JFET produce the following:
Open Gate: Measured Voltages at the drain and source are similar.
Open Drain or Source: Measured Voltages at the drain and source reflect the fact that $I_D = 0$A.
Shorts: Measured Voltages at the two shorted terminals are identical.

PROBLEMS

SECTION 20–2   The Junction Field Effect Transistor

1. An N-channel JFET has a maximum pinch-off voltage of $V_p = 5$V. Find the drain-source pinch-off voltage ($V_{DS_{po}}$) if
   (a) $V_{GS} = 0$V   (b) $V_{GS} = -2$V   (c) $V_{GS} = -4$V

2. Find the drain-source resistance (R_{DS}) for an N-channel JFET that has the following parameters:
   (a) $V_p = 4$V and $I_{DSS} = 20$mA   (b) $V_p = 2$V and $I_{DSS} = 10$mA
   (c) $V_p = 10$V and $I_{DSS} = 40$mA

3. Indicate the **OHMIC** and **ACTIVE** regions for the I_D-V_{DS} curve in Figure 20–32 (p. 676).

FIGURE 20–32

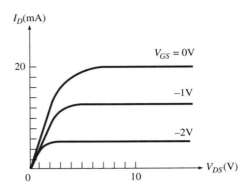

4. Figure 20–32 shows a JFET drain-source family of curves.
   (a) What is I_{DSS}?   (b) What is V_p?
   (c) Find $I_{D_{po}}$ and $V_{DS_{po}}$ for each curve shown.   (d) Find R_{DS} for each curve shown.

SECTION 20–3   Active Region Drain Current vs. Gate Voltage

5. Figure 20–33 shows a JFET I_D-V_{GS} curve.
   (a) What is I_{DSS}?   (b) What is $V_{GS(off)}$?
   (c) What is the pinch-off voltage for this device?

FIGURE 20–33

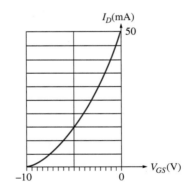

6. Use Figure 20–33 to find $I_{D_{po}}$ and $V_{DS_{po}}$ if
   (a) $V_{GS} = 0V$ (b) $V_{GS} = -1V$ (c) $V_{GS} = -3V$
7. Use Figure 20–33 to construct the I_D-V_{DS} family of curves when $V_{GS} = 0V, -1V, -3V$
8. Find R_{DS} for each of the curves you drew for Problem 7.
9. Describe pinch-off and **CUT-OFF.** What is the difference between V_p and $V_{GS(off)}$?

SECTION 20–4   Basic N-Channel JFET Biasing

Where appropriate, use the Universal N-channel (Figure 20–20, p. 663) or P-channel (Figure 20–24, p. 666) I-V curve to find the bias values in the following problems.

10. For the JFET in Figure 20–34, $V_{GS(off)} = -5V$ and $I_{DSS} = 10mA$. If $V_{DD} = +12V$, $V_{GS} = -2V$, $R_D = 2K\Omega$
    (a) Assume the transistor is **ACTIVE,** and find in the following order

    V_{GS}

    I_D

    V_D

    (b) What is the state of the transistor?

FIGURE 20–34

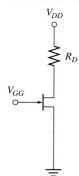

V_{DD}

R_D

V_{GG}

11. For the JFET in Figure 20–34, $V_{GS(off)} = -5V$ and $I_{DSS} = 10mA$. If $V_{DD} = +15V$, $V_{GS} = -1V$, $R_D = 1\ K\Omega$

   (a) Assume the transistor is **ACTIVE,** and find in the following order

      V_{GS}

      I_D

      V_D

   (b) What is the state of the transistor?

12. For the JFET in Figure 20–34, $V_{GS(off)} = -5V$ and $I_{DSS} = 10mA$. If $V_{DD} = +10V$, $V_{GS} = -2V$, $R_D = 5K\Omega$

   (a) Assume the transistor is **ACTIVE,** and find in the following order

      V_{GS}

      I_D

      V_D

   (b) What is the state of the transistor?

   (c) If the JFET is in the **OHMIC** region, what are $V_{DS_{po}}$, $I_{D_{po}}$, and R_{DS}?

   (d) What are the actual values of I_D and V_D?

13. For the JFET in Figure 20–34, $V_{GS(off)} = -5V$ and $I_{DSS} = 10mA$. If $V_{GS} = -2V$, and $R_D = 2K\Omega$, find the value of the drain supply voltage (V_{DD}) that will put the JFET at the edge of the **OHMIC** region.

14. For the JFET in Figure 20–34, $V_{GS(off)} = -5V$ and $I_{DSS} = 10mA$. If $V_{GS} = -2V$ and $V_{DD} = +10V$, find the value of the drain resistor (R_D) that will put the JFET at the edge of the **OHMIC** region.

15. Figure 20–35 shows a JFET inverter circuit, where $V_{GS(off)} = -3V$ and $I_{DSS} = 20mA$.

   (a) What state is the JFET in when $V_G = -5V$?   **(b)** What is V_o when $V_G = -5V$?

   (c) What state is the JFET in when $V_G = 0V$?   **(d)** What is R_{DS} when $V_G = 0V$?

   (e) What is V_o when $V_G = 0V$?   **(f)** Sketch V_o.

FIGURE 20–35

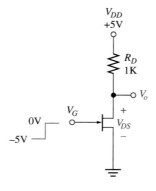

V_{DD}
$+5V$

R_D
$1K$

V_o

V_G

$0V$

$-5V$

$+$
V_{DS}
$-$

FIGURE 20–36

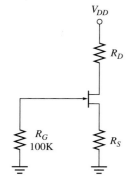

V_{DD}

R_D

R_G
$100K$

R_S

SECTION 20–5   Single-Source JFET Biasing (Graphical)

16. For the circuit in Figure 20–36

$$V_{GS(off)} = -5V, I_{DSS} = 5mA, V_{DD} = +10V, R_D = 2K\Omega, R_S = 1K\Omega.$$

   (a) Sketch the I_D-V_{GS} curve.

   (b) Assume the transistor is **ACTIVE,** and find in the following order

      V_{GS}

      I_D

      V_S

      V_D

   (c) What is the state of the transistor?

17. For the circuit in Figure 20–36 (p. 677)

$$V_{GS(off)} = -5V, I_{DSS} = 5\text{mA}, V_{DD} = +10V, R_D = 7\text{K}\Omega, \text{ and } R_S = 2\text{K}\Omega.$$

(a) Sketch the I_D-V_{GS} curve.

(b) Assume the transistor is **ACTIVE,** and find in the following order

V_{GS}

I_D

V_S

V_D

(c) What is the state of the transistor?

(d) If the JFET is in the **OHMIC** region, what are $V_{DS_{po}}$, $I_{DS_{po}}$, and R_{DS}?

(e) What are the actual values of I_D, V_S, and V_D?

18. For the circuit in Figure 20–36 (p. 677)

$$V_{GS(off)} = -5V, I_{DSS} = 5\text{mA}, R_D = 5\text{K}\Omega, \text{ and } R_S = 1\text{K}\Omega.$$

Find the drain supply voltage (V_{DD}) that puts the JFET at the edge of the **OHMIC** region.

19. For the circuit in Figure 20.36 (p. 677)

$$V_{GS(off)} = -5V, I_{DSS} = 5\text{mA}, V_{DD} = +10V, \text{ and } R_S = 1\text{K}\Omega.$$

Find the drain resistor (R_D) that puts the JFET at the edge of the **OHMIC** region.

SECTION 20–6 Single-Source JFET Biasing (Analytical)

20. Repeat Problem 16 (p. 677) using the analytical method.

21. Repeat Problem 17 using the analytical method.

SECTION 20–7 Voltage Divider Biasing

FIGURE 20–37

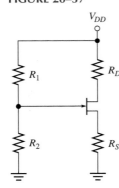

22. For the circuit in Figure 20-37

$$V_{GS(off)} = -3V, I_{DSS} = 10\text{mA}, V_{DD} = +10V, R_1 = 190\text{K}\Omega,$$
$$R_2 = 10\text{K}\Omega, R_D = 3\text{K}\Omega, R_S = 2\text{K}\Omega$$

(a) Assume the transistor is **ACTIVE,** and find in the following order

V_G

V_{GS}

I_D

V_S

V_D

(b) What is the state of the transistor?

23. The JFET in Figure 20–37 is in the **OHMIC** region.

(a) Redraw the circuit, replacing the JFET with R_{DS}. Label V_D and V_S.

(b) Find V_S.   **(c)** Find V_D.   **(d)** Find V_{DS}.

24. For the circuit in Figure 20–37

$$V_{GS(off)} = -3V, I_{DSS} = 10\text{mA}, V_{DD} = +10V, R_1 = 190\text{K}\Omega,$$
$$R_2 = 10\text{K}\Omega, R_D = 10\text{K}\Omega, R_S = 2\text{K}\Omega$$

(a) Assume the transistor is **ACTIVE,** and find in the following order

V_G

V_{GS}

I_D

V_S

V_D

(b) What is the state of the transistor?

    (c) If the JFET is in the **OHMIC** region, what are $V_{DS_{po}}$, $I_{D_{po}}$, R_{DS}?

    (d) Find the actual values of I_D, V_S, and V_D.

25. For the circuit in Figure 20–38, draw both the JFET I-V curve and the gate load-line equation. Is there a problem with this circuit?

26. Using the analytical method, repeat Problem 22.

27. Using the analytical method, repeat Problem 24.

FIGURE 20–38

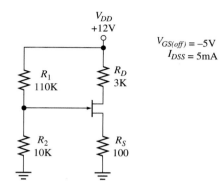

SECTION 20–8 The P-Channel JFET

28. For the circuit in Figure 20–39, $V_{GS(off)} = 3V$, $I_{DSS} = 10mA$.

    (a) Sketch the I_D-V_{GS} curve.

    (b) Assume the transistor is **ACTIVE,** and find in the following order

        V_{GS}

        I_D

        V_D

    (c) What is the state of the transistor?

29. For the circuit in Figure 20–40, $V_{GS(off)} = 3V$, $I_{DSS} = 10mA$.

    (a) Sketch the I_D-V_{GS} curve.

    (b) Assume the transistor is **ACTIVE,** and find in the following order

        V_{GS}

        I_D

        V_D

    (c) What is the state of the transistor?

    (d) Find the actual values of I_D, V_S, V_D.

FIGURE 20–39

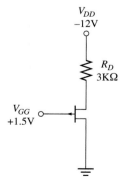

FIGURE 20–40

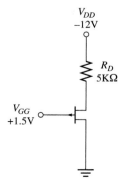

30. For the circuit in Figure 20–41, $V_{GS(off)} = 5\text{V}$, $I_{DSS} = 5\text{mA}$.

    (a) Sketch the I_D-V_{GS} curve.

    (b) Assume the transistor is **ACTIVE,** and find in the following order

$$V_{GS}$$
$$I_D$$
$$V_S$$
$$V_D$$

    (c) What is the state of the transistor?

31. Using the analytical method, repeat Problem 30.

32. For the circuit in Figure 20–42, $V_{GS(off)} = 5\text{V}$, $I_{DSS} = 10\text{mA}$.

    (a) Sketch the I_D-V_{GS} curve.

    (b) Assume the transistor is **ACTIVE,** and find in the following order

$$V_{GS}$$
$$I_D$$
$$V_S$$
$$V_D$$

    (c) What is the state of the transistor?

33. Repeat Problem 32 using the analytical method.

FIGURE 20–41

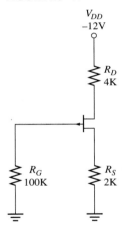

FIGURE 20–42

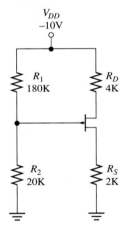

SECTION 20–9   JFET Data Sheets and Computer Models

34. Use Figure 20–25 (p. 669) to find the maximum values for $V_{GS(off)}$ (don't forget the sign) and I_{DSS} for the following JFETs:

    (a) 2N4220A   **(b)** 2N5361   **(c)** 2N5364   **(d)** 2N3909   **(e)** 2N5269

35. Find R_{DS} for the following JFETs (use the maximum values given in Figure 20–25, p. 669):

    (a) 2N4220A   **(b)** 2N5361   **(c)** 2N5364   **(d)** 2N3909   **(e)** 2N5269

36. For the 2N5361 (Figure 20–25, p. 669)

    (a) Draw the maximum and minimum gate I-V curves.

    (b) If the gate bias circuitry results in $I_D = -V_{GS}/0.5\text{K}$, draw this load-line on your I-V curve.

    (c) What is the range of values for V_{GS}?

    (d) What is the range of values for I_D?

37. Refer to Figures 20–25 (p. 669) and 20–27 (p. 671). You want to model the 2N5361 JFET. What are VTO and β?

SECTION 20–10 Troubleshooting the JFET

38. An N-Channel JFET shows a high drain-to-source resistance. Is there a problem with the JFET? Why?

39. An N-Channel JFET shows a low resistance in both directions between the drain and the gate. Is there a problem with this JFET? Why?

40. If the connection and resistors are good, determine the likely cause of the DC Troubleshooting results shown in Figure 20–43.

41. If the connection and resistors are good, determine the likely cause of the DC Troubleshooting results shown in Figure 20–44.

FIGURE 20–43

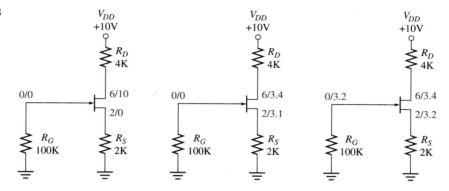

FIGURE 20–44

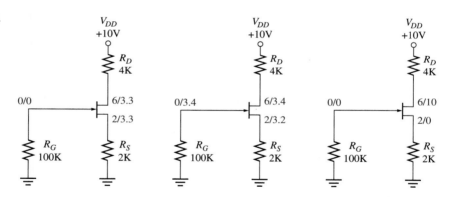

Computer Problems

42. Examine the JFET models available in the computer program you are using. For each model, determine I_{DSS}.

43. Simulate the circuit of Problem 10 (p. 676), and find all voltages and currents.

44. Simulate the circuit of Problem 16 (p. 677), and find all voltages and currents.

45. Simulate the circuit of Problem 22 (p. 678), and find all voltages and currents.

46. Simulate the circuit of Problem 28 (p. 679), and find all voltages and currents.

47. Simulate the circuit of Problem 30, and find all voltages and currents.

48. Simulate the circuit of Problem 32, and find all voltages and currents.

49. Simulate the circuit of Problem 16 (p. 677). Find the voltages at the top of R_G, the top of R_S, and the bottom of R_D after

    (a) You open the gate.   **(b)** You open the source.   **(c)** You open the drain.

50. Simulate the circuit of Problem 16 (p. 677). Find the voltages at the top of R_G, the top of R_S, and the bottom of R_D after

    (a) You short the gate to the source.   **(b)** You short the gate to the drain.

    (c) You short the drain to the source.

CHAPTER

21

THE MOSFET

C**HAPTER** **OBJECTIVES**

To determine the DC behavior of the D-MOSFET.

To find the bias levels in an N-channel D-MOSFET circuit.

To find the bias levels in a P-channel D-MOSFET circuit.

To determine the DC behavior of the E-MOSFET.

To find the bias levels in an N-channel E-MOSFET circuit.

To find the bias levels in a P-channel E-MOSFET circuit.

To read MOSFET data sheets.

To build a CMOS inverter.

To DC Troubleshoot MOSFET circuits.

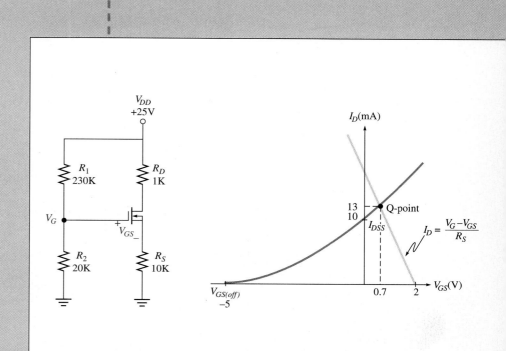

21–1 ■ INTRODUCTION

In addition to the JFET, there are two other types of FETs that you might see in electronic circuits. The first of these has the rather long name of *depletion-enhancement metal oxide semiconductor field effect transistor* (whew!), which we abbreviate as **D-MOSFET.**

Fortunately for the student, the D-MOSFET has the same electrical behavior as the JFET, with one minor variation. Like the JFET, the D-MOSFET is also a device that is normally **ON;** the conducting channel is cut off by a voltage applied at the gate. In the D-MOSFET, however, the channel can also be increased (enhanced) by a correctly applied gate voltage.

The second of these new FETs is the enhancement mode MOSFET, or **E-MOSFET.** In this device, if no gate voltage is applied, there is no conducting channel. Like the BJT, the E-MOSFET is normally **OFF.** It requires applied gate voltage to create a conducting channel in an E-MOSFET.

You might note, to your relief, that even though we cover two FET types here, this chapter is shorter than the previous one. The biasing circuits of the MOSFETs are identical to the biasing circuits of the JFET so there is not much new to learn.

21–2 ■ THE N-CHANNEL D-MOSFET

Structure and Function

Figure 21–1 shows the construction and two circuit symbols for the N-channel D-MOSFET (shortly, we'll explain why there are two circuit symbols for this device). The P-type substrate acts as a platform upon which the MOSFET is built and which also acts as a reservoir of charges. You can see the two deep wells of N material that have been diffused into the P-type substrate to form the drain and source of the MOSFET. There is also a channel of N-type material *between* the drain and source. As with the JFET, then, because there is a conducting channel between the drain and source, the D-MOSFET is normally **ON.**

FIGURE 21–1
*The N-channel
D-MOSFET*

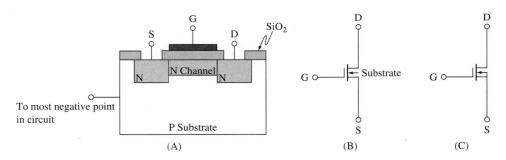

The gate lead doesn't actually touch any semiconductor material. In fact, there is an insulator (made of silicon dioxide [SiO_2], a glass-like material) between the metal terminal of the gate and the N-channel. This construction gives us a metal terminal, an oxide insulator, and the semiconductor material creating the source-drain channel, hence MOSFET. (Because the gate is insulated, some manufacturers refer to this device as an insulated gate FET, or IGFET.) Due to the insulation at the gate, there is even less gate current in the MOSFET than there is in the JFET.

As with any FET, PN junctions must never be forward biased if the device is to operate properly. In the MOSFET the only PN junctions are between the source and the substrate and between the drain and the substrate. To keep these junctions reverse biased, we apply a positive voltage to the N-type drain. In addition, we want the P-type substrate connected to the most negative voltage in the circuit. This is why some MOSFETs have four external terminals (Figure 21–1B).

In most MOSFET circuits the source is also connected to the most negative point (or ground). In this case we can connect the substrate to the source. Many MOSFETs have the substrate internally connected to the source and have only three external terminals (Figure 21–1C).

When $V_{GS} = 0V$ and we apply a positive V_{DS}, we get a current (I_{DSS}) in the channel between the source and the drain (Figure 21–2A). We now apply a negative voltage to the gate. The electric field created by this field pushes electrons out of the channel region down into the substrate. The channel narrows, causing the current to decrease (Figure 21–2B). Further increasing the negative voltage at the gate pushes even more electrons away, narrowing the channel and decreasing the current. Finally, we reach a voltage ($V_{GS(off)}$) where the region between the source and the drain is completely *depleted* of electrons and the drain current falls to 0A (Figure 21–2C).

FIGURE 21–2

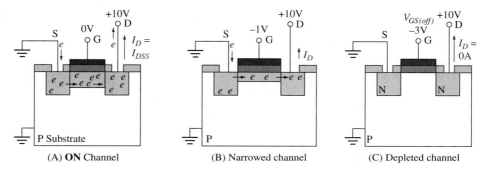

(A) **ON** Channel   (B) Narrowed channel   (C) Depleted channel

To the outside world the I-V behavior of the MOSFET looks similar to the I-V behavior of the JFET. In fact, for negative gate voltages the gate I-V curve of the MOSFET is the same as the gate I-V curve of the JFET.

$$\textbf{D-MOSFET: } I_D = I_{DSS}\left(1 - \frac{V_{GS}}{V_{GS(off)}}\right)^2$$

The difference between the D-MOSFET and the JFET is that the D-MOSFET will work with positive gate voltages. Even though the substrate is P-type material, there are still a significant number of electrons wandering aimlessly around in the substrate. If we apply a positive voltage to the gate, these electrons are pulled towards the gate, *enhancing* the channel. As the channel width increases, the drain current increases.

We are actually being a bit unfair to the JFET. It too can be operated in the enhanced region. However, to ensure that the gate-source PN junction is not forward biased, V_{GS} for the JFET must be less than 0.7V. The gate I-V curve for the N-channel D-MOSFET is shown in Figure 21–3A. Note that we include the depletion region ($V_{GS} < 0V$) and the enhancement region ($V_{GS} > 0V$). Even with the differences between the two devices, biasing the D-MOSFET uses exactly the same circuitry and analysis methods as biasing the JFET.

FIGURE 21–3
D-MOSFET I-V curves

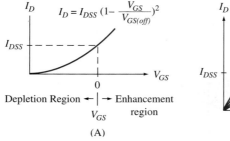

(A)

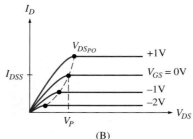

(B)

The drain I-V family of curves for the D-MOSFET is shown in Figure 21–3B (p. 686). You can see that the maximum drain current is now greater than I_{DSS}. The drain pinch-off voltage is still found from

$$V_{DS_{po}} = V_p + V_{GS}$$

EXAMPLE 21–1
Biasing the D-MOSFET, Depletion Region

Find the bias levels for the circuit shown in Figure 21–4.

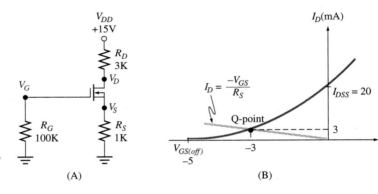

FIGURE 21–4
Basic D-MOSFET biasing circuit

(A)　　　　　　　　　　　　　(B)

Solution　We first note that

$$V_G = 0V$$

so

$$V_S = V_G - V_{GS} = -V_{GS}$$

This leads to

$$I_D = \frac{-V_{GS}}{R_S} = \frac{-V_{GS}}{1K}$$

We now plot this gate load-line relation along with the I-V curve (Figure 21–4B) to find

$$V_{GS} = -3V$$

and

$$I_D = 3mA$$

To check the state of the transistor, we find $V_{DS_{po}}$:

$$V_{DS_{po}} = V_p + V_{GS} = 5 + (-3) = 2V$$

The drain-source voltage is

$$V_{DS} = V_{DD} - (R_D + R_S)I_D = 15 - (3K + 1K) \times 3mA = 3V$$

Because $V_{DS} > V_{DS_{po}}$, the transistor is **ACTIVE.**

DRILL EXERCISE

Redraw the circuit of Figure 21–4, changing the source resistor to $R_S = 2K\Omega$. Find the bias levels.

Answer　$V_{GS} = -3.5$ V, $I_D = 1.8$ mA

EXAMPLE 21–2
Biasing the D-MOSFET, Enhancement Region

Find the bias levels for the circuit in Figure 21–5. Note that to bias a D-MOSFET in the enhancement region, we usually use a voltage divider at the gate.

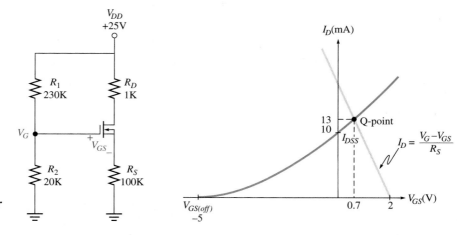

FIGURE 21–5
Voltage divider D-MOSFET

Solution   We first note that

$$V_G = \frac{R_2}{R_1 + R_2} V_{DD} = \frac{20K}{230K + 20K} 25 = 2V$$

Because

$$V_S = V_G - V_{GS}$$

we get

$$I_D = \frac{V_S}{R_S} = \frac{V_G}{R_S} - \frac{V_{GS}}{R_S}$$

so

$$I_D = \frac{2}{0.1K} - \frac{V_{GS}}{0.1K} = 20mA - \frac{V_{GS}}{0.1K}$$

We now plot this gate load-line along with the I-V curve (Figure 21–5B) to find

$$V_{GS} = 0.7V$$

and

$$I_D = 20mA - \frac{0.7}{0.1K} = 13mA$$

To check the state of the transistor, we find $V_{DS_{po}}$:

$$V_{DS_{po}} = V_p + V_{GS} = 5 + 0.7 = 5.7V$$

Note that the pinch-off voltage is greater than V_p in the enhancement region. The drain-source voltage is

$$V_{DS} = V_{DD} - (R_D + R_S)I_D = 25 - (1K + 0.1K) \times 13mA = 10.7V$$

Because $V_{DS} > V_{DS_{po}}$, the transistor is **ACTIVE.**

Redraw the circuit of Figure 21–5 (p. 687), changing the source resistor to $R_S = 200\Omega$. Find the bias levels.

Answer $V_{GS} = 0V, I_D = 10mA$

EXAMPLE 21–3
D-MOSFET
Biasing, Analytical
Method

The analytical method presented in the last chapter for the N-channel JFET also applies to the D-MOSFET. Use this method to find the bias level in Example 21–2 (p. 687).

Solution The quadratic equation

$$aV^2_{GS} + bV_{GS} + c = 0$$

has these parameters:

$$a = \frac{1}{V^2_{GS(off)}}$$

$$b = \frac{-2}{V_{GS(off)}} + \frac{1}{R_S I_{DSS}}$$

$$c = 1 - \frac{V_G}{R_S I_{DSS}}$$

The equation is solved as

$$V_{GS} = \frac{-b + \sqrt{b^2 - 4ac}}{2a}$$

For our circuit

$$V_G = \frac{R_2}{R_1 + R_2} V_{DD} = \frac{20K}{230K + 20K} 25 = 2V$$

$$a = \frac{1}{V^2_{GS(off)}} = \frac{1}{(-5)^2} = 0.04$$

$$b = \frac{-2}{V_{GS(off)}} + \frac{1}{R_S I_{DSS}} = \frac{-2}{-5} + \frac{1}{0.1K \times 10mA} = 1.4$$

$$c = 1 - \frac{V_G}{R_S I_{DSS}} = 1 - \frac{2}{0.1K \times 10mA} = -1$$

so

$$V_{GS} = \frac{-b + \sqrt{b^2 - 4ac}}{2a} = 0.7V$$

and

$$I_D = \frac{V_G - V_{GS}}{R_S} = \frac{2 - 0.7}{0.1K} = 13mA$$

The Almost Universal N-Channel D-MOSFET I-V Curve

In Figure 21–6 we present an N-channel D-MOSFET gate I-V curve that you can use to determine bias values for V_{GS} and I_D. For a given circuit, make a copy of this figure, insert the appropriate values for $V_{GS(off)}$ and I_{DSS}, and then draw the gate load-line. Because the range of the enhancement region depends on the particular D-MOSFET you are using, we cannot present a truly universal curve. Or, in some cases you may find that V_G for your circuit is off the scale of this figure. In such cases, just extend the horizontal axis.

FIGURE 21–6

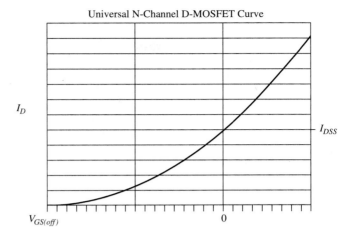

Universal N-Channel D-MOSFET Curve

REVIEW QUESTIONS

1. Show the construction of an N-channel D-MOSFET, and explain its operation.

2. Draw two circuit symbols for the N-channel D-MOSFET. Why are there two symbols?

3. Draw the gate I-V curve for this device. Identify the depletion and enhancement regions.

4. How does this curve differ from the JFET I-V curve?

5. Draw the drain I-V family of curves. For each curve in the family, indicate $V_{DS_{po}}$.

6. Assume that there is a voltage divider at the gate. Sketch the gate I-V curve and two gate load-lines, one that puts the Q-point in the depletion region and another that puts the Q-point in the enhancement region.

7. What are the formulas for a, b, and c in the quadratic formula for V_{GS}?

8. Write the formula for V_{GS}.

21–3 ■ THE P-CHANNEL D-MOSFET

Structure and Function

Figure 21–7 shows the construction and the two circuit symbols for the P-channel D-MOSFET. The N-type substrate acts as a platform upon which the MOSFET is built and also acts as a reservoir of charges. In this MOSFET the P-type wells form the drain and source. There is a conducting channel of P-type material between the drain and source. The I-V behavior of the P-channel D-MOSFET looks similar to the I-V behavior of the P-channel JFET. In fact, for positive gate voltages the gate I-V curve of this MOSFET is the same as the gate I-V curve of the P-channel JFET.

FIGURE 21–7
P-channel D-MOSFET

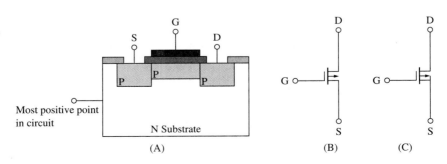

The difference between the D-MOSFET and the JFET is that the P-channel D-MOSFET will work with negative gate voltages. Even though the substrate is N-type material, there are still a significant number of positive charges (holes) in the substrate. If we apply a negative voltage to the gate, these holes are pulled towards the gate, enhancing the channel. As the channel width increases, the drain current increases.

The gate I-V curve for the P-channel D-MOSFET is shown in Figure 21–8. Note that we include the depletion region ($0 < V_{GS} < V_{GS(off)}$) and the enhancement region ($V_{GS} < 0$). Biasing the P-channel D-MOSFET uses the same circuitry and analysis methods as biasing the P-channel JFET.

FIGURE 21–8
P-channel D-MOSFET
I-V curve

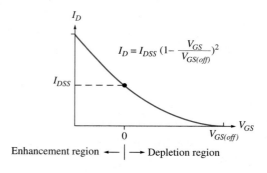

EXAMPLE 21–4
Biasing the
P-Channel
D-MOSFET

Find the bias levels for the circuit in Figure 21–9. Note that to bias a D-MOSFET in the enhancement region, we usually use a voltage divider at the gate.

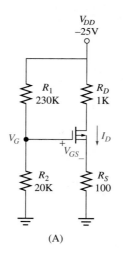

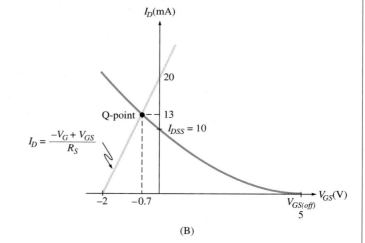

FIGURE 21–9   (A)   (B)

Solution   We first note that

$$V_G = \frac{R_2}{R_1 + R_2} V_{DD} = \frac{20K}{230K + 20K}(-25) = -2V$$

Now

$$V_S = V_G - V_{GS}$$

and because of the direction of I_D

$$I_D = \frac{-V_S}{R_S} = \frac{-V_G}{R_S} + \frac{V_{GS}}{R_S} = \frac{2}{0.1K} + \frac{V_{GS}}{0.1K}$$

We plot this gate load-line along with the I-V curve (Figure 21–9B) to find

$$V_{GS} = -0.7\text{V (enhancement region)}$$

and

$$I_D = \frac{-V_G + V_{GS}}{R_S} = \frac{-(-2) - 0.7}{0.1\text{K}} = 13\text{mA}$$

DRILL EXERCISE

Redraw the circuit of Figure 21–9, changing the source resistor to $R_S = 200\Omega$. Find the bias levels.

Answer $V_{GS} = 0\text{V}, I_D = 10\text{mA}$

EXAMPLE 21–5
P-Channel
D-MOSFET
Biasing, Analytical
Method

The analytical method presented in the last chapter also applies to the P-channel D-MOSFET. Use the equation to find the bias level of Example 21–4.

Solution We can still solve the quadratic equation:

$$aV^2_{GS} + bV_{GS} + c = 0$$

to find

$$V_{GS} = \frac{-b - \sqrt{b^2 - 4ac}}{2a}$$

where for the P-channel D-MOSFET

$$a = \frac{1}{V^2_{GS(off)}}$$

$$b = \frac{-2}{V_{GS(off)}} - \frac{1}{R_S I_{DSS}}$$

and

$$c = 1 + \frac{V_G}{R_S I_{DSS}}$$

We first find V_G:

$$V_G = \frac{R_2}{R_1 + R_2} V_{DD} = \frac{20\text{K}}{230\text{K} + 20\text{K}} (-25) = -2\text{V}$$

We now find a, b, and c:

$$a = \frac{1}{V^2_{GS(off)}} = \frac{1}{5^2} = 0.04$$

$$b = \frac{-2}{V_{GS(off)}} - \frac{1}{R_S I_{DSS}} = \frac{-2}{5} - \frac{1}{0.1\text{K} \times 10\text{mA}} = -1.4$$

and

$$c = 1 + \frac{V_G}{R_S I_{DSS}} = 1 + \frac{-2}{1} = -1\text{V}$$

V_{GS} is found from

$$V_{GS} = \frac{-b - \sqrt{b^2 - 4ac}}{2a} = -0.7\text{V}$$

and the drain current is

$$I_D = \frac{-V_G + V_{GS}}{R_S} = \frac{(2 - 0.7)}{0.1\text{K}} = 13\text{mA}$$

The Almost Universal P-Channel D-MOSFET I-V Curve

In Figure 21–10 we present a P-channel D-MOSFET gate I-V curve that you can use to determine bias values for V_{GS} and I_D. For a given circuit, make a copy of this figure, insert the appropriate values for $V_{GS(off)}$ and I_{DSS}, then draw the gate load-line.

FIGURE 21–10

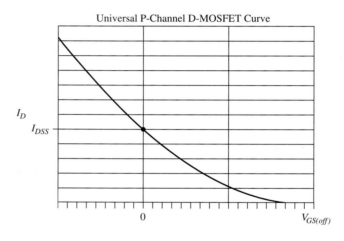

Universal P-Channel D-MOSFET Curve

REVIEW QUESTIONS

1. Show the construction of a P-channel D-MOSFET, and explain its operation.
2. Draw two circuit symbols for the P-channel D-MOSFET. Why are there two symbols?
3. Draw the gate I-V curve for this device. Identify the depletion and enhancement regions.
4. How does this curve differ from the P-channel JFET I-V curve?
5. Assume that there is a voltage divider in the gate. Sketch the I-V curve and two gate load-lines, one that puts the Q-point in the depletion region and another that puts the Q-point in the enhancement region.
6. What are the formulas for a, b, and c in the quadratic formula for V_{GS}?
7. Write the formula for V_{GS}.

21–4 ■ THE N-CHANNEL E-MOSFET

Structure and Function

Figure 21–11 shows the construction and circuit symbols for the N-channel enhancement-mode MOSFET, or **E-MOSFET.** As with the D-MOSFET, this device comes with either four external terminals or three external terminals (substrate internally connected to source). You can see that there is no conducting channel between the drain and the source. This means if no gate voltage is applied, there is no drain current. Like the BJT, the E-MOSFET is normally **OFF.**

FIGURE 21–11
N-channel E-MOSFET

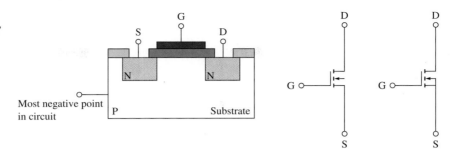

When a positive voltage is applied to the gate, electrons are pulled up from the substrate, and an N-channel starts to form (Figure 21–12A). The channel does not form uniformly but rather begins to form on the drain side. As the gate voltage increases, the channel length also increases. Finally, the gate voltage increases to the point where the channel reaches from the drain to the source, and conduction begins (Figure 21–12B). This voltage is known as the **gate threshold voltage** ($V_{GS(th)}$).

FIGURE 21–12
N-channel E-MOSFET
I-V curve

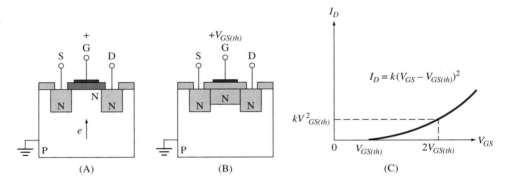

For gate voltages greater than $V_{GS(th)}$, the gate I-V curve looks the same as the other FET curves (Figure 21–12C). Because a positive voltage is required at the gate, bias circuits for the E-MOSFET usually contain voltage dividers. We will show you how feedback can be used to bias the E-MOSFET in Problem 22 at the end of the chapter.

EXAMPLE 21–6
N-Channel
E-MOSFET Biasing

Find the bias levels for the circuit in Figure 21–13.

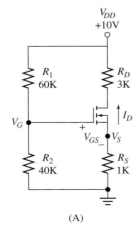

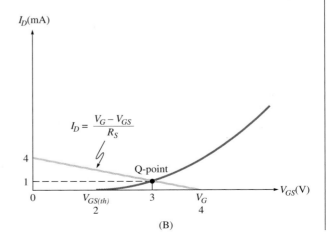

FIGURE 21–13
E-MOSFET biasing

(A)

(B)

Solution   We first find V_G:

$$V_G = \frac{R_2}{R_1 + R_2} V_{DD} = \frac{40K}{60K + 40K} 10 = 4V$$

The gate load-line is given by

$$I_D = \frac{V_G}{R_S} - \frac{V_{GS}}{R_S}$$

which leads to the two points:

$$V_{GS} = V_G = 4V \rightarrow I_D = 0A$$

and

$$V_{GS} = 0V \rightarrow I_D = \frac{V_G}{R_S} = \frac{4}{1K} = 4mA$$

The load-line for our circuit is plotted in Figure 21–13B (p. 693). We see the Q-points are

$$V_{GS} = 3V \quad \text{and} \quad I_D = 1mA$$

DRILL EXERCISE   Redraw the circuit of Figure 21–13 (p. 693), changing the source resistor to 200Ω. Find the bias levels.

Answer $V_{GS} = 3.5V, I_D = 2.5mA$

Analytical N-Channel E-MOSFET Biasing

The I-V curve for the E-MOSFET can also be represented with a parabola. The formula used here is slightly different from the JFET and the D-MOSFET:

$$I_D = k(V_{GS} - V_{GS(th)})^2$$

where k has units of A/V^2.

For now we will assume that k is given. In Section 21–7 (p. 703) we will discuss how to find this parameter. We can still find V_{GS} from the quadratic formula:

$$aV^2_{GS} + bV_{GS} + c = 0$$

where

$$a = 1$$
$$b = -2V_{GS(th)} + \frac{1}{kR_S}$$
$$c = V^2_{GS(th)} - \frac{V_G}{kR_S}$$

The formula for V_{GS} is

$$V_{GS} = \frac{-b + \sqrt{b^2 - 4c}}{2}$$

EXAMPLE 21–7
N-Channel
E-MOSFET Biasing,
Analytical Method

Use the quadratic formula to find the bias levels for the circuit in Example 21–6. For the FET in this circuit

$$V_{GS(th)} = 2V \quad \text{and} \quad k = 0.001 A/V^2$$

Solution We use $V_G = 4V$ found in Example 21–6 to find

$$a = 1$$

$$b = -2V_{GS(th)} + \frac{1}{kR_S} = -2 \times 2 + \frac{1}{0.001 \times 1K} = -3$$

$$c = V^2_{GS(th)} - \frac{V_G}{kR_S} = 2^2 - \frac{4}{0.001 \times 1K} = 0$$

Finally, V_{GS} is found from

$$V_{GS} = \frac{-b + \sqrt{b^2 - 4c}}{2} = \frac{-(-3) + 3}{2} = 3V$$

The drain current is found from

$$I_D = \frac{V_G - V_{GS}}{R_S} = 1mA$$

DRILL EXERCISE

Redraw the circuit of Figure 21–13 (p. 693), changing the source resistor to 200Ω. Find the bias levels.

Answer $V_{GS} = 3.5V$, $I_D = 2.5mA$

The Almost Universal N-Channel E-MOSFET I-V Curve

The I-V curve in Figure 21–14 covers the range from $V_{GS(th)}$ to $3V_{GS(th)}$. Note that to save room, the horizontal axis begins at $V_{GS(th)}$. Also note that when

$$V_{GS} = 2V_{GS(th)}$$

the drain current is

$$I_D = kV^2_{GS}$$

Copy this graph and use it to determine bias levels for the N-channel E-MOSFET.

FIGURE 21–14

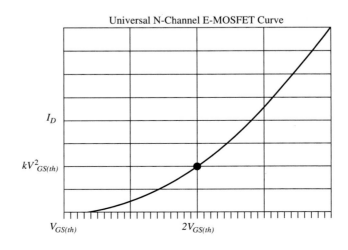

Universal N-Channel E-MOSFET Curve

1. Show the construction of an N-channel E-MOSFET, and explain its operation. Is this device normally **ON** or **OFF**?

2. Draw two circuit symbols for the N-channel E-MOSFET. Why are there two symbols?

3. Draw the gate I-V curve for this device.

4. Sketch the I-V curve and a gate load-line. Identify the Q-point.

5. What are the formulas for a, b, and c in the quadratic formula for V_{GS}?

6. Write the formula for V_{GS}.

7. What is the formula for I_D?

21–5 ▪ THE P-CHANNEL E-MOSFET

Structure and Function

Figure 21–15 shows the construction and circuit symbols for the P-channel enhancement-mode MOSFET. When a negative voltage exceeding $V_{GS(th)}$ is applied to the gate, holes are pulled up from the substrate, forming a P-channel. The P-channel gate I-V curve is shown in Figure 21–16.

FIGURE 21–15
P-channel E-MOSFET

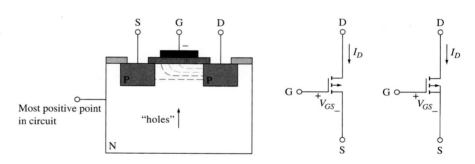

FIGURE 21–16
P-channel E-MOSFET I-V curve

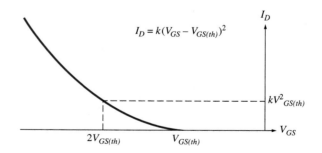

$$I_D = k(V_{GS} - V_{GS(th)})^2$$

EXAMPLE 21–8
P-Channel E-MOSFET Biasing

Find the bias levels for the circuit in Figure 21–17. Assume that $V_{GS(th)} = -2V$ and $k = 0.002 A/V^2$

Solution   We first find V_G:

$$V_G = \frac{R_2}{R_1 + R_2} V_{DD} = \frac{40K}{80K + 40K} (-12) = -4V$$

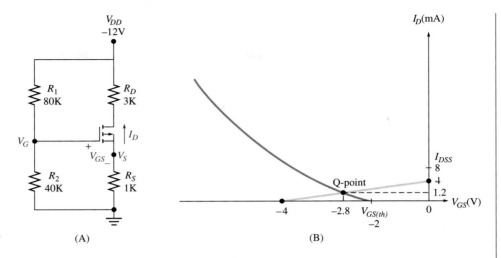

FIGURE 21–17   (A)   (B)

The gate load-line is now given by

$$I_D = \frac{-V_S}{R_S} = \frac{-V_G}{R_S} + \frac{V_{GS}}{R_S}$$

which leads to the two points:

$$V_{GS} = V_G = -4\text{V} \rightarrow I_D = 0\text{A}$$

and

$$V_{GS} = 0\text{V} \rightarrow I_D = \frac{-V_G}{R_S} = \frac{4}{1\text{K}} = 4\text{mA}$$

The load-line is plotted in Figure 21–17B. We see the Q-points are

$$V_{GS} = -2.8\text{V} \qquad \text{and} \qquad I_D = 1.2\text{mA}$$

DRILL EXERCISE   Redraw the circuit of Figure 21–17, changing the source resistor to 200Ω. Find the bias levels.

Answer $V_{GS} = -3.3\text{V}, I_D = 3.5\text{mA}$

Analytical P-Channel E-MOSFET Biasing

Here, the quadratic formula for V_{GS} is

$$aV^2_{GS} + bV_{GS} + c = 0$$

where

$$a = 1$$

$$b = -2V_{GS(th)} - \frac{1}{kR_S}$$

$$c = V^2_{GS(th)} + \frac{V_G}{kR_S}$$

The formula for V_{GS} is

$$V_{GS} = \frac{-b - \sqrt{b^2 - 4c}}{2}$$

EXAMPLE 21–9

N-Channel E-MOSFET Biasing, Analytical Method

Use the quadratic formula to find the bias levels for the circuit in Example 21–8 (p. 696). For the FET in this circuit

$$V_{GS(th)} = -2V \qquad \text{and} \qquad k = 0.002 A/V^2$$

Solution We use $V_G = -4V$ found in Example 21–8 to find

$$a = 1$$

$$b = -2V_{GS(th)} - \frac{1}{kR_S} = -2 \times (-2) - \frac{1}{0.002 \times 1K} = 3.5$$

$$c = V^2_{GS(th)} + \frac{V_G}{kR_S} = -2^2 + \frac{-4}{0.002 \times 1K} = 2$$

Finally V_{GS} is found from

$$V_{GS} = \frac{-b - \sqrt{b^2 - 4c}}{2} = -2.8V$$

The drain current is found from

$$I_D = -\frac{V_G - V_{GS}}{R_S} = 1.2mA$$

DRILL EXERCISE

Redraw the circuit of Figure 21–17 (p. 697), changing the source resistor to 200Ω. Find the bias levels.

Answer $V_{GS} = -3.3V$, $I_D = 3.5mA$

The Almost Universal P-Channel E-MOSFET I-V Curve

The I-V curve given in Figure 21–18 covers the range from $V_{GS(th)}$ to $3V_{GS(th)}$. Note that to save room, the horizontal axis begins at $V_{GS(th)}$. Also note that when

$$V_{GS} = 2V_{GS(th)}$$

the drain current is

$$I_D = kV^2_{GS(th)}$$

Copy this graph, and use it to determine bias levels for the P-channel E-MOSFET.

FIGURE 21–18

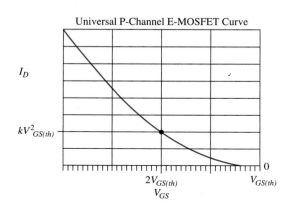

Universal P-Channel E-MOSFET Curve

21–6 ■ MOSFET Data Sheets and Computer Models

Data Sheets

Figure 21–19 (p. 700) shows a data sheet for an N-channel D-MOSFET. This data sheet presents information similar to that given for the JFETs. You can see that in the ON CHARACTERISTICS, two currents are given. The Zero-Gate-Voltage Drain Current (I_{DSS}) is the same current defined for the JFET when the gate voltage is 0V. For this device I_{DSS} varies from 2 to 6mA. The On-State Drain Current ($I_{D(on)}$) is the current when the gate voltage is positive. This is the enhanced region for the N-channel D-MOSFET. When $V_{GS} = +3.5$V, this current can vary from 9 to 18mA.

Figure 21–20 (p. 701) shows a data sheet for an N-channel E-MOSFET. From the ON CHARACTERISTICS, we can see that this device has a gate threshold voltage ($V_{GS(th)}$) that can vary from 1 to 5V. Note that this data sheet gives I_{DSS}; however, because there is almost no current in this device when the gate voltage is 0V, this parameter is not as useful as it is for the depletion devices, the JFET and the D-MOSFET.

Because the E-MOSFET only operates in the enhanced region, the important parameter here is the On-State Drain Current ($I_{D(on)}$). For this device $I_{D(on)} = 3.0$mA when $V_{GS} = 10$V.

To use the analytical method or the universal E-MOSFET curves, we need to know the parameter k. As you can see (or, more to the point, as you cannot see), this parameter is not given in the data sheet. To determine k, we can use the data given for On-State Drain Current in Figure 2–20 (p. 701). We know that

$$I_D = k(V_{GS} - V_{GS(th)})^2$$

so

$$k = \frac{I_D}{(V_{GS} - V_{GS(th)})^2}$$

The data sheet tells us that for this E-MOSFET

$$V_{GS(th)} = 1.0\text{V}$$

and

$$I_{D(on)} = 3.0\text{mA when } V_{GS} = 10\text{V}$$

Using this information, we find

$$k = \frac{I_D}{(V_{GS} - V_{GS(th)})^2} = \frac{3\text{mA}}{(10 - 1)^2} = 0.000037\text{A/V}^2$$

Note that we have used the minimum values for all specifications because that is all that is available for $I_{D(on)}$.

FIGURE 21–19

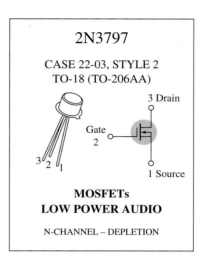

2N3797

CASE 22-03, STYLE 2
TO-18 (TO-206AA)

3 Drain

Gate
2

1 Source

MOSFETs
LOW POWER AUDIO

N-CHANNEL – DEPLETION

MAXIMUM RATINGS

Rating	Symbol	Value	Unit
Drain-Source Voltage 2N3797	V_{DS}	20	Vdc
Gate-Source Voltage	V_{GS}	±10	Vdc
Drain Current	I_D	20	mAdc
Total Device Dissipation @ $T_A = 25°C$ Derate above 25°C	P_D	200 1.14	mW mW/°C
Junction Temperature Range	T_J	+175	°C
Storage Channel Temperature Range	T_{stg}	−65 to +200	°C

ELECTRICAL CHARACTERISTICS ($T_A = 25°C$ unless otherwise noted)

Characteristic	Symbol	Min	Typ	Max	Unit
OFF CHARACTERISTICS					
Drain Source Breakdown Voltage ($V_{GS} = -7.0$ V, $I_D = 5.0$ μA) 2N3797	$V_{(BR)DSX}$	20	25	—	Vdc
Gate Reverse Current (1) ($V_{GS} = -10$ V, $V_{DS} = 0$) ($V_{GS} = -10$ V, $V_{DS} = 0$, $T_A = 150°C$)	I_{GSS}	— —	— —	1.0 200	pAdc
Gate Source Cutoff Voltage ($I_D = 2.0$ μA, $V_{DS} = 10$ V) 2N3797	$V_{GS(off)}$	—	−5.0	−7.0	Vdc
Drain-Gate Reverse Current (1) ($V_{DG} = 10$ V, $I_S = 0$)	I_{DGO}	—	—	1.0	pAdc
ON CHARACTERISTICS					
Zero-Gate Voltage Drain Current ($V_{DS} = 10$ V, $V_{GS} = 0$) 2N3797	I_{DSS}	2.0	2.9	6.0	mAdc
On-State Drain Current ($V_{DS} = 10$ V, $V_{GS} = +3.5$ V) 2N3797	$I_{D(on)}$	9.0	14	18	mAdc
SMALL-SIGNAL CHARACTERISTICS					
Forward Transfer Admittance ($V_{DS} = 10$ V, $V_{GS} = 0$, f = 1.0 kHz) 2N3797	$|y_{fs}|$	1500	2300	3000	μmhos
($V_{DS} = 10$ V, $V_{GS} = 0$, f = 1.0 MHz) 2N3797		1500	—	—	
Output Admittance ($I_{DS} = 10$ V, $V_{GS} = 0$, f = 1.0 kHz) 2N3797	$|y_{os}|$	—	27	60	μmhos
Input Capacitance ($V_{DS} = 10$ V, $V_{GS} = 0$, f = 1.0 MHz) 2N3797	C_{iss}	—	6.0	8.0	pF
Reverse Transfer Capacitance ($V_{DS} = 10$ V, $V_{GS} = 0$, f = 1.0 MHz)	C_{rss}	—	0.5	0.8	pF
FUNCTIONAL CHARACTERISTICS					
Noise Figure ($V_{DS} = 10$ V, $V_{GS} = 0$, f = 1.0 kHz, $R_S = 3$ megohms)	NF	—	3.8	—	dB

(1) This value of current includes both the FET leakage current as well as the leakage current associated with the test socket and fixture
when measured under best attainable conditions.

FIGURE 21–20

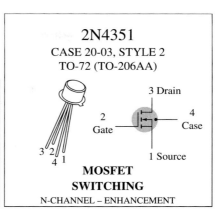

2N4351
CASE 20-03, STYLE 2
TO-72 (TO-206AA)

3 Drain

2
Gate

4
Case

1 Source

**MOSFET
SWITCHING**

N-CHANNEL – ENHANCEMENT

MAXIMUM RATINGS

Rating	Symbol	Value	Unit
Drain-Source Voltage	V_{DS}	25	Vdc
Drain-Gate Voltage	V_{DG}	30	Vdc
Gate-Source Voltage*	V_{GS}	30	Vdc
Drain Current	I_D	30	mAdc
Total Device Dissipation @ $T_A = 25°C$ Derate above 25°C	P_D	300 1.7	mW mW/°C
Junction Temperature Range	T_J	175	°C
Storage Temperature Range	T_{stg}	−65 to +175	°C

*Transient potentials of $\pm$ 75 Volt will not cause gate-oxide failure.

ELECTRICAL CHARACTERISTICS ($T_A = 25°C$ unless otherwise noted.)

Characteristic	Symbol	Min	Max	Unit
OFF CHARACTERISTICS				
Drain-Source Breakdown Voltage ($I_D = 10$ μA, $V_{GS} = 0$)	$V_{(BR)DSX}$	25	—	Vdc
Zero-Gate-Voltage Drain Current ($V_{DS} = 10$ V, $V_{GS} = 0$) $T_A = 25°C$ $T_A = 150°C$	I_{DSS}	— —	10 10	nAdc μAdc
Gate Reverse Current ($V_{GS} = \pm 15$ Vdc, $V_{DS} = 0$)	I_{GSS}	—	$\pm$ 10	pAdc
ON CHARACTERISTICS				
Gate Threshold Voltage ($V_{DS} = 10$ V, $I_D = 10$ μA)	$V_{GS(Th)}$	1.0	.5	Vdc
Drain-Source On-Voltage ($I_D = 2.0$ mA, $V_{GS} = 10$V)	$V_{DS(on)}$	—	1.0	V
On-State Drain Current ($V_{GS} = 10$ V, $V_{DS} = 10$V)	$I_{D(on)}$	3.0	—	mAdc
SMALL-SIGNAL CHARACTERISTICS				
Forward Transfer Admittance ($V_{DS} = 10$ V, $I_D = 2.0$ mA, $f = 1.0$ kHz)	$|y_{fs}|$	1000	—	μmho
Input Capacitance ($V_{DS} = 10$ V, $V_{GS} = 0$, $f = 140$ kHz)	C_{iss}	—	5.0	pF
Reverse Transfer Capacitance ($V_{DS} = 0$, $V_{GS} = 0$, $f = 140$ kHz)	C_{rss}	—	1.3	pF
Drain-Substrate Capacitance ($V_{D(SUB)} = 10$ V, $f = 140$ kHz)	$C_{d(sub)}$	—	5.0	pF
Drain-Source Resistance ($V_{GS} = 10$ V, $I_D = 0$, $f = 1.0$ kHz)	$r_{ds(on)}$	—	300	ohms

SWITCHING CHARACTERISTICS

		Symbol	Min	Max	Unit
Turn-On Delay (Fig. 5)		t_{d1}		45	ns
Rise Time (Fig. 6)	$I_D = 2.0$ mAdc, $V_{DS} = 10$ Vdc,	t_r		65	ns
Turn-Off Delay (Fig. 7)	($V_{GS} = 10$ Vdc) (See Figure 9; Times Circuit Determined)	t_{d2}		60	ns
Fall Time (Fig. 8)		t_f		100	ns

Computer Models

D-MOSFET Figure 21–21 (p. 702) shows the D-MOSFET computer model used by Electronic Workbench. The Threshold Voltage (*VTO*) is the same as the gate cut-off voltage ($V_{GS(off)}$). That is

$$VTO = V_{GS(off)}$$

FIGURE 21–21

3-Terminal Depletion N-MOSFET Model "2N3796" Library "default"			
VTO	Threshold voltage	−3	V
β	Transconductance coefficient	0.0005	A/V
λm	Channel-length modulation	0.0025	1/V
φ	Surface potential	0.75	V
δ	Bulk-threshold parameter	1e-05	V**0.5
rd	Drain ohmic resistance	21	Ω
rs	Source ohmic resistance	21	Ω
Is	Bulk-junction saturation current	8.33e-15	A
Cgb	Gate-bulk capacitance	3.9e-08	F
Cgd	Zero-bias gate-drain junction capacitance	5e-09	F
Cgs	Zero-bias gate-source junction capacitance	6e-09	F
Cbd	Zero-bias bulk-drain junction capacitance	1.43e-11	F
Cbs	Zero-bias bulk-source junction capacitance	1.71e-11	F
φB	Bulk-junction potential	0.8	V

As with the JFET, the computer model uses the parameter β (remember not to confuse this use of the symbol β with the BJT current gain). Note that the actual unit of β is A/V^2. The relation between I_{DSS} and β is given by

$$\beta = \frac{I_{DSS}}{V_{GS(off)}^2}$$

The term *transconductance coefficient* may also cause some confusion. In the next chapter we will introduce the small-signal AC parameter that relates the output current to the input voltage. This parameter is known as the transconductance (g_m) and bears no direct relation to the transconductance coefficient used in this computer program.

E-MOSFET Figure 21–22 shows the E-MOSFET computer model used by Electronic Workbench. Comparing the computer model to the parameters we have used for the E-MOSFET, you can see that

$$VTO = V_{GS(th)}$$

and

$$\beta = k$$

FIGURE 21–22

3-Terminal Enhancement N-MOSFET Model "IRF511" Library "default"			
VTO	Threshold voltage	3	V
β	Transconductance coefficient	0.0408	A/V
λm	Channel-length modulation	0.00075	1/V
φ	Surface potential	0.75	V
δ	Bulk-threshold parameter	9.08e-07	V**0.5
rd	Drain ohmic resistance	0.084	Ω
rs	Source ohmic resistance	0.084	Ω
Is	Bulk-junction saturation current	3.75e-15	A
Cgb	Gate-bulk capacitance	9.49e-07	F
Cgd	Zero-bias gate-drain junction capacitance	2.5e-07	F
Cgs	Zero-bias gate-source junction capacitance	3e-07	F
Cbd	Zero-bias bulk-drain junction capacitance	4.57e-10	F
Cbs	Zero-bias bulk-source junction capacitance	5.49e-10	F
φB	Bulk-junction potential	0.8	V

21–7 ■ THE COMPLEMENTARY MOSFET (CMOS)—AN INVERTER

An N-channel E-MOSFET and a P-channel E-MOSFET can be combined on a chip to form a *complementary* pair (Figure 21–23) known as a **CMOS.** The two MOSFETs are constructed so that their device parameters are the same. That is

$$|V_{GS(th)}|_N = |V_{GS(th)}|_P \text{ and } k_N = k_P$$

FIGURE 21–23
CMOS

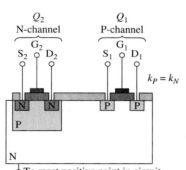

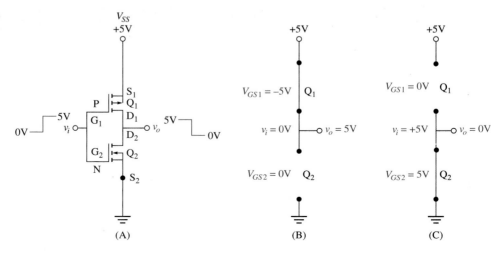

The CMOS is used to build amplifiers and logic gates. We will discuss these applications in the next chapter. For now we will just consider the basic CMOS inverter.

In the CMOS inverter (Figure 21–24A), the two drains of the FETs are tied together, as are the two gates. Typically, the supply voltage and ground are connected to the two sources, the input is applied to the gate, and the output is taken from the drain.

FIGURE 21–24
CMOS inverter

Remember, an E-MOSFET is normally **OFF.** To turn it **ON,** you need to apply a positive gate-source voltage to the N-channel device and a negative gate-source voltage to the P-channel device. This means that in a typical CMOS application, one of the MOSFETs will be **OFF.** Because of this, the current draw on the supply source is very small, a distinct advantage in portable devices.

In this circuit we have connected a +5V supply to the source of the P-channel MOSFET (Q_1) and have grounded the source of the N-channel MOSFET (Q_2). Let's see what happens when we apply an input of 0V and an input of +5V. Assume that the threshold voltage is +3V for the N-channel MOSFET and −3V for the P-channel MOSFET. First note that

$$\textbf{P-channel: } V_{GS_1} = V_i - V_{DD} = V_i - 5 \qquad V_{GS(th)} = -3\text{V}$$

and

$$\textbf{N-channel: } V_{GS_2} = V_i - 0 = V_i \qquad V_{GS(th)} = +3\text{V}$$

When $V_i = 0$V

$$\textbf{P-channel: } V_{GS_1} = 0 - 5 = -5\text{V} \rightarrow Q_1 \text{ is } \textbf{ON}$$
$$\textbf{N-channel: } V_{GS_2} = V_i = 0\text{V} \rightarrow Q_2 \text{ is } \textbf{OFF}$$

Figure 21–24B (p. 703) shows the ideal case where Q_1 is **ON** and Q_2 is **OFF** (we have ignored the **ON** resistance of the real MOSFET). The output is connected to the supply voltage, so

$$V_i = 0\text{V} \rightarrow V_o = V_{DD} = +5\text{V}$$

When the input is +5V

$$\textbf{P-channel: } V_{GS_1} = 5 - 5 = 0\text{V} \rightarrow Q_1 \text{ is } \textbf{OFF}$$
$$\textbf{N-channel: } V_{GS_2} = V_i = 5\text{V} \rightarrow Q_2 \text{ is } \textbf{ON}$$

Figure 21–24C (p. 703) shows the ideal case where Q_1 is **OFF** and Q_2 is **ON.** The output is connected to ground, so

$$V_i = 5\text{V} \rightarrow V_o = 0\text{V}$$

In this simple CMOS circuit, a HIGH input produces a LOW output; and a LOW input produces a HIGH output. The basic CMOS circuit acts as an *inverter.*

REVIEW QUESTIONS

1. How is the CMOS constructed?
2. What can you say about the relation between the N-channel and P-channel parameters?
3. Why is one of the MOSFETs in a CMOS usually **OFF?**
4. Build a CMOS inverter, and explain its operation.

21–8 ■ OTHER MOSFETs

The major advantage of the FET over the BJT is its high input resistance. For this reason, many integrated circuit amplifiers use FETs in the input circuit. Two disadvantages of the FET are:

Low current capability—approximately 1/10 of the current handling capability of the BJT.

Poor high-frequency response—the reverse-bias gate PN junctions of the JFET and the insulated gate of the MOSFETs act as capacitors that limit frequency response.

Clever designers have found ways to build FETs that overcome these problems.

Power FETs

Through special construction techniques an FET can be built with channel geometries that allow the FET to handle large currents. The VMOS has a deep vertical gate, with the source at the top and the drain at the bottom of the device (Figure 21–25). While this construction allows for large drain currents, the V-shaped gate results in a low breakdown voltage. To avoid this problem, manufacturers have produced MOSFETs with U-shaped gates.

Another vertical MOSFET (source at the top—drain at the bottom) is the double-diffused MOSFET (DMOS). This device, which has an extra semiconductor layer, also provides for a large drain current.

FIGURE 21–25
VMOS

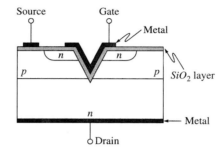

Dual-Gate MOSFETs

The dual-gate MOSFET controls the channel with two gates that are connected in series, putting the gate capacitances in series. As you might remember, combining two capacitors in series lowers the total capacitance, which will improve the frequency response.

The MESFET

The gate in the MESFET is constructed in a manner much like the Schottky diode (Chapter 5). That is, the gate here is a metal-semiconductor interface, which reduces the gate capacitance, improving the frequency response.

REVIEW QUESTIONS

1. What are two disadvantages of the FET?
2. What advantage does the VMOS have? What disadvantage?
3. Why does the dual-gate MOSFET have a frequency advantage?
4. What device has a similar construction to the MESFET? Why is the MESFET used?

21–9 ■ DC Troubleshooting MOSFET Circuits

You must exercise great care when handling MOSFETs; the static electrical charge on your fingers can damage many of these devices. You must, therefore, make sure that you are grounded (in an electrical not a psychological sense) when you work with MOSFETs. For this reason, we do not recommend testing MOSFETs out of the circuit, unless you have been explicitly trained to do so. We will limit our discussion here to in-circuit troubleshooting.

DC Troubleshooting the D-MOSFET

Figures 21–26 and 21–27 (p. 706) show the results of DC Troubleshooting an N-channel D-MOSFET circuit. As with the JFET, it is unlikely that you will be able to determine the Estimated Voltages at the drain and source. For this reason we show only the Measured Voltages at these terminals. We begin by considering the results of open failures.

FIGURE 21–26
Open failures

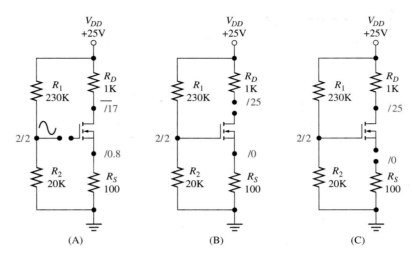

(A)   (B)   (C)

FIGURE 21–27
Short failures

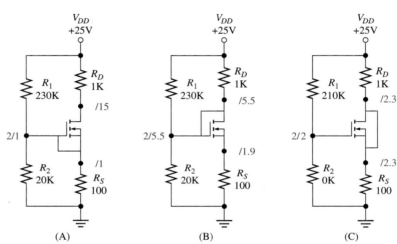

(A)   (B)   (C)

Open Failures

You can see from Figure 21–26 that the Measured Voltage at the gate always equals the Estimated Voltage at the gate. The reason for this is that even in a properly operating MOSFET, there is no gate current. Therefore, we look only at the other two measurements to determine what the problem might be.

Open Gate It is next to impossible to determine an open gate failure (Figure 21–26A) by measuring the drain and source DC voltages because the D-MOSFET (unlike the JFET) can be operated in the depletion or the enhancement mode. A D-MOSFET designed to operate in the depletion mode will draw more drain current with an open gate. A D-MOSFET designed to operate in the enhancement mode will draw less drain current with an open gate.

The only sure way to know that the gate has failed open is to inject an AC signal at the gate. If the drain voltage remains constant, the gate has failed open.

Open Drain or Open Source If either the drain or source is open (Figures 21–26B and C), the drain current will fall to 0A. We would then measure the supply voltage at the bottom of R_D and 0V at the top of R_S.

Short Failures

As Figure 21–27 shows, when an internal short occurs between two terminals, the Measured Voltages at these terminals are the same.

DC Troubleshooting the E-MOSFET

Figures 21–28 and 21–29 show the results of DC Troubleshooting an N-channel E-MOSFET circuit.

FIGURE 21–28

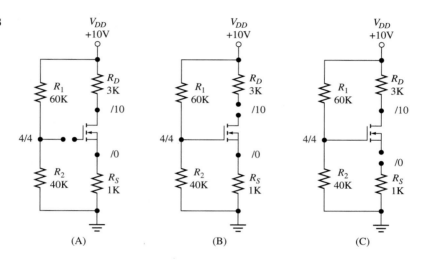

(A)   (B)   (C)

FIGURE 21–29

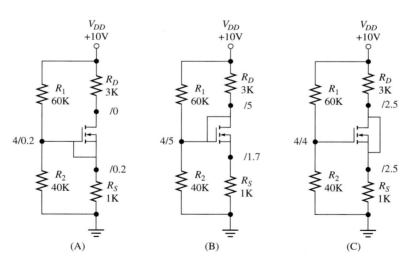

(A)   (B)   (C)

Open Failures You can see from Figure 21–28 that the Measured Voltages at all terminals are the same for an open gate (A), open drain (B), or open source (C). You already know that an open drain or source will cause I_D to be 0A. Because this is an enhancement device, if there is an open at the gate, there will be no gate voltage to create the channel. This also results in zero drain current. Therefore, DC Troubleshooting can tell you that one of the E-MOSFET terminals has opened up but cannot tell you which terminal has failed.

Short Failures As Figure 21–29 shows, when an internal short occurs between two terminals, the Measured Voltages at these terminals are the same.

1. Why must MOSFETs be handled with care?

2. What do you expect if a D-MOSFET has an internal open at the drain? The source? The gate? Explain your answers.

3. What do you expect if a D-MOSFET has an internal short between the gate and the source? The gate and the drain? The drain and the source?

4. What do you expect if an E-MOSFET has an internal open at the drain? The source? The gate? Explain your answers.

5. What do you expect if an E-MOSFET has an internal short between the gate and the source? The gate and the drain? The drain and the source?

SUMMARY

- The MOSFET is constructed with an insulated gate, resulting in near zero gate current.
- The D-MOSFET is normally **ON**. When the gate voltage is 0V, the drain current is I_{DSS}.
- A negative gate voltage is applied to the N-channel D-MOSFET to cut off the channel. Drain current is 0A when $V_{GS} = V_{GS(off)}$. A positive gate voltage is applied to enhance the channel, increasing the drain current.
- A positive gate voltage is applied to the P-channel D-MOSFET to cut off the channel. A negative voltage is applied to the gate to enhance the channel.
- The D-MOSFET I-V curve is approximated by a parabola.

$$I_D = I_{DSS}\left(1 - \frac{V_{GS}}{V_{GS(off)}}\right)^2$$

- When the gate is grounded, the bias values are found by drawing one of the following gate load-lines on the same graph as the MOSFET I-V curve.

$$I_D = \frac{-V_{GS}}{R_S} \qquad \textbf{N-Channel}$$

$$I_D = \frac{V_{GS}}{R_S} \qquad \textbf{P-Channel}$$

- When a voltage divider is used at the gate, the gate load-line equation is

$$I_D = \frac{V_G}{R_S} - \frac{V_{GS}}{R_S} \qquad \textbf{N-channel}$$

$$I_D = \frac{-V_G}{R_S} + \frac{V_{GS}}{R_S} \qquad \textbf{P-channel}$$

- Bias values can be calculated from the quadratic equation solution, where for the N-channel D-MOSFET

$$a = \frac{1}{V^2_{GS(off)}}$$

$$b = \frac{-2}{V_{GS(off)}} + \frac{1}{R_S I_{DSS}}$$

$$c = 1 - \frac{V_G}{R_S I_{DSS}}$$

$$V_{GS} = \frac{-b + \sqrt{b^2 - 4ac}}{2a}$$

and for the P-channel D-MOSFET

$$a = \frac{1}{V^2_{GS(off)}}$$

$$b = \frac{-2}{V_{GS(off)}} - \frac{1}{R_S I_{DSS}}$$

$$c = 1 + \frac{V_G}{R_S I_{DSS}}$$

$$V_{GS} = \frac{-b - \sqrt{b^2 - 4ac}}{2a}$$

- The E-MOSFET is normally **OFF.** A gate voltage, positive for the N-channel and negative for the P-channel, is applied to enhance the conducting channel.
- Conduction in the E-MOSFET begins when the magnitude of V_{GS} exceeds $V_{GS(th)}$. The current in an E-MOSFET is found from

$$I_D = k(V_{GS} - V_{GS(th)})^2$$

- Graphical determination of the bias levels in the E-MOSFET follows the same procedure as used for the D-MOSFET.
- Bias values can be calculated from solving the quadratic equation. For the N-channel E-MOSFET

$$a = 1$$

$$b = -2V_{GS(th)} + \frac{1}{kR_S}$$

$$c = V^2_{GS(th)} - \frac{V_G}{kR_S}$$

$$V_{GS} = \frac{-b + \sqrt{b^2 - 4c}}{2}$$

For the P-channel E-MOSFET

$$a = 1$$

$$b = -2V_{GS(th)} - \frac{1}{kR_S}$$

$$c = V^2_{GS(th)} + \frac{V_G}{kR_S}$$

$$V_{GS} = \frac{-b - \sqrt{b^2 - 4c}}{2}$$

- The CMOS is a complementary combination of an N- and a P-channel E-MOSFET.
- The VMOS is used in power applications.
- The dual-gate MOSFET is used in high-frequency applications.
- MOSFETs can be damaged by the static electricity on the hands; handle them with care.
- An open failure at the gate of D-MOSFET cannot be easily determined from DC Troubleshooting. An AC signal injected at the gate will not show up at the drain.
- An open failure at the drain or the source will result in zero drain current.
- A short failure between any two MOSFET terminals will lead to the same Measured Voltages at the two terminals.
- An open failure at any terminal of the E-MOSFET will result in zero drain current.

PROBLEMS *Where appropriate, use the Almost Universal I-V curves of Figures 21–6 (p. 689), 21–10 (p. 692), 21–14 (p. 695), or 21–18 (p. 698) to find the bias values in the following problems.*

SECTION 21–2   The N-channel D-MOSFET

FIGURE 21–30

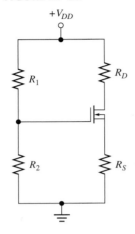

1. For the circuit in Figure 21–30, assume $V_{GS(off)} = -4V$ and $I_{DSS} = 10mA$. If

$$V_{DD} = +10V, R_1 = 90K\Omega, R_2 = 10K\Omega, R_D = 3K\Omega, R_S = 2K\Omega$$

   (a) Find in the following order
   V_G
   V_{GS}
   I_D
   V_S
   V_D

   (b) Is the transistor operating in the depletion or enhancement mode?

2. For the circuit in Figure 21–30, assume $V_{GS(off)} = -10V$ and $I_{DSS} = 20mA$. If

$$V_{DD} = +20V, R_1 = 190K\Omega, R_2 = 10K\Omega, R_D = 3K\Omega, R_S = 2K\Omega$$

   Repeat Problem 1.

3. For the circuit in Figure 21–30, assume $V_{GS(off)} = -4V$ and $I_{DSS} = 5mA$. If

$$V_{DD} = +10V, R_1 = 70K\Omega, R_2 = 30K\Omega, R_D = 1K\Omega, R_S = 500\Omega$$

   (a) Find in the following order
   V_G
   V_{GS}
   I_D
   V_S
   V_D

   (b) Is the transistor operating in the depletion or enhancement mode?

4. For the circuit in Figure 21–30, assume $V_{GS(off)} = -5V$ and $I_{DSS} = 15mA$. Assume that

$$V_{DD} = +20V, R_1 = 170K\Omega, R_2 = 30K\Omega, R_D = 500\Omega, R_S = 100K\Omega$$

   Repeat Problem 3.

FIGURE 21–31

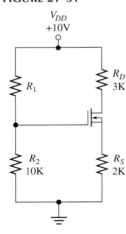

5. In Figure 21–31, $V_{GS(off)} = -4V$, and $I_{DSS} = 10mA$. Choose R_1 so that the Q-points are $V_{GS} = 0V$ and $I_D = I_{DSS} = 10mA$. Proceed as follows:

   (a) Draw the gate I-V curve.

   (b) The slope of the gate load-line is $-1/R_S$. Draw a gate load-line with this slope through the desired Q-point.

   (c) The point at which the load-line crosses the V_{GS} axis is the required gate voltage (V_G).

   (d) Find R_1 to produce V_G.

6. Repeat Problem 1 using the analytical method.

7. Repeat Problem 2 using the analytical method.

8. Repeat Problem 3 using the analytical method.

9. Repeat Problem 4 using the analytical method.

SECTION 21–3   The P-Channel D-MOSFET

10. For the circuit in Figure 21–32, assume $V_{GS(off)} = 4V$ and $I_{DSS} = 10mA$. Find in the following order
    V_G
    V_{GS}
    I_D
    V_S
    V_D

FIGURE 21–32 **FIGURE 21–33**

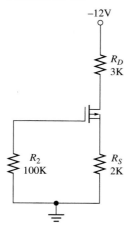 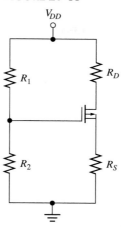

11. For the circuit in Figure 21–33, assume $V_{GS(off)} = 4V$ and $I_{DSS} = 10mA$. Assume that

$$V_{DD} = -10V, R_1 = 90K\Omega, R_2 = 10K\Omega, R_D = 3K\Omega, R_S = 2K\Omega$$

Find in the following order

V_G

V_{GS}

I_D

V_S

V_D

12. For the circuit in Figure 21–33, assume $V_{GS(off)} = 4V$ and $I_{DSS} = 5mA$. Assume that

$$V_{DD} = -10V, R_1 = 70K\Omega, R_2 = 30K\Omega, R_D = 1K\Omega, R_S = 500\Omega$$

Repeat Problem 11.

13. For the circuit in Figure 21–33, assume $V_{GS(off)} = 5V$ and $I_{DSS} = 15mA$. Assume that

$$V_{DD} = -10V, R_1 = 190K\Omega, R_2 = 30K\Omega, R_D = 500\Omega, R_S = 100\Omega$$

Repeat Problem 11.

14. Repeat Problem 10 using the analytical method.

15. Repeat Problem 11 using the analytical method.

16. Repeat Problem 12 using the analytical method.

17. Repeat Problem 13 using the analytical method.

SECTION 21–4   The N-Channel E-MOSFET

FIGURE 21–34

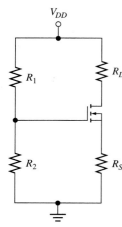

18. For the circuit in Figure 21–34, assume $V_{GS(th)} = 2V$ and $k = 0.00125A/V^2$. Assume that

$$V_{DD} = +12V, R_1 = 20K\Omega, R_2 = 10K\Omega, R_D = 3K\Omega, R_S = 1K\Omega$$

Find in the following order

V_G

V_{GS}

I_D

V_S

V_D

19. For the circuit in Figure 21–34, assume $V_{GS(th)} = 1V$ and $k = 0.000125A/V^2$. Assume that

$$V_{DD} = +10V, R_1 = 70K\Omega, R_2 = 30K\Omega, R_D = 2K\Omega, R_S = 500\Omega$$

Repeat Problem 18.

20. Repeat Problem 18 using the analytical method.

FIGURE 21–35

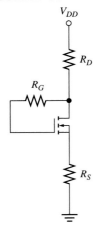

21. Repeat Problem 19 (p. 711) using the analytical method.

22. Figure 21–35 shows how to bias an E-MOSFET with a feedback resistor between the drain and the gate.

(a) What is the current in R_G? (b) What is the voltage across R_G?

(c) Does $V_G = V_D$? Why? (d) What is the current in R_D? R_S?

(e) Write a relation between V_{DS} and I_D.

(f) Rearrange the equation just found to find I_D as a function of V_{DS}.

(g) Does $V_{DS} = V_{GS}$? Why?

(h) Replace V_{DS} in the equation you wrote for part (f) with V_{GS}.

(i) Sketch a typical E-MOSFET gate I-V curve. On the same sketch, draw the load-line you found in part (h). Indicate the Q-point for this circuit.

23. The E-MOSFET in Figure 21–35 has the parameters

$$V_{GS(th)} = 2\text{V}, \ k = 0.000625\text{A/V}^2$$

If $V_{DD} = +20$ V, $R_G = 100\text{K}\Omega$, $R_D = 2\text{K}\Omega$, $R_S = 1\text{K}\Omega$, find V_{GS}, and I_D.

SECTION 21–5   The P-Channel E-MOSFET

FIGURE 21–36

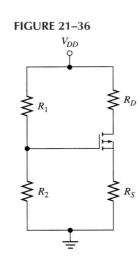

24. For the circuit in Figure 21–36, assume $V_{GS(th)} = -2$V and $k = 0.00125\text{A/V}^2$. Assume that

$$V_{DD} = -12\text{V}, \ R_1 = 20\text{K}\Omega, \ R_2 = 10\text{K}\Omega, \ R_D = 3\text{K}\Omega, \ R_S = 1\text{K}\Omega$$

Find in the following order

V_G

V_{GS}

I_D

V_S

V_D

25. For the circuit in Figure 21–36, assume $V_{GS(th)} = -1$V and $k = 0.000125\text{A/V}^2$. Assume that

$$V_{DD} = -10\text{V}, \ R_1 = 70\text{K}\Omega, \ R_2 = 30\text{K}\Omega, \ R_D = 2\text{K}\Omega, \ R_S = 500\Omega$$

Repeat Problem 24.

26. Repeat Problem 24 using the analytical method.

27. Repeat Problem 25 using the analytical method.

SECTION 21–6   MOSFET Data Sheets and Computer Models

28. Refer to Figure 21–19 (p. 700) to find the range of $V_{GS(off)}$ (typical to maximum) and I_{DSS} (typical to maximum) for the 2N3797.

(a) Draw the typical and maximum gate I-V curves for this device.

(b) Assume that the gate load-line is given by $I_D = -V_{GS}/1\text{K}$; draw the load-line on top of the I-V curves. What is the range of values for V_{GS} and I_D?

29. Find the computer parameter β for typical and maximum values given in the data sheet for the 2N3797.

30. The data sheet for the E-MOSFET (Figure 21–20, p. 701) shows the parameter I_{DSS}. Why isn't this value used in the bias calculations for an E-MOSFET?

31. Draw the gate I-V curve for the 2N4351 (Figure 21–20, p. 701).

32. Find the computer parameter (β) for the 2N4351.

SECTION 21–7   The Complementary MOSFET (CMOS)

33. For the CMOS inverter in Figure 21–37 (p. 713)

(a) Which MOSFET is **ON** when the input is 0V? What is V_o?

(b) Which MOSFET is **ON** when the input is +5V? What is V_o?

FIGURE 21–37 **FIGURE 21–38** **FIGURE 21–39**

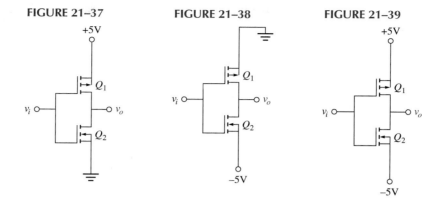

34. For the CMOS inverter in Figure 21–38
   (a) Which MOSFET is **ON** when the input is 0V? What is V_o?
   (b) Which MOSFET is **ON** when the input is −5V? What is V_o?
35. For the CMOS inverter in Figure 21–39
   (a) Which MOSFET is **ON** when the input is +5V? What is V_o?
   (b) Which MOSFET is **ON** when the input is −5V? What is V_o?
   (c) Which MOSFET is **ON** when the input is 0V? Can you use the ideal MOSFET model in which the **ON** resistance is 0Ω? If each **ON** MOSFET is represented with the resistance r_{DS}, what is V_o?

SECTION 21–9 DC Troubleshooting MOSFET Circuits

36. If the connections and resistors are good, determine the likely cause of the DC Troubleshooting results shown in Figure 21–40.

FIGURE 21–40

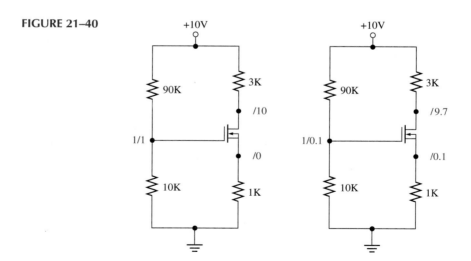

37. If the connections and resistors are good, determine the likely cause of the DC Troubleshooting results shown in Figure 21–41 (p. 714).

FIGURE 21–41

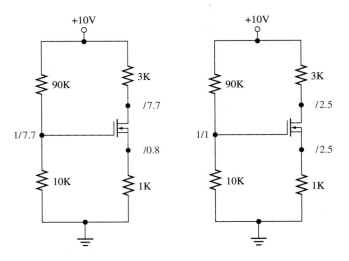

38. If the connections and resistors are good, determine the likely cause of the DC Troubleshooting results shown in Figure 21–42.

FIGURE 21–42

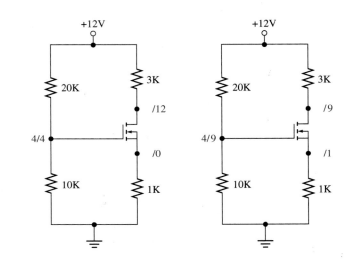

39. If the connections and resistors are good, determine the likely cause of the DC Troubleshooting results shown in Figure 21–43.

FIGURE 21–43

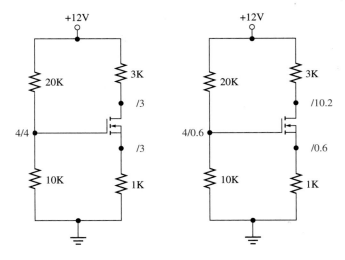

Computer Problems

40. Examine the D-MOSFET models available in the computer program you are using. For each model, determine I_{DSS}.

41. Simulate the circuit of Problem 1 (p. 710), and find all voltages and currents.

42. Simulate the circuit of Problem 3 (p. 710), and find all voltages and currents.

43. Simulate the circuit of Problem 10 (p. 710), and find all voltages and currents.

44. Simulate the circuit of Problem 12 (p. 711), and find all voltages and currents.

45. Simulate the circuit of Problem 1 (p. 710). Find the voltages at the top of R_G, the top of R_2, and the bottom of R_D after

    (a) You open the gate.   **(b)** You open the source.   **(c)** You open the drain.

46. Simulate the circuit of Problem 1 (p. 710). Find the voltages at the top of R_G, the top of R_2, and the bottom of R_D after

    (a) You short the gate to the source.   **(b)** You short the gate to the drain.

    (c) You short the drain to the source.

47. Simulate the circuit of Problem 18 (p. 711). Find the voltages at the top of R_G, the top of R_2, and the bottom of R_D after

    (a) You open the gate.   **(b)** You open the source.   **(c)** You open the drain.

48. Simulate the circuit of Problem 18 (p. 711). Find the voltages at the top of R_G, the top of R_2, and the bottom of R_D after

    (a) You short the gate to the source.   **(b)** You short the gate to the drain.

    (c) You short the drain to the source.

CHAPTER

22

FET AC BEHAVIOR AND APPLICATIONS

C HAPTER OBJECTIVES

To determine the AC model for FETs.

To determine g_m for the JFET and D-MOSFET.

To determine g_m for the E-MOSFET.

To analyze the FET common-source amplifier by comparison
to the BJT common-emitter amplifier.

To analyze the FET common-drain amplifier (FET follower)
by comparison to the BJT common-collector (emitter-
follower) amplifier.

To analyze the FET common-gate amplifier by comparison
to the BJT common-base amplifier.

To analyze the FET differential amplifier by comparison to
the BJT differential amplifier.

To use the FET as a voltage variable resistor (VVR).

To build CMOS logic circuits.

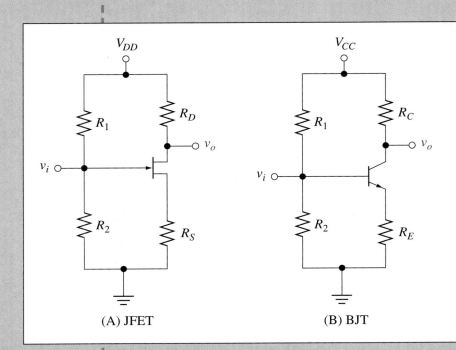

(A) JFET   (B) BJT

22–1 ■ INTRODUCTION

You might have noticed that the title of this chapter refers to FET AC behavior, not JFET, D-MOSFET, or E-MOSFET AC behavior. *All* FETs exhibit essentially the same AC behavior. You have already finished the tough part, determining the DC bias levels in the various FET device circuits. As we will show you, if you know the AC response of a BJT circuit, you know the AC response of the equivalent FET circuit.

22–2 ■ FET AC MODEL

The AC model we use for the FET looks very similar to the AC model for the BJT; the modeling details are given in a Special Topic section at the end of this chapter. For comparison, we show you the BJT AC model in Figure 22–1A and FET AC model in Figure 22–1B. The important AC model variables and parameters are tabulated below:

BJT AC MODEL	FET AC MODEL
$i_b = \dfrac{i_c}{\beta}$	$i_g = 0\text{A}$
$i_c \approx i_e$	$i_s = i_d$
$\hat{v}_e = v_b$	$\hat{v}_s = v_g$
AC emitter resistance $= r_e$	AC source resistance $= \dfrac{1}{g_m}$

A comparison of these model parameters shows two important differences:

☐ The gate current of the FET is 0A, which means the gate resistance of the FET is infinite. Because of this, the input resistance of FET circuits is usually higher than the input resistance of equivalent BJT circuits.

☐ To follow standard FET modeling practices, we use a transconductance (g_m) to represent the AC source resistance.

FIGURE 22–1
BJT-FET AC model comparison

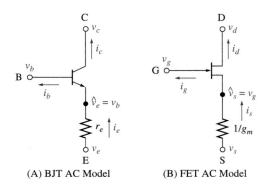

(A) BJT AC Model   (B) FET AC Model

It is the second difference that makes FET AC analysis slightly more difficult than BJT analysis. We remind you that we found the AC emitter resistance of the BJT emitter by determining the slope of its I-V curve (Figure 22–2A):

$$m = \frac{\Delta i_e}{\Delta v_{be}}$$

FIGURE 22–2

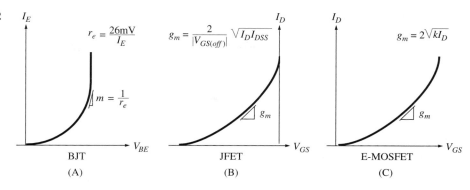

where

$$m = \frac{1}{r_e}$$

and the AC emitter resistance is found from the simple formula:

$$\textbf{BJT: } r_e = \frac{26\text{mV}}{I_E}$$

The AC source resistance of the FET is found by determining the slope of the gate I-V curves: JFET or D-MOSFET (Figure 22–2B) and E-MOSFET (Figure 22–2C). The slope of the FET curves is known as the *transconductance* (g_m). (Do not confuse the AC transconductance we find here with the transconductance parameter (k) used in computer programs. Although the names are similar, these are two different parameters and are not directly related.)

Determining g_m for the JFET and D-MOSFET

For the JFET or the D-MOSFET (Figure 22–2B), we know that

$$I_D = I_{DSS}\left(1 - \frac{V_{GS}}{V_{GS(off)}}\right)^2$$

The slope of this curve is given by

$$g_m = \frac{\Delta I_D}{\Delta V_{GS}}$$

We can use the DC bias level for either I_D or V_{GS} to derive a formula for g_m:

$$g_m = \frac{2}{|V_{GS(off)}|}\sqrt{I_D I_{DSS}}$$

or

$$g_m = 2\left|\frac{I_{DSS}}{V_{GS(off)}}\right|\left(1 - \frac{V_{GS}}{V_{GS(off)}}\right)$$

Note the use of the absolute values. Whether the FET is N-channel or P-channel, g_m is always positive.

You will sometimes see the parameter g_{mo} used in determining the transconductance. This value is found when $V_{GS} = 0V$:

$$V_{GS} = 0V \rightarrow g_m = g_{mo} = 2\left|\frac{I_{DSS}}{V_{GS(off)}}\right|$$

We can then express the general transconductance as

$$g_m = g_{mo}\left(1 - \frac{V_{GS}}{V_{GS(off)}}\right)$$

Because we find both I_D and V_{GS} when we do bias calculations, you can use either the I_D or the V_{GS} based formula. The choice is a personal one.

EXAMPLE 22–1
Finding g_m
for the JFET

An N-channel JFET has the following parameters:

$$V_{GS(off)} = -4V \text{ and } I_{DSS} = 10mA$$

The bias circuitry produces the DC bias levels:

$$I_D = 2.5mA \text{ and } V_{GS} = -2V$$

(a) Find g_m using I_D.   **(b)** Find g_m using V_{GS}.

Solution
(a) We first find g_m using I_D:

$$g_m = \frac{2}{|V_{GS(off)}|}\sqrt{I_D I_{DSS}} = \frac{2}{4}\sqrt{(2.5 \times 10^{-3})(10 \times 10^{-3})}$$
$$= 0.5\sqrt{25 \times 10^{-6}} = 2.5mS$$

Note that g_m has the unit of conductance, the siemen (S).
(b) We now find g_m using V_{GS}:

$$g_m = 2\left|\frac{I_{DSS}}{V_{GS(off)}}\right|\left(1 - \frac{V_{GS}}{V_{GS(off)}}\right) = \frac{2(10 \times 10^{-3})}{4}\left(1 - \frac{-2}{-4}\right) = 2.5mS$$

DRILL EXERCISE

Find g_m in terms of I_{DSS} and $V_{GS(off)}$ if the bias circuitry results in $I_D = \dfrac{I_{DSS}}{4}$.

Answer $g_m = \left|\dfrac{I_{DSS}}{V_{GS(off)}}\right|$

Determining g_m for the E-MOSFET

The gate I-V curve for the E-MOSFET (Figure 22–2C) is described by

$$I_D = k(V_{GS} - V_{GS(th)})^2$$

The slope of this curve is given by either

$$g_m = 2\sqrt{kI_D}$$

or

$$g_m = 2k\,|V_{GS} - V_{GS(th)}|$$

Again, note the use of the absolute value; whether we use an N-channel or a P-channel E-MOSFET, g_m is always positive.

EXAMPLE 22–2
Finding g_m
for the E-MOSFET

An N-channel E-MOSFET has the following parameters:

$$V_{GS(th)} = 2\text{V} \qquad \text{and} \qquad k = 2 \times 10^{-3}\text{A/V}^2$$

(a) Find g_m if the bias circuitry produces the DC bias level $I_D = 2.5\text{mA}$.
(b) Find g_m if the bias circuitry produces the DC bias level $V_{GS} = 3.1\text{V}$.

Solution
(a) Using I_D, we find g_m from

$$g_m = 2\sqrt{k\,I_D} = 2\sqrt{(2 \times 10^{-3})(2.5 \times 10^{-3})} = 2\sqrt{5 \times 10^{-6}} = 4.47\text{mS}$$

(b) Using V_{GS}, we find g_m from

$$g_m = 2k\,|V_{GS} - V_{GS(th)}| = 2(2 \times 10^{-3}) \times |3.1 - 2| = 4.4\text{mS}$$

DRILL EXERCISE

Find g_m in terms of k and $V_{GS(th)}$ if the bias circuitry results in $V_{GS} = 2V_{GS(th)}$.

Answer $g_m = 2k\,|V_{GS(th)}|$

We have stated that AC emitter resistance (r_e) and the AC source resistance $\left(\dfrac{1}{g_m}\right)$ are equivalent parameters in transistor equations. There is a difference in typical values for these parameters, however. In the previous example, a drain current of 2.5mA led to

$$g_m = 4.47\text{mS}$$

so

$$\frac{1}{g_m} = \frac{1}{4.47 \times 10^{-3}} = 224\Omega$$

For a BJT with an emitter current of 2.5mA, we get

$$r_e = \frac{26\text{mV}}{I_E} = \frac{26\text{mV}}{2.5\text{mA}} = 10.4\Omega$$

For the same bias currents, the AC source resistance of the FET is 20 times greater than the AC emitter resistance of the BJT.

FET Data Sheets and AC Parameters

The data sheets for the 2N3797 D-MOSFET and the 2N4351 E-MOSFET are shown again in Figures 22–3 (p. 722) and 22–4 (p. 723). The parameter that we label the transconductance (g_m) is labeled the *Forward Transfer Admittance* (y_{fs}) by the manufacturer. That is

Transconductance: $g_m = y_{fs}$

For the 2N3797, g_m can vary from 1500 to 3000μS. For the 2N4351, g_m has a minimum value of 1000μS.

As we did with the BJT, we have assumed until now that the FET drain I-V curve is flat in the **ACTIVE** region. In fact, the drain current will continue to rise as the drain-

FIGURE 22–3

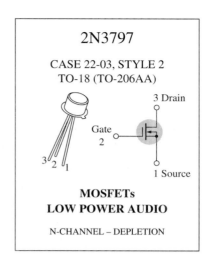

2N3797

CASE 22-03, STYLE 2
TO-18 (TO-206AA)

3 Drain

Gate
2

1 Source

MOSFETs
LOW POWER AUDIO

N-CHANNEL – DEPLETION

MAXIMUM RATINGS

Rating	Symbol	Value	Unit
Drain-Source Voltage 2N3797	V_{DS}	20	Vdc
Gate-Source Voltage	V_{GS}	±10	Vdc
Drain Current	I_D	20	mAdc
Total Device Dissipation @ $T_A = 25°C$ Derate above 25°C	P_D	200 1.14	mW mW/°C
Junction Temperature Range	T_J	+175	°C
Storage Channel Temperature Range	T_{stg}	−65 to +200	°C

ELECTRICAL CHARACTERISTICS ($T_A = 25°C$ unless otherwise noted)

Characteristic	Symbol	Min	Typ	Max	Unit
OFF CHARACTERISTICS					
Drain Source Breakdown Voltage ($V_{GS} = -7.0$ V, $I_D = 5.0$ μA) 2N3797	$V_{(BR)DSX}$	20	25	—	Vdc
Gate Reverse Current (1) ($V_{GS} = -10$ V, $V_{DS} = 0$) ($V_{GS} = -10$ V, $V_{DS} = 0$, $T_A = 150°C$)	I_{GSS}	— —	— —	1.0 200	pAdc
Gate Source Cutoff Voltage ($I_D = 2.0$ μA, $V_{DS} = 10$ V) 2N3797	$V_{GS(off)}$	—	−5.0	−7.0	Vdc
Drain-Gate Reverse Current (1) ($V_{DG} = 10$ V, $I_S = 0$)	I_{DGO}	—	—	1.0	pAdc
ON CHARACTERISTICS					
Zero-Gate Voltage Drain Current ($V_{DS} = 10$ V, $V_{GS} = 0$) 2N3797	I_{DSS}	2.0	2.9	6.0	mAdc
On-State Drain Current ($V_{DS} = 10$ V, $V_{GS} = +3.5$ V) 2N3797	$I_{D(on)}$	9.0	14	18	mAdc
SMALL-SIGNAL CHARACTERISTICS					

Forward Transfer Admittance
($V_{DS} = 10$ V, $V_{GS} = 0$, $f = 1.0$ kHz)
     2N3797
($V_{DS} = 10$ V, $V_{GS} = 0$, $f = 1.0$ MHz)
     2N3797
	$|y_{fs}|$	1500
		
		1500
			2300
			
			—
				3000
				
				—
					μmhos

Output Admittance ($I_{DS} = 10$ V, $V_{GS} = 0$, $f = 1.0$ kHz) 2N3797	$|y_{os}|$	—	27	60	μmhos
Input Capacitance ($V_{DS} = 10$ V, $V_{GS} = 0$, $f = 1.0$ MHz) 2N3797	C_{iss}	—	6.0	8.0	pF
Reverse Transfer Capacitance ($V_{DS} = 10$ V, $V_{GS} = 0$, $f = 1.0$ MHz)	C_{rss}	—	0.5	0.3	pF
FUNCTIONAL CHARACTERISTICS					
Noise Figure ($V_{DS} = 10$ V, $V_{GS} = 0$, $f = 1.0$ kHz, $R_S = 3$ megohms)	NF	—	3.8	—	dB

(1) This value of current includes both the FET leakage current as well as the leakage current associated with the test socket and fixture when measured under best attainable conditions.

FIGURE 22–4

MAXIMUM RATINGS

Rating	Symbol	Value	Unit
Drain-Source Voltage	V_{DS}	25	Vdc
Drain-Gate Voltage	V_{DG}	30	Vdc
Gate-Source Voltage*	V_{GS}	30	Vdc
Drain Current	I_D	30	mAdc
Total Device Dissipation @ T_A = 25°C Derate above 25°C	P_D	300 1.7	mW mW/°C
Junction Temperature Range	T_J	175	°C
Storage Temperature Range	T_{stg}	−65 to −175	°C

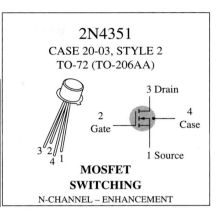

2N4351
CASE 20-03, STYLE 2
TO-72 (TO-206AA)

MOSFET
SWITCHING
N-CHANNEL – ENHANCEMENT

*Transient potentials of ± 75 Volt will not cause gate-oxide failure.

ELECTRICAL CHARACTERISTICS (T_A = 25°C unless otherwise noted.)

Characteristic	Symbol	Min	Max	Unit
OFF CHARACTERISTICS				
Drain-Source Breakdown Voltage (I_D = 10 μA, V_{GS} = 0)	$V_{(BR)DSX}$	25	—	Vdc
Zero-Gate-Voltage Drain Current (V_{DS} = 10 V, V_{GS} = 0) T_A = 25°C T_A = 150°C	I_{DSS}	— —	10 10	nAdc μAdc
Gate Reverse Current (V_{GS} = ± 15 Vdc, V_{DS} = 0)	I_{GSS}	—	± 10	pAdc
ON CHARACTERISTICS				
Gate threshold Voltage (V_{DS} = 10 V, I_D = 10 μA)	$V_{GS(Th)}$	1.0	.5	Vdc
Drain-Source On-Voltage (I_D = 2.0 mA, V_{GS} = 10V)	$V_{DS(on)}$	—	1.0	V
On-State Drain Current (V_{GS} = 10 V, V_{DS} = 10V)	$I_{D(on)}$	3.0	—	mAdc
SMALL-SIGNAL CHARACTERISTICS				
Forward Transfer Admittance (V_{DS} = 10 V, I_D = 2.0 mA, f = 1.0 kHz)	$|y_{fs}|$	1000	—	μmho
Input Capacitance (V_{DS} = 10 V, V_{GS} = 0, f = 140 kHz)	C_{iss}	—	5.0	pF
Reverse Transfer Capacitance (V_{DS} = 0, V_{GS} = 0, f = 140 kHz)	C_{rss}	—	1.3	pF
Drain-Substrate Capacitance ($V_{D(SUB)}$ = 10 V, f = 140 kHz)	$C_{d(sub)}$	—	5.0	pF
Drain-Source Resistance (V_{GS} = 10 V, I_D = 0, f = 1.0 kHz)	$r_{ds(on)}$	—	300	ohms

SWITCHING CHARACTERISTICS

		Symbol	Min	Max	Unit
Turn-On Delay (Fig. 5)	I_D = 2.0 mAdc, V_{DS} = 10 Vdc, (V_{GS} = 10 Vdc) (See Figure 9; Times Circuit Determined)	t_{dt1}	—	45	ns
Rise Time (Fig. 6)		t_r	—	65	ns
Turn-Off Delay (Fig. 7)		t_{d2}	—	60	ns
Fall Time (Fig. 8)		t_f	—	100	ns

FIGURE 22–5
OHMIC and ACTIVE region resistances

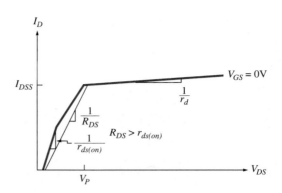

source voltage is increased (Figure 22–5). This rise reflects the fact that the FET has a finite output resistance (r_d). Note that r_d is the inverse of the slope of the I-V curve in the **ACTIVE** region. The data sheets in Figures 22–3 (p. 722) and 22–4 (p. 723) give us the actual slope, rather than its inverse. The manufacturer labels this parameter the Output Admittance (y_{os}). The relation between these two parameters is

$$\text{FET Output Resistance: } r_d = \frac{1}{y_{os}}$$

For the 2N3797, the output admittance can vary from 27 to 60μS. This means the output resistance of the 2N3797 can vary from 16 to 37KΩ.

Because the output resistance of an FET appears approximately in parallel with the external drain resistance (R_D), the total drain resistance of an FET amplifier is

$$R_{Dtotal} \approx r_d \| R_D$$

Some FETs have output resistances as low as 2KΩ. What would happen if this FET is used with a drain resistor of $R_D = 2$KΩ? The total drain resistance would be 1KΩ, resulting in a gain that is one half of what you would expect.

FET OHMIC Region Resistance

In previous chapters we described the **OHMIC** region of the drain I-V curve with a single straight line represented by $R_{DS} = V_{DS_{po}}/I_{DS_{po}}$. A more accurate model in this region is shown in Figure 22–5.

The rise in current is steepest when both I_D and V_{DS} are small. The slope of this initial rise is labeled $1/r_{ds(on)}$. The value we calculate for R_{DS} (shown by the dashed line in Figure 22–5) is always larger than the value of $r_{ds(on)}$. For the 2N4351, $r_{ds(on)}$ has a maximum value of 300Ω. Note that this value is found at low drain current.

REVIEW QUESTIONS

1. Draw the BJT and FET AC models. Compare the two models.
2. What is the slope of the BJT I-V curve? What is the formula for r_e?
3. Write the gate I-V curve equation for the JFET or D-MOSFET.
4. What is the slope of the gate I-V curve?
5. Write the two equations that can be used to find g_m.
6. Write the gate I-V curve equation for the E-MOSFET.
7. Write the two equations that can be used to find g_m.
8. Where do you find g_m on a data sheet?
9. What does r_d represent? How do you find its value from a data sheet?
10. What does $r_{ds(on)}$ represent? How does it compare to R_{DS}?

22–3 ■ THE COMMON-SOURCE FET AMPLIFIER

Figure 22–1 (p. 718) shows that except for different labels, the AC model for the FET is the same as the AC model for the BJT. This means that FET circuits must have the same AC response as equivalent BJT circuits.

Common Source Box Model Parameters

Figure 22–6A shows the basic FET amplifier, the common-source configuration. Figure 22–6B shows the equivalent BJT common-emitter amplifier. You already know how to find the Box Model parameters for the BJT circuit. We find the Box Model parameters for the FET circuit by direct substitution into the BJT equations. The results are given in this table:

BJT	FET
r_e	$\dfrac{1}{g_m}$
$R_B = R_1 \| R_2$	$R_G = R_1 \| R_2$
R_C	R_D
$R_{Etotal} = r_e + R_E$	$R_{Stotal} = \dfrac{1}{g_m} + R_S$
$R_{base} = \beta R_{Etotal}$	$R_{gate} \rightarrow \infty \ (i_g = 0A)$

Box Model Parameters

BJT	FET
$A = -\dfrac{R_C}{R_{Etotal}}$	$A = -\dfrac{R_D}{R_{Stotal}}$
$R_o = R_C$	$R_o = R_D$
$R_i = R_B \| R_{base}$	$R_i = R_G \ (i_g = 0A)$

FIGURE 22–6

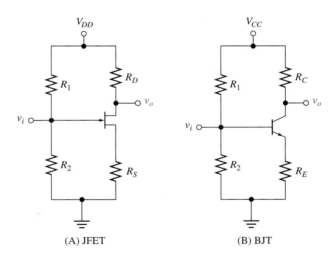

(A) JFET (B) BJT

While we can find the FET Box Model parameters by simple substitution with the BJT parameters, there are numerical differences.

☐ The input resistance of the FET amplifier is R_G, which is greater than the BJT input resistance of $R_B \| R_{base}$. The higher input resistance of an FET is one of its major advantages over the BJT.

☐ As shown in the previous section, the AC source resistance of the FET is significantly greater than the AC emitter resistance of the BJT. This results in a larger R_{Stotal} than R_{Etotal}, which in turn leads to a smaller gain for the FET.

Typically, then, FET amplifiers have a higher input resistance and a lower gain than equivalent BJT amplifiers.

EXAMPLE 22–3
The Common-Source Amplifier

For the common-source JFET circuit shown in Figure 22–7, $V_{GS(off)} = -5V$, $I_{DSS} = 10mA$.

(a) Find I_D and V_{GS}.

(b) Find g_m.

(c) Find the Box Model parameters.

(d) Find the system gain.

Note that because the subscript "S" is used for the FET, we do not use v_s and R_s to represent the input source and its resistance. Rather, we now use the notation v_{signal} and R_v.

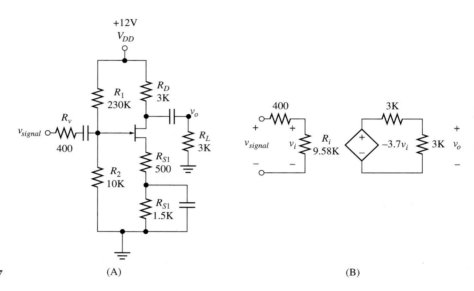

FIGURE 22–7

(A)   (B)

Solution   You might first notice that the source resistor is partially bypassed. This is fairly common in discrete FET circuits. For DC calculations we use

$$R_{S_{DC}} = R_{S1} + R_{S2} = 2K\Omega$$

Because R_{S2} is bypassed, for AC calculations we use

$$R_{S_{AC}} = R_{S1} = 0.5K\Omega$$

and

$$R_{Stotal} = \frac{1}{g_m} + R_{S_{AC}}$$

(a) DC Calculations

When we open the capacitors, the DC source resistance is

$$R_{S_{DC}} = 0.5K + 1.5K = 2K\Omega$$

With the capacitors opened we get the same bias circuit shown in Example 20–10 of Chapter 20 (Figure 20–19, p. 679). The Q-point values we found there were

$$I_D = 1.72mA \qquad \text{and} \qquad V_{GS} = -2.93V$$

(b) g_m Calculations

$$g_m = \frac{2}{|V_{GS(off)}|} \sqrt{I_D I_{DSS}} = \frac{2}{5} \sqrt{1.72mA \times 10mA} = 1.66mS$$

(c) Box Model Parameters

We first find R_G and R_{Stotal}:

$$R_G = R_1 \| R_2 = 230\text{K} \| 10\text{K} = 9.58\text{K}\Omega$$

and

$$R_{Stotal} = \frac{1}{g_m} + R_{S_{AC}} = \frac{1}{1.66 \times 10^{-3}} + 0.5\text{K} = 1.1\text{K}\Omega$$

The Box Model parameters are found from

$$A = -\frac{R_D}{R_{Stotal}} = -\frac{3\text{K}}{1.1\text{K}} = -2.73$$

$$R_o = R_D = 3\text{K}\Omega$$

$$R_i = R_G = 9.58\text{K}\Omega$$

(d) System Gain

The Box Model, along with the source and load, is shown in Figure 22–7B. To find the system gain, we first find v_i:

$$v_i = \frac{R_i}{R_i + R_v} v_{signal} = \frac{9.58\text{K}}{9.58\text{K} + 0.4\text{K}} v_{signal} = 0.96 v_{signal}$$

We now find v_o:

$$v_o = \frac{R_L}{R_L + R_o} A v_i = \frac{3\text{K}}{3\text{K} + 3\text{K}} \times -2.73 \times 0.96 v_{signal} = -1.31 v_{signal}$$

so

$$A_{sys} = \frac{v_o}{v_{signal}} = -1.31$$

DRILL EXERCISE

Redraw the circuit of Figure 22–7. Assume the Q-point drain current is now $I_D = 2.1\text{mA}$. Find g_m, the Box Model parameters, and the system gain.

Answer $g_m = 2.47\text{mS}, A = -3.32, R_o = 3\text{K}\Omega, R_i = 9.6\text{K}\Omega,$
$A_{sys} = -0.864$

REVIEW QUESTIONS

1. Draw the AC model for the FET.
2. Draw the common-source circuit with R_S not bypassed. What is R_{Stotal} for this circuit?
3. What are the Box Model parameters?
4. If R_S is completely bypassed, what is R_{Stotal}?
5. Draw the common-source circuit with R_S partially bypassed. What is R_{Stotal} now?

22–4 ■ ADDITIONAL FET CIRCUITS

We do not want to belabor the point by showing you an example of how the Box Model parameters for each FET circuit is derived from the equivalent BJT circuit. Rather, Figure 22–8 shows the remaining basic transistor amplifiers and the model parameters for each. The results are tabulated below. Note that

$$R_{Etotal} = r_e + R_E \qquad \text{(not bypassed)}$$

$$R_{Stotal} = \frac{1}{g_m} + R_S \qquad \text{(not bypassed)}$$

FIGURE 22–8
FET-BJT amplifier comparison

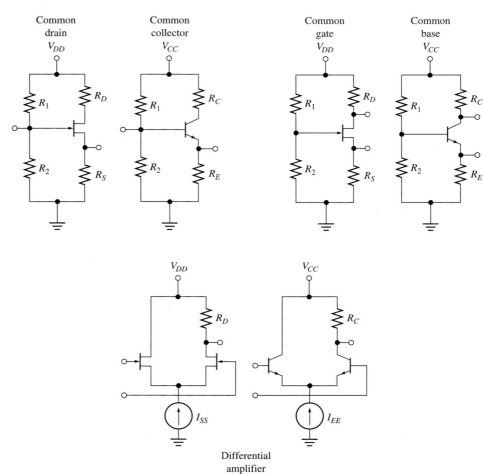

Differential amplifier

BJT	FET
Common Collector (Emitter Follower)*	**Common Drain (FET Follower)***
$A = \dfrac{R_E}{R_{Etotal}}$	$A = \dfrac{R_S}{R_{Stotal}}$
$R_i \approx R_B \| \beta R_{Etotal}$	$R_i = R_G$
$R_o \approx r_e \| R_E$	$R_o = \dfrac{1}{g_m} \| R_S$

Common Base	Common Gate
$A = \dfrac{R_C}{r_e}$	$A = \dfrac{R_D}{1/g_m} = g_m R_D$
$R_i = r_e \| R_E$	$R_i = \dfrac{1}{g_m} \| R_S$
$R_o = R_C$	$R_o = R_D$
Differential Amplifier	**Differential Amplifier**
$A = \dfrac{R_C}{2r_e}$	$A = \dfrac{R_D}{2/g_m} = \dfrac{g_m R_D}{2}$
$R_i = 2\beta r_e$	$R_i \to \infty$ (open circuit)
$R_o = R_C$	$R_o = R_D$

NOTE: Input and output resistance of the emitter follower are approximate because these parameters are affected by the load and signal source resistances. The values are exact for the FET follower, due to zero gate current.

EXAMPLE 22–4
The FET Follower

For the circuit shown in Figure 22–9, find

(a) The Q-point voltage and current. **(b)** g_m.
(c) The Box Model parameters. **(d)** The system gain.

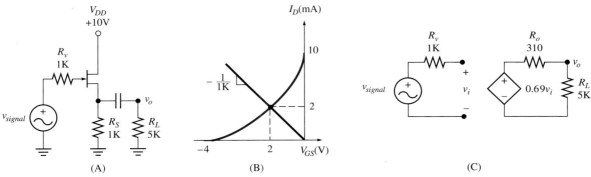

FIGURE 22–9

(A)   (B)   (C)

Solution

(a) DC Bias Calculations. Note that we have applied the input signal directly to the gate of the JFET. For DC calculations we set the input to 0V. Because there is no gate current, the DC gate voltage is

$$V_G = 0\text{V}$$

Therefore, the gate load-line is

$$I_D = \frac{V_S}{R_S} = \frac{-V_{GS}}{1\text{K}}$$

Plotting this load-line curve on the gate I-V curve (Figure 22–9B) gives us

$$V_{GS} = -2\text{V}$$

and

$$I_D = \frac{-V_{GS}}{R_S} = \frac{2}{1K} = 2mA$$

(b) We find g_m from

$$g_m = \frac{2}{|V_{GS(off)}|} \sqrt{I_D I_{DSS}} = \frac{2}{4} \sqrt{2mA \times 10mA} = 2.24mS$$

(c) Box Model Parameters

$$R_{Stotal} = \frac{1}{g_m} + R_S = \frac{1}{2.24 \times 10^{-3}} + 1K = 0.446K + 1K = 1.45K\Omega$$

$$A = \frac{R_S}{R_{Stotal}} = \frac{1K}{1.45K} = 0.69$$

Note: Because $\dfrac{1}{g_m}$ is, in general, larger than r_e, the FET follower gain tends to be lower than the emitter follower gain. Because there is no gate current and no R_G in this circuit

$$R_i \to \infty$$

That is, the FET presents an open circuit to the signal source. The output resistance is

$$R_o = \frac{1}{g_m} \| R_S = 0.45K \| 1K = 0.31K\Omega$$

(d) The Box Model, along with the signal source and the load, are shown in Figure 22–9C (p. 729). Note the open circuit on the input side, which results in

$$v_i = v_{signal}$$

This is the major advantage of the FET follower. It has such a high input resistance that the resistance of the signal source has a negligible effect. On the output side

$$v_o = \frac{R_L}{R_L + R_o} A v_i = \frac{5K}{5K + 0.31K} \times 0.69 v_{signal} = 0.65 v_{signal}$$

so

$$A_{sys} = \frac{v_o}{v_{signal}} = 0.65$$

DRILL EXERCISE

Redraw the circuit of Figure 22–9 (p. 729), changing the supply voltage to $V_{DD} = 24V$. Find the DC bias levels (g_m), the Box Model parameters, and the system gain.

Answer $V_{GS} = -2V$, $I_D = 2mA$, $g_m = 2.24mS$
$A = 0.69$, $R_i \to \infty$, $R_o = 0.31K\Omega$, $A_{sys} = 0.65$

REVIEW QUESTIONS

1. Draw the FET follower circuit. What are its Box Model parameters?
2. What advantage does the FET follower have over the emitter follower? What disadvantage?
3. Draw the common-gate circuit. What are its Box Model parameters?
4. Draw the FET differential amplifier circuit. What are its Box Model parameters?

22–5 ■ FET-BJT CIRCUITS

We have seen that the FET has a very high input resistance but a low gain. For this reason, FETs and BJTs are often combined, the FET providing the necessary input resistance, and the BJTs supplying the gain. A simple combined circuit is shown in Figure 22–10.

To analyze this circuit we proceed as follows:

☐ Determine the DC bias levels for each transistor.

☐ Find the Box Model for each transistor stage.

☐ Connect the Box Models together to find the overall system gain.

FIGURE 22–10

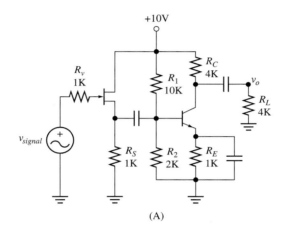

(A)

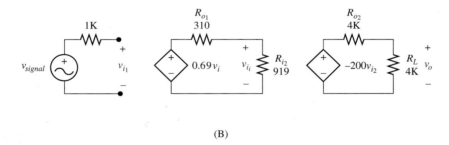

(B)

The input stage in this circuit is the same FET follower than we analyzed in Example 22–4 (p. 729), where we found

$$V_{GS} = -2V$$

$$I_D = 2mA$$

$$g_m = 2.24mS$$

The Box Model parameters are (we use the subscript "1" to indicate that we are working with the first stage)

$$A_1 = 0.69$$

$$R_{i_1} \to \infty$$

$$R_{o_1} = 0.31K\Omega$$

The BJT (stage 2) is analyzed as follows:

$$V_B = \frac{R_2}{R_1 + R_2} V_{CC} = \frac{2K}{10K + 2K} 12 \doteq 2V$$

$$V_E = V_B - 0.7 = 1.3\text{V}$$

$$I_E = \frac{V_E}{R_E} = \frac{1.3}{1\text{K}} = 1.3\text{mA}$$

So the AC emitter resistance is

$$r_e = \frac{26\text{mV}}{I_E} = \frac{26\text{mV}}{1.3\text{mA}} = 20\Omega$$

The Box Model parameters for this BJT common-emitter amplifier are

$$A_2 = -\frac{R_C}{R_{Etotal}} = -\frac{R_C}{r_e} = -\frac{4\text{K}}{0.02\text{K}} = -200$$

$$R_{i_2} = R_B \| \beta R_{Etotal} = 1.7\text{K} \| 100 \times 0.02\text{K} = 0.919\text{K}\Omega$$

$$R_{o_2} = R_C = 4\text{K}\Omega$$

The two Box Models are shown in Figure 22–10B (p. 731). Note that R_{i_2} is the load for the first stage; and the output voltage of the first stage is the input voltage to the second stage. We find the system gain by working from the left to the right:

$$v_{i_1} = v_{signal}$$

$$v_{i_2} = v_{o_1} = \frac{R_{i_2}}{R_{i_2} + R_{o_1}} A_1 v_{i_1} = \frac{0.919\text{K}}{0.919\text{K} + 0.31\text{K}} \times 0.69 v_{signal} = 0.516 v_{signal}$$

Finally

$$v_o = \frac{R_L}{R_L + R_{o_2}} A_2 v_{i_2} = \frac{4\text{K}}{4\text{K} + 4\text{K}} \times -200 \times 0.516 v_{signal} = -51.6 v_{signal}$$

and the overall system gain is

$$A = \frac{v_o}{v_{signal}} = -51.6$$

DRILL EXERCISE

Redraw the circuit of Figure 22–10 (p. 731), changing the supply voltage to 24V. Noting that this change does not affect the FET parameters, find the new DC bias levels for the BJT, r_e, the BJT Box Model Parameters, and the overall system gain.

Answer $I_C = 3.3\text{mA}, r_e = 8\Omega, A_2 = -500, R_{i_2} = 0.5\text{K}\Omega, A_{sys} = -109$

REVIEW QUESTIONS

1. What advantage does the FET have over the BJT?
2. What advantage does the BJT have over the FET?
3. When an amplifier circuit contains both FETs and BJTs, how can we use Box Models to find the overall system gain?

22–6 ■ THE VOLTAGE VARIABLE RESISTOR

Figure 22–11 shows a family of drain I-V curves in the **OHMIC** region of an FET. In this region the FET acts as a resistor rather than an amplifier. The resistance in this region is the inverse of the slope of the I-V curve.

FIGURE 22–11
FET as voltage variable resistor

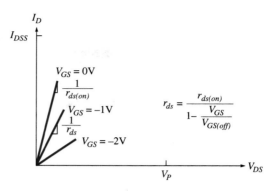

Resistance of FET

As you can see, the resistance of the FET depends on the applied gate-source voltage. Therefore, when V_{DS} is small, the FET acts a resistor whose value is set by V_{GS}. In the **OHMIC** region, the FET is used as a **voltage variable resistor.** The parameter $r_{ds(on)}$ is the FET resistance (depletion devices only) when V_{DS} is small and $V_{GS} = 0V$. The general **OHMIC** region resistance has the following dependence on V_{GS}:

$$r_{ds} = \frac{r_{ds(on)}}{1 - \dfrac{V_{GS}}{V_{GS(off)}}}$$

EXAMPLE 22–5
The FET as a Voltage Variable Resistor

Figure 22–12A shows a voltage divider circuit in which one of the resistors is replaced with an FET where $r_{ds(on)} = 300\Omega$ and $V_{GS(off)} = -4V$. Assuming the JFET is in its **OHMIC** region, find V_o if

(a) $V_G = 0V$ **(b)** $V_G = -1V$ **(c)** $V_G = -2V$

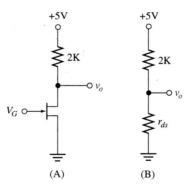

FIGURE 22–12   (A)   (B)

Solution   Because the JFET is operating in its **OHMIC** region, it acts as a resistor (Figure 22–12B). The output voltage, which is the drain-source voltage of the JFET, is given by

$$V_o = \frac{r_{ds}}{r_{ds} + 2K} \, 5$$

Because the source is grounded, the gate-source voltage is

$$V_{GS} = V_G$$

(a) For $V_G = 0\text{V}$

$$r_{ds} = r_{ds(on)} = 300\Omega$$

so

$$V_o = \frac{r_{ds}}{r_{ds} + 2\text{K}} 5 = \frac{0.3\text{K}}{0.3\text{K} + 2\text{K}} 5 = 0.652\text{V}$$

(b) For $V_G = -1\text{V}$

$$r_{ds} = \frac{r_{ds(on)}}{1 - \dfrac{V_{GS}}{V_{GS(off)}}} = \frac{300}{1 - \dfrac{-1}{-4}} = 400\Omega$$

so

$$V_o = \frac{r_{ds}}{r_{ds} + 2\text{K}} 5 = \frac{0.4\text{K}}{0.4\text{K} + 2\text{K}} 5 = 0.833\text{V}$$

(c) For $V_G = -2\text{V}$

$$r_{ds} = \frac{r_{ds(on)}}{1 - \dfrac{V_{GS}}{V_{GS(off)}}} = \frac{300}{1 - \dfrac{-2}{-4}} = 600\Omega$$

so

$$V_o = \frac{r_{ds}}{r_{ds} + 2\text{K}} 5 = \frac{0.6\text{K}}{0.6\text{K} + 2\text{K}} 5 = 1.15\text{V}$$

Automatic Gain Control (AGC)

You are at home late at night watching an old romantic movie. The hero is whispering sweet nothings into the ear of his beloved. You strain to hear the soft murmurings—**SUDDENLY CRAZY PHIL IS ON THE AIR** selling used cars at a decibel level that would do a rock band proud. When the ringing in your ears subsides, you decide that your next TV will have automatic gain control. That is, the TV circuitry will decrease the gain when the incoming commercial audio signal is louder than the program audio.

Figure 22–13A shows an amplifier circuit (we will discuss this circuit in detail in the next chapter) in which the amplifier gain is given by

$$A \approx \frac{R_2}{R_1}$$

Note that when the input voltage goes positive, the output voltage goes positive. In Figure 22–13B we have added an FET to the circuit. Because the gate voltage is positive, we use a P-channel JFET in this circuit.

If the FET operates in the **OHMIC** region, the gain is now given by

$$A \approx \frac{R_2}{R_1 + r_{ds}}$$

The amplifier output sets the gate-source voltage of the FET. As the amplifier output increases, r_{ds} increases, and the gain goes down. This circuit has automatic gain control.

FIGURE 22–13
Automatic gain control

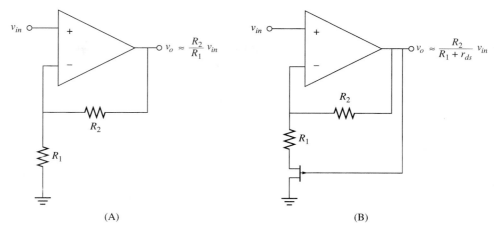

(A)   (B)

22–7 ■ CMOS Logic Circuits

Inverter

Figure 22–14A shows the CMOS inverter that we introduced in the last chapter. When the input is LOW ($V_i = 0V$)

$$V_{GS_1} = 5V \text{ and } V_{GS_2} = 0V$$

FIGURE 22–14
CMOS inverter

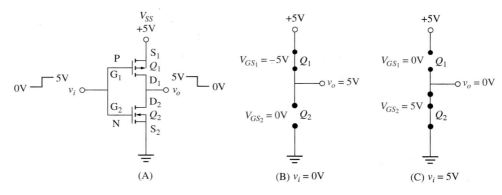

(A)   (B) $v_i = 0V$ (C) $v_i = 5V$

Therefore, the P-channel Q_1 is **ON** and the N-channel Q_2 is **OFF.** As Figure 22–14B shows, the output now is $+5V$ (HIGH).

     When the input is HIGH ($V_i = +5V$)

$$V_{GS_1} = 0V \text{ and } V_{GS_2} = 5V$$

Therefore, the P-channel Q_1 is **OFF** and the N-channel Q_2 is **ON.** As Figure 22–14C shows, the output now is 0V (LOW).

Because a HIGH input produces a LOW output, and a LOW input produces a HIGH output, this circuit is an inverter.

NAND Gate

The CMOS circuit in Figure 22–15 has two inputs, V_A and V_B. There are, therefore, four possible input combinations

V_A LOW (0V)	V_B LOW (0V)
V_A LOW (0V)	V_B HIGH (+5V)
V_A HIGH (+5V)	V_B LOW (0V)
V_A HIGH (+5V)	V_B HIGH (+5V)

FIGURE 22–15
CMOS NAND gate

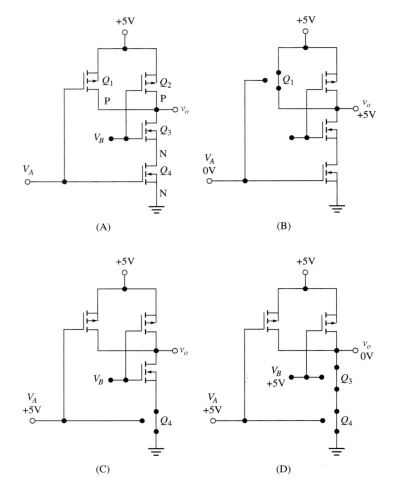

In the following analysis the two P-channel E-MOSFET gate-source voltages are given by

$$V_{GS_1} = V_A - 5 \text{ and } V_{GS_2} = V_B - 5$$

Remember that a P-channel E-MOSFET turns **ON** when its gate-source voltage is negative. The two N-channel E-MOSFET gate-source voltages are given by

$$V_{GS_4} = V_A \text{ and } V_{GS_2} = V_B - V_{DS_4}$$

An N-channel E-MOSFET turns **ON** when its gate-source voltage is positive.

$$\text{When } V_A = 0\text{V (LOW)}$$
$$V_{GS_1} = -5\text{V}$$

Because the gate-source voltage is negative, the P-channel Q_1 turns **ON** and can be approximated with a drain-source short (Figure 22–15B). In this case the output is connected to the supply and we get

$$V_A = 0\text{V (LOW)} \rightarrow V_o = +5\text{V (HIGH)}.$$

When $V_B = 0\text{V}$, the output is connected to the supply through Q_2. Therefore, if either input is LOW, the output is HIGH.

What happens when both inputs are HIGH? In this case, both Q_1 and Q_2 are **OFF.** When V_A goes HIGH, Q_4 turns **ON,** connecting the source of Q_3 to ground (Figure 22–15C). Because the gate-source voltage of Q_3 is now $+5\text{V}$, Q_3 also turns **ON.** The output is effectively connected to ground (Figure 22–15D), and we get

$$V_A = +5\text{V (HIGH) and } V_B = +5\text{V (HIGH)} \rightarrow V_o = 0\text{V (LOW)}$$

This CMOS circuit produces the following:

V_A	V_B	V_o
LOW	LOW	HIGH
LOW	HIGH	HIGH
HIGH	LOW	HIGH
HIGH	HIGH	LOW

Because the output is LOW only when both inputs are HIGH, the circuit in Figure 22–14 is a CMOS NAND gate.

NOR Gate

Figure 22–16A (p. 738) shows another arrangement of CMOS devices. In this circuit, if V_A is HIGH, Q_3 turns **ON** and the output is connected to ground (Figure 22–16B, p. 738). If V_B is HIGH, Q_4 turns **ON,** again connecting the output to ground. Therefore, if either V_A or V_B is HIGH, the output goes LOW.

What happens if we set both V_A and V_B LOW, that is $V_A = V_B = 0\text{V}$? In this case, both of the N-channel devices are **OFF.** Because $V_{GS_1} = -5\text{V}$, Q_1 will turn **ON** (Figure 22–16C, p. 738). When Q_1 turns **ON,** the source of Q_2 is connected to the supply voltage. This results in $V_{GS_2} = -5\text{V}$, so Q_2 also turns **ON.** The output is connected to the supply (Figure 22–16D, p. 738) and equals $+5\text{V}$ (HIGH).

This CMOS circuit produces the following:

V_A	V_B	V_o
LOW	LOW	HIGH
LOW	HIGH	LOW
HIGH	LOW	LOW
HIGH	HIGH	LOW

Because the output is HIGH only when both inputs are LOW, the circuit in Figure 22–16 is a CMOS NOR gate.

FIGURE 22–16
CMOS NOR gate

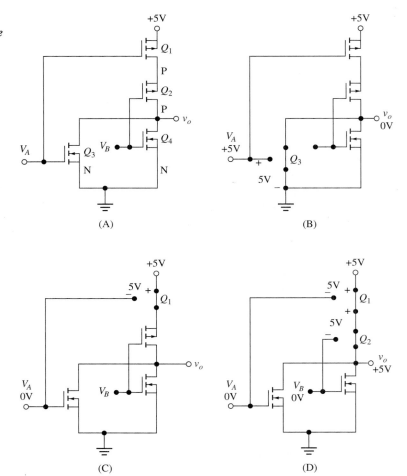

(A)

(B)

(C)

(D)

**REVIEW
QUESTIONS**

1. Draw a CMOS inverter and explain its operation.
2. Which FET in a CMOS NAND gate turns **ON** when V_A is LOW? When V_B is LOW?
3. Why do both inputs in a CMOS NAND gate need to be HIGH to produce a LOW output?
4. Which FET in a CMOS NOR gate turns **ON** when V_A is HIGH? When V_B is HIGH?
5. Why do both inputs in a CMOS NOR gate need to be LOW to produce a HIGH output?

SUMMARY

■ The AC model for the FET is the same as the AC model for the BJT; the AC source resistance $(1/g_m)$ is analogous to the AC emitter resistance (r_e).

■ The AC transconductance (g_m) is found from

$$\text{JFET or D-MOSFET: } g_m = \frac{2}{|V_{GS(off)}|}\sqrt{I_D I_{DSS}} = 2\left|\frac{I_{DSS}}{V_{GS(off)}}\right|\left(1 - \frac{V_{GS}}{V_{GS(off)}}\right)$$

$$\text{E-MOSFET: } \quad g_m = 2\sqrt{k\,I_{DSS}} = 2k\,(V_{GS} - V_{GS(off)})$$

■ The FET common-source amplifier is analogous to the BJT common-emitter amplifier. Its Box Model parameters are given by

$$A = -\frac{R_D}{R_{Stotal}}$$

$$R_i = R_G$$

$$R_o = R_D$$

■ The FET common-drain amplifier (FET follower) is analogous to the BJT common-collector amplifier (emitter follower). Its Box Model parameters are given by

$$A = \frac{R_S}{R_{Stotal}}$$

$$R_i = R_G$$

$$R_o = \frac{1}{g_m} \| R_S$$

■ The FET common-source and common-drain amplifiers have a higher input resistance and a lower gain than their BJT equivalents.

■ The FET common-gate amplifier is analogous to the BJT common-base amplifier. Its Box Model parameters are given by

$$A = g_m R_D$$

$$R_i = \frac{1}{g_m} \| R_S$$

$$R_o = R_D$$

■ The FET differential amplifier is analogous to the BJT differential amplifier. Its Box Model parameters are

$$A = \frac{g_m R_D}{2}$$

$$R_i \to \infty$$

$$R_o = R_D$$

■ When operated in the **OHMIC** region, the FET acts as a voltage variable resistor. This property can be used to build automatic gain control (AGC) circuits.

■ CMOS devices can be used to build logical inverters, NAND gates, and NOR gates.

SPECIAL TOPIC **FET MODELING**

BJT Models

Figure 22–17A shows the AC model we have been using for the BJT. We can change the current-controlled voltage source into a voltage-controlled current source by noting that

$$v_{be} = r_e i_e = r_e(i_b + \beta i_b) = (\beta + 1)r_e i_b$$

We solve for i_b:

$$i_b = \frac{v_{be}}{(\beta + 1)r_e}$$

FIGURE 22–17

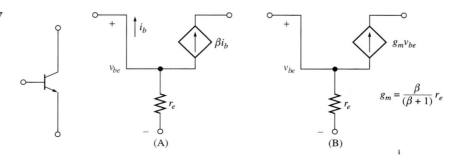

(A)   (B)

Therefore, we can substitute for the i_b term at the controlled source:

$$\beta i_b = \frac{\beta}{(\beta + 1)r_e}\, v_{be}$$

To simplify, define

$$g_m \equiv \frac{\beta}{(\beta + 1)r_e}$$

The controlled current source is now controlled by a voltage:

$$i_c = \beta i_b = g_m v_{be}$$

as shown in Figure 22–17B (p. 739).

FET Models

The standard AC model for the FET is shown in Figure 22–18A. Note the open circuit between the gate and the source. Also note the voltage-controlled voltage source that represents the dependence of the drain current on the gate-source voltage, that is

$$i_d = g_m v_{gs}$$

FIGURE 22–18

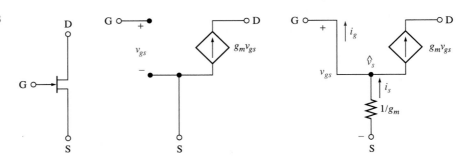

To make the FET AC model look like the BJT AC model, we can develop the FET AC model shown in Figure 22–18B. We know that the gate current in a FET is 0A. Although it may not appear to be true, the gate current in our new model is, indeed, 0A. To show this, we solve for i_g by summing the currents at the $\hat{v}_s$ node:

$$i_s = i_g + g_m v_{gs}$$

Noting that $\hat{v}_s = v_g$, we get from Ohm's Law

$$i_s = \frac{v_{gs}}{1/g_m} = g_m v_{gs}$$

We substitute into the current equation:

$$g_m v_{gs} = i_g + g_m v_{gs}$$

which leads to

$$i_g = 0\text{A}$$

verifying that the FET AC model shown in Figure 22–18B is valid.

Now, comparing the BJT AC model (Figure 22–17B) to the FET AC model (Figure 22–18B), we see that the only difference is the labeling of the voltages, currents, and resistor. This justifies determining the Box Model parameters of FET amplifiers by direct substitution into the equivalent BJT amplifier Box Model parameters.

PROBLEMS

SECTION 22–2 FET AC Model

1. An N-Channel JFET has the parameters $V_{GS(off)} = -3V$, $I_{DSS} = 5mA$. Find g_m if
   (a) $I_D = 2mA$ (b) $V_{GS} = -1V$

2. An N-Channel E-MOSFET has the parameters $V_{GS(th)} = 1V$, $k = 0.002A/V^2$. Find g_m if
   (a) $I_D = 2mA$ (b) $V_{GS} = 3V$

3. The AC transconductance (g_m) is the slope of the gate I-V curve at the Q-point. The I-V curve of a P-Channel D-MOSFET is given in Figure 22–19. Use a scale to find g_m for
   (a) $V_{GS} = 0V$ (b) $V_{GS} = 3V$

FIGURE 22–19

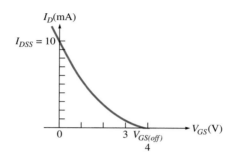

4. Use the maximum values given in the data sheet for the 2N3797 (Figure 22–3) to sketch the gate I-V curve for this device.

5. Use the data sheet for the 2N3797 (Figure 22–3, p. 722) to answer the following:
   (a) What is the typical value for g_m (y_{fs})?
   (b) Find g_m for the typical values given for I_{DSS} and $V_{GS(off)}$. Note that $V_{GS} = 0V$ for these values.
   (c) Compare the results of a and b.

SECTION 22–3 The Common-Source FET Amplifier

FIGURE 22–20

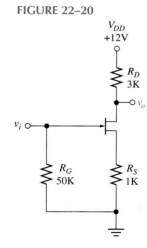

6. Given the common-source amplifier in Figure 22–20, assume

$$V_{GS(off)} = -4V \text{ and } I_{DSS} = 10mA$$

   (a) Find the DC bias levels.   (b) Find g_m.
   (c) Draw the AC model for this circuit. Be sure to label $\hat{v}_s$ (don't forget to turn off all DC voltage sources).
   (d) Find in the following order
   $$v_g$$
   $$\hat{v}_s$$
   $$i_d$$
   $$v_d$$
   $$v_o$$
   (e) Find the voltage gain of the common-source amplifier.

7. For the circuit of Figure 22–21 (p. 742), assume

$$v_{GS(off)} = -4V \text{ and } I_{DSS} = 10mA$$

   (a) Find the DC bias levels.   (b) Find g_m.
   (c) Find the Box Model parameters.   (d) Find the overall system gain.

8. For the circuit of Figure 22–22 (p. 742), assume $V_{GS(off)} = -4V$ and $I_{DSS} = 10mA$.
   (a) Find the DC bias levels.   (b) Find g_m.
   (c) Find the Box Model parameters.   (d) Find the overall system gain.

FIGURE 22–21   **FIGURE 22–22**

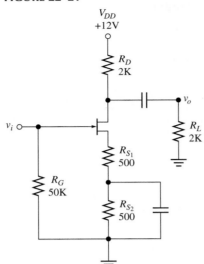

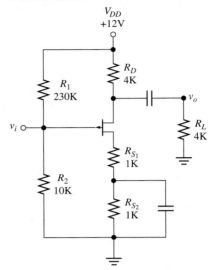

9. For the circuit of Figure 22–23, assume $V_{gs(th)} = 2V$ and $k = 0.0001 A/V^2$.
   (a) Find the DC bias levels.   (b) Find g_m.
   (c) Find the Box Model parameters.   (d) Find the overall system gain.

FIGURE 22–23

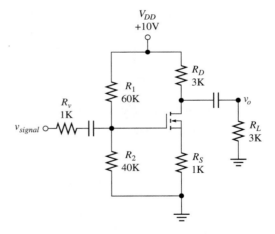

SECTION 22–4   Additional FET Circuits

10. For the circuit of Figure 22–24, assume $V_{GS(off)} = -4V$ and $I_{DSS} = 10mA$.
    (a) What type of amplifier is this?   (b) Find the DC bias levels.
    (c) Find g_m.   (d) Find the Box Model parameters.
    (e) Find the overall system gain.

11. For the circuit of Figure 22–25, assume $V_{gs(th)} = 2V$ and $k = 0.0001 A/V^2$.
    (a) Find the DC bias levels.   (b) Find g_m.
    (c) Find the Box Model parameters.   (d) Find the overall system gain.

12. For the circuit of Figure 22–26, assume $V_{GS(off)} = -4V$ and $I_{DSS} = 10$ mA.
    (a) Find the DC bias levels of each transistor. (Hint: First find I_D for each transistor; then use the
        I-V curve to find V_{GS}).
    (b) Find g_m.   (c) Find the Box Model parameters.
    (d) Find the overall system gain.

FIGURE 22–24

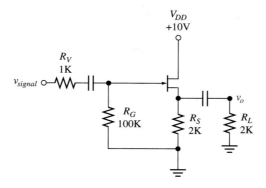

FIGURE 22–25

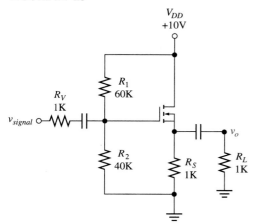

FIGURE 22–26

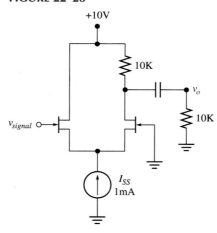

13. Match the FET amplifier with its equivalent BJT amplifier:

FET	BJT
(a) Common Gate	**1.** Common Emitter
(b) Common Source	**2.** Common Collector
(c) Common Drain	**3.** Common Base

SECTION 22–5   The Voltage Variable Resistor

FIGURE 22–27

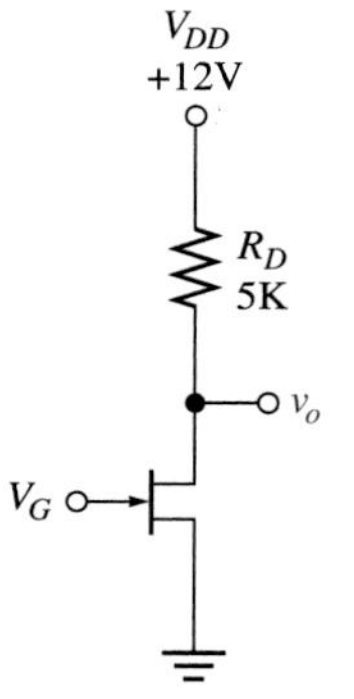

14. In what state must an FET operate when it is used as a voltage variable resistor?

15. An FET has the following parameters:

$$r_{ds(on)} = 200\Omega, \; I_{DSS} = 20\text{mA}, \; V_{GS(off)} = -5\text{V}$$

Find r_{ds} if

   (a) $V_{GS} = 0\text{V}$.   **(b)** $V_{GS} = -1\text{V}$.   **(c)** $V_{GS} = -2\text{V}$.

16. In Figure 22–27, $V_{GS} = -5\text{V}$ and $I_{DSS} = 20$ mA. Find V_o if

   (a) $V_{GS} = 0\text{V}$.   **(b)** $V_{GS} = -1\text{V}$.   **(c)** $V_{GS} = -2\text{V}$.

17. We want V_o in Figure 22–27 to equal 1V.

   (a) Find the required r_{ds}.   **(b)** Find the required V_G.

18. In Figure 22–27, $V_{GS} = -5\text{V}$ and $I_{DSS} = 20\text{mA}$.

   (a) Find r_{ds} and V_o if $V_{GS} = -4\text{V}$.   **(b)** The result you just found is wrong. Why?

SECTION 22–6   CMOS Logic Circuits

19. Figure 22–28 shows a CMOS NAND gate. For the following combinations of input voltages, determine which FETs are **ON** and which FETs are **OFF.**

    (a) $V_A = 0V, V_B = 0V$ **(b)** $V_A = +5V, V_B = 0V$

    (c) $V_A = 0V, V_B = +5V$ **(d)** $V_A = +5V, V_B = +5V$

20. Figure 22–29 shows a CMOS NOR gate. For the following combinations of input voltages, determine which FETs are **ON** and which FETs are **OFF.**

    (a) $V_A = 0V, V_B = 0V$ **(b)** $V_A = +5V, V_B = 0V$

    (c) $V_A = 0V, V_B = +5V$ **(d)** $V_A = +5V, V_B = +5V$

FIGURE 22–28   **FIGURE 22–29**

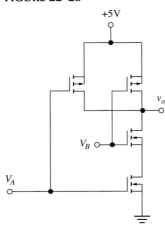
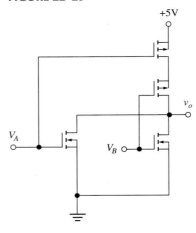

Computer Problems

For the problems below, most computer models for the FET require the transconductance parameter (β) *rather than* I_{DSS}. *The relation between these two parameters is*

$$\text{JFET and D-MOSFET: } \beta = \frac{I_{DSS}}{V^2_{GS(off)}}$$

$$\text{E-MOSFET: } \beta = k$$

21. For the computer program you are using

    (a) List the FET devices available in your program's library.

    (b) For each JFET and D-MOSFET in your library, find I_{DSS} from the value given for the transconductance parameter (β).

22. Simulate the circuit of Figure 22–21 (p. 742) (set the capacitors to 30μF).

    (a) Find the DC bias levels and the system gain.

    (b) Determine the maximum symmetrical output swing. You can do this by applying a 1KHz input signal at varying amplitudes.

23. Repeat Problem 22 for the circuit of Figure 22–22 (p. 742).

24. Repeat Problem 22 for the circuit of Figure 22–23 (p. 742).

25. Repeat Problem 22 for the circuit of Figure 22–24.

26. Repeat Problem 22 for the circuit of Figure 22–25.

27. Repeat Problem 22 for the circuit of Figure 22–26.

28. Using an operational amplifier library model, simulate the automatic gain control circuit of Figure 22–13B (p. 735). Let $R_2 = 20K\Omega, R_1 = 1.7K\Omega$. Find V_o and $A = V_o/V_i$ for $V_i = 0.1, 0.25, 0.5, 0.75, 1, 1.5V$.

OPERATIONAL AMPLIFIERS

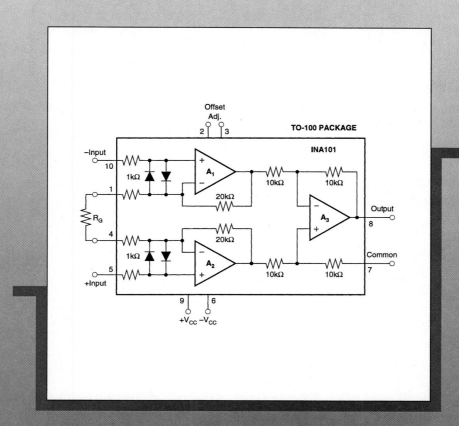

Chapter

23

The Operational Amplifier (Op-Amp)

C

C HAPTER OBJECTIVES

To determine how negative feedback acts to minimize an error voltage.

To determine the properties of the Op-Amp.

To determine the output of the non-inverting Op-Amp by talking around the loop.

To determine the output of the inverting Op-Amp by talking around the loop.

To model the Op-Amp as an ideal device.

To use the ideal Op-Amp model and node analysis to find the closed-loop gain of the non-inverting and inverting amplifiers.

To determine how real Op-Amp deviations from the ideal affect the output.

To troubleshoot Op-Amp circuits.

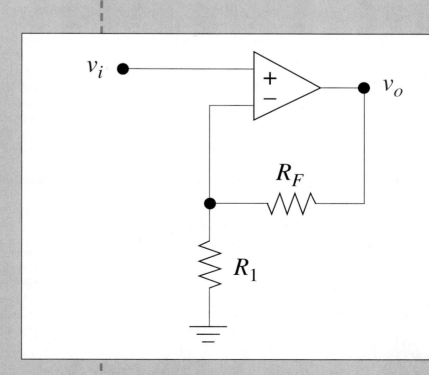

23–1 ■ INTRODUCTION

In years past we built all electronic devices with the types of discrete transistor circuits discussed in the previous chapters. The stereo amplifier circuit shown in Figure 23–1 is an example of such a device. (Don't worry—you are not being asked to analyze this circuit, although you may recognize many of the sub-circuits.)

FIGURE 23–1
Discrete amplifier

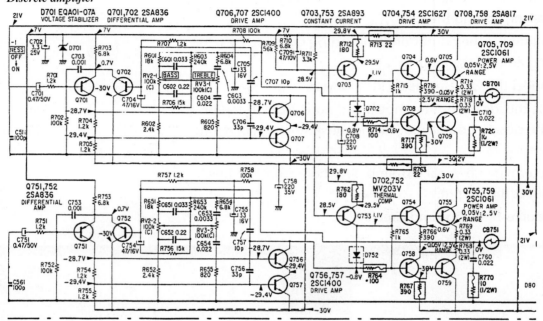

These days, in addition to using discrete transistors, we most often build devices with special-purpose integrated circuits, or chips. Integrated circuits use technologies that combine anywhere from several to many hundreds of transistors on a single chip. Figure 23–2 shows an integrated circuit used to construct an FM amplifier.

If you compare the integrated circuit of Figure 23–2 to the discrete circuit of Figure 23–1, you can see the simplicity of integrated circuits. In particular, all the necessary bias circuitry is incorporated in the integrated circuit. The user simply provides the required supply voltages. It is also much easier to troubleshoot an integrated circuit than a discrete circuit. A few measurements at the external pins of the chip will tell you whether it is working; repairing the circuit is a simple matter of replacing the bad chip with a good one.

On the other hand, discrete components also have an advantage. They allow very flexible design, whereas special purpose integrated circuits do what they do, and nothing else.

23–2 ■ THE OPERATIONAL AMPLIFIER (OP-AMP)

A circuit design tool that is intermediate between discrete components and special-purpose integrated circuits is provided by integrated circuit **operational amplifiers,** or **Op-Amps.** The Op-Amp is a very high-gain voltage amplifier that we combine with resistors, capacitors, and even transistors to construct a wide variety of electronic devices.

The Op-Amp combines a differential amplifier input stage (Chapter 17), a high-gain common-emitter amplifier stage (Chapter 10), and a push-pull output stage (Chapter 19)

FIGURE 23–2
Integrated circuit amplifier

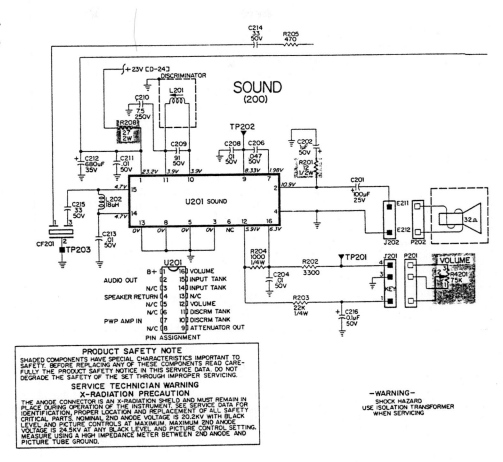

on a single chip (Figure 23–3). The engineer that designs the Op-Amp chip must be very familiar with the semiconductor physics and the Op-Amp's circuit behavior. Fortunately for most of us who use the Op-Amp, we can get away with much simpler models for this device.

FIGURE 23–3
Operational amplifier

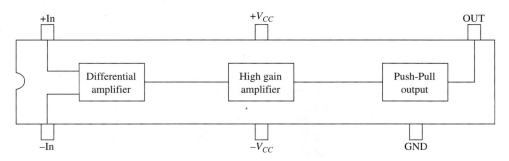

The circuit diagram for the Op-Amp is a triangle with two inputs and, usually, one output (Figure 23–4A, p. 750). Most Op-Amps are supplied by a positive and negative voltage of approximately ±9V to ±15V. A single supply can also be used.

Although the input terminals are labeled as + and −, this does not mean you have to apply positive voltages to the + terminal and negative voltages to the − terminal. Any voltage can be applied to either terminal. The true meaning of the input terminal labels is that a positive voltage applied to the + terminal drives the output voltage towards +15V; a positive voltage applied to the − terminal drives the output voltage towards −15V.

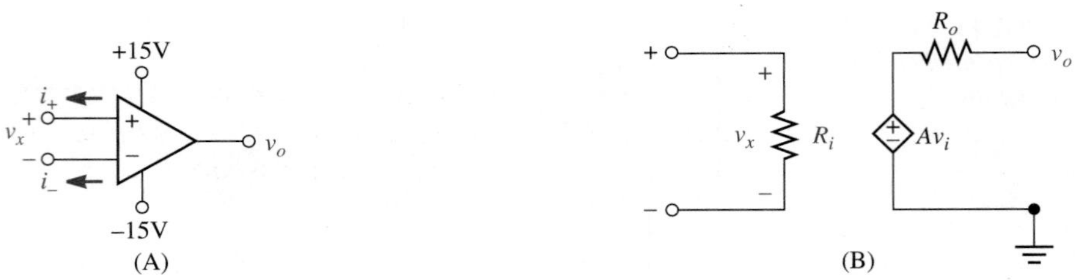

FIGURE 23–4
Op-Amp symbol and Box Model

Figure 23–4B shows the Box Model for the Op-Amp. It is similar to the Box Model we have used to represent other amplifiers; however, the Op-Amp has two inputs. The approximate model parameter values for the 741 Op-Amp are

$$A \approx 1.5 \times 10^5$$
$$R_i \approx 2M\Omega$$
$$R_o \approx 100\Omega$$

You can see that the Op-Amp has a very high gain, high input resistance, and low output resistance.

Note that the Op-Amp amplifies the voltage *difference* across its two input terminals, that is

$$v_o = A(v_+ - v_-)$$

The Op-Amp is almost always used in a feedback configuration. That is, we take some of the output and feed it back around to the input. Therefore, we will first examine some feedback concepts.

REVIEW QUESTIONS

1. Draw the circuit symbol for the Op-Amp.
2. What three stages are usually found in an Op-Amp?
3. Draw the Box Model for the Op-Amp. Describe the Box Model parameters (which are large; which are small?)
4. What does the Op-Amp amplify?

23–3 ■ NEGATIVE FEEDBACK

The basic negative feedback amplifier is shown schematically in Figure 23–5. The circle with the capital sigma (Σ) inside it is a summing junction. In negative feedback the output of the summing junction is the difference between its two inputs. In this network the output voltage is directly connected to the summing junction. This is known as **unity feedback.**

FIGURE 23–5
Unity feedback

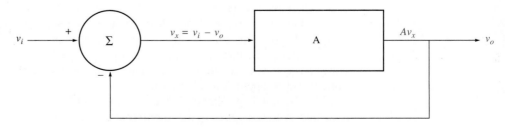

Talking Around the Loop

Feedback systems are also known as **closed-loop systems.** We can determine the output of the closed-loop amplifier of Figure 23–5 by tracing the signals around the loop.

The output of the summer is known as the **error signal.** It is given by

$$v_x = v_i - v_o$$

The box labeled A is the **forward gain,** also known as the **open-loop gain.** The output voltage is given by

$$v_o = Av_x$$

We can see the effect of negative feedback by talking around the loop. As an example, we will use

$$A = 10$$

Let's first assume that the output is 0V. We now apply an input, for example

$$v_i = 1V$$

At this time $v_o = 0V$, so

$$v_x = v_i - v_o = 1 - 0 = 1V$$

Therefore, the output tries to reach

$$v_o = Av_x = 10 \times 1 = 10V$$

The key word here is "tries." The output never actually reaches 10V. To go from 0V to 10V, the output has to first go through 0.25, 0.50, 0.75V, and so on. In the real world, voltage can never change instantaneously. Let's see what happens when the output reaches 0.50V.

When

$$v_o = 0.50V$$

the error signal equals

$$v_x = v_i - v_o = 1 - 0.50 = 0.50V$$

so the output is now headed for

$$v_o = 10v_x = 5V$$

We see that the negative feedback has reduced v_x, so v_o now is heading for a lower voltage. Let's continue by assuming that the output has reached 0.75V. Now

$$v_x = v_i - v_o = 1 - 0.75 = 0.25V$$

and

$$v_o = 10v_x = 2.50V$$

Again, negative feedback has reduced v_x and v_o.

How long does this continue? Let's see what happens when the output reaches 0.909V. For

$$v_o = 0.909V$$

the error signal is

$$v_x = v_i - v_o = 1 - 0.909 = 0.091V$$

and the output voltage is

$$v_o = 10v_x = 0.91V$$

We see that the feedback system has reached equilibrium (within round-off error). When we assume that the output is 0.909V and talk around the loop, we find the output is still approximately 0.909V.

Negative feedback stabilizes the output at a value that is consistent around the loop. An amplifier uses negative feedback if the results of talking around the loop produce the following:

☐ Increasing the input voltage increases the error signal.

☐ Increasing the error signal increases the output voltage.

☐ Increasing the output voltage decreases the error signal.

Is there an easier way to find the true output voltage? Luckily, the answer is yes. Now that you understand how negative feedback works, we can return to the two original equations:

$$\text{Error voltage} \qquad v_x = v_i - v_o$$
$$\text{Output voltage} \qquad v_o = Av_x$$

If we put these two equations together, we get

$$v_o = Av_x = A(v_i - v_o)$$

so

$$v_o = Av_i - Av_o$$

Note that v_o appears on both sides of the equation. This is typical in feedback systems. Moving v_o to the left side of the equation, we get

$$v_o + Av_o = Av_i$$

or

$$(1 + A)v_o = Av_i$$

and

$$v_o = \frac{A}{1 + A}v_i$$

The total system gain, which we will call the **closed-loop gain (A_{CL})** is given by

$$A_{CL} = \frac{v_o}{v_i} = \frac{A}{1 + A}$$

For the example we worked, $A = 10$ and $v_i = 1$V. The closed-loop equations applied to our example gives us

$$v_o = \frac{10}{1 + 10} \, v_i = \frac{10}{11} \times 1 = 0.909V$$

and

$$A_{CL} = \frac{A}{1 + A} = \frac{10}{1 + 10} = 0.909.$$

The forward gain is 10 in this example, but the closed-loop gain is just 0.909. This is a very general property of negative feedback systems:

The gain of the closed-loop system is always less than the forward gain.

Feedback reduces the gain but can improve the stability and frequency bandwidth of the system.

The General Negative Feedback Amplifier

Figure 23–6 shows a more general feedback system. In this case, the output is multiplied by the gain B before it is connected to the summing junction. The gain B is known as the **feedback fraction.**

FIGURE 23–6
General feedback

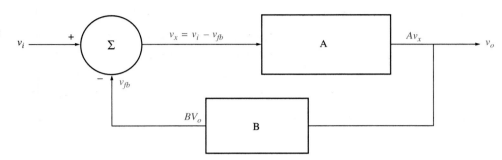

Most often, feedback gain is provided by resistor and capacitor circuits. Because passive circuits always have a gain that is less than 1, the feedback fraction is usually less than 1. What effect would this have on our previous example? We are feeding back only a fraction of the output voltage now. Therefore, the error voltage cannot be reduced as fast. This means that the output voltage must reach a higher level to bring the system to equilibrium.

We can formalize the above explanation by deriving the feedback equation for the general system. Because

$$v_{fb} = Bv_o$$

The error voltage is now

$$v_x = v_i - v_{fb} = v_i - Bv_o$$

and

$$v_o = Av_x$$

so

$$v_o = A(v_i - Bv_o) = Av_i - ABv_o$$

Completing the analysis

$$v_o + ABv_o = Av_i$$

so

$$(1 + AB)v_o = Av_i$$

and we get the general negative feedback equations:

$$v_o = \frac{A}{1 + AB} v_i$$

and

$$A_{CL} = \frac{A}{1 + AB}$$

EXAMPLE 23–1
Closed-Loop Gain

For the feedback amplifier of Figure 23–6

$$A = 100 \text{ and } V_i = 0.1V$$

Find the output voltage and closed-loop gain if

(a) $B = 0$ (open-loop, or forward, gain) **(b)** $B = 1$ (unity feedback)
(c) $B = 0.5$ **(d)** $B = 0.05$ **(e)** $B = -0.01$ (note the negative sign)

Solution For the values given for the forward gain and input voltage, the feedback amplifier will produce

$$V_o = \frac{A}{1 + AB} V_i = \frac{100}{1 + 100B} 0.1$$

and a closed-loop gain of

$$A_{CL} = \frac{A}{1 + AB} = \frac{100}{1 + 100B}$$

(a) For $B = 0$

$$V_o = \frac{A}{1 + AB} V_i = \frac{100}{1 + 100B} 0.1 = \frac{100}{1 + 0} 0.1 = 10V$$

and

$$A_{CL} = \frac{A}{1 + AB} = \frac{100}{1 + 100B} = \frac{100}{1 + 0} = 100 = A$$

(b) For $B = 1$

$$V_o = \frac{A}{1 + AB} V_i = \frac{100}{1 + 100B} 0.1 = \frac{100}{1 + 100} 0.1 = 0.099V$$

and

$$A_{CL} = \frac{A}{1 + AB} = \frac{100}{1 + 100B} = \frac{100}{1 + 100} = 0.99$$

(c) For $B = 0.5$

$$V_o = \frac{A}{1 + AB} V_i = \frac{100}{1 + 100B} 0.1 = \frac{100}{1 + 50} 0.1 = 0.196V$$

and

$$A_{CL} = \frac{A}{1 + AB} = \frac{100}{1 + 100B} = \frac{100}{1 + 50} = 1.96$$

(d) For $B = 0.05$

$$V_o = \frac{A}{1 + AB} V_i = \frac{100}{1 + 100B} 0.1 = \frac{100}{1 + 5} 0.1 = 1.67V$$

and

$$A_{CL} = \frac{A}{1 + AB} = \frac{100}{1 + 100B} = \frac{100}{1 + 5} = 16.7$$

(e) For $B = -0.01$

$$V_o = \frac{A}{1 + AB} V_i = \frac{100}{1 + 100B} 0.1 = \frac{100}{1 + (-1)} 0.1 = \frac{100}{0} 0.01 = ?$$

and

$$A_{CL} = \frac{A}{1 + AB} = \frac{100}{1 + 100B} = \frac{100}{1 + (-1)} = \frac{100}{0} = ?$$

You can get into trouble in a negative feedback system when $AB = -1$. We will discuss this further in Chapter 28.

High Forward Gain

In Op-Amp circuits, the forward gain (A) is provided by the Op-Amp. As previously stated, Op-Amp gains are very high, for example $A \approx 1 \times 10^5$. This leads to two important results:

1. If A is very large, then $AB > 1$. Therefore

$$1 + AB \approx AB$$

and

$$v_o = \frac{A}{1 + AB} v_i \approx \frac{A}{AB} v_i$$

so

$$v_o \approx \frac{1}{B} v_i$$

and the closed-loop gain is

$$A_{CL} \approx \frac{1}{B}$$

If the forward gain (A) is large, the output voltage depends only on B! In an Op-Amp circuit, B is set by the external resistors and capacitors that we connect to the Op-Amp.

2. Let's now look at the relationship between v_x and v_o:

$$v_o = Av_x$$

In theory, v_o can reach any value. However, in real life there are limits on the output voltage. For example, if we supply an Op-Amp with $\pm 15\text{V}$, the output voltage magnitude must be less than 15V. The maximum output voltage that is available from an Op-Amp is known as the **saturation voltage.**

Consider an Op-Amp with a gain of 1×10^5 and a saturation voltage of $\pm 14\text{V}$. The output voltage is given by

$$v_o = 1 \times 10^5 \, v_x$$

If $v_x = 0.1\text{mV}$, then $v_o = 10\text{V}$. This is no problem, but if $v_x = 1\text{V}$, then $v_o = 1 \times 10^5\text{V}$! Do any of you believe this? Remember, v_o can not be greater than the supply voltage. In linear circuits, feedback acts to keep v_x small enough so that v_o will always be less than the saturation voltage.

We have shown that feedback acts to keep v_x small. To see how small, we ask ourselves the following: If v_o cannot be larger than 14V and $A = 1 \times 10^5$, what is largest possible value of v_x? We use the simple expression above to find

$$v_x = \frac{v_o}{A} = \frac{14}{1 \times 10^5} = 0.00014\text{V}$$

This is the largest that v_x can be. Therefore, feedback must keep v_x below 0.00014V. Rather than pick nits, we can simply say that in Op-Amp circuits, feedback is used to make

$$v_x \approx 0\text{V}$$

REVIEW QUESTIONS

1. Draw a unity feedback system.
2. What is the error voltage?
3. How does the output voltage depend on the error voltage?
4. How does the output voltage depend on the input voltage?
5. What is the closed-loop gain?
6. Draw a general feedback system.
7. What is the feedback fraction?
8. What is the error voltage?
9. How does the output voltage depend on the input voltage?
10. What is the closed-loop gain?
11. If the forward gain is very high, how can we approximate the output voltage?
12. Explain why the error voltage in an Op-Amp feedback circuit is approximately 0V.

23–4 ■ THE NON-INVERTING AMPLIFIER— TALKING AROUND THE LOOP

We have just seen that negative feedback is used to keep v_x very close to zero. For the Op-Amp (Figure 23–4, p. 750), v_x is the difference between the two input terminal voltages. That is,

$$v_x = v_+ - v_- = 0\text{V}$$

Feedback in linear Op-Amp circuits, therefore, must keep

$$v_- = v_+$$

for the circuit to operate properly.

One more fact, and then we can proceed with Op-Amp circuit analysis. We have stated that the input resistance of an Op-Amp is very high. In fact, we can buy Op-Amps that have input resistances from a few megohms to thousands of megohms. Because the input resistance is so high, we usually assume that the input current to the Op-Amp is zero (Figure 23–4, p. 750). That is

$$i_- = i_+ = 0\text{A}$$

The Unity-Gain Amplifier (Buffer Amplifier)

FIGURE 23–7
Op-Amp buffer

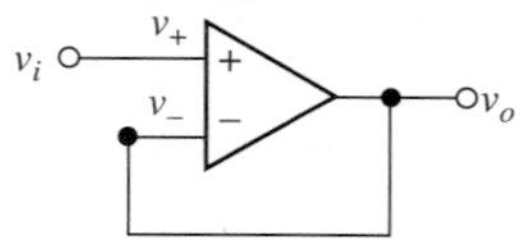

The simplest Op-Amp circuit is shown in Figure 23–7. The input is connected to the positive terminal of the Op-Amp. The output is directly fed back to the negative terminal. This is an electronic example of a unity feedback system.

We analyze this Op-Amp circuit in exactly the same way that we analyzed the unity feedback system in Section 23–3. We can talk around the loop in the same way. We note that

$$v_+ = v_i$$

and because the output is directly connected to the negative terminal

$$v_- = v_o$$

Therefore, the error signal is

$$v_x = v_+ - v_- = v_i - v_o$$

We first assume that v_o is 0V and apply an input v_i. The output tries to reach $v_o = Av_i$. As the output increases, the error voltage (v_x) decreases. Finally, v_x decreases to a value that allows the output to actually reach $v_o = Av_x$, so equilibrium is reached.

Because the open loop gain for the Op-Amp is so large, we can assume the feedback system will reach equilibrium when $v_x \approx 0$. For this to be true we need

$$v_x = v_i - v_o = 0V$$

In other words, we reach equilibrium when

$$v_o = v_i$$

That is, this Op-Amp circuit has unity closed-loop gain:

$$A_{CL} = \frac{v_o}{v_i} = 1.$$

We can also use the unity feedback equations derived in the last section:

$$v_o = \frac{A}{1 + A} v_i$$

and

$$A_{CL} = \frac{A}{1 + A}$$

For A very large, $1 + A \approx A$, so

$$v_o \approx \frac{A}{A} v_i = v_i$$

and

$$A_{CL} = \frac{A}{1 + A} \approx \frac{A}{A} = 1$$

This unity gain Op-Amp circuit serves the same purpose as the emitter and FET followers. It provides a buffer between a source and a load. The Op-Amp circuit is a better buffer than the simple transistor circuits for two reasons: the Op-Amp buffer has a true gain of 1; and the input resistance of the Op-Amp buffer is effectively infinite.

EXAMPLE 23–2 **The Unity-Gain** **(Buffer) Amplifier**	**(a)** An electrode used to record neural activity has an internal resistance of 1MΩ. Figure 23–8A (p. 758) shows this electrode connected directly to a voltage amplifier that has an input resistance of 100KΩ. Find the amplifier input voltage. **(b)** An Op-Amp buffer is placed between the electrode and the voltage amplifier (Figure 23–8B (p. 758). What is the amplifier input voltage?

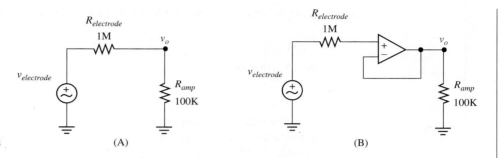

FIGURE 23–8

(A)   (B)

Solution

(a) The circuit in Figure 23–8A forms a voltage divider, so

$$v_a = \frac{100K}{100K + 1M} v_{electrode} = 0.0909 v_{electrode}$$

A mismatch in the signal source resistance and the amplifier input resistance results in only 9% of the signal reaching the amplifier.

(b) We already know that the output of the Op-Amp is not affected by the load and is given by

$$v_a = v_+$$

Because there is no current into the Op-Amp, there is no current in the source resistor ($R_{electrode}$), so

$$v_{source\ resistor} = 0V$$

This means that the Op-Amp input voltage equals the source voltage:

$$v_+ = v_{electrode}$$

and the amplifier voltage is now

$$v_a = v_+ = v_{electrode}$$

The Op-Amp buffer has effectively isolated the signal source from the load.

The Non-Inverting Amplifier

Consider now the Op-Amp circuit shown in Figure 23–9A. The input voltage is applied to the positive terminal of the Op-Amp. The output voltage is fed back to the negative terminal through R_F and R_1. What is v_o?

FIGURE 23–9
Non-inverting amplifier

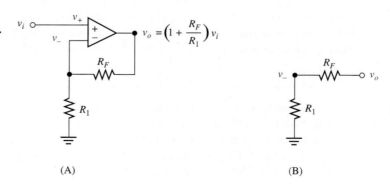

(A)   (B)

We can see from the circuit that once again

$$v_+ = v_i$$

If the circuit is operating properly, the feedback circuit of R_1 and R_F must assure us that $v_- = v_+$. Therefore,

$$v_- = v_+ = v_i$$

But where does v_- come from? Because $i_- = 0A$, R_1 and R_F form a voltage divider with v_o supplying the driving voltage. We show this voltage divider in Figure 23–9B:

$$v_- = \frac{R_1}{R_1 + R_F} v_o$$

Because $v_- = v_i$, v_o must take on a value that satisfies

$$v_i = \frac{R_1}{R_1 + R_F} v_o$$

We can find v_o by inverting this relationship:

$$v_o = \frac{R_1 + R_F}{R_1} v_i$$

or, more commonly

$$v_o = \left(1 + \frac{R_F}{R_1} \right) v_i$$

The closed-loop gain is, therefore

$$A_{CL} = \frac{v_o}{v_i} = 1 + \frac{R_F}{R_1}$$

This circuit is known as the **non-inverting amplifier.** Note that the gain of the closed loop system is set only by the feedback resistors.

EXAMPLE 23–3
The Non-Inverting Amplifier

For the Op-Amp circuit in Figure 23–10, find the system gain if

(a) $R_1 = R_F = 1K\Omega$ **(b)** $R_1 = 1K\Omega$ and $R_F = 10K\Omega$

FIGURE 23–10

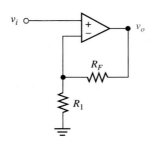

Solution

(a) For $R_1 = R_F = 1\text{K}\Omega$

$$A_{CL} = 1 + \frac{R_F}{R_1} = 1 + \frac{1\text{K}}{1\text{K}} = 2$$

(b) For $R_1 = 1\text{K}\Omega$ and $R_F = 10\text{K}\Omega$

$$A_{CL} = 1 + \frac{R_F}{R_1} = 1 + \frac{10\text{K}}{1\text{K}} = 11$$

Note the simplicity of the Op-Amp amplifier, compared to the transistor amplifiers discussed earlier.

EXAMPLE 23–4
Non-Inverting Amplifier Design

For the circuit of Figure 23–10 (p. 759), if $R_1 = 2\text{K}\Omega$, find R_F so that you get a closed-loop gain of $A_{CL} = 20$.

Solution   We know that

$$A_{CL} = 1 + \frac{R_F}{R_1}$$

Using the given information, we get

$$20 = 1 + \frac{R_F}{2\text{K}}$$

We rearrange this to get

$$\frac{R_F}{2\text{K}} = 19 \rightarrow R_F = 38\text{K}\Omega$$

DRILL EXERCISE

If $R_F = 50\text{K}\Omega$, find R_1 so that $A_{CL} = 26$.

Answer $R_1 = 2\text{K}\Omega$

REVIEW QUESTIONS

1. What is the relation between the voltages at the two Op-Amp terminals? Why?
2. What do we assume the Op-Amp input currents are? Why?
3. Draw the unity-gain buffer amplifier. What is the output voltage?
4. Why do we use unity-gain buffer amplifiers?
5. Draw the non-inverting amplifier.
6. What is v_- in this circuit?
7. Does $v_- = v_i$? Why?
8. What is the closed-loop gain for the non-inverting amplifier?

23–5 ▪ THE INVERTING AMPLIFIER—TALKING AROUND THE LOOP

The circuit in Figure 23–11A is similar to the non-inverting amplifier we just analyzed. Now, however, the positive terminal is grounded. The input is brought in at the negative terminal. We still feed the output back to the input by the same combination of R_1 and R_F.

FIGURE 23–11
Non-inverting amplifier

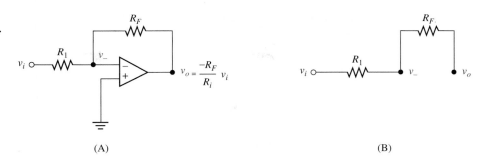

(A)   (B)

To simplify the schematic in Figure 23–11A, we have turned the Op-Amp upside-down (this is the usual way to show the Op-Amp). The positive terminal is now on the bottom.

In this circuit, the positive terminal of the Op-Amp is grounded. This means that

$$v_+ = 0V$$

and

$$v_x = v_+ - v_- = -v_-$$

so

$$v_o = Av_x = -Av_-$$

If we apply a positive voltage at the input, then v_- will also go positive. The above equation tells us that when v_- is positive, v_o heads for a large negative value. Because R_F connects v_o back to the negative terminal, when v_o goes negative, v_- will also go negative. So, we have the following process occurring:

☐ v_i going positive leads to v_- going positive.

☐ v_- going positive leads to v_o going negative.

☐ v_o going negative leads to v_- going negative.

This process establishes that this circuit is a negative feedback amplifier.

We already know that a negative feedback system stabilizes the error voltage (v_x) at a small value. In the case of the Op-Amp, the high forward gain stabilizes v_x at approximately 0V. Because in this circuit

$$v_x = -v_-$$

we get

$$v_x = 0V \rightarrow v_- = 0V$$

As you know, any point in a circuit that is wired to ground is held at 0V. When any point in a circuit is held to 0V because of the circuit connections, we say that point acts as a *virtual ground*. The − terminal in this Op-Amp circuit, therefore, acts as a virtual ground. Feedback in this circuit must adjust the output voltage until $v_- = 0V$. We have isolated R_1 and R_F and the appropriate voltages in Figure 23–11B. Solving for v_-, we get

$$v_- = \frac{R_F}{R_1 + R_F} v_i + \frac{R_1}{R_1 + R_F} v_o$$

Now, because

$$v_- = 0V$$

we have

$$0 = \frac{R_F}{R_1 + R_F} v_i + \frac{R_1}{R_1 + R_F} v_o$$

Multiplying through by $(R_1 + R_F)$

$$0 = R_F v_i + R_1 v_o$$

Solving for v_o

$$v_o = -\frac{R_F}{R_1} v_i$$

So, the closed-loop gain is

$$A_{CL} = -\frac{R_F}{R_1}$$

In the inverting amplifier the closed-loop gain is set by the simple ratio of R_F/R_1. Note the negative sign of the gain, indicating that this circuit inverts the input. That is, when the input goes positive, the output goes negative.

EXAMPLE 23–5
The Inverting Amplifier

Find the closed-loop gain for the Op-Amp circuit shown in Figure 23–12 if

(a) $R_1 = 1K\Omega$ and $R_F = 20K\Omega$ **(b)** $R_1 = 2K\Omega$ and $R_F = 1K\Omega$

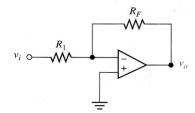

FIGURE 23–12

Solution
(a) For $R_1 = 1K\Omega$ and $R_F = 20K\Omega$

$$A_{CL} = -\frac{R_F}{R_1} = \frac{-20}{1K} = -20$$

(b) For $R_1 = 2K\Omega$ and $R_F = 1K\Omega$

$$A_{CL} = -\frac{R_F}{R_1} = -\frac{1K}{2K} = -0.5$$

Note that the gain for the inverting amplifier can be less than 1.

EXAMPLE 23–6
The Inverting Amplifier, Design

You have the following resistor values available:

$$1K\Omega, \ 5K\Omega, \ 10K\Omega, \ \text{and } 20K\Omega$$

Design an Op-Amp circuit with a system gain of -4.

Solution   Because the closed-loop gain of the inverting amplifier is

$$A_{CL} = -\frac{R_F}{R_1}$$

we need to use resistors that have a ratio of 4:1. The two resistors that satisfy this requirement are

$$R_F = 20\text{K}\Omega$$

and

$$R_1 = 5\text{K}\Omega$$

DRILL EXERCISE

Use the resistors of Example 23–4 (p. 760) to construct an Op-Amp circuit with a gain of −2.

Answer $R_F = 20\text{K}\Omega$ and $R_1 = 10\text{K}\Omega$, or $R_F = 10\text{K}\Omega$ and $R_1 = 5\text{ K}\Omega$

REVIEW QUESTIONS

1. Draw the inverting Op-Amp amplifier circuit.
2. What does v_+ equal? What does v_x equal?
3. What does v_- equal in terms of v_i and v_o?
4. What is a virtual ground?
5. What does v_o equal?
6. What is the system gain?
7. What is the meaning of the negative sign in the system gain?

23–6 ■ THE IDEAL OP-AMP

Talking around the loop of an Op-Amp circuit will let you know whether or not you are working with a negative feedback system. We have also used this technique in the previous two sections to analyze some simple Op-Amp circuits. In this section, we show you a more formal method for analyzing general Op-Amp circuits.

You can determine the output of an Op-Amp circuit by applying node analysis to the *ideal* Op-Amp. The real Op-Amp has a very high gain, a very high input resistance, and a very low output resistance. The ideal Op-Amp parameters are derived by taking the real Op-Amp parameters to their extreme, as shown below:

Real Op-Amp	Ideal Op-Amp
$R_i = 2\text{M}\Omega$	$R_i \to \infty$ (open circuit)
$A = 1 \times 10^5$	$A \to \infty$
$R_o = 100\Omega$	$R_o = 0\Omega$

Because the output resistance of the Op-Amp is small, in the ideal case we set it to zero. The consequence of this is shown in Figure 23–13 (p. 764). Here we have connected a load to the Op-Amp and solve for the load voltage:

$$v_L = \frac{R_L}{r_o + R_L} Av_x$$

As r_o goes to 0Ω

$$v_L = \frac{R_L}{R_L} Av_x = Av_x$$

FIGURE 23–13
Op-Amp output resistance

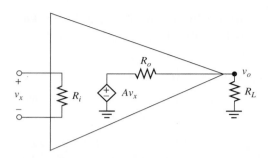

What is important here is that the load voltage does not depend on R_L. This means that when we analyze an Op-Amp circuit, we do not have to be concerned about what is connected to the Op-Amp. As you remember, this was not the case with BJT and FET circuits.

The other ideal Op-Amp properties have already been discussed. Infinite input resistance means that the ideal Op-Amp has zero input current. The high gain (approaching infinity for the ideal case) means that we can set the error voltage to 0V. In summary

Ideal Op-Amp $r_o = 0\Omega \rightarrow$ Ignore Op-Amp load
$R_i \rightarrow \infty \ \rightarrow i_+ = i_- = 0$A
$A \rightarrow \infty \ \rightarrow v_x = 0$V or $v_- = v_+$

The Non-Inverting Amplifier

We are now ready to proceed. We first reexamine the non-inverting amplifier, repeated in Figure 23–14. We can write a node equation at the junction of R_F and R_1 (the assignment of current direction is arbitrary):

$$i_F + i_1 = i_-$$

FIGURE 23–14

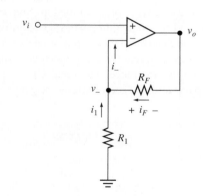

We know that for the ideal Op-Amp $i_- = 0$A, so

$$i_F + i_1 = 0$$

We also know that for this ideal Op-Amp circuit:

$$v_- = v_+ = v_i$$

Now we can write the currents as

$$i_1 = \frac{v_-}{R_1} = \frac{v_i}{R_1}$$

and

$$i_F = \frac{v_- - v_o}{R_F} = \frac{v_i - v_o}{R_F}$$

The current equation at the − terminal can be written as

$$i_1 + i_F = \frac{v_i}{R_1} + \frac{v_i - v_o}{R_F} = 0$$

so

$$\frac{v_i}{R_1} + \frac{v_i}{R_F} = \frac{v_o}{R_F}$$

We rearrange this to get

$$\frac{v_o}{R_F} = \left(\frac{1}{R_1} + \frac{1}{R_F}\right) v_i = \frac{R_1 + R_F}{R_1 R_F} v_i$$

and we find

$$v_o = \frac{R_1 + R_F}{R_1} v_i = \left(1 + \frac{R_F}{R_1}\right) v_i$$

so

Non-Inverting Amplifier: $v_o = \left(1 + \dfrac{R_F}{R_1}\right) v_i$

The closed-loop gain for the non-inverting Op-Amp is

Non-Inverting Amplifier: $A_{CL} = 1 + \dfrac{R_F}{R_1}$

What are the overall input and output resistances of the Op-Amp circuit when we include the external resistors? We already know that the ideal Op-Amp has zero output resistance. It turns out that even for the real Op-Amp, feedback reduces the already small Op-Amp output resistance to just a few ohms.

Because there is no current into the + terminal of the Op-Amp, the input resistance of the total non-inverting Op-Amp circuit (the resistance "seen" by the source) is infinite. This is one of the main advantages of the non-inverting amplifiers.

Non-Inverting Amplifier: $R_{in_{CL}} \to \infty$ (open circuit)
$$R_o \approx 0\Omega$$

EXAMPLE 23–7
The Non-Inverting Amplifier

The source in the Op-Amp circuit in Figure 23–15 (p. 766) has an internal resistance of $10K\Omega$. Find the closed-loop gain.

Solution Because of the low output resistance of the Op-Amp circuit (0Ω in the ideal case), we do not need to worry about the load resistor. How does the source resistance affect the results? To find out, let's find v_+:

$$v_+ = v_{source} - R_{source} i_+$$

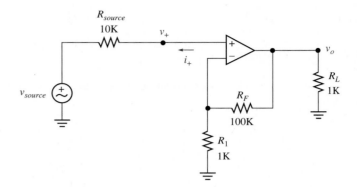

FIGURE 23–15

Because of the infinite input resistance of the Op-Amp, $i_+ = 0A$, and

$$v_+ = v_{source}$$

That is, for the non-inverting Op-Amp we can ignore the internal resistance of the source. The closed-loop gain is still

$$A_{CL} = 1 + \frac{R_F}{R_1}$$

so

$$A_{CL} = 1 + \frac{R_F}{R_1} = 1 + \frac{100K}{1K} = 101$$

The Inverting Amplifier

Let's now use node analysis on the inverting amplifier in Figure 23–16. Note again that the positive terminal is down. Because the positive terminal is grounded

$$v_+ = 0V$$

FIGURE 23–16

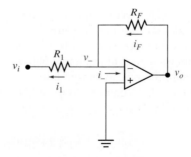

Because $v_- = v_+$

$$v_- = 0V \text{ (virtual ground)}$$

We now write the node equation at the negative terminal:

$$i_F = i_1 + i_-$$

As before, $i_- = 0A$. Here, though

$$i_F = \frac{v_- - v_o}{R_F} = \frac{-v_o}{R_F}$$

and

$$i_1 = \frac{v_i - v_-}{R_1} = \frac{v_i}{R_1}$$

so

$$-\frac{v_o}{R_F} = \frac{v_i}{R_1}$$

and we get the final results for the inverting amplifier:

> **Inverting amplifier:** $v_o = -\dfrac{R_F}{R_1} v_i$

and

> **Inverting amplifier:** $A_{CL} = -\dfrac{R_F}{R_1}$

The output resistance of the inverting Op-Amp circuit is still approximately 0Ω, so we can also ignore any load connected to the Op-Amp. However, the input resistance here must be taken into account. Referring to Figure 23–16, we see that

$$i_i = \frac{v_i - v_-}{R_1} = \frac{v_i}{R_1}$$

In effect, the source sees R_1 connected to the virtual ground. The input resistance of the inverting Op-Amp is, therefore

> **Inverting amplifier:** $R_{in} = \dfrac{v_i}{i_i} = R_1$

The resistance seen by the source in this amplifier is set by R_1.

EXAMPLE 23–8
The Inverting Op-Amp Amplifier

Figure 23–17 shows an inverting Op-Amp amplifier. Find the closed-loop gain if

(a) $R_{source} = 0\Omega$ **(b)** $R_{source} = 1K\Omega$

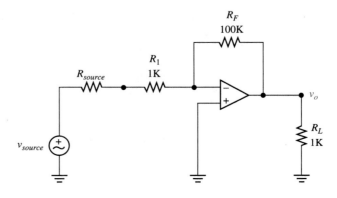

FIGURE 23–17

Solution We first note that once again the load resistor does not affect the gain of this circuit.

(a) If $R_{source} = 0\Omega$, then this is the same circuit as in Figure 23–16 (p. 766). The closed-loop gain is, therefore

$$A_{CL} = -\frac{R_F}{R_1} = -\frac{100K}{1K} = -100$$

(b) For $R_{source} = 1K\Omega$, the source resistance is in series with R_1, so adds to it. The gain formula becomes

$$A_{CL} = -\frac{R_F}{R_{source} + R_1} = -\frac{100K}{1K + 1K} = -50$$

You can see that we have lost half of the gain. The low input resistance of the inverting Op-Amp amplifier is a drawback to this design.

DRILL EXERCISE   Find the closed-loop gain in Figure 23–17 (p. 767) if $R_{source} = 3K\Omega$.

Answer $A_{CL} = -25$

REVIEW QUESTIONS

1. What are the parameters of the ideal Op-Amp?
2. What are the results of $A \rightarrow \infty$, $R_i \rightarrow \infty$, $R_o = 0\Omega$?
3. Draw the non-inverting Op-Amp circuit, and write the node equation at the negative terminal.
4. What is the non-inverting Op-Amp closed-loop gain?
5. Draw the inverting Op-Amp circuit, and write the node equation at the negative terminal.
6. What is the inverting Op-Amp closed-loop gain?
7. What is the resistance seen by the source in the inverting amplifier?

23–7 ■ THE REAL OP-AMP

The real Op-Amp produces essentially the same results as the ideal Op-Amp. There are, however, some limitations of the real Op-Amp that you should consider. This doesn't mean that you need to perform more detailed analyses or need a computer simulation to work with real Op-Amps. It just means that you should be aware of the ways that the real Op-Amp deviates from the ideal and how those differences might affect your Op-Amp behavior.

We will first discuss the most important parameters of the real Op-Amp, then look at a data sheet to familiarize you with the remaining parameters. The parameters that are usually of most concern are these:

Input Bias Current

Input Offset Voltage

Bandwidth (gain-bandwidth product)

Rise Time

Slew Rate

Input Bias Current

We have stated that there is no current into the terminals of the ideal Op-Amp. However, as you know from your study of transistors, bias currents are required for the proper AC functioning of these devices. In fact, real Op-Amps do have a small bias current at each terminal, a current in the range of μA to pA.

Consider the Op-Amp circuit in Figure 23–18A, where we model the input bias currents with DC current sources (I_{BIAS}) added to the ideal Op-Amp. Note that we have grounded the inputs to the Op-Amp and would normally expect to see 0V at the output. Let's apply our analysis technique to this circuit, noting that because we have explicitly modeled the bias currents, we still assume that no current enters the ideal portion of the Op-Amp model. Because the positive terminal is grounded

$$v_+ = 0V$$

FIGURE 23–18
Bias currents

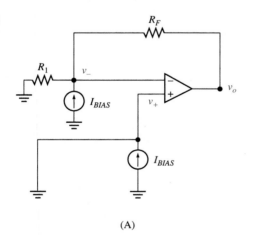

(A)

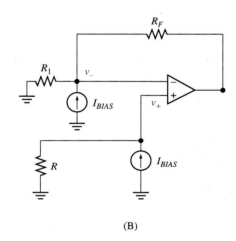

(B)

So

$$v_- = v_+ = 0V$$

Because the positive terminal is grounded, the current source there has no affect. Node analysis at the negative terminal gives us

$$\frac{v_- - v_o}{R_F} = I_{BIAS} + \frac{-v_-}{R_1}$$

Because $v_- = 0V$, this reduces to

$$-\frac{v_o}{R_F} = I_{BIAS}$$

Solving for the output voltage

$$v_o = -R_F I_{BIAS}$$

Even though I_{BIAS} is small, when it is multiplied by the large feedback resistor, we can get a significant output voltage even when there is no input.

To correct for this, we usually connect a resistor (R) to the + terminal (Figure 23–18B), where

$$R = R_1 \| R_F$$

To see the affect of this additional resistor, let's analyze Figure 23–18B. The key here is that at the positive terminal, all of the bias current must go through R (no current enters the ideal portion of the Op-Amp), so

$$v_+ = RI_{BIAS}$$

Now

$$v_- = v_+ = RI_{BIAS}$$

The node equation at the negative terminal gives us

$$\frac{RI_{BIAS} - v_o}{R_F} = I_{BIAS} - \frac{RI_{BIAS}}{R_1}$$

Rearranging and factoring, we get

$$v_o = \left(\frac{R}{R_1} + \frac{R}{R_F} - 1\right)I_{BIAS}$$

Because we have set

$$R = R_1 \| R_F = \frac{R_1 R_F}{R_1 + R_F}$$

we get

$$v_o = \left(\frac{R_F}{R_1 + R_F} + \frac{R_1}{R_1 + R_F} - 1\right)I_{BIAS}$$

Combining the fractions

$$v_o = \left(\frac{R_1 + R_F}{R_1 + R_F} - 1\right)I_{BIAS} = (1 - 1)I_{BIAS} = 0V$$

This is the reason for adding R to Op-Amp circuits—to eliminate effect of I_{BIAS}. How does R affect the amplification of an Op-Amp circuit? As we show in the next example, it has no affect.

EXAMPLE 23–9
The Non-Inverting Amplifier

Figure 23–19 shows an Op-Amp circuit with R added to correct for input bias current affects. Find the closed-loop gain (you can still treat the Op-Amp as ideal).

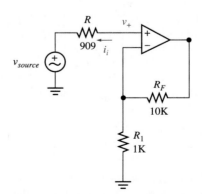

FIGURE 23–19

Solution   We already know the closed-loop gain of the non-inverting amplifier when the input is directly connected to the + terminal. In this circuit

$$v_+ = v_{source} - Ri_i$$

Because we still assume that there is no signal current into the Op-Amp (we can ignore the bias current), we find

$$v_+ = v_{source} - R \times 0 = v_{source}$$

Because there is no signal current into the Op-Amp, there is no voltage drop across R and the system gain is still

$$A_{CL} = 1 + \frac{R_F}{R_1} = 1 + \frac{10K}{1K} = 11$$

The resistor R, therefore, has no affect on the gain of this amplifier, or on the inverting amplifier, for that matter.

Input Offset Voltage

Even when the bias current correcting resistor R is used, we might still see a few millivolts at the output when the inputs are grounded. To correct this problem, we can apply a small DC voltage at the input (positive or negative, as required) to drive the output voltage to 0V. The voltage we need to apply at the input to zero the output is known as the **input offset voltage.**

Most of today's Op-Amps provide additional external pins at which the input offset voltage can be applied. Figure 23–20 shows these pins and the manufacturer's recommended set-up for the 741 Op-Amp. When the circuit is built, set the inputs to 0V, and turn the potentiometer until the output measures 0V.

FIGURE 23–20

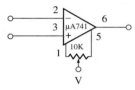

Bandwidth (Gain-Bandwidth Product)

We will discuss the frequency behavior of electronic devices in Chapter 27. For now it is enough to know that all electronic devices only work over a limited range of frequencies known as the **bandwidth (BW).** The bandwidth of an Op-Amp amplifier depends on the closed-loop gain of the Op-Amp circuit. Usually the Op-Amp bandwidth is given for the unity-gain buffer circuit. For example, the unity-gain bandwidth of the 741 Op-Amp is approximately 1MHz. Beyond this frequency, the gain will fall off very rapidly.

The bandwidth of the general Op-Amp circuit depends on the closed-loop gain of the circuit, where

$$A_{CL} = 1 + \frac{R_F}{R_1}$$

We also use this gain to find the bandwidth of the inverting amplifier.

The bandwidth of the general amplifier is found from the parameter the **gain-bandwidth product (GBW).** This product is formed by multiplying the closed-loop gain of an amplifier by the amplifier bandwidth at that gain. For example, for a unity gain 741 Op-Amp

$$GBW = A_{CL_{unity\,gain}} \times BW_{unity\,gain} = 1 \times 1.5\text{MHz} = 1.5\text{MHz}$$

Without proof, we state the following:

The gain-bandwidth product of an Op-Amp amplifier is constant.

This means that for the 741 Op-Amp (whatever the actual closed-loop gain is)

$$GBW = 1.5\text{MHz}$$

The actual bandwidth at a given closed-loop gain can be found dividing the gain-bandwidth product by the closed-loop gain:

$$BW = \frac{GBW}{A_{CL}}$$

EXAMPLE 23–10
Bandwidth and Gain-Bandwidth Product

The 741 Op-Amp has a gain-bandwidth product of $GBW = 1.5\text{MHz}$. Find the operating bandwidth for the following closed-loop gains.

(a) $A_{CL} = 1$ (unity-gain buffer amplifier)　　**(b)** $A_{CL} = 10$　　**(c)** $A_{CL} = 100$

Solution　　The bandwidth is found from

$$BW = \frac{GBW}{A_{CL}}$$

(a) For $A = 1$

$$BW = \frac{GBW}{A_{CL}} = \frac{1.5\text{MHz}}{1} = 1.5\text{MHz}$$

(b) For $A = 10$

$$BW = \frac{GBW}{A_{CL}} = \frac{1.5\text{MHz}}{10} = 150\text{KHz}$$

(c) For $A = 100$

$$BW = \frac{GBW}{A_{CL}} = \frac{1.5\text{MHz}}{100} = 15\text{KHz}$$

As the closed-loop gain is increased, the bandwidth decreases.

Rise Time and Slew Rate

The bandwidth of a system tells us how the system responds to sinusoidal inputs. The **rise time** (T_R) is a measure of how fast the system responds to a transient input. The rise time of a system is inversely proportional to the system bandwidth; that is, as the bandwidth increases, the rise time decreases. For the Op-Amp the relation between these two parameters is given by

$$BW = \frac{0.35}{T_R} \quad \text{or} \quad T_R = \frac{0.35}{BW}$$

This means that for the unity gain 741 Op-Amp circuit we expect a rise time of approximately

$$T_R = \frac{0.35}{1.5\text{MHz}} = 0.233\mu\text{s}$$

Figure 23–21 (p. 773) shows the expected response of the unity gain Op-Amp amplifier to a step input. The rise time for this circuit tells us that the output should reach its final value in approximately $0.4\mu\text{s}$. This is what we would actually see if the input were limited to a few millivolts. If the input is larger than a few millivolts, what we actually see at the output is a much slower rise to the final value. A large step input saturates the input differential amplifier, so the system responds in a non-linear manner.

　　The time it takes the Op-Amp output voltage to actually reach its final value depends on the **slew rate** of the Op-Amp. The slew rate is the slope of the straight line seen in the actual output shown in Figure 23–21 (p. 773). For the μA741 the unity-gain slew rate is approximately $0.5\text{V}/\mu\text{s}$. Therefore, it actually takes $2\mu\text{s}$ for the output to reach 1V.

　　Be aware that rise time and slew rate are specified for the unity gain circuit. As the gain of the Op-Amp circuit increases, both of these response times also increase. If you are amplifying small magnitude signals, choose an Op-Amp according to its rise time. If you are amplifying large magnitude signals choose an Op-Amp that has a fast enough slew rate.

FIGURE 23–21

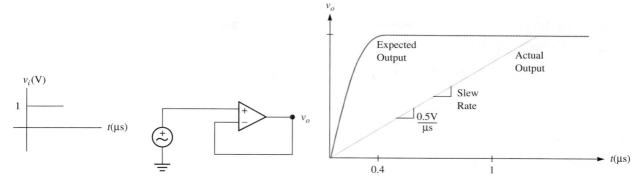

Op-Amp Data Sheets

Figure 23–22 (p. 774–75) shows data sheets for the 741 Op-Amp. We have already discussed in detail many of these data entries. We finish off this section with a brief description of the remaining entries. You can see that there are two sets of data given: one for room temperature (25°) and another for a range of temperatures. We will concentrate, for the most part, on the first set of data.

Average Input Offset Voltage Drift: The input offset voltage of the Op-Amp will change with temperature. For this device, the input offset voltage will change $15\mu V$ for each 1°C change in temperature.

Input Offset Current: The input stage of an Op-Amp is a differential amplifier. If the transistors that make up each side of the differential amplifier are perfectly matched, the input bias current for each terminal would be the same. Because it is impossible to perfectly match the transistors, the bias current will be slightly different.

The input offset current is a measure of the difference in the two input bias currents. When you zero out the output voltage, you are actually canceling the affects of both the input offset voltage and the input offset current.

Average Input Offset Current Drift: This is the change expected in the input offset current as the temperature changes.

Input Resistance: This is the input resistance of the Op-Amp, considered to be infinity for the ideal Op-Amp. Note that the input resistance of the A741 Op-Amp is a few megohms. Op-Amps with much higher input resistance can also be found. Even though the input resistance of the 741 Op-Amp is only a few megohms, when used with feedback, the input resistance of the non-inverting amplifier appears much greater. That is, with feedback the actual input resistance $(R_{in_{CL}})$ is given by

$$R_{in_{CL}} = A_{CL} \times R_i$$

where R_i is the Op-Amp input resistance.

Output Voltage Swing: This parameter gives us the saturation voltage for the Op-Amp. In this device, the output voltage can reach a magnitude of 15 or 16V when the supply voltage is 20V.

Output Short Circuit Current: Also known as "Watch out, ham-handed Henry is connecting the output terminal to ground again!" This is the current that flows when the output is short-circuited. Most Op-Amps have internal circuitry to limit the short circuit current, preventing damage to the push-pull output stage.

Common-Mode Rejection Ratio (CMRR): This information is located in the middle of the second set of data. As you recall from our discussion of differential amplifiers in Chapter 17, the common-mode rejection ratio is a comparison of the differential-mode and the common-mode gains. Ideally, the common-mode gain is 0, and the *CMRR* approaches infinity. For this Op-Amp

$$CMRR_{dB} \approx 100dB$$

FIGURE 23–22

LM741/LM741A/LM741C/LM741E Operational Amplifier

General Description

The LM741 series are general purpose operational amplifiers which feature improved performance over industry standards like the LM709. They are direct, plug-in replacements for the 709C, LM201, MC1439 and 748 in most applications.

The amplifiers offer many features which make their application nearly foolproof: overload protection on the input and output, no latch-up when the common mode range is exceeded, as well as freedom from oscillations.

The LM741C/LM741E are identical to the LM741/LM741A except that the LM741C/LM741E have their performance guaranteed over a 0°C to +70°C temperature range, instead of −55°C to +125°C.

Electrical Characteristics (Note 3)

Parameter	Conditions	LM741A/LM741E			LM741			LM741C			Units
		Min	Typ	Max	Min	Typ	Max	Min	Typ	Max	
Input Offset Voltage	$T_A = 25°C$ $R_S \leq 10\,k\Omega$ $R_S \leq 50\Omega$		0.8	3.0		1.0	5.0		2.0	6.0	mV mV
	$T_{AMIN} \leq T_A \leq T_{AMAX}$ $R_S \leq 50\Omega$ $R_S \leq 10\,k\Omega$			4.0			6.0			7.5	mV mV
Average Input Offset Voltage Drift				15							µV/°C
Input Offset Voltage Adjustment Range	$T_A = 25°C$, $V_S = \pm20V$	±10				±15			±15		mV
Input Offset Current	$T_A = 25°C$		3.0	30		20	200		20	200	nA
	$T_{AMIN} \leq T_A \leq T_{AMAX}$			70		85	500			300	nA
Average Input Offset Current Drift				0.5							nA/°C
Input Bias Current	$T_A = 25°C$		30	80		80	500		80	500	nA
	$T_{AMIN} \leq T_A \leq T_{AMAX}$			0.210			1.5			0.8	µA
Input Resistance	$T_A = 25°C$, $V_S = \pm20V$	1.0	6.0		0.3	2.0		0.3	2.0		MΩ
	$T_{AMIN} \leq T_A \leq T_{AMAX}$, $V_S = \pm20V$	0.5									MΩ
Input Voltage Range	$T_A = 25°C$							±12	±13		V
	$T_{AMIN} \leq T_A \leq T_{AMAX}$				±12	±13					V
Large Signal Voltage Gain	$T_A = 25°C$, $R_L \geq 2\,k\Omega$ $V_S = \pm20V$, $V_O = \pm15V$ $V_S = \pm15V$, $V_O = \pm10V$	50			50	200		20	200		V/mV V/mV
	$T_{AMIN} \leq T_A \leq T_{AMAX}$, $R_L \geq 2\,k\Omega$, $V_S = \pm20V$, $V_O = \pm15V$ $V_S = \pm15V$, $V_O = \pm10V$ $V_S = \pm5V$, $V_O = \pm2V$	32 10			25			15			V/mV V/mV V/mV

FIGURE 23–22, *continued*

Electrical Characteristics (Note 3) (Continued)

Parameter	Conditions	LM741A/LM741E			LM741			LM741C			Units
		Min	Typ	Max	Min	Typ	Max	Min	Typ	Max	
Output Voltage Swing	$V_S = \pm20V$ $R_L \geq 10\ k\Omega$ $R_L \geq 2\ k\Omega$	±16 ±15									V V
	$V_S = \pm15V$ $R_L \geq 10\ k\Omega$ $R_L \geq 2\ k\Omega$				±12 ±10	±14 ±13		±12 ±10	±14 ±13		V V
Output Short Circuit Current	$T_A = 25°C$ $T_{AMIN} \leq T_A \leq T_{AMAX}$	10 10	25	35 40		25			25		mA mA
Common-Mode Rejection Ratio	$T_{AMIN} \leq T_A \leq T_{AMAX}$ $R_S \leq 10\ k\Omega, V_{CM} = \pm12V$ $R_S \leq 50\Omega, V_{CM} = \pm12V$	80	95		70	90		70	90		dB dB
Supply Voltage Rejection Ratio	$T_{AMIN} \leq T_A \leq T_{AMAX},$ $V_S = \pm20V$ to $V_S = \pm5V$ $R_S \leq 50\Omega$ $R_S \leq 10\ k\Omega$	86	96		77	96		77	96		dB dB
Transient Response Rise Time Overshoot	$T_A = 25°C$, Unity Gain		0.25 6.0	0.8 20		0.3 5			0.3 5		μs %
Bandwidth (Note 4)	$T_A = 25°C$	0.437	1.5								MHz
Slew Rate	$T_A = 25°C$, Unity Gain	0.3	0.7			0.5			0.5		$V/\mu s$
Supply Current	$T_A = 25°C$					1.7	2.8		1.7	2.8	mA
Power Consumption	$T_A = 25°C$ $V_S = \pm20V$ $V_S = \pm15V$		80	150		50	85		50	85	mW mW
LM741A	$V_S = \pm20V$ $T_A = T_{AMIN}$ $T_A = T_{AMAX}$			165 135							mW mW
LM741E	$V_S = \pm20V$ $T_A = T_{AMIN}$ $T_A = T_{AMAX}$			150 150							mW mW
LM741	$V_S = \pm15V$ $T_A = T_{AMIN}$ $T_A = T_{AMAX}$					60 45	100 75				mW mW

Note 1: For operation at elevated temperatures, these devices must be derated based on thermal resistance, and T_j max. (listed under "Absolute Maximum Ratings"). $T_j = T_A + (\theta_{jA}\ P_D)$.

Thermal Resistance	Cerdip (J)	DIP (N)	HO8 (H)	SO-8 (M)
θ_{jA} (Junction to Ambient)	100°C/W	100°C/W	170°C/W	195°C/W
θ_{jC} (Junction to Case)	N/A	N/A	25°C/W	N/A

Note 2: For supply voltages less than $\pm15V$, the absolute maximum input voltage is equal to the supply voltage.

Note 3: Unless otherwise specified, these specifications apply for $V_S = \pm15V$, $-55°C \leq T_A \leq +125°C$ (LM741/LM741A). For the LM741C/LM741E, these specifications are limited to $0°C \leq T_A \leq +70°C$.

Note 4: Calculated value from: BW (MHz) = 0.35/Rise Time(μs).

Note 5: For military specifications see RETS741X for LM741 and RETS741AX for LM741A.

Note 6: Human body model, 1.5 kΩ in series with 100 pF.

The decibel measure comes from

$$CMRR_{db} = 20 \log \frac{A_{differential}}{A_{common\,mode}}$$

Because the differential-mode gain here is $A_{differential} \approx 1.5 \times 10^5$, the common-mode gain is approximately

$$A_{common\,mode} \approx 1.5$$

Because the *CMRR* is so large, we can safely ignore the common-mode gain when we use most Op-Amps.

Supply Voltage Rejection Ratio: This is the change expected in the input offset voltage as the power supply voltage changes.

Power Consumption: Power dissipated by the Op-Amp. Note the low power consumption.

REVIEW QUESTIONS

1. What is the input bias current? How can we compensate for its affects?
2. What is the input offset voltage?
3. What information do you get from the bandwidth?
4. What is the gain-bandwidth product? How can it be used to find the bandwidth of a closed-loop amplifier?
5. What is the rise time, and how is it related to the bandwidth?
6. What is the slew rate? When is the slew rate limit a concern?
7. What is meant by drift in an Op-Amp parameter?
8. What is the input offset current?
9. How is the input resistance of the non-inverting Op-Amp circuit affected by feedback?
10. How is the common-mode rejection ratio defined? What is the *CMRR* for the 741? What is the common-mode gain for this Op-Amp?

23–8 ■ INSTRUMENTATION AMPLIFIERS

Another type of integrated circuit amplifier is the instrumentation amplifier. Unlike the very high-gain Op-Amp, the instrumentation amplifier has either a fixed gain, or a gain that can be set over a limited range.

Figure 23–23 shows the circuit description for the 1NA101 instrumentation amplifier. As with the Op-Amp, there is a differential input (pins 10 and 5), a single output (pin 8), positive and negative supplies (pins 9 and 6), and terminals for offset adjustment (pins 2 and 3). The gain of this particular instrumentation amplifier is set by connecting an external resistor between pins 1 and 4. The gain is given by

$$A = 1 + \frac{40\text{K}}{R_G}$$

The gain can be varied from 1 (R_G is removed) to 1000 ($R \approx 40\Omega$).

The instrumentation amplifier is used only to amplify signals. The operational amplifier, on the other hand, has many uses. The advantages of the instrumentation amplifier is that it is optimized to perform only the one function of voltage amplification.

FIGURE 23–23

INA101

High Accuracy
INSTRUMENTATION AMPLIFIER

FEATURES

● **LOW DRIFT: 0.25μV/°C max**

● **LOW OFFSET VOLTAGE: 25μV max**

● **LOW NONLINEARITY: 0.002%**

● **LOW NOISE: 13nV/√Hz**

● **HIGH CMR: 106dB AT 60Hz**

● **HIGH INPUT IMPEDANCE: $10^{10}\Omega$**

● **14-PIN PLASTIC AND CERAMIC DIP SOL-16, TO-100 PACKAGES**

APPLICATIONS

● **STRAIN GAGES**

● **THERMOCOUPLES**

● **RTDs**

● **REMOTE TRANSDUCERS**

● **LOW-LEVEL SIGNALS**

● **MEDICAL INSTRUMENTATION**

DESCRIPTION

The INA101 is a high accuracy instrumentation amplifier designed for low-level signal amplification and general purpose data acquisition. Three precision op amps and laser-trimmed metal film resistors are integrated on a single monolithic integrated circuit.

The INA101 is packaged in TO-100 metal, 14-pin plastic and ceramic DIP, and SOL-16 surface-mount packages. Commercial, industrial and military temperature range models are available.

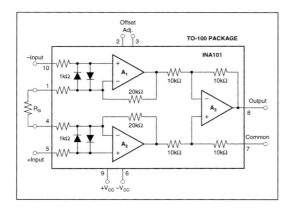

International Airport Industrial Park • Mailing Address: PO Box 11400 • Tucson, AZ 85734 • Street Address: 6730 S. Tucson Blvd. • Tucson, AZ 85706
Tel: (602) 746-1111 • Twx: 910-952-1111 • Cable: BBRCORP • Telex: 066-6491 • FAX: (602) 889-1510 • Immediate Product Info: (800) 548-6132

PDS-454J

23–9 ■ TROUBLESHOOTING OP-AMP CIRCUITS

Troubleshooting integrated circuits, including Op-Amp circuits, is quite straight-forward. If all connections are good and the Op-Amp is getting the appropriate input, then an inappropriate output indicates a bad Op-Amp. Replace the bad Op-Amp with another one. The following examples demonstrate what you might see if there are connection failures in the non-inverting Op-Amp circuit.

EXAMPLE 23–11
Op-Amp Troubleshooting

Consider the non-inverting Op-Amp circuit and the outputs shown in Figure 23–24. Note that we also show the supply voltages. If the Op-Amp is good, determine the likely cause of the failures shown.

(A)

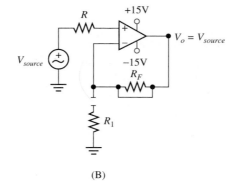

(B)

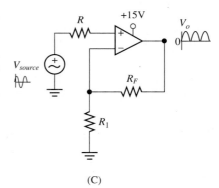

(C)

FIGURE 23–24

Solution

(a) Because we see $+14V$ at the output when the input is turned off, the likely problem is either a failed open feedback resistor or a bad connection between the feedback resistor and the Op-Amp. In this case, the very large Op-Amp gain will cause the output to saturate. The feedback resistor and its connections to the Op-Amp can be checked with an ohmmeter. **Be sure to disconnect the supply voltages.**

(b) Here the output voltage equals the input voltage; that is $A_{CL} = 1$. This will happen if R_F has failed short (an unlikely problem) or (more likely) R_1 has failed open or has bad connections. Turn off the input voltage and the supply voltages, and check the resistors with an ohmmeter.

(c) The large positive DC offset voltage indicates that the negative supply voltage is not properly connected to the Op-Amp. Use a voltmeter to check the supply voltages at the appropriate pins.

REVIEW QUESTIONS

1. If all connections and resistors in an Op-Amp circuit are good, what is the cause of an unexpected output?
2. What will happen if the feedback resistor or its connections are bad?
3. What should you check if you get no output when the input signal is applied?
4. What will you see in the output if R_1 (refer to Figure 23–24) or its connections are bad?
5. What will you see in the output if the negative supply voltage is not properly connected to the Op-Amp?
6. What will you see in the output if the positive supply voltage is not properly connected to the Op-Amp?

SUMMARY

- The operational amplifier (Op-Amp) is a high-gain integrated circuit amplifier.
- The Op-Amp amplifies the difference in the voltage between the $+$ and $-$ terminals.
- The Op-Amp is usually used with negative feedback supplied by external resistors. This feedback acts to keep the voltage difference at the input terminals close to 0V.
- The ideal Op-Amp has infinite gain, infinite input resistance, and 0Ω output resistance.
- Ideal Op-Amp circuits can be analyzed in the following manner. Note that because of the 0Ω output resistance, the Op-Amp output does not depend on the load.

    Find v_+
    Set $v_- = v_+$
    Write the node equation at the $-$ terminal ($i_- = i_+ = 0A$)
    Solve for v_o.

- The closed-loop gain of the non-inverting Op-Amp circuit is $A_{CL} = 1 + \dfrac{R_F}{R_1}$.

- The input resistance of the non-inverting Op-Amp circuit approaches infinity.
- The closed-loop gain of the inverting Op-Amp circuit is

$$A_{CL} = -\frac{R_F}{R_1}$$

- The input resistance of the inverting Op-Amp circuit is R_1.
- The most important deviations from the ideal Op-Amp are
    Input Bias Current
    Input Offset Voltage
    Bandwidth $\left(BW = \dfrac{GBW}{A_{CL}} \right)$
    Slew Rate

■ The instrumentation amplifier is an integrated circuit amplifier that is available with a fixed or limited range of gains.

■ If all connections and external resistors are good, an inappropriate Op-Amp output indicates that the Op-Amp is bad. Replace it.

PROBLEMS

SECTION 23–2 The Operational Amplifier

1. An operational amplifier is usually composed of three stages. What are those stages, and what are their functions?

SECTION 23–3 Negative Feedback

2. For the negative feedback system in Figure 23–25, if $v_{in} = 2V$
   (a) Find v_o. **(b)** Find v_x. **(c)** Find A_{CL}.

FIGURE 23–25

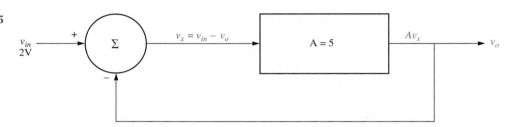

3. Find the forward gain in Figure 23–25 required to produce an output voltage of $v_o = 1V$.
4. What is the highest closed-loop gain that can be achieved in Figure 23–25? Why?
5. For the negative feedback system in Figure 23–26, if $v_{in} = 2V$
   (a) Find v_o. **(b)** Find v_x. **(c)** Find A_{CL}.

FIGURE 23–26

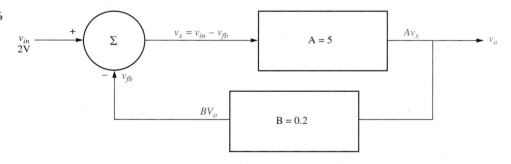

6. Choose a new forward gain (A) in Figure 23–26 to produce a closed-loop gain of $A_{CL} = 4.5$.
7. In Figure 23–26
   (a) Find the closed-loop gain (A_{CL}) if the forward gain is set to $A = 50$.
   (b) Find the closed-loop gain (A_{CL}) if the forward gain is set to $A = 100$.
   (c) Find the closed-loop gain (A_{CL}) if the forward gain is set to $A = 500$.
   (d) What is the maximum possible closed-loop gain when $B = 0.2$?
   (e) How is this maximum closed-loop gain related to the feedback fraction B?
8. An amplifier has an output that is given by $v_o = Av_x$. If the saturation voltage is $\pm10V$, what is the maximum allowable v_x if
   (a) $A = 10$. **(b)** $A = 1000$.

SECTION 23–4   The Non-Inverting Amplifier—Talking Around the Loop

9. Draw the non-inverting Op-Amp amplifier with an input v_i.

   (a) What procedure would you follow to determine that this is a negative feedback amplifier?

   (b) What is the differential input voltage (v_x)? With negative feedback, what does this voltage approximately equal?

   (c) What is the relation between v_+ and v_-?   **(d)** How does v_- depend on v_o?

   (e) What is the closed-loop gain of this amplifier?

SECTION 23–5   The Inverting Amplifier—Talking Around the Loop

10. Draw the inverting Op-Amp amplifier with an input v_i.

   (a) What procedure would you follow to determine that this is a negative feedback amplifier?

   (b) What is the differential input voltage (v_x)? With negative feedback, what does this voltage approximately equal?

   (c) What is the relation between v_+ and v_-?   **(d)** How does v_- depend on v_o?

   (e) What is the closed-loop gain of this amplifier?

SECTION 23–6   The Ideal Op-Amp

11. For the Op-Amp circuit in Figure 23–27

   (a) Find v_- as a function of v_o.   **(b)** Find v_+ as a function of v_o.

   (c) Find v_o as a function of v_i.   **(d)** What is the closed-loop gain (A_{CL})?

12. The circuit in Figure 23–28 isolates the feedback circuitry of the non-inverting Op-Amp amplifier.

   (a) Find i; then use the result to find v_-.   **(b)** Use node analysis to find v_-.

   (c) Use superposition and voltage dividers to find v_-.

FIGURE 23–27

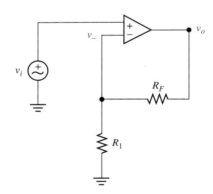

FIGURE 23–28

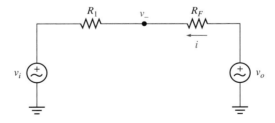

13. For the circuit in Figure 23–29 (p. 782)

   (a) Find v_+.   **(b)** Find v_-.

   (c) Write a node equation at v_- and find v_o.   **(d)** What is A_{CL}?

14. Why doesn't the load resistor in Figure 23–29 (p. 782) affect the closed-loop gain?

15. The non-inverting Op-Amp in Figure 23–29 (p. 782) has the resistors $R_1 = 1\text{K}\Omega$ and $R_F = 10\text{K}\Omega$.

   (a) What is the closed-loop gain?   **(b)** If $v_i = 0.5\text{V}$, what is v_o?

   (c) If $v_i = -0.5\text{V}$, what is v_o?

FIGURE 23–29

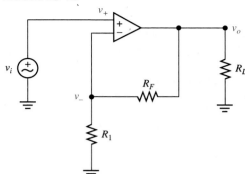

FIGURE 23–30

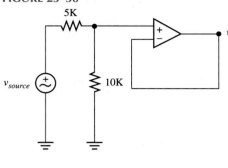

16. The non-inverting Op-Amp in Figure 23–29 has the resistors $R_1 = 1K\Omega$ and $R_F = 10K\Omega$. The Op-Amp also has $\pm15V$ supply voltages.

(a) Use the closed-loop gain to find v_o if $v_i = 2V$.   (b) Why is the answer to (a) wrong?

17. Find v_o in Figure 23–30.

18. For the circuit in Figure 23–31

(a) Find v_+.   (b) Find v_-.   (c) Write a node equation at v_-, and find v_o.

(d) Find the closed-loop gain.

19. If the Op-Amp in Figure 23–31 has $\pm12V$ supplies, what is the maximum allowable input voltage?

FIGURE 23–31

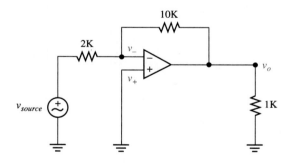

20. The inverting amplifier circuit in Figure 23–32 includes a source resistance (R_S).

(a) Find the output voltage and closed-loop gain if $R_S = 0\Omega$.

(b) Find the output voltage and closed-loop gain if $R_S = 2K\Omega$.

(c) How does the source resistor affect the closed-loop gain?

(d) What can you place between the source and the amplifier to eliminate the affect of the source resistor?

21. For the circuit in Figure 23–33

(a) Find v_+.   (b) Find v_-.   (c) Write a node equation at v_- and find v_o.

22. For the two Op-Amp circuits in Figure 23–34

(a) Find v_{o1}. Note that you do not need to worry about the second Op-Amp.

(b) Find v_o in terms of v_{o1}. Note that here you do not need to worry about the first Op-Amp.

(c) Find v_o as a function of v_i.

FIGURE 23–32

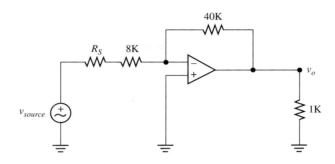

FIGURE 23–33

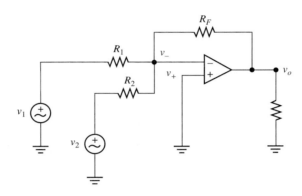

FIGURE 23–34

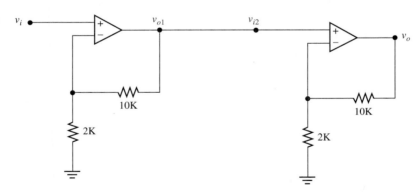

SECTION 23–7   Real Op-Amps

23. The circuit in Figure 23–35 (p. 784) models I_{BIAS} with current sources. If $I_{BIAS} = 1\mu A$

   (a) Remove R, and find the output voltage.

   (b) Replace R, and choose a value for it to set the output voltage to 0V.

24. You are building an amplifier to have a gain of 100 with 741 Op-Amps that have a gain-bandwidth product of 1.5MHz.

   (a) If you use a single Op-Amp, what will be the actual bandwidth of your circuit?

   (b) You now build the amplifier with two Op-Amp circuits, each with a gain of 10. What is the actual bandwidth for each Op-Amp now?

25. What is the rise time for the amplifier in Problem 24a?

26. What is the rise time for each of the Op-Amps in Problem 24b?

27. An Op-Amp has a $CMRR_{dB}$ of 80 dB and a forward gain of 1×10^6. What is the common-mode gain for this device?

FIGURE 23–35

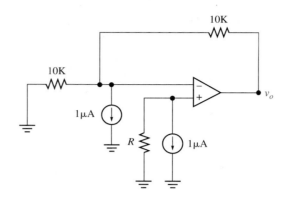

FIGURE 23–36

INSTRUMENTATION AMPLIFIERS										Boldface = NEW
						Input Parameters				
Description	**Model**	**Gain Range**	**Gain Error G = 100 25°C, max(%)**	**Gain Drift G = 100 (ppm/°C)**	**Non-Linearity G = 100 max(%)**	**CMR[5] min(dB)**	**Offset Voltage vs Temp max(µV/°C)**	**Dynamic Response G = 100 −3dB BW (kHz)**	**Temp Range[1]**	**Pkg**	**Page No.**
Very High Accuracy	INA114	1–10,000[2]	0.5	25	±0.002	96	±0.25 + 5/G	10	Ind	DIP, SOIC	4.74
	INA115	1–10,000[2]	0.5	25	±0.002	96	±0.25 + 5/G	10	Ind	SOIC	4.87
	INA131	100	0.024	10	±0.002	110	±0.25	70	Ind	DIP, SOIC	4.135
	INA120	1, 10, 100, 1000	0.5	30	±0.01	96	±0.25 ± 10/G	20	Ind	DIP	4.125
	INA104	1–1000[2]	0.15	22[3]	±0.003	96	±0.25 ± 10/G	25	Ind	DIP	A
	INA101	1–1000[2]	0.1	22[3]	±0.003	96	±0.25 ± 10/G	25	Ind, Mil	DIP TO-100, DIP	4.4
Electrometer Input I$_8$ = 10fA typ	**INA116**	**1–10,000[2]**	**0.5**	**25**	**±0.005**	**90**	**±5 ± 20/G**	**70**	**Ind**	**DIP/SOIC**	**4.98**
Dual	**INA2128**	**1–10,000[2]**	**0.5**	**10**	**±0.002**	**97**	**±0.75 ± 20G**	**200**	**XInd**	**DIP/SOIC**	**4.145**
	INA2141	**10, 100**	**0.15**	**10**	**±0.002**	**97**	**1[7]**	**200**	**XInd**	**DIP/SOIC**	**4.147**
Low Quies-cent Power	INA118	1–10,000	0.5	25	±0.002	97	±0.5 + 20/G	100	Ind	DIP, SOIC	4.114
	INA102	1, 10, 100 1000	0.15	15	±0.02	90	±2 ± 5/G	3	Com, Ind	DIP, SOIC	4.10
Low Noise, Low Distortion	INA103	1–1000[2]	0.1	25	±0.004	100	±0.5 + 10/G type	800	Ind	DIP	4.22
		1–1000[2]	0.25	25	±0.010	90	±0.5 + 20/G type	800	Com	DIP, SOIC	
Fast Settling FET Input	INA110	1, 10, 100, 200, 500	0.1	20	±0.01	96	±2 + 50/G	470	Ind	DIP, SOIC	4.52
	INA111	1–10,000[2]	0.5	25	±0.01	96	±5 + 100/G	450	Ind	DIP, SOIC	4.63
Unity-Gain Difference Amp	INA105	1V/V, fixed	0.01[3]	5	±0.001[3]	86[4]	10	1000[3]	Ind	TO-99, DIP SOIC	4.34
	3627	1V/V, fixed	0.01[3]	5	±0.001[3]	100	20	800[3]	Ind	TO-99	A
Gain of 10 Diff. Amp	INA106	10V/V, fixed	0.025	10	±0.001	100[4]	0.20	500	Ind	DIP, SOIC	4.46
High Com. Mode Volt. Diff. Amp (200VDC CMV)	INA117	1V/V, fixed	0.02	10	±0.001	86[4]	20	200[3]	Ind	TO-99, DIP SOIC	4.100
4–20mA Loop Receiver	RCV420	4–20mA in 0–5V Out	0.05	25	±0.002	86	25[6]	150	Com, Ind	DIP	4.190

NOTES: (1) Com = 0°C to +70°C, Ind = −25°C to +85°C, Mil = −55°C to +125°C. (2) Set with external resistor. (3) Unity-gain. (4) No source imbalance.
(5) DC to 60Hz, Gain = 10, (or specified gain of device). (6) RTO. (7) G = 100.
"A" indicates a product that is not included in the 1995 Data Books—contact factory for data sheet.

SECTION 23–8   Instrumentation Amplifiers

Figure 23–36 shows data for a number of instrumentation amplifiers. Use this data to answer the following.

28. You need an amplifier with a gain of 500. Which instrumentation amplifiers can you use?

29. You need to provide amplification of 1000 to a 400 KHz signal. Which instrumentation amplifiers can you use?

30. What is the gain-bandwidth product of the 1NA120 instrumentation amplifier?

SECTION 23–9   Troubleshooting Op-Amp Circuits

31. What is the likely cause of the faulty output shown in Figure 23–37?

32. What is the likely cause of the faulty output shown in Figure 23–38?

33. What is the likely cause of the faulty output shown in Figure 23–39?

FIGURE 23–37

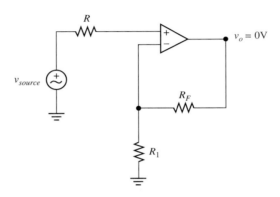

FIGURE 23–38

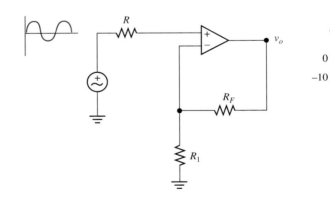

FIGURE 23–39

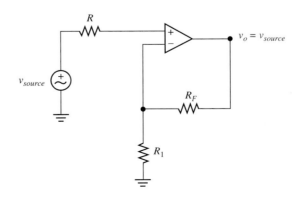

FIGURE 23–40

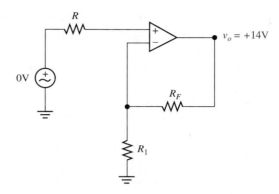

34. What is the likely cause of the faulty output shown in Figure 23–40?

Computer Problems

35. If your computer program has a library of Op-Amps

(a) What Op-Amp model parameters are used? Compare them to the parameters given in a typical data sheet.

(b) What Op-Amp models are in your library? List their parameter values.

36. Figure 23–41 shows a non-inverting Op-Amp amplifier where the Op-Amp has been replaced with its circuit model. Simulate this circuit, apply a 1V input signal, and find the output voltage.

FIGURE 23–41

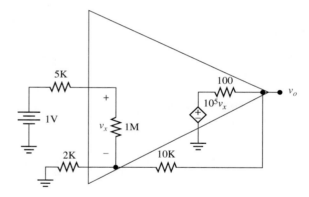

37. Simulate the circuit of Figure 23–29 (p. 782) with $R_F = 10K\Omega$ and $R_1 = 2K\Omega$. Apply a 1 V input, and find the output as you vary R_L from 100KΩ down to 1Ω.

38. Simulate the circuit of Figure 23–31 (p. 782). Apply a 1V input, and find the output as you vary R_L from 100KΩ down to 1Ω.

39. Simulate the circuit of Figure 23–34 (p. 783). Apply a 1mV input, and find the output voltage.

Chapter

24

Operational Amplifier Circuits

Chapter Outline

C HAPTER OBJECTIVES

To build an inverting summing amplifier.

To build a non-inverting summing amplifier.

To build a differential amplifier.

To build an electronic differentiator.

To build an electronic integrator.

To build an improved voltage regulator.

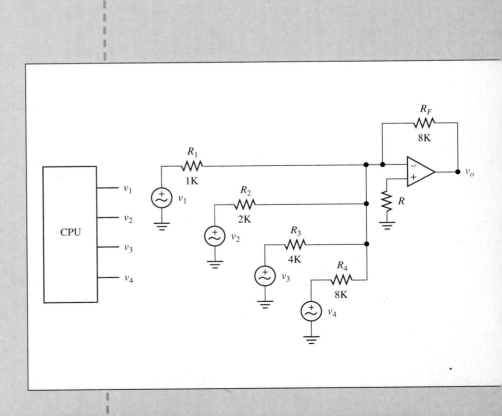

24–1 ■ INTRODUCTION

In the preceding chapter we introduced you to the ideal Op-Amp and the basic Op-Amp circuits. In this chapter, we will show you some typical uses of the Op-Amp. We will just be scratching the surface with these circuits; there are a mindboggling number of circuits that use Op-Amps. You can find most of these in the many excellent Op-Amp handbooks that are available. We first review the two basic circuits: the non-inverting and the inverting amplifier.

Non-Inverting Amplifier

The basic non-inverting amplifier is shown in Figure 24–1. According to standard practice, we draw the Op-Amp with the +terminal down for most Op-Amp circuits. The output of the non-inverting Op-Amp is given by

$$v_o = \left(1 + \frac{R_F}{R_1}\right)v_i$$

FIGURE 24–1
Non-inverting amplifier

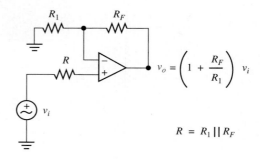

$$v_o = \left(1 + \frac{R_F}{R_1}\right)v_i$$

$$R = R_1 \,||\, R_F$$

The resistor *(R)* has no effect on the signal behavior of this circuit but is included to eliminate the offset voltage due to input bias currents. Its value is given by

$$R = R_1 \,||\, R_F$$

The major advantage of the non-inverting amplifier is its near infinite input resistance. As we shall see, this advantage is lost in some non-inverting circuits.

Unity-Gain Buffer Amplifier

Figure 24–2 shows the unity-gain Op-Amp buffer amplifier. Here

$$v_o = v_i$$

The Op-Amp buffer provides better source-load isolation than the transistor buffers (emitter follower and FET follower) because it has near infinite input resistance and near 0Ω output resistance.

FIGURE 24–2
Buffer amplifier

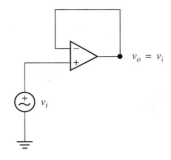

$$v_o = v_i$$

Inverting Amplifier

Figure 24–3 shows the basic inverting amplifier. The output is given by

$$v_o = -\frac{R_F}{R_1}v_o$$

Again, the resistor ($R = R_1 \| R_F$) is there to eliminate the effect of the input bias currents. Because the gain formula for the inverting amplifier is simpler, it is easier to design with than the non-inverting amplifier.

FIGURE 24–3
Inverting amplifier

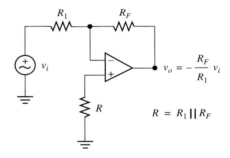

$$v_o = -\frac{R_F}{R_1}v_i$$

$$R = R_1 \| R_F$$

The negative sign is not really a problem because Op-Amps are cheap enough that we can use several of them to build an amplifier. One inverting amplifier feeding a second inverting amplifier will produce a non-inverting output. In fact, many actual Op-Amp designs use several Op-Amp stages, a process made easier by the fact that manufacturers produce integrated circuits with four Op-Amps in a single package (Figure 24–4).

FIGURE 24–4
Quad Op-Amp chip

CONNECTION DIAGRAM

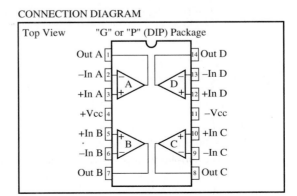

The main disadvantage of the inverting Op-Amp is that, because of the virtual ground at the $-$terminal, the resistance seen by the signal source is just R_1. If we desire a high input resistance, we can still maintain the simplicity of inverting amplifier design by preceding the amplifier with a unity-gain buffer, as in Figure 24–5.

As we add more elements to Op-Amp circuits, the analysis procedure that we introduced in the previous chapter still (actually always) works. We repeat it here.

FIGURE 24–5

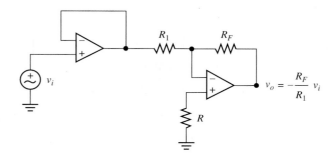

Op-Amp Analysis Procedure

☐ Find v_+ (sometimes we assume we know v_+, then find it at the end of the analysis).

☐ Set $v_- = v_+$.

☐ Write a node equation at the $-$terminal.

☐ Solve for v_o.

Troubleshooting

As we discussed in Chapter 23, if all circuit connections and passive components (resistors and capacitors) are good, an incorrect output indicates that the Op-Amp is bad. As we introduce new Op-Amp circuits, we will discuss the results of bad connections or bad passive components.

REVIEW QUESTIONS

1. Draw the non-inverting amplifier, and write the formula for the output voltage.
2. Draw the unity-gain buffer amplifier. What are the advantages of this buffer?
3. Draw the inverting amplifier, and write the formula for its output voltage. What is the disadvantage of the inverting amplifier?
4. What is the procedure for analyzing an Op-Amp circuit?

24–2 ■ SUMMING AMPLIFIERS

Inverting Summer

One of the advantages of the Op-Amp is that we can use it to combine several signals in a number of interesting ways. At some point you will recognize these rather common circuits and be able to write the output formulae from memory. For now, we will use the procedure given previously. Note that when there are several inputs, we cannot find a system gain. Rather, we find how the output depends on each of the several inputs.

FIGURE 24–6
Inverting summing amplifier

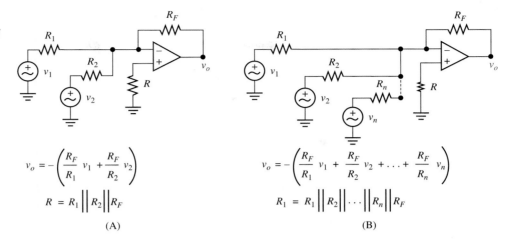

(A)

(B)

Let's look at the multiple input circuits in Figure 24–6. We first analyze the two-input circuit of Figure 24–6A. Because the positive terminal is grounded through R, and because $i_+ = 0A$, there is still a virtual ground at the $-$terminal. That is

$$v_+ = 0V \rightarrow v_- = 0V$$

We now write the node equation at the negative terminal ($i_- = 0A$):

$$\frac{v_1}{R_1} + \frac{v_2}{R_2} = \frac{-v_o}{R_F}$$

so

$$\frac{v_1}{R_1} + \frac{v_2}{R_2} = -\frac{v_o}{R_F}$$

Rearranging, we get

$$\frac{v_o}{R_F} = -\left(\frac{v_1}{R_1} + \frac{v_2}{R_2}\right)$$

Finally, we get

$$v_o = -\left(\frac{R_F}{R_1} v_1 + \frac{R_F}{R_2} v_2\right)$$

We have built a **weighted summer** (this circuit also inverts the inputs). To eliminate the bias current effect, we set

$$R = R_1 \| R_2 \| R_F$$

If

$$R_F = R_1 = R_2$$

then

$$v_o = -(v_1 + v_2)$$

and the circuit is a simple summer.
If

$$R_2 = R_1$$

then

$$v_o = -\frac{R_F}{R_1}(v_1 + v_2)$$

and we have built a summer with a gain of R_F/R_1.
If, for example

$$R_F = 10K\Omega, R_1 = 5K\Omega, \text{ and } R_2 = 10K\Omega$$

then the output voltage is

$$v_o = -(2v_1 + v_2)$$

This is what we mean by a weighted summer. In this case, we add twice as much of the signal represented by v_1 to the signal represented by v_2. Note that to cancel the effect of the input bias currents we now set

$$R = R_1\|R_2\|R_F = 2.5K\Omega$$

We can extend this analysis to the more general circuit shown in Figure 24–6B (p. 793). Here the only difference is that there are more currents to consider in the node equation at the negative terminal:

$$\frac{v_1}{R_1} + \frac{v_2}{R_2} + \ldots + \frac{v_n}{R_n} = -\frac{v_o}{R_F}$$

Solving for v_o

$$v_o = -\left(\frac{R_F}{R_1}v_1 + \frac{R_F}{R_2}v_2 + \ldots + \frac{R_F}{R_n}v_n\right)$$

In general, to eliminate the offset voltage due to the bias currents, we want the total resistance connected to the +terminal to equal the total resistance (including R_F) connected to the −terminal

$$R = R_1\|R_2 \ldots R_n\|R_F$$

EXAMPLE 24–1
**The Inverting
Summer**

Figure 24–7 shows an inverting summer that combines three input voltages.

(a) Find the output voltage.
(b) Choose R to eliminate the DC offset due to the bias currents.

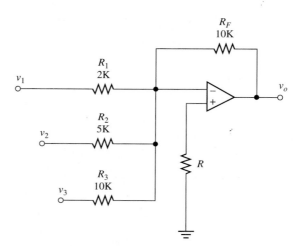

FIGURE 24–7

Solution

(a) The output of the summer is found from

$$v_o = -\left(\frac{R_F}{R_1}v_1 + \frac{R_F}{R_2}v_2 + \frac{R_F}{R_3}v_3\right) = -\left(\frac{10K}{2K}v_1 + \frac{10K}{5K}v_2 + \frac{10K}{10K}v_3\right)$$

So the output is

$$v_o = -(5v_1 + 2v_2 + v_3)$$

The output is a weighted sum of the three inputs.

(b) The resistance that we need is found from

$$R = R_1\|R_2\|R_3\|R_F = 2K\|5K\|10K\|10K = 1.11K\Omega$$

DRILL EXERCISE

Redraw the inverting summer of Figure 24–7 with the feedback resistor set instead to $R_F = 20K\Omega$. Find v_o and R.

Answer $v_o = -(10v_1 + 4v_2 + 2v_3)$ and $R_F = 1.18K\Omega$

EXAMPLE 24–2
Digital-to-Analog
Converter (DAC)

Figure 24–8 shows how we can use the inverting summer to build a 4-bit digital-to-analog converter. The voltages v_1 through v_4 are provided by the output bus of a 4-bit computer and represent either a logical HIGH (1) or a logical LOW (0). The sequence of voltages represent a 4-bit binary number, with v_1 being the most significant bit (MSB) and v_4 being the least significant bit (LSB).

(a) Find v_o as a function of v_1 through v_4.
(b) Assume that a logic HIGH is represented by 1V and a logic LOW is represented by 0V. Find the output voltage if

$$(v_1 \; v_2 \; v_3 \; v_4) = (1 \; 0 \; 1 \; 1)$$

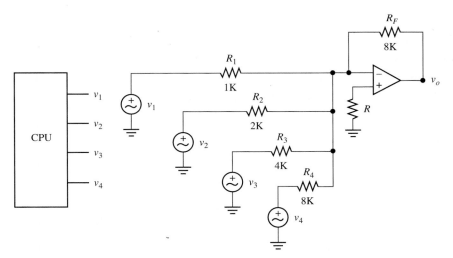

FIGURE 24–8
Digital-to-analog
converter

Solution

(a) From the formula for the weighted summer, we find

$$v_o = -\left(\frac{R_F}{R_1}v_1 + \frac{R_F}{R_2}v_2 + \frac{R_F}{R_3}v_3 + \frac{R_F}{R_4}v_4\right) = -(8v_1 + 4v_2 + 2v_3 + v_4)$$

(b) For $(v_1\ v_2\ v_3\ v_4) = (1\ 0\ 1\ 1)$, the output voltage is

$$v_o = -(8v_1 + 4v_2 + 2v_3 + v_4) = -(8\times1 + 4\times0 + 2\times1 + 1\times1) = -11\text{V}$$

DRILL EXERCISE Find the output of the DAC in Figure 24–8 (p. 795) if $(v_1\ v_2\ v_3\ v_4) = (1\ 1\ 0\ 1)$

Answer $v_o = -13\text{V}$

Inverting Amplifier with Buffers

Each signal generator in the DAC of Figure 24–9 sees a different resistance. That is, when we have multiple inputs, we cannot talk about a single amplifier input resistance. In the case of the inverting summer, because the −terminal is a virtual ground,

v_1 sees an input resistance of $R_1 = 1\text{K}\Omega$

v_2 sees an input resistance of $R_2 = 2\text{K}\Omega$

v_3 sees an input resistance of $R_3 = 4\text{K}\Omega$

v_4 sees an input resistance of $R_4 = 8\text{K}\Omega$

If these resistances are too small for the circuit, we can insert unity-gain Op-Amp buffers between each signal source and the amplifier (Figure 24–9).

FIGURE 24–9

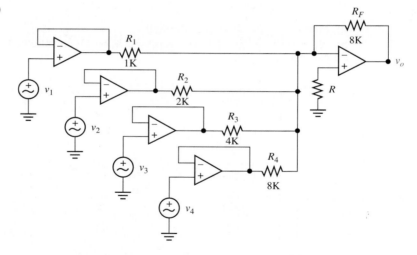

Non-Inverting Summer

Before we consider the non-inverting summer, let's examine the circuit of Figure 24–10. For Op-Amp circuits that have several resistors connected to the +terminal, it is easier to assume first that we know v_+, find the output, and then find the actual value of v_+. That is, if we know v_+

$$v_- = v_+$$

The node equation at the −terminal gives us

$$-\frac{v_+}{R_A} = \frac{v_+ - v_o}{R_F}$$

FIGURE 24–10
*Non-inverting amplifier
with voltage divider input*

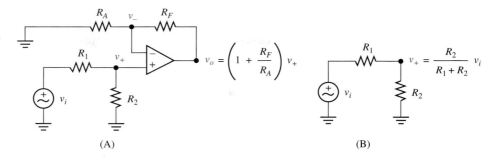

(A)   (B)

Solving for the output voltage

$$v_o = \left(1 + \frac{R_F}{R_A}\right)v_+$$

You might recognize this as the standard formula for the non-inverting gain, with v_+ as the input voltage. To find v_+ we recognize that because there is no current into the Op-Amp, R_1 and R_2 form a voltage divider (Figure 24–10B).

$$v_+ = \frac{R_2}{R_1 + R_2}\, v_i$$

We now have the complete picture:

$$v_o = \left(1 + \frac{R_F}{R_A}\right)\frac{R_2}{R_1 + R_2}v_i$$

This non-inverting amplifier results in a pretty messy formula. To eliminate the bias current effect, we would usually set

$$R_1\|R_2 = R_A\|R_F$$

For once the student gets a break here. If, *and only if,* the above relation is true, then the nasty formula for the output voltage here reduces to

$$R_1\|R_2 = R_A\|R_F \rightarrow v_o = \frac{R_F}{R_1}v_i$$

Now, that's better. This formula is as simple as the inverting amplifier formula. Note that the resistance seen by the source is no longer infinite. Source current now has a path to ground through R_2, making the input resistance here

$$R_{in} = R_1 + R_2$$

EXAMPLE 24–3
**Non-Inverting
Amplifier**

Find the output voltage for the circuit in Figure 24–11 (p. 798) if

(a) $R_2 = 1K\Omega$ **(b)** $R_2 = 4K\Omega$

Solution
(a) For $R_2 = 1K\Omega$

$$v_o = \left(1 + \frac{R_F}{R_A}\right)\frac{R_2}{R_1 + R_2}v_i = \left(1 + \frac{4K}{1K}\right)\frac{1K}{1K + 1K}v_i = 5 \times \frac{1}{2}v_i = 2.5v_i$$

Note that $R_1\|R_2 \neq R_A\|R_F$ here, and we are stuck with the long formula.

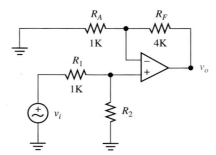

FIGURE 24–11

(b) For $R_2 = 4K\Omega$

$$v_o = \left(1 + \frac{R_F}{R_A}\right) \frac{R_2}{R_1 + R_2} v_i = \left(1 + \frac{4K}{1K}\right) \frac{4K}{1K + 4K} v_i = 5 \times \frac{4}{5} v_i = 4v_i$$

In this circuit $R_1 \| R_2 = R_A \| R_F$, so we could have used the simple formula:

$$v_o = \frac{R_F}{R_1} v_i = \frac{4K}{1K} v_i = 4v_i$$

The non-inverting summing amplifier is shown in Figure 24–12. The general formula for the output voltage in this circuit is quite hairy, so we will not give it to you. However, if R_A is chosen to eliminate bias current effects, that is, if

$$R_A \| R_F = R_1 \| R_2 \ldots \| R_n$$

then the non-inverting output is given by the simple formula:

$$v_o = \left(\frac{R_F}{R_1} v_1 + \frac{R_F}{R_2} v_2 + \ldots + \frac{R_F}{R_n} v_n\right)$$

FIGURE 24–12
Non-inverting summing amplifier

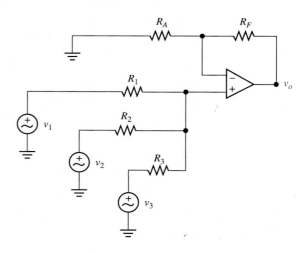

EXAMPLE 24–4
The Non-Inverting Summing Amplifier

Find the output voltage in the circuit of Figure 24–13.

Solution   We first find

$$R_A \| R_F = 5K \| 100K = 4.8K\Omega$$

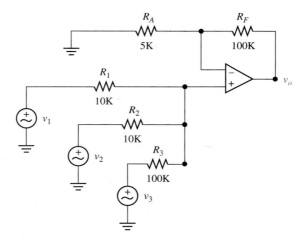

FIGURE 24–13

and

$$R_1 \| R_2 \| R_3 = 10K \| 10K \| 100K = 4.8K\Omega$$

Because $R_A \| R_F = R_1 \| R_2 \| R_3$, we can use the simple formula to find v_o:

$$v_o = \left(\frac{R_F}{R_1} v_1 + \frac{R_F}{R_2} v_2 + \frac{R_F}{R_3} v_3 \right) = 10v_1 + 10v_2 + v_3$$

Troubleshooting

If the connections of the feedback resistor are bad, or if the feedback resistor has failed open, the feedback pathway is effectively removed. This means that the Op-Amp is operating with its internal gain of approximately 1×10^5. With such a large gain, the output will be pegged at the saturation voltage, which is approximately equal to the supply voltage (Figure 24–14A).

If the Measured Voltage at the output is significantly different from the Expected Voltage (Figure 24–14B), most likely one of the signal sources is not properly connected or the resistor for the signal source has failed open. You may be able to tell from the expected gains and the given inputs which source is not properly connected. Otherwise, turn on the signal sources one at a time. When you get no output, you have found the bad connection.

FIGURE 24–14
Bad connections

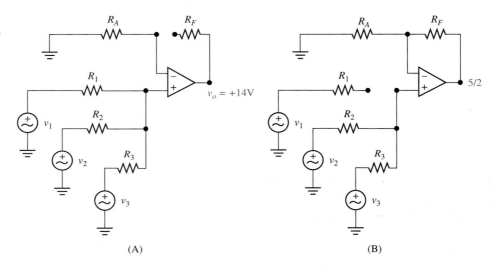

(A)   (B)

1. Draw the inverting summer, and write the formula for the output voltage.
2. Discuss the resistances seen by the sources in the inverting summer.
3. What resistance should we connect from the +terminal to ground to eliminate the effect of input bias currents?
4. For the 4-bit digital-to-analog converter, what are the ratios needed between R_F and the other resistors?
5. Draw a non-inverting amplifier with a voltage divider input. What is the output voltage?
6. What is the resistance seen by the source in the above circuit?
7. Draw the general non-inverting summer, and write the formula for the output voltage. What assumption have you made to write this formula?
8. What output do you expect if R_F or its connections are bad?
9. How can you tell if one of the input sources is not properly connected to a summer?

24–3 ■ DIFFERENCE (DIFFERENTIAL) AMPLIFIER

What happens if we connect inputs to both the positive and negative terminals of the Op-Amp? Let's first examine the circuit in Figure 24–15A.

We begin by recognizing that

$$v_+ = v_2$$

FIGURE 24–15
Difference amplifier

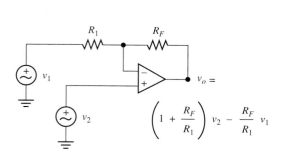

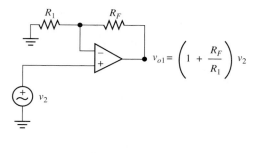

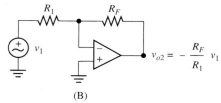

(A) (B)

so

$$v_- = v_+ = v_2$$

We now write the node equation at the negative terminal:

$$\frac{v_1 - v_2}{R_1} = \frac{v_2 - v_o}{R_F}$$

so

$$\frac{v_o}{R_F} = \left(\frac{1}{R_F} + \frac{1}{R_1}\right)v_2 - \frac{v_1}{R_1}$$

Solving for v_o:

$$v_o = \left(1 + \frac{R_F}{R_1}\right)v_2 - \frac{R_F}{R_1}v_1$$

This circuit takes the difference in the two input voltages. Note however, that the two input voltages are multiplied by slightly different gains.

Although we found the output here with our standard analysis procedure, it is often easier to use superposition to find the output of Op-Amp circuits. This procedure is demonstrated in Figure 24–15B. If we turn off v_2, we get the basic non-inverting amplifier and can write the output by inspection:

$$v_{o_2} = -\frac{R_F}{R_1}v_1$$

We now turn off v_1 and turn on v_2 to get the basic inverting amplifier. The output here is

$$v_{o_1} = \left(1 + \frac{R_F}{R_1}\right)v_2$$

The total output is the sum of v_{o_1} and v_{o_2}:

$$v_o = v_{o_1} + v_{o_2} = \left(1 + \frac{R_F}{R_1}\right)v_2 - \frac{R_F}{R_1}v_1$$

Differential Amplifier

To get a true differential amplifier, we can use the circuit of Figure 24–16. Using superposition is the fastest way to find the output. When we turn off v_1, we get the voltage divider non-inverting amplifier. Because we have purposely set the resistances at the positive and negative terminals to be equal, we can use the short formula:

$$v_{o_1} = \frac{R_F}{R_1}v_2$$

If we now turn off v_2 and turn on v_1, we get the inverting amplifier output:

$$v_{o_2} = -\frac{R_F}{R_1}v_1$$

Combining these answers we get the total output

$$v_o = \frac{R_F}{R_1}v_2 - \frac{R_F}{R_F}v_1 = \frac{R_F}{R_1}(v_2 - v_1)$$

FIGURE 24–16

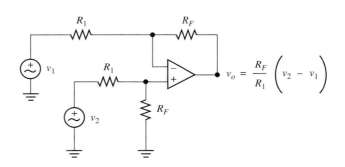

The Bridge Amplifier

As you might have seen in a circuits course, the bridge circuit shown in Figure 24–17A can be used to determine an unknown resistance. The circuit is in *balance* ($R_{unknown} = R_{pot}$) when $v_o = 0V$.

Figure 24–17B shows the implementation of this circuit with an Op-Amp, where R_{pot} represents a precision-calibrated potentiometer. The voltage applied to this circuit is a known DC reference voltage. In fact, this circuit is the same as the difference circuit of Figure 24–15, where

$$v_1 = v_2 = V_{ref}$$

FIGURE 24–17
Bridge amplifier

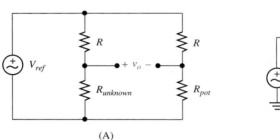

(A)

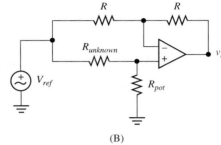

(B)

Therefore, we can use the previous result to find that

$$v_o = \left(1 + \frac{R}{R}\right)\frac{R_{pot}}{R_{unknown} + R_{pot}} V_{ref} - \frac{R}{R} V_{ref} = \frac{2R_{pot}}{R_{unknown} + R_{pot}} V_{ref} - V_{ref}$$

Factoring, we get

$$v_o = \left(\frac{2R_{pot}}{R_{unknown} + R_{pot}} - 1\right)V_{ref}$$

To find the unknown resistance, we adjust the potentiometer until the output voltage is 0V. An examination of the above formula shows that

$$v_o = 0V \rightarrow R_{unknown} = R_{pot}$$

EXAMPLE 24–5
Bridge Amplifier

The bridge amplifier can also be used to determine temperature, light level, strain gauge response, or the response of any sensor that works by changing its resistance. As an example, let's assume that we wish to measure changes in temperature in a fluid bath.

One possible circuit arrangement is shown in Figure 24–18, where the sensor is placed in the feedback pathway. Resistors that equal the nominal resistance of the sensor make up the rest of the circuit. In practice, if the nominal resistance of the sensor is small, then the sensor is placed in series with a fixed resistor, and R represents the series combination of the sensor resistor and the fixed resistor. As the temperature changes, the changes in the sensor resistance are represented with ΔR.

(a) Assume that when the temperature sensor is held at its reference temperature, $\Delta R = 0\Omega$. Find v_o.
(b) Find the output voltage when the temperature changes.
(c) Assume that the nominal resistance of the sensor is 100Ω and its resistance changes $-1\Omega/°C$. If $V_{ref} = 10V$, find how much the output voltage changes for each degree change in temperature.

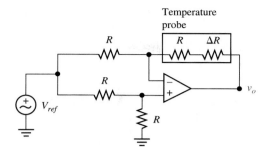

FIGURE 24–18

Solution We can use the results from the difference amplifier of Figure 24–15 (p. 800), where in this case

$$v_1 = v_2 = V_{ref}$$

(a) When $\Delta R = 0\Omega$, the bridge is in balance. We find

$$v_o = \left(1 + \frac{R}{R}\right)\frac{R}{R + R}\ V_{ref} - \frac{R}{R}V_{ref} = V_{ref} - V_{ref} = 0\text{V}$$

At the nominal temperature, there is no output voltage. Therefore, any changes in the output voltage must reflect changes in temperature.

(b) When we include ΔR in the equations, we get

$$v_o = \left(1 + \frac{R + \Delta R}{R}\right)\frac{R}{R + R}\ V_{ref} - \frac{R + \Delta R}{R}\ V_{ref}$$

So

$$v_o = \left(1 + 1 + \frac{\Delta R}{R}\right)\frac{V_{ref}}{2} - \left(1 + \frac{\Delta R}{R}\right)V_{ref}$$

Combining fractions and factoring, we get

$$v_o = \left(\frac{1}{2} + \frac{1}{2} + \frac{\Delta R}{2R} - 1 - \frac{\Delta R}{R}\right)V_{ref}$$

And we get the final result

$$v_o = -\frac{\Delta R}{2R}\ V_{ref}$$

Changes in the sensor resistance are reflected in changes in the output voltage. By calibrating the voltage changes with known temperature changes, we now have an effective electronic temperature probe.

(c) We know that $R = 100\Omega$, $V_{ref} = 10\text{V}$, and $\Delta R = -1\Omega/°\text{C}$. Putting these all together gives us

$$v_o = -\frac{\Delta R}{2R}\ V_{ref} = -\frac{-1}{200}\ 10 = 50\text{mV}/°\text{C}$$

That is, for each $1°$ increase in temperature, the output will increase by 50mV.

Troubleshooting

We troubleshoot the difference amplifier the same way we troubleshoot the summing amplifier. If the connections of the feedback resistor are bad or the feedback resistor has failed open, the output will be pegged at the saturation voltage.

If you suspect that one of the signal sources is not properly connected or that the resistor or that signal source has failed open, turn on the signal sources one at a time. When you get no output, you have found the bad connection.

REVIEW QUESTIONS

1. What are two methods you can use to find the output of a differential amplifier?
2. Draw the basic differential amplifier. What is its output?
3. Draw the differential amplifier with a voltage divider at the +terminal. What is its output?
4. Draw a bridge amplifier, and discuss what we mean by a balanced bridge amplifier. What is the output voltage here?
5. How can the bridge amplifier be used to determine an unknown resistance?
6. How can the bridge amplifier be used to determine temperature?
7. What do you expect to see in the output if the feedback resistor or its connections have failed open?

24–4 ■ DIFFERENTIATION AND INTEGRATION

Those of you who have had a calculus course know how differentiation and integration are defined. For those who haven't had such a course, you can still come to appreciate a plain word interpretation of these mathematical functions.

Differentiation

There are two ways to think about differentiation. Consider the graph in Figure 24–19A. The slope of the straight line is given by

$$r_e = \frac{\Delta y}{\Delta x}$$

For the straight line the slope is constant over the entire data range. That is, no matter how large or small Δx is, the slope is the same.

FIGURE 24–19

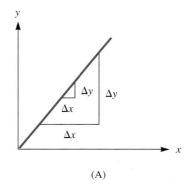

(A)

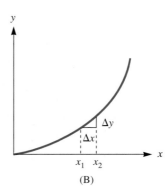

(B)

Consider now the graph in Figure 24–19B. Because this graph is not a straight line, you cannot find a slope with the above expression. However, if we let Δx get very small, then we can approximate the curve between x_1 and x_2 as a straight line. Mathematically, when Δx and Δy get very small, we write

$$slope = \frac{\Delta y}{\Delta x} \rightarrow slope = \frac{dy}{dx}$$

We say that the slope of y is given by the *differentiation* of y with respect to x, and dy/dx is known as the *derivative* of y.

When we gave you the formula for the transconductance (g_m) of the FET, we used the derivative of I_D. Here, we derive g_m only to show you an example of the use of derivatives; this is not a calculus textbook and we do not necessarily expect you to follow the math. The gate I-V curve of the FET is described by

$$I_D = I_{DSS}\left(1 - \frac{V_{GS}}{V_{GS(off)}}\right)^2$$

The slope of this curve is

$$slope = g_m = \frac{dI_D}{dV_{GS}} = -2\frac{I_{DSS}}{V_{GS(off)}}\left(1 - \frac{V_{GS}}{V_{GS(off)}}\right)$$

After a bit of fancy footwork, this reduces to

$$g_m = \frac{-2}{V_{GS(off)}}\sqrt{I_D I_{DSS}}$$

There is another, more practical interpretation of differentiation. You are stopped by a cop and asked if you know how fast you were going. Trying to impress the officer, you state that because it took you 5 seconds to cover the last 500 feet, your speed was just

$$speed = \frac{500\text{ft}}{5\text{s}} = 100\text{ft/s}$$

So you tell the officer "I was only going 100 feet per second." Unfortunately for you, you have been stopped by a cop who knows her math, and, while handing you your well-earned ticket, she lets you know that 100ft/s is approximately 70mi/hr.

The point of this little vignette is that you calculated your speed as

$$speed = \frac{\Delta distance}{\Delta time}$$

If you repeated this calculation over a very small change in time, this equation reduces to

$$v = \frac{dx}{dt}$$

where v is speed, x is distance, and t is time. A derivative, then, is also a measure of how fast the variable of interest is changing.

In Chapter 4 we reminded you that the I-V characteristic of the capacitor is given by

$$i_C = C\frac{\Delta v_C}{\Delta t}$$

We can use derivatives to express this relation as

$$i_C = C\frac{dv_C}{dt}$$

Now, consider the Op-Amp circuit in Figure 24–20 (p. 806). To find the output voltage, we first note that

$$v_- = v_+ = 0\text{V}$$

We now write the node equation at the −terminal:

$$i_C = \frac{0 - v_o}{R_F} = -\frac{v_o}{R_F}$$

Noting that $v_C = v_i - 0 = v_i$ and using the capacitor I-V relation, we can write

$$C\frac{dv_i}{dt} = -\frac{v_o}{R_F}$$

Rearranging, we get

$$v_o = -R_F C \frac{dv_i}{dt}$$

and we have built an electronic **differentiator.** The output of the Op-Amp gives us the rate of change of the input. Note that the unit of $R_F C$ is seconds, so the output is a true voltage. If the input voltage represents the distance traveled, the output voltage represents the speed (velocity) of travel; if the input voltage represents the speed (e.g. from an electronic speedometer), the Op-Amp output gives us the acceleration.

A Note of Caution

Figure 24–20 is a helpful illustration, but we never actually built it. The problem is that differentiation enhances high-frequency signals, and most noise is high frequency. For those of you with some calculus background consider the voltage

$$v = 10 \sin 10t + 0.1 \sin 1000t$$

FIGURE 24–20
Differentiator

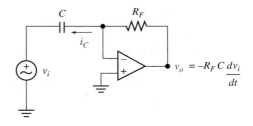

where the low-frequency voltage is the signal and the high-frequency voltage is the noise. We rate how "clean" a signal is by the signal-to-noise ratio *(S/N)*. Here

$$S/N = \frac{10}{0.1} = 100$$

That is, the signal is 100 times stronger than the noise. Now, let's differentiate the voltage:

$$\frac{dv}{dt} = 100 \cos 10t + 100 \cos 10t$$

and the signal-to-noise ratio has become

$$S/N = \frac{100}{100} = 1$$

After differentiation, the noise is now as large as the signal! We can build differentiators if we first filter the input signal to reduce the high-frequency noise. We will show you this in a later chapter.

EXAMPLE 24–6
Differentiation

For the Op-Amp circuit in Figure 24–21

(a) Find the expression for the output voltage.
(b) Find the output voltage for the given input.
(c) What real Op-Amp limitation must you be concerned about in this circuit?

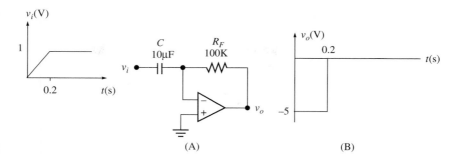

FIGURE 24–21

(A)   (B)

Solution

(a) For this differentiator the output voltage is given by

$$v_o = -R_F C \frac{dv_i}{dt} = -100 \times 10^3 \times 10^{-6} \frac{dv_i}{dt} = -\frac{dv_i}{dt}$$

(b) Because the input voltage is a straight line between 0.0 and 0.2s, we get

$$v_o = -\frac{dv_i}{dt} = -\frac{1.0 - 0.0}{0.2 - 0.0} = -\frac{1.0}{0.2} = -5.0\text{V for } 0.0 < t < 0.2\text{s}$$

For $t > 0.2$s, the input is constant; that is, the input has 0 slope, so

$$v_o = \frac{dv_i}{dt} = 0\text{V for } t > 0.2\text{s}$$

The output is shown in Figure 24–21B.

(c) We must ensure that the *slew rate* of the Op-Amp is fast enough that the output reaches −5V in an acceptable amount of time.

DRILL EXERCISE

For the circuit of Figure 24–21, find the output voltage between 0.0s and 0.2s, if

(a) $R_F = 10\text{K}\Omega$
(b) $R_F = 1\text{M}\Omega$ (Assume that the Op-Amp has a saturation voltage of ±15V.)

Answer   (a) $v_o = -0.5$V
             (b) $v_o \approx -15$V (output voltage cannot reach −50V)

Integration

There are two interpretations of **integration.** Integrating a curve between two points will give us the area under the curve. We can also find the volume of solids by integration. Of more interest to us here is that integration can also be thought of as the inverse of differentiation. That is, if we can find velocity by differentiating distance

$$v = \frac{dx}{dt}$$

we can find the distance traveled over time by integrating the velocity over the time traveled:

$$x = \int_0^t V(t)\, dt$$

Using this principle is how we navigate all of our airplanes, satellites, ships, etc. The output of velocity sensors are integrated to provide position information.

FIGURE 24–22
Integrator

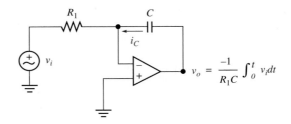

Figure 24–22 shows how we can construct an electronic integrator with the Op-Amp. Noting that the voltage at the negative terminal is 0V, we write the node equation

$$\frac{v_i - 0}{R_1} = i_C$$

Because the capacitor voltage is now

$$v_C = 0 - v_o = -v_o$$

the capacitor current is

$$i_C = C\frac{dv_C}{dt} = -C\frac{dv_o}{dt}$$

And the node equation becomes

$$\frac{v_i}{R_1} = -C\frac{dv_o}{dt}$$

So

$$\frac{dv_o}{dt} = -\frac{1}{R_1 C}v_i$$

To find the output voltage, we integrate both sides of the equation to get

$$v_o = -\frac{1}{R_1 C}\int_0^t v_i\,dt$$

Integration does not have the noise problem associated with differentiation, so this circuit is actually used. In fact, the Op-Amp integrator is the basic element in the analog computer (if you can still find one).

EXAMPLE 24–7
Op-Amp
Integration

In Figure 24–23, given the Op-Amp integrator and the square wave input shown, find the output voltage.

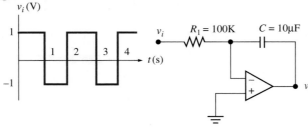

(A)

(B)

FIGURE 24–23

Solution The output voltage of this circuit is given by

$$v_o = -\frac{1}{R_1 C}\int_0^t v_i\, dt = -\int_0^t v_i\, dt$$

When we integrate a constant, we get a straight line. Therefore, the integration of the square wave results in the triangle wave shown in Figure 24–23B. Once again, in the real world the slew rate of your Op-Amp must be fast enough to support the ramp output shown.

DRILL EXERCISE

Draw the output of the Op-Amp in Figure 24–23 if $C = 1\mu F$.

Answer The output is still a triangle wave as shown in Figure 24–23B. Now, however, the peak voltages is $-10V$.

Troubleshooting

For the differentiator: If the feedback resistor has failed open or its connections are bad, the output will be close to the supply voltage. If the capacitor has failed open, there will be no output. If the capacitor has failed short, the signal source is connected directly to the −terminal, so the output will saturate.

   For the integrator: If the resistor has failed open, there will be no appreciable output. If the capacitor has failed open, the feedback path is removed, so the output will saturate at the supply voltage. If the capacitor fails short, the output will be directly connected to the −terminal and will therefore be held at 0V.

REVIEW QUESTIONS

1. Give two interpretations of differentiation.
2. Draw an Op-Amp differentiator, and write the formula for the output voltage.
3. Why do we avoid differentiation?
4. Give two interpretations of integration.
5. Draw an Op-Amp integrator, and write the formula for the output voltage.
6. When we integrate or differentiate, what practical Op-Amp parameter must we consider? Why?
7. What output voltage will you see if the capacitor in the differentiator has failed open? Failed short?
8. What output voltage will you see if the feedback capacitor in the integrator has failed open? Failed short?

24–5 ■ IMPROVED VOLTAGE REGULATION

In Chapter 5 we showed you how the Zener diode could be used to provide a well-regulated DC voltage. We give you a quick review here with Figure 24–24A (p. 810). When there is no load, $R_L \to \infty$:

$$I_{Z_{\text{no-load}}} = \frac{15 - V_Z}{R} = \frac{15 - 5}{5K} = 2mA$$

This no-load Zener current is the amount of current that can be supplied to a load before the diode stops Zenering and we lose voltage regulation. As the load resistor decreases, we draw increasing current away from the Zener. The minimum load that we can supply with a constant 5V is

$$R_{L_{min}} = \frac{V_Z}{I_{Z_{no\text{-}load}}} = \frac{5}{2\text{mA}} = 2.5\text{K}\Omega$$

To improve the regulation of the Zener circuit we can place a non-inverting Op-Amp amplifier between the Zener and the load. For example, in Figure 24–24B we have used a unity-gain buffer. Because there is almost no current into the Op-Amp, no current is drawn away from the Zener. The load voltage is, therefore, independent of the load and is

$$V_L = V_Z = 5\text{V}$$

FIGURE 24–24
Improved Zener voltage regulation

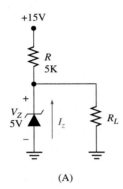
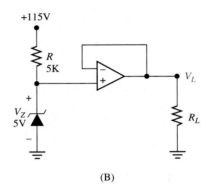

(A) (B)

EXAMPLE 24–8
Buffered Zener with Voltage Divider

Figure 24–25 shows a Zener circuit connected to a voltage divider circuit, which in turn is connected to a unity-gain buffer.

(a) What is the no-load Zener current?
(b) How much current is drawn from the Zener by the voltage divider?
(c) What is the load voltage?

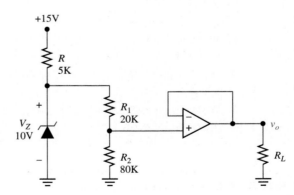

FIGURE 24–25

Solution
(a) The no-load Zener current is given by

$$I_{Z_{no\text{-}load}} = \frac{15 - 10}{5\text{K}} = 1\text{mA}$$

This is the maximum current that can be supplied to the Zener load, the voltage divider.

(b) The current drawn from the Zener is the current in the series combination of R_1 and R_2:

$$I_{R_1+R_2} = \frac{V_Z}{R_1 + R_2} = \frac{10}{20K + 80K} = 0.1mA$$

Because the current drawn from the Zener is one-tenth of the no-load current, we have a well-regulated voltage supply.

(c) The load voltage is equal to the voltage at the +terminal, so

$$V_L = v_+ = \frac{R_2}{R_1 + R_2} V_Z = \frac{80K}{20K + 80K} 10 = 8V$$

With the aid of the Op-Amp buffer, we can produce any regulated load voltage that we desire (as long as it is smaller than V_Z).

DRILL EXERCISE Choose values for R_1 and R_2 in Figure 24–25 to produce a load voltage of 3V. Make sure that the series combination of these two resistors is still 100KΩ.

Answer $R_1 = 70K\Omega, R_2 = 30K\Omega$

Troubleshooting

If you are not getting the expected output voltage, check the voltage at the Zener diode and the voltage divider. If the Zener diode has failed open, you will see a voltage significantly different from V_Z. If R has failed open, then there will be no Zener current and no Zener voltage. If R has failed short, you will see the supply voltage at the Zener diode. If the voltage at the Zener diode is correct but the voltage at the voltage divider output is not, the problem is with one of the resistors in the voltage divider.

REVIEW QUESTIONS

1. What is the maximum current that can be drawn from a Zener diode?
2. What is the limitation on a load resistor being supplied by a Zener diode?
3. How can we use an Op-Amp to provide improved Zener regulation?
4. How can we provide a regulated voltage that is smaller than the Zener diode voltage?
5. What will you find if the Zener diode has failed open?
6. What will you find if the Zener supply resistor has failed open? Failed short?

SUMMARY

■ The Op-Amp analysis procedure is

☐ Find v_+ (or assume we know v_+, then find it at the end of the analysis).
☐ Set $v_- = v_+$.
☐ Write a node equation at the −terminal.
☐ Solve for v_o.

■ The inverting output is

$$v_o = -\frac{R_F}{R_1}$$

- The inverting summer output is

$$v_o = -\left(\frac{R_F}{R_1}v_1 + \frac{R_F}{R_2}v_2 + \ldots + \frac{R_F}{R_n}v_n\right)$$

- The non-inverting output is

$$v_o = \left(1 + \frac{R_F}{R_A}\right)v_1$$

- To eliminate the bias current effect, the total resistance at the $+$terminal must equal the total resistance at the $-$terminal.
- If the above is true, that is, if

$$R_1 \| R_2 \ldots \| R_n = R_A \| R_F$$

then the output of the non-inverting summer is

$$v_o = \frac{R_F}{R_1}v_1 + \frac{R_F}{R_2}v_2 + \ldots + \frac{R_F}{R_n}v_n$$

- We build a difference amplifier by applying an input voltage at both Op-Amp terminals. A voltage divider is used at the $+$terminal to ensure that the positive and negative gains are the same.
- We can build an electronic differentiator by placing a capacitor between the signal source and the Op-Amp.
- We can build an electronic integrator by placing a capacitor in the feedback path of the inverting Op-Amp circuit.
- We can improve Zener voltage regulation by placing a unity-gain buffer between the Zener diode and the load.

PROBLEMS

SECTION 24–1   Introduction

1. For the inverting amplifier in Figure 24–26, let $R = 0\Omega$ and find in the following order

   v_+

   v_-

   v_o

   i_i

   i_f

   $R_i = \dfrac{v_i}{i_i}$

2. For the inverting amplifier in Figure 24–26, let $R = 4\text{K}\Omega$ and find in the following order

   v_+

   v_-

   v_o

   i_i

   i_f

   $R_i = \dfrac{v_i}{i_i}$

FIGURE 24–26

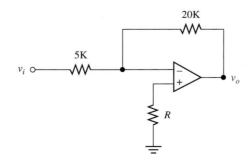

3. The input source in Figure 24–26 has an internal resistance of 5KΩ.

   (a) Redraw the circuit to include this resistor, and find v_o.

   (b) How can you prevent the internal resistance of the source from affecting the output voltage?

4. For the non-inverting amplifier in Figure 24–27, let $R = 0\Omega$, and find in the following order

   v_+

   v_-

   v_o

   i_i

   i_f

   $R_i = \dfrac{v_i}{i_i}$

FIGURE 24–27

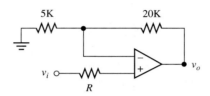

5. For the non-inverting amplifier in Figure 24–27, let $R = 4\ K\Omega$, and find in the following order

   v_+

   v_-

   v_o

   i_i

   i_f

   $R_i = \dfrac{v_i}{i_i}$

6. Figure 24–28 shows a unity-gain buffer amplifier feeding an inverting amplifier.

   (a) Find v_{o_1}. Note that you can ignore the second Op-Amp here.

   (b) Find v_o. Consider the output voltage of the first Op-Amp as the input to the second Op-Amp.

   (c) What resistance is seen by the signal source?

7. Figure 24–29 shows a three Op-Amp circuit. To find the output of each Op-Amp, you can ignore the Op-Amp to which it is connected.

   (a) Find v_{o_1}.   (b) Find v_{o_2}.   (c) Find v_o.

FIGURE 24–28

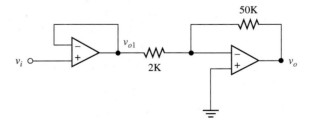

FIGURE 24–29

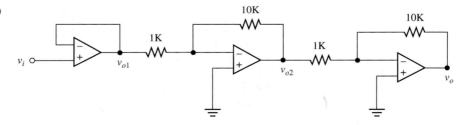

SECTION 24–2   Summing Amplifiers

8. For the inverting summer in Figure 24–30, find v_o if
   (a) $v_2 = 0V$.   **(b)** $v_1 = 0V$.   **(c)** Both input sources are turned on.
   (d) Find R to eliminate the effect of the bias currents.
9. (a) Find R_1 and R_2 in Figure 24–30 so that $v_o = -(10v_1 + v_2)$
   (b) Find R to eliminate the effect of the bias currents.

FIGURE 24–30

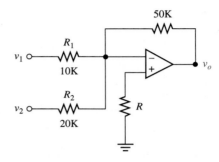

10. Figure 24–31 shows another way to combine two signals.
   (a) Find v_{o_1} (ignore the second Op-Amp).
   (b) Find v_{o_2} (treat v_{o_1} as an input to the second Op-Amp).
   (c) What have we built with this circuit?

FIGURE 24–31

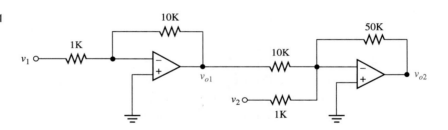

11. For Figure 24–32, let $R = 10K\Omega$, and find in the following order
   v_+
   v_-
   v_o (Short form does not apply here.)
12. For Figure 24–32, let $R = 5K\Omega$, and find in the following order
   v_+
   v_-
   v_o (Does the short form apply here?)

FIGURE 24–32

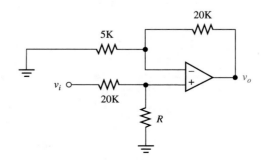

13. For Figure 24–33, let $R = 10\text{K}\Omega$

    (a) Does $R \parallel R_F = R_1 \parallel R_2$?   **(b)** Find v_o.

14. Use Figure 24–33 to design a circuit so that

$$v_o = 5v_1 + 10v_2$$

    (Hint: Choose R_1 and R_2 to get the gains; then choose R so that $R \parallel R_F = R_1 \parallel R_2$).

FIGURE 24–33

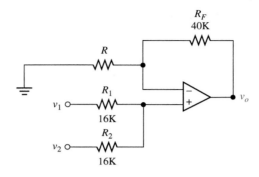

15. For the circuit in Figure 24–34, let $R = 2\text{K}\Omega$, and find v_o if

    (a) $v_2 = 0\text{V}$.   **(b)** $v_1 = 0\text{V}$ (Use the long form here).   **(c)** Both inputs are on.

16. For the circuit in Figure 24–34, let $R = 10\text{K}\Omega$ and find v_o if

    (a) $v_2 = 0\text{V}$.   **(b)** $v_1 = 0\text{V}$ (Use the short form here).   **(c)** Both inputs are on.

FIGURE 24–34

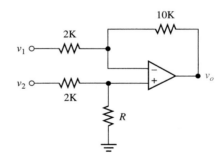

17. The Op-Amp can be used to construct a **current source** by placing the load resistor in the feedback path. Consider the basic inverting Op-Amp circuit shown in Figure 24–35.

    (a) Let $v_i = 1\text{V}$ and $R_1 = 1\text{K}\Omega$. Find the current in R_1.

    (b) Where does all of this current go?   **(c)** What is the current in R_L?

    (d) Does this current depend on R_L?

    (e) What have we built if the current delivered to the load does not depend on the load resistance?

    (f) How is the load current determined in this circuit?

FIGURE 24–35

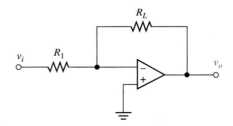

18. Find the output in Figure 24–36 as follows:

    (a) Find the Thevenin equivalent of the input circuit (left of dashed line).

    (b) Replace the input circuit with its Thevenin equivalent, and find v_o.

19. Find v_o in terms of v_i for the Op-Amp amplifier of Figure 24–37. Note that the input voltage can be written as

$$v_i = v_1 - v_2$$

    (a) Ground v_2, and find v_o in terms of v_1.   **(b)** Ground v_1, and find v_o in terms of v_2.

    (c) Find the total output in terms of v_1 and v_2.   **(d)** Find v_o in terms of v_i.

FIGURE 24–36

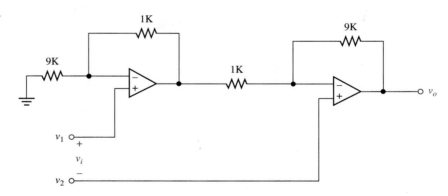

FIGURE 24–37

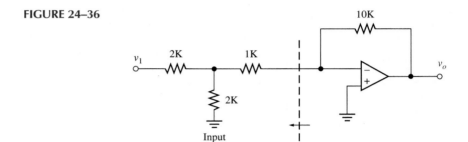

SECTION 24–3   Difference (Differential) Amplifier

20. Find R in Figure 24–34 (p. 815) to produce a differential amplifier where $v_o = 5(v_2 - v_1)$.

21. The input sources to the differential amplifier you just built have internal resistances in the kilohm range. Build a circuit that will eliminate the effect of these resistances.

22. For the bridge circuit in Figure 24–38

    (a) Find v_o if $\Delta R = 0\Omega$.   **(b)** Find v_o if $\Delta R = 10\Omega$.

FIGURE 24–38

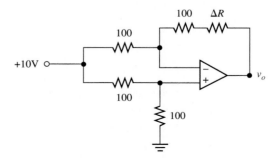

SECTION 24–4   Differentiation and Integration

23. For the differentiator in Figure 24–39
   (a) Find the general expression for v_o.   (b) Find v_o if $v_i = \cos 10t$.
   (c) Find v_o if $v_i = \sin 10t$.   (d) Find v_o if $v_i = e^{-2t}$

24. For the integrator in Figure 24–40
   (a) Find the general expression for v_o.   (b) Find v_o if $v_i = \cos 10t$.
   (c) Find v_o if $v_i = \sin 10t$.   (d) Find v_o if $v_i = e^{-2t}$.

25. The Op-Amp in Figure 24–40 has a saturation voltage of ± 10V.
   (a) Find v_o if $v_i = 1$V.   (b) Find v_o if $v_i = 5$V.

FIGURE 24–39

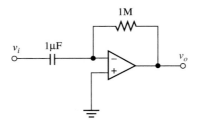

FIGURE 24–40

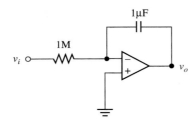

SECTION 24–5   Improved Voltage Regulation

26. For the Zener circuit of Figure 24–41A
   (a) Find the Zener no-load current.   (b) What is the minimum allowable R_L?
   (c) Discuss the effect of the buffer amplifier in Figure 24–37B.

FIGURE 24–41

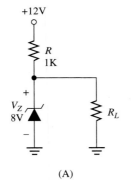

(A)

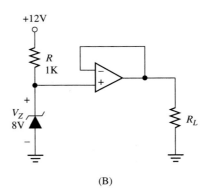

(B)

27. For the circuit of Figure 24–42 (p. 818)
   (a) Find the Zener no-load current.
   (b) Find the current in the series combination of $R_1 + R_2$. Compare this current to the Zener no-load current.
   (c) Find v_+.   (d) Find v_o.

FIGURE 24–42

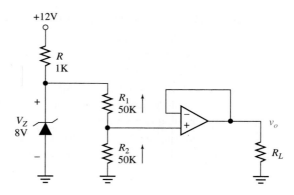

Troubleshooting Problems

28. Figure 24–43 shows Op-Amp circuits with the Estimated and Measured Voltages shown at the +terminal and the output. In each case, describe a problem that could cause the discrepancies.

FIGURE 24–43

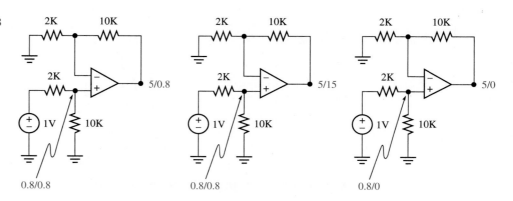

29. Figure 24–44 shows Op-Amp circuits with the Estimated and Measured Voltages shown at the output. In each case, describe a problem that could cause the discrepancies.

FIGURE 24–44

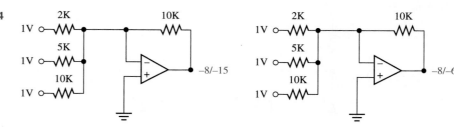

FIGURE 24–45

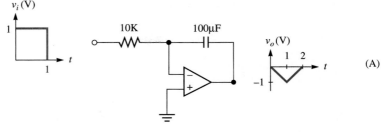

30. Figure 24–45A shows an integrator circuit with a pulse input and its triangle output. Figures 24–45B and C show two unexpected outputs. In each case, describe a problem that could cause these outputs.

Computer Problems

31. Use the computer to answer Problem 5 (p. 813).

32. Simulate the circuit of Figure 24–30 (p. 814) (with R = 0Ω).

    (a) Set $v_1 = 0.5V$, $v_2 = 0V$, and find the output voltage.

    (b) Set $v_1 = 0V$, let v_2 be a 1V 1KHz sinewave, and find the output voltage.

    (c) Set $v_1 = 0.5V$, let v_2 be a 1V 1KHz sinewave, and find the output voltage.

33. Simulate the circuit of Figure 24–31 (p. 814), and repeat the procedure of Problem 10.

34. Simulate the circuit of Figure 24–38 (p. 816). Find the output voltage as you vary ΔR from 0 to 100Ω. Plot v_o vs. ΔR.

35. Simulate the integrator of Figure 24–40 (p. 817). Apply the various inputs available in your program and record the output voltage.

Chapter

25

Non–Linear Operational Amplifier Circuits

To build and troubleshoot inverting Op-Amp rectifiers.

To build and troubleshoot non-inverting Op-Amp rectifiers.

To build and troubleshoot improved Op-Amp limiter circuits.

To determine the behavior of the open-loop Op-Amp.

To build zero-crossing detectors.

To build and troubleshoot comparators (level detectors).

To build and troubleshoot the Schmitt trigger.

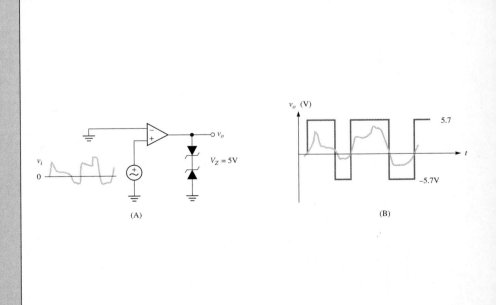

(A)

(B)

25–1 ■ INTRODUCTION

In the preceding chapter, we showed you how to use the Op-Amp to build amplifiers, summers, differencers, integrators, and differentiators. These circuits all share one thing in common—they are linear applications of the Op-Amp. In a linear circuit a sine wave input produces a sine wave output (Figure 25–1A). In a non-linear circuit a sine wave input produces a non-sine wave output (Figure 25–1B). Figure 25–2 shows the input-output responses of the specific non-linear circuits we will discuss in this chapter. We begin by showing you how the Op-Amp can be used to produce improved versions of the diode rectifier and limiter circuits. We will then show you how we can take advantage of the high open-loop gain of the Op-Amp to construct zero-crossing detectors, comparators (level detectors), and the Schmitt trigger.

FIGURE 25–1

FIGURE 25–2

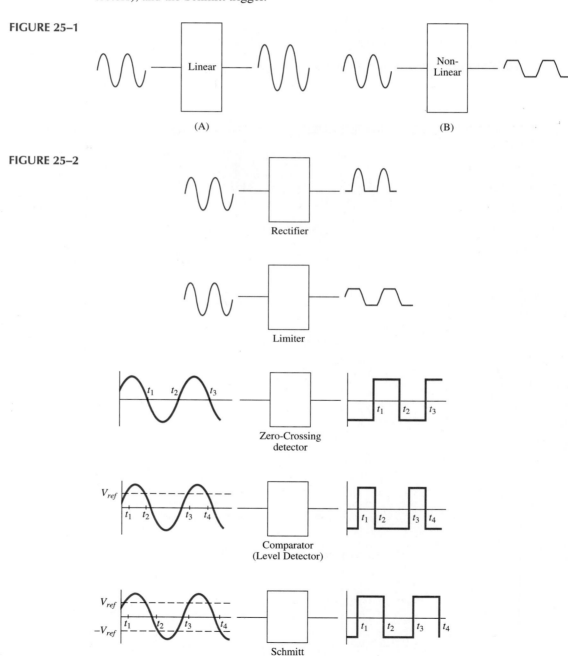

The circuits covered in this chapter are far from an exhaustive list of the non-linear Op-Amp circuits that you may encounter. Our goal is to give you a firm enough grounding that you can confidently read the many fine Op-Amp applications books that are available.

25–2 ▪ RECTIFIERS

The behavior of real diode rectifiers depends on signal source and load resistances. The Op-Amp can be used to improve on these circuits by isolating the signal source and the load. We can also construct an Op-Amp rectifier to eliminate the voltage loss due to diode turn-on voltage. To keep this chapter short and relatively simple, we discuss only the half-wave rectifier.

Simple Inverting Rectifier

Figure 25–3 shows how we can use a diode in the feedback path of the basic inverting amplifier to produce a rectifier. The details of this circuit can be determined by examining it when the diode is **OFF** and **ON.**

FIGURE 25–3
Inverting rectifier

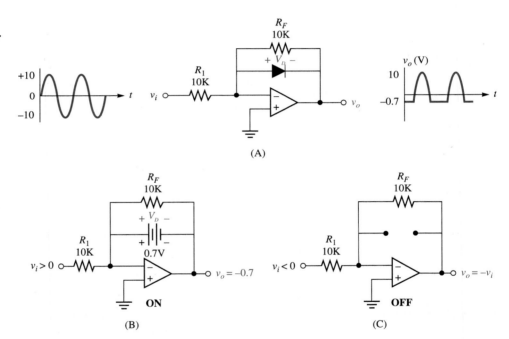

When the input goes positive, current is drawn towards the supply, turning the diode **ON** (Figure 25–3B). As before, when the feedback path is closed, whether through R_F or the diode,

$$v_- = v_+$$

and because the +terminal is grounded

$$v_- = v_+ = 0V$$

When the diode is **ON** it acts as a 0.7V battery. The output voltage then is fixed at

$$v_o = 0 - V_D = -0.7V$$

When the input voltage goes negative, current goes from the source towards the output. This is the reverse direction for the diode, so when the input goes negative, the diode is

OFF (Figure 25–3C). When the diode is **OFF,** the circuit behaves as a basic inverting amplifier and the output is given by

$$v_o = -\frac{R_F}{R_1}v_i = -\frac{10K}{10K}v_i = -v_i$$

Because the input is negative, the output goes positive. Combining these results, we get the output shown in Figure 25–3A. We have built an inverting half-wave rectifier.

EXAMPLE 25–1
Simple Inverting Rectifier

Find the output voltage of the inverting rectifier in Figure 25–4. Include the diode turn-on voltage of 0.7V.

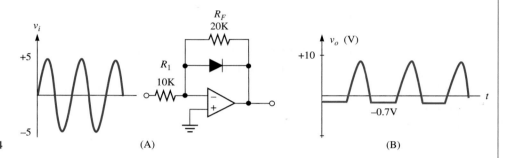

FIGURE 25–4

(A) (B)

Solution   When the input goes positive, current is drawn towards the source so the diode is **ON** and the output is held at

$$v_o = -0.7V$$

When the input goes negative, the diode is **OFF** and we can use the inverting amplifier formula:

$$v_o = -\frac{R_F}{R_1}v_i = -\frac{20K}{10K}v_i = -2v_i$$

Note that we can use the Op-Amp to amplify the signal. The response is shown in Figure 25–4B.

DRILL EXERCISE

Reverse the diode in Figure 25–4. When is the diode **ON? OFF?** What is the output voltage?

Answer   The diode is **ON** when the input is negative and **OFF** when the input is positive. The output voltage is an "upside down" version of the output in Figure 25–4B.

Precision Rectifiers

The simple inverting rectifier of Figure 25–4 isolates the load from the source and, if desired, also provides gain. However, it has the same disadvantage as the simple diode rectifier: the output voltage is affected by the diode turn-on voltage. We can build rectifiers, known as **precision rectifiers,** that eliminate the diode turn-on voltage.

Non-Inverting Precision Rectifier

A non-inverting precision rectifier is shown in Figure 25–5A. Note that in this circuit we do not take the load voltage directly from the Op-Amp output. The load voltage is actually the voltage at the −terminal.

FIGURE 25–5
Non-inverting rectifier

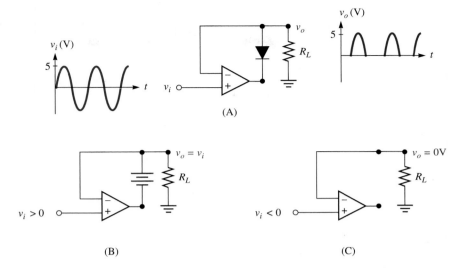

(A)

(B)   (C)

When the input voltage goes positive, the Op-Amp output voltage goes positive, turning the diode **ON** (Figure 25–5B). This closes the loop, and we know that when the loop is closed

$$v_- = v_+$$

Therefore, when the input goes positive

$$v_o = v_- = v_+ = v_i$$

and the output voltage tracks the input voltage.

When the input voltage goes negative, the output voltage goes negative, turning the diode **OFF.** When the diode is **OFF,** the feedback pathway is effectively removed (Figure 25–5C). Because the current into the Op-Amp is near 0A, there is no current to the load, so the output voltage goes to 0V.

The total output voltage is shown in Figure 25–5A. Note that the diode turn-on voltage is not seen in the output. There are two disadvantages to this circuit:

1. When the feedback pathway is removed, the differential input voltage is no longer 0V. In fact, v_x can get large enough to damage Op-Amps that do not have internal protective circuitry.

2. When the feedback pathway is removed, the output voltage is pegged at the saturation voltage. In a real Op-Amp, it takes a finite time to come out of **SATURATION.** This circuit may not be able to follow a high-frequency input signal.

EXAMPLE 25–2

Non-Inverting Rectifier

(a) Find the output of the non-inverting rectifier shown in Figure 25–6A.
(b) If the Op-Amp saturation voltage is ±14V, what is the peak inverse voltage *(PIV)* the diode will experience?

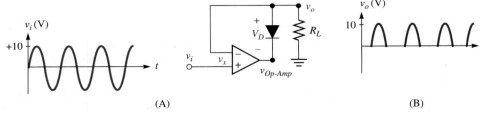

(A)   (B)

FIGURE 25–6

Solution

(a) Because this is a precision rectifier, we get the true half-wave rectified signal shown in Figure 25–6B (p. 825).

(b) The diode voltage is given by

$$v_D = v_o - v_{Op\text{-}Amp}$$

When the diode is **OFF,** we know that $v_o = 0V$. Also, because the loop opens when a negative voltage is applied to the input (Figure 25–5C, p. 825), the Op-Amp output will be pegged at its negative saturation voltage. Therefore, the peak inverse will be

$$PIV = v_{Op\text{-}Amp} - v_o = -14 - 0 = -14V$$

Inverting Precision Rectifier

Figure 25–7A shows an inverting Op-Amp precision rectifier. In this circuit there are two feedback pathways. Because one of the two diodes is always **ON,** there is always an intact feedback loop. This prevents the Op-Amp from saturating, avoiding the disadvantages of the non-inverting rectifier.

FIGURE 25–7
Precision rectifier

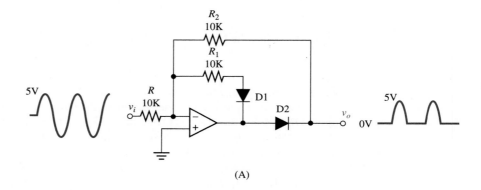

(A)

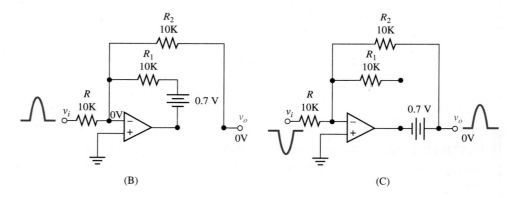

(B)   (C)

When the input voltage swings positive, the Op-Amp output voltage swings negative. This applies a negative voltage at the cathode of D1 and the anode of D2. This means that D1 is **ON** and D2 is **OFF** (Figure 25–7B). With D2 **OFF,** the output is connected through R_2 to the virtual ground at the −terminal. Therefore, the output voltage is held at 0V.

When the input voltage swings negative, the Op-Amp output voltage swings positive. This puts a positive voltage at the cathode of D1 and the anode of D2: D1 is

OFF and D2 is **ON** (Figure 25–7C). The output voltage now is given by the inverting gain formula:

$$v_o = -\frac{R_2}{R} v_i = -\frac{10K}{10K} = -v_i$$

Because the input is on its negative half-cycle, the output will be a positive half-wave rectified signal (Figure 25–7A). Although we have set the gain to 1, it is also possible to use this circuit to amplify the input voltage.

EXAMPLE 25–3
Inverting Precision Rectifier

We can also take the output from between R_1 and D1 of the inverting precision rectifier (Figure 25–8).

(a) When the input voltage swings positive, what happens to the Op-Amp output voltage?
(b) During the positive half-cycle of the input, which diode is **ON?** Which diode is **OFF?**
(c) What is v_o during the positive half-cycle of the input?
(d) When the input voltage swings negative, what happens to the output voltage?
(e) During the negative half-cycle of the input, which diode is **ON?** Which diode **OFF?**
(f) What is v_o during the negative half-cycle of the input?
(g) Sketch the input and output voltage waveforms on the same graph.

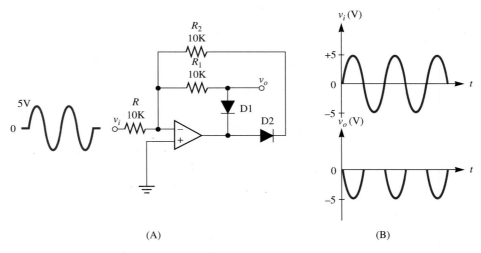

FIGURE 25–8

(A)   (B)

Solution
(a) Because this is an inverting amplifier, when the input voltage swings positive, the Op-Amp output voltage swings negative.
(b) Therefore, a negative voltage is applied to the cathode of D1 and the anode of D2. This means that D1 is **ON** and D2 is **OFF.**
(c) With D1 **ON,** the output voltage is given by the inverting gain formula:

$$v_o = -\frac{10K}{10K} v_i = -v_i$$

Because v_i is positive, the output voltage will swing negative.
(d) When the input swings negative, the Op-Amp output swings positive.

(e) A positive voltage is applied to the cathode of D1 and the anode of D2; D1 is now **OFF** and D2 is now **ON.**

(f) The output is connected to the virtual ground through R_1, so $v_o = 0$V. Note that a feedback pathway is still intact (R_2 and D2).

(g) The input and output voltages are shown in Figure 25–8B (p. 827).

Troubleshooting

Possible element failures and the resulting Op-Amp output are listed below. Because resistors rarely fail short, we exclude this possibility.

Inverting Rectifier

R_1 *failed open:* There is no input to Op-Amp, so the output voltage is 0V.

R_F *failed open* (Figure 25–9A): When the diode is **OFF,** the feedback path is removed, so the Op-Amp output will saturate.

Diode failed short: Op-Amp output is connected directly to the −terminal, so output voltage is 0V, independent of the input.

Diode failed open (Figure 25–9B): The circuit acts as simple inverting amplifier, so $v_o = -v_i$.

Non-inverting Rectifier

Diode failed short: The output is directly connected to −terminal, so circuit acts as unity-gain buffer, and $v_o = v_i$.

Diode failed open (Figure 25–9C): The feedback pathway is disabled, so v_{Op-Amp} will bounce between + and − supply voltage, and v_o will remain fixed at approximately 0V.

FIGURE 25–9

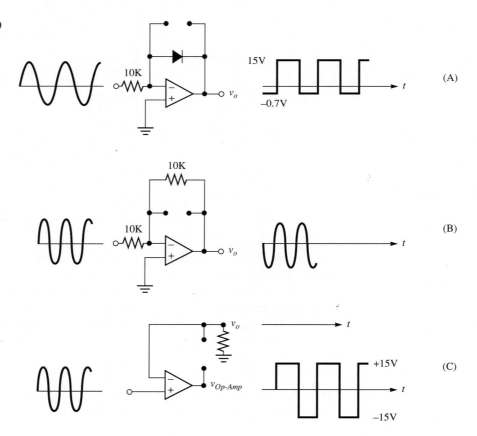

1. Draw the inverting rectifier circuit. If the input is a 1V sine wave, draw the input and output voltage on the same graph.

2. What is a disadvantage of this circuit?

3. Reverse the diode in your circuit, and repeat Question 1.

4. Draw the non-inverting rectifier circuit. If the input is a 1V sine wave, draw the input and output voltage on the same graph.

5. What are some disadvantages of this circuit?

6. Describe the expected output voltages of the inverting rectifier if the resistors fail open.

7. Describe the expected output voltages of the non-inverting rectifier if the diode fails open.

25–3 ▪ LIMITERS

Op-Amp Limiter Circuit

Figure 25–10 shows an Op-Amp limiter circuit. Let's see what happens when the input goes positive. Remember, a Zener diode operates in one of three states: **OFF** (open circuit), **ON** (0.7V battery), and Zenering (battery equal to V_Z).

FIGURE 25–10
Zener diode limiting

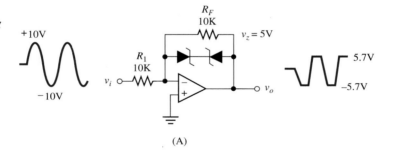

(A)

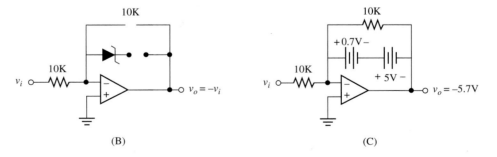

(B)   (C)

When the input goes positive, current is drawn towards the source. This is the forward direction for Z1 and the reverse direction for Z2. As the input begins to increase from 0V, there is not enough reverse-bias voltage across Z2 to put it into breakdown, so it acts as a regular reverse-biased diode and is open. Because Z2 is open, the diodes are removed from the feedback pathway (Figure 25–10B) and

$$v_o = -\frac{R_F}{R_1} v_i = -\frac{10K}{10K} v_i = -v_i$$

As the input continues to increase, the reverse-bias voltage on Z2 reaches its breakdown voltage, and the diode acts as a battery. Because the current in Z2 is heading towards the supply, Z1 is forward biased and turns **ON.** The feedback pathway now effectively contains two batteries in series (Figure 25–10C, p. 829). Because $v_- = 0V$, the output voltage becomes

$$v_o = v_- - 0.7 - V_Z = 0.0 - 0.7 - 5.0 = -5.7V$$

As the input voltage increases further, the output remains fixed at $-5.7V$.

The circuit behaves in a similar manner when the input goes negative. At first the output voltage is an inverted replica of the input voltage. When the input voltage drops below $-5.7V$, the output voltage stays fixed at $-5.7V$. We have built a symmetrical limiter. That is, the output voltage is limited to $\pm 5.7V$. In general, the output voltage is limited to

$$v_{o_{\text{lim}}} = \pm(0.7 + V_Z)$$

EXAMPLE 25–4
Op-Amp Limiter

Find the output voltage for the limiter shown in Figure 25–11A.

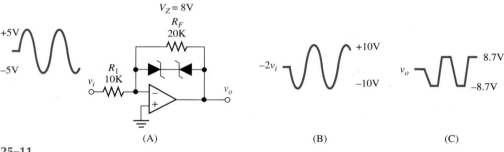

(A) (B) (C)

FIGURE 25–11

Solution Because the gain here is greater than 1, we first plot the output as though the Zener diodes are not there:

$$v_o = -\frac{R_F}{R_1} v_i = -\frac{20K}{10K} v_i = -2v_i$$

and we get the 10V sine wave shown in Figure 25–11B. Now when the output tries to increase beyond the limiting voltage of $0.7 + V_Z$, a Zener diode goes into breakdown, limiting the voltage to

$$v_{o_{\text{lim}}} = \pm(0.7 + V_Z) = \pm 8.7V$$

The actual output is shown in Figure 25–11C.

Troubleshooting

R_1 *failed open:* No signal reaches Op-Amp, so output is fixed at 0V.

R_F *failed open* (Figure 25–12A): The gain becomes infinite, and the Op-Amp output tries to reach the supply voltage. The Zener diodes will still limit the output, but the output will look like a square wave that bounces between $\pm(0.7 + V_Z)$.

A Zener fails open: Zener diodes are removed from feedback pathway, and the circuit acts as simple inverting amplifier with no limiting.

A Zener fails short (Figure 25–12B): Limiting will occur on one side only.

FIGURE 25–12

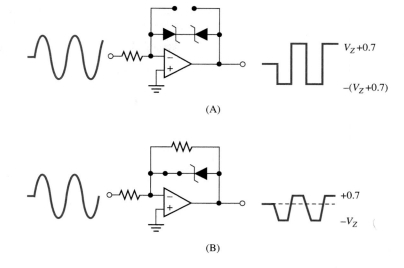

(A)

(B)

25–4 ■ ZERO-CROSSING DETECTOR

To this point all of the Op-Amp circuits we have shown you use negative feedback for proper operation. But we also build Op-Amp circuits that take advantage of the very high open-loop gain. These circuits work without negative feedback and depend on the saturation of the Op-Amp output voltage. The simplest of these circuits is shown in Figure 25–13.

In the circuit of Figure 25–13, we have grounded the −terminal and applied the signal source directly to the +terminal. This is as simple an Op-Amp circuit as we can build.

FIGURE 25–13
Zero-crossing detector

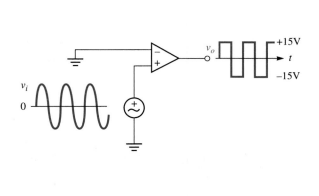

(A)

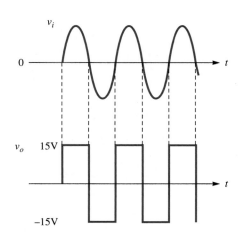

(B)

Note that there is no feedback pathway here. Without feedback, the differential voltage between the + and −terminals is not driven to 0V. That is, without feedback

$$v_+ \neq v_-$$

Also, without feedback the output voltage is given by

$$v_o = A(v_+ - v_-)$$

Because the gain *(A)* is so high, as soon as v_+ is a few microvolts greater than v_-, the output will jump to approximately the positive supply voltage. As soon as v_+ drops a few microvolts below v_-, the output will jump to approximately the negative supply voltage. That is

$$v_+ > v_- \rightarrow v_o \approx +V_{supply}$$
$$v_+ < v_- \rightarrow v_o \approx -V_{supply}$$

For the circuit of Figure 25–13 (p. 831)

$$v_- = 0V$$

Therefore, as soon as the input signal goes positive, the output jumps to approximately +15V. When the input goes negative, the output jumps to approximately −15V. The slew rate of the Op-Amp limits how fast the output voltage can actually jump.

We can use this simple circuit in two ways. First, as you can see from Figure 25–13 (p. 831), the circuit produces a square wave output from a sine wave input. That is, the circuit can be used as a square wave generator.

Figure 25–13B (p. 831) illustrates the second use. We have drawn the input and output voltages on the same graph. You can see that every time the input crosses 0V going positive, the output jumps to +15V. Therefore, every time we see this +15V jump in the output, we know the input signal has crossed 0V. This circuit is a **zero-crossing detector.**

An example of the use of a zero-crossing detector is shown in Figure 25–14A. The *RC* combination produces a series of voltage spikes each time the Op-Amp output voltage changes from + to − or from − to +. The diode allows only the positive-going spikes to pass. The input signal and the resultant spike train are shown in Figure 25–14B. Every time we see a voltage spike, we know the input has crossed 0V in a positive direction. These voltage spikes can be used for such things as triggering an oscilloscope, a recording device, or an enabling input to a computer.

One common modification to the zero-crossing detector is to limit the output voltage. Figure 25–15 shows how to use back-to-back Zener diodes to do this. The output now jumps between $\pm(0.7 + V_Z)$.

FIGURE 25–14

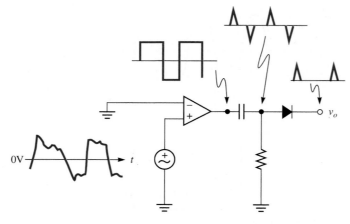

(A)

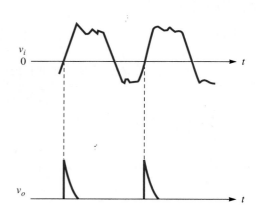

(B)

FIGURE 25–15

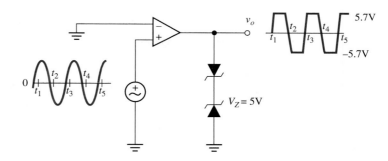

EXAMPLE 25–5
Zero-Crossing Detector

On the same graph, draw the input signal and the output voltage for the zero-crossing detector in Figure 25–16A.

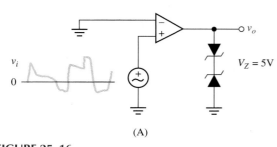

(A)

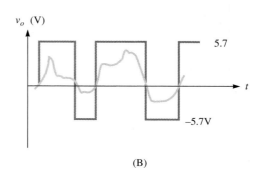

(B)

FIGURE 25–16

Solution The Zener diodes limit the output voltage to ±5.7V, as shown in Figure 25–16B.

The zero-crossing detector is a special type of comparator. We discuss the comparator and its troubleshooting in the next section.

REVIEW QUESTIONS

1. What can you say about v_+ and v_- if there is no negative feedback in an Op-Amp circuit?
2. What will the Op-Amp output voltage be if $v_+ > v_-$? If $v_+ < v_-$?
3. What is the limiting factor in how fast the Op-Amp output voltage can jump?
4. Draw the zero-crossing detector. On the same graph, draw an arbitrary input and the resultant output.
5. Draw the zero-crossing detector with Zener diode limiting. On the same graph, draw an arbitrary input and the resultant output.

25–5 ■ COMPARATORS (LEVEL DETECTORS)

The zero-crossing detector is a special type of **comparator.** Comparators are also known as **level detectors.** In the zero-crossing detector, the Op-Amp output voltage changes every time the input voltage goes through 0V. In the comparator the Op-Amp output voltage changes every time the input voltage goes through a preset reference voltage. The comparator can be constructed in either a non-inverting or inverting mode.

The Non-Inverting Comparator

In the circuit in Figure 25–17, the reference battery sets the voltage at the −terminal. That is

$$v_- = V_{ref} = 5V$$

Now, we know that when

$$v_+ > v_- \rightarrow v_o \approx +15V$$

And when

$$v_+ < v_- \rightarrow v_o \approx -15V$$

In other word, whenever the input voltage is greater than the reference voltage, the output goes to the negative voltage supply. Whenever the input voltage is less than the reference voltage, the output goes to the positive supply voltage. The input and output signals for this non-inverting comparator are shown on the same graph in Figure 25–17B.

FIGURE 25–17
Non-inverting comparator

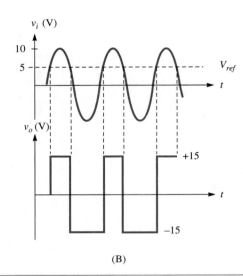

(A)   (B)

EXAMPLE 25–6
Non-Inverting Comparator

Figure 25–18A shows that a signal that we are trying to record is corrupted by noise. Because we are only interested in recording the signal spikes that appear above the noise, design a system to accomplish this task.

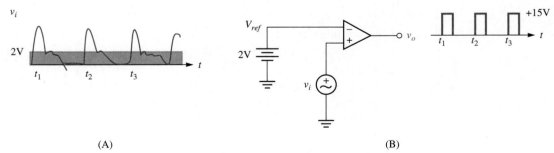

(A)   (B)

FIGURE 25–18

Solution We will record the signal with a computer that will only record when its enable input goes positive. Because the noise in Figure 25–18A has a maximum magnitude of 2V, we can use the comparator in Figure 25–18B. The comparator will produce a positive output voltage whenever the input signal increases above 2V.

Voltage Divider Reference Voltage for Comparator

As Figure 25–19 shows, we do not need a battery to produce a reference voltage. We can use a voltage divider connected to the positive supply voltage (for a positive reference) or to the negative supply voltage (for a negative reference voltage). The reference voltage is given by

$$V_{ref} = \frac{R_2}{R_1 + R_2} \; V_{supply} = \frac{50K}{50K + 100K} \; 15 = 5V$$

FIGURE 25–19

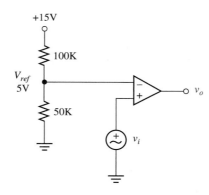

The Inverting Comparator

If we want the comparator output voltage to go negative when the reference voltage is exceeded, then we connect the reference voltage to the +terminal. The signal is brought in at the −terminal. Figure 25–20 shows an inverting comparator with the reference voltage set by a voltage divider. Note that the comparator, non-inverting or inverting, becomes a zero-crossing detector if the reference voltage is set to 0V.

FIGURE 25–20

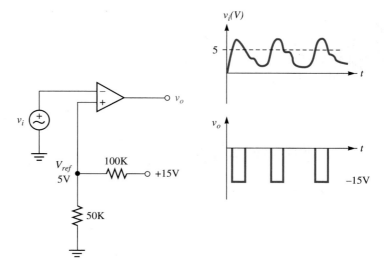

EXAMPLE 25–7 Voltage Level Indicator	In Example 6–9 (p. 169) of Chapter 6, we showed you in general terms how to use level detectors and LEDs to build a voltage level indicator. We repeat this circuit in Figure 25–21A (p. 836). Build the first level detector—that is, build a level detector that will output +5V when the input voltage passes +2V.

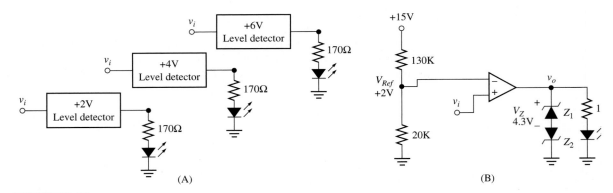

FIGURE 25–21

Solution Figure 25–21B shows the level detector circuit. We connect the non-inverting comparator to a Zener diode limiting circuit, which drives the LED. Assuming that we are using ±15V supplies, the reference voltage is

$$V_{ref} = \frac{20K}{20K + 130K} \, 15 = +2V$$

When the input passes +2V, the Op-Amp saturates near +15V. The Zener Z1 now will be in breakdown, and Z2 will be **ON.** The output of the limiter is then

$$v_o = V_Z + 0.7 = 4.3 + 0.7 = 5V$$

This +5V output biases the LED in its forward direction, and the LED lights up.
    When the input is less than +2V, the Op-Amp will saturate near −15V, and the limiter will produce an output of −5V. This negative output voltage will reverse bias the LED, and it will not light up.

FIGURE 25–22

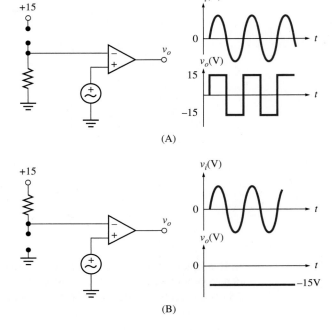

Troubleshooting

If the Op-Amp is good and all components are properly connected, the only problems could be with the resistors in the voltage divider.

R_1 *open* (Figure 25–22A): $v_- = 0V$ and the circuit behaves as a zero-crossing detector.

R_2 *short* (Figure 25–22B): here $v_- = +15V$, so the output saturates at $-15V$.

REVIEW QUESTIONS

1. Draw a non-inverting comparator with a battery reference voltage.
2. Explain the operation of this comparator, and demonstrate your answer by drawing a sample input voltage and resultant output voltage on the same graph.
3. How can we eliminate the battery?
4. When does a comparator become a zero-crossing detector?
5. Draw the inverting comparator with a voltage divider reference voltage. Draw a sample input voltage and the resultant output voltage on the same graph.
6. What will the output of the inverting comparator be if R_1 fails open? If R_2 fails open?

25–6 ▪ SCHMITT TRIGGER

The comparator has some problems when it is used with very fast signals, such as the one in Figure 25–23A. For one thing, it takes a finite amount of time for the output of the Op-Amp to saturate at the supply voltage. If the signal exceeds the reference level for only a very short time, the Op-Amp may not have time to respond (Figure 25–23B). For another, even if the Op-Amp can respond, the output pulse produced can be so narrow that it is not useful (Figure 25–23C). The problems with the comparator can be overcome with a slightly more complicated circuit known as a **Schmitt trigger.** This circuit comes in several configurations; we will discuss only the basic inverting Schmitt trigger (Figure 25–24A, p. 838).

FIGURE 25–23

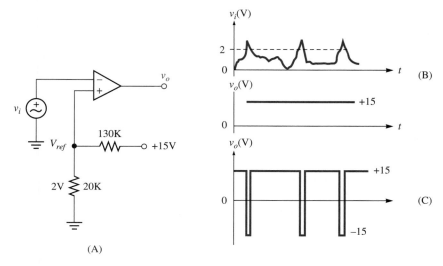

Positive Feedback

This Schmitt trigger looks very similar to the circuit of Figure 25–23. There is, however, a very important difference. The drive for the voltage divider is not the supply voltage. It is the Op-Amp output voltage. Note that this circuit connects the output voltage back to the *posi-*

FIGURE 25–24
Schmitt trigger

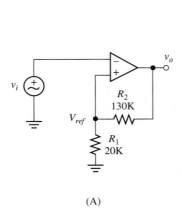

(A)

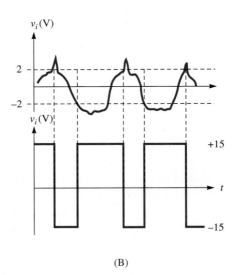

(B)

tive terminal. You might remember that we talked around the loop to describe the operation of the negative feedback loop that most Op-Amp circuits use. It went something like this:

☐ A positive voltage applied to the −terminal leads to a negative output voltage.

☐ The negative output voltage is fed back to the −terminal.

☐ This feedback negative voltage at the −terminal leads to a positive output voltage.

The hallmark of a negative feedback system is that an increase in the output voltage leads to a decrease in the output voltage, until equilibrium is reached. The Op-Amp negative feedback system reaches equilibrium when $v_+ = v_-$.

On the other hand, talking around the loop for the Schmitt trigger leads to the following:

☐ Assume that a positive voltage is applied to the −terminal. This will lead to a negative output voltage.

☐ The negative output voltage is fed back through R_1 and R_2 to the +terminal.

☐ This feedback negative voltage applied to the +terminal leads to a negative output voltage.

Unlike a negative feedback system, once the output starts to go negative the feedback will make it go even more negative, until the output saturates at close to the supply voltage. This is a **positive feedback system.**

The positive feedback loop takes care of the problem shown in Figure 25–23B (p. 837). The Schmitt trigger's positive feedback saturates the output voltage as soon as the input signal is slightly greater than the reference voltage providing a much more rapid response than the comparator.

The Schmitt trigger also provides a longer output pulse than the comparator. This is a bit tricky to see, but we will break the process down into bite-sized chunks. In Figure 25–25A we assume that the output voltage is saturated at +15V. Therefore, the trigger reference voltage is

$$V_{ref} = \frac{R_2}{R_1 + R_2} V_o = \frac{20K}{20K + 130K}(+15) = +2V$$

As soon as the input passes +2V, the output will jump down to −15V (this is an inverting circuit). When the input voltage goes from +15V to −15V, the reference voltage also changes! This is the key to the working of the Schmitt trigger. The reference voltage is now

$$V_{ref} = \frac{R_2}{R_1 + R_2} V_o = \frac{20K}{20K + 130K}(-15) = -2V$$

FIGURE 25–25

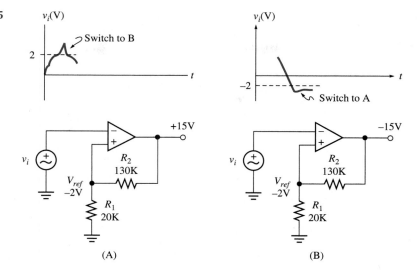

Note what has happened here. The reference voltage has been reduced to -2V. That is, as long as the input stays above -2V, the $-$terminal voltage will be greater than the $+$terminal voltage and the output will, therefore, stay pegged at -15V. The output pulse has been lengthened.

The input voltage now decreases (Figure 25–25B). When the input drops below -2V, the $+$terminal is at a greater voltage than the $-$terminal. At this point the Op-Amp output goes back to $+15$V, and the reference voltage goes to $+2$V. As long as the input remains below $+2$V, the output will stay at $+15$V. When the input signal once again passes $+2$V, the output voltage swings negative. The input and output signals for this Schmitt trigger are shown in Figure 25–24B.

EXAMPLE 25–8
The Schmitt Trigger

For the Schmitt trigger in Figure 25–26A, draw the input and output voltages on the same graph.

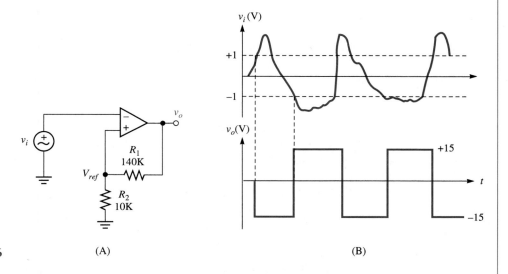

FIGURE 25–26　　　(A)　　　　　　　　　　　　　　　　　　　(B)

Solution　　If we assume that the output is at $+15$V, the reference voltage is

$$V_{ref} = \frac{R_2}{R_1 + R_2} V_o = \frac{10\text{K}}{10\text{K} + 140\text{K}} (+15) = +1\text{V}$$

Therefore, the circuit will trigger when the input goes above $+1V$ and falls below $-1V$. The input and output signals are shown in Figure 25–26B (p. 839).

Both the comparator and the Schmitt trigger provide an output that changes states from a positive voltage to a negative voltage, or vice versa, when preset input reference levels are passed. While similar, there is an important functional difference in the two circuits.

The reference level that triggers the comparator to go from a negative to a positive output is the same level that triggers the comparator to go from a positive to a negative output. For example, let's say that the reference level for a non-inverting comparator is $+2V$. When the input exceeds $+2V$, the output will go positive. When the input drops below this same $+2V$, the output goes negative.

The Schmitt trigger, on the other hand, has one reference level for a negative-to-positive output change and a separate reference level for a positive-to-negative output change. For example, a general Schmitt trigger may be designed to go positive when the input passes $+2V$ but not to go negative until the input drops below $+1V$. When a device has one trigger level for a positive output change and a different trigger level for a negative output change, the characteristic is known as **hysteresis.**

A final word on Schmitt triggers. The basic model uses the Op-Amp saturation voltage as a reference for the feedback voltage divider. It may have occurred to you that the saturation voltage for the particular Op-Amp you are using is not precisely known. Therefore, you cannot set the trigger level precisely. We can fix this problem by connecting a limiter circuit to the Op-Amp output.

Figure 25–27 shows how we can use Zener diodes to provide the reference voltage for the feedback circuit. When the Op-Amp output swings to its positive saturation volt-

FIGURE 25–27

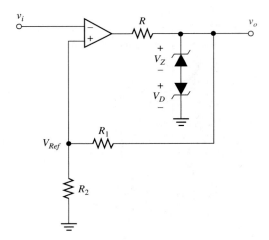

age, Z1 is reverse biased and Z2 is forward biased. Assuming the diode **ON** voltage is 0.7V, the reference voltage now is

$$V_{ref} = V_Z + 0.7 = +5.7V$$

When the Op-Amp output swings to its negative saturation voltage, the reference voltage becomes $V_{ref} = -5.7V$.

The resistor (R) that is connected between the Op-Amp output and the Zener diodes limits the current into the Op-Amp. This is done to assure that there will be enough current available to keep the appropriate Zener in breakdown.

Troubleshooting

Signal source not connected (Figure 25–28A): Output voltage is fixed at positive or negative supply voltage.

R_1 *open* (Figure 25–28B): The circuit behaves as zero-crossing detector.

R_2 *open* (Figure 25–28C): Output voltage is fixed at positive or negative supply voltage.

FIGURE 25–28

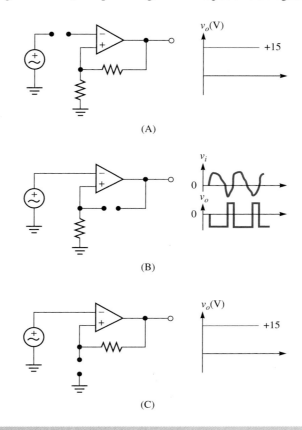

(A)

(B)

(C)

REVIEW QUESTIONS

1. What are two disadvantages of the comparator?
2. Draw the basic inverting Schmitt trigger.
3. What type of feedback is used in this circuit? Describe the effect of this feedback.
4. What determines the reference voltage in the Schmitt trigger?
5. How does the reference voltage in the Schmitt trigger differ from the reference voltage in the comparator?
6. Draw a typical set of input and output signals for the Schmitt trigger.
7. Is there a problem with using the Op-Amp saturation voltage to provide the reference voltage? How do you fix this problem?
8. Discuss the output voltage of the Schmitt trigger if the signal source is not connected. If R_1 is open. If R_2 is open.

SUMMARY

- The simple inverting rectifier can produce an amplified half-wave rectified voltage. The diode voltage drop is seen at the output.
- Precision rectifiers eliminate the diode voltage drop.
- The non-inverting precision rectifier will saturate when the diode is turned **OFF.** The **OFF** diode effectively removes the feedback pathway.

- The inverting precision rectifier uses two feedback pathways, one of which is always intact. Either a negative-going or a positive-going half-wave rectified voltage can be obtained from this rectifier.
- The Zener diode limiter has an output that must lie between $\pm(0.7 + V_Z)$.
- The zero-crossing detector comes in inverting and non-inverting forms. The output jumps to the supply voltage every time the input signal crosses 0V.
- The comparator compares the input signal to a reference voltage. The output jumps every time the input signal increases above or decreases below the reference signal.
- The reference voltage for the comparator can be derived from a voltage divider connected to the supply voltage.
- The Schmitt trigger has a faster response and produces a longer output pulse than the comparator. The reference voltage is derived from a voltage divider connected to the Op-Amp output. When the Op-Amp output changes sign, so does the reference voltage.

PROBLEMS

SECTION 25–2 Rectifiers

1. For the rectifier circuit of Figure 25–29
   (a) Draw the circuit when D1 is **ON.**   (b) Draw the circuit when D1 is **OFF.**
   (c) For a 5V 1KHz input signal, draw the input and output signals on the same graph. Do not forget to include the diode voltage drop.
2. Let $R_F = 20K\Omega$ in Figure 25–29, and repeat Problem 25–1c.
3. Let $R_F = 50K\Omega$ in Figure 25–29. Assume that the Op-Amp has a saturation voltage of $\pm 15V$.
   (a) What is the maximum Op-Amp output voltage?   (b) Repeat Problem 25–1c.

FIGURE 25–29

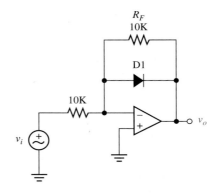

4. For the rectifier circuit of Figure 25–30
   (a) Draw the circuit when D1 is **ON.**   (b) Draw the circuit when D1 is **OFF.**
   (c) For a 5V 1KHz input signal, draw the input and output signals on the same graph. Do not forget to include the diode voltage drop.
5. Let $R_F = 20K\Omega$ in Figure 25–30, and repeat Problem 25–4c.

FIGURE 25–30

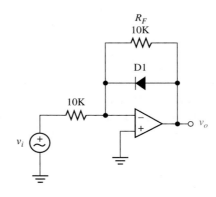

6. You have the following resistors available:

$$1K\Omega,\ 2.2K\Omega,\ 10K\Omega,\ 47K\Omega$$

   (a) Build a half-wave rectifier that will produce positive half-waves and will amplify the input voltage by approximately 20.

   (b) Assume your input is a 100mV 1KHz voltage. Sketch the input and actual output voltage on the same graph.

7. For the rectifier circuit of Figure 25–31.

   (a) Draw the circuit when D1 is **ON.**   (b) Draw the circuit when D1 is **OFF.**

   (c) For a 5V 1KHz input signal, draw the input and output signals on the same graph. Do not forget to include the diode voltage drop.

FIGURE 25–31

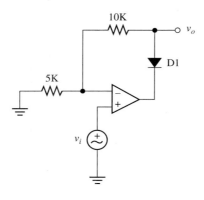

8. For the rectifier circuit of Figure 25–32.

   (a) Draw the circuit when D1 is **ON.**   (b) Draw the circuit when D1 is **OFF.**

   (c) For a 5V 1KHz input signal, draw the input and output signals on the same graph. Do not forget to include the diode voltage drop.

FIGURE 25–32

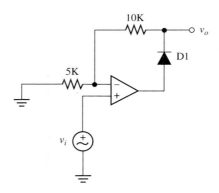

9. For the inverting precision rectifier in Figure 25–33 (p. 844)

   (a) Which diode is **ON** when the input is positive? Which diode is **OFF?**

   (b) Which diode is **ON** when the input is negative? Which diode is **OFF?**

   (c) When D1 is **ON,** what is the gain of the amplifier?

   (d) When D2 is **ON,** what is the gain of the amplifier?

10. For the circuit of Figure 25–33 (p. 844)

$$v_i = 2V\ 1KHz,\ R = 1K\Omega,\ R_1 = 5K\Omega,\ R_2 = 5K\Omega$$

   Sketch v_i, v_{o_1}, and v_{o_2} on the same graph.

FIGURE 25–33

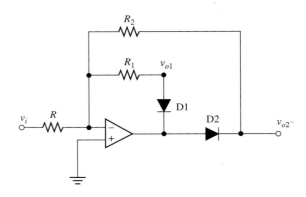

11. For the circuit of Figure 25–33

$$v_i = 2V\ 1KHz,\ R = 1K\Omega,\ R_1 = 2K\Omega,\ R_2 = 5K\Omega$$

Sketch v_i, v_{o_1}, and v_{o_2} on the same graph. Note that the feedback resistors have different values in this problem.

SECTION 25–3   Limiters

12. **(a)** What are the three possible states in which a Zener diode operates?

 (b) Draw the circuit of Figure 25–34 if Z2 is **OFF**. What is v_o?

 (c) Draw the circuit if Z1 is **ON** and Z2 is in breakdown. What is v_o?

 (d) Draw the circuit if Z2 is **ON** and Z1 is in breakdown. What is v_o?

 (e) For a 5V 1KHz input signal, draw the input and output signals on the same graph.

13. Let $R_F = 50K\Omega$ in Figure 25–34 and repeat Problem 12(e).

14. Build a limiter that will have a gain of -5 and an output voltage limit of $\pm 7V$. Be sure to indicate the resistor and Zener values that you use.

FIGURE 25–34

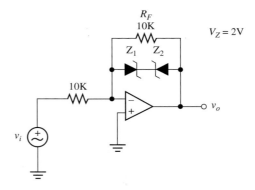

SECTION 25–4   Zero-Crossing Detector

15. If the input signal to the circuit in Figure 25–35 is a 1V 1KHZ signal, draw the input and output voltages on the same graph. Assume that the Op-Amp saturation voltage is $\pm 13V$.

FIGURE 25–35

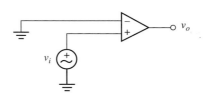

16. If the input signal to the circuit in Figure 25–36 is a 1V 1KHz signal, draw the input and output voltages on the same graph. Assume that the Op-Amp saturation voltage is ±15V.

17. If the input signal to the circuit in Figure 25–36 is a 2V 1KHz signal, draw the input and output voltages on the same graph. Assume that the Op-Amp saturation voltage is ±15V.

FIGURE 25–36

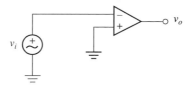

18. The Op-Amp in the circuit of Figure 25–37 has a saturation voltage of ±15V. This circuit approximates a zero-crossing detector but retains a feedback pathway.

  (a) What is the output voltage of this non-inverting amplifier?

  (b) What is the maximum possible output voltage?

  (c) At what input voltage does the output saturate?

  (d) Sketch the output for a 15mV 1KHz input voltage.

  (e) Sketch the output for a 60mV 1KHz input voltage.

  (f) Sketch the output for a 5V 1KHz input voltage.

FIGURE 25–37

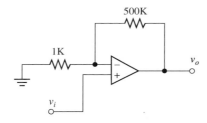

SECTION 25–5 Comparators (Level Detectors)

19. If the input signal to the circuit in Figure 25–38 is a 2V 1KHz signal, draw the input and output voltages on the same graph. Assume the Op-Amp has a saturation voltage of ±12V.

FIGURE 25–38

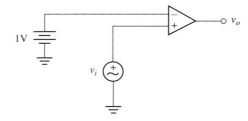

20. Assume the Op-Amp in Figure 25–39 has a saturation voltage of ±14V. If the input signal to the circuit is a 2V 1KHz signal, draw the input and output voltages on the same graph.

FIGURE 25–39

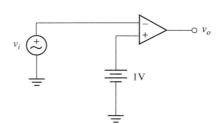

21. For the comparator of Figure 25–40, the Op-Amp saturation voltage is $\pm 15V$.

    (a) What is V_{ref}?   **(b)** Draw the input and output signals on the same graph.

22. Modify the circuit of Figure 25–40 so that the output voltage will be limited to $\pm 2V$. Be sure to indicate device values.

FIGURE 25–40

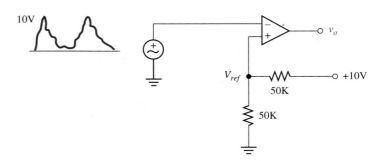

23. You are recording a signal that is corrupted by noise that has a 2V amplitude (Figure 25–41).

    (a) Design a comparator that will produce a positive pulse whenever the signal rises above the noise. Assume that you have $\pm 15V$ supplies available. Be sure to indicate the resistor values that you use.

    (b) Assume your Op-Amp has a saturation voltage of ± 14. Sketch the input and output voltages on the same graph.

FIGURE 25–41

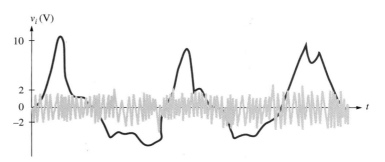

SECTION 25–6   Schmitt Trigger

24. For the Schmitt trigger of Figure 25–42, the Op-Amp saturation voltage is $\pm 15V$.

    (a) What is V_{ref} when the output is saturated at $+15V$?

    (b) What is V_{ref} when the output is saturated at $-15V$?

    (c) Draw the input and output signals on the same graph.

FIGURE 25–42

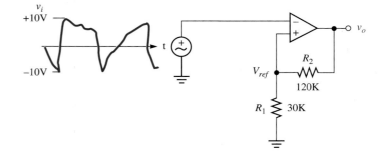

25. For the Schmitt trigger in Figure 25–42, with an Op-Amp saturation voltage $\pm 15V$, choose a new R_1 so that the trigger voltages will be $\pm 7.5V$.

26. For the Schmitt trigger in Figure 25–42, with an Op-Amp saturation voltage of $\pm 12V$, choose a new R_2 so that the trigger voltages will be $\pm 4V$.

27. For the Schmitt trigger in Figure 25–42, find V_{ref} if the Op-Amp saturation voltage is
    (a) $\pm 10V$ **(b)** $\pm 15V$ **(c)** $\pm 20V$

28. Modify the Schmitt trigger circuit in Figure 25–42 so the output voltage will not exceed $\pm 6V$. Change the resistors to maintain $V_{ref} = \pm 3V$ when the Op-Amp saturation voltage is $\pm 15V$.

Troubleshooting Problems

29. What are the likely causes of the problems seen in the rectifier circuit of Figure 25–43?

30. What are the likely causes of the problems seen in the limiter circuit of Figure 25–44?

31. What are the likely causes of the problems seen in the comparator circuit of Figure 25–45 (p. 848)?

32. What are the likely causes of the problems seen in the Schmitt trigger circuit of Figure 25–46 (p. 848)?

FIGURE 25–43

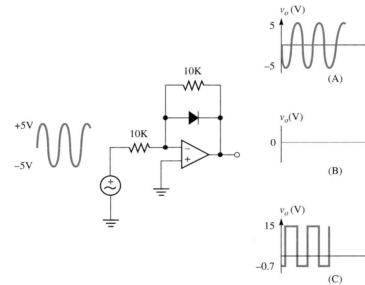

FIGURE 25–44

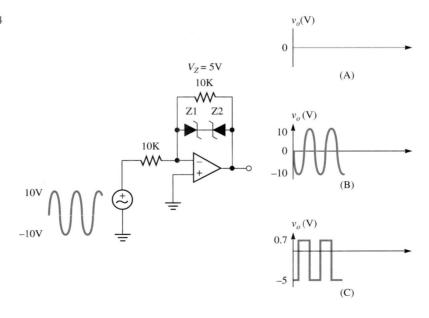

FIGURE 25–45

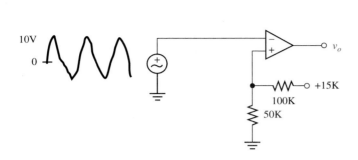

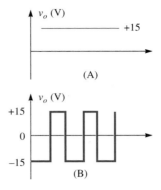

FIGURE 25–46

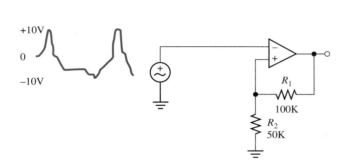

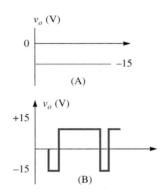

33. Referring to Figure 25–14 (p. 832), what is the likely cause or causes of the following:

    (a) The output of the Op-Amp is zero volts.   **(b)** The Op-Amp output is good, but $v_o = 0V$

34. Referring to Figure 25–24 (p. 838), what is the likely cause of the following:

    (a) $v_o = +15V$ at all times.

    (b) The Op-Amp output changes states when the input signal crosses 0V.

    (c) $v_o = -15V$ as soon as the input is applied.

Computer Problems

35. Simulate the circuit of Figure 25–29 (p. 842).

    (a) Find the output for a 5V 1KHz sine wave.

    (b) Find the output for a 5V 1KHz triangular wave.

    (c) Find the output for a 0.5V 1KHz sine wave. Explain your results.

36. Simulate the circuit of Figure 25–31 (p. 843).

    (a) Find the output for a 5V 1KHz sine wave.

    (b) Find the Op-Amp output for a 5V 1KHz sine wave.   **(c)** Compare your answers.

37. Simulate the circuit of Figure 25–33 (p. 844).

    (a) Repeat Problem 10 (p. 843).   **(b)** Repeat Problem 11 (p. 844), and discuss the results.

38. Simulate the circuit of Figure 25–38 (p. 845).

    (a) Find the output for a 5V 1KHz input signal.

    (b) Find the output for a 0.5V 1KHz input signal. Explain your results.

39. Simulate the circuit you built for Problem 24 (p. 846). Use a 10V input (sine, square, or triangular wave) to demonstrate that your circuit will perform as desired.

40. Simulate the circuit of Figure 25–42 (p. 846). Let the input be a 10V 1KHz sine wave, and find and plot v_o and V_{ref} on the same graph as v_i.

ADVANCED TOPICS RESPONSE AND FILTERS

PART

5

FREQUENCY RESPONSE

FILTERS

FEEDBACK AMPLIFIERS

SIGNAL GENERATORS

THYRISTORS

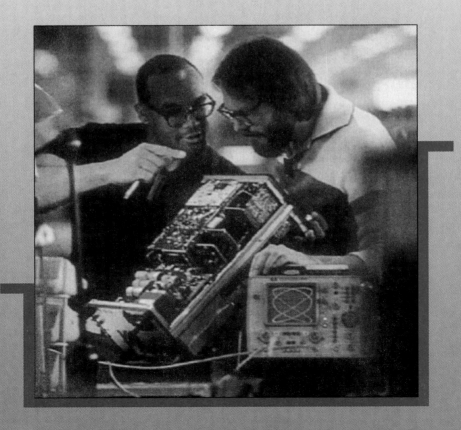

CHAPTER

26

FREQUENCY RESPONSE

CHAPTER OBJECTIVES

To determine the half-power frequencies and bandwidth of an amplifier.

To represent the frequency response of an amplifier with the Bode gain plot.

To determine low-frequency amplifier behavior due to coupling capacitors.

To determine low-frequency amplifier behavior due to bypass capacitors.

To determine the high-frequency behavior of amplifiers.

To find the Miller capacitance, and use it to determine the high-frequency 3-dB frequency.

To determine the frequency response of multistage amplifiers.

To determine BJT and FET internal capacitances from data sheet information.

26–1 ■ INTRODUCTION

We haven't been completely honest with you. When we have encountered a capacitor in an amplifier circuit, we have assumed that it was open for DC analysis and short for AC analysis. This is an accurate statement for DC, but not for AC.

While it is true that we can approximate the capacitor as a short circuit when the signal frequency is high, two questions might have occurred to you:

(1) How high is high (does the capacitor short at 100Hz, 1KHz, 100MHz)?

(2) How does the capacitor behave between DC and the frequency at which it shorts, and how is the amplifier affected by the capacitor behavior?

Figure 26–1 shows the answer to the last question. This figure shows the response of a typical amplifier to an AC input signal that is fixed at 10mV amplitude but varies in frequency from 1Hz to 1000Hz. (The vertical scales for the input and output voltages are different). You can see that at 1Hz, there is almost no output. As the input frequency increases, the output increases, finally reaching 100mV. After that, it is interesting to note that as the input frequency is increased further, the output decreases.

FIGURE 26–1

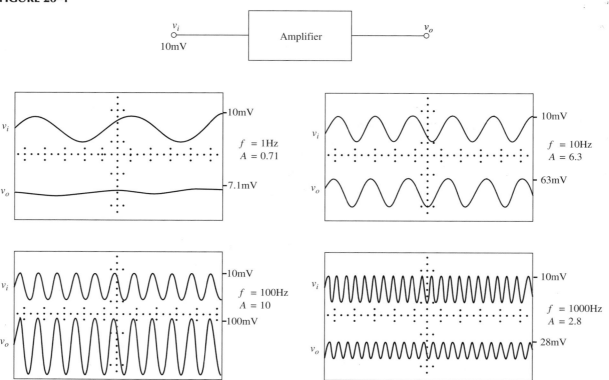

As we will show you in this chapter, the initial increase in the output amplitude is due to shorting of the external coupling and bypass capacitors. The decrease at higher frequency is due to the shorting of internal PN-junction capacitances. Given these processes, it is important for you to know the frequency limitations of the devices and instruments that you use. If you try to use audio amplifier components to build an amplifier for satellite signals, you will be very disappointed (and so will your boss). Audio amplifiers work in the KHz range; satellite communication can take place in the GHz range.

26–2 ■ THE GAIN RESPONSE

Amplifier Gain

We can describe the frequency behavior of an amplifier by examining amplifier gain as a function of frequency. It is important to note here that we ignore any minus signs when we plot the gain. For now we are only interested in the magnitude of the gain. For the amplifier in Figure 26–1, we get

$$A = \frac{v_o}{v_i} = \frac{7.1\text{mV}}{10\text{mV}} = 0.71 \quad @ \quad f = 1\text{Hz}$$

$$A = \frac{v_o}{v_i} = \frac{63\text{mV}}{10\text{mV}} = 6.30 \quad @ \quad f = 10\text{Hz}$$

$$A = \frac{v_o}{v_i} = \frac{100\text{mV}}{10\text{mV}} = 10.0 \quad @ \quad f = 100\text{Hz}$$

$$A = \frac{v_o}{v_i} = \frac{28\text{mV}}{10\text{mV}} = 2.80 \quad @ \quad f = 1000\text{Hz}$$

You may have noticed in Figure 26–1 that the output signal not only has a different amplitude than the input signal, there is also a shift along the time (horizontal) axis. This shift is known as a **phase shift.** At this point it is not critical to dwell on the phase shift. Instead, we will concentrate on the change in signal amplitudes.

We can picture the frequency response of an amplifier by plotting the amplifier gain as a function of input frequency. Figure 26–2 shows the gain plot for the amplifier of Figure 26–1. This plot clearly shows the increase in gain, the maximum gain, and the decrease in gain as input frequency increases. As the vertical dashed lines in the figure show, we can separate the graph into a low-frequency region, a mid-frequency region, and a high-frequency region.

FIGURE 26–2

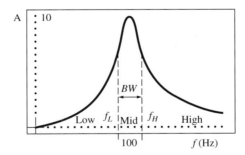

In all of our previous chapters on amplifiers, we assumed the transistors and Op-Amps have been operating in the mid-frequency (or mid-band) range. This is the range where the external coupling and bypass capacitors have shorted. The high-frequency drop-off is due to internal capacitances in the electronic devices. We will discuss these capacitors later in this chapter.

Bandwidth

The useful range of frequencies for any amplifier is expressed as the *bandwidth (BW)*, where (Figure 26–2)

$$BW = f_H - f_L$$

FIGURE 26–3

When you buy an amplifier for your stereo, one of the most important specifications you should be interested in is the amplifier's bandwidth. Because we can hear from approximately 20Hz to 20KHz, you want to make sure that the bandwidth of the stereo encompasses this audible range. Anybody who has listened to the public address system in a bus or train station understands the problem of limited bandwidth (Figure 26–3).

How do we decide where the mid-band frequency range begins and ends? The accepted standard is that the bandwidth is given by the range of frequencies that produces at least 50% of the maximum power available at the output. The edge frequencies f_L and f_H are known as the **half-power frequencies;** we find them as follows. The output voltage is given by

$$v_o = Av_i$$

Because the power to a resistive load is given by

$$P_L = \frac{v^2_o}{R_L}$$

and because

$$v_o = Av_i$$

the power to the load is

$$P_L = \frac{(Av_i)^2}{R_L}$$

The maximum power to the load is, therefore

$$P_{L_{max}} = \frac{(A_{max}v_i)^2}{R_L}$$

The bandwidth frequencies, f_H and f_L, are the frequencies at which the power is half the maximum power. That is, the half-power frequencies are found when

$$P_L = \frac{1}{2}P_{L_{max}}$$

In terms of the amplifier gain, the half-power frequencies are found when

$$\frac{(Av_i)^2}{R_L} = \frac{1}{2}\frac{(A_{max}v_i)^2}{R_L}$$

This reduces to

$$A^2 = \frac{1}{2}A^2_{\text{max}}$$

So, we get

$$A = \frac{1}{\sqrt{2}}A_{\text{max}} \quad \text{when} \quad \begin{matrix} f = f_H \\ \text{or} \\ f = f_L \end{matrix}$$

EXAMPLE 26–1

Half-Power Frequencies— Band-Pass Response

In the RLC circuit of Figure 26–4, the output we are interested in is the voltage across the resistor. The frequency response of this circuit is also shown. Find f_H, f_L, BW.

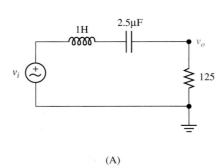

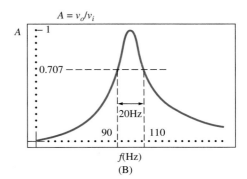

FIGURE 26–4   (A)   (B)

Solution   The maximum gain is

$$A_{\text{max}} = 1$$

To find the bandwidth, we first find the frequencies at which the gain is

$$A = \frac{1}{\sqrt{2}}A_{\text{max}} = \frac{1}{\sqrt{2}}1 = 0.707$$

From the figure, we see that

$$f_H = 110\text{Hz}$$
$$f_L = 90\text{Hz}$$

So, the bandwidth is

$$BW = f_H - f_L = 110 - 90 = 20\text{Hz}$$

In this circuit, low-frequency and high-frequency signals are greatly reduced. Only a band of frequencies from 90 to 110 Hz are passed through to the output. Because this circuit response has a useful band of frequencies from 90 to 110Hz, it is known as a **band-pass circuit.**

EXAMPLE 26–2

Half-Power Frequency—Low-Pass Response

In the RC circuit of Figure 26–5 (p. 856), the output we are interested in is the voltage across the capacitor. The frequency response of this circuit is also shown. Find f_H, f_L, BW.

Solution   The maximum gain is

$$A_{\text{max}} = 1$$

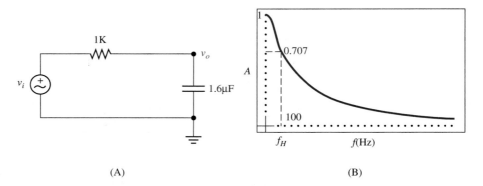

FIGURE 26–5   (A)   (B)

To find the bandwidth, we first find the frequencies at which the gain is

$$A = \frac{1}{\sqrt{2}} A_{\max} = \frac{1}{\sqrt{2}} 1 = 0.707$$

From the figure, we see that

$$f_H = 100\text{Hz}$$

In this response the maximum output occurs at 0Hz, so there is no f_L. The bandwidth is

$$BW = f_H - f_L = 100\text{Hz}$$

In this circuit, low-frequency signals are passed through to the output. High-frequency signals are greatly reduced. This is known as a **low-pass response.**

EXAMPLE 26–3
Half-Power
Frequency—High-
Pass Response

In the RC circuit of Figure 26–6, the output we are interested in is the voltage across the resistor. The frequency response of this circuit is also shown. Find f_H, f_L, BW.

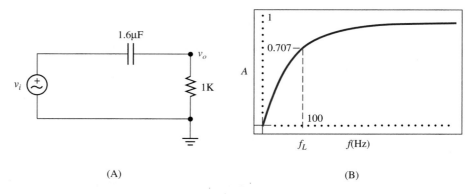

FIGURE 26–6   (A)   (B)

Solution   The maximum gain is

$$A_{\max} = 1$$

To find the bandwidth, we first find the frequencies at which the gain is

$$A = \frac{1}{\sqrt{2}} A_{\max} = \frac{1}{\sqrt{2}} 1 = 0.707$$

From the figure, we see that

$$f_L = 100\text{Hz}$$

In this response, the maximum output occurs at high frequency. Because there is no drop-off at high frequency, $f_H \rightarrow \infty$. Bandwidth here is infinite.

In this circuit, high-frequency signals are passed through to the output. Low-frequency signals are greatly reduced. This is known as a **high-pass response.**

REVIEW QUESTIONS

1. Define the bandwidth of an amplifier?
2. Why do we need to know an amplifier's bandwidth?
3. How do we find the edge frequencies $(f_L$ and $f_H)$? What are these frequencies called?
4. Draw a typical band-pass response, and indicate the half-power frequencies.
5. Draw a typical low-pass response, and indicate the half-power frequency. Is it f_L or f_H?
6. Draw a typical high-pass response, and indicate the half-power frequency. Is it f_L or f_H?

26–3 ■ THE BODE GAIN PLOT

The examples used in the previous section are not very realistic. For example, the bandwidth in Example 26–1 (p. 855) was purposely kept small (20Hz) so that the frequency response could be easily plotted. Real amplifiers usually have a much greater bandwidth; audio amplifiers have bandwidths of approximately 20KHz. Consider the problem of plotting a gain response over the range of frequencies 10Hz to 100KHz. If we use a frequency scale in which 1 inch = 10Hz, we would need a piece of paper 833 ft. long to reach 100KHz!

We can solve this problem by plotting the gain versus log f. The logarithm is used to compress the frequency scale. The relation between f and log f is shown in the following table:

f	$\log f$
0Hz	(Note that we cannot represent 0Hz on a log scale.)
1Hz	0
10Hz	1
100Hz	2
1KHz	3
10KHz	4
100KHz	5
1MHz	6

You can see that for every 10-fold increase in frequency, there is an increase of only 1 in log f. This is similar to the Richter scale used to measure earthquakes; a magnitude 5 quake is ten times stronger than a magnitude 4 shaker.

Two measures are commonly used to relate the frequencies on the log f scale: the decade, and the octave. The first, the **decade,** is a 10-fold increase in frequency. That is, if f_2 is a decade above f_1, then

$$decade: \quad f_2 = 10f_1$$

For example, 100Hz is a decade above 10Hz; 1KHz is a decade above 100Hz. And if f_2 is a decade above f_1, then f_1 is a decade *below* f_2. For example, 40Hz is a decade below 400Hz.

EXAMPLE 26–4

Log *f* Scale-
Decades

(a) What frequency is a decade above 25KHz?
(b) What frequency is a decade below 25KHz?
(c) What frequency is two decades above 25KHz?

Solution

(a) If f_2 is a decade above f_1, then

$$f_2 = 10f_1$$

so a decade above 25KHz is

$$f_2 = 10f_1 = 10 \times 25\text{KHz} = 250\text{KHz}$$

(b) If f_1 is a decade below f_2, then

$$f_1 = \frac{f_2}{10} = \frac{25\text{KHz}}{10} = 2.5\text{KHz}$$

(c) We have already found that one decade above 25KHz is 250KHz. If we go up another decade, we get

$$f_3 = 10f_2 = 10(10f_1) = 10 \times 250\text{KHz} = 2.5\text{MHz}$$

The second commonly used log *f* measure is the **octave.** An octave is a doubling in frequency. That is,

$$octave: \quad f_2 = 2f_1$$

For example, 40Hz is one octave above 20Hz; 100Hz is an octave below 200Hz.

You may be aware that the term octave implies eight of something. Why is this term used for a doubling of frequency? The answer comes from the world of music. In the standard music scale (diatonic scale), there are eight notes in every doubling of frequency.

EXAMPLE 26–5

Log *f* Scale-
Octaves

(a) What frequency is an octave above 440Hz?
(b) What frequency is an octave below 20KHz?
(c) What frequency is two octaves above 440Hz?

Solution

(a) If f_2 is an octave above f_1, then

$$f_2 = 2f_1 = 2 \times 440\text{Hz} = 880\text{Hz}$$

(b) If f_1 is an octave below f_2, then

$$f_1 = \frac{f_2}{2} = \frac{20\text{KHz}}{2} = 10\text{KHz}$$

(c) We have already found that one octave above 440Hz is 880Hz. If we go up another octave, we get

$$f_3 = 2f_2 = 2(2f_1) = 2 \times 880\text{Hz} = 1760\text{Hz}$$

Figure 26–7A shows the gain response of an audio amplifier plotted versus a log frequency scale. In this type of plot we can comfortably plot the entire frequency range in a limited space. For clarity, we show the actual frequency below the log measure. For ease of

FIGURE 26–7

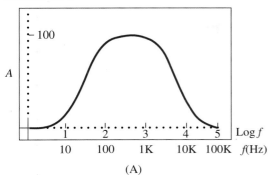

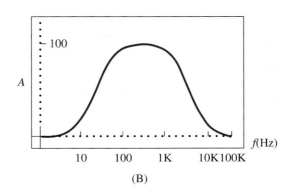

(A)   (B)

reading these plots, we usually label just the actual frequencies, omitting the log numbers (Figure 26–7B).

We also find it convenient to compress the gain scale, so we most commonly plot the log of the gain, rather than the actual gain. In fact, we usually multiply the log of the gain by 20. That is, we plot

$$20 \log A \text{ versus } \log f$$

As discussed in Chapter 17, the units of this new measure of gain are *decibels* (or dB) and we abbreviate $20 \log A$ as A_{dB}. The relation between the actual gain measure and the decibel measure is given by

$$A_{dB} = 20 \log A$$

A comparison of actual gain and A_{dB} is given in the table below.

A	A_{dB}
0.1	−20dB
0.707	−3dB
1.0	0dB
10	20dB
100	40dB
1000	60dB

Note a few interesting points from this table. Gains less than 1 show up as negative decibels. A gain of 1 is 0dB; that is, 0dB represents unity gain. For every ten-fold increase in gain there is a 20dB increase.

An Historical Note: The decibel, named after Alexander Graham Bell, was originally used to describe the relation between input and output acoustic power. That is, the original definition of the decibel was

$$dB = 10 \log \left(\frac{P_o}{P_i} \right)$$

We still use this definition when we want to relate output and input electrical powers. As you know, electrical power is related to the square of the voltage. That is

$$P \alpha v^2$$

If we replace the power in the decibel equation with squared voltages, we get

$$dB = 10 \log \left(\frac{P_o}{P_i} \right) = 10 \log \left(\frac{v^2_o}{v^2_i} \right)$$

Because $\log x^2 = 2 \log x$, we get the voltage ratio decibel measure

$$dB = 20 \log \left(\frac{v_o}{v_i}\right) = 20 \log A$$

where the voltage ratio is the gain $A = v_o/v_i$.

The band-pass response of Figure 26–7 is replotted in Figure 26–8. This time we use the dB measure for the gain. This type of plot is known as the **Bode gain plot.** One of the advantages of the Bode plot is that it produces a curve that is essentially composed of straight lines. You can see this when you compare the band-pass responses shown in Figure 26–4B and Figure 26–8.

FIGURE 26–8

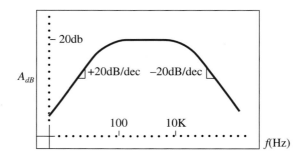

To find the half-power frequencies from the regular gain plot, we look for the frequencies where

$$A = \frac{1}{\sqrt{2}} A_{max}$$

To find these frequencies on a Bode Plot, we take the logarithm of this equation (remember, $\log [a \times b] = \log a + \log b$):

$$20 \log A = 20 \log \left(\frac{1}{\sqrt{2}} A_{max}\right) = 20\left(\log \frac{1}{\sqrt{2}} + \log A_{max}\right)$$

so

$$20 \log A = 20 \log A_{max} + 20 \log \frac{1}{\sqrt{2}}$$

The half-power frequencies are found on the Bode plot where

$$A_{dB} = A_{dB_{max}} - 3\text{dB}$$

That is, we look for the frequencies where the Bode gain is 3dB lower than the maximum gain. We label this gain as $A_{3_{dB}}$. The half-power frequencies are more commonly known as the **3-dB frequencies.**

EXAMPLE 26–6
3-dB
Frequencies—
Band-Pass
Response

The Bode gain plot of the output of a two-capacitor circuit is shown in Figure 26–9. Find f_H, f_L, BW.

Solution The maximum gain is

$$A_{dB_{max}} = -6.5\text{dB} \text{ (Note that this means the actual gain is less than 1.)}$$

To find the bandwidth, we first find the frequencies at which the gain is

$$A_{dB} = -6.5 - 3 = -9.5\text{dB}$$

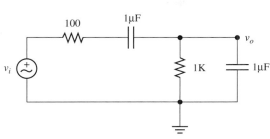

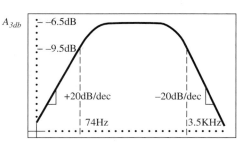

FIGURE 26–9

From the figure, we see that

$$f_H = 3.5\text{KHz}$$

$$f_L = 74\text{Hz}$$

So, the bandwidth is

$$BW = f_H - f_L = 3.5\text{KHz} - 74\text{Hz} \approx 3.5\text{KHz}$$

EXAMPLE 26–7
3-dB Frequency—
Low-Pass
Response

In the RC circuit in Figure 26–10, the output we are interested in is the voltage across the capacitor. The frequency response of this circuit is also shown. Find f_H, f_L, BW.

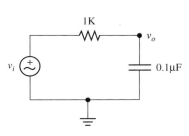

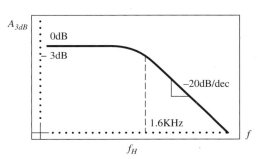

FIGURE 26–10

Solution The maximum gain is

$$A_{dB_{max}} = 0\text{dB (the actual gain is unity)}$$

To find the bandwidth, we first find the frequencies at which the gain is

$$A_{dB} = 0 - 3 = -3\text{dB}$$

From the figure, we see that

$$f_H = 1.6\text{KHz}$$

In this response the maximum output occurs at low frequency (we cannot show 0Hz on a Bode plot). In this case there is no f_L. The bandwidth here is

$$BW = f_H - f_L = f_H = 1.6\text{KHz}$$

EXAMPLE 26–8
3-dB Frequency—
High-Pass
Response

In the RC circuit in Figure 26–11 (p. 862), the output we are interested in is the voltage across the resistor. The frequency response of this circuit is also shown. Find f_H, f_L, BW.

Solution The maximum gain is

$$A_{dB_{max}} = 0\text{dB}$$

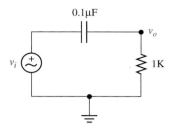

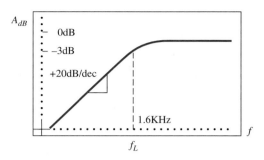

FIGURE 26–11

To find the bandwidth we first find the frequencies at which the gain is

$$A_{dB} = 0 - 3 = -3dB$$

From the figure we see that

$$f_L = 1.6KHz$$

In this response the maximum output occurs at high frequency. Because there is no drop-off at high frequency, $f_H \to \infty$. Bandwidth here is infinite.

REVIEW QUESTIONS

1. What is the problem with plotting frequency response versus f?

2. For each 10-fold increase in f, what is the change in $\log f$? Find $\log f$ if $f = 20\text{Hz}$. 200Hz. 2000Hz.

3. What is a decade?

4. What is an octave?

5. How is the decibel measure of gain defined?

6. For each 10-fold increase in A, how much does A_{dB} increase? Find A_{dB} if $A = 0.707$. $A = 1$. $A = 20$. $A = 200$.

7. What is the actual gain if $A_{dB} = 0dB$?

8. How do we find the 3-dB frequencies and the bandwidth from a Bode gain plot?

26–4 ■ LOW-FREQUENCY RESPONSE—COUPLING CAPACITORS

An AC amplifier is shown in general form in Figure 26–12. As before, the AC input signal is coupled to the amplifier via a capacitor. Likewise, the load is capacitively coupled to the amplifier output. The amplifier in the figure can represent a BJT common-emitter amplifier, an FET common-source amplifier, or an Op-Amp amplifier (inverting or non-inverting). We can model any one of these amplifiers with the Box Model shown in Figure 26–12B. If you put your hand over the right side of the Box Model, you will see that the input side is a single-capacitor circuit. This is also true of the output side. To determine the frequency behavior of a capacitor-coupled AC amplifier, we need only analyze two single-capacitor circuits.

We begin by again examining the single capacitor-single resistor circuit shown in Figure 26–13A. If you compare this circuit to the circuit in Example 26–6, you will see that this circuit exhibits a high-pass response, as shown in Figure 26–13B.

FIGURE 26–12

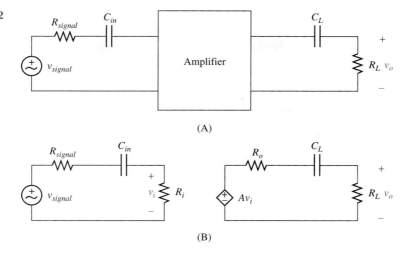

(A)

(B)

FIGURE 26–13

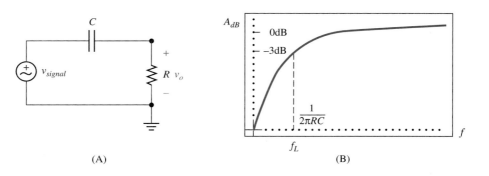

(A)

(B)

Capacitor Reactance

You already know that at DC the capacitor acts as an open circuit, so the circuit has no output voltage. Because the capacitor acts as a short circuit at high frequency, $v_o = v_{signal}$. Therefore

At DC	$A = 0$
At high frequency	$A = 1$

In other words, the apparent "resistance" of the capacitor depends on the input frequency.

At DC this capacitor "resistance" is infinite; at high frequency the capacitor resistance goes to 0Ω. The apparent resistance of the capacitor depends on the input frequency. We call this apparent resistance the **reactance** of the capacitor. The symbol for reactance is X. The reactance for the capacitor is given by

$$X_C = \frac{1}{2\pi f C}$$

The unit of reactance is also the ohm. Note that as $f \rightarrow 0$, $X_C \rightarrow \infty$; as $f \rightarrow \infty$, $X_C \rightarrow 0\Omega$.

As the input frequency increases, the reactance of the capacitor starts at infinity (open circuit), then decreases. As the capacitor reactance decreases, more of the input signal gets to the load.

The 3-dB Frequency

We do not want to go into any mathematical detail here. Rather, we will just state that the 3-dB frequency (f_L) occurs when the capacitor reactance is equal to the total resistance seen by the capacitor. For the simple circuit in Figure 26–13 (p. 863), f_L occurs when

$$X_C = R \rightarrow \frac{1}{2\pi f_L C} = R$$

So

$$f_L = \frac{1}{2\pi RC}$$

EXAMPLE 26–9
Simple RC Circuit

For the RC circuit in Figure 23–14, sketch the Bode gain plot. Show the maximum gain $(A_{dB_{max}})$, f_L, and the gain at f_L (A_{3dB}).

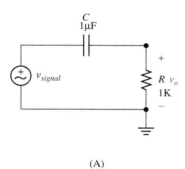

FIGURE 26–14   (A)   (B)

Solution   We find the 3-dB frequency from

$$f_L = \frac{1}{2\pi RC} = \frac{1}{2\pi (1K) \times 1\mu F} = 159 Hz$$

The maximum gain is 1, so for the Bode plot we get

$$A_{dB_{max}} = 20 \log A = 20 \log 1 = 0 dB$$

The gain at the 3-dB frequency of 159Hz is given by

$$A_{dB} = A_{dB_{max}} - 3 = 0 - 3 = -3 dB.$$

The resulting Bode gain plot is shown in Figure 23–14B.

DRILL EXERCISE

Find f_L if the circuit elements in Figure 23–14 have the following values: $R = 10K\Omega$ and $C = 0.22\mu F$

Answer $f_L = 72.3 Hz$

Two-Resistor Circuit

Figure 26–15 shows a circuit that has two resistors and a single capacitor, a circuit that is similar to the circuits in the Box Model. The output here is the voltage across R_2. We find the maximum gain when the frequency is high and the capacitor shorts, resulting in a voltage divider circuit:

$$A_{max} = \frac{v_o}{v_i} = \frac{R_2}{R_1 + R_2}$$

To find the 3-dB frequency we note that the total resistance seen by the capacitor is

$$R_1 + R_2$$

and find f_H from

$$f_H = \frac{1}{2\pi(R_1 + R_2)C}$$

FIGURE 26–15

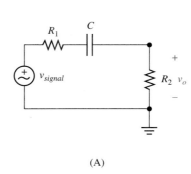

(A)

$$A_{max} = \frac{R_2}{R_1 + R_2}$$

$$f_L = \frac{1}{2\pi(R_1 + R_2)C}$$

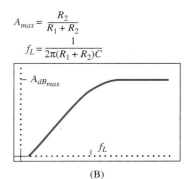

(B)

EXAMPLE 26–10
Two-Resistor, Single Capacitor Circuit

Sketch the Bode plot for the circuit in Figure 26–16A.

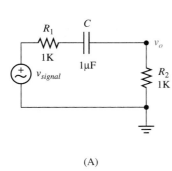

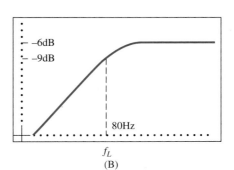

FIGURE 26–16   (A)   (B)

Solution

The maximum gain is found when the capacitor is shorted. The resultant voltage divider gives us

$$A_{max} = \frac{v_o}{v_i} = \frac{R_2}{R_1 + R_2} = \frac{1\text{K}}{1\text{K} + 1\text{K}} = 0.5$$

In decibels, this becomes

$$A_{dB_{max}} = 20 \log A_{max} = -6\text{dB}$$

The 3-dB frequency is found from

$$f_L = \frac{1}{2\pi(R_1 + R_2)C} = \frac{1}{2\pi(1\text{K} + 1\text{K}) \times 1\mu\text{F}} = 79.6\text{Hz}$$

The gain at this frequency is given by

$$A_{3dB} = A_{dB_{max}} - 3 = -6 - 3 = -9\text{dB}$$

The Bode plot for this circuit is shown in Figure 26–16B (p. 865).

DRILL EXERCISE   Find the maximum gain (A_{dB}), f_L, and the 3-dB gain for the circuit in Figure 26–16 (p. 865), given the following circuit elements: $R_1 = 1\text{K}\Omega$, $R_2 = 5\text{K}\Omega$, $C = 4.7\mu\text{F}$

Answer $A_{dB_{max}} = -1.58\text{dB}$, $f_L = 5.64\text{Hz}$, $A_{3dB} = -4.58\text{dB}$

The Box Model

We now return to the Box Model, shown again in Figure 26–17. The maximum gain is still found at high frequency, when the capacitors short. This gives us the system gain we have found in previous chapters:

$$A_{max} = A_{sys} = \left(\frac{R_i}{R_i + R_{signal}}\right)\left(\frac{R_L}{R_L + R_o}\right)A$$

More complicated is finding the overall f_L. Both the left and right parts of the Box Model have their own 3-dB frequencies:

On the left:
$$f_1 = \frac{1}{2\pi(R_{signal} + R_i)C_{in}}$$
On the right:
$$f_2 = \frac{1}{2\pi(R_L + R_o)C_L}$$

The overall 3-dB frequency actually depends on both of these values. In most cases either f_1 or f_2 is substantially greater than the other value. We can then approximate the overall 3-dB frequency as

$$f_L \approx \text{the larger of } f_1 \text{ or } f_2$$

This value for f_L is usually close enough for our purposes. For those of you who prefer exact values, the 3-dB frequency is actually given by

$$f_L = \sqrt{f_1^2 + f_2^2}$$

FIGURE 26–17

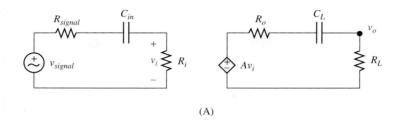

(A)

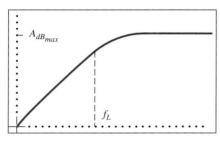

$$A_{max} = \frac{R_i}{R_i + R_{signal}} \times \frac{R_L}{R_L + R_o} \times A$$

$$f_1 = \frac{1}{2\pi(R_i + R_{signal})\,C_{in}}$$

$$f_2 = \frac{1}{2\pi(R_L + R_o)\,C_L}$$

$$f_L = \text{greater of } f_1, f_2$$

(B)

EXAMPLE 26–11
Common-Emitter Amplifier

The DC bias current for the common-emitter amplifier shown in Figure 26–18 is $I_E = 1mA$. The 2N2222 has a current gain of $\beta = 255$.

(a) Find the AC emitter resistance.   **(b)** Find the Box Model parameters.
(c) Find the maximum gain (capacitors are shorted). This is the gain A_{sys} found in earlier chapters. Express the gain in decibels.
(d) Find the two individual 3-dB frequencies (f_1 and f_2) and the total 3-dB frequency (f_L).
(e) Sketch the Bode gain plot for this amplifier.

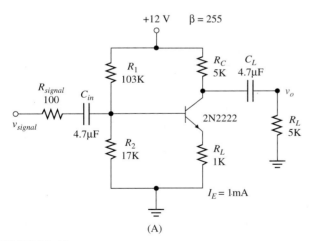

(A)

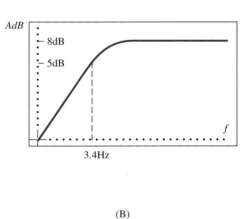

(B)

FIGURE 26–18

Solution
(a) The AC emitter resistance is found from

$$r_E = \frac{26mV}{I_E} = \frac{26mV}{1mA} = 26\Omega$$

(b) The Box Model parameters are ($\beta = 255$)

$$R_{Etotal} = r_e + R_E = 0.026K + 1K \approx 1K$$

$$R_i = R_B \parallel \beta\, R_{Etotal} = 14.6K \parallel 255K = 13.8K\Omega$$

$$R_o = R_C = 5K\Omega$$

$$A = -\frac{R_C}{R_{Etotal}} = -\frac{5K}{1K} = -5$$

(c) The system gain is found from

$$A_{sys} = \left(\frac{R_i}{R_i + R_{signal}}\right)\left(\frac{R_L}{R_L + R_o}\right)A = \left(\frac{13.8K}{13.8K + 0.1K}\right)\left(\frac{5K}{5K + 5K}\right)(-5) = -2.48$$

Ignoring the negative sign, the gain in decibels is

$$A_{dB_{max}} = 20 \log 2.48 = 7.89dB$$

(d) The individual 3-dB frequencies are

$$f_1 = \frac{1}{2\pi(R_i + R_{signal})\,C_{in}} = \frac{1}{2\pi(13.8K) \times 4.7\mu F} = 2.45Hz$$

and

$$f_2 = \frac{1}{2\pi(R_L + R_o)C_L} = \frac{1}{2\pi(10K) \times 4.7\mu F} = 3.39Hz$$

The overall 3-dB frequency is approximately the larger of the two individual ones, because

$$f_2 > f_1 \rightarrow f_L \approx f_2 \approx 3.4Hz$$

The Bode plot is shown in Figure 26–18B (p. 867).

DRILL EXERCISE   Find f_1, f_2, and f_L if the capacitors in Figure 26–18 are changed to $C_{in} = C_L = 1\mu F$

Answer $f_1 = 11.4Hz$, $f_2 = 15.9Hz$, and $f_L \approx 16Hz$

EXAMPLE 26–12
Common-Source
Amplifier

The DC bias current for the common-source amplifier shown in Figure 26–19 is $I_D = 2mA$.

(a) Find the AC transconductance $\left(g_m = \dfrac{2}{|V_{GS(off)}|}\sqrt{I_D\,I_{DSS}}\right)$.

(b) Find the Box Model parameters.

(c) Find the maximum gain (capacitors are shorted). This is the gain A_{sys} found in earlier chapters. Express the gain in decibels.

(d) Find the two individual 3-dB frequencies (f_1 and f_2) and the overall 3-dB frequency (f_L).

(e) Sketch the Bode gain plot for this amplifier.

Solution

(a) The AC transconductance is found from

$$g_m = \frac{2}{|V_{GS(off)}|}\sqrt{I_D\,I_{DSS}} = \frac{2}{3}\sqrt{2mA \times 4mA} = 1.89mS$$

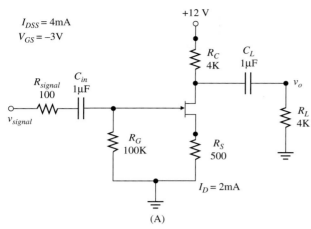

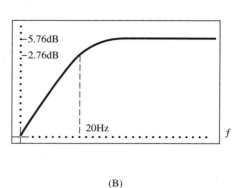

(A)

(B)

FIGURE 26–19

(b) The Box Model parameters are

$$R_{S_{total}} = \frac{1}{g_m} + R_S = 530 + 500 = 1.03K\Omega$$

$$R_i = R_G = 100K\Omega$$

$$R_o = R_D = 4K\Omega$$

$$A = -\frac{R_D}{R_{S_{total}}} = -\frac{4K}{1.03K} = -3.8$$

(c) The maximum gain is found from

$$A_{sys} = \left(\frac{R_i}{R_i + R_{signal}}\right)\left(\frac{R_L}{R_L + R_o}\right)A$$

so

$$A_{sys} = \left(\frac{100K}{100K + 0.1K}\right)\left(\frac{4K}{4K + 4K}\right)(-3.88) = -1.94$$

Ignoring the negative sign, the gain in decibels is

$$A_{dB_{max}} = 20 \log 1.94 = 5.76dB$$

(d) The individual 3-dB frequencies are

$$f_1 = \frac{1}{2\pi(R_i + R_{signal})C_{in}} = \frac{1}{2\pi(100K) \times 1\mu F} = 1.59Hz$$

and

$$f_2 = \frac{1}{2\pi(R_L + R_o)C_L} = \frac{1}{2\pi(8K) \times 1\mu F} = 19.9Hz$$

The overall 3-dB frequency is approximately the larger of the two individual ones, so

$$f_2 > f_1 \rightarrow f_L \approx f_2 \approx 20Hz$$

The Bode plot is shown in Figure 26–19B.

Find $f_1, f_2,$ and f_L if the capacitors in Figure 26–19 (p. 869) are changed to $C_{in} = 0.1\mu F$, and $C_L = 10\mu F$

Answer $f_1 \approx 16\text{Hz}, f_2 \approx 2\text{Hz}, \text{ and } f_L \approx 16\text{Hz}$

Op-Amp Amplifiers

We do not usually construct the Box Model for Op-Amps. Rather, we have shown you how to find the system gain using the ideal Op-Amp analysis. To complete a frequency response analysis, we need to know the input and output resistances of the Op-Amp circuit, as well as the load resistance.

EXAMPLE 26–13
Inverting Op-Amp
AC Amplifier

For the inverting amplifier in Figure 26–20, we know that

$$A_{sys} = -\frac{R_F}{R_1} = -\frac{10\text{K}}{1\text{K}} = -10 \rightarrow A_{dB_{max}} = 20\text{dB}$$

$$R_i = R_1 = 1\text{K}\Omega$$

$$R_o = 0\Omega$$

Find f_1, f_2, f_L and sketch the Bode gain plot.

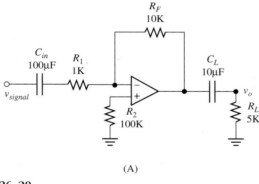

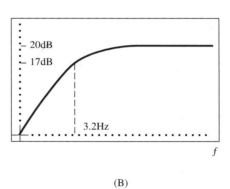

(A)   (B)

FIGURE 26–20

Solution We find the individual 3-dB frequencies as follows (note that there is no R_{signal} here):

$$f_1 = \frac{1}{2\pi(R_{signal} + R_i)C_{in}} = \frac{1}{2\pi(1\text{K}) \times 100\mu F} = 1.59\text{Hz}$$

$$f_2 = \frac{1}{2\pi(R_o + R_L)C_o} = \frac{1}{2\pi(5\text{K}) \times 10\mu F} = 3.18\text{Hz}$$

Because $f_2 > f_1$

$$f_L \approx 3.2\text{Hz}$$

The Bode gain plot for this circuit is shown in Figure 26–20B.

EXAMPLE 26–14
Non-Inverting Op-Amp AC Amplifier

For the non-inverting amplifier in Figure 26–21, we know that

$$A_{sys} = 1 + \frac{R_F}{R_1} = 1 + \frac{10K}{1K} = 11 \rightarrow A_{dB_{max}} = 20.8dB$$

$$R_i = R_2 = 100K\Omega \text{ (Remember, no current enters the Op-Amp.)}$$

$$R_o = 0\Omega$$

Find f_1, f_2, f_L and sketch the Bode gain plot.

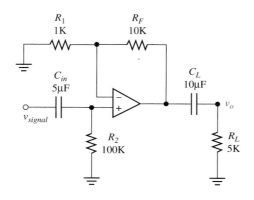

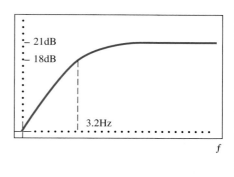

FIGURE 26–21

Solution   We find the individual 3-dB frequencies as follows:

$$f_1 = \frac{1}{2\pi(R_{signal} + R_i)C_{in}} = \frac{1}{2\pi(100K) \times 5\mu F} = 0.318Hz$$

$$f_2 = \frac{1}{2\pi(R_o + R_L)C_o} = \frac{1}{2\pi(5K) \times 10\mu F} = 3.18Hz$$

Because $f_2 > f_1$

$$f_L \approx 3.2Hz$$

The Bode gain plot for this circuit is shown in Figure 26–21B.

DRILL EXERCISE

Find f_1, f_2, and f_L if the capacitors in Figure 26–21 are changed to $C_{in} = 0.47\mu F$, $C_L = 0.22\mu F$.

Answer $f_1 \approx 3.4Hz, f_2 \approx 145Hz, f_L \approx 145Hz$

REVIEW QUESTIONS

1. What is the 3-dB frequency for a simple RC circuit?
2. Draw a generic Box Model with a signal source resistor and a load resistor. What is the system gain when the capacitors are shorted? What is the 3-dB frequency for the left side of the circuit? The right side? What is the overall 3-dB frequency?
3. What are f_1, f_2, f_L, and the system gain for the common-emitter amplifier?
4. What are f_1, f_2, f_L, and the system gain for the common-source amplifier?
5. What are f_1, f_2, f_L, and the system gain for the inverting Op-Amp amplifier?
6. What are f_1, f_2, f_L, and the system gain for the non-inverting Op-Amp amplifier?

26–5 ■ LOW-FREQUENCY RESPONSE—BYPASS CAPACITORS

To increase the system gain of the common-emitter amplifier, we can place a bypass capacitor across the emitter resistor. Similarly, we can increase the gain of the common-source amplifier with a source resistor bypass capacitor. With three capacitors in the amplifier circuit, the frequency response can get quite complicated. To explain the effect of this third capacitor, we look at the circuit shown in Figure 26–22. This is the common-emitter circuit of Figure 26–18, with an added emitter bypass capacitor.

FIGURE 26–22

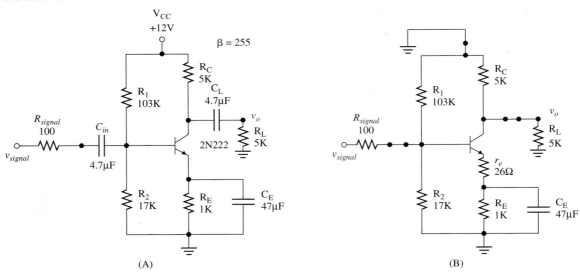

(A)   (B)

As usual, the two coupling capacitors short before the bypass capacitor does. That is, C_E is open when we determine the individual 3-dB frequencies due to C_{in} and C_L:

$$\text{At low frequency: } R_{Etotal} = r_e + R_E = 0.026K + 1K \approx 1K\Omega$$

Because

$$R_B = 103K \parallel 17K = 14.6K\Omega$$

and the current gain is $\beta = 255$, we get the input resistance

$$R_i = R_B \parallel \beta R_{Etotal} = 14.6K \parallel 255K = 13.8K\Omega$$

and

$$f_1 = \frac{1}{2\pi(R_{signal} + R_i)C_{in}} = \frac{1}{2\pi(13.8K) \times 4.7\mu F} = 2.45 \text{Hz}$$

$$f_2 = \frac{1}{2\pi(R_o + R_L)C_L} = \frac{1}{2\pi(10K) \times 4.7\mu F} = 3.39 \text{Hz}$$

Now comes the tough part. As Figure 26–22B shows, when we consider C_E, the coupling capacitors have already become shorts. Rather than try a complicated analysis, let's use what we already know.

C_E **Open:** When C_E is open,

$$R_{Etotal} = r_e + R_E \approx 1K\Omega$$

And the system gain is

$$A_{sys} \approx -2.5 \rightarrow A_{dB} \approx 8 \text{dB}$$

C_E **Short:** When C_E shorts   *i.e at high frequency.*

$$R_{Etotal} = r_e = 0.026\text{K}\Omega$$

The input resistance is now

$$R_i = R_B \parallel \beta r_e = 14.6\text{K} \parallel (255 \times 0.026) = 4.56\text{K}\Omega$$

Note that the bypass capacitor reduces the input resistance. The system gain is

$$A_{sys} = -94 \rightarrow A_{dB} = 39.5\text{dB}$$

Figure 26–23 shows the Bode plot for this circuit. We see the gain rise as the coupling capacitors short. We then see the flat gain response at 8dB before C_E shorts. Finally, we get the maximum system gain of approximately 40dB after C_E has shorted completely.

FIGURE 26–23

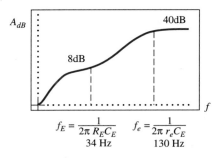

$$f_E = \frac{1}{2\pi R_E C_E}$$
34 Hz

$$f_e = \frac{1}{2\pi r_e C_E}$$
130 Hz

The increase in gain at f_E occurs because the reactance of C_E begins to decrease compared to the resistance of R_E. At this frequency

$$X_{C_E} = R_E \rightarrow \frac{1}{2\pi f_E C_E} = R_E$$

So we find f_E from

$$f_E = \frac{1}{2\pi R_E C_E}$$

For this circuit

$$f_E = \frac{1}{2\pi R_E C_E} = \frac{1}{2\pi(1\text{K}) \times 4.7\mu\text{F}} \approx 34\text{Hz}$$

When an emitter resistor bypass capacitor is used, it usually sets the overall 3-dB frequency (f_L). The resistance seen by C_E, which we will call R_{C_E}, is the resistance seen looking into the emitter. This resistance is given by the rather complicated formula:

$$R_{C_E} = \left(\frac{R_{signal} \parallel R_B}{\beta} + r_e \right) \parallel R_E$$

Luckily for us, this reduces to approximately

$$R_{C_E} \approx r_e$$

so

$$f_L \approx \frac{1}{2\pi r_e C_E}$$

For our circuit

$$f_L = \frac{1}{2\pi r_e C_E} = \frac{1}{2\pi(0.026\text{K}) \times 47\mu\text{F}} = 130\text{Hz}$$

EXAMPLE 26–15
Source Bypass Capacitor

Figure 26–24 repeats the common-source amplifier of Example 26–12, now with a bypass capacitor connected across R_S. Use the results from that example, and find the system gain when C_S shorts, the new f_L, and the FET equivalent of f_E (which we call f_S). Sketch the resultant Bode plot.

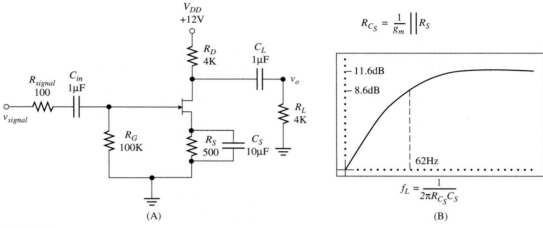

FIGURE 26–24

Solution From Example 26–12 (p. 868), we know the following:

C_S Open

$g_m = 1.89\text{mS}$

$f_1 \approx 1.6\text{Hz}$

$f_2 \approx 20\text{Hz}$

$A_{sys} = -1.5$

C_S Short

The maximum system gain occurs when C_S has shorted. This gain is given by

$$A_{sys} = \left(\frac{R_i}{R_i + R_{signal}}\right)\left(\frac{R_L}{R_L + R_o}\right)(-g_m R_D)$$

For our circuit, the maximum system gain is

$$A_{sys} = -3.8 \rightarrow A_{dB_{max}} = 11.6\text{dB}$$

The bypass capacitor starts to short when its reactance drops below R_S. The frequency at which this occurs is given by

$$f_S = \frac{1}{2\pi R_S C_S}$$

For our circuit

$$f_S = \frac{1}{2\pi R_S C_S} = \frac{1}{2\pi(0.5\text{K}) \times 10\mu\text{F}} = 31.8\text{Hz}$$

The total resistance seen by C_S is

$$R_{C_S} = \frac{1}{g_m} \parallel R_S$$

The values for $1/g_m$ and R_S are generally the same order of magnitude, so we use both in our calculations. For our circuit

$$R_{C_S} = \frac{1}{g_m} \parallel R_S = 530 \parallel 500 = 257\Omega$$

The overall 3-dB frequency is set by C_S and is given by

$$f_L = \frac{1}{2\pi R_{C_S} C_S}$$

For our circuit, we get

$$f_L = \frac{1}{2\pi R_{C_S} C_S} = \frac{1}{2\pi (0.257\text{K}) \times 10\mu\text{F}} = 61.9\text{Hz}$$

The Bode plot for this circuit is shown in Figure 26–24B. You can see that the plateau between f_S and f_L is not evident here. This is because f_S and f_2 are close to each other.

DRILL EXERCISE

For the circuit in Figure 26–24, change the bypass capacitor to $C_S = 4.7\mu\text{F}$ and find f_S and f_L.

Answer $f_S \approx 68\text{Hz}, f_L \approx 132\text{Hz}$

REVIEW QUESTIONS

1. What effect does a bypass capacitor have on the system gain?
2. When does a bypass capacitor begin to short?
3. What are the formulas for f_E and f_S?
4. What is the resistance seen by C_E? What is f_L?
5. What is the resistance seen by C_S? What is f_L?
6. Assume that f_E is much greater than f_1 and f_2. Sketch a Bode plot that shows the plateau for frequencies less than f_E, the rise at f_E, the leveling off at f_L, and the maximum gain.

26–6 ■ High-Frequency Response—The Miller Effect

In Figure 26–1 (p. 852) we showed you that the typical amplifier output increases with increasing frequency until a maximum output is obtained. This increase in system gain is due to the external coupling and bypass capacitors shorting as frequency increases. As frequency was further increased, the output started to decrease, due to the shorting of internal capacitances.

The internal capacitances in electronic devices are due to PN-junction physics that are too complicated to go into here. You should be aware, though, that there is a capacitance that is associated with every PN junction. This capacitance gets larger as the reverse bias on the PN junction increases.

The BJT has a base-collector and a base-emitter junction, so the transistor has an internal capacitance between the base and collector and another internal capacitance between the base and the emitter. These two capacitances are shown schematically in Figure 26–25. The figure also shows the Electronic Workbench model parameters for a typical BJT. We have checked the two junction capacitances for the device, C_e and C_c.

The actual value of the junction capacitances depends on the DC bias levels of the transistor. The values given for C_e and C_c in Figure 26–25 apply only for zero voltage bias. Be aware that when we use these values for our calculations, we get approximate answers. We can obtain more accuracy with a full computer simulation for your particular transistor.

FIGURE 26–25

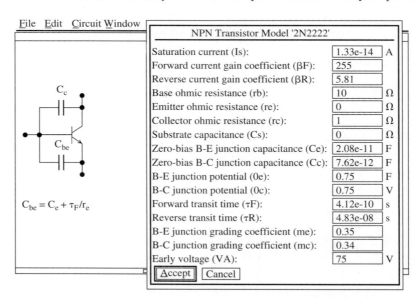

File Edit Circuit Window		

NPN Transistor Model '2N2222'		
Saturation current (Is):	1.33e-14	A
Forward current gain coefficient (βF):	255	
Reverse current gain coefficient (βR):	5.81	
Base ohmic resistance (rb):	10	Ω
Emitter ohmic resistance (re):	0	Ω
Collector ohmic resistance (rc):	1	Ω
Substrate capacitance (Cs):	0	Ω
Zero-bias B-E junction capacitance (Ce):	2.08e-11	F
Zero-bias B-C junction capacitance (Cc):	7.62e-12	F
B-E junction potential (0e):	0.75	F
B-C junction potential (0c):	0.75	V
Forward transit time (τF):	4.12e-10	s
Reverse transit time (τR):	4.83e-08	s
B-E junction grading coefficient (me):	0.35	
B-C junction grading coefficient (mc):	0.34	
Early voltage (VA):	75	V

[Accept] [Cancel]

In addition to the junction capacitance in a BJT, we also need to model the fact that it takes a finite amount of *time* to move charges into and out of the base. This is modeled in the computer program with the **forward transit time parameter** (τ_F), which is also checked in Figure 26–25. To get a circuit model, we can represent this effect with a capacitor that has the value

$$\text{transit time capacitance} = \frac{\tau_F}{r_e}$$

This capacitance appears in parallel with B-E junction capacitance to give us a total base-emitter capacitance of

$$C_{be} = C_e + \frac{\tau_F}{r_e}$$

FETs are constructed with several reverse-biased junctions. Depending on the FET type, there can be several junction capacitances that we must account for in FET amplifiers. We will limit our discussion to the JFET; MOSFETs have additional capacitances between the FET terminals and the substrate. Figure 26–26 shows a JFET and its gate-to-drain and gate-to-source capacitances. JFETs do not have a τ_F because there is no gate current in these devices.

Amplifiers with Bypass Capacitors

To show you how internal capacitances affect amplifier behavior, we have shown the AC model for the BJT common-emitter amplifier of Figure 26–22 in Figure 26–27A. Here, we are only concerned with the high-frequency behavior, so we have shorted the external cou-

FIGURE 26–26

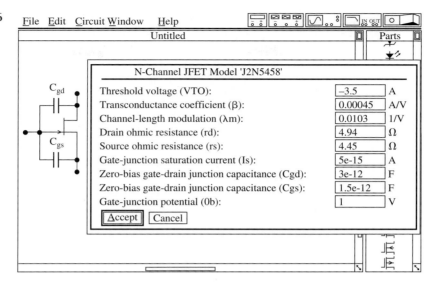

FIGURE 26–27

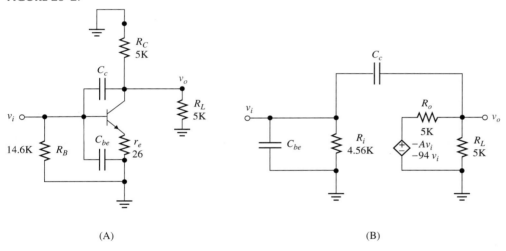

(A)   (B)

pling and bypass capacitors, plus the DC supply voltage. We also show the C_{be} and C_c capacitors.

In Figure 26–27B we show where these capacitances show up in the Box Model. This looks like a difficult circuit to analyze, but we have our tricks.

The Miller Capacitance

In Figure 26–28A (p. 878) we have isolated the B-C junction capacitor. We know that before the internal capacitors come into play, the output voltage is given approximately by

$$v_o \approx A_{sys} v_i$$

This relation is approximate because v_o is really equal to $A_{sys} v_{signal}$. The approximation is accurate if R_{signal} is small compared to R_i, which is the usual case.

The magnitude of the current in the capacitor is given by

$$i_{cap} = \frac{v_{cap}}{X_{C_C}}$$

FIGURE 26–28

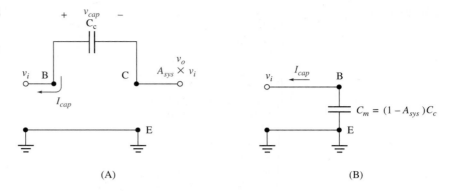

(A)   (B)

Now, because

$$v_{cap} = v_i - v_o$$

and

$$v_o = A_{sys}v_i$$

the current is

$$i_{cap} = \frac{v_i - A_{sys}v_i}{X_{C_c}} = \frac{1 - A_{sys}}{X_{C_c}}v_i$$

Rearranging this equation, we can find the relation between v_i and i_{cap}:

$$v_i = \frac{X_{C_c}}{1 - A_{sys}}i_{cap}$$

In Figure 26–28B, we show a capacitor (C_m) connected from the base to the emitter. The idea here is to use C_m to model the relation between v_i and i_{cap}. To accomplish this, we compare the V-I relation in Figure 26–28B to the one in Figure 26–28A. In the simpler circuit

$$v_i = X_{c_m}i_{cap}$$

For the two circuits to give the same V-I relation at the base:

$$X_{c_m} = \frac{X_{C_c}}{1 - A_{sys}}$$

Because a capacitor's reactance is given by $X = 1/(2\pi fC)$, we get

$$\frac{1}{2\pi fC_m} = \frac{1}{2\pi C_c(1 - A_{sys})}$$

Therefore, for the two circuits to give the same V-I relation at the base

$$C_m = (1 - A_{sys})C_c$$

This relation is known as the *Miller effect,* and C_m is the **Miller capacitance.** The important result here is that a capacitor connected between the input and output of an amplifier looks approximately A_{sys} times larger at the input.

In Figure 26–29A we have redrawn the input side of our BJT amplifier and accounted for C_c by inserting the Miller capacitance between the base and the emitter. As Figure 26–29B shows, the capacitors in the base combine to give us a single capacitor circuit. The total capacitance is given by

$$C_t = C_{be} + C_m$$

FIGURE 26–29

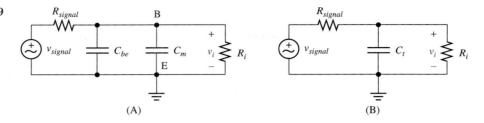

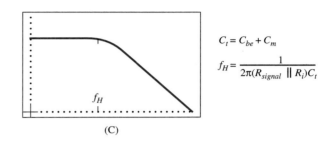

$$C_t = C_{be} + C_m$$

$$f_H = \frac{1}{2\pi(R_{signal} \parallel R_i)C_t}$$

The total resistance seen by the capacitor is

$$R_t = R_{signal} \parallel R_i$$

Don't forget that when there is an emitter bypass resistor, $R_i = R_B \parallel \beta r_e$. The 3-dB frequency for this part of the frequency response is given by

$$f_H = \frac{1}{2\pi R_t C_t} = \frac{1}{2\pi(R_{signal} \parallel R_i)C_t}$$

A typical high-frequency Bode gain plot is shown in Figure 26–29C.

You should note that if an ideal voltage source is directly connected to the base, $R_{signal} = 0\Omega$. In this case, R_t is also 0Ω, and because the high-frequency response never drops off, there is no f_H.

EXAMPLE 26–16
Complete Common-Emitter Frequency Response

Figure 26–30 (p. 880) repeats the circuit of Figure 26–22. Find the complete frequency response for this amplifier. Use the transistor data given in Figure 26–25 (p. 876) for the 2N2222. Sketch the complete Bode gain response.

Solution The device parameters for the 2N2222 are

$$\beta = 255$$
$$C_e = 20.8\text{pF}$$
$$C_c = 7.62\text{pF}$$
$$\tau_F = 412\text{ps}$$

We have already found that for this circuit, the AC emitter resistance is

$$r_e = 26\Omega$$

The input resistance $(C_{Eshorted})$ is

$$R_i = 4.56\text{K}\Omega$$

The maximum gain is

$$A_{sys} = -94$$

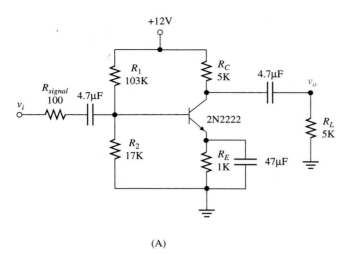

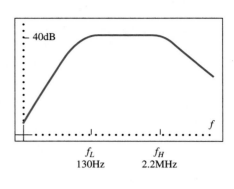

(A)

(B)

FIGURE 26–30

We now determine C_{be} and C_m:

$$C_{be} = C_e + \frac{\tau_F}{r_e} = 20.8\text{pF} + \frac{412\text{ps}}{26} = 36.6\text{pF}$$

The Miller capacitance is found from

$$C_m = (1 - A_{sys})C_c = (1 + 94)\,7.62\text{pF} = 724\text{pF}$$

As is usual, C_m is much greater than C_{be}. The total capacitance is

$$C_t = C_{be} + C_m = 36.6 + 724 = 761\text{pF}$$

The resistance seen by this capacitance is

$$R_t = R_{signal} \parallel R_i = 100 \parallel 4.56\text{K} = 98\Omega$$

As previously stated, the resistance seen by the C_t is usually determined by R_{signal}. The high-frequency 3-dB point is

$$f_H = \frac{1}{2\pi R_t C_t} = \frac{1}{2\pi(98) \times 724\text{pF}} = 2.2\text{MHz}$$

Figure 26–30B shows the complete Bode magnitude response for this common-emitter amplifier.

DRILL EXERCISE

Find C_m, C_{be}, C_t, and f_H for the circuit of Figure 26–22 given the following data $C_e = 10\text{pF}$, $C_c = 2\text{pF}$, $\tau_F = 100\text{ps}$.

Answer $C_m = 190\text{pF}$, $C_{be} = 13.8\text{pF}$, $C_t = 204\text{pF}$, $f_H = 7.96\text{MHz}$

EXAMPLE 26–17

Complete Common-Source Frequency Response

Find the total frequency response for the common-source amplifier in Figure 26–31A (a repeat of Figure 26–24).

Use the results from Example 26–15 (p. 874) and the device data:

$$C_{gd} = 3\text{pF}$$
$$C_{gs} = 1.5\text{pF}$$

to find and sketch the complete Bode magnitude response.

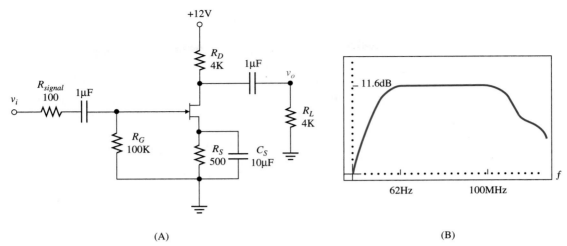

FIGURE 26–31

Solution We have already found that for this circuit

$$R_i = 100\text{K}\Omega$$

$$A_{sys} = -3.8$$

The Miller capacitance at the gate is given by

$$C_m = (1 - A_{sys})C_{gd}$$

For our circuit

$$C_m = (1 - A_{sys})\,C_{gd} = (1 + 3.8)\ 3\text{pF} = 14.4\text{pF}$$

For the FET the total input capacitance is

$$C_t = C_{gs} + C_m$$

For our circuit

$$C_t = C_{gs} + C_m = 1.5\text{pF} + 14.4\text{pF} = 15.9\text{pF}$$

The resistance seen by this capacitance is

$$R_t = R_{signal} \parallel R_i = 100 \parallel 100\text{K} = 100\Omega$$

As previously stated, the resistance seen by the C_t is usually determined by R_{signal}. The high-frequency 3-dB point is

$$f_H = \frac{1}{2\pi R_t C_t} = \frac{1}{2\pi(100) \times 15.9\text{pF}} = 100\text{MHz}$$

The complete Bode magnitude response for this common-emitter amplifier is shown in Figure 26–31B.

DRILL EXERCISE   Find C_m, C_t, and f_H for the circuit of Figure 26–24 (p. 874), given the following data $C_{gd} = 5\text{nF}$, $C_{gs} = 6\text{nF}$

Answer $C_m = 24\text{nF}$, $C_t = 30\text{nF}$, $f_H = 53\text{KHz}$

Amplifiers without Bypass Capacitors

When the emitter or drain resistor is not bypassed, the high-frequency 3-dB frequency depends primarily on C_{be} for the BJT and C_{gs} for the FET. Because these capacitors are small, the bandwidth of amplifiers without bypass capacitors is very large.

The calculation of the high-frequency 3-dB point is too complicated to go into now. For this reason we will just show you the results of computer simulations: Figure 26–32 shows the frequency response of a common-emitter amplifier without a bypass capacitor. Figure 26–32B shows that, as expected, the gain of the common emitter is much lower when a bypass capacitor is not used. Along with the lower gain, the bandwidth is increased from 2.2MHz to 45MHz.

FIGURE 26–32

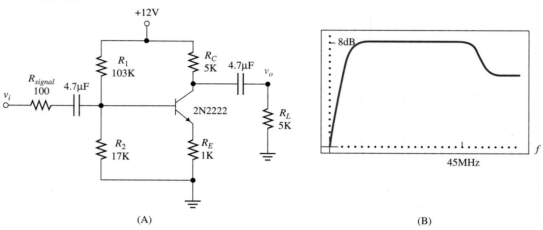

(A)   (B)

Figure 26–33 shows the frequency response of an emitter follower. Figure 26–33B shows that the emitter follower has a low gain ($A \approx 1$) and a very high bandwidth of 580MHz.

The equivalent FET amplifiers show a similar increase in bandwidth. These amplifiers are rarely used by themselves. They are usually combined with other amplifiers that have a smaller bandwidth and so will not affect the overall bandwidth of the amplifier.

FIGURE 26–33

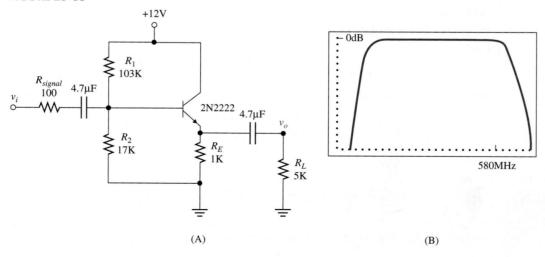

(A)   (B)

REVIEW QUESTIONS

1. What is the cause of the high-frequency drop-off that we see in amplifier responses?
2. What does the forward transit time measure? How can we model this effect?
3. What is the total base capacitance?
4. What does the Miller capacitance model? What is the formula for C_m?
5. Why does using A_{sys} in the Miller formula give an approximate answer?
6. What is the total resistance seen by the base capacitance? What is a good approximation to this resistance?
7. What is f_H for the BJT?
8. Sketch a typical complete Bode gain response for the common-emitter amplifier.
9. What is the Miller capacitance for the JFET?
10. What is the total gate capacitance?
11. What is the resistance seen by this capacitance?
12. What is f_H for the JFET common-source amplifier?
13. Sketch a complete Bode magnitude response for the common-source amplifier.
14. What happens to the bandwidth of an amplifier that is built without bypass capacitors?

26–7 ■ MULTISTAGE AMPLIFIERS

Most practical and commercial amplifiers use several stages to amplify the input signal. When we connect several stages together, we can significantly affect the frequency behavior of the amplifier. To give you some insight into multistage frequency response, we will show you some computer simulations of the following two-stage amplifiers: common emitter-to-common emitter, emitter follower-to-common emitter, and common emitter-to-common base (cascode).

Common Emitter-to-Common Emitter (CE-CE)

Figure 26–34A (p. 884) shows a two-stage amplifier in which a common-emitter amplifier (Q_1) drives another common-emitter amplifier (Q_2). When one amplifier drives a second amplifier, we say the amplifiers are **cascaded amplifiers.**

The frequency response for the cascaded common-emitter amplifiers is shown in Figure 26–34B (p. 884). Because each stage provides a high gain, we get a very high gain from this circuit (80dB = 10,000). This circuit has two major disadvantages: (1) Because the emitter resistor of the first stage is bypassed, the input resistance is low; (2) The bandwidth is very limited ($BW = 520$KHz).

Emitter Follower-to-Common Emitter (EF-CE)

Figure 26–35A (p. 884) shows a two-stage amplifier in which the first stage is an emitter follower. Figure 26–35B (p. 884) shows the frequency response for this amplifier. You can see that the overall gain is lower than the CE-CE amplifier. This is expected because the first stage has a gain of 1. On the plus side, the bandwidth is now 6.8MHz. This circuit has an additional advantage: Because the first-stage emitter resistor is not bypassed, the input resistance of the amplifier is greater than the CE-CE amplifier.

FIGURE 26–34

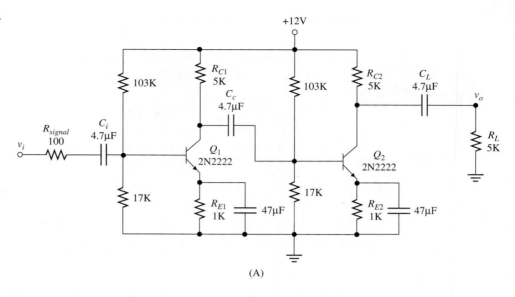

(A)

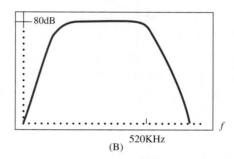

(B)

FIGURE 26–35

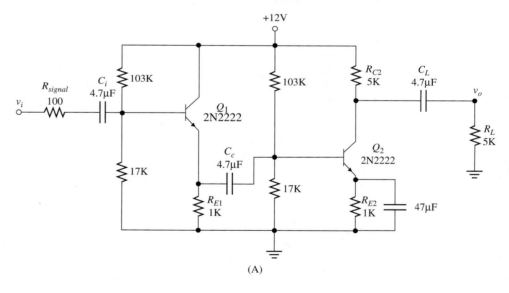

(A)

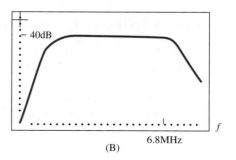

(B)

Cascode Amplifier (Common Base-to-Common Emitter)

Finally, in Figure 26–36A we show you the *cascode connection* first discussed in Chapter 16, Sections 14 and 15. Because the collector junction capacitance of Q_2 is not connected back to the input (base of Q_1), the Miller effect here is not operative.

The frequency response of this circuit is shown in Figure 26–36B. As you can see, this cascode amplifier achieves the same gain as the EF-CE cascade amplifier; however, the cascode amplifier has the largest bandwidth of all of the amplifiers we have discussed. The cascode does have a disadvantage. As you may remember, it has a very low input resistance.

FIGURE 26–36

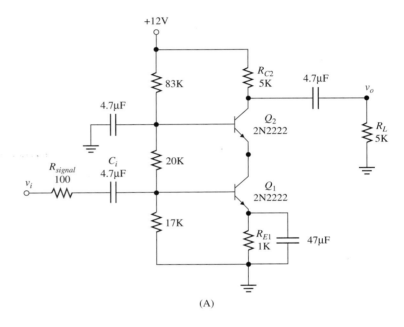

(A)

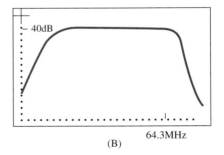

(B)

REVIEW QUESTIONS	
	1. Discuss the gain and bandwidth (large? small?) of the CE-CE amplifier.
	2. What is the disadvantage of the CE-CE amplifier?
	3. Compare EF-CE amplifier to the CE-CE amplifier. What is the advantage of the EF-CE amplifier?
	4. Compare the cascode amplifier to the other multistage amplifiers. What is the disadvantage of the cascode amplifier?

26–8 ■ TRANSISTOR DATA SHEETS

Electronic computer programs were originally constructed to help the chip designer build integrated circuits. For this reason, many of the transistor parameters in these programs are based on the internal physics of the device.

Manufacturer's data sheets, on the other hand, are provided to help the user of the device, not the designer. Manufacturers derive transistor models on the basis of laboratory tests they perform. This gives rise to a different set of parameters to those required by computer programs.

BJT Data Sheet

Figure 26–37 shows a page from a transistor manufacture data book. The most typical high-frequency parameters that are given by manufacturers are C_{ob} and f_T. The base-collector junction capacitance is represented by C_{ob}, therefore

$$C_c = C_{ob}$$

The base-emitter junction capacitance and the forward transit time are not given. Instead, the parameter f_T, known as the *current gain-bandwidth product,* provides the same information.

Figure 26–38 (p. 888) shows the Bode plot for the current gain of a transistor (β, or h_{fe}). Although we have considered h_{fe} as a constant, the current gain will actually start to drop as the base capacitance begins to short. The parameter f_T is the frequency at which the current gain falls to one (0dB); hence, f_T is known as the *unity-gain frequency.*

Because of the test set-up for measuring f_T, the total base capacitance here is

$$C_t = C_{be} + C_{ob}$$

The frequency at which gain of h_{fe} drops to one is given by

$$f_T = \frac{1}{2\pi r_e C_t}$$

Note in Figure 26–37 that f_T is found at a specific collector bias current. Use this current to find r_e. The total base capacitance is

$$C_t = C_{be} + C_{ob} = \frac{1}{2\pi r_e f_T}$$

so

$$C_{be} = \frac{1}{2\pi r_e f_T} - C_{ob}$$

This is the capacitance we need to perform a hand analysis. To use a computer program, we need to find C_e and τ_F. We know that

$$C_{be} = C_e + \frac{\tau_F}{r_e}$$

We can solve for τ_F:

$$\tau_F = r_e(C_{be} - C_e)$$

In some cases, the manufacturer might also give you the base-emitter junction capacitance (C_{ib}), which is the same parameter as C_e. If it is not given, assume that $C_e = 0$.

PNP Transistors

National Semiconductor

General Purpose Amplifiers and Switches

Type No.	Case Style	V_{CBO} (V) Min	V_{CEO} (V) Min	V_{EBO} (V) Min	I_{CES}* I_{CBO} @ (nA) Max	V_{CB} (V)	h_{FE} Min Max	@ I_C (mA) & V_{CE} (V)	$V_{CE(SAT)}$ (V) Max	$V_{BE(SAT)}$ (V) Min Max	@ I_C (mA)	C_{OB} (pF) Max	f_T (MHz) Min Max	@ I_C (mA)	t_{OFF} (ns) Max	NF (dB) Max	Test Conditions	Process No.
2N2904	TO-5	60	40	5	20	50	20 40 120 35 25 20	500 10 150 10 10 10 1 10 0.1 10	0.4 1.6	1.3 2.6	150 500	8	200	50	100		(Note 2)	63
2N2904A	TO-5	60	60	5	10	50	40 40 40 120 40 40	500 10 150 10 10 10 1 10 0.1 10	0.4 1.6	1.3 2.6	150 500	8	200	50	100		(Note 2)	63
2N2905 also Avail. JAN/TX/V Versions	TO-5	60	40	5	20	50	30 100 300 75 50 35	500 10 150 10 10 10 1 10 0.1 10	0.4 1.6	1.3 2.6	150 500	8	200	50	100		(Note 2)	63
2N2905A also Avail. JAN/TX/V Versions	TO-5	60	60	5	10	50	50 100 300 100 100 75	500 10 150 10 10 10 1 10 0.1 10	0.4 1.6	1.3 2.6	150 500	8	200	50	100		(Note 2)	63
2N2906	TO-18	60	40	5	20	50	20 40 120 35 25 20	500 10 150 10 10 10 1 10 0.1 10	0.4 1.6	1.3 2.6	150 500	8	200	50	100		(Note 2)	63

FIGURE 26–37

FIGURE 26–38

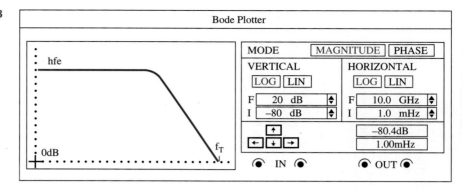

EXAMPLE 26–18
BJT High-Frequency Parameters

Use the data given in Figure 26–37 (p. 887) for the 2N2904 to derive C_e and τ_F.

Solution From the data sheet

$$C_c = C_{ob} = 8pF$$

The unity-gain frequency is

$$f_T = 200MHz \quad @ \quad I_C = 50mA$$

Because $I_E \approx I_C$, the AC emitter resistance is

$$r_e = \frac{26mV}{50mA} = 0.52\Omega$$

The total base capacitance is

$$C_t = C_{be} + C_{ob} = \frac{1}{2\pi r_e f_T} = \frac{1}{2\pi(0.52) \times 200MHz} = 1530pF$$

Subtracting C_{ob}:

$$C_{be} = C_t - C_{ob} = 1530pF - 8pF \approx 1520pF$$

Because C_{ib} is not given, we find the forward transit time from

$$\tau_F = r_e C_{be} = 0.52 \times 1520pF = 790ps$$

JFET Data Sheet

Figure 26–39 shows a manufacturer's data sheet for several JFETs. The gate-source junction capacitance is labeled C_{iss}, so

$$C_{gs} = C_{iss}$$

The gate-drain junction capacitance is labeled C_{rss}, so

$$C_{gd} = C_{rss}$$

FIGURE 26–39

Case 318-02
TO-236AA
(SOT-23)

PIN 1. DRAIN
2. SOURCE
3. GATE

P-CHANNEL JFETs TO-236AA (SOT-23) Case 318-02, Style 10

Device	$|Y_{fs}|$ mmho Min	$|Y_{os}|$ μmho Max	C_{iss} pF Max	C_{rss} pF Max	$V_{(BR)GSS}$ V Min	$V_{GS(off)}$ V Min	$V_{GS(off)}$ V Max	I_{DSS} mA Min	I_{DSS} mA Max
MMBF5460	1	75	7	2	40	0.75	6	1	5
MMBF5463	1	75	7	2	60	0.75	6	1	5
MMBF5462	2	75	7	2	40	1.8	9	4	16
MMBF5461	1.5	75	7	2	40	1	7.5	2	9

High-Frequency Amplifiers

PIN 1. DRAIN
2. SOURCE
3. GATE

N-CHANNEL JFETs TO-226AA (TO-92) Case 29-02, Style 5

Device	$|Y_{fs}|$ mmho Min	@ f MHz	$|Y_{os}|$ μmho Max	@ f MHz	C_{iss} pF Max	C_{rss} pF Max	NF dB Max	@ RG = 1K f MHz	$V_{(BR)GSS}$ V Min	$V_{GS(off)}$ V Min	$V_{GS(off)}$ V Max	I_{DSS} mA Min	I_{DSS} mA Max
2N5669	1.6	100	100	100	7	3	2.5	100	25	1	6	4	10
MPF108	1.6	100	200	100	6.5	2.5	3	100	25	0.5	8	1.5	24
MPF102	1.6	100	200	100	7	3	—	—	25	—	8	2	20
2N3819#	1.6	100	—	—	—	4	—	8	25	—	8	2	20
2N5668	1	100	50	100	7	3	2.5	100	25	0.2	4	1	5
2N5484	2.5	100	75	100	5	1	3	100	25	0.3	3	1	5
2N5670	2.5	100	150	100	7	3	2.5	100	25	2	8	8	20
2N5246	2.5	400	100	400	4.5	1	—	—	30	0.5	4	1.5	7
MPF4223	—	—	200	200	6	2	5	200	30	0.1	8	3	18
2N5485	3	400	100	400	5	1	4	400	25	5	4	4	10
MPF3823	—	—	—	—	6	2	2.5	100	30	—	8	4	20
2N5486	3.5	400	100	400	5	1	4	400	25	2	6	8	20
MPF4416	4	400	100	400	4	0.8	4	400	30	2	6	5	15
MPF4416A	4	400	100	400	4	0.8	4	400	35	2.5	6	5	15
2N5245##	4	400	100	400	4.5	1	4	400	30	1	6	5	15
2N5247##	4	400	150	400	4.5	1	4	400	30	1.5	8	8	24
J308	12 Typ	100	250 Typ	100	7.5	2.5	1.5 Typ	450	25	1	6.5	12	60
J309	12 Typ	100	25 Typ	100	7.5	2.5	1.5 Type	450	25	1	4	12	30
J310	12 Typ	100	25 Typ	100	7.5	2.5	1.5 Typ	450	25	2	6.5	24	60

***JAN, JTX, JTXV   #Case Style 22   ##Case Style 23

1. What high-frequency BJT parameters are typically given in a manufacturer's data sheet?

2. How is the base-collector junction capacitor labeled in a data sheet?

3. What is the name of f_T?

4. Given f_T, how do you find C_t? C_{be}?

5. How is the JFET gate-source junction capacitance labeled in a data sheet?

6. How is the JFET gate-drain junction capacitance labeled in a data sheet?

26–9 ■ OP-AMP FREQUENCY RESPONSE

The Op-Amp is an inherently wide-band, high-gain amplifier that is intended for use in feedback configuration. While you might think that having a wide bandwidth is always a good thing, as we shall show you in the next chapter, a large bandwidth can cause stability problems in feedback amplifiers. To narrow the bandwidth of some Op-Amps, the user connects an external compensation capacitor to appropriate pins of the Op-Amp (Figure 26–40).

Some Op-Amps (the very popular 741 for example) have a compensation capacitor integrated onto the chip (Figure 26–41A). Problem is, proper compensation requires capacitors in the μF range, but we can only integrate capacitors in the pF range. To solve this problem, we use the Miller effect, placing the pF capacitor between the input and output of the very high-gain amplifier stage.

FIGURE 26–40

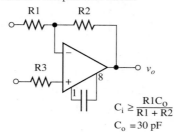

Standard Compensation Circuit

$$C_i \geq \frac{R1C_0}{R1 + R2}$$
$$C_o = 30 \text{ pF}$$

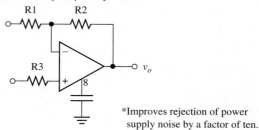

Alternate* Frequency Compensation

*Improves rejection of power supply noise by a factor of ten.

FIGURE 26–41

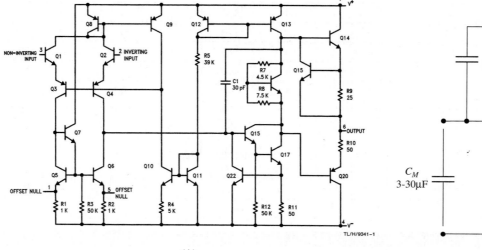

TL/H/9341–1

(A)

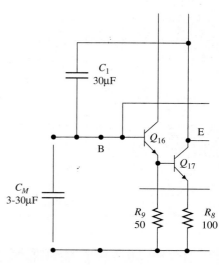

(B)

As Figure 26–41B shows, the integrated capacitor is connected between the base and the emitter of the common-emitter amplifier stage of the 741. The Miller effect tells us that a capacitor connected between input and output acts as though there is a large capacitor at the input:

$$C_m = (1 - A) \, C_c$$

Because the gain is between 10^5 and 10^6, the integrated compensation capacitor of 30pF acts as the much larger input capacitance of

$$C_m = 3 \text{ to } 30\mu\text{F}$$

The Gain-Bandwidth Product

As discussed in Chapter 23, Section 7, the product of the gain and the bandwidth *(GBW)* is constant. As the gain decreases, the bandwidth increases; as the gain increases, the bandwidth decreases. For a given gain, the bandwidth of an Op-Amp amplifier is given by

$$BW = \frac{GWB}{A_{sys}}$$

The 741 Op-Amp has a bandwidth of 1.5MHz when the Op-Amp is configured as a unity-gain buffer. Therefore, the gain bandwidth product for the 741 is

$$GBW_{741} = 1 \times 1.5\text{MHz} = 1.5\text{MHz}$$

The bandwidth of a 741 amplifier with a gain of 10 is

$$BW_{741} = \frac{GBW_{741}}{A_{sys}} = \frac{1.5\text{MHz}}{10} = 150\text{KHz}$$

If you want a gain of 100 with the 741, then you will have a bandwidth of only

$$BW_{741} = \frac{GBW_{741}}{A_{sys}} = \frac{1.5\text{MHz}}{100} = 15\text{KHz}$$

This bandwidth does not cover the audio frequency range, which goes up to 20KHz. You either have to lower the gain (use several stages to achieve the required gain) or use an Op-Amp with a larger unity-gain frequency.

REVIEW QUESTIONS

1. What problem can be caused by a large bandwidth in a feedback amplifier?
2. How did the first integrated circuit Op-Amp narrow the amplifier bandwidth?
3. What is the limit on the size of a capacitor that can be integrated on a chip?
4. What effect can be used to increase the apparent size of this capacitor?

SUMMARY

■ An amplifier's frequency response is shown by plotting the amplifier gain versus the input frequency.

■ The useful frequency range of an amplifier is known as the bandwidth, where

$$BW = f_H - f_L$$

■ f_H and f_L are the half-power frequencies, found when $A = 0.707 A_{max}$.

■ The Bode plot gives us the gain in decibels versus the log of the input frequency.

■ The decibel measure for gain is

$$A_{dB} = 20 \log A$$

- On a Bode plot, f_H and f_L are known as the 3-dB frequencies and are found when $A_{dB} = A_{dB_{max}} - 3\text{dB}$.
- Each coupling capacitor in a BJT or FET circuit produces a low-frequency 3-dB frequency of

$$f_L = \frac{1}{2\pi RC}$$

where R is the total resistance seen by the coupling capacitor C.

- As a bypass capacitor begins to short, the gain increases; the gain reaches a maximum when the bypass capacitor has shorted completely.
- The frequency at which the bypass capacitor begins to short and the gain begins to increase is labeled f_E (BJT) or f_S (FET) and is given by

$$f_E = \frac{1}{2\pi R_E C_E} \qquad f_S = \frac{1}{2\pi R_S C_S}$$

- The frequency at which the bypass capacitor is completely shorted provides the f_L due to this capacitor

$$\textbf{BJT:} \; f_L \approx \frac{1}{2\pi r_e C_E} \qquad \textbf{FET:} \; f_L = \frac{1}{2\pi \left(\dfrac{1}{g_m} \| R_S \right) C_S}$$

- The BJT has internal capacitances between the base and collector and between the base and emitter. Because of these capacitances, the gain drops off at high frequencies.
- The base-collector junction capacitance is labeled C_c in computer programs and C_{ob} in data sheets.
- There are two capacitors between the base and the emitter. The junction capacitance is labeled C_e in computer programs and C_{ib} in data sheets. There is also a capacitance due to the transit time (τ_F). The total base-emitter capacitance is given by

$$C_{be} = C_e + \frac{\tau_F}{r_e}$$

- Because the base-collector capacitor connects the output back to the input, the Miller effect produces the large effective capacitance in the base:

$$C_m = (1 - A_{sys}) \, C_c$$

- The high-frequency 3-dB point is given by

$$f_H = \frac{1}{2\pi R_t C_t}$$

where $C_t = C_{be} + C_m$ and $R_t = R_{signal} \| R_i$

- For the FET, the gate-drain junction capacitance is labeled C_{gd} in computer programs and C_{iss} in data sheets. The gate-source capacitance is labeled C_{gs} in computer programs and C_{rss} in data sheets. The high-frequency 3-dB point is

$$f_H = \frac{1}{2\pi R_t C_t}$$

where $C_t = C_{gs} + C_m, \quad C_m = (1 - A_{sys}) \, C_{gd}, \quad R_t = R_{signal} \| R_G$

- The gain of the common emitter-to-common emitter multistage amplifier is very high. The bandwidth is small, and the input resistance is low.
- The gain of the emitter follower-to-common emitter amplifier is the same as a single-stage common-emitter amplifier. The bandwidth is much greater and the input resistance is higher than the cascaded common-emitter amplifier.
- The gain of the cascode amplifier (common emitter-to-common base) is the same as the EF-CE amplifier. It has a higher bandwidth but a very low input resistance.
- Wide-band high-gain Op-Amps used in feedback configuration can become unstable. To correct for this, either an external or internal capacitor is used to reduce the bandwidth. Because we can

only integrate capacitors in the pF range, the Miller effect is used to increase the apparent value of the integrated capacitor into the μF range.

■ The gain-bandwidth product of an Op-Amp amplifier is a constant. As the gain increases, the bandwidth decreases.

PROBLEMS

SECTION 26–2   The Gain Response

1. Find f_H, f_L, BW for the frequency response shown in Figure 26–42.
2. Find f_H, f_L, BW for the frequency response shown in Figure 26–43.
3. Find f_H, f_L, BW for the frequency response shown in Figure 26–44.

FIGURE 26–42

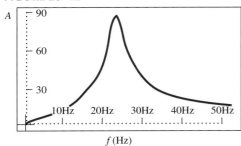

FIGURE 26–43

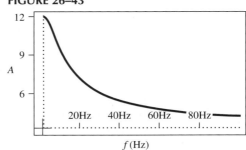

FIGURE 26–44

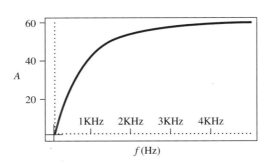

SECTION 26–3   The Bode Gain Plot

4. Find the logarithm of the following frequencies

$$f = 1, 2, 3, 4, 5, 6, 7, 8, 9, 10, 20, 30, 40, 50, 60, 70, 80, 90, 100.$$

5. Use the results of the previous problem to draw a horizontal scale of $\log f$. Be sure the distance from $\log f = 0$ to $\log f = 1$ is the same as the distance from $\log f = 1$ to $\log f = 2$.

6. Given the following frequencies

$$10\text{Hz}, 25\text{Hz}, 50\text{Hz}, 100\text{Hz}, 2.5\text{KHz}, 5\text{KHz}$$

(a) Which frequencies are a decade apart?   (b) Which frequencies are two decades apart?

7. Given the following frequencies

$$30\text{Hz}, 100\text{Hz}, 120\text{Hz}, 200\text{Hz}, 240\text{Hz}, 800\text{Hz}$$

(a) Which frequencies are an octave apart?   (b) Which frequencies are two octaves apart?

8. Find A_{3dB} for the following gains: $A = 0.1, 0.5, 0.707, 1, 25, 150$.

9. Find f_H, f_L, BW for the Bode gain plot in Figure 26–45 (p. 894).

FIGURE 26–45

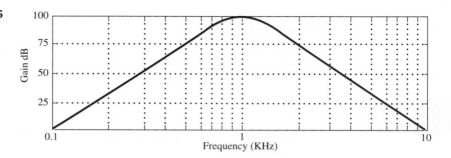

10. Find f_H, f_L, BW for the Bode gain plot in Figure 26–46.

FIGURE 26–46

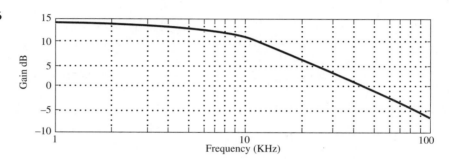

11. Find f_H, f_L, BW for the Bode gain plot in Figure 26–47.

FIGURE 26–47

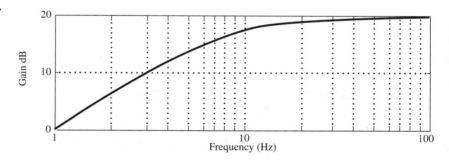

SECTION 26–4 Low-Frequency Response—Coupling Capacitors

FIGURE 26–48

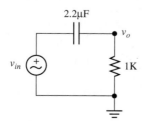

12. For the RC circuit of Figure 26–48, find A_{max} and f_L. Sketch the Bode gain plot for this circuit.

13. Choose a new capacitor in Figure 26–48 so that $f_L = 540\text{KHz}$.

14. For the RC circuit of Figure 26–49, find A_{max} and f_L. Sketch the Bode gain plot for this circuit.

15. The bias current for the common-emitter amplifier in Figure 26–50 is $I_E = 1\text{mA}$.

    (a) Find r_e.   **(b)** Find the Box Model parameters, as well as A_{sys}.

    (c) Find the 3-dB frequency for the input and output coupling capacitors.

    (d) Find the overall 3-dB frequency.   **(e)** Sketch the Bode gain plot.

16. Choose a new set of coupling capacitors in Figure 26–50 so that $f_L = 30\text{Hz}$.

FIGURE 26–49

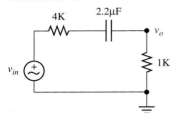

FIGURE 26–50

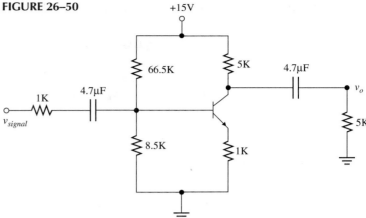

17. The bias current for the common-source amplifier shown in Figure 26–51 is $I_D = 1\text{mA}$.

    (a) Find g_m.   **(b)** Find the Box Model parameters, as well as A_{sys}.

    (c) Find the 3-dB frequency for the input and output coupling capacitors.

    (d) Find the overall 3-dB frequency.   **(e)** Sketch the Bode gain plot.

18. Choose a new set of capacitors in Figure 26–51 so that $f_L = 20\text{Hz}$.

FIGURE 26–51

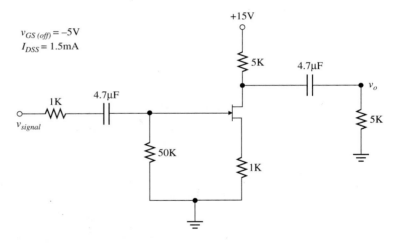

19. Find the individual and overall 3-dB frequencies for the Op-Amp AC amplifier in Figure 26–52. Sketch the Bode gain plot.

20. Change the output capacitor in Figure 26–52 so that $f_L = 20\text{Hz}$.

21. Find the individual and overall 3-dB frequencies for the Op-Amp AC amplifier in Figure 26–53. Sketch the Bode gain plot.

22. Change the output capacitor in Figure 26–53 so that $f_L = 100\text{Hz}$.

FIGURE 26–52

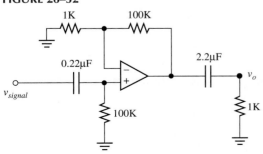

FIGURE 26–53

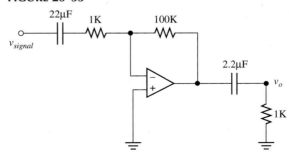

SECTION 26–5   Low-Frequency Response—Bypass Capacitors

23. The bias current for the common-emitter amplifier in Figure 26–54 is $I_E = 1$mA.

    (a) Find r_e.   **(b)** Find the Box Model parameters as well as A_{sys}.

    (c) Find the 3-dB frequency for the input and output coupling capacitors.

    (d) Find f_E, the 3-dB frequency due to C_E, and the overall 3-dB frequency.

    (e) Sketch the Bode gain plot.

24. Choose a new bypass capacitor in Figure 26–54 so that $f_L = 250$Hz.

FIGURE 26–54

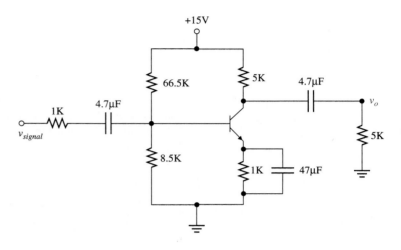

25. The bias current for the common-source amplifier in Figure 26–55 is $I_D = 1$mA.

    (a) Find g_m.   **(b)** Find the Box Model parameters, as well as A_{sys}.

    (c) Find the 3-dB frequency for the input and output coupling capacitors.

    (d) Find f_S, the 3-dB frequency due to C_S, and the overall 3-dB frequency.

    (e) Sketch the Bode gain plot.

26. Choose a new bypass capacitor in Figure 26–55 so that $f_L = 30$Hz.

FIGURE 26–55

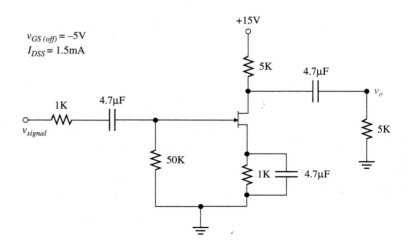

SECTION 26–6   High-Frequency Response—The Miller Effect

27. A BJT has the following parameters:

$$C_{ob} = 8\text{pF}, \ C_{ib} = 2\text{pF}, \text{ and } f_T = 200\text{MHz} @ I_C = 50\text{mA}$$

If $A_{sys} = -200$, find C_m, C_{be}, and C_r.

28. Find the Miller capacitance seen at the input in Figure 26–56.

FIGURE 26–56

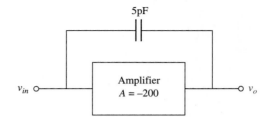

29. For the BJT in Figure 26–54 and Problem 23, $C_e = 5pF$, $C_c = 3pF$, $\tau_F = 500ps$.
   (a) Find f_H.   **(b)** Sketch the complete Bode gain plot.
30. For the BJT in Figure 26–54 and Problem 23, $C_e = 7pF$, $C_c = 10pF$, $\tau_F = 400ps$.
   (a) Find f_H.   **(b)** Sketch the complete Bode gain plot.
31. For the FET in Figure 26–55 and Problem 25, $C_{gd} = 5pF$, $C_{gs} = 1pF$.
   (a) Find f_H.   **(b)** Sketch the complete Bode gain plot.
32. For the FET in Figure 26–55 and Problem 25, $C_{gd} = 3pF$, $C_{gs} = 2pF$.
   (a) Find f_H.   **(b)** Sketch the complete Bode gain plot.

Computer Problems

33. List the BJT models available in your computer program. Next to each model, write the values given for C_c, C_e, τ_F.
34. If $I_E = 1mA$, find C_{be} for each value of C_e and τ_F listed in Problem 33.
35. List the JFET models available in your computer program. Next to each model, write the values given for C_{gs} and C_{gd}.
36. Simulate the circuit of Figure 26–54 with the device parameters given in Problem 29. Show the complete Bode gain plot.
   (a) What is the mid-band gain?   **(b)** What is f_L?   **(c)** What is f_H?
37. Repeat Problem 36 using different BJT devices from your library. Discuss the results.
38. Simulate the circuit of Figure 26–55 with the device parameters given in Problem 31. Show the Bode gain plot.
   (a) What is the mid-band gain?   **(b)** What is f_L?   **(c)** What is f_H?
39. Repeat Problem 38 using different JFET devices from your library. Discuss your results.
40. For the following Op-Amps, simulate the circuit of Figure 26–52 (p. 895), and sketch the complete Bode plot:
   (a) An ideal Op-Amp.   **(b)** The 741 Op-Amp.   **(c)** Another Op-Amp from your library.
   (d) Compare A_{sys}, f_L, and f_H for the three different Op-Amps.
41. The gain-bandwidth product (the unity-gain frequency) for the 741 Op-Amp is $GBW = 1.5MHz$. Simulate the basic non-inverting 741 amplifier with $R_1 = 1K\Omega$ (You can eliminate the coupling capacitors for this problem).
   (a) Find the gain, the bandwidth, and GBW when $R_F = 2K\Omega$.
   (b) Find the gain, the bandwidth, and GBW when $R_F = 20K\Omega$.
   (c) Find the gain, the bandwidth, and GBW when $R_F = 200K\Omega$.
   (d) Is the gain-bandwidth product a constant?

Chapter

27

Active Filters

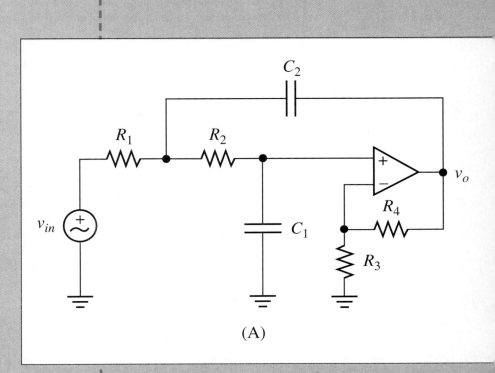

(A)

27–1 ■ INTRODUCTION

In the last chapter we found the bandwidth of amplifiers to determine the useful frequency range of the device. The bandwidth of these amplifiers was limited at the low-frequency end by coupling and bypass capacitors. At the high-frequency end, internal capacitances determined f_H. When we want to ensure that most of the input signal is amplified, we look for a very wide-band amplifier response.

However, very often we want to build circuits that have limited bandwidth. These circuits, known as **filters,** allow us to selectively amplify certain frequency ranges and not others. Consider, for example, the receiver in Figure 27–1A. The arrows in the figure point to the two tone controls, the treble control and the bass control. These tone controls are actually filters. The treble control filters out low-frequency audio, increasing the high-frequency content of the signal. That is, the treble control is a high-pass filter. If you are listening to a soprano—or Little Richard—you would increase your amplifier's treble response.

FIGURE 27–1

(A) Tone controls   BASS   TREBLE

60Hz   400Hz   2.4KHz   16KHz
150Hz   1KHz   6KHz
(B) Equalizer

The bass control, on the other hand, filters out high-frequency audio, increasing the low-frequency content of the signal. That is, the bass control is a low-pass filter. If you want to hear the kettle drums of classical music or the gut-pounding beat of rock and roll, you want to increase your amplifier's bass response.

Equalizers (Figure 27–1B) are devices that allow you to further tailor the frequency response of your amplifier. The equalizer in Figure 27–1B has slide switches that are labeled with the frequencies 60Hz, 150Hz, 400Hz, 1KHz, 2.4KHz, 6KHz, and 16KHz. Each one of these switches controls a filter that passes a small band of frequencies around the labeled frequency. An equalizer, then, is a set of band-pass filters.

Filters are an integral part of all electronic devices that are used in communication and signal transmission. This includes such devices as radio, TV, telephone, cellular phone, satellite communication, remote control, feedback control, voice recognition and synthesis, and CAT and MRI scans. Because filters are so widely used, it is important to understand their basic behaviors.[1]

In the old days we built filters using resistors, capacitors, and inductors. Because these filters use passive components, they are now known as **passive filters.** Passive filters are still used today in high-power and high-frequency situations. Figure 27–2A shows examples of commercially-available passive filters used to remove noise from power lines.

One of the disadvantages of passive filters is that the gain of the filter is always less than 1. That is, the output voltage is always smaller than the input voltage. This is due to voltage losses in the passive elements of the filter. To compensate for the loss in a passive filter, we could always cascade the filter with an amplifier; however, by using Op-Amps we

---

[1]Many of the applications mentioned here can use digital technology to do the filtering. Even though digital implementation of a filter uses different electronics than described in this chapter, the basic principles of digital and analog filters are the same.

FIGURE 27–2
Filters

S, V, W, AND VS, VV, VW SERIES FILTERS

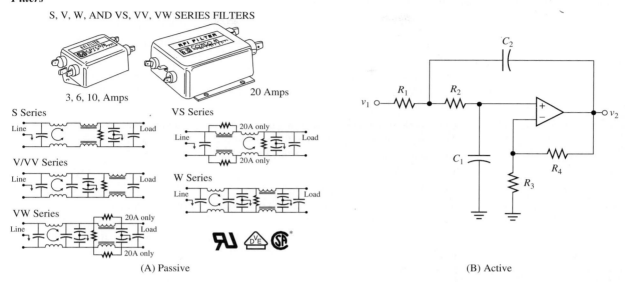

3, 6, 10, Amps

20 Amps

S Series

VS Series

V/VV Series

W Series

VW Series

(A) Passive

(B) Active

can construct a combined filter-amplifier. In fact, one of the most popular uses of the Op-Amp is to build filter circuits. Because the Op-Amp is an active device, these filters are known as **active filters** (Figure 27–2B).

Active filters have several major advantages over passive filters: (1) Active filters can provide amplification of the signal as well as filtering; (2) Active filters are not affected by the load; and (3) Active filters are constructed with resistors and capacitors only. Inductors are not used in active filters.

27–2 ▪ LOW-PASS FILTERS

The Ideal Low-Pass Filter

We have already discussed the general concepts of the frequency response. To apply these ideas to filters, we begin by defining the ideal filter. Figure 27–3A shows the frequency response (linear, not Bode) of an ideal low-pass filter. The filter passes all frequencies less than the *cut-off frequency* of 1KHz. That is, it blocks all frequencies greater than 1KHz, so

FIGURE 27–3
Low-pass filters

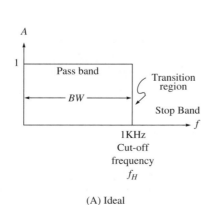

(A) Ideal

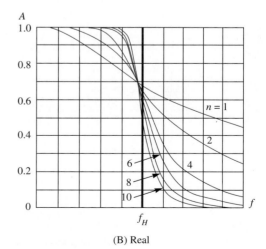

(B) Real

no frequency above 1KHz is seen in the output of this ideal low-pass filter. It may be helpful to define some terms we use when we discuss filters:

Pass-Band: the range of frequencies seen in the filter output. For our ideal filter, the pass-band is 0 to 1KHz. The pass-band has the same meaning as the bandwidth *(BW)* of the filter.

Stop-Band: the range of frequencies blocked by the filter. These frequencies are not seen in the filter output. The stop-band for our filter is all frequencies greater than 1KHz.

Cut-Off Frequency: the frequency that determines the pass-band (1KHz in our filter). The cut-off frequency of a real filter is the 3-dB frequency of that filter. For the low-pass filter, the cut-off frequency is the f_H described in Chapter 26.

Transition Region: the range of frequencies between the pass-band and the stop-band. In the ideal low-pass filter, at the cut-off frequency the pass-band transitions immediately into the stop-band. That is, the slope of the frequency response is infinite (vertical) at the cut-off frequency. In real filters, it takes a while for the filter response to go from the pass-band to the stop band. We also look at the slope, called the **roll-off,** of the frequency response at the cut-off frequency. The steeper the roll-off of an actual filter, the more it behaves as an ideal filter.

The two important parameters that we see in the ideal low-pass filter is that the filter gain is constant (flat) in the pass-band, and the transition from pass-band to stop-band is immediate. Figure 27–3B (p. 901) shows the linear frequency plots of the responses of several real low-pass filters. The number *n* is the *order* of the filter and is determined by the number of capacitors and inductors (if any) in the circuit. You can see that real filters do not have absolutely flat gains in the pass-band or an immediate transition between the pass- and stop-bands.

The Passive First-Order Low-Pass Filter

Figure 27–4A shows a simple passive RC circuit. Because the circuit contains a single capacitor, it is a *first-order circuit.* As we determined in Chapter 26, this circuit has the low-pass response shown in Figure 27–4B. For comparison, we added a colored line to show the ideal low-pass filter response.

FIGURE 27–4
Passive low-pass filter

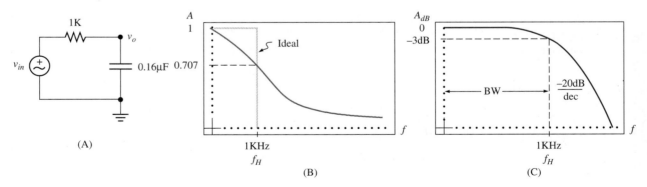

You can see that the real filter does not have a completely flat response in the pass-band. More importantly, the transition region between the pass- and stop-bands is quite large. Because there is no abrupt change in the response of the real filter, there is no clear cut-off frequency.

By convention we define the cut-off frequencies of a filter as the half-power (3-dB) frequencies, f_H and f_L. For this filter we get maximum gain at low frequency, where the capacitor is open. When the capacitor opens, $v_o = v_{signal}$, so

$$A_{max} = 1$$

The cut-off frequency for this filter is

$$f_H = \frac{1}{2\pi RC} = \frac{1}{2\pi \times 1K \times 0.16\mu F} \approx 1KHz$$

For low-pass filters, the bandwidth (pass-band) is equal to f_H; that is

$$\boxed{BW = f_H}$$

so, for our filter

$$BW = f_H \approx 1KHz$$

The Bode gain plot for the RC filter is shown in Figure 27–4C. You can see that on the Bode plot the gain of this filter is fairly flat in the pass-band. The slope, or roll-off, of the first-order filter is 20dB/dec. From now on we will usually use the Bode plot to show filter behavior.

EXAMPLE 27–1
First-Order Passive RC Filter

(a) Find the maximum gain and bandwidth of the RC filter in Figure 27–5A.
(b) In Figure 27–5B we have connected a load to the filter. Find the gain and bandwidth.
(c) Compare the above results.

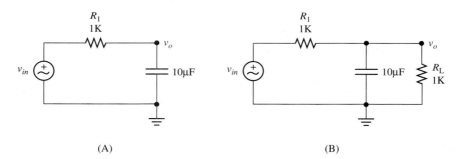

FIGURE 27–5   (A)   (B)

Solution

(a) For the simple RC circuit in Figure 27–5A, we know that

$$A_{max} = 1$$

and

$$BW = f_H = \frac{1}{2\pi R_1 C} = \frac{1}{2\pi \times 1K \times 10\mu F} \approx 16Hz$$

(b) To find the maximum gain at low frequency, we open the capacitor. The resulting circuit is a voltage divider, so

$$A_{max} = \frac{R_L}{R_L + R_1} = \frac{1K}{1K + 1K} = 0.5$$

In the last chapter we saw that f_H is given by

$$BW = f_H = \frac{1}{2\pi(R_1 \| R_L) \times C} = \frac{1}{2\pi \times 0.5K \times 10\mu F} \approx 32Hz$$

(c) The addition of a load to the filter has drastically altered the filter response. This is one of the problems with passive filters. They must be designed to a specific load.

REVIEW QUESTIONS

1. Draw the ideal low-pass filter response. Indicate the pass-band, the stop-band, the cut-off frequency.
2. For the low-pass filter, what is the relation between bandwidth, pass-band, cut-off frequency, and f_H?
3. How many capacitors are there in a first-order filter?
4. Draw a simple RC low-pass filter, and show its frequency response.
5. Compare this response to the ideal response.
6. Where does the maximum gain occur? What is it?
7. How is the cut-off frequency defined for the real filter? What is it?
8. What happens to a passive filter when you connect a load to it?

27-3 ■ THE LOW-PASS ACTIVE FILTER

First-Order Filters

Figure 27–6A shows a first-order active filter built with an Op-Amp in the non-inverting mode. It is also possible to build inverting active filters, but here we will consider just the non-inverting filters.

FIGURE 27–6
Active low-pass filter

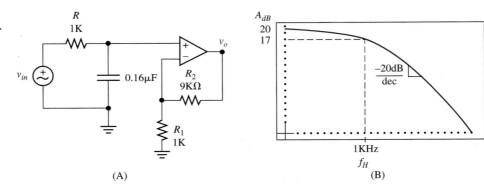

(A)

(B)

Because there is no current into the Op-Amp, the RC part of this circuit is identical to the passive low-pass filter we just discussed. The Op-Amp merely amplifies the output of the RC filter. At low frequency the capacitor is open, and the circuit acts as the basic non-inverting amplifier. Remembering that R has no effect on the gain, we get

$$A_{max} = 1 + \frac{R_2}{R_1} = 1 + \frac{9K}{1K} = 10$$

so

$$A_{dB_{max}} = 20dB$$

Bandwidth and cut-off frequency in this circuit are still determined by R and C:

$$BW = f_H = \frac{1}{2\pi RC}$$

For our circuit

$$BW = f_H = \frac{1}{2\pi RC} = \frac{1}{2\pi \times 1\,K \times 0.16\mu F} \approx 1\,KHz$$

The Bode plot for this filter is shown in Figure 27–6B. We see that one advantage of the active filter is that we can amplify as well as filter the input signal. Note that the first-order filter has a 20dB/dec roll-off.

EXAMPLE 27–2
First-Order Low-Pass Active Filter

(a) Find the maximum gain and the bandwidth for the filter in Figure 27–7A, and sketch the Bode plot.

(b) Compare this filter to the one in Example 27–1 (p. 903).

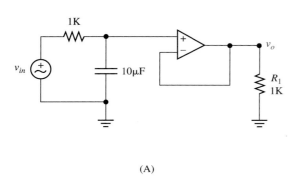

(A)

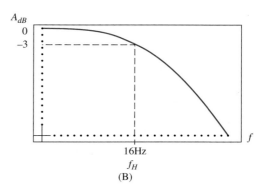

(B)

FIGURE 27–7

Solution

(a) The Op-Amp in this circuit is connected as a unity-gain buffer. Therefore, the Op-Amp output is identical to v_+ (the voltage across the capacitor), so

$$A_{\max} = 1$$

or

$$A_{dB_{\max}} = 0dB$$

and

$$BW = f_H = \frac{1}{2\pi RC} = \frac{1}{2\pi \times 1K \times 10\mu F} \approx 16Hz$$

The Bode plot for this filter is shown in Figure 27–7B.

(b) This filter has exactly the same response as the unloaded RC filter in Example 27–1. The difference here is that the load resistor does not affect the behavior of the active filter. We did not even consider the load resistor when we found the gain and bandwidth.

We have already stated that one advantage of an active filter is that we can also amplify the signal. As you can see from this example, we do not always use this advantage. Another advantage of the active filter is that the filter behavior does not depend on the filter load.

DRILL EXERCISE

Redraw the circuit of Figure 27–7A, changing the element values to $R = 10\Omega$ and $C = 1\mu F$. Find the maximum gain in decibels and the bandwidth of this filter.

Answer $A_{dB_{\max}} = 0dB$, $BW = f_H \approx 16KHz$

Second-Order Filters

The roll-off of 20dB/dec in the first-order filter is not very steep; remember the greater the roll-off, the closer our filter approaches the ideal. We can increase the roll-off by increasing the order of the filter. For active filters, the order of a filter is determined by the number of capacitors (n) built into it. The final roll-off of a low-pass active filter is given by

$$\textbf{final roll-off} = 20n\text{dB/dec}$$

All second-order filters, therefore, have a final roll-off of 40dB/dec.

When we use second-order and higher order filters, we can produce some interesting frequency responses. Consider the two linear frequency responses shown in Figure 27–8. By properly selecting the resistor and capacitor values of an active filter, we can produce either of these two responses.

FIGURE 27–8

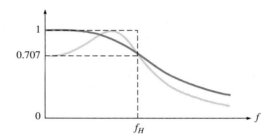

The frequency response in dark blue gives us a reasonably flat gain in the pass-band. This is the response of a **Butterworth filter.** The frequency response in light blue looks a bit strange. Rather than a flat gain response in the pass-band, the gain response goes up and down. Why would we use this type of filter, known as a **Chebychev filter?** The answer is in the roll-off at the cut-off frequency. Although both filters eventually provide the same roll-off, you can see that right at the cut-off frequency, the Chebychev response drops a lot faster than the Butterworth response. The trade-off here is rapid roll-off versus flat gain.

For this discussion, because we limit ourselves to filters that provide a maximally flat gain response in the pass-band, we will only discuss Butterworth Filters. The most common type of second-order low-pass filter is shown in Figure 27–9A. This filter is known as a **Sallen-Key filter,** after its original designers. This type of filter is also sometimes called a *VCVS* (voltage controlled–voltage source) filter. Note that for ease of drawing the circuit, the −terminal is down.

The gain in this filter is provided by R_3 and R_4. The bandwidth of the filter is determined by R_1, R_2 C_1, and C_2. Circuit equations for this circuit are quite involved. You don't have to crank out a lot of equations, though, to get a general understanding of the behavior of this filter or any other. All you have to do is to look at the circuit when the frequency is very low and when the frequency is very high.

When the frequency is low, all capacitors are open, as shown in Figure 27–9B. When the capacitors are open, the circuit becomes a basic non-inverting amplifier (R_1 and R_2 do not affect the gain), so

$$A_{\text{max}} = 1 + \frac{R_4}{R_3}$$

At high frequency the capacitors short, as shown in Figure 27–9C. When C_1 shorts, the input +terminal is connected to ground so there is no output. In this filter, then, we get maximum gain at low frequency and no gain at high frequency. This circuit must be a low-pass filter.

FIGURE 27–9
Sallen-Key low-pass filter

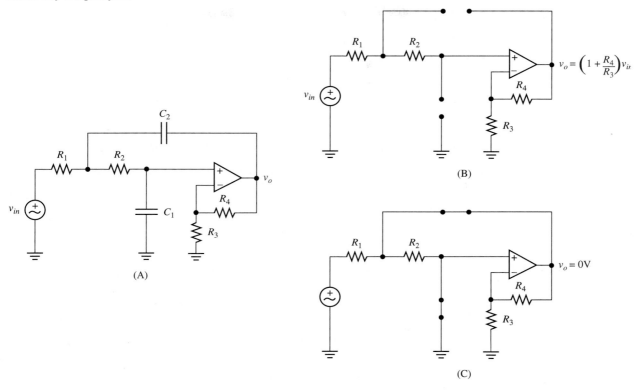

We have already found the maximum gain of the Sallen-Key filter. The exact formula for the bandwidth of the Sallen-Key filter is very hairy, but we can (luckily) use a simpler formula to approximate the bandwidth:

$$BW = f_H \approx \frac{1}{2\pi\sqrt{R_1 R_2 C_1 C_2}}$$

EXAMPLE 27–3
Sallen-Key Filter

A Butterworth active filter is shown in Figure 27–10A. Find the maximum gain of the filter and its bandwidth. Sketch the Bode gain plot for this filter.

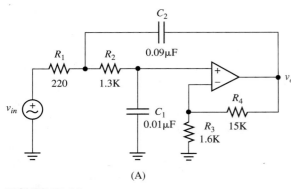

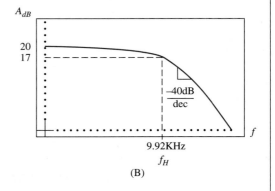

FIGURE 27–10

Solution   The maximum gain of the filter occurs at low frequency when the capacitors open:

$$A_{max} = 1 + \frac{R_4}{R_3} = 1 + \frac{15K}{1.6K} = 10$$

so the gain in decibels is

$$A_{dB_{max}} = 20dB$$

The bandwidth is approximately

$$BW = f_H \approx \frac{1}{2\pi\sqrt{R_1 R_2 C_1 C_2}} = \frac{1}{2\pi\sqrt{220 \times 1.3K \times 0.09\mu F \times 0.01\mu F}} = 9.92KHz$$

Because the filter is a Butterworth design, we know that the gain in the pass-band is reasonably flat. The Bode plot for this filter is shown in Figure 27–10C (p. 907). The roll-off for this second-order filter is 40dB/dec.

DRILL EXERCISE

Draw a Sallen-Key filter with all resistors set to 1KΩ and both capacitors set to 1μF. Find the maximum gain in decibels and the bandwidth. Note that this filter is *not* a Butterworth filter; however, the formulas we have for gain and bandwidth still apply.

Answer $A_{dB_{max}} = 6dB$, $BW \approx 160Hz$

REVIEW QUESTIONS

1. What are the advantages of an active filter? (See Section 27–1, p. 900)
2. Draw a first-order active filter.
3. Where does the maximum gain of a low-pass filter occur? What is this gain for the first-order filter?
4. What is the bandwidth of the first-order filter?
5. What is the roll-off of the first-order filter?
6. How does a load affect the behavior of the active low-pass filter?
7. What is the final roll-off of a second-order filter?
8. What is the characteristic of the Butterworth filter?
9. Draw the second-order low-pass Sallen-Key circuit. What is the maximum gain? What is the bandwidth?

27–4 ■ THE HIGH-PASS ACTIVE FILTER

Figure 27–11A shows the ideal high-pass response. Here we see that the cut-off frequency is again 1KHz. Now, however, only frequencies greater than 1KHz are seen in the output. That is, the cut-off frequency here is the f_L we found in Chapter 26. For the high-pass filter the pass-band (bandwidth) is the range of frequencies greater than the cut-off frequency; the pass-band of the high-pass filter is infinite. Note that in the high-pass filter, the bandwidth is not equal to the cut-off frequency. As in the ideal low-pass filter, the transition between stop-band and pass-band is immediate.

Figure 27–11B shows the Bode plots of the responses of several real high-pass filters. Again note that the real response is not an exact match to the ideal response; the pass-

FIGURE 27–11
High-pass filter

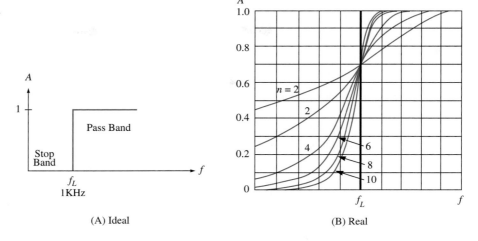

(A) Ideal   (B) Real

band is not completely flat, and the transition region is finite. The higher the filter order, the smaller the transition region.

First-Order Filters

In Chapter 26 we showed you that the resistor voltage in an RC circuit has a high-pass response. Figure 27–12A shows how we connect an RC circuit to an Op-Amp to produce an active high-pass filter. Because there is a single capacitor, this is a first-order filter.

At low frequency the capacitor is open, so the input signal cannot get the output. That is, there is zero output. At high frequency the capacitor shorts and the circuit acts as the basic non-inverting amplifier. Again, R has no affect on the gain and we get

$$A_{max} = 1 + \frac{R_2}{R_1} = 1 + \frac{3K}{1K} = 4$$

so

$$A_{dB_{max}} = 12dB$$

The cut-off frequency (f_L) in this circuit is determined by R and C:

$$f_L = \frac{1}{2\pi RC}$$

FIGURE 27–12
High-pass filter

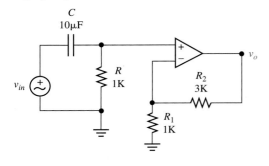

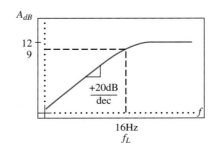

(A)   (B)

For our circuit

$$f_L = \frac{1}{2\pi RC} = \frac{1}{2\pi(1\text{K} \times 10\mu\text{F})} \approx 16\text{Hz}$$

The Bode plot for this filter is shown in Figure 27–12B (p. 909). The first-order filter has a positive slope of 20dB/dec. Even though the slope is positive, we still use the term roll-off to describe the slope.

Note: The bandwidth of a high-pass active filter is not infinite. We saw in Chapter 26 that the Op-Amp itself has a limited frequency range. That is, the Op-Amp, separate from the filter, has its own upper cut-off frequency given by

$$f_{H_{Op\text{-}Amp}} = \frac{unity\ gain\ bandwidth}{1 + \dfrac{R_2}{R_1}}$$

If the filter here is built with a 741 Op-Amp, we get

$$f_{H_{Op\text{-}Amp}} = \frac{unity\ gain\ bandwidth}{1 + \dfrac{R_2}{R_1}} = \frac{1 \times 10^6}{1 + \dfrac{3\text{K}}{1\text{K}}} = 250\text{KHz}$$

This filter will actually pass signals from 16Hz to 250KHz. In fact, then, there is no such thing as a true high-pass active filter.

EXAMPLE 27–4
First-Order High-Pass Active Filter

(a) Find the maximum gain and the cut-off frequency for the filter in Figure 27–13A.
(b) Sketch the Bode plot.

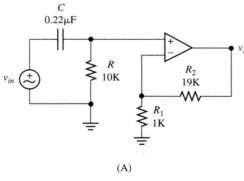

(A)

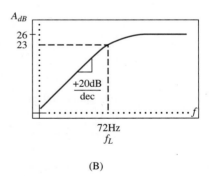

(B)

FIGURE 27–13

Solution

The maximum gain occurs at high frequencies and is

$$A_{max} = 1 + \frac{R_2}{R_1} = 1 + \frac{19\text{K}}{1\text{K}} = 20$$

or

$$A_{dB_{max}} = 26\text{dB}$$

The cut-off frequency is

$$f_L = \frac{1}{2\pi RC} = \frac{1}{2\pi \times 10\text{K} \times 0.22\mu\text{F}} \approx 72\text{Hz}$$

The Bode plot for this filter is shown in Figure 27–13B. Again note that the active filter frequency response does not depend on the filter load.

DRILL EXERCISE Redraw the circuit of Figure 27–13A, changing the element values to $R_2 = 1K\Omega$, $R_1 = 1K\Omega$, $R = 1K\Omega$, and $C = 0.22\mu F$. Find the maximum gain in decibels and f_L.

Answer $A_{dB_{max}} = 6dB, f_L = 723Hz$

Second-Order Filters

As with low-pass filters, we can increase the roll-off of the high-pass filter by increasing the order of the filter. For example, second-order filters have a final roll-off of 40dB/dec.

The Sallen-Key high-pass filter is shown in Figure 27–14A. If you compare it to the low-pass filter of Figure 27–9A (p. 907), you can see that we have just interchanged the resistors and capacitors (R_4 and R_3 set the gain and do not affect f_L). To get a basic understanding of the high-pass circuit, we examine the circuit at low and high frequencies.

When the frequency is low, all capacitors are open, as shown in Figure 27–14B. When the capacitors are open, the input signal cannot reach the Op-Amp, so there is no output.

FIGURE 27–14
Sallen-Key high-pass filter

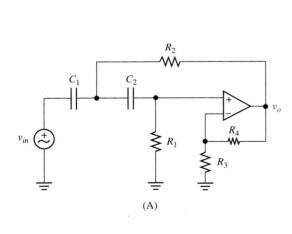

(A)

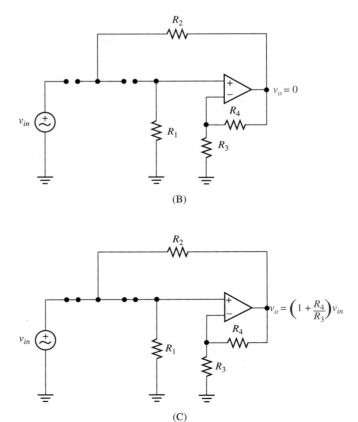

(B)

(C)

When the frequency is high, the capacitors short, as shown in Figure 27–14C (p. 911). The circuit is now the basic non-inverting amplifier (R_1 and R_2 do not affect the gain), so

$$A_{max} = 1 + \frac{R_4}{R_3}$$

The gain in this circuit is low at low frequencies and high at high frequencies; therefore, this is a high-pass filter.

Again, the exact formula for the cut-off frequency of the Sallen-Key filter is very hairy. Rather we can approximate f_L:

$$f_L \approx \frac{1}{2\pi\sqrt{R_1 R_2 C_1 C_2}}$$

EXAMPLE 27–5
Sallen-Key High-Pass Filter

A Butterworth active filter is shown in Figure 27–15A. Find the maximum gain of the filter and its cut-off frequency. Sketch the Bode gain plot for this filter.

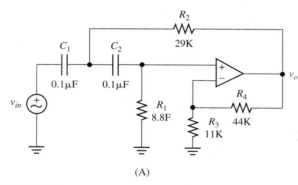

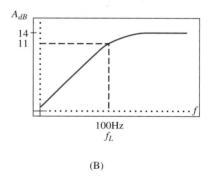

(A)

(B)

FIGURE 27–15

Solution The maximum gain of the filter occurs at high frequency where the capacitors short:

$$A_{max} = 1 + \frac{R_4}{R_3} = 1 + \frac{44K}{11K} = 5$$

so the gain in decibels is

$$A_{dB_{max}} = 14dB$$

The cut-off frequency is approximated by

$$f_L \approx \frac{1}{2\pi\sqrt{R_1 R_2 C_1 C_2}} = \frac{1}{2\pi\sqrt{8.8K \times 29K \times 0.1\mu F \times 0.1\mu F}} \approx 100Hz$$

Because the filter is a Butterworth design, we know that the gain in the pass-band is reasonably flat. The Bode plot for this filter is shown in Figure 27–15B.

DRILL EXERCISE   Draw a Sallen-Key filter with all resistors set to 10KΩ and both capacitors set to 0.01 μF. Find the maximum gain in decibels and the cut-off frequency. Note that this filter is *not* a Butterworth filter; however, the formulas we have for gain and cut-off frequency still apply.

Answer $A_{dB_{max}} = 6dB, f_L = 1.6KHz$

REVIEW QUESTIONS
1. Draw a first-order active high-pass filter.
2. Where does the maximum gain of a high-pass filter occur? What is this gain for the first-order filter?
3. What is the cut-off frequency of the first-order filter?
4. What is the roll-off of the first-order filter?
5. Why is there no such thing as true high-pass active filter? How do you find the upper frequency limitation of an Op-Amp amplifier?
6. What is the final roll-off of a second-order filter?
7. Draw the second-order high-pass Sallen-Key circuit. What is the maximum gain? What is the cut-off frequency?

27–5 ■ THE BAND-PASS ACTIVE FILTER

Figure 27–16A shows an ideal band-pass filter that passes frequencies between 10KHz and 20KHz. The band-pass filter has two cut-off frequencies, equivalent to the f_L and f_H of the previous chapter. Figure 27–16B shows Bode plots for several real band-pass filters.

FIGURE 27–16
Band-pass filter

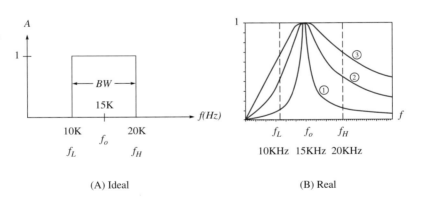

(A) Ideal   (B) Real

In addition to the filter gain and bandwidth already described, for the band-pass filter we also want to know these characteristics:

Center Frequency (f_o): Also known as the **resonant frequency,** f_o is the frequency in the center of the band-pass response. In Figure 27–16B the center frequency for all of the filter responses shown is 15KHz. The frequencies labeled on the equalizer of Figure 27–1B (p. 900) are the center frequencies for each of the band-pass filters used in the device.

Q: Q is a measure of the **selectivity** of the band-pass filter. A highly selective filter—a high Q filter—passes only a small band of frequencies about the center frequency. Consider the three filter responses shown in Figure 27–16B. You can see that filter number 1 has the narrowest bandwidth; this filter will pass fewer frequencies than the other two filters. Filter number 1 has the highest Q; filter number 3 has the lowest Q.

Figure 27–17 shows the responses of three different filters, each with the same bandwidth of 0.5KHz but with different center frequencies. Note that for ease in reading, this figure is not drawn to scale. Here the bandwidth does not give us a good comparison between the filters.

FIGURE 27–17

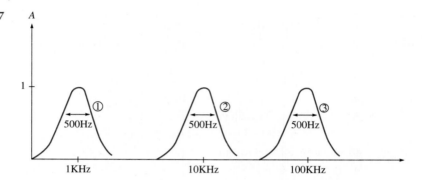

For filter number 1, the bandwidth of 0.5KHz is one-half of the center frequency; that is the range of frequencies passed is 50% of the center frequency. For filter number 2, the bandwidth of 0.5KHz implies that the range of frequencies passed is 5% of the center frequency. For filter number 3, the bandwidth of 0.5KHz tells us that the range of frequencies passed is only 0.5% of the center frequency. Filter 3 here is the most selective and will have the highest Q.

The selectivity of a band-pass filter increases with decreasing bandwidth and increases with increasing center frequency. The formula for Q is

$$Q = \frac{f_o}{BW}$$

Relating f_L and f_H to *BW* and f_o

We know how to find the bandwidth if we know f_L and f_H:

$$BW = f_H - f_L$$

We can also find the center frequency from

$$f_o = \sqrt{f_H f_L}$$

Interestingly, the center frequency is not half-way between the cut-off frequencies. However, the band-pass Bode plot is symmetrical about f_o. Remember, the Bode plot is gain versus log f, so on a Bode plot

$$\log f_o = \frac{\log f_H + \log f_L}{2}$$

That is, $\log f_o$ is half-way between $\log f_L$ and $\log f_H$.

If we know the bandwidth and the center frequency, we can find the cut-off frequencies. The exact formula is somewhat involved

$$f_H = \sqrt{\frac{BW^2}{4} + f_o{}^2} + \frac{BW}{2}$$

and

$$f_L = \sqrt{\frac{BW^2}{4} + f_o{}^2} - \frac{BW}{2}$$

Fortunately, if the filter has a reasonably high Q response (small BW compared to f_o), we can approximate the cut-off frequencies as

$$f_H \approx f_o + \frac{BW}{2}$$

and

$$f_L \approx f_o - \frac{BW}{2}$$

High Pass–Low Pass Combination

We can build a band-pass filter by combining a high-pass filter with a low-pass filter, as shown in Figure 27–18 (the order of the two filters doesn't matter). The high-pass filter blocks frequencies below f_L, and the low-pass filter blocks frequencies above f_H.

FIGURE 27–18

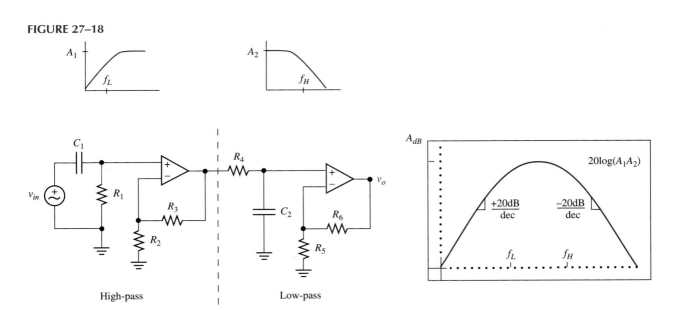

Each Op-Amp filter can be separately analyzed to find its gain and cut-off frequency. The filter's total gain is the product of the individual gains, and the filter cut-off frequencies of the complete filter are given by the individual f_L and f_H. When we combine the two individual filters, we get the complete band-pass response shown.

Even though there are two capacitors in the band-pass filter ($n = 2$), the low-frequency and high-frequency roll-offs are only 20dB/dec. This happens because one capacitor contributes to the low-frequency roll-off and the other contributes to the high-frequency roll-off. This band-pass filter, therefore, is considered to be a first-order band-pass filter. For band-pass filters

$$band\text{-}pass\ filter\ order = \frac{n}{2}$$

and the high- and low-frequency roll-offs have the slope

$$band\text{-}pass\ roll\text{-}off = 10n$$

EXAMPLE 27–6

High Pass–Low Pass Combination

For the circuit of Figure 27–19A, find

(a) The gain and f_L for the high-pass filter.
(b) The gain and f_H for the low-pass filter.
(c) The total gain and bandwidth of this band-pass filter.
(d) The center frequency f_o and the Q of the filter.
(e) Draw the Bode plot for the complete filter.

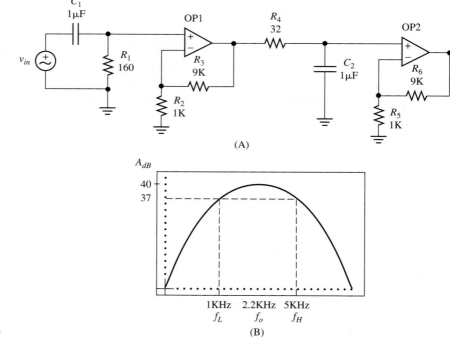

(A)

(B)

FIGURE 27–19

Solution

(a) OP1 is connected as a first-order high-pass filter. Its gain and f_L are

$$A_1 = 1 + \frac{R_3}{R_2} = 1 + \frac{9K}{1K} = 10$$

and

$$f_L = \frac{1}{2\pi R_1 C_1} = \frac{1}{2\pi \times 160 \times 1\mu F} \approx 1KHz$$

(b) OP2 is connected as a first-order low-pass filter. Its gain and f_H are

$$A_2 = 1 + \frac{R_6}{R_5} = 1 + \frac{9K}{1K} = 10$$

and

$$f_H = \frac{1}{2\pi R_4 C_2} = \frac{1}{2\pi \times 32 \times 1\mu F} \approx 5KHz$$

(c) The total gain and bandwidth are

$$A_{total} = A_1 A_2 = 100$$

or

$$A_{dB_{total}} = 40dB$$

The bandwidth is

$$BW = f_H - f_L = 5KHz - 1KHz = 4KHz$$

(d) The center frequency and Q are

$$f_o = \sqrt{f_H f_L} = \sqrt{5K \times 1K} = 2.24KHz$$

and

$$Q = \frac{f_o}{BW} = \frac{2.24K}{4K} = 0.56$$

Note that because the bandwidth is greater than the center frequency, the Q is very small. This particular filter is not highly selective.

(e) The Bode plot for the complete filter is shown in Figure 27–19B.

DRILL EXERCISE

Redraw the circuit of Figure 27–19 with the following parameter values:

$$R_1 = 500\Omega, R_2 = 2K\Omega, R_3 = 4K\Omega, R_4 = 250\Omega, R_5 = 1K\Omega, R_6 = 1K\Omega$$
$$C_1 = C_2 = 0.1\mu F$$

Find $A_{dB}, f_L, f_H, BW, f_o, Q$

Answer $A_{dB_{total}} = 15dB, f_L = 3.2KHz, f_H = 6.4KHz, BW = 3KHz, f_o = 4.5KHz,$ $Q = 1.5$

Sallen-Key Band-Pass Filter

Figure 27–20A (p. 918) shows a Sallen-Key band-pass filter. To understand its operation, let's examine the circuit at high and low frequency. At low frequency, both capacitors are open (Figure 27–20B, p. 918). When C_2 is open, the signal cannot reach the Op-Amp, so we get zero output.

At high frequency, both capacitors are short (Figure 27–20C, p. 918). When C_1 shorts, the input is shorted to ground and, again, the output is zero. We see that at both low and high frequency, there is no output. Therefore, this circuit must amplify a middle range of frequencies. It is a band-pass filter.

The formulas for the filter parameters of this band-pass filter are quite complicated. To simplify matters, we will assume that

$$C_1 = C_2 = C$$

FIGURE 27–20
Sallen-Key band-pass filter

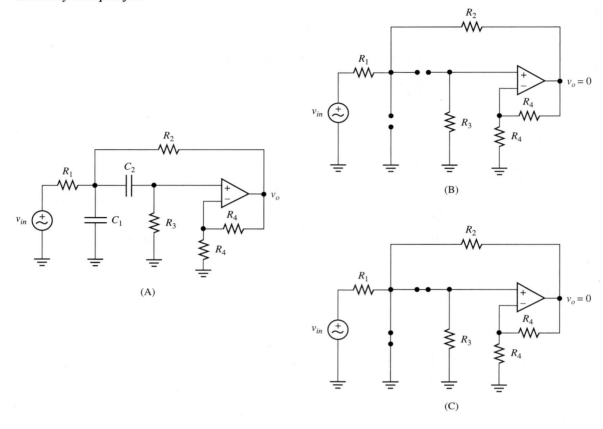

(A)

(B)

(C)

Also note that we use the same resistors in the negative feedback loop. Even with these simplifications, the formulas are still nasty. The filter parameters that we find first are the center frequency, bandwidth, and gain.

$$f_o = \frac{1}{2\pi C \sqrt{R_3(R_1 \parallel R_2)}}$$

$$BW = \frac{\dfrac{1}{R_1} - \dfrac{1}{R_2} + \dfrac{2}{R_3}}{2\pi C}$$

$$A_{max} = \frac{1}{\pi C R_1 BW}$$

Because this type of filter is usually high Q (although in practice we don't exceed $Q = 10$), we find the cut-off frequencies from

$$f_H \approx f_o + \frac{BW}{2}$$

and

$$f_L \approx f_o - \frac{BW}{2}$$

EXAMPLE 27–7
Sallen-Key Band-Pass Filter

For the band-pass filter in Figure 27–21A, find f_o, BW, A, f_H, f_L. Draw the Bode plot for the filter.

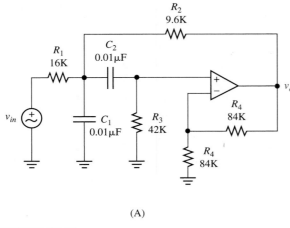

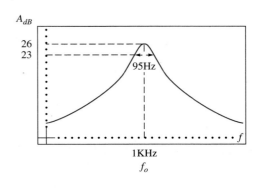

(A)

(B)

FIGURE 27–21

Solution   We use the formulas to find

$$f_o = \frac{1}{2\pi C \sqrt{R_3(R_1 \| R_2)}} = \frac{1}{2\pi \times 0.01\mu F \times \sqrt{42K \,(16K \| 9.6K)}} = 1KHz$$

$$BW = \frac{\dfrac{1}{R_1} - \dfrac{1}{R_2} + \dfrac{2}{R_3}}{2\pi C} = \frac{\dfrac{1}{16K} - \dfrac{1}{9.6K} + \dfrac{2}{42K}}{2\pi \times 0.01\mu F} = 95Hz$$

$$A_{max} = \frac{1}{\pi C R_1 BW} = \frac{1}{\pi \times 0.01\mu F \times 16K \times 95} = 21$$

so

$$A_{dB_{max}} = 26dB$$

The cut-off frequencies are found from

$$f_H \approx f_o + \frac{BW}{2} = 1000 + \frac{95}{2} = 1048Hz$$

and

$$f_L \approx f_o - \frac{BW}{2} = 1000 - \frac{95}{2} = 953Hz$$

The Bode plot for this filter is shown in Figure 27–21B.

DRILL EXERCISE

Redraw the circuit of Figure 27–21, changing the capacitor values to $C = 0.1\mu F$. Find f_o, BW, $A_{dB_{max}}$, f_H, f_L.

Answer $f_o = 100Hz$, $BW = 9.5Hz$, $A_{db_{max}} = 20dB$, $f_H = 105Hz$, $f_L = 96Hz$

REVIEW QUESTIONS

1. Draw an ideal band-pass frequency response.

2. What is the center frequency of the filter? What does Q measure, and how is it found?

3. Draw several real band-pass Bode plots that all have the same f_o. Discuss the Q of these filters.

4. Draw several real band-pass Bode plots that all have the same BW but different f_o. Discuss the Q of these filters.

5. How is the center frequency related to the cut-off frequencies?

6. For a high Q filter, how can you approximate the cut-off frequencies, given the center frequency and the bandwidth?

7. Draw a band-pass filter built with a high- and a low-pass filter. What is the total gain of the filter?

8. Which filter sets f_H? f_L?

9. Draw the Sallen-Key band-pass filter.

10. What are the formulas for f_o, BW, and A?

11. How do you find f_H and f_L for the Sallen-Key filter?

27–6 ■ HIGHER ORDER FILTERS

The roll-off achieved with a first-order filter is 20dB/dec; the roll-off achieved with a second-order filter is 40dB/dec. We often need to build filters with much steeper roll-offs. There are two basic ways to build a higher-order filter:

☐ Combine several lower order filters.

☐ Build a single amplifier high-order filter.

FIGURE 27–22
Third-order low-pass filter

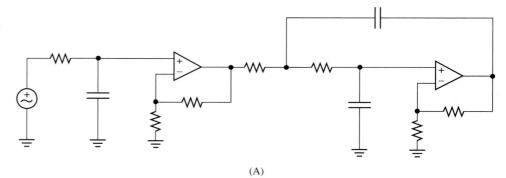

(A)

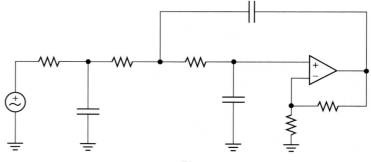

(B)

Figure 27–22A shows how we can construct a third-order low-pass filter by connecting a first-order filter to a second-order circuit. If we want a fourth-order filter, we would connect two second-order filters together. Figure 27–22B shows how we can construct a third-order filter with a single Op-Amp and three capacitors. A fourth-order filter would require four capacitors, and so on.

The formulas we have given you for filter parameters apply only to the first- and second-order filters. You can consult active filter or Op-Amp handbooks to get the appropriate formulas for higher order filters.

27–7 ■ TROUBLESHOOTING ACTIVE FILTERS

Troubleshooting active filters can be a bit tricky. You can determine failures of resistors or the Op-Amp (the most likely failure in an active filter) by tracing the AC voltages in the circuit. Failures of the capacitors can be more difficult to detect.

To test the resistors and the Op-Amp in an active filter, apply an input signal at a frequency that is well within the pass-band of the filter. In the pass-band, capacitors will either act as opens or shorts, resulting in an Op-Amp circuit that effectively contains only resistors.

Low-Pass Active Filter

Consider for example, the low-pass filter ($f_H = 1$ KHz) shown in Figure 27–23A. If we apply a 100Hz signal to the input, we will be well within the pass-band of the filter. For the low-pass filter, capacitors act as opens for frequencies within the pass-band (Figure 27–23B).

FIGURE 27–23

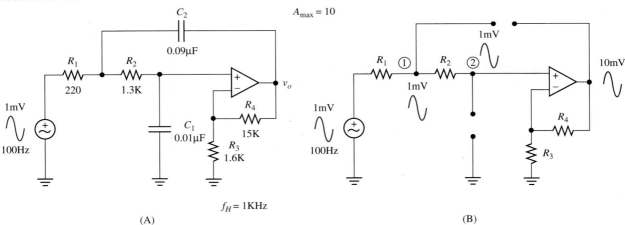

(A)

(B)

Figure 27–23B shows the AC signals we would expect to see if the resistors and Op-Amp are good. Likely failures of the elements would lead to

Open R_1: no signal at point 1.

Open R_2: either no signal or noise at point 2.

Open R_3: output signal equals input signal (unity gain).

Open R_4: output saturates at supply voltage.

Failed Op-Amp: no output.

We now assume the resistors and the Op-Amp are good and look to the capacitors.

C_1 *or* C_2 *short:* Filter acts as though it sees a high-frequency input (when capacitors normally short). Because this is a low-pass filter, there will be no appreciable output.

C_1 *or* C_2 *open:* This is the tricky situation. Removing one of the capacitors still leaves you with an active filter. However, the order of the filter is reduced by 1, so we end up with a first-order active filter. The only way to know that a capacitor has failed open is to run a full frequency analysis of the filter and compare the measured roll-off to the expected roll-off (Figure 27–24). Even if you know that a capacitor has failed, you will probably not know which specific one is the problem. You will have to individually test each capacitor until you find the culprit.

FIGURE 27–24

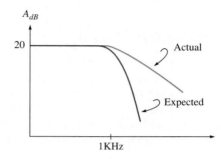

High-Pass Active Filter

If we apply 1 KHz signal to the input of the high-pass filter in Figure 27–25A, we will be well within the pass-band of the filter. For the high-pass filter, capacitors act as shorts for frequencies within the pass-band (Figure 27–25B). This figure also shows the AC signals we would expect to see if the resistors and Op-Amp are good. Likely failures of the elements would lead to

Open R_1*:* no noticeable difference. C_2 is effectively removed from the circuit, reducing the order of the filter and decreasing the roll-off.

Open R_2*:* no noticeable difference. C_1 and C_2 are now in series and combine, again reducing the order of the filter.

Open R_3*:* Output signal equals input signal (unity gain).

Open R_4*:* Output saturates at supply voltage.

Failed Op-Amp: no output.

We now assume the resistors and the Op-Amp are good and look to the capacitors.

C_1 *or* C_2 *short:* reduces the order of the filter.

C_1 *or* C_2 *open:* signal does not reach Op-Amp input so there is no output.

FIGURE 27–25

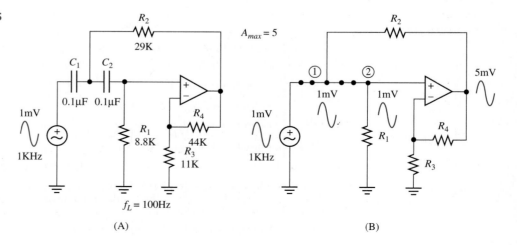

(A)   (B)

Band-Pass Active Filter

The band-pass filter in Figure 27–26A has $f_o = 1KHz$ and $BW = 95Hz$. We apply an input at the center frequency of 1KHz. At the center frequency, C_1 acts as an open and C_2 acts as a short, as shown in Figure 27–26B. Also shown are the AC signals we would expect to see if the resistors and Op-Amp are good. Likely failures of the elements would lead to

Open R_1: no signal at point 1.

Open R_2: may be a change in the gain (a drop in the output voltage).

Open R_3: significant drop in gain.

Open R_4 in feedback path: output saturates at supply voltage.

Failed Op-Amp: no output.

We now assume the resistors and the Op-Amp are good and look to the capacitors. Because the capacitors have different effects in a band-pass filter, we look at them individually:

C_1 Short: The input signal is shorted to ground, so there is no output.

C_2 Short: The filter becomes a low-pass filter.

C_1 Open: The filter becomes a high-pass filter.

C_2 Open: No signal gets to Op-Amp input, so there is no output.

FIGURE 27–26

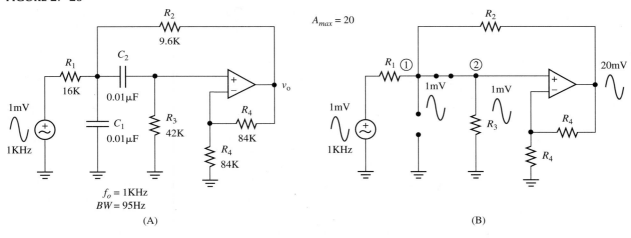

$$f_o = 1KHz$$
$$BW = 95Hz$$

(A)

(B)

EXAMPLE 27–8
Troubleshooting the Low-Pass Filter

A low-pass second-order filter is shown in Figure 27–27A (p. 924). Below on the left is a list of troubleshooting results. Match the problem with the likely cause listed on the right.

(a) No signal at point 1.
(b) Signal at point 2, no signal at output.
(c) $v_o = +15V$.
(d) See gain plot in Figure 27–27B.
(e) 0V at point 2.

1. R_4 open.
2. R_1 open.
3. C_2 open.
4. C_1 short.
5. Op-Amp failed.

Solution

a-2: No signal at point 1 indicates that R_1 has failed open.

b-5: The signal gets to the Op-Amp, but there is no output. The Op-Amp is bad.

c-1: Saturation of the Op-Amp voltage at the supply voltage indicates that the negative feedback has failed. R_4 is probably open.

d-3: The actual roll-off indicates a first-order filter response. A capacitor must have failed open.

e-4: 0V at point 2 results from C_1 shorting to ground.

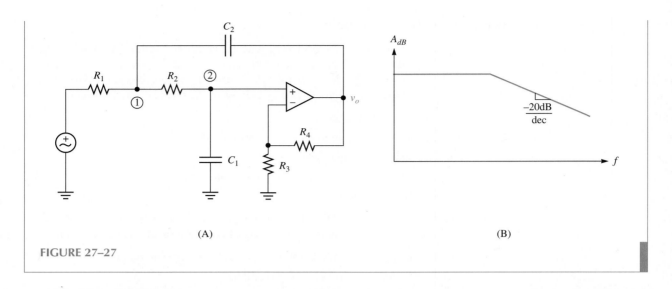

FIGURE 27–27

SUMMARY

- We have defined the following filter parameters:

  □ Pass-band: range of frequencies passed by filter (same as bandwidth BW).

  □ Stop-band: range of frequencies blocked by filter.

  □ Roll-off: the slope in the transition region between the pass- and stop-bands.

  □ f_L and f_H: the low- and high-frequency filter cut-off frequencies.

  □ Center frequency: frequency (f_o) in the middle of the pass-band of a band-pass filter.

  □ Q: a measure of selectivity of a band-pass filter: $Q = \dfrac{f_o}{BW}$

- Ideal filters have perfectly flat pass-bands and infinite roll-off (abrupt change between pass- and stop-bands).

- Real first-order low-pass active filters can be built by connecting an RC filter to an Op-Amp amplifier. First-order low-pass filters have a roll-off of 20dB/dec.

- The Sallen-Key circuit is a common way to build second-order active filters containing two capacitors and a single Op-Amp. The second-order low-pass filter has a 40dB/dec roll-off.

- The Butterworth filter is a design that produces the flattest possible pass-band gain achievable with a real filter.

- Real first-order high-pass active filters can be built by connecting an RC filter to an Op-Amp amplifier. First-order high-pass filters have a roll-off of 20dB/dec.

- Interchanging the capacitor and filter resistors of the low-pass Sallen-Key filter produces a high-pass filter. The second-order high-pass filter has a 40dB/dec roll-off.

- A band-pass filter can be constructed by connecting a high-pass to a low-pass filter. The band-pass filter cut-off frequencies are f_L and f_H of the individual filters. The bandwidth and center frequency are found from

$$BW = f_H - f_L$$
$$f_o = \sqrt{f_H f_L}$$

- For high-Q band-pass filters, we can find the cut-off frequencies from the center frequency and the bandwidth:

$$f_H \approx f_o + \frac{BW}{2}$$

$$f_L \approx f_o - \frac{BW}{2}$$

- A single Op-Amp band-pass filter can be constructed with a Sallen-Key circuit. These tend to be high-Q filters.

- Higher-order low-pass and high-pass filters can be constructed by connecting first- and second-order filters or a single Op-Amp and n capacitors. In either case, the order of the filter is equal to the number of capacitors in the circuit. The final roll-off of a filter equals $20n$dB/dec.
- The order of a band-pass filter is $\frac{n}{2}$. The final roll-offs of a band-pass filter equals $10n$dB/dec.
- A bad resistor or bad Op-Amp can usually be found by applying an input signal at a frequency well within the pass-band of the filter. Tracing the AC signals should locate the problem.
- In a low-pass filter, a failed short capacitor will produce zero output.
- A failed open capacitor will reduce the filter order by one. The filter frequency response will show a smaller roll-off than expected.
- In a high-pass filter, a failed open capacitor will produce zero output.
- A failed short capacitor will reduce the filter order by 1. The filter frequency response will show a smaller roll-off than expected.
- In a band-pass filter, a failed short capacitor will either produce zero output or turn the filter into a low-pass filter, depending on which capacitor shorts.
- In a band-pass filter, a failed open capacitor will either produce zero output or turn the filter into a high-pass filter, depending on which capacitor opens.

PROBLEMS

SECTION 27–2 Low-Pass Filters

1. Draw the frequency response for an ideal low-pass filter that has a pass-band of 5KHz and a pass-band gain of 10. On your sketch identify the bandwidth, pass-band, stop-band, and f_H.

FIGURE 27–28

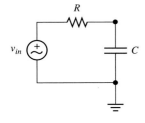

2. Sketch the Bode plot for the RC circuit in Figure 27–28 if $R = 1K\Omega$ and the capacitor has the following values. In each case, find f_H.
   (a) $C = 0.1\mu F$.   (b) $C = 1\mu F$.   (c) $C = 10\mu F$.
   (d) How does the bandwidth of the low-pass filter depend on C?

3. Sketch the Bode plot for the RC circuit in Figure 27–28 if $C = 1\mu F$ and the resistor has the following values. In each case, find f_H.
   (a) $R = 100\Omega$.   (b) $R = 1K\Omega$.   (c) $R = 10K\Omega$.
   (d) How does the bandwidth of the low-pass filter depend on R?

4. In Figure 27–28, $R = 100\Omega$ and $C = 0.01\mu F$.
   (a) Find the pass-band gain and f_H.
   (b) Now connect a 200Ω load to the RC filter, and find the pass-band gain and f_H.
   (c) Compare the results of parts (a) and (b).

SECTION 27–3 The Low-Pass Active Filter

5. Figure 27–29 shows an active first-order low-pass filter. Sketch the Bode plot for each set of parameter values given below. In each case be sure to label the pass-band gain (f_H) and the final roll-off.
   (a) $R = 100\Omega$, $C = 0.47\mu F$, $R_1 = 1K\Omega$, $R_2 = 1K\Omega$
   (b) $R = 1K\Omega$, $C = 0.22\mu F$, $R_1 = 1K\Omega$, $R_2 = 7K\Omega$
   (c) $R = 10K\Omega$, $C = 0.1\mu F$, R_1 has been removed, and R_2 replaced with a short circuit.

6. Design a first-order low-pass filter that has a gain of 26dB and a bandwidth of 500Hz. In your design, let $R = R_1 \| R_2$, and choose C to meet the bandwidth requirement.

FIGURE 27–29

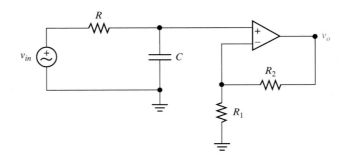

FIGURE 27–30

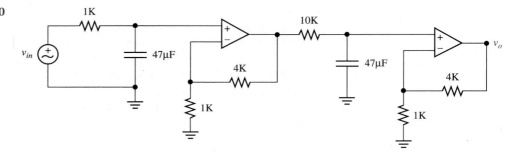

7. In Figure 27–30, we have connected two first-order filters together.
    (a) Find the gain and f_H for each filter.   **(b)** What is the overall gain in decibels?
    (c) What is the overall f_H?   **(d)** What is the overall filter order?
    (e) What is the final roll-off?
8. Figure 27–31 shows a Sallen-Key second-order low-pass filter.
    (a) Redraw the circuit with the capacitors open, and find the gain.
    (b) Find the pass-band gain of the original circuit. Compare with part (a).
    (c) Find the bandwidth.
    (d) Assuming the filter is a Butterworth filter, sketch the Bode plot.
    (e) What is the order of this filter?   **(f)** What is the final roll-off?
9. Connect a 1KΩ load to the filter of Figure 27–31.
    (a) Now what are the gain and bandwidth of the filter?
    (b) What effect does the load have on the filter response?
10. Connect a signal source with an internal resistance of 100Ω to the Sallen-Key filter of Figure 27–31.
    (a) Find the gain and bandwidth (remember that the signal source resistance is in series with R_1).
    (b) Compare this result to the original filter response found in Problem 8.
    (c) It might interest you to know that the filter is no longer a Butterworth filter. How can we use an additional Op-Amp to prevent the signal-source resistance from affecting the filter?

FIGURE 27–31

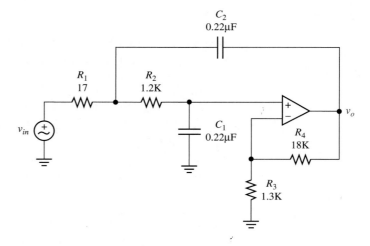

SECTION 27–4   The High-Pass Active Filter

11. Draw an ideal high-pass frequency response. Label the pass-band gain, the pass-band, the stop-band, and f_L. What is the bandwidth of the high-pass filter?
12. Sketch the Bode plot for the RC circuit in Figure 27–32 (p. 927) if $R = 1$KΩ and the capacitor has the following values. In each case, find f_L.
    (a) $C = 0.1\mu F.$ **(b)** $C = 1\mu F.$ **(c)** $C = 10\mu F.$
    (d) How does f_L of the high-pass filter depend on C?

FIGURE 27–32

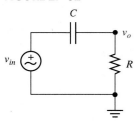

13. Sketch the Bode plot for the RC circuit in Figure 27–32 if $C = 1\,\mu F$ and the resistor has the following values. In each case, find f_L.

    (a) $R = 100\,\Omega$ **(b)** $R = 1K\Omega$ **(c)** $R = 10K\Omega$

    (d) How does f_L of the high-pass filter depend on R?

14. Figure 27–33 shows a Sallen-Key second-order high-pass filter.

    (a) Find the gain and f_L.

    (b) Assuming the filter is a Butterworth filter, sketch the Bode plot.

    (c) What is the order of this filter?

    (d) What is the final roll-off?

FIGURE 27–33

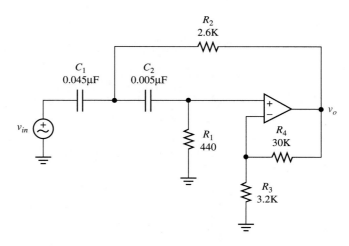

SECTION 27–5   The Band-Pass Active Filter

15. Draw an ideal band-pass response. Label the pass-band, the pass-band gain, the stop-bands, the bandwidth, and the cut-off frequencies.

16. Figure 27–34 shows several band-pass responses. For each filter response find Q, f_L, and f_H.

FIGURE 27–34

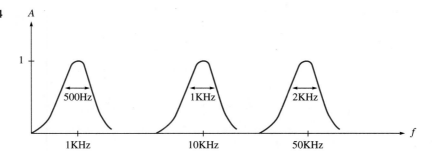

17. Figure 27–35 (p. 928) shows a high-pass filter connected to a low-pass filter.

    (a) Find the gain and cut-off frequency for the low-pass filter.

    (b) Find the gain and cut-off frequency for the high-pass filter.

    (c) What is the overall gain, f_L, and f_H of the filter?

    (d) Sketch the overall Bode response for the filter.

    (e) What are the bandwidth, center frequency, and Q of the filter?

18. Use the circuit configuration of Figure 27–35 (p. 928) to design a band-pass filter with an overall gain of 40dB and with cut-off frequencies of $f_L = 20Hz$ and $f_H = 20KHz$. What are the bandwidth, center frequency, and Q of your filter?

FIGURE 27–35

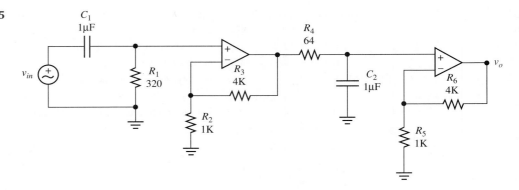

19. Figure 27–36 shows a Sallen-Key band-pass filter.
    (a) Redraw the circuit with C_1 open. Discuss the filter response.
    (b) Redraw the circuit with C_1 short. Discuss the filter response.
    (c) Redraw the circuit with C_2 open. Discuss the filter response.
    (d) Redraw the circuit with C_2 short. Discuss the filter response.

20. For the Sallen-Key band-pass filter of Figure 27–36
    (a) Find the center frequency, bandwidth, and gain.   (b) Find Q, f_L, and f_H.
    (c) Sketch the Bode plot.   (d) What are the roll-offs here? Why?

FIGURE 27–36

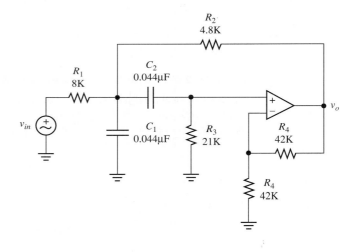

SECTION 27–7   Troubleshooting Active Filters

21. The following problems have been found when troubleshooting the low-pass filter of Figure 27–37. What is the likely cause of each problem?
    (a) No signal at point 1.   (b) Signal at point 2 but no signal at output.
    (c) The measured roll-off is 20dB/dec.   (d) 0V at point 2.

22. The following problems have been found when troubleshooting the high-pass filter of Figure 27–38. What is the likely cause of the problem?
    (a) The measured roll-off is 20dB/dec.   (b) Output voltage equals input voltage.
    (c) No signal at point 1.   (d) +15V at output.

23. The following problems have been found when troubleshooting the band-pass filter of Figure 27–39. What is the likely cause of each problem?
    (a) Circuit acts as a low-pass filter.   (b) Circuit acts as a high-pass filter.
    (c) 0V at point 2.   (d) 0V at point 1.

FIGURE 27–37

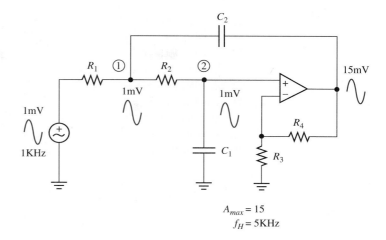

$A_{max} = 15$
$f_H = 5\text{KHz}$

FIGURE 27–38

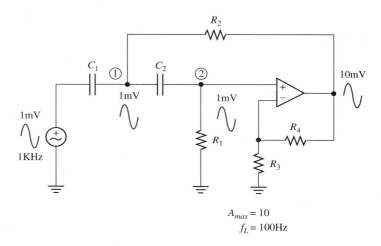

$A_{max} = 10$
$f_L = 100\text{Hz}$

FIGURE 27–39

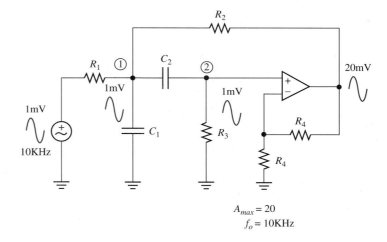

$A_{max} = 20$
$f_o = 10\text{KHz}$

Computer Problems

Unless told otherwise, use the ideal Op-Amp model if it is available in your computer program.

24. Simulate the circuits of Problem 2 (p. 925).

25. Simulate the circuits of Problem 3 (p. 925).

26. Simulate the circuits of Problem 4 (p. 925).

27. Simulate the circuits of Problem 5 (p. 925).

28. Simulate the circuit of Problem 7 (p. 926).

29. Simulate the circuit of Problem 8 (p. 926).

30. Simulate the circuit of Problem 10 (p. 926).

31. Simulate the circuit of Problem 12 (p. 926).

32. Simulate the circuit of Problem 14 (p. 927).

33. Simulate the circuit of Problem 14 with a 741 Op-Amp. Why is this Bode plot different from the one you got in Problem 32?

34. Simulate the circuit of Problem 17 (p. 927).

35. Simulate the circuit of Problem 20 (p. 928).

CHAPTER

28

FEEDBACK AMPLIFIERS

To determine the system gain and advantages of negative feedback.

To determine the system gain and properties of a positive feedback amplifier.

To determine the stability of positive and negative feedback amplifiers.

To determine the system gain of a voltage feedback amplifier.

To determine the system gain of a current feedback amplifier.

To determine the input and output resistances of a voltage feedback amplifier.

To determine the input and output resistances of a current feedback amplifier.

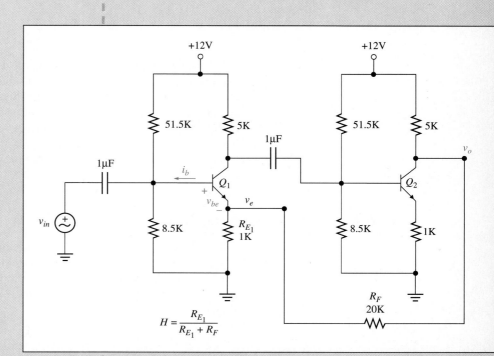

$$H = \frac{R_{E_1}}{R_{E_1} + R_F}$$

28–1 ■ INTRODUCTION

Feedback has been used for centuries to control the behavior of devices such as water clocks and millstones. In more recent history, feedback was used to regulate the speed of steam engines. Today, feedback systems are used in almost all electrical and mechanical devices. The telephone company in the earlier part of this century was the first to use electronic feedback amplifiers to reduce distortion. Feedback amplifiers also made effective radio communication possible.

Consider the cruise control available in most automobiles built in recent years (Figure 28–1). A sensor measures the actual speed of the car, and the sensor output is compared to the speed setting chosen by the driver. If the car slows down (for example, in going up a hill), a signal is sent to the carburetor or fuel injector to increase the gas flow. The car's speed increases until it matches the setting speed. Likewise, if the car speeds up (going down a hill), gas flow is reduced. You can easily make up what you pay extra for cruise control with the money you save in speeding tickets. In other words, feedback can keep you out of traffic school.

FIGURE 28–1
Cruise control

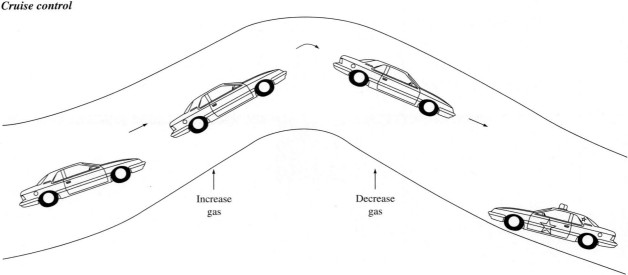

Increase
gas

Decrease
gas

While the principle of using feedback to control a system's output is very basic, many surprises lurk for the unwary. For instance, the governor used in the feedback control of a steam engine can cause instability, leading to pretty wild oscillations of the engine. An unstable controller in a satellite intended to orbit the earth could cause that satellite to go boldly (no split infinitives for us) where no satellite has gone before. The need to control instability, also present in electronic feedback amplifiers, led to the formal study of feedback. In this chapter we present some of the results of these studies.

28–2 ■ NEGATIVE FEEDBACK BASICS

Figure 28–2 shows the feedback block diagram that we first introduced in Chapter 23. The letter *S* is used to represent the various signals (either voltage or current) in the system. The circle with the Greek Σ is a comparator (or differencer) that subtracts the feedback signal from the input signal to produce an error signal. Negative feedback works to minimize the error signal.

FIGURE 28–2
Negative feedback amplifier

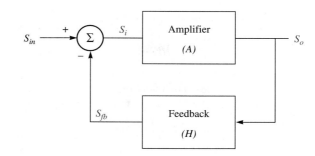

Terms and Formulas

Before we derive the negative feedback relations, we want to define some terms

- ☐ Forward-path gain = A
- ☐ Feedback gain (or feedback factor) = H
- ☐ Loop gain = AH
- ☐ Closed-loop gain (system gain) = A_{sys}

We now trace the signals around the loop. The comparator output is

$$S_i = S_{in} - S_{fb}$$

The output of the amplifier in the forward path is

$$S_o = AS_i = A(S_{in} - S_{fb})$$

The feedback signal is found by applying the feedback factor to the output signal:

$$S_{fb} = HS_o$$

This leads to

$$S_o = A(S_{in} - HS_o) = AS_{in} - AHS_o$$

Note that the output signal appears on both sides of the equation. This is typical in feedback systems. Solving for the output

$$S_o = \frac{A}{1 + AH} S_{in}$$

The total system gain (closed-loop gain) is

$$A_{sys} = \frac{S_o}{S_{in}} = \frac{A}{1 + AH}$$

One of the properties of negative feedback is that it reduces the overall gain. We used this property to our advantage with Op-Amp circuits, using feedback to reduce the very high, and unusable, gain.

EXAMPLE 28–1
Closed-Loop Gain

Find the loop gain *(AH)* and the system gain, given the following data:

(a) $A = 2, H = 1$ **(b)** $A = 100, H = 0.01$ **(c)** $A = 100, H = 0.5$

Solution
(a) The loop gain is

$$AH = 2 \times 1 = 2$$

The system gain is

$$A_{sys} = \frac{A}{1 + AH} = \frac{2}{1 + 2} = 0.667$$

(b) The loop gain is

$$AH = 100 \times 0.01 = 1$$

The system gain is

$$A_{sys} = \frac{A}{1 + AH} = \frac{100}{1 + 1} = 50$$

(c) The loop gain is

$$AH = 100 \times 0.5 = 50$$

The system gain is

$$A_{sys} = \frac{A}{1 + AH} = \frac{100}{1 + 50} = 1.96$$

Note that in this example the loop gain is very large, so

$$A_{sys} = \frac{A}{1 + AH} \approx \frac{A}{AH} = \frac{1}{H}$$

DRILL EXERCISE   Find the loop gain and the system gain in Figure 28–2 (p. 935) if $A = 25$, $H = 1$.

Answer $AH = 25$, $A_{sys} = 0.962$

Amplifier Parameter Variations and Non-Linearities

One of the primary advantages of feedback is that parameter variations or non-linearities in the forward-path amplifier *(A)* do not affect the overall output. The overall system gain of a feedback amplifier is

$$A_{sys} = \frac{A}{1 + AH}$$

If the loop gain is large

$$AH \gg 1$$

and the system gain becomes

$$A_{sys} \approx \frac{1}{H}$$

In general, the forward path contains the various types of amplifiers we have been discussing. The feedback path *(H)* is usually constructed with a few resistors. Because we can use precision resistors, it is usually much easier to control the exact value of H than the exact value of A.

For a large loop gain, the overall system is controlled by the gain of the feedback element. Because the forward gain *(A)* does not show up in the final formula, variations in A do not affect the system gain. In Example 28–1 Part C (p. 935) we found

$$A = 100, H = 0.5 \rightarrow A_{sys} = 1.96$$

Let's now assume that component aging has caused the amplifier gain to decrease by 50%. The system gain is now

$$A_{sys} = \frac{50}{1 + 50 \times 0.5} = 1.92$$

Because of feedback, a 50% variation in the forward-path gain results in a 2% variation in the system gain.

The results we just found also apply to an amplifier that is not linear. In a non-linear amplifier, the amplifier gain will change as the input magnitude changes. For example, at a 1mV input the gain might be 100, but at 1V the gain is only 80. Whether the forward-path amplifier gain changes due to parameter variations or non-linearities, a high loop gain results in a system gain that depends only on H.

REVIEW QUESTIONS

1. Draw a block diagram of a negative feedback system. Identify the comparator, the forward-path gain, and the feedback gain. What is the loop gain?

2. Talk around the loop of your diagram to determine how negative feedback works.

3. What is the overall system gain?

4. If the loop gain is large, what is the overall system gain?

5. What is the advantage of using negative feedback?

28–3 ■ POSITIVE FEEDBACK BASICS

As Figure 28–3 shows the block diagram of a positive feedback system. In this case, the feedback signal is *added* to the input signal. We can find the system gain from

$$S_i = S_{in} + S_{fb} = S_{in} + HS_o$$

FIGURE 28–3
Positive feedback amplifier

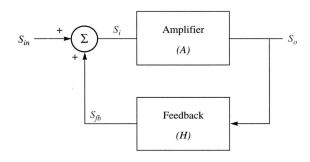

The output is

$$S_o = AS_i$$

so

$$S_o = AS_i = AS_{in} + AHS_o$$

Once again we see that the output appears on both sides of the equation. Solving for the output and the system gain, we get

$$S_o = \frac{A}{1 - AH} S_{in}$$

and

$$A_{sys} = \frac{S_o}{S_{in}} = \frac{A}{1 - AH}$$

In positive feedback the denominator of the system gain is found by *subtracting* the loop gain from 1. Let's find the system gain of an amplifier with positive feedback, where the amplifier gain and feedback fraction are

$$A = 2, \quad H = 0.4$$

The system gain due to positive feedback is

$$A_{sys} = \frac{A}{1 - AH} = \frac{2}{1 - 2 \times 0.4} = 10$$

You can see that where negative feedback reduces the overall gain, positive feedback increases the gain.

This fact was made use of by radio pioneer E.H. Armstrong. He used positive feedback, known as **regeneration,** to increase the gain of the very weak vacuum tube amplifiers available in his day. His discovery of regeneration made radio communication commercially viable.

You might have noticed that the loop gain in our positive feedback example was less than 1. To get an appreciation for how a positive feedback system is greatly influenced by the loop gain, let's see what happens if we apply a short duration pulse to the input and vary the loop gain.

In Figure 28–4 we have applied a pulse input to three positive feedback loops having loop gains less than 1, equal to 1, and greater than 1. To simplify the analysis we will assume that it takes a finite time for the pulse to work its way around the loop. That is, when we first apply the pulse, there is as yet no feedback signal. The input then goes to zero, and we trace each successive pulse around the loop.

Stable Response

We begin with Figure 28–4A, which has a loop gain less than 1 ($AH = 0.5$). The pulse train shown at the output comes about as follows:

    1V pulse applied to input

    1V pulse at S_i because there is no feedback yet

    2V pulse at output

    0.5V pulse at S_{fb} ($S_{fb} = 0.25 \, S_o$)

    0.5V pulse at S_i because input has gone to 0V

    1V pulse at output

    0.25V pulse at S_i

    0.5V pulse at output

    1.25V pulse at S_i

    0.25V pulse at output

As you see, the output starts at 2V and decreases towards 0V. This is known as a **stable response.** We jerk the system a bit, and it settles back down to rest. If you use positive feedback to build a regenerative amplifier, the loop gain must be less than 1.

FIGURE 28–4
Stability

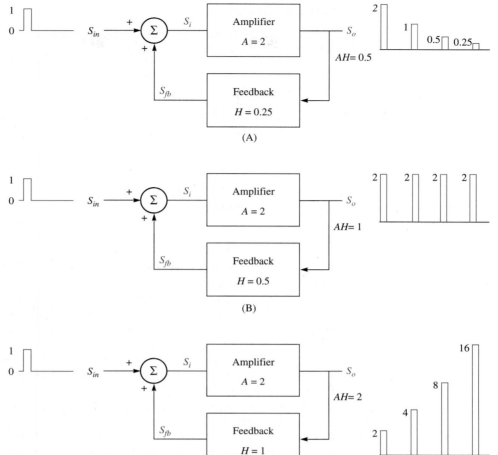

(A)

(B)

(C)

Sustained Pulse

In Figure 28–4B, the loop gain is set to 1. The resulting output sequence now is

> 1V pulse applied to input
> 1V pulse at S_i
> **2V pulse at output**
> 1V pulse at S_{fb} ($S_{fb} = 0.5S_o$)
> 1V pulse at S_i because input has gone to 0V
> **2V pulse at output**
> 1V pulse at S_i
> **2V pulse at output**
> 1V pulse at S_i
> **2V pulse at output**

As you see, if the loop gain is 1, the positive feedback loop produces a sustained pulse train output. This is useful if we want to build oscillators; however, sustained oscillations like this will produce unwanted hum in an amplifier.

Unstable Response

Finally, Figure 28–4C (p. 940) shows a positive feedback loop with a loop gain greater than 1 ($AH = 2$). The pulse train sequence here is

1V pulse applied to input

1V pulse at S_i

2V pulse at output

2V pulse at S_{fb} ($S_{fb} = 1S_o$)

2V pulse at S_i because input has gone to 0V

4V pulse at output

4V pulse at S_i

8V pulse at output

8V pulse at S_i

16V pulse at output

We see here that if the loop gain in a positive feedback system is greater than 1, any small input will cause the output to grow uncontrollably. This is an **unstable response.** In real devices, the output cannot reach infinity but will instead saturate at or near the supply voltage. You have all probably experienced an unstable positive feedback loop. The most common example is what happens when a microphone is placed too close to the speaker. The result is a lot of loud noise and everybody screaming "Turn it down!"

Have you ever seen a parent yell at a child to stop crying, only to have the child cry even louder, resulting in the parent yelling louder, etc? You have witnessed an unstable biological positive-feedback system.

We want amplifiers to be stable. However, there are circuits (such as the Schmitt trigger we described in Chapter 25) that use unstable behavior to advantage. The unstable positive-feedback loop of the Schmitt trigger takes a small positive output and feeds it back to the +terminal. This voltage is amplified to create a larger positive output voltage, which again is fed back to the +terminal. The output voltage, therefore, grows very rapidly until the amplifier saturates.

EXAMPLE 28–2
Positive Feedback

Given the following feedback factors, draw the expected output of the positive feedback system in Figure 28–5. In each case, determine whether the response is stable or unstable. If the response is stable, find the system gain.

(a) $H = 0.1$ **(b)** $H = 0.5$

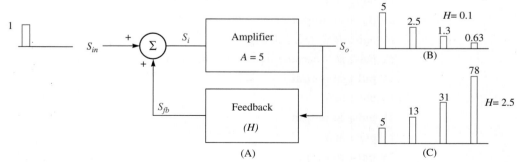

FIGURE 28–5

Solution
(a) For $H = 0.1$, the loop gain is

$$AH = 0.5$$

Because the loop gain is less than 1, the system is stable. The pulse sequence at the output is found from

$$S_o = AS_i = AS_{in} = 5$$
$$S_{fb} = HS_o = 0.5$$
$$S_o = 2.5$$
$$S_{fb} = 0.25$$
$$S_o = 1.25$$
$$S_{fb} = 0.125$$
$$S_o = 0.625$$

This pulse train output is shown in Figure 28–5B. The system gain for this stable system is

$$A_{sys} = \frac{A}{1 - AH} = \frac{5}{0.5} = 10$$

(b) For $H = 0.5$, the loop gain is $AH = 2.5$. Because the loop gain is greater than 1, the system is unstable. The pulse sequence at the output is found from

$$S_o = AS_i = AS_{in} = 5$$
$$S_{fb} = HS_o = 2.5$$
$$S_o = 12.5$$
$$S_{fb} = 6.25$$
$$S_o = 31.3$$
$$S_{fb} = 15.6$$
$$S_o = 78$$

The pulse train for this unstable system is shown in Figure 28–5C.

REVIEW QUESTIONS

1. Draw a block diagram of a positive feedback system.
2. How does a positive feedback system differ from a negative feedback system?
3. What is the system gain? How does positive feedback affect the overall gain?
4. How is stability of a positive feedback system determined?
5. Draw a pulse train output that demonstrates the stability of a positive feedback system with loop gain less than 1.
6. Draw a pulse train output that demonstrates the oscillations of a positive feedback system with loop gain equal to 1.
7. Draw a pulse train output that demonstrates the instability of a positive feedback system with loop gain greater than 1.

28–4 ▪ STABILITY OF NEGATIVE FEEDBACK AMPLIFIERS

The preceding two sections may have given you the impression that only positive feedback systems can become unstable. Unfortunately, negative feedback amplifiers can also go unstable. To see how this can occur, we need to take another look at the input-output behavior of amplifiers.

In Chapter 26 we described the frequency behavior of an amplifier as a ratio of the input and output magnitudes. Figure 28–6 (p. 942) shows how the output amplitude varies

FIGURE 28–6
Time delay and phase angle

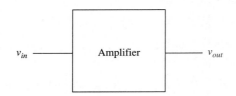

$$\theta = 360°\frac{t_d}{T}$$

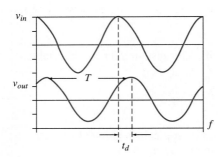

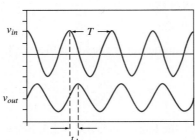

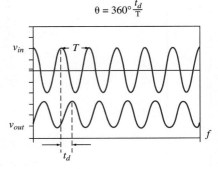

with frequency. Figure 28–6 also shows that the input and output sine waves are not lined up. In fact, the output sine wave peaks later than the input sine wave. In this particular amplifier as frequency increases, the time delay between the input and output sine waves increases.

Rather than talk about time delays in the output, we talk about the **phase angle** between the output and input sine waves. The phase angle (θ) is defined as

$$\theta = -360°\frac{t_d}{T}$$

where t_d is the time delay and T is the period of the sine wave. This relation results from the fact that one cycle of a sine wave covers 360° and that it takes T seconds to complete that one cycle. A negative phase indicates that the output peaks after the input.

EXAMPLE 28–3
Phase Angle and Time Delay

(a) A 1KHz sine wave input produces an output that peaks 0.25ms after the input. Find the phase angle.
(b) A 100Hz signal input produces an output that has a phase angle of −30°. Find the time delay in the output.

Solution
(a) A 1KHz signal has a period of

$$T = \frac{1}{f} = \frac{1}{1K} = 1\text{ms}$$

A time delay of 0.25ms results in a phase angle of

$$\theta = -360\frac{t_d}{T} = -360 \times \frac{0.25}{1} = -90°$$

(b) A 100Hz signal has a period of

$$T = \frac{1}{f} = \frac{1}{100} = 10\text{ms}$$

We need to rearrange the formula to find the time delay

$$t_d = T\frac{\theta}{-360} = 0.01 \times \frac{-30}{-360} = 0.833\text{ms}$$

The phase angle gives us information that does not depend on the frequency (the period) of the sine wave. For example, a phase angle of $-90°$ produces a sine wave that is shifted by one-fourth of a cycle. The following table shows some relations between phase angle and the delay in the sine wave:

Phase Angle (°) θ	Delay t_d
0	0
-45	1/8 cycle
-90	1/4 cycle
-135	3/8 cycle
-180	1/2 cycle

Of primary concern to us here is the phase angle of $-180°$. In Figure 28–7, we show an input-output pair that has a phase angle of $-180°$. You can see that the output is actually the negative of the input. A phase angle of $-180°$—or $+180°$, for that matter—produces the same result as multiplying the input by -1. That is, the phase angle between v_{in} and v_o is $-180°$, then

$$v_o = -v_{in} \quad \text{for} \quad \theta = -180°$$

FIGURE 28–7

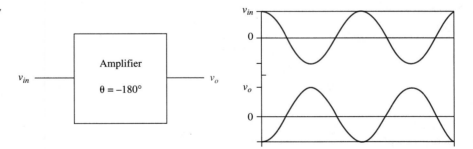

Figure 28–8 shows a negative feedback system in which the phase angle produced by the amplifier is $-180°$. Because of this phase angle, the output of the feedback amplifier is the negative of the input to the amplifier:

$$S_{fb} = -H_{so}$$

FIGURE 28–8

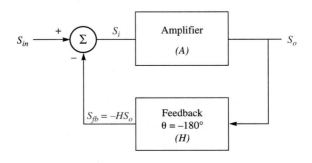

To find the overall system gain, we can replace H with $-H$ in the standard negative feedback formula to get

$$A_{sys} = \frac{A}{1 + (-AH)} = \frac{A}{1 - AH}$$

If you compare this denominator with the denominator of a positive feedback system, you will see that they are the same. That is

**If the loop gain has a phase angle of $-180°$, a negative
feedback system becomes a positive feedback system.**

As with any positive feedback system, if loop gain equals 1, then we will get sustained oscillations; if the loop gain is greater than 1, the system will be unstable. Summarizing:

☐ If $\theta = -180°$ and $AH < 1$ the amplifier is stable.

☐ If $\theta = -180°$ and $AH = 1$ the amplifier will oscillate.

☐ If $\theta = -180°$ and $AH > 1$ the amplifier is unstable.

You may remember from an AC circuits class that a capacitor can produce up to $-90°$ of phase angle. Therefore, an amplifier needs three or more capacitors before its phase angle can actually reach $-180°$. Unfortunately, this capacitor count includes all of the internal capacitances of the transistors making up the amplifier.

For example, Figure 28–9A shows the Bode gain and phase angle plots for an Op-Amp without internal compensation. Note that for the first time we are showing a plot of the phase angle. The details of the phase plot are not important. Rather we want to know the amplifier gain when the phase angle is $-180°$. Looking up the dashed line from $\theta = -180°$, we see that the gain is

$$A_{dB} = 120dB \rightarrow A = 10^6 \text{ at } \theta = -180°$$

FIGURE 28–9
Operation amplifier Bode plots

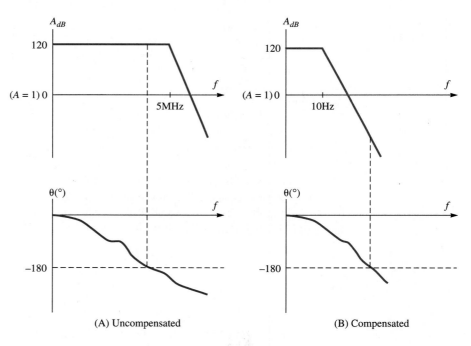

(A) Uncompensated   (B) Compensated

If we close a negative feedback loop around this Op-Amp, we will create an unstable amplifier that will be useless to us.

The internal compensation capacitor in the 741 gives us the gain and phase plots shown in Figure 28–9B. You can see that by knocking the gain down at low frequency, we have produced an amplifier with a gain less than 1 (0dB) when the phase angle hits $-180°$. A negative feedback loop around the compensated Op-Amp will remain stable.

1. Draw a 1KHz sine wave. Under this sine wave, draw another that has a time delay. On this plot, indicate the period and the time delay.

2. What is the relation between time delay and phase angle?

3. What part of a cycle is a phase angle of $-135°$?

4. How else can we represent a phase angle of $-180°$?

5. If the loop gain in a negative feedback system has a phase angle of $-180°$, what type of feedback system does it become?

6. Discuss the stability of a negative feedback amplifier.

7. How much phase angle can be contributed by a single capacitor?

8. Why is internal compensation used in an Op-Amp?

28–5 ■ VOLTAGE FEEDBACK

Four Feedback Possibilities

To this point we have discussed feedback with the general signal S. The two signals that we deal with in electronic amplifiers are voltage and current. The output S_o can be either a voltage or a current. The signal we feed back (S_{fb}) can also be either a voltage or a current. This gives us four different possibilities for a feedback amplifier. The four feedback possibilities are listed below, each with a descriptive name. Be aware that every textbook has a different name for these feedback types.

S_o	S_{fb}	Feedback Type
voltage	voltage	voltage out–voltage feedback
voltage	current	voltage out–current feedback
current	current	current out–current feedback
current	voltage	current out–voltage feedback

Note: S_{in}, S_i, and S_{fb} must all be the same—all voltages or all currents.

To simplify matters, in the following discussion we will consider in depth only the two feedback amplifiers that have a voltage output. We will then give a brief description of the two current output amplifiers.

The trickiest thing about using the feedback formula for gain (and the resistance formulas we will be giving you) is knowing which type of feedback is used in an amplifier. We will try to show you the difference by looking at some examples.

Figure 28–10A (p. 946) shows the basic non-inverting amplifier. Remember that for the Op-Amp

$$v_o = Av_i$$

and

$$v_i = v_+ - v_-$$

Comparing these relations to the closed-loop diagram for a negative feedback system, we see that

$$S_o = v_o$$
$$S_{in} = v_+$$
$$S_{fb} = v_-$$
$$S_i = v_+ - v_-$$

FIGURE 28–10

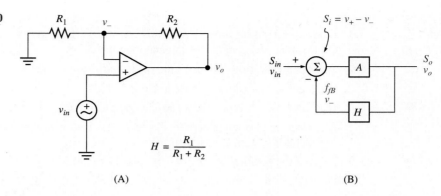

(A)   (B)

Figure 28–10B shows this non-inverting amplifier modeled as a negative feedback system. The forward-path gain is provided by the Op-Amp and is approximately

$$A = 1 \times 10^6$$

Because there is no current into the Op-Amp, R_1 and R_2 form a voltage divider that gives us the feedback factor:

$$v_- = Hv_o = \frac{R_1}{R_1 + R_2} \, v_o$$

so

$$H = \frac{R_1}{R_1 + R_2}$$

Because the forward-path gain is so large

$$A_{sys} = \frac{A}{1 + AH} \approx \frac{1}{H} = \frac{R_1 + R_2}{R_1} = 1 + \frac{R_2}{R_1}$$

This feedback analysis has given us the non-inverting gain that we are familiar with.

EXAMPLE 28–4
Feedback Analysis of Non-Inverting Amplifier

The Op-Amp in Figure 28–11 has a gain of $A = 1 \times 10^4$. Find the feedback factor and use feedback analysis to find the system gain.

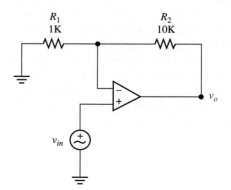

FIGURE 28–11

Solution The feedback factor is found from the voltage divider relation:

$$v_- = \frac{R_1}{R_1 + R_2} v_o = \frac{1K}{1K + 10K} v_o = \frac{1}{11} v_o$$

so

$$H = \frac{1}{11}$$

The system gain is

$$A_{sys} = \frac{A}{1 + AH} = \frac{10^4}{1 + \dfrac{1}{11} \times 1 \times 10^4} = 11$$

BJT Feedback Analysis

In Figure 28–12 we show a two-stage BJT amplifier with a feedback connection from the collector of Q_2 to the emitter of Q_1. BJT feedback amplifiers use both voltage and current feedback. How do we know what type of feedback we are dealing with here?

FIGURE 28–12
Voltage out–voltage back feedback amplifier

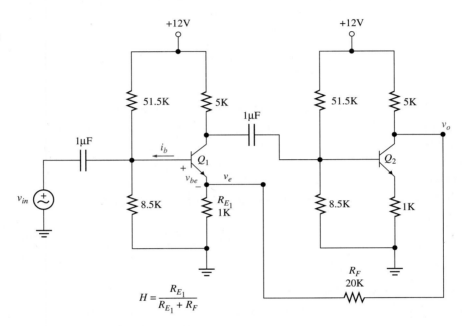

$$H = \frac{R_{E_1}}{R_{E_1} + R_F}$$

We know that in a feedback system the forward-path amplifier amplifies a *difference* in the input and feedback signals. For instance, we normally think of the BJT as amplifying the input current (βi_b). If the base current can be found as difference in an input current and a feedback current, then the amplifier is using current feedback. A quick glance at the circuit in Figure 28–12 will tell you that the base current here depends only on the input. This cannot be a current feedback amplifier.

From Chapter 13 you know that we can also model the BJT as a device that amplifies the base-emitter voltage $(g_m v_{be})$. (This is equivalent, by the way, to the amplification in an FET $[g_m v_{gs}]$.) In this situation

$$S_{in} = v_b$$
$$S_{fb} = v_e$$

and

$$S_i = v_{be} = v_b - v_e$$

Because the emitter in this circuit is connected to the output, this is a voltage out–voltage feedback amplifier.

The forward-path amplifier gain is found by removing R_F (actually, we disconnect R_F from the emitter, and connect it to ground):

$$A = A_{Q_1} A_{Q_2} = 13$$

To find the feedback factor, we need to find v_e as a function of v_o. We do this by assuming the R_F and R_{E_1} form a voltage divider, giving us the approximate relation

$$v_e \approx \frac{R_{E_1}}{R_{E_1} + R_F} v_o$$

so

$$H = \frac{R_{E_1}}{R_{E_1} + R_F}$$

For our amplifier

$$H = \frac{R_{E_1}}{R_{E_1} + R_F} = \frac{1K}{1K + 20K} = 0.0476$$

The system gain is, therefore

$$A_{sys} = \frac{A}{1 + AH} = \frac{13}{1 + 0.0476 \times 13} = 8.03$$

EXAMPLE 28–5
BJT Voltage Out–Voltage Feedback Amplifier

Change the feedback resistor in Figure 28–12 (p. 947) to

$$R_F = 5K$$

and find the system gain.

Solution   We find the feedback factor from

$$H = \frac{R_{E_1}}{R_{E_1} + R_F} = \frac{1K}{1K + 5K} = 0.167$$

The system gain is

$$A_{sys} = \frac{A}{1 + AH} = \frac{13}{1 + 0.167 \times 13} = 4.1$$

As the feedback resistor is reduced, the feedback voltage at the emitter increases. Increasing the feedback voltage reduces the gain.

REVIEW QUESTIONS

1. What types of feedback are there?
2. How can you tell if the feedback signal is a voltage or a current?
3. Draw a non-inverting Op-Amp amplifier. Identify S_{in}, S_i, S_{fb}, and A. Find H and A_{sys}.
4. For a voltage out–voltage feedback BJT amplifier, identify S_{in}, S_i, S_{fb}, and A.
5. Assume feedback is provided by R_F connected between the output and the emitter. Find H.

28–6 ■ VOLTAGE OUT–CURRENT FEEDBACK

The analysis of current feedback systems requires you to get used to looking at amplifiers in a new light. To this point we have always assumed that the input to the amplifier was a voltage, and, so, the gain was a ratio of output voltage to input voltage.

FIGURE 28–13
Voltage out–current back feedback amplifier

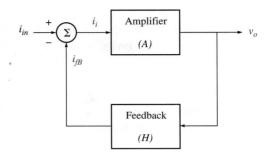

However, if a system uses current feedback (Figure 28–13), then we *must* consider the input to be a current. The forward-path and overall system gain now will be the ratios of output voltages to input currents and has the units of resistance

$$A_i = \frac{v_o}{i_i}$$

and

$$A_{sys} = \frac{v_o}{i_{in}}$$

The feedback factor relates a current to a voltage and, so, has units of conductance

$$H = \frac{i_{fb}}{v_o}$$

The system gain is still given by

$$A_{sys} = \frac{A_i}{1 + A_i H}$$

Figure 28–14A shows the basic inverting Op-Amp amplifier and also shows the Box Model for the Op-Amp. In this circuit, because the +terminal is grounded

$$v_+ = 0\text{V}$$

FIGURE 28–14

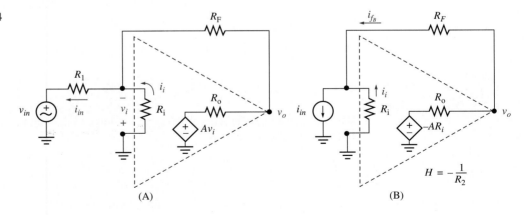

the error voltage is

$$v_i = v_+ - v_- = -v_-$$

In this case, the error voltage is *not* found as a difference in input and feedback voltages. This cannot be a voltage feedback amplifier. If it is not voltage feedback, it must be current feedback.

Because we are dealing with a voltage out–current feedback amplifier, we must use current as an input, as shown in Figure 28–14B (p. 949). After we solve the feedback equations to find v_o/i_{in}, we will show you how to convert this to the more standard v_o/v_{in}.

We choose the current directions in Figure 28–14B so that

$$i_i = i_{in} - i_{fb}$$

This confirms that the error signal is the difference in the input and feedback signals. Figure 28–14B also shows that we can use Ohm's law to find

$$v_i = -R_i i_i$$

The minus sign is due to the direction of i_i. So, the controlled source can be modeled with

$$Av_i = -AR_i i_i \text{ (Remember, this is an inverting amplifier.)}$$

That is, the forward-path gain is now

$$A_i = \frac{v_o}{i_i} = -AR_i$$

Note that we are not performing an ideal Op-Amp analysis; here we use the input current and input resistance.

We now know the forward-path gain and need to find the feedback factor. We know that feedback will minimize v_i, so $v_- \approx v_+$. Therefore, when we find i_{fb}, we can assume the voltage at the −terminal is near enough to zero that

$$i_{fb} \approx -\frac{v_o}{R_F}$$

This leads to the feedback factor:

$$H = \frac{i_{fb}}{v_o} \approx -\frac{1}{R_F}$$

The system gain is

$$A_{i_{sys}} = \frac{A_i}{1 + A_i H} = \frac{-AR_i}{1 + \dfrac{A_i R_i}{R_F}}$$

Because the loop gain (AR_i) is very large, the system gain becomes

$$A_{i_{sys}} = \frac{1}{H} = -R_F$$

We have just found the ratio of the output voltage to the input current. We can use Ohm's law to find the voltage gain:

$$A_{sys} = \frac{v_o}{v_{in}}$$

By definition the input voltage and current are related by

$$v_{in} = R_{in} i_{in}$$

where R_{in} is the input resistance of the closed loop system. Substituting this relation into the system gain equation gives us

$$A_{sys} = \frac{v_o}{v_{in}} = \frac{v_o}{R_{in} i_{in}} = \frac{1}{R_{in}} \frac{v_o}{i_{in}}$$

So the system voltage gain is

$$A_{sys} = \frac{1}{R_{in}} A_{i_{sys}}$$

Because we know that the input resistance in the inverting Op-Amp amplifier of Figure 28–14A (p. 949) is

$$R_{in} = R_1$$

we find the system voltage gain as

$$A_{sys} = \frac{1}{R_{in}} A_{i_{sys}} = -\frac{R_F}{R_1}$$

EXAMPLE 28–6

Voltage Out–Current Feedback Amplifier

The Op-Amp shown in Figure 28–15 has the parameters

$$A = 10^6 \text{ and } R_i = 1\text{MHz}$$

Find A_i, H, $A_{i_{sys}}$, R_{in}, A_{sys}.

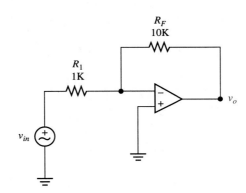

FIGURE 28–15

Solution   The forward-path gain for this voltage out–current feedback amplifier is

$$A_i = -AR_i = -1 \times 10^6 \times 1\text{K} = 1 \times 10^9 \text{V/A}$$

The feedback factor is found from

$$H = -\frac{1}{R_F} = -\frac{1}{10\text{K}}$$

Because the forward-path gain is so large, the system gain for the current feedback amplifier is

$$A_{i_{sys}} = \frac{1}{H} = -10\text{K}$$

The input resistance of the inverting amplifier is

$$R_{in} = R_1 = 1\text{K}$$

Finally, the voltage gain for this system is

$$A_{sys} = \frac{1}{R_{in}} A_{i_{sys}} = -\frac{10\text{K}}{1\text{K}} = -10$$

We have seen that we need to know the input resistance of the closed-loop system to convert $A_{i_{sys}}$ to a voltage system gain. While finding the input resistance of the inverting Op-Amp is very simple, finding the input resistance of a feedback transistor circuit requires more work. We will return to current feedback amplifiers after we discuss input and output resistances of feedback amplifiers.

REVIEW QUESTIONS

1. In a current feedback system what type of electrical signals are S_{in}, S_{fb}, S_i?
2. Because the output signal is still a voltage, what are the units of A_i, H, $A_{i_{sys}}$?
3. How can we convert $A_{i_{sys}}$ to the voltage gain A_{sys}?
4. Draw an inverting Op-Amp amplifier. What are A_i, H, $A_{i_{sys}}$, R_{in}, A_{sys}?

28–7 ■ INPUT AND OUTPUT RESISTANCE—VOLTAGE FEEDBACK

When we close a loop around an amplifier, we change the amplifier's input and output resistances. Box Models of the feedback system will help us find these resistances. In Figure 28–16 the forward-path amplifier is represented with a complete Box Model. To simplify the analysis, we have modeled the feedback element *(H)* with a simplified Box Model that assumes an infinite input resistance and zero output resistance.

FIGURE 28–16
Box Models for voltage out–voltage back amplifier

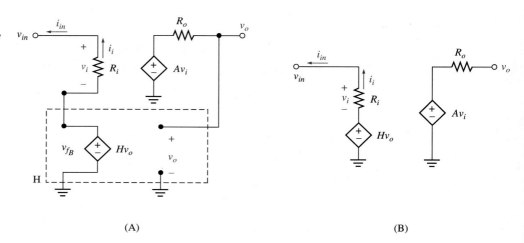

(A)

(B)

Note that the comparator is not explicitly seen in Figure 28–16A. We will show you how this circuit accomplishes the necessary subtraction of the feedback voltage from the input voltage. In Figure 28–16B, we have redrawn the feedback system to make it easier to visualize the following relations.

System Gain

We first find the system gain to verify that this model is correct. At the output

$$v_o = Av_i$$

The error voltage appears across R_i, the input resistance of the forward-path amplifier, and is given by

$$v_i = v_{in} - Hv_o$$

Combining these equations, we get

$$v_o = Av_{in} - AHv_o$$

Solving for the output voltage

$$v_o = \frac{A}{1 + AH} \, v_{in}$$

so

$$A_{sys} = \frac{v_o}{v_{in}} = \frac{A}{1 + AH}$$

This confirms the usefulness of this model to analyze feedback amplifiers.

Input Resistance

Remember, resistance is a measure of the relation between a current and a voltage. We can see in Figure 28–16 (p. 952) that for voltage feedback, the system input current is the same as the input current to the forward-path amplifier:

$$i_{in} = i_i$$

From Ohm's law

$$i_{in} = i_i = \frac{v_i}{R_i}$$

We know that when we close the loop on a voltage feedback system, the error voltage is driven towards 0V. As v_i is decreased by the voltage feedback, the input current decreases. Because the effect of voltage feedback is to reduce the input current, we can say that

Voltage feedback increases the system input resistance.

We can derive a formula for the input resistance of a voltage feedback amplifier as follows:

$$i_{in} = i_i = \frac{v_i}{R_i} = \frac{v_{in} - Hv_o}{R_i}$$

We already know that

$$v_o = \frac{A}{1 + AH} \, v_{in}$$

So we can get rid of the v_o term in the current equation to get

$$i_{in} = \frac{v_{in} - \left(\dfrac{AH}{1 + AH}\right)v_{in}}{R_i}$$

After combining terms, this reduces to

$$i_{in} = \frac{v_{in}}{(1 + AH)R_i}$$

The closed-loop input resistance is found from

$$R_{in_{closed\,loop}} = \frac{v_{in}}{i_{in}} = (1 + AH)R_i$$

This confirms that the input impedance is increased with voltage feedback.

Output Resistance

To find the output resistance of an amplifier, we turn off the input, apply a voltage to the output, and solve for the output current. That is

$$R_{out} = \frac{v_o}{i_o} \text{ when } v_{in} = 0V$$

When there is no feedback, $v_i = 0V$. This turns off the voltage source on the output side of the Box Model, so

Open loop: $R_{out} = R_o$

When we connect the feedback loop, v_i becomes

$$v_i = v_{in} - Hv_o = -Hv_o$$

The output current is now

$$i_o = \frac{v_o - Av_i}{R_o} = \frac{v_o - A(-Hv_o)}{R_o}$$

or

$$i_o = \frac{1 + AH}{R_o} v_o$$

Solving for the closed-loop output resistance, we get

$$R_{out_{closed \, loop}} = \frac{v_o}{i_o} = \frac{R_o}{1 + AH}$$

If we are feeding back a fraction of the output voltage, output resistance decreases.

**EXAMPLE 28–7
Input and Output
Resistance—
Voltage Feedback**

The Op-Amp in the non-inverting amplifier of Figure 28–17 has the following parameters:

$$A = 1 \times 10^6$$
$$R_i = 2M\Omega$$
$$R_o = 100\Omega$$

Find H, A_{sys}, AH, R_{in}, R_{out}.

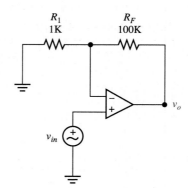

FIGURE 28–17

Solution The feedback factor for the non-inverting amplifier is

$$H = \frac{R_1}{R_1 + R_F} = \frac{1K}{1K + 100K} = \frac{1}{101}$$

The loop gain is

$$AH = \frac{1 \times 10^6}{101} \approx 1 \times 10^4$$

Because the loop gain is so large

$$A_{sys} = \frac{1}{H} = 101$$

The closed-loop input resistance is found from

$$R_{in_{closed\ loop}} = (1 + AH)R_i = 1 \times 10^4 \times 2M = 2 \times 10^{10}\Omega!$$

We see that feedback has greatly increased the already large input resistance of the Op-Amp.

Remember that in the ideal Op-Amp analysis we assumed that the input resistance was infinity. You can now see why that assumption works so well in the real world. (By the way, while the input resistance of the non-inverting Op-Amp amplifier is very high, it never gets as high as we have indicated here; other internal factors act to limit this resistance.) The output resistance is found from

$$R_{out_{closed\ loop}} = \frac{R_o}{1 + AH} = \frac{100}{1 \times 10^4} = 0.01\Omega$$

Our ideal Op-Amp analysis assumed that the output resistance of the amplifier was 0Ω. You can see how good this approximation is for the feedback amplifier.

REVIEW QUESTIONS

1. Use a Box Model to represent the forward-path amplifier and the feedback element. Connect these two together to form a voltage out–voltage feedback amplifier.

2. Write the formulas for the closed-loop gain, input resistance, and output resistance.

3. Discuss the effect of voltage feedback on system gain, input resistance, and output resistance.

28–8 ■ INPUT AND OUTPUT RESISTANCE—CURRENT FEEDBACK

Figure 28–18A (p. 956) shows how we can model voltage out–current feedback with Box Models. This time the output voltage source depends on i_i. Also, the feedback element is a current source (H has units of conductance here); note how this source is connected to the input of the forward-path amplifier. For easier analysis, we have redrawn the feedback system in Figure 28–18B (p. 956). For this circuit you can see that

$$i_i = i_{in} - i_{fb} = i_{in} - Hv_o$$

which provides us with the error signal that we need.

System Gain

From Figure 28–18B (p. 956), we see that

$$v_o = A_i i_i = A_i(i_{in} - Hv_o)$$

FIGURE 28–18
Box Models for voltage out–current back feedback amplifier

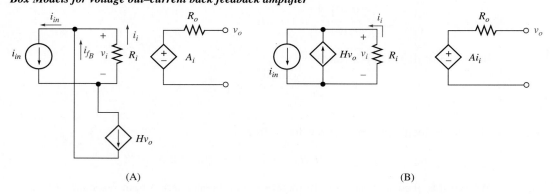

(A)   (B)

Solving for the output voltage, we get the system gain:

$$A_{i_{sys}} = \frac{v_o}{i_{in}} = \frac{A_i}{1 + A_i H}$$

Again note that the system gain here is the ratio of the output voltage to the input current.

Input Resistance

Because current feedback increases the current in R_i, we can say that

Current feedback decreases input resistance.

The input resistance of the current feedback amplifier is found from

$$R_{in} = \frac{v_{in}}{i_{in}}$$

Noting that the input voltage here is the same as the error voltage:

$$v_{in} = v_i = R_i i_i$$

and using the fact that

$$i_i = i_{in} - H v_o$$

we get

$$v_{in} = v_i = R_i(i_{in} - H v_o)$$

The output voltage is related to the input current by the system gain:

$$v_o = \frac{A_i}{1 + A_i H} i_{in}$$

Substituting in the previous equation and rearranging gives us

$$v_{in} = \frac{R_i}{1 + A_i H} i_{in}$$

So the closed-loop input resistance is

$$R_{in_{closed\ loop}} = \frac{v_{in}}{i_{in}} = \frac{R_i}{1 + A_i H}$$

Because current feedback pumps more current into the amplifier input, current feedback reduces the input resistance of the amplifier.

Output Resistance

To save time, we will just tell you that the output resistance of the voltage out–current feedback amplifier is still

$$R_{out} = \frac{R_o}{1 + A_i H}$$

As long as we are picking off the output voltage to produce the feedback signal—either voltage or current—the output resistance of the closed-loop amplifier is reduced.

EXAMPLE 28–8
Input and Output Resistance— Current Feedback

The two-stage BJT amplifier shown in Figure 28–19 uses current feedback to the base of Q_1. Note that because this is a current feedback amplifier, we must assume that the input is a current. If we open the loop by removing R_F, then the forward-path gain, input resistance, and output resistance are

$$A_i = -18\text{KV/A}$$
$$R_i = 6.8\text{K}\Omega$$
$$R_o = 25\Omega$$

Find H, $A_i H$, $A_{i_{sys}}$, R_{in}, R_{out}, A_{sys}.

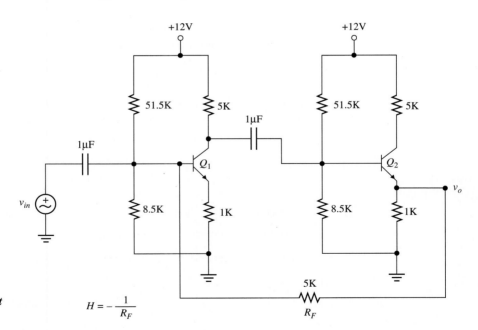

FIGURE 28–19
Voltage out–current back amplifier

$$H = -\frac{1}{R_F}$$

Solution

Because we approximate the feedback current as

$$i_{fb} \approx -\frac{v_o}{R_F} = -\frac{v_o}{5\text{K}}$$

The feedback factor is

$$H = -\frac{v_o}{i_{fb}} \approx -\frac{1}{5\text{K}}$$

The loop gain is

$$A_i H = (-18\text{K})\left(-\frac{1}{5\text{K}}\right) = 3.6$$

The system gain for this current feedback amplifier is

$$A_{i_{sys}} = \frac{A_i}{1 + A_i H} = \frac{-18\text{K}}{1 + 3.6} = -3910\text{V/A}$$

The closed-loop input resistance is found from

$$R_{in_{closed\ loop}} = \frac{R_i}{1 + A_i H} = \frac{6.8\text{K}}{1 + 3.6} = 1.48\Omega$$

The closed-loop output resistance is

$$R_{out_{closed\ loop}} = \frac{R_o}{1 + AH} = \frac{25}{1 + 3.6} = 5.43\Omega$$

The system voltage gain (A_{sys}) is found from

$$A_{sys} = \frac{1}{R_{in}}A_{i_{sys}} = \frac{-3910}{1.48\text{K}} = -2.64$$

How Do I Know if a Transistor Amplifier Is Voltage or Current Feedback?

☐ If the feedback element connects the output back to the emitter of the input transistor, you have a voltage out–voltage feedback amplifier.

☐ If the feedback element connects the output back to the base of the input transistor, you have a voltage out–current feedback amplifier.

REVIEW QUESTIONS

1. Use a Box Model to represent the forward-path amplifier and the feedback element in a voltage out–current feedback amplifier.

2. Write the formulas for the closed-loop gain, input resistance, and output resistance.

3. Discuss the effect of current feedback on system gain, input resistance, and output resistance.

4. How can you find the closed-loop voltage gain from a voltage out–current feedback analysis?

5. How can you tell if a transistor amplifier is a current feedback or a voltage feedback amplifier?

28–9 ■ CURRENT-OUT FEEDBACK AMPLIFIERS

To give you a more complete picture of feedback amplifiers, we will briefly describe amplifiers that feed back a fraction of the output current. In the voltage feedback amplifiers of the previous sections, the feedback pick-off point was at the output. If the feedback pick-off point is not at the output, you are dealing with a current feedback amplifier. Because you are not familiar with finding output currents (we have concentrated in this text on output voltages), we will just discuss these amplifiers in general terms.

Current Out–Current Feedback Amplifier

Figure 28–20 is a repeat of Figure 28–19; however, we now consider the collector voltage to be the output voltage. Because the feedback pick-off point is not at the output but is instead at the emitter, this cannot be a voltage-out amplifier. It must, therefore, be a current-output amplifier.

FIGURE 28–20

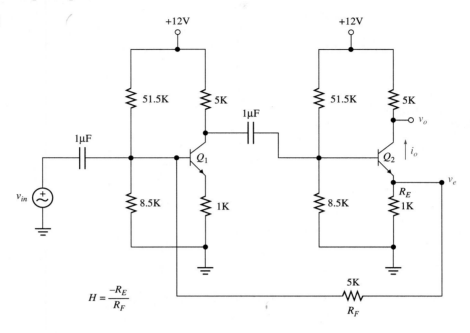

$$H = \frac{-R_E}{R_F}$$

We already know that because the feedback element connects to the transistor base, this is a current feedback amplifier. Just by changing the output from the emitter to the collector, we have created a current out–current feedback amplifier. The forward-path and system gains are ratios of output currents to input currents.

It is instructive to find the feedback factor here, but after we do this, we will just tabulate the feedback parameters for you. In the last section, we found the feedback factor for this amplifier when the emitter voltage was the output voltage. Now, the emitter voltage can be found from the approximation (ignoring the current into R_F)

$$v_e \approx R_E i_o$$

The feedback current is now

$$i_{fb} = -\frac{v_e}{R_F} = -\frac{R_E i_o}{R_F}$$

So the feedback factor is

$$H = -\frac{R_E}{R_F}$$

The closed loop current out–current feedback parameters are listed below:

$$A_{sys} = \frac{i_o}{i_{in}} = \frac{A}{1 + AH}$$

$$R_{in} = \frac{R_i}{1 + AH}$$

$$R_{out} = (1 + AH)R_o$$

Because this is still a current feedback amplifier, the input resistance decreases. The output resistance increases in a current-out feedback amplifier.

Current Out–Voltage Feedback Amplifier

Figure 28–21 is a repeat of the BJT amplifier shown in Figure 28–12. Now, however, we are taking the output voltage off the emitter. This means that the output voltage is not directly connected to the feedback path. This must be a current out–voltage feedback amplifier. The amplifier gains, therefore, are ratios of output voltage to input current.

FIGURE 28–21

Current out–voltage back amplifier

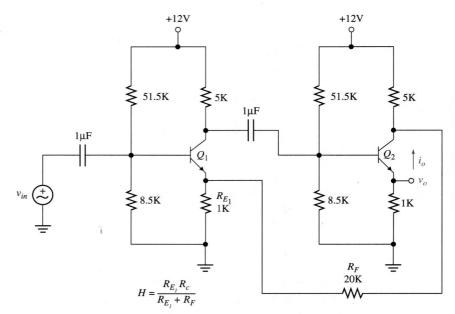

$$H = \frac{R_{E_1} R_c}{R_{E_1} + R_F}$$

We can approximate the feedback factor by assuming that all of the output current goes through R_C:

$$v_c \approx R_C i_o$$

so

$$v_{fb} = \frac{R_E}{R_E + R_f} v_c = \frac{R_E R_C}{R_E + R_F} i_o$$

so the feedback factor is

$$H = \frac{R_E R_C}{R_E + R_F}$$

The closed loop current out–voltage feedback parameters are listed below:

$$A_{sys} = \frac{i_o}{v_{in}} = \frac{A}{1 + AH}$$

$$R_{in} = (1 + AH)R_{in}$$

$$R_{out} = (1 + AH)R_o$$

Because this is still a voltage feedback amplifier, the input resistance increases. Again, the output resistance increases in a current-out feedback amplifier.

1. How do you know if you are dealing with a voltage-out or current-out amplifier?
2. How do you know if you are dealing with a voltage or current feedback amplifier?
3. List the four possible feedback amplifier types.
4. What is the feedback type if the feedback pick-off point is at the emitter and the output is at the collector? What is v_e?
5. What is the feedback type if the feedback pick-off point is at the collector and the output is at the emitter? What is v_c?
6. List the parameters for the current out–current feedback amplifier. Indicate which resistances increase or decrease.
7. List the parameters for the current out–voltage feedback amplifier. Indicate which resistances increase or decrease.

28–10 ■ TROUBLESHOOTING

There is not too much new to learn about troubleshooting feedback amplifiers. You have already seen how to troubleshoot Op-Amp circuits. And DC Troubleshooting is *still* the first procedure to use in a transistor feedback circuit that is not working properly. The only difference in applying DC Troubleshooting is that finding the Estimated Voltages for a feedback circuit can be quite difficult. Because of the feedback, bias levels in one transistor affect the bias levels in the other transistors. There are several alternatives to having to calculate the Estimated Voltages:

☐ You refer to a manufacturer's schematic that gives the Expected Voltages for the bias levels in the circuit.

☐ You simulate the amplifier on a computer to find the bias levels.

☐ You must work without the Expected Voltages. In this case, assume all transistors are operating somewhere in the middle of the **ACTIVE** region. If the Measured Voltages indicate that a transistor is operating at or near **CUT-OFF** or **SATURATION,** there is a problem with that transistor or its bias circuitry.

If DC Troubleshooting doesn't locate the problem, then apply to the input a small AC voltage at a frequency in the middle of the pass-band of the amplifier. Compare the actual output to the expected output. Two problems that are specific to feedback amplifiers are

1. Losing the feedback loop ($H \rightarrow 0$).
2. Degradation in the forward-path amplifier (A decreases dramatically).

If you remember some of the properties of a feedback amplifier, you will understand how these two problems can affect the output of the amplifier.

Loosing the Feedback Loop (R_F Fails Open)

We know that feedback reduces the system gain and reduces distortion caused by nonlinearities in A. If the feedback resistor fails open, the feedback path is removed (see Figures 28–20, p. 959, and 28–21, p. 960). Figure 28–22A (p. 962) shows the expected AC output for the feedback amplifier under study. If R_F fails open, we can expect the following:

1. An increase in the measured gain, resulting in an increased output (Figure 28–22B, p. 962). If, as is usually the case, the forward-path gain is very large, losing the feedback path will result in the saturation of the forward-path amplifier. The output will either remain fixed at the supply voltage (Figure 28–22C) or will be clipped (Figure 28–22D).

FIGURE 28–22

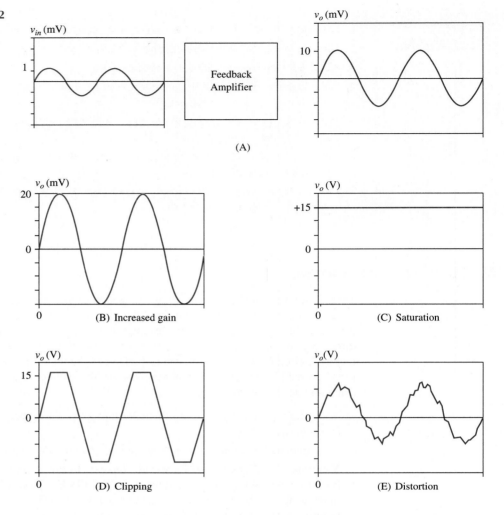

(A)

(B) Increased gain

(C) Saturation

(D) Clipping

(E) Distortion

2. An increase in the distortion of the output signal. That is, the output signal will not look like a perfect sine wave (Figure 28–22E).

Degradation in the Forward-Path Amplifier

The most likely cause of a problem with a feedback amplifier is that a transistor in the forward-path degrades in performance. This will result in a reduced forward-path gain, which will result in a decrease in the closed-loop gain. Note that if the forward-path gain is normally very high, then only a major degradation in the gain will be reflected in the output. The results of a reduced gain can be seen in

1. A reduced output voltage (Figure 28–23B).

2. Increased distortion in the output (Figure 28–23C). This happens because feedback reduces forward-path distortion only if the loop gain is high enough that

$$A_{sys} \approx \frac{1}{H}$$

As the forward-path gain drops, the system gain becomes dependent on A, and distortion due to A will show up in the output.

FIGURE 28–23
Forward-gain degradation

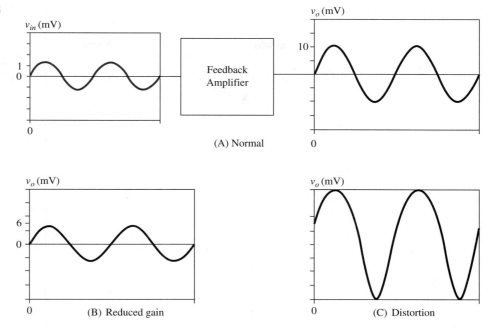

(A) Normal

(B) Reduced gain

(C) Distortion

A Note of Caution

How can you perform an in-circuit test to check the forward-path amplifier? It might seem reasonable to remove the feedback path, then measure the AC signals in the forward path. The problem with this is that even a properly functioning amplifier might well show inter-stage distortion in the forward path. After all, the feedback is there to reduce naturally-occurring distortion. In addition, removing the feedback path might lead to saturation in the forward path, making your forward-path measurements useless.

We recommend the following procedure for troubleshooting many feedback amplifiers. With the feedback loop intact, use an oscilloscope to measure the feedback signal and the output signal. If the ratio of these two signals approximates the expected value for H, the feedback path is probably intact, which tells you that the problem is with the forward-path amplifier. If possible, remove the transistors in the forward path and test them in isolation. If the transistors prove good, a failed capacitor is the next likeliest cause of the trouble. Remove and check them.

REVIEW QUESTIONS

1. What are the ways you can perform DC Troubleshooting of feedback amplifiers when you cannot easily calculate the Expected Voltages?

2. What are two likely problems you might encounter in troubleshooting a feedback amplifier?

3. What are some of the possible results of losing the feedback path?

4. What are some of the possible results of degradation in the forward-path amplifier?

5. Why is opening the feedback loop to test the forward-path amplifier not advisable?

6. Describe a useful procedure for testing feedback amplifiers.

SUMMARY

■ In a negative feedback amplifier, a portion of the output HS_o is subtracted from the input to produce an error signal (S_i). The error signal is multiplied by the forward-path gain to produce the final output $(S_o = AS_i)$.

■ The system gain is

$$A_{sys} = \frac{A}{1 + AH}$$

■ If the forward-path gain is large, the output is set by the feedback fraction:

$$\text{For } A \text{ large } A_{sys} \approx \frac{1}{H}$$

■ A large forward-path gain makes the output insensitive to parameter changes or non-linearities in the forward-path amplifier.
■ In a positive feedback system, the feedback signal is added to the input signal.
■ A positive feedback system is stable if the loop gain *(AH)* is less than 1.
■ A positive feedback system will oscillate if the loop gain equals 1.
■ A positive feedback system will be unstable if the loop gain is greater than 1.
■ The output of an amplifier is delayed with respect to the input (t_d). This delay is measured with the phase angle of the output, where

$$\theta = -\frac{t_d}{T} \times 360°$$

■ A negative feedback system is stable only if the loop gain is less than 1 when the phase angle of the loop gain equals ±180°.
■ In voltage out–voltage feedback, a portion of the output voltage is fed back as a voltage to produce the closed-loop gain $\frac{v_o}{v_{in}}$.
■ In voltage out–current feedback, a portion of the output voltage is fed back as a current to produce the closed-loop gain $\frac{v_o}{i_{in}}$.
■ In current out–current feedback, a portion of the output current is fed back as a current to produce the closed-loop gain $\frac{i_o}{i_{in}}$.
■ In current out–voltage feedback, a portion of the output current is fed back as a voltage to produce the closed-loop gain $\frac{i_o}{v_{in}}$.
■ The non-inverting Op-Amp amplifier is a voltage out–voltage feedback amplifier.
■ The inverting Op-Amp amplifier is a voltage out–current feedback amplifier.
■ If the feedback element in a transistor amplifier returns to the emitter, it is a voltage feedback amplifier.
■ If the feedback element in a transistor amplifier returns to the base, it is a current feedback amplifier.
■ If the feedback point for a transistor amplifier is the output, then it is a voltage-out amplifier.
■ If the feedback point for a transistor amplifier is not the output, then it is a current-out amplifier.
■ The closed-loop gain (A_{sys}), input resistance (R_{in}), and output resistance (R_o) have the following relations to the forward-path gain *(A)*, the open input resistance (R_i), and the open-loop output resistance (R_o).

Feedback Type	A_{sys}	R_{in}	R_{out}
Voltage out–Voltage feedback	$\frac{A}{1 + AH}$	$(1 + AH)R_i$	$\frac{R_o}{1 + AH}$
Voltage out–Current feedback	$\frac{A}{1 + AH}$	$\frac{R_i}{1 + AH}$	$\frac{R_o}{1 + AH}$
Current out–Current feedback	$\frac{A}{1 + AH}$	$\frac{R_i}{1 + AH}$	$(1 + AH)R_o$
Current out–Voltage feedback	$\frac{A}{1 + AH}$	$(1 + AH)R_i$	$(1 + AH)R_o$

- An open feedback pathway will result in increased gain (most likely to the point of saturation) and output distortion.
- A degraded forward-path amplifier will result in a decreased gain and increased output distortion.

PROBLEMS

SECTION 28–2 Negative Feedback Basics

1. A block diagram representation of a negative feedback system is shown in Figure 28–24.
   (a) Find S_{fb} in terms of S_o.   (b) Find S_i in terms of S_{in} and S_{fb}.
   (c) Find S_o in terms of S_i.   (d) Find S_o in terms of S_{in}.
   (e) Find the overall system gain.

FIGURE 28–24

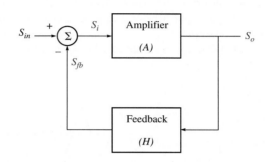

2. Find the closed-loop gain for the feedback system shown in Figure 28–24 if
   (a) $A = 100, H = 0.25$.   (b) $A = 5, H = 2$.
   (c) $A = -10, H = -2$.

3. For the forward gains given in Problem 2a
   (a) Assume the forward-path gain in Problem 2a decreases by 50%. Find the percentage change in the system gain.
   (b) Assume the forward-path gain in Problem 2a increases by 50%. Find the percentage change in the system gain.
   (c) Discuss the effect of forward-path gain changes on the overall system gain.

4. For the forward gain given in Problem 2b
   (a) Assume the forward-path gain in Problem 2b decreases by 50%. Find the percentage change in the system gain.
   (b) Assume the forward-path gain in Problem 2b increases by 50%. Find the percentage change in the system gain.
   (c) Discuss the effect of forward-path gain changes on the overall system gain.

SECTION 28–3 Positive Feedback Basics

5. Figure 28–25 shows a positive feedback system. Repeat Problem 1 for this system. Are there any limitations to the use of the system gain formula for positive feedback systems?

FIGURE 28–25

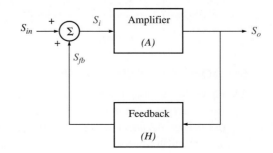

6. Apply a short duration 1V pulse to the input of the positive feedback system in Figure 28–25 (p. 965). Plot the resulting output pulse train if

   (a) $A = 10, H = 0.05$ (b) $A = 10, H = 0.1$

   (c) $A = 10, H = 1$

   (d) Use these results to discuss the stability of positive feedback systems.

7. The positive feedback system shown in Figure 28–25 (p. 965) has a forward-path gain of $A = 25$. What is the maximum feedback gain (H) allowed if you want a stable system? What feedback gain do you need if you want to build an oscillator?

SECTION 28–4   Stability of Negative Feedback Amplifiers

8. Figure 28–26 shows the input and output waveforms for an amplifier driven at several frequencies. At each frequency find the amplifier gain and the amplifier phase.

9. An amplifier has a gain of 10 and a phase of 180°; draw the amplifier output if the input is $5\cos\omega t$.

10. When does a negative feedback amplifier act as a positive feedback amplifier? Why?

FIGURE 28–26

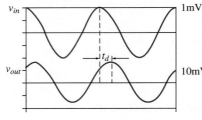

$f = 100\text{Hz}$
$t_d = 0.08\text{s}$

$f = 1\text{KHz}$
$t_d = 4\text{ms}$

$f = 5\text{KHz}$
$t_d = 0.88\text{ms}$

11. Figure 28–27 shows the Bode gain and phase plots for several forward-path amplifiers. If $H = 1$ (so $AH = A$), determine which of these amplifiers will produce a stable feedback response. Explain your answers.

12. Draw a block diagram of a negative feedback system that shows the correct signals—S_{in}, S_{fb}, S_i, S_o— for

   (a) Voltage out–voltage feedback.   (b) Voltage out–current feedback.

   (c) Current out–current feedback.   (d) Current out–voltage feedback.

FIGURE 28–27

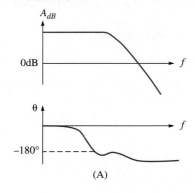
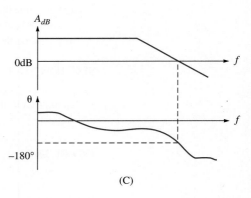

(A)   (B)   (C)

SECTION 28–5   Voltage Feedback

13. Figure 28–28 shows the non-inverting Op-Amp amplifier.

   (a) Identify the feedback type used here.

   (b) Redraw the circuit, and indicate the input signal, the feedback signal, the error signal, and the output signal.

   (c) Find the feedback factor (H).

FIGURE 28–28

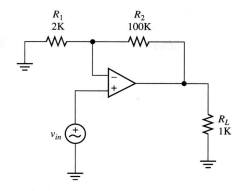

14. Figure 28–29 shows a BJT feedback amplifier.
    (a) Identify the feedback type used here.
    (b) Redraw the circuit, and indicate the input signal, the feedback signal, the error signal, and the output signal.
    (c) Find the feedback factor *(H)*.
    (d) If the forward-path gain is $A = 10$, find the system gain.

FIGURE 28–29

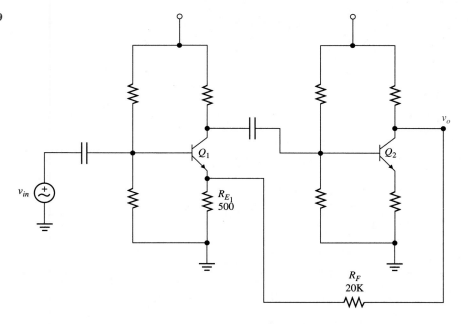

SECTION 28–6   Voltage Out–Current Feedback

15. Figure 28–30 (p. 968) shows an inverting operational amplifier.
    (a) Identify the feedback type used here.
    (b) Redraw the circuit, and indicate the input signal, the feedback signal, the error signal, and the output signal.
    (c) Find the feedback factor *(H)*.

16. Figure 28–31 (p. 968) shows a BJT amplifier.
    (a) Identify the feedback type used here.
    (b) Redraw the circuit, and indicate the input signal, the feedback signal, the error signal, and the output signal.
    (c) Find the feedback factor *(H)*.
    (d) If the forward gain is 10, find the system gain.

FIGURE 28–30

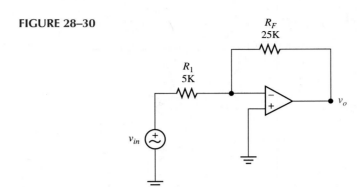

FIGURE 28–31

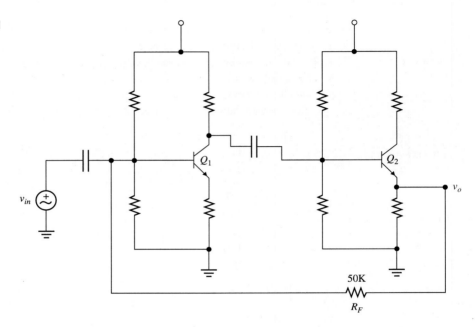

SECTION 28–7　Input and Output Resistance—Voltage Feedback

17. How does voltage feedback affect the input resistance of a feedback amplifier?

18. Find the input and output resistances for the Op-Amp amplifier of Figure 28–28 (p. 967) if $A = 10^5$, $R_i = 4M\Omega$, $R_o = 50\Omega$.

19. Find the input and output resistances for the BJT amplifier of Figure 28–29 (p. 967) if $A = 10^5$, $R_i = 4K\Omega$, $R_o = 5K\Omega$.

SECTION 28–8　Input and Output Resistance—Current Feedback

20. How does current feedback affect the input resistance of a feedback amplifier?

21. Find the input and output resistances for the BJT amplifier of Figure 28–31 if $A = 10$, $R_i = 4M\Omega$, $R_o = 50\Omega$.

22. Find the input and output resistances for the current out–voltage feedback BJT amplifier of Figure 28–32 if $A = 10^5$, $R_i = 4K\Omega$, $R_o = 50K\Omega$.

SECTION 28–10　Troubleshooting Problems

23. Discuss what you might see if the feedback pathway in a feedback amplifier fails.

24. Discuss what you might see if the forward-path amplifier of a feedback amplifier degrades (A decreases).

25. Figure 28–33 shows the expected output from a feedback amplifier and several outputs caused by a failure in the amplifier. For each output, identify a likely cause, or causes, of the problem.

FIGURE 28–32

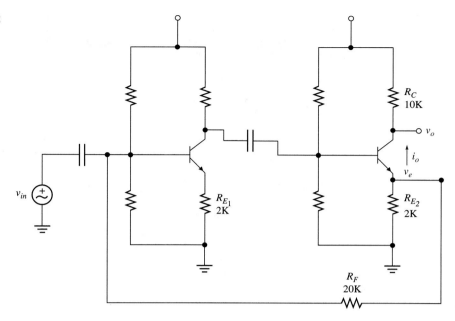

FIGURE 28–33

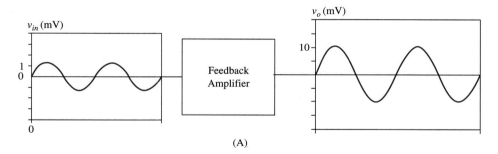

(A)

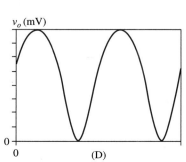

(B)

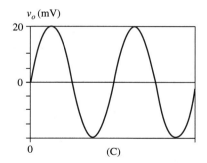

(C)

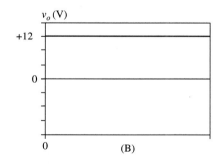

(D)

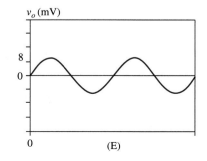

(E)

Computer Problems

26. Simulate the BJT amplifier of Figure 28–12 (p. 947). If you have a Bode plotter or routine in your program, use it to answer the following questions. Remember to convert decibels to actual gain. If a Bode routine is not available, apply a 1mV, 1KHz signal to the input. Find the appropriate output, and divide by 1mV to find the gain.

    (a) Find $\dfrac{v_o}{v_{in}}$. This is A_{sys}.

    (b) Find $\dfrac{v_e}{v_o}$. This will give you the feedback gain *(H)*.

    (c) To properly open the loop, disconnect R_F from the emitter of Q_1 and reconnect it to ground. Find $\dfrac{v_o}{v_{in}}$. This will give you the forward-path gain *(A)*.

    (d) Does $A_{sys} = \dfrac{A}{1 + AH}$?

27. You can find the input resistance by placing a 1Ω resistor in series with the input source. The voltage across this resistor is a direct measure of the input current.

    (a) Find R_{in} for the closed-loop system.

    (b) Open the loop as described in Problem 26c, and find the open-loop input resistance. This is R_i.

    (c) Does $R_{in} = \dfrac{R_i}{1 + AH}$? Use the values for A and H found in the previous problem.

28. To simulate a failed feedback path, remove R_F from the circuit. What happens to the gain?

29. To simulate a degraded forward-path amplifier, lower β in your computer model. Keep lowering β until you notice a significant change in the closed-loop gain.

CHAPTER

29

SIGNAL GENERATORS

CHAPTER OBJECTIVES

To use a Class C tuned amplifier to amplify a sinusoidal input.

To build an LC Colpitts sine wave oscillator.

To build an LC Hartley sine wave oscillator.

To build an RC phase shift sine wave oscillator.

To build a Wien bridge sine wave oscillator.

To build square wave generators.

To build triangle wave generators.

To use integrated circuit signal generators.

To troubleshoot signal generators.

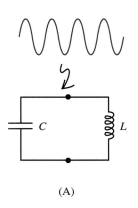

(A)

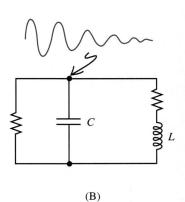

(B)

29–1 ■ INTRODUCTION

Figure 29–1 shows a type of function generator available in most labs. This device allows you to test circuits with the following inputs:

☐ Sine wave

☐ Square wave

☐ Triangle wave

The reason we want to have these signals available is that they are the most common signals used in electronic circuits. Some of the applications of these signals are

☐ **Sine wave:** radio, TV, satellite communication, AC-circuit response testing

☐ **Square wave:** digital clock pulses, switching circuits, DC-circuit testing

☐ **Triangle wave:** raster displays (TV, oscilloscope)

FIGURE 29–1
Function generator

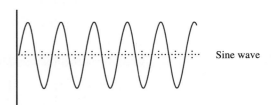

Sine wave

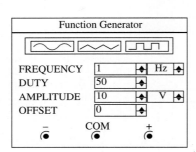

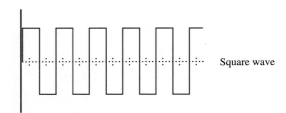

Square wave

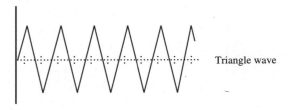

Triangle wave

29–2 ■ THE CLASS C TUNED AMPLIFIER

Consider the jack-in-the-box (the toy, not the fast food restaurant) in Figure 29–2A. When you turn the crank, the lid opens, and Jack pops out and bounces up and down. Why does Jack oscillate? The answer tells us how we can build an electronic oscillator.

Jack is constructed by connecting his head (a mass) to a spring. The spring is compressed when Jack is unceremoniously stuffed into his box. When the lid opens, the following sequence of events occurs:

1. The spring expands and pushes Jack's head up. That is, the energy stored in the spring is used to raise the mass (weight is *mass × gravity*) of Jack's head. When the spring is fully extended, no more energy is stored in the spring.

FIGURE 29–2

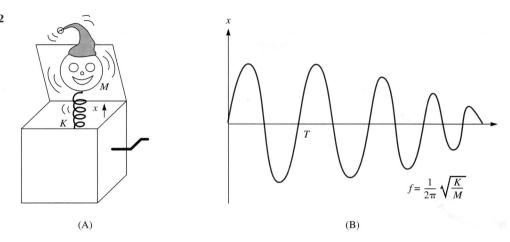

(A)   (B)

2. Because Jack's head has been raised, potential energy is now stored in the mass of Jack's head. The weight of Jack's head causes Jack to drop, recompressing the spring. When the spring is compressed, the potential energy stored in Jack's head is gone, transferred back to the spring.

3. The process now repeats itself. The compressed spring raises the head. The raised head compresses the spring. The result is a sinusoidal oscillation of the head. The frequency of oscillation is

$$f_o = \frac{1}{2\pi} \sqrt{\frac{K}{M}}$$

where M is the mass of the head and K is the elasticity of the spring.

4. Internal friction in the spring and the friction of the air combine to decrease the magnitude of the oscillations until Jack stops bobbing and comes to rest (Figure 29–2B).

The principle behind Jack's oscillations is that energy is transferred from the spring to the mass and back again. If there were no friction in the system, Jack would bounce up and down forever.

The electrical equivalent of Jack is the inductor-capacitor (LC) circuit in Figure 29–3A. The capacitor stores energy by storing electrical charge; the inductor stores energy in its magnetic field. Let's assume that we have stored some charge on the capacitor, then connected the capacitor to the inductor (Figure 29–3B). The following occurs:

1. The capacitor discharges current into the inductor. This reduces the charge stored on the capacitor and, so, reduces the capacitor voltage.

2. The current discharged into the inductor creates a magnetic field around the inductor. As the capacitor releases its stored energy, the inductor increases its stored energy (the energy stored in its magnetic field).

FIGURE 29–3
LC tuned (tank) circuit

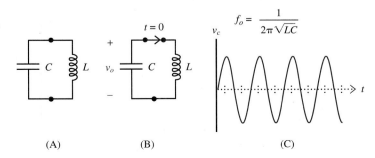

(A)   (B)   (C)

3. As the capacitor discharges, current in the circuit decreases. As the current decreases, the magnetic field around the inductor collapses. This magnetic field collapse induces a current in the coil.

4. The current produced by the coil recharges the capacitor and the process begins again.

In this process, energy is transferred back and forth between the capacitor and inductor. The result is a sinusoidal oscillation of the circuit voltage (Figure 29–3C, p. 975). The frequency of this oscillation is given by

$$f_o = \frac{1}{2\pi\sqrt{LC}}$$

The LC circuit *resonates* at the frequency of oscillation. *Resonance* means the LC circuit responds best to signals with a frequency of f_o. This is similar to making one tuning fork vibrate by holding another vibrating tuning fork, tuned to the same frequency, next to it.

The LC circuit is often called a **tuned circuit,** or a **tank circuit.** The tuned circuit can be used to amplify signals at f_o, or to generate sine waves at this frequency.

EXAMPLE 29–1
The LC Tuned Circuit

Find the frequency of oscillation for the following LC combinations:

(a) $L = \dfrac{1}{2\pi}H, \quad C = \dfrac{1}{2\pi}F$ **(b)** $L = 4.3\text{mH}, \quad C = 20\text{pF}$

(c) $L = 2.4\mu\text{H}, \quad C = 1\text{pF}$

Solution   Using the formula for the oscillation frequency of an LC circuit, we get

(a) $f_o = \dfrac{1}{2\pi\sqrt{LC}} = \dfrac{1}{2\pi\sqrt{\dfrac{1}{2\pi} \times \dfrac{1}{2\pi}}} = 1\text{Hz}$

(b) $f_o = \dfrac{1}{2\pi\sqrt{LC}} = \dfrac{1}{2\pi\sqrt{4.3\text{mH} \times 20\text{pF}}} = 543\text{KHz}$

This frequency is at the low end of the AM radio band.

(c) $f_o = \dfrac{1}{2\pi\sqrt{LC}} = \dfrac{1}{2\pi\sqrt{2.4\mu\text{H} \times 1\text{pF}}} = 103\text{MHz}$

This frequency is in the middle of the FM radio band.

EXAMPLE 29–2
Designing the LC Tuned Circuit

Find the capacitor required to build an LC oscillator that uses an inductance of $L = 1\text{mH}$ to produce sine waves of the following frequencies:

(a) $f_o = 90\text{MHz}$ **(b)** $f_o = 1\text{GHz}$ $(1\text{GHz} = 1 \times 10^{12}\text{Hz})$

Solution   We rearrange the frequency equation to find that

$$C = \frac{1}{L(2\pi f_o)^2}$$

(a) $C = \dfrac{1}{L(2\pi f_o)^2} = \dfrac{1}{1\text{mH} \times (2\pi \times 90\text{MHz})^2} = 0.00313\text{pF}$

(b) $C = \dfrac{1}{L(2\pi f_o)^2} = \dfrac{1}{1\text{mH} \times (2\pi \times 1\text{GHz})^2} = 2.53 \times 10^{-11}\text{pF!}$

It would be very easy to build sine wave generators if we had ideal inductors, capacitors, and connecting wires (Figure 29–4A). Unfortunately, all real devices have resistance: the wires in the inductor have a finite resistance; the insulator in a capacitor is not a true open circuit. As the energy is transferred between the capacitor and inductor, energy will be lost in the form of heat to the resistances in the LC circuit. This energy loss will cause the sinusoidal oscillations to die out, resulting in a damped sinusoid that goes to 0V (Figure 29–4B).

FIGURE 29–4

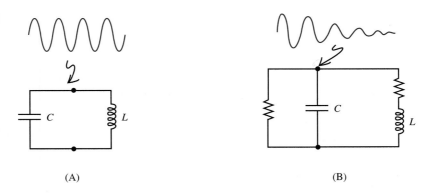

(A) (B)

To maintain the oscillations in a real LC circuit, we have to provide some means to inject additional energy into the circuit. This additional energy will compensate for the energy lost to the resistance of the tuned circuit. Timing the injection of this energy into the tuned circuit is critical. We want to make sure that we apply the external signal so that it reinforces the natural oscillation. For example, we do *not* want to inject a negative signal into the LC circuit when the circuit response is going positive. This will cancel the oscillations.

The Class C Tuned Amplifier

Figure 29–5A shows a BJT amplifier with an LC tuned circuit in the collector. The transistor is not used as a linear amplifier. Rather it is biased in **CUT-OFF,** as you can see from the negative base supply voltage. The transistor will only conduct when the input signal is large enough to overcome the negative bias. This is known as **Class C** operation. In Class C operation, a transistor is on for only a portion of the input signal. Because the Class C amplifier is biased in **CUT-OFF,** there is no drain on the supply voltages, and we get a very efficient amplifier.

FIGURE 29–5
Class C amplifier

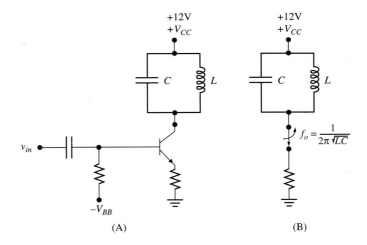

(A) (B)

Whenever the transistor is turned on, it effectively connects the LC circuit to ground. This results in the LC tank circuit being energized by the supply voltage (Figure 29–5B). To sustain the oscillations of the LC circuit, we want to pulse the transistor on and off at the resonant frequency of the LC circuit (f_o). Any other switching frequency will not sustain the LC circuit oscillations.

FIGURE 29–6

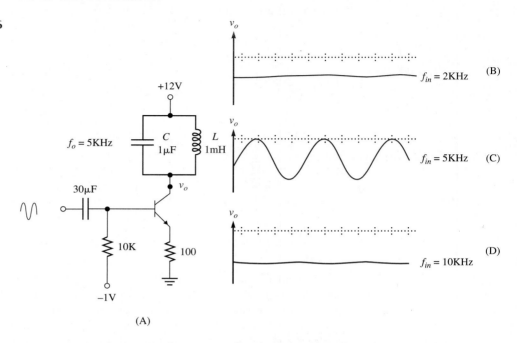

In Figure 29–6A we apply a sinusoidal input to the base of the transistor. Whenever the input sinusoid plus the negative bias results in a base voltage greater than 0.7V, the transistor turns on. Figures 29–6B–D show the output of this amplifier for three different input frequencies. You can see that we only get a significant sinusoidal output when the input frequency equals the resonant frequency of the tuned circuit:

$$\textbf{Tuned circuit: } f_o = \frac{1}{2\pi\sqrt{LC}} = \frac{1}{2\pi\sqrt{1\text{mH} \times 1\mu\text{F}}} \approx 5\text{KHz}$$

Because a small input sinusoid has produced a large output sinusoid, we have built an amplifier tuned to f_o.

The tuned amplifier will actually produce a sinusoidal output for any input as long as the input frequency is f_o. Figure 29–7 shows the response of our tuned amplifier to a pulse train input. In effect, we have created a pulse-to-sine wave converter.

FIGURE 29–7
Pulse-to-sine wave converter

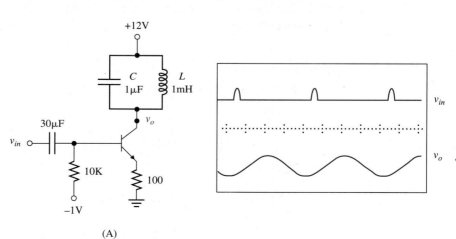

1. What names are given to an LC circuit?
2. What is the resonant frequency of an LC circuit? What is the formula for the resonant frequency?
3. Describe how the LC circuit produces oscillations.
4. Where is the tuned Class C amplifier biased? What is its mode of operation?
5. What does the transistor do in this circuit?
6. What input frequency causes an oscillation in the tank circuit?
7. How can the tuned amplifier be used to convert any type of waveform into a sinusoid?

29–3 ■ LC OSCILLATORS

In the tuned amplifier we produced a sinusoidal output, but we needed a separate input signal to generate this output. We can build a sinusoidal oscillator that does not require any external signal. We do this by using a portion of the tuned circuit output itself to provide the input signal needed.

When you switch on an oscillator circuit, the electrical transients caused by the switching provide the energy to kick-off the oscillations. A fraction of the output voltage is fed back to an amplifier. The amplifier output energizes the tank circuit, providing for sustained oscillations (Figure 29–8).

FIGURE 29–8
Sustaining oscillations

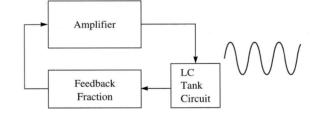

The Colpitts Oscillator

One of the most common ways to obtain a fraction of the LC voltage is to split the tank capacitor into two capacitors. Figure 29–9 shows an LC oscillator built with two capacitors in series: the output voltage is taken across C_1; the feedback voltage is taken across C_2. The opposite polarities of v_o and v_{fb} are necessary for proper timing, or *phasing,* in the feedback loop. Oscillators using this type of tank circuit are known as **Colpitts oscillators.**

FIGURE 29–9
Colpitts oscillator

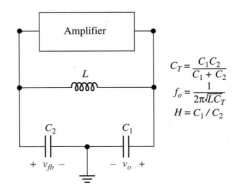

$$C_T = \frac{C_1 C_2}{C_1 + C_2}$$

$$f_o = \frac{1}{2\pi\sqrt{LC_T}}$$

$$H = C_1 / C_2$$

Because the two capacitors are in series, the total capacitance here is

$$C_T = \frac{C_1 C_2}{C_1 + C_2}$$

The frequency of oscillation (the resonant frequency) of this oscillator is

$$f_o = \frac{1}{2\pi\sqrt{LC_T}}$$

The relation between v_o and v_{fb} is given by

$$v_{fb} = -\frac{C_1}{C_2}v_o$$

You are shown how to derive this formula in Problem 4 at the end of the chapter.
The feedback gain *(H)* also known as the **feedback factor,** is defined by

$$v_{fb} = Hv_o \rightarrow H = \frac{v_{fb}}{v_o}$$

Because the standard practice is to use a positive number for the feedback fraction, for the Colpitts oscillator

$$H = \frac{C_1}{C_2}$$

EXAMPLE 29–3
The Colpitts Oscillator

Find the resonant frequency and the feedback factor for a Colpitts oscillator with the following components:

$$L = 1\text{mH}, \; C_1 = 1\mu\text{F}, \; C_2 = 5\mu\text{F}$$

Solution   The total capacitance of the tank circuit is

$$C_T = \frac{C_1 C_2}{C_1 + C_2} = \frac{1\mu\text{F} \times 5\mu\text{F}}{1\mu\text{F} + 5\mu\text{F}} = 0.833\mu\text{F}$$

The frequency of oscillation (the resonant frequency) of this oscillator is

$$f_o = \frac{1}{2\pi\sqrt{LC_T}} = \frac{1}{2\pi\sqrt{1\text{mH} \times 0.833\mu\text{F}}} = 5.51\text{KHz}$$

The feedback factor is

$$H = \frac{C_1}{C_2} = \frac{1\mu\text{F}}{5\mu\text{F}} = 0.2$$

DRILL EXERCISE

Find f_o and H if C_1 in the previous example is changed to $C_1 = 2\mu\text{F}$

Answer $f_o = 4.21\text{KHz}, H = 0.4$

EXAMPLE 29–4
Colpitts Design

A 1mH inductor is available. Choose the capacitor values in a Colpitts oscillator so $f_o = 1MHz$ and $H = 0.25$.

Solution We first find the required C_T by rearranging the resonant frequency formula:

$$C_T = \frac{1}{L(2\pi f_o)^2} = \frac{1}{1mH(2\pi \times 1MHz)^2} = 25.3pF$$

This gives us one relation for C_1 and C_2:

$$C_T = \frac{C_1 C_2}{C_1 + C_2} = 25.3pF$$

To get the given feedback fraction, we need

$$H = \frac{C_1}{C_2} = 0.25$$

so

$$C_2 = 4C_1$$

Substituting this relation into the total capacitance formula gives us

$$\frac{C_1 C_2}{C_1 + C_2} = \frac{4C_1{}^2}{5C_1} = 0.8C_1 = 25.3pF$$

And we find

$$C_1 = \frac{25.3pF}{0.8} = 31.6pF$$

The other capacitor is given by

$$C_2 = 4C_1 = 126pF$$

DRILL EXERCISE

Repeat the previous example if the inductor is 10mH.

Answer $C_1 = 3.16pF$, $C_2 = 12.6pF$

FIGURE 29–10

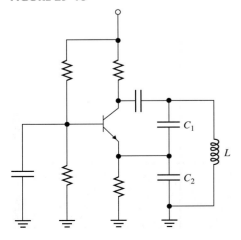
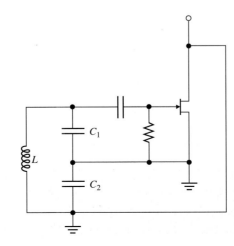
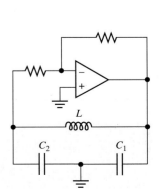

Any of the amplifiers we have described in this textbook can be used to construct a Colpitts Oscillator. Figure 29–10 shows this oscillator built with a BJT amplifier, an FET amplifier, and an Op-Amp amplifier. **Note:** All of the coupling capacitors in these circuits are much larger than the tank circuit capacitors. This prevents the coupling capacitors from affecting the behavior of the circuit at the oscillation frequency.

The Hartley Oscillator

Figure 29–11 shows how we can use two inductors in series to produce the feedback voltage. This type of circuit is known as a **Hartley oscillator.** The formulas for the Hartley oscillator are

$$L_T = L_1 + L_2$$

$$f_o = \frac{1}{2\pi\sqrt{L_T C}}$$

$$H = \frac{L_2}{L_1}$$

FIGURE 29–11
Hartley oscillator

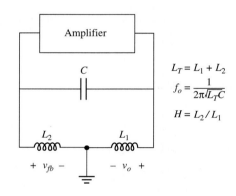

$$L_T = L_1 + L_2$$
$$f_o = \frac{1}{2\pi\sqrt{L_T C}}$$
$$H = L_2 / L_1$$

EXAMPLE 29–5
The Hartley Oscillator

Find the resonant frequency and the feedback factor for a Hartley oscillator with the following components:

$$L_1 = 3\text{mH}, L_2 = 1\text{mH}, C = 10\mu\text{F}$$

Solution The total inductance the tank circuit is

$$L_T = L_1 + L_2 = 3\text{mH} + 1\text{mH} = 4\text{mH}$$

The frequency of oscillation (the resonant frequency) of this oscillator is

$$f_o = \frac{1}{2\pi\sqrt{L_T C}} = \frac{1}{2\pi\sqrt{4\text{mH} \times 10\mu\text{F}}} = 796\text{Hz}$$

The feedback factor is

$$H = \frac{L_2}{L_1} = \frac{1\text{mH}}{3\text{mH}} = 0.333$$

DRILL EXERCISE

Find f_o and H if L_2 in Example 29.5 is changed to $L_2 = 0.5\text{mH}$.

Answer $f_o = 851\text{Hz}, H = 0.167$

EXAMPLE 29–6
Hartley Design

A 1pF capacitor is available. Choose the inductor values in a Hartley oscillator so that $f_o = 1\text{MHz}$ and $H = 0.2$.

Solution   We first find the required L_T by rearranging the resonant frequency formula:

$$L_T = L_1 + L_2 = \frac{1}{C(2\pi f_o)^2} = \frac{1}{1\text{pF}\,(2\pi \times 1\text{MHz})^2} = 25.3\text{mH}$$

To get the given feedback factor, we need

$$H = \frac{L_2}{L_1} = 0.2$$

so

$$L_2 = 0.2L_1$$

Substituting this relation into the total inductance formula gives us

$$L_T = 1.2L_1 = 25.3\text{mH}$$

so

$$L_1 = \frac{25.3\text{mH}}{1.2} = 21.1\text{mH}$$

The other inductor is

$$L_2 = 0.2L_1 = 4.22\text{mH}$$

DRILL EXERCISE

Repeat the Example 29–6 if the capacitor is 10pF.

Answer $L_1 = 2.11\text{mH}, \quad L_2 = 0.422\text{mH}$

Figure 29–12 shows Hartley oscillators built with a BJT amplifier, an FET amplifier, and an Op-Amp amplifier. We offer the following little trick to help you remember which circuit goes with which name

Colpitts → **C**apacitor is split

Hartley → **H**enry → Inductor is split

FIGURE 29–12

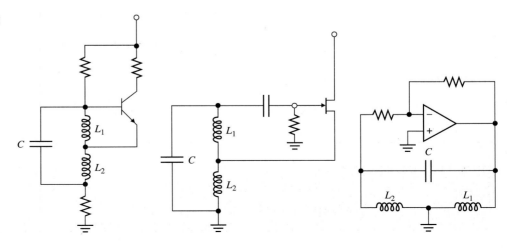

Crystal Oscillator

Very often a piezo-electric crystal (usually made from quartz) will be used in an oscillator circuit (Figure 29–13). A piezo-electric crystal can be cut in such a way that it will respond, or resonate, at an exactly controlled frequency. That is, when electrically excited, the crystal will vibrate at f_o. Also, when the crystal vibrates, it produces a sinusoidal electrical response at a frequency of f_o.

FIGURE 29–13
Crystal oscillator

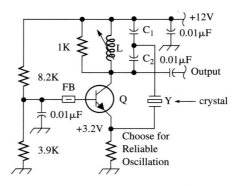

$X_L = 300\Omega$ at desired frequency
C1 = 3X to 4X C2
Q = 2N2222 or 3904 below 30MHz
Y = Specify series-resonant when ordering.

Oscillators built only with capacitors and inductors will often fluctuate slightly in frequency. This fluctuation is known as **frequency drift.** This occurs because of minor fluctuations in the tank circuit capacitance or inductance due to factors such as temperature and age. The resonant frequency of a crystal does not change; it can only vibrate at f_o. Crystal oscillators produce an extremely stable output frequency.

Variable Frequency Oscillators

Figure 29–14 shows examples of oscillators that can produce several frequencies. The first is known as a **variable frequency oscillator** (VFO). This particular circuit is an FET Hartley oscillator. The capacitors in the tank circuit can be varied to vary the oscillation frequency.

The second circuit is known as a **voltage-controlled oscillator** (VCO). The VCO shown is a Colpitts oscillator that uses both a variable capacitor and varactor diodes in the tuning circuit. As described in Chapter 6, the capacitance of the varactor diodes changes as the diode voltage changes. The output frequency of this oscillator, therefore, depends on the value of the frequency control voltage.

REVIEW QUESTIONS

1. How is an LC oscillator constructed?
2. Draw the tank circuit of a Colpitts oscillator.
3. What is the total capacitance, the resonant frequency, and the feedback factor of a Colpitts oscillator?
4. Draw the tank circuit of a Hartley oscillator.
5. What is the total inductance, the resonant frequency, and the feedback factor of a Hartley oscillator?
6. What is a piezo-electric crystal, and why is it used in oscillators?
7. What is a VFO? What controls the frequency?
8. What is a VCO? What controls the frequency?

FIGURE 29–14
Variable frequency oscillators

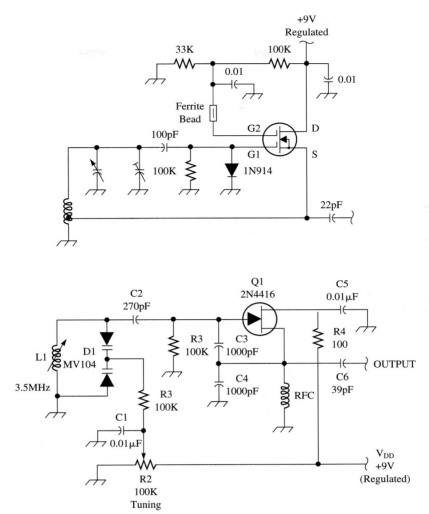

29–4 ■ RC OSCILLATORS

In Chapter 28 we explained how a positive feedback system can oscillate. We show this again in Figure 29–15 (p. 986), where a single-input pulse can produce any one of the outputs shown. If the loop gain *(AH)* is less than 1, each time the pulse travels around the loop it decreases in size, so the circuit is stable (Figure 29–15B). If the loop gain is greater than 1, each time the pulse travels around the loop it increases in size, so the circuit is unstable (Figure 29–15C). If the loop gain equals 1, the pulse doesn't change, so we get a sustained pulse train (Figure 29–15D). That is, if the loop gain in a positive feedback system is +1, the system will oscillate.

In general, the loop gain varies with frequency, as discussed in the last chapter. To determine the system behavior, we examine the magnitude of the loop gain ($|AH|$) and the phase of the loop gain (θ_{AH}).

For oscillation to occur in a positive feedback system we need

> **Positive Feedback Oscillator:** $AH = +1 \rightarrow |AH| = 1$ and $\theta_{AH} = 0°$

If the loop gain satisfies the above criteria at some frequency (f_o) the system will oscillate sinusoidally at the frequency f_o.

FIGURE 29–15
Positive feedback

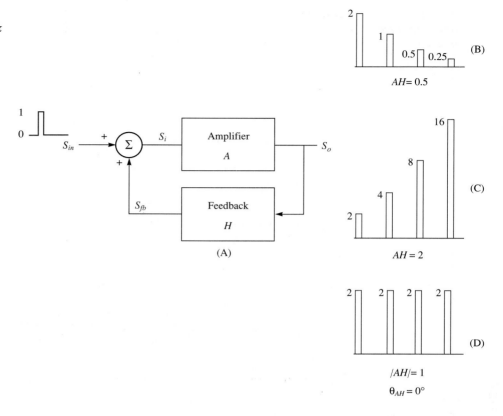

Figure 29–16 shows the more common negative feedback system. The criteria for oscillation is now modified to

Negative Feedback Oscillator: $AH = -1 \rightarrow \left| AH \right| = 1$ and $\theta_{AH} = \pm 180°$

Note that either $+$ or $-180°$ is the equivalent of a minus sign. A negative feedback system will sinusoidally oscillate at the frequency where the loop gain is 1 and the phase is 180°.

FIGURE 29–16
Negative feedback

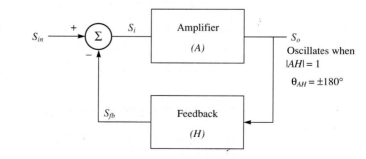

The RC Phase Shift Oscillator

Figure 29–17 shows an Op-Amp sinusoidal generator that uses an RC ladder network in the negative feedback loop. You might remember from Chapter 28 that a single RC circuit produces from 0 to $-90°$ of phase lag. Note that $-90°$ is only approached but never

FIGURE 29–17
RC phase shift oscillator

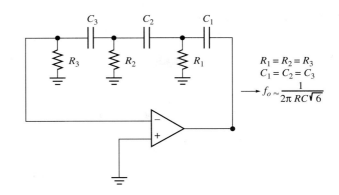

$$R_1 = R_2 = R_3$$
$$C_1 = C_2 = C_3$$
$$\rightarrow f_o \approx \frac{1}{2\pi\,RC\sqrt{6}}$$

reached. Because we need to reach exactly $-180°$, we need to use three RC stages, each RC stage contributing approximately $-60°$. The frequency at which the three-stage RC network reaches $-180°$ is the resonant frequency of this oscillator. In the case where $R_1 = R_2 = R_3 = R$, and $C_1 = C_2 = C_3 = C$, the resonant frequency is given approximately by

$$\textbf{RC Oscillator: } f_o \approx \frac{1}{2\pi RC\sqrt{6}}$$

EXAMPLE 29–7
The RC Phase Shift Oscillator

(a) Find the resonant frequency of an RC oscillator if $R = 1\text{K}\Omega$ and $C = 10\mu\text{F}$.
(b) Using 5pF capacitors, find R to produce a resonant frequency of $f_o = 800\text{KHz}$.

Solution
(a) For the given resistor and capacitor values

$$f_o \approx \frac{1}{2\pi RC\sqrt{6}} = \frac{1}{2\pi \times 1\text{K} \times 10\mu\text{F} \times \sqrt{6}} = 6.5\text{Hz}$$

(b) We first rearrange the frequency formula to solve for the resistance

$$R = \frac{1}{2\pi f_o C\sqrt{6}}$$

Using the values given, we find

$$R = \frac{1}{2\pi f_o C\sqrt{6}} = \frac{1}{2\pi \times 800\text{K} \times 5\text{pF} \times \sqrt{6}} = 16.2\text{K}\Omega$$

The Wien Bridge Oscillator

Figure 29–18 (p. 988) shows another way we can use RC circuits to produce a sinewave generator. This circuit is known as a **Wien bridge oscillator.** In the Wien bridge oscillator, the feedback voltage is derived from the output voltage through the voltage divider formed by Z_1 and Z_2. Because the amplifier does not invert, and because v_{fb} has the same polarity as v_o, this is a *positive* feedback circuit.

Although the resistors and capacitors do not need to have the same values, it is much easier to design the circuit when

$$R_1 = R_2 = R \quad \text{and} \quad C_1 = C_2 = C$$

FIGURE 29–18
Wien bridge oscillator

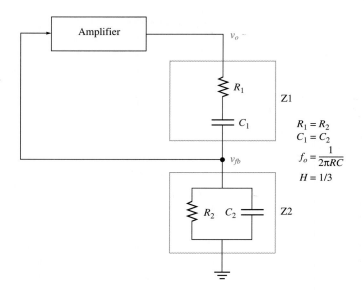

Given these relations, the center frequency, feedback factor, and amplifier gain (these are too complicated to derive) are

$$\text{Wien Bridge Oscillator: } f_o = \frac{1}{2\pi RC}$$

$$H = \frac{1}{3}$$

$$A = 3$$

It is interesting to note that when we use identical resistors and capacitors, the feedback factor is a fixed $H = \frac{1}{3}$. Because the loop gain AH must equal 1, the amplifier gain needs to be $A = 3$.

Figure 29–19 shows how we construct a Wien bridge oscillator with an Op-Amp. The Op-Amp and the resistors forming the negative feedback loop (R_3 and R_4) give us the non-inverting amplifier with a gain of

$$A = 1 + \frac{R_3}{R_4}$$

FIGURE 29–19

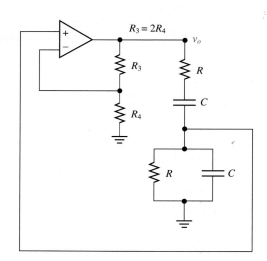

To get an amplifier gain of 3, we need to set

$$R_3 = 2R_4.$$

EXAMPLE 29–8
The Wien Bridge Oscillator

In the Wien bridge oscillator of Figure 29–19, $R = 1\text{K}\Omega$, $C = 1000\text{pF}$, and $R_4 = 2\text{ K}\Omega$.

(a) Find the resonant frequency and feedback factor for this oscillator.
(b) Find R_3 to ensure oscillation.

Solution
(a) For the Wien bridge oscillator

$$f_o = \frac{1}{2\pi RC} = \frac{1}{2\pi \times 1\text{K} \times 1000\text{pF}} = 159\text{KHz}$$

and

$$H = \frac{1}{3}$$

(b) To find R_3 we know that to get a gain of $A = 3$

$$R_3 = 2R_4 = 2 \times 2\text{K} = 4\text{K}\Omega$$

REVIEW QUESTIONS

1. Under what conditions will a positive feedback amplifier oscillate?
2. Under what conditions will a negative feedback oscillator oscillate?
3. Does the RC phase shift oscillator use positive or negative feedback?
4. How much phase can a single RC circuit produce? Why do we need a three-RC network for this oscillator?
5. What is the resonant frequency for this oscillator? What assumptions have you made?
6. Does the Wien bridge oscillator use positive or negative feedback?
7. What are the resonant frequency, feedback factor, and amplifier gain for the Wien bridge oscillator? What assumptions have you made?
8. What is the relation between the two resistors forming the non-inverting Op-Amp amplifier in the Wien bridge oscillator?

29–5 ■ SQUARE WAVE GENERATOR

Figure 29–20A (p. 990) shows a general pulse waveform. The voltage swings between a constant $+10\text{V}$ and a constant -5V; repeating every 10s. In this waveform, the positive part of the pulse lasts for 4s, while the negative part lasts for 6s. Figure 29–20B (p. 990) shows a very special pulse wave known as a **square wave.** In the square wave, the positive and negative values are the same, and the positive and negative sections last for the same amount of time.

The basic parameters of pulse waves are shown in Figure 29–21A (p. 990):

□ $+A1$: The amplitude of positive part of wave.
□ $-A_2$: The amplitude of negative part of wave.

FIGURE 29–20
Pulse and square waves

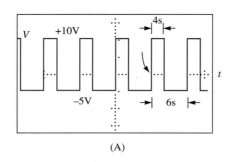

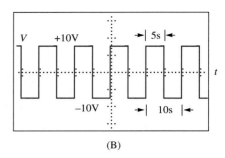

(A)

(B)

FIGURE 29–21

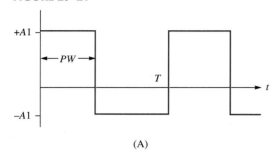

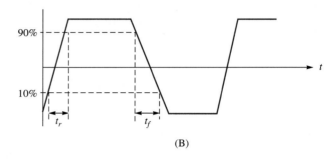

(A)

(B)

☐ *PW:* the pulse width of the signal. This is measured on the positive side

☐ *T:* the period of the signal. The pulse repeats itself every *T* seconds

☐ f_o: the frequency of the signal: $f_o = \dfrac{1}{T}$

☐ duty cycle: ratio of the on-time (positive part) to the total period *(T)*. Duty cycle is usually expressed as a percent; for example, the duty cycle of a square wave is 50%. That is, the square wave is on (positive) half the time and off (negative) half the time. The formula is

$$duty\ cycle = 100\,\frac{PW}{T}$$

In the real world it is impossible to instantaneously switch from one state to another. If you examine a pulse on an oscilloscope that has its time scale adjusted to measure very small times, what appears to be vertical lines between the on and off states will actually look like the pulse shown in Figure 29–21B. The additional parameters that describe pulses are

☐ t_r: The **rise time,** or *on-time,* is the time it takes for the pulse to go positive.

☐ t_f: The **fall time,** or *off-time,* is the time it takes for the pulse to go negative (or to 0V).

Rise and fall times are determined by the time it takes to go from 10% to 90% of the total change in amplitude.

When you use a pulse generator, you want to make sure that the rise and fall times are small enough that they won't affect circuit behavior. This is very important in digital and computer circuits that operate at high speeds.

EXAMPLE 29–9
Pulse Wave
Parameters

What are the parameters of the pulse wave shown in Figure 29–22?

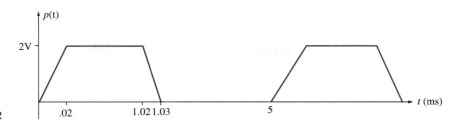

FIGURE 29–22

Solution Except for the duty cycle, the pulse wave parameters can be read directly from the plot:

$+A1 = 2V$

$-A2 = 0V$

$PW = 1ms$

$T = 5ms$

$t_r = 0.02ms$

$t_f = 0.01ms$

The duty cycle is calculated from

$$duty\ cycle = 100\frac{PW}{T} = 100\frac{1ms}{5ms} = 20\%$$

A pulse wave can be generated by repetitively switching DC sources on and off. Figure 29–23 shows the voltage that results when we periodically switch between $+10V$ and $-5V$. Pulse generator circuits use transistors or Op-Amps to provide the necessary switching. Whatever device is used, it must be able to switch fast enough to provide the required rise and fall times.

FIGURE 29–23

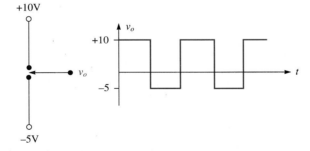

Figure 29–24 (p. 992) shows an Op-Amp circuit that produces a square wave output. This circuit is also known as an **astable** or **free-running multivibrator** (no jokes here). You can see that we have both positive and negative feedback in this circuit. Let's first look at the voltage divider connecting the output back to the positive terminal of the Op-Amp (Figure 29–25A (p. 992)). If you take a quick look back at Figure 25–24 in Chapter 25 (p. 838), you will see that this part of the circuit is actually a Schmitt trigger.

To review how the Schmitt trigger works, let's first assume that the output voltage is pinned at the positive saturation voltage, $+12V$ here. The reference voltage at the +terminal is given by

$$v_{ref} = \frac{R_2}{R_1 + R_2}v_o$$

FIGURE 29–24
Square wave generator

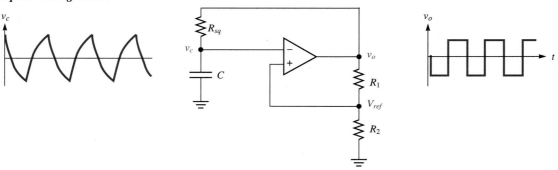

FIGURE 29–25

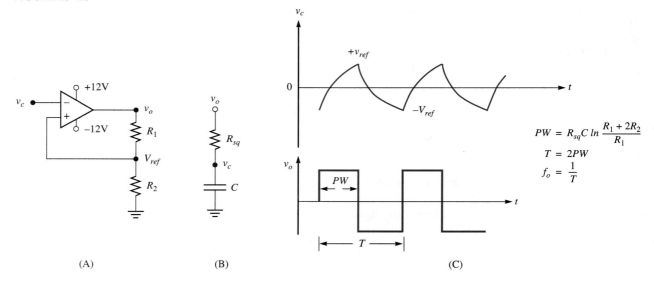

(A)   (B)   (C)

As an example, let's assume that $R_1 = 8$KHz and $R_2 = 4$KHz. This gives us

$$v_o = +12\text{V} \rightarrow v_{ref} = \frac{R_2}{R_1 + R_2} v_o = \frac{4\text{K}}{4\text{K} + 8\text{K}} 12 = +4\text{V}$$

Now we apply an arbitrary voltage to the −terminal. When this voltage increases beyond the reference voltage, the voltage at the −terminal is greater than the reference voltage. The output voltage jumps down to −12V. At this point, the reference voltage changes to

$$v_o = -12\text{V} \rightarrow v_{ref} = -4\text{V}$$

When the input voltage drops below −4V, the +terminal is now more positive than the −terminal and the output jumps to +12V.

Figure 29–25B shows the RC charging circuit. Let's see what happens to the capacitor voltage (v_{fb}). When $v_o = +12$V, the capacitor starts to charge towards +12V. At some point, the capacitor voltage will have increased to +4V, causing the Schmitt trigger to change states, setting $v_o = -12$V.

With the voltage driving the RC circuit now at −12V, the capacitor will discharge and then recharge in the opposite direction, heading for −12V. When the capacitor voltage hits −4V, the Schmitt trigger is tripped and the output jumps to +12V.

This process repeats itself periodically, producing the capacitor and output voltages shown in Figure 29–25C.

The combination of a charging capacitor and some form of comparator that triggers jumps in the output voltage is the basis for all multivibrator designs. For this square wave generator, the pulse width is given by (this result is derived in Problem 21, p. 1005, at the end of the chapter).

$$PW = R_{sq}C\ln\frac{R_1 + 2R_2}{R_1}$$

Because the duty cycle of all square wave generators is 50%

$$T = 2PW \text{ and } f_o = \frac{1}{T}$$

EXAMPLE 29–10

Square Wave Generator

The square wave generator of Figure 29–24 has the following parameters:

$$V_{supply} = 15\text{V}, R_1 = 4\text{K}\Omega, R_2 = 1\text{K}\Omega, R_{sq} = 1\text{K}\Omega, C = 5\mu\text{F}.$$

Find v_{ref}, PW, T, f_o.

Solution   The reference voltage is found from

$$v_{ref} = \frac{R_2}{R_1 + R_2}V_{supply} = \frac{1\text{K}}{4\text{K} + 1\text{K}}15 = 3\text{V}$$

The pulse width is found from

$$PW = R_{sq}C\ln\frac{R_1 + 2R_2}{R_1} = (1\text{K} \times 5\mu\text{F})\ln\frac{4\text{K} + 2 \times 1\text{K}}{4\text{K}} = 2\text{ms}$$

The period and frequency are given by

$$T = 2PW = 4\text{ms} \text{ and } f_o = \frac{1}{T} = 250\text{Hz}$$

EXAMPLE 29–11

Designing a Square Wave Generator

Using the same voltage supply and resistors of Example 29–10, find the capacitance needed to construct a square wave generator with a frequency of $f_o = 20$KHz.

Solution   We work backwards to find C. The given frequency means that we will have a period and pulse width of

$$T = \frac{1}{f_o} = 0.05\text{ms}$$

$$PW = \frac{T}{2} = 0.025\text{ms}$$

We now invert the pulse width formula to find

$$C = \frac{PW}{R_{sq}\ln\frac{R_1 + 2R_2}{R_1}} = \frac{0.025\text{ms}}{1 \times 1\text{K} \times \ln\frac{6\text{K}}{4\text{K}}} = 0.0617\mu\text{F}$$

DRILL EXERCISE   Find C to produce a frequency of $f_o = 1\text{MHz}$

Answer $C = 1230\text{pF}$

REVIEW QUESTIONS

1. Draw a general pulse wave, and describe its parameters.
2. What are other names for pulse generators?
3. Draw the Op-Amp square wave generator. What circuit is formed by the positive feedback loop?
4. How is the reference voltage found?
5. What is the function of the RC circuit?
6. Describe how the circuit works.
7. What are the formulas for *PW, T,* and f_o?

29–6 ■ TRIANGLE WAVE GENERATOR

Figure 29–26 shows the inverting Op-Amp integrator discussed in Chapter 24. Because the integral of a constant is a straight line, also known as a **ramp,** we get the output voltage shown. Note that because this is an inverter, we apply a negative input voltage to get a positive output voltage. The output voltage increases linearly until the maximum Op-Amp output voltage is reached. At this point, the output remains constant.

The slope of the ramp is given by

$$m = \frac{-V_{in}}{RC} \text{ V/s}$$

The slope is positive because V_{in} is negative. We can control the slope by changing the capacitor, the resistor, or the driving voltage.

FIGURE 29–26
Ramp generator

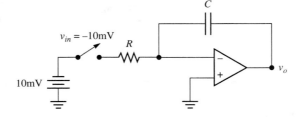 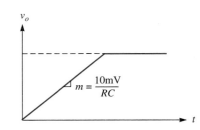

EXAMPLE 29–12
Ramp Generator

The ramp generator circuit in Figure 29–26 has the following parameters:

$$V_{in} = -10\text{mV}$$
$$R = 10\text{K}\Omega$$
$$C = 10\mu\text{F}$$

Find the slope of the ramp.

Solution   The slope of the ramp is

$$m = \frac{-V_{in}}{RC} \text{ V/s} = \frac{10\text{mV}}{10\text{K} \times 10\mu\text{F}} = 0.1\text{V/s}$$

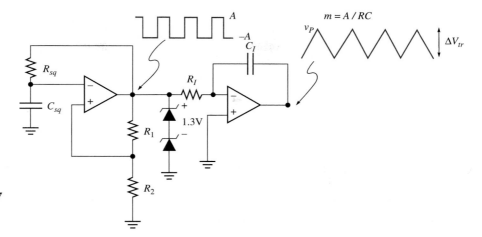

FIGURE 29–27
Triangle wave generator

In Figure 29–27 we use a square wave generator, such as the circuit described in the previous section, to drive the integrator. The response is the triangle wave shown. When the square wave is negative, we get a positive-going ramp with the slope:

$$m_+ = \frac{A}{R_I C_I}$$

The ramp will increase until the square wave generator goes positive. At this point, the ramp is negative going with the slope:

$$m_- = \frac{-A}{R_I C_I}$$

EXAMPLE 29–13
Triangle Wave Generator

The circuit in Figure 29–27 has the following parameters:

> square wave generator: $A = 0.1\text{V}$, $T = 20\text{ms}$
> integrator: $R_{int} = 1\text{K}\Omega$, $C_{int} = 1\mu\text{F}$

Find the slope of the triangle wave and its peak value.

Solution
(a) The slope of the positive-going ramp is

$$m_+ = \frac{A}{R_{int}C_{int}} = \frac{0.1}{1\text{K} \times 10\mu\text{F}} = 0.1\text{V/ms (Note the use of ms here.)}$$

The negative slope is, therefore

$$m_- = -0.1\text{V/ms}$$

Because the period of the square wave is 20ms, the positive-going ramp rises for 10ms. The positive-going ramp voltage will rise a total of

$$\Delta V_{tr} = m\frac{T}{2} = 0.1\text{V/ms} \times 10\text{ms} = 1\text{V}$$

Because of the symmetry of the triangle wave, the peak voltage is

$$V_p = \frac{\Delta V_{tr}}{2} = 0.5\text{V}$$

DRILL EXERCISE

Using the same square wave generator parameters, change the integrator parameters to $R_{int} = 2K\Omega$ and $C_{int} = 2\mu F$. Find the triangle wave positive slope and positive peak.

Answer $m_+ = 0.5V/ms$, $V_P = 2.5V$

Triangle Wave Generator

Figure 29–28 shows a triangle wave generator circuit that uses a single capacitor. The circuit is composed of an integrator (OP_1) and a Schmitt trigger (OP_2); note that the integrator has the +terminal down and the Schmitt trigger has the +terminal up. The circuit works as follows:

1. The voltage at the positive terminal of OP_2 depends on both the integrator output and the Schmitt trigger output.

2. Assume that the integrator has no output yet and the Schmitt trigger output is negative (saturated due to the positive feedback in the Schmitt trigger). This results in a negative voltage at the +terminal of OP_2.

3. The negative Schmitt trigger output voltage is also fed around to the integrator, which produces a positive-going ramp. The positive-going output of the integrator reduces the negative voltage at the +terminal.

4. At some point the integrator output ramp increases to a value where the voltage at the +terminal of OP_2 goes positive. When this happens, the Schmitt trigger output jumps to its positive voltage.

5. The positive Schmitt trigger output creates a negative-going ramp from the integrator. This negative-going ramp will eventually create a negative voltage at the +terminal of the trigger, causing the trigger output to swing negative.

This process repeats itself, and we get the triangle wave shown at the output of OP_1.

FIGURE 29–28
Triangle wave generator

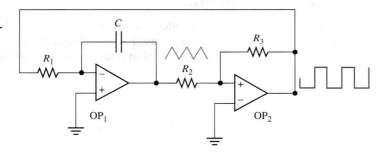

The Sawtooth Generator

Figure 29–29A shows a variation of the triangle wave, known as the **sawtooth wave.** This is the signal that is used to drive the electron beam across the screen of oscilloscopes and TVs. The beam is pulled to the right during the ramp and snaps back when the sawtooth jumps down zero, forming a raster display.

Sawtooth circuitry is a bit too involved to go into here, but Figure 29–29B shows conceptually how we produce this waveshape. We start with a triangle wave generator that has a switch across the integrating capacitor. Every time the ramp reaches its peak value, we momentarily close the switch; this discharges the capacitor and sets the output voltage to 0. When the switch reopens, the capacitor charges again, and we get another ramp.

FIGURE 29–29
Sawtooth generator

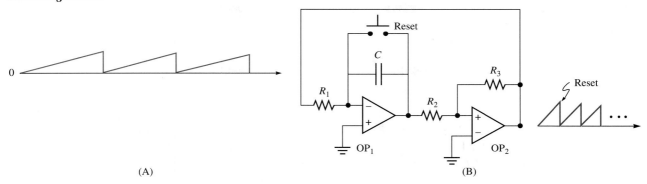

(A)

(B)

1. Draw the inverting Op-Amp integrator. What is the output of this integrator when the input is $-V_P$ volts? What is the slope of the output?

2. Does the ramp rise forever? Why?

3. Draw the output of an integrator if it is driven by a square wave of amplitude A and period T. What is the peak value of the triangle wave?

4. Describe how the triangle wave generator of Figure 29–28 works.

5. Draw a sawtooth wave.

6. How do we generate sawtooth waves from a triangle wave generator?

29–7 ■ INTEGRATED CIRCUIT SIGNAL GENERATORS

The 555 Pulse Generator

The most commonly used integrated circuit signal generator is the 555 timer. This chip is a pulse wave generator that can be used several different ways. Figure 29–30A (p. 998) shows the external connections needed to turn this chip into an astable multivibrator. The output of the 555 will be the pulse wave shown. The pulse width is determined approximately by

$$PW \approx 0.7 \ (R_1 + R_2) \ C$$

The period of the pulse is given approximately by

$$T \approx 0.7 \ (R_1 + 2R_2) \ C$$

the 555 IC is commonly used to provide the clock pulses needed in digital circuitry.

Figure 29–30B (p. 998) shows another use of the 555. In this circuit an external signal is used to drive the chip. Connected as shown, the 555 will produce a single-pulse output in response to each pulse that comes in. This type of circuit is known as a **monostable** (or **one-shot**) **multivibrator.** The output pulse width of a monostable multivibrator is independent of the input; it is given by

$$PW_{mono} = 1.1RC$$

FIGURE 29–30
Integrated circuit timer (555)

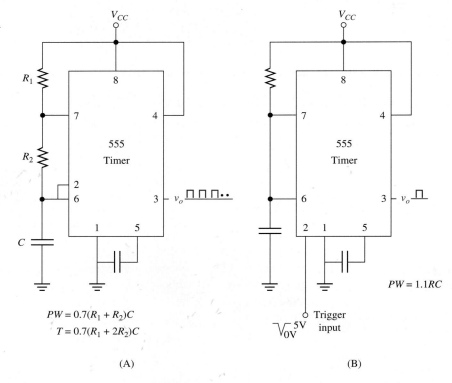

$$PW = 0.7(R_1 + R_2)C$$
$$T = 0.7(R_1 + 2R_2)C$$

(A)

$$PW = 1.1RC$$

(B)

EXAMPLE 29–14
The 555 Astable Multivibrator

The 555 circuit of Figure 29–30A has the following element values:

$$R_1 = 1K\Omega, \quad R_2 = 4K\Omega, \quad C = 10\mu F.$$

Find the pulse width and the pulse frequency.

Solution   The pulse width is

$$PW = 0.7(R_1 + R_2)C = 0.7(1K + 4K) \times 10\mu F = 35ms$$

The period of the pulse train is

$$T = 0.7(R_1 + 2R_2)C = 0.7\ (1K + 2 \times 4K)10\mu F = 63ms$$

The frequency, therefore, is

$$f_o = \frac{1}{T} = \frac{1}{63ms} = 15.9Hz$$

Function Generator Integrated Circuit

The chip shown in Figure 29–31 is an integrated circuit oscillator that can produce a sine wave and a cosine wave. The manufacturer shows the external connections required to produce sine waves from as low as 0.002Hz up to 20KHz.

FIGURE 29–31

PRECISION QUADRATURE OSCILLATOR

DESCRIPTION

The Model 4423 is a precision quadrature oscillator. It has two outputs 90 degrees out of phase with each other, thus providing sine and cosine wave outputs available at the same time. The 4423 is resistor programmable and is easy to use. It has low distortion (0.2% max up to 5 kHz) and excellent frequency and amplitude stability.

The Model 4423 also includes an uncommitted operational amplifier which may be used as a buffer, a level shifter or as an independent operational amplifier. The 4423 is packaged in a versatile, small, low cost DIP package.

EXTERNAL CONNECTIONS

1. 20 kHz Quadrature Oscillator

The 4423 does not require any external component to obtain a 20 kHz quadrature oscillator. The connection diagram is as shown in Figure 5.

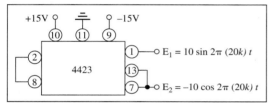

FIGURE 5.

2. Resistor Programmable Quadrature Oscillator

For resistor programmable frequencies in the 2 kHz to 20 kHz frequency range, the connection diagram is shown in Figure 6. Note that only two resistors of equal value are required. The resistor R can be expressed by,

$$R = \frac{3.785f}{42.05 - 2f} \quad \begin{array}{l} \text{, R in k}\Omega \\ \text{f in kHz} \end{array}$$

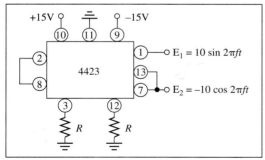

FIGURE 6.

3. Quadrature Oscillator Programmable to 0.002 Hz

For oscillator frequencies below 2000 Hz, use of two capacitors of equal value and two resistors of equal value as shown in Figure 7 is recommended. Connections shown in Figure 7 can be used to get oscillator frequency in the 0.002 Hz to 20 kHz range.

The frequency f can be expressed by:

$$f = \frac{42.05\,R}{(C + 0.001)\,(3.785 + 2R)}$$

where, f is in Hz
C is in μF
and R is in kΩ

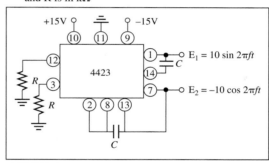

FIGURE 7.

For best results, the capacitor values shown in Table 1 should be selected with respect to their frequency ranges.

	f	20 kHz to 2 kHz	2 kHz to 200 Hz	200 Hz to 20 Hz
	C	0	0.01μF	0.1μF
f	20 Hz to 2 Hz	2 Hz to 0.2 Hz	0.2 Hz to 0.02 Hz	0.02 Hz to 0.002 Hz
C	1μF	10μF	100μF	1000μF

TABLE 1.

After selecting the capacitor for a particular frequency the value of the required resistor can be obtained by using the resistor selection curve shown in Figure 8 or by the expression:

$$R = \frac{3.785f\,(C + 0.001)}{42.05 - 2f\,(C + 0.001)}$$

where
R is in kΩ
f is in Hz
and C is in μF

1. What are the operating modes of the 555 IC (p. 999)?
2. What is an astable multivibrator?
3. How are the pulse width and period of the astable multivibrator found?
4. What is a monostable multivibrator?
5. How is the pulse width of the monostable multivibrator found?

29–8 ■ TROUBLESHOOTING OSCILLATORS

It is very easy to tell when an oscillator is not working—there is no signal output. It is not so easy to locate the problem. All oscillators use some form of feedback. A failure, or even a degradation, of any element in the loop, resistor, capacitor, inductor, transistor, or Op-Amp, can cause an oscillator not to oscillate.

It is a bit easier to locate the problem if the oscillator is producing an output, even though it isn't the output you expected. Here are two possible faulty oscillator outputs and their probable causes:

Oscillator output voltage is too small: The likely problem here is a degradation of the active device, the BJT, FET, or Op-Amp. Remove and check these components.

Oscillator output has the wrong frequency: Look for a problem with components that directly affect the frequency, e.g., the L or C in a tank circuit, the Rs and Cs in a phase shift oscillator.

The following examples show how to troubleshoot various oscillators when there is no signal output. The first step, as always in electronic circuits, is to perform DC Troubleshooting. Remember, for DC all capacitors are open and all inductors are short. DC Troubleshooting can find a shorted (or leaky) capacitor and an open inductor. However, it will not tell you that a capacitor has failed open or an inductor failed short.

The next step, if possible, is to separately check the forward and feedback pathways. We do this by disconnecting the loop and injecting signals as shown in Figure 29–32.

FIGURE 29–32

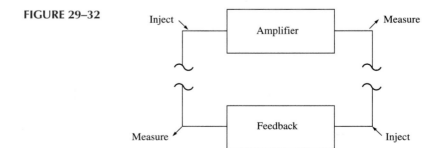

EXAMPLE 29–15
Troubleshooting LC Oscillators

The Hartley oscillator in Figure 29–33A is not oscillating.

(a) Find the likely causes of the DC Troubleshooting results shown in Figure 29–33B. (Note that for DC all capacitors are opened and all inductors are shorted.)
(b) DC Troubleshooting shows no problems. Troubleshoot the circuit.

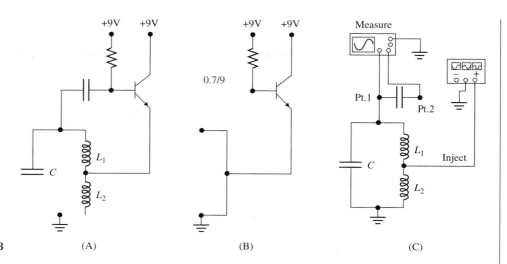

FIGURE 29–33   (A)   (B)   (C)

Solution

(a) The fact that the base is at 9V indicates that the transistor base has failed open.

(b) Remove the transistor to open the feedback loop in this oscillator. Using a function generator, inject a sinusoidal signal as shown in Figure 29–33C. Using an oscilloscope, observe the voltages at the indicated points as you change the input frequency.

　　If the tank circuit is good, the observed output at point #1 should be maximum when the input frequency equals the tank circuit resonant frequency (f_o). If the response does not peak at the expected f_o, remove and check the tank circuit components.

　　If the tank circuit response is what you expected but the signal does not show up at point #2, the coupling capacitor has probably failed open.

EXAMPLE 29–16
Troubleshooting
RC Phase Shift
Oscillators

The Op-Amp in the RC phase shift oscillator shown in Figure 29–34A is known to be good but there is no output. Troubleshoot this circuit.

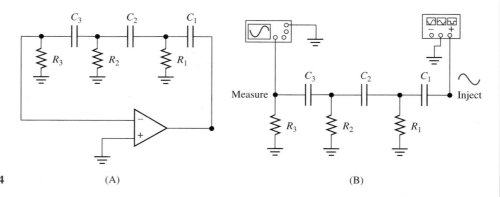

FIGURE 29–34   (A)   (B)

Solution   Because we know the Op-Amp is good, we will break the loop and trace signals in the feedback path. We inject the test signal into the RC feedback network and observe the response (Figure 29–34B). At the expected f_o of the oscillator, the RC response should be 180° out of phase with the input test signal. Note that it is easier to check for proper phase than proper magnitude here. If the two signals are not properly phased, then a component in the RC feedback network has failed or has significantly degraded.

EXAMPLE 29–17
Troubleshooting Square Wave Generators

The square wave generator shown in Figure 29–35A produces only a constant output. Locate the problem.

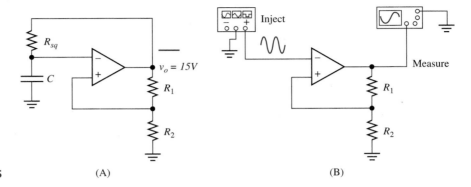

FIGURE 29–35   (A)   (B)

Solution   The square wave generator is built with two feedback loops. In this circuit we disconnect the negative feedback pathway. This leaves us with the Schmitt trigger circuit shown in Figure 29–35B. We now apply a sinusoidal input of sufficient magnitude that we should see the Op-Amp output periodically change states.

   If the output remains fixed at a DC level, the Schmitt trigger is not working. The problem is either with the Op-Amp or the resistors in the positive feedback path.

   If the Schmitt trigger is good, there must be a problem in the RC charging circuit in the negative feedback pathway. Check these components.

SUMMARY

■ An LC circuit, known as a tank or tuned circuit, will oscillate at a resonant frequency of

$$f_o = \frac{1}{2\pi\sqrt{LC}}$$

■ A Class C amplifier is biased in **CUT-OFF,** resulting in a very efficient amplifier.

■ A Class C tuned amplifier has an LC tank circuit in the output circuit. It will produce a sinusoidal output at f_o when the input signal has the same frequency.

■ A real LC circuit will produce oscillations that decrease with time. This is due to the resistances of these devices.

■ An LC oscillator can be built by feeding a portion of the tank circuit output back around to an amplifier. The amplifier will then supply the energy needed to counteract the resistance in an LC tank circuit.

■ A Colpitts oscillator splits the tank circuit capacitance into two capacitors. The two capacitors form a voltage divider; the voltage divider output is fed back to the amplifier. The total tank capacitance is

$$C_T = \frac{C_1 C_2}{C_1 + C_2}$$

■ A Hartley oscillator splits the tank circuit inductance into two inductors. The two inductors form a voltage divider; the voltage divider output is fed back to the amplifier. The total tank inductance is

$$L_T = L_1 + L_2$$

■ Piezo-electric crystals are often used in oscillators to prevent frequency drift.

■ A positive feedback loop will oscillate at the frequency where the loop gain is

$$AH = +1 \rightarrow |AH| = 1 \quad \text{and} \quad \theta_{AH} = 0°$$

■ A negative feedback loop will oscillate at the frequency where the loop gain is

$$AH = -1 \rightarrow |AH| = 1 \quad \text{and} \quad \theta_{AH} = \pm 180°$$

■ The RC phase shift oscillator uses a negative feedback loop where the feedback pathway is composed of three RC stages. Each stage contributes 60° of phase lag. If all of the Rs are the same and all of the Cs are the same

$$f_o = \frac{1}{2\pi RC\sqrt{6}}$$

■ The Wien bridge oscillator uses a positive feedback loop formed by a series RC combination and a parallel RC combination. If the Rs are the same and the Cs are the same, the resonant frequency of the Wien bridge oscillator is

$$f_o = \frac{1}{2\pi RC}$$

■ The feedback factor for the Wien bridge oscillator is one-third; the forward gain is 3.
■ A pulse wave is described by

$+A1$: the positive constant value

$-A2$: the negative constant value

PW: the pulse width of the positive part of the wave

T: the period

Duty Cycle $= 100\dfrac{PW}{T}$: ratio of the on-time to the total period

$f_o = 1/T$: the frequency

t_r *rise time (on-time):* the time it takes to rise from 10% to 90% of the total amplitude change.

t_f *fall time (off-time):* the time it takes to fall from 90% to 10% of the total amplitude change.

■ A square wave is a pulse train where $+A1 = -A2$ and *duty cycle = 50%*.
■ Square wave generation is accomplished by triggering a Schmitt trigger with an RC charging circuit.
■ A square wave generator is an astable, or free-running, multivibrator.
■ A triangle wave is produced by integrating a square wave.
■ A sawtooth wave is generated by resetting the integrator in a triangle wave generator at the peak of the positive-going ramp.
■ The 555 is an integrated circuit device that can be set up as an astable multivibrator or a monostable (one-shot) multivibrator.
■ Weak oscillations are probably due to degradation of the active device.
■ An oscillator that produces the wrong frequency has probably suffered degradation of a component in the tank circuit, or RC feedback circuit.
■ An oscillator that produces no oscillations is first checked with DC Troubleshooting.
■ The integrity of the forward and feedback paths can be checked by opening the loop and injecting signals at the appropriate locations.

PROBLEMS

SECTION 29–2 The Class C Tuned Amplifier

1. An LC tank circuit has capacitance and inductance values of $C = 0.22\text{pF}$ and $L = 2\text{mH}$. Find the resonant frequency f_o.

2. An LC tank circuit has an inductance of $L = 0.05\text{mH}$. Find the tank capacitance needed to produce oscillations with a frequency of $f_o = 2.5\text{MHz}$.

3. An LC tank circuit has a capacitance of $C = 0.47\text{pF}$. Find the tank inductance needed to produce oscillations with a frequency of $f_o = 10\text{MHz}$.

FIGURE 29–36

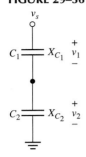

SECTION 29–3   LC Oscillators

4. Figure 29–36 shows a voltage divider formed by two capacitors, represented by their reactances X_{C_1} and X_{C_2}, where $X_C = \dfrac{-1}{2\pi fC}$. A voltage divider formed by reactances is analyzed the same way we analyze a voltage divider formed by resistors—just use X instead of R.

   (a) Find v_1 in terms of v_s.   (b) Find v_2 in terms of v_s.   (c) Find v_2 in terms of v_1.

5. A Colpitts oscillator has the following parameters:

$$C_1 = 2\text{pF}, C_2 = 10\text{pF}, L = 1\text{mH}.$$

   (a) What is the resonant frequency (f_o)?   (b) What is the feedback factor *(H)?*

6. A Colpitts oscillator is designed to produce a resonant frequency of 700KHz. If the inductor used is 1mH and $C_1 = 300\mu\text{F}$, find the required C_2.

7. A Colpitts oscillator is designed to produce a resonant frequency of 100MHz. If the inductor used is 0.1mH, and we want a feedback factor of $H = 0.2$, find the required capacitors.

FIGURE 29–37

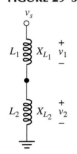

8. Figure 29–37 shows a voltage divider formed by two inductors, represented by their reactances X_{L_1} and X_{L_2}, where $X_L = 2\pi fL$. A voltage divider formed by reactances is analyzed the same way we analyze a voltage divider formed by resistors—just use X instead of R.

   (a) Find v_1 in terms of v_s.   (b) Find v_2 in terms of v_s.   (c) Find v_2 in terms of v_1.

9. A Hartley oscillator has the following parameters: $L_1 = 0.8\text{mH}, L_2 = 0.2\text{mH}, C = 5\mu\text{F}$.

   (a) What is the resonant frequency f_o?   (b) What is the feedback factor H?

10. A Hartley oscillator is designed to produce a resonant frequency of 1400KHz. If the capacitor used is 1μF and $L_1 = 100\text{mH}$, find the required L_2.

11. A Hartley oscillator is designed to produce a resonant frequency of 92MHz. If the capacitor used is 0.22μF and we want a feedback factor of $H = 0.2$, find the required inductors.

SECTION 29–4   RC Oscillators

12. A phase shift oscillator is constructed with

$$R = 2\text{K}\Omega \quad \text{and} \quad C = 4.7\mu\text{F}$$

   Find the resonant frequency of this oscillator.

13. What R do you need to construct a phase shift oscillator that has a resonant frequency of $f_o = 1200\text{KHz}$ and capacitors of $C = 0.0022\mu\text{F}$?

14. What C do you need to construct a phase shift oscillator that has a resonant frequency of $f_o = 92\text{MHz}$ and resistors of $R = 100\Omega$?

15. A Wien bridge oscillator is constructed with $R_1 = R_2 = 20\text{K}\Omega$ and $C_1 = C_2 = 5\text{pF}$. What is the resonant frequency of this oscillator?

16. You have the following resistors available:

$$10\Omega, 470\Omega, 600\Omega, 940\Omega, 1\text{K}\Omega, \text{ and } 1.2\text{K}\Omega$$

   Which combination of resistors can be used in the negative feedback loop of the Wien bridge oscillator? Why?

17. A Wien bridge oscillator is supposed to have a resonant frequency of $f_o = 1200\text{KHz}$. If you have 0.0022μF capacitors, what resistors do you need to complete the positive feedback loop?

18. A Wien bridge oscillator is supposed to have a resonant frequency of $f_o = 92\text{MHz}$. If you have 100Ω resistors, what capacitors do you need to complete the positive feedback loop?

SECTION 29–5   Square Wave Generator

19. For the pulse train shown in Figure 29–38

   (a) Find $+A1$ and $-A2$.   (b) Find *PW, T, duty cycle.*   (c) Find t_r and t_f.

20. The maximum output of the Op-Amp in the square wave generator shown in Figure 29–39 is ±12V.

   (a) What is v_{ref} when the Op-Amp output is +12V?

   (b) What is the output pulse width?

FIGURE 29–38

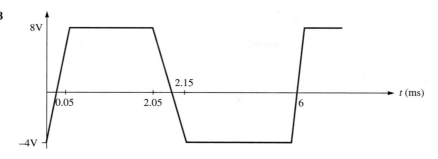

FIGURE 29–39

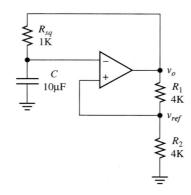

(c) What is the output period?

(d) Draw the capacitor and output voltages.

21. In this problem you are going to derive the pulse width formula for the square wave generator of Figure 29–39.

(a) Find v_{ref} in terms of R_1, R_2, and V_o. (Hint: This part of the circuit is a voltage divider.)

(b) You might recognize the formula for the feedback voltage (which is a capacitor voltage) as

$$v_C = (-v_{ref} - V_o)e^{-t/R_{sq}C} + V_o$$

The Schmitt trigger will fire when $v_C = v_{ref}$. The time at which this occurs is the pulse width we are looking for. Equate v_C to v_{ref}, and solve for t. Remember that if $e^x = y$, then $x = \ln y$.

SECTION 29–6   Triangle Wave Generator

22. Draw the output of the integrator in Figure 29–40 if the maximum Op-Amp output voltage is $\pm 12V$, and we switch in one of the following inputs. In each case, find the slope of your ramp.

(a) $V_{in} = -10\text{mV}$ **(b)** $V_{in} = -0.5V$ **(c)** $V_{in} = +10\text{mV}$

FIGURE 29–40

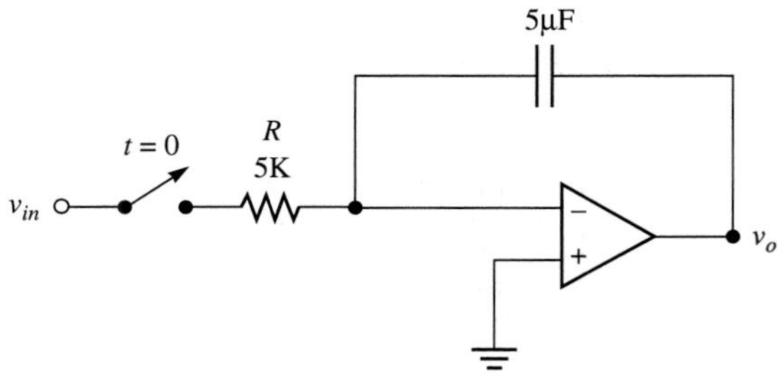

FIGURE 29–41

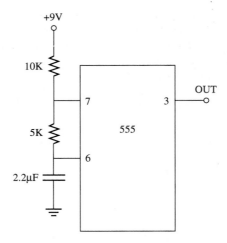

SECTION 29–7   Integrated Circuit Signal Generators

23. Draw the output of the 555 circuit shown in Figure 29–41. What are *PW, T,* and f_o?

Troubleshooting Problems

24. Find the likely causes of the following problems that might be found in the Colpitts oscillator shown in Figure 29–42.

   (a) There is no AC output. DC Troubleshooting finds that $V_B = 9V$, $V_C = 9V$, $V_E = 0V$.

   (b) A faint oscillation at the correct frequency is seen at the output.

   (c) The output oscillates but at a frequency higher than expected.

   (d) No oscillations, but the transistor is good. You disconnect the tank circuit, inject a signal at v_o, and observe no signal at v_{fb}.

25. An RC phase shift oscillator is not working. You have disconnected the phase shift network from the Op-Amp and injected a sinusoidal signal (at the resonant frequency) into the RC network. What are likely causes of the following observed outputs?

   (a) The phase shift is observed to be only 120°.

   (b) There is no signal at the output of the RC network.

FIGURE 29–42

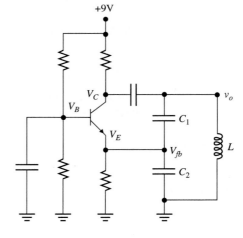

Computer Problems

Oscillators can be very difficult to simulate. These circuits do not have any inputs (they are often set off by transient electrical surges that occur when the switch is thrown); also, the numerical routines used in computer programs often cannot converge to an answer in feedback (especially positive feedback) circuits.

Some computer programs allow you to set up an initial voltage on a capacitor. These programs can sometimes be used to analyze oscillators. Because you might not have this type of program, we will limit these problems to analyzing pieces of an oscillator circuit.

26. Simulate the Class C tuned amplifier of Figure 29–6 (p. 978). Observe the output for the following *sinusoidal* inputs:

    (a) 2V, $f = 5$KHz   (b) 2V, $f = 12$KHz   (c) 2V, $f = 2$KHz

27. Simulate the Class C tuned amplifier of Figure 29–6 (p. 978). Observe the output for the following *pulse* inputs:

    (a) 2V, $f = 5$KHz, *duty cycle* $= 5\%$.   (b) 2V, $f = 12$KHz, *duty cycle* $= 5\%$.

    (c) 2V, $f = 2$KHz, *duty cycle* $= 5\%$.

28. Simulate the three-stage RC network of the RC phase shift oscillator with $R = 1$KΩ and $C = 5\mu$F (Figure 29.17, p. 987).

    (a) Find the frequency response of this circuit. Remember, the input is connected to the first capacitor, and the output is taken across the last resistor.

    (b) At what frequency does the phase pass through $-180°$?

    (c) Eliminate the second capacitor, and repeat the simulation. Does the phase pass through $-180°$? Why?

29. Simulate the series-parallel RC combinations of the Wien bridge amplifier with $R = 1$KΩ and $C = 5\mu$F (Figure 29.18, p. 988).

    (a) Find the frequency response of this circuit. Remember, the input is connected to the series RC combination. The output is taken from the connecting point between the series RC and the parallel RC combinations.

    (b) At what frequency does the phase pass through $0°$?

30. Simulate just the RC portion of the square wave generator shown in Figure 29–39 (p. 1005). Drive the resistor with a 12V square wave of period $T = 40$ms. This represents the changes seen in the Op-Amp output. The reference voltage for this square wave generator is $v_{ref} = \pm6$V. Observe the capacitor voltage, and determine how long it takes to go from -12 to $+6$V or from $+12$ to -6V. This time is the pulse width of the square wave generator.

31. Simulate the integrator of Figure 29–40 (p. 1005) (you don't need the switch). Use a square wave input of varying frequencies, and observe the ramp. At each frequency, record the slope of the ramp. Be sure that one of your runs uses a low enough frequency that you see the Op-Amp output saturate.

CHAPTER

30

ELECTRONIC SWITCHES (THYRISTORS)

CHAPTER OBJECTIVES

To determine the switching behavior and use of the four-layer diode.

To determine the switching behavior and use of the DIAC.

To determine the switching behavior of the silicon-controlled rectifier (SCR).

To use the SCR in time-delay circuits.

To determine the conduction angle in AC SCR circuits.

To build an SCR relaxation oscillator.

To determine the switching behavior of the TRIAC.

To determine the switching properties of the unijunction transistor (UJT).

To set the trigger voltage for a programmable unijunction transistor (PUT).

To briefly describe the switching behavior of the gate turn-off thyristor (GTO), the silicon bilateral switch (SBS), the silicon-controlled switch (SCS), and optoisolators.

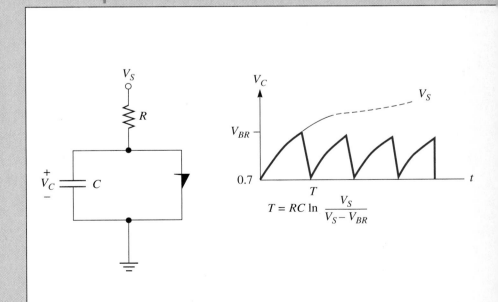

30–1 ▪ INTRODUCTION

The diode acts as an electronic switch that is controlled by its anode-to-cathode voltage. Set the diode voltage to 0.7V (for silicon), and the diode turns **ON.** The diode will be **OFF** for diode voltages less than this value. We have also seen that the BJT and FET transistors can be used as switches (operating between **CUT-OFF** and **SATURATION**).

The devices described in this chapter act as switches, but they also provide many more on-off properties than the diode or transistor. These on-off devices all come under the common heading of **thyristors.** The advantage of thyristors is their ability to handle large currents, currents that reach up to 100A.

We will discuss the following thyristors and related devices in this chapter:

☐ Four-layer diode

☐ DIAC

☐ Silicon-controlled switch (SCR)

☐ TRIAC

☐ Unijunction transistor (UJT)

☐ Programmable unijunction transistor (PUT)

The I-V characteristics of thyristors look very strange to the uninitiated. Because of this, we will use the metaphor of a line of people trying to pass through a door to explain the switching properties of these devices. To introduce this concept, we begin by revisiting the standard silicon PN-junction diode.

In Figure 30–1 folks have politely lined up in front of the door. You are at the head of the line. This particular door opens only in one direction (rectification), has no latch on it, and is spring-loaded so that it will close on its own.

To open the door you must exert enough pressure to counteract the spring. Once you overcome the spring, the door opens and you pass through. As long as the folks that follow you maintain this level of pressure on the door, everybody can pass through.

The one-way aspect of the door represents the one-way conduction of the diode. The pressure you apply to open the door is similar to the 0.7V you need to turn on a diode. The people passing through the door represent the diode current.

FIGURE 30–1

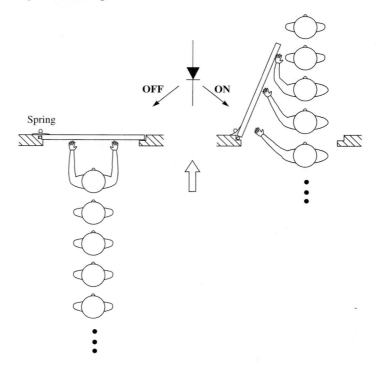

30–2 ■ THE FOUR-LAYER DIODE AND THE DIAC

The Four-Layer Diode

The **four-layer diode,** is shown in Figure 30–2A. This device has many other names: *PNPN diode, Shockley diode,* and *reverse-blocking diode thyristor.* It is a PNPN device that has the I-V characteristic shown in Figure 30–2B. As with the regular diode, current passes only from cathode (K) to anode (A). Before looking at the consequences of the I-V characteristic, let's go back to our door metaphor—with one major change.

FIGURE 30–2
The four-layer diode

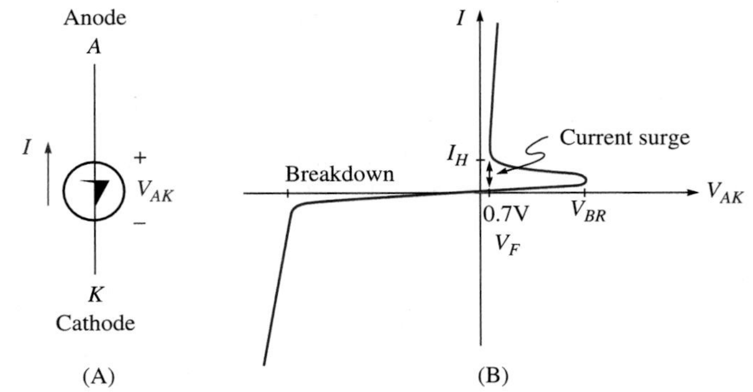

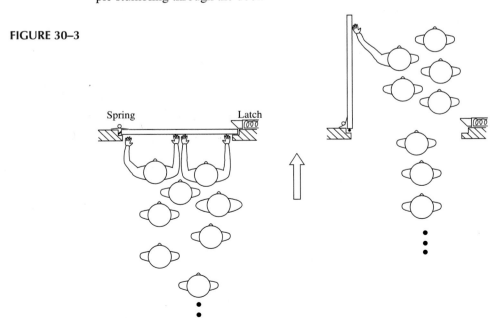

This time, in addition to the spring that automatically closes the door, there is also a latch with a very strong spring that holds the door closed (Figure 30–3). Now if you apply the same pressure as before, the door will not open; the latch prevents it. To open the door you must apply enough pressure to overcome the latch. To do this, you enlist the aid of your friends standing behind you to increase the pressure on the door.

When you exceed the pressure of the latch, the door will pop open to let you pass. As soon as the latch is overcome, the only thing holding the door in position is the door spring. Opening the door requires a lot of pressure; keeping the door open requires very little pressure. As a result, it is highly likely that because there are several of you applying pressure to overcome the latch, when the door suddenly opens, there will be an initial surge of people stumbling through the door.

FIGURE 30–3

Back to the four-layer diode I-V characteristic of Figure 30–2B (p. 1011). Let's start at the origin and slowly increase the diode voltage. The initial slope of the curve is very small, indicating that the diode has very little current in it—the diode is effectively **OFF.** Even when we hit 0.7V, the diode is still **OFF.** We need to increase the voltage to overcome the internal "latch" of the diode; the voltage required for this is known as the **breakover voltage,** or V_{BR}, sometimes labeled $V_{(BO)}$.

As soon as the diode voltage reaches V_{BR}, the diode turns **ON** and begins to conduct. As soon as it begins to conduct (as soon as the "latch" is overcome), the diode behaves as a regular diode and the diode voltage drops to V_F, which is approximately 0.7V.

The I-V characteristic also indicates that as soon as the diode turns **ON** there is a surge of current (comparable to you and your friends surging through the suddenly-open door). This surge of current can be used to trigger other electronic devices. When the current in it drops below a minimum value, known as the holding current (I_H), the four-layer diode turns **OFF.**

Figure 30–4 shows a four-layer diode relaxation oscillator that produces a sawtooth-like output. This circuit works as follows:

1. The capacitor is initially uncharged, so the diode voltage is initially 0V. The diode is **OFF** at this time.

2. The switch is closed. The capacitor voltage starts to charge towards the supply voltage. This explains the initial rise in the output waveform. The formula for a charging capacitor that is initially uncharged is

$$v_C(t) = V_S(1 - e^{-t/RC})$$

3. At time T, the capacitor voltage reaches V_{BR}; that is, at $t = T$

$$v_C(T) = V_{BR} = V_S(1 - e^{-T/RC})$$

This causes the four-layer diode to turn **ON.** Because any **ON** diode acts essentially as a short circuit, the capacitor will discharge through the diode. The discharge time will depend on the total resistance in the diode-capacitor circuit. The faster the discharge, the greater the capacitor discharge current. (To protect the capacitor from excess current, this circuit is usually built with an additional resistor in series with the capacitor.)

4. When the capacitor current drops below I_H—that is, when the capacitor voltage drops below 0.7V—the diode turns **OFF.** The discharged capacitor now recharges towards the supply voltage, and the process repeats.

FIGURE 30–4
Four-layer diode
relaxation oscillator

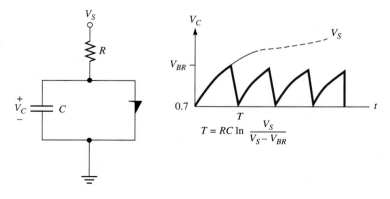

This waveform actually goes from 0.7V to V_{BR}. If we ignore the 0.7V, we get the following approximate formula for the pulse period T:

$$T \approx RC\ln\frac{V_S}{V_S - V_{BR}}$$

| **EXAMPLE 30–1**
The Four-Layer
Diode Relaxation
Oscillator | For the circuit of Figure 30–4, find the pulse period and pulse train frequency, given the following data:

$$V_S = 20V, \; V_{BR} = 10V, \; R = 1K\Omega, \; C = 10\mu F$$

Solution   The period is given by

$$T = RC \ln \frac{V_S}{V_S - V_{BR}} = 1K \times 10\mu F \ln \frac{20}{20 - 10} = 6.93 ms$$

The frequency is found from

$$f = \frac{1}{T} = \frac{1}{6.93 ms} = 144 Hz$$ |

| **DRILL EXERCISE** | Find the period and frequency for the circuit in Figure 30–4 given the following data:
$V_S = 100V, \; V_{BR} = 25V, \; R = 2K\Omega, \; C = 5\mu F.$

Answer $T = 2.88 ms$ and $f = 347 Hz$ |

The DIAC

The DIAC (Figure 30–5) acts as a bidirectional four-layer diode. That is, it is a door that swings both ways. When the breakover voltage is exceeded in either direction, the DIAC conducts. Because the DIAC conducts in both directions, it is often used in AC circuits.

Figure 30–6 shows data on two commercial DIACs. The parameter $V_{(BO)}$ is the breakover voltage for the devices and ranges from 27 to 45V. The maximum current that DIACs can handle is given by the parameter I_{TRM} and is 2A for both of these devices.

The DIAC I-V characteristic is not exactly symmetric; there is usually a difference between the breakover voltage in one direction and the breakover voltage in the other direction. The symmetry parameter gives us a measure of this difference.

FIGURE 30–5
The DIAC

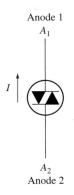
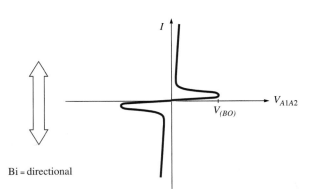

FIGURE 30–6

Pkg.	$V_{(BO)}$ (V)		$V_{(BO)}$ Symmetry (V)	Dynamic Breakback Voltage (V)	$I_{(BO)}$ (μA)	$I_{(BO)}$ (A)
	Min.	Max.				
DO-35	27	37	3	5	50	2.0
DO-35	35	45	3	5	50	2.0

30–3 ■ THE SILICON-CONTROLLED RECTIFIER (SCR)

The **silicon-controlled rectifier** (SCR) (Figure 30–7A) is one of the most common thyristors in use. Note that this is a three-terminal device, the third terminal being the gate (G).

To more fully understand the operation of the SCR, let's return to our door example. Consider the type of entry door where you have to be "buzzed in" (Figure 30–7B). The latch of this door is electrically controlled. This is in addition to the door-closing spring. If you and your friends push very hard on the door, you can probably pop the latch. This is analogous to the SCR with no input at the gate. If you apply enough voltage between the anode and cathode, the SCR will turn **ON.** Without a gate input, the SCR behaves the same as the four-layer diode.

FIGURE 30–7
The silicon-controlled rectifier (SCR)

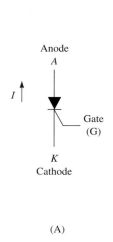

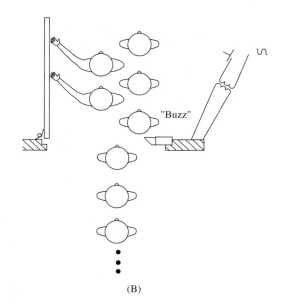

(A) (B)

Normally, you and your friends would pass through the door by having someone send a signal (the buzz) that electrically releases the latch. Once the latch has been released, to keep the door open you only need to apply enough pressure to counteract the door spring.

Back to the SCR. When you apply enough gate current (or gate voltage) to the SCR, the internal latch is released and the SCR conducts. In Figure 30–8A you see the SCR I-V curve; note its similarity to the I-V curve of the four-layer diode. One way to think of how an SCR works is that as we increase the gate current, we reduce the forward breakover voltage to the point where the SCR turns on (Figure 30–8B). The parameters for the SCR are

V_{GT}: the gate voltage needed to trigger the SCR, usually less than 2V.

V_{BR}: forward breakover voltage. The voltage at which the SCR turns on without a gate trigger.

FIGURE 30–8
SCR I-V curve

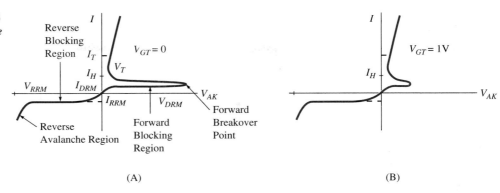

(A)   (B)

V_{TM}: the anode-to-cathode voltage when the SCR is **ON,** typically 1.5 to 2V.

V_{DRM}: the maximum anode-to-cathode voltage the SCR can withstand when it is **OFF,** varies from 30 to 600V.

I_H: holding current. The minimum current necessary to keep the SCR conducting.

Forward blocking region: the region of the I-V curve where forward voltage is applied but there is no conduction, that is, $(I_{SCR} < I_H)$.

Reverse blocking region: the region of the I-V curve where the SCR is reverse biased and there is no conduction (similar to a regular diode).

FIGURE 30–9

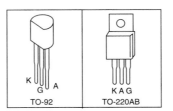

$I_{T\,RMS}$ (A)	V_{DRM} & V_{RRM} (V)	Package	I_{DRM} & I_{GT} (mA) Min.	I_{DRM} & I_{GT} (mA) Max.	I_{RRM} @25°C (mA)	V_{TM} @25°C (V)	V_{GT} @25°C (V)	$P_{G(AV)}$ (W)	t_{gt} (μsec.)
		SCRs—Sensitive Gate (0.8-10.0 Amp)							
	200	TO-92 (Isolated)	—	.200	.001	1.7	0.8	0.1	2.2
	400	TO-92 (Isolated)	—	.200	.001	1.7	0.8	0.1	2.2
X	600	TO-92 (Isolated)	—	.200	.002	1.7	0.8	0.1	2.2
	30	TO-92 (Isolated)	—	.200	.001	1.7	0.8	0.1	2.2
	200	TO-92 (Isolated)	—	.200	.001	1.7	0.8	0.1	2.2
	50	TO-92 (Isolated)	—	.200	.001	1.5	0.8	0.1	2.2
	100	TO-92 (Isolated)	—	.200	.001	1.5	0.8	0.1	2.2
1.5	200	TO-92 (Isolated)	—	.200	.001	1.5	0.8	0.1	2.2
	400	TO-92 (Isolated)	—	.200	.001	1.5	0.8	0.1	2.2
	600	TO-92 (Isolated)	—	.200	.002	1.5	0.8	0.1	2.2
	100	TO-92 (Isolated)	1	10	.01	1.6	1.5	0.3	2.0
1.0	200	TO-92 (Isolated)	1	10	.01	1.6	1.5	0.3	2.0
	400	TO-92 (Isolated)	1	10	.01	1.6	1.5	0.3	2.0
	600	TO-92 (Isolated)	1	10	.01	1.6	1.5	0.3	2.0
	200	TO-220AB (Isolated)	1	15	.01	1.6	1.5	0.5	2.0
6.0	400	TO-220AB (Isolated)	1	15	.01	1.6	1.5	0.5	2.0
	600	TO-220AB (Isolated)	1	15	.01	1.6	1.5	0.5	2.0
	200	TO-220AB (Isolated)	1	15	.01	1.6	1.5	0.5	2.0
8.0	400	TO-220AB (Isolated)	1	15	.01	1.6	1.5	0.5	2.0
	600	TO-220AB (Isolated)	1	15	.01	1.6	1.5	0.5	2.0
	200	TO-220AB (Non-Isolated)	1	15	.01	1.6	1.5	0.5	2.0
	400	TO-220AB (Non-Isolated)	1	15	.01	1.6	1.5	0.5	2.0
	600	TO-220AB (Non-Isolated)	1	15	.01	1.6	1.5	0.5	2.0
10.0	400	TO-220AB (Isolated)	1	15	.01	1.6	1.5	0.5	2.0
	600	TO-220AB (Isolated)	1	15	.01	1.6	1.5	0.5	2.0
12.0	400	TO-220AB (Non-Isolated)	1	20	.01	1.6	1.5	0.5	2.0
	600	TO-220AB (Non-Isolated)	1	20	.01	1.6	1.5	0.5	2.0
15.0	400	TO-220AB (Isolated)	1	30	.01	1.6	1.5	0.6	2.0
	600	TO-220AB (Isolated)	1	30	.01	1.6	1.5	0.6	2.0
20.0	400	TO-220AB (Isolated)	1	30	.01	1.6	1.5	0.5	2.0
	600	TO-220AB (Isolated)	1	30	.01	1.6	1.5	0.5	2.0
25.0	400	TO-220AB (Isolated)	1	30	.01	1.6	1.5	0.5	2.0
	600	TO-220AB (Isolated)	1	30	.01	1.6	1.5	0.5	2.0

Figure 30–9 (p. 1015) shows the data for some commercial SCRs. The first column shows how much anode-to-cathode current (I_T) each device can handle. Note that because this device is usually used in AC power circuits, this value is given in RMS.

The other parameter of interest is the gate voltage (V_{GT}) required to turn on the SCR, V_{GT}, which ranges from 0.8 to 2V.

Turning the SCR **OFF**

An NPN BJT that is **OFF** can be turned on by applying a positive base voltage. An NPN BJT that is **ON** can be turned off by applying a negative voltage at the base. The SCR, on the other hand, cannot be turned off from the gate. Once the latch on our door has been released, there is no way to signal it to reclose the door. The only way to close the door is to stop the flow of people through it. The only way to turn off an SCR is to stop the current in it. As Figure 30–10 shows, there are several general ways this is achieved:

1. Figure 30–10A: Disconnect the SCR from its supply voltage or ground.
2. Figure 30–10B: Divert the current with a short circuit around the SCR.
3. Figure 30–10C: Force current into the SCR in the **OFF** direction. This is known as *forced commutation*.

FIGURE 30–10
Turning the SCR OFF

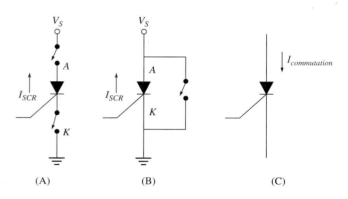

(A)　　　　　　(B)　　　　　　(C)

EXAMPLE 30–2
SCR Indicator Circuit

Mary suspects the cookie jar is being raided while she is at work. To confirm her suspicions, she designs the circuit shown in Figure 30–11.

(a) Describe how this circuit works.
(b) The gate turn-on voltage is $V_{GT} = 0.8$V, the anode-to-cathode **ON** voltage of the SCR is $V_{AK} = 0.6$V, and the LED turn-on voltage is $V_{LED} = 2$V. Find the gate current and the LED current.

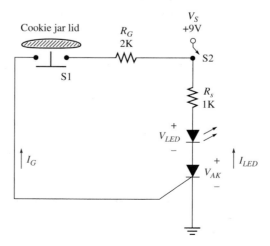

FIGURE 30–11

Solution

(a) When Mary leaves for work in the morning, she closes switch S2. Switch S1 is normally closed; therefore, when the cookie lid is on the jar, S1 is open. With S1 open, there is no voltage at the gate. Therefore, the SCR does not conduct, and there is no current in the LED.

When the cookie monster opens the lid, S1 closes, triggering the SCR **ON**. The LED lights and will stay lit, even if the lid is replaced and S1 opens again. The LED will stay lit until Mary opens switch S2.

(b) To trigger the SCR, the gate voltage needed is 0.8V. The gate current, therefore, is

$$I_G = \frac{V_S - V_G}{R_G} = \frac{9 - 0.8}{2K} = 4.1mA$$

The LED current is

$$I_{LED} = \frac{V_S - V_{LED} - V_{AK}}{R_S} = \frac{9 - 2 - 0.6}{1K} = 6.4mA$$

EXAMPLE 30–3
SCR Relaxation Oscillator

Describe the operation of the circuit in Figure 30–12A.

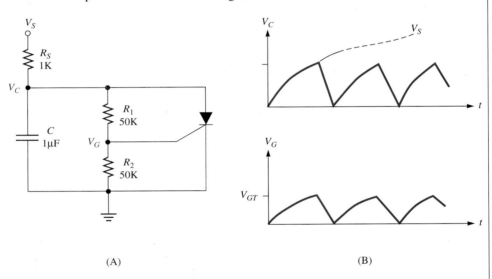

FIGURE 30–12
SCR relaxation oscillator

(A)　　　　　　　　　　　　(B)

Solution The capacitor voltage supplies both the anode voltage and, through the voltage divider, the gate voltage to the SCR. Initially, the SCR is open, and the capacitor charges through R_S. Because R_1 and R_2 are very large, they have a minimal effect on the charging of the capacitor.

As the capacitor voltage increases, the anode and gate voltage of the SCR increase. At some point, the gate voltage (V_G) reaches V_{GT}, the value necessary to trigger the SCR. When the SCR turns **ON**, it provides a short circuit path that rapidly discharges the capacitor. As the capacitor discharges, the anode voltage and the SCR current drop. When the SCR current falls below the holding current (this happens very quickly), the SCR turns **OFF,** and the process repeats itself. The capacitor and gate voltages are shown in Figure 30–12B.

False Triggering

Circuits that use SCRs for switching often operate at high voltages. This can create problems with circuits similar to Figure 30–12. The SCR in this circuit will trigger when the gate reaches a few volts. In everyday use, it is not unusual to find noise or surges of a few volts on supply lines. Such noises or surges could accidentally trigger the SCR. For this reason, many SCR trigger circuits use a four-layer diode in the gate circuit. Because it takes a large voltage to trigger a four-layer diode, surges of a few volts will not accidentally trigger the diode.

EXAMPLE 30–4
Four-Layer Diode Triggering of SCR

The circuit in Figure 30–13 uses both a four-layer diode and an SCR. The breakover voltage for the diode is 25V. Determine the operation of this circuit.

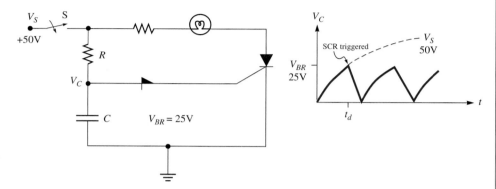

FIGURE 30–13
SCR relaxation oscillator with DIAC trigger

Solution The SCR is initially **OFF,** so the bulb is dark. When the switch is closed, the capacitor starts to charge towards the 50V supply. The capacitor voltage eventually reaches the breakover voltage of the four-layer diode, which turns **ON,** sending a current surge into the gate of the SCR. The SCR turns **ON,** and the bulb lights.

This is a **time-delay circuit:** the bulb doesn't turn on until t_d seconds after the switch is closed. We find the delay time by finding when the capacitor voltage equals the four-layer diode breakover voltage (we ignore the small SCR gate voltage):

$$v_C(t_d) = V_{BR} = V_S(1 - e^{-t_d/RC})$$

Solving for t_d we get

$$t_d = RC\ln\frac{V_S}{V_S - V_{BR}}$$

The delay time for this circuit is given by

$$t_d = RC\ln\frac{V_S}{V_S - V_{BR}} = RC\ln\frac{50}{50 - 25} = RC\ln 2 \approx 0.7RC$$

REVIEW QUESTIONS

1. Draw the circuit symbol for the SCR, and sketch its I-V curve when $V_G = 0$V.
2. Sketch the I-V curve as the gate voltage rises. What is the effect of increasing the gate voltage?
3. Describe the operation of the SCR.
4. How can we turn off an SCR?
5. What is false triggering? What can cause it?
6. How can we use the four-layer diode to prevent false triggering?

30–4 ■ AC SCR APPLICATIONS

Thyristors are most often used to control power distribution in AC circuits. Applications range from motor controllers to light dimmers. Strange as it might sound, SCRs can be used to control the speed of DC motors when the supply voltage is AC.

DC Motor Control

The speed of a DC motor is set by the DC voltage that drives the motor. Rather than use a DC supply, with the attendant problems of DC drift and speed variation, we can actually drive a DC motor with a pulse train (Figure 30–14). The motor does not respond to the on-off nature of the pulses. Rather the DC motor speed depends on the average value of the driving voltage. The average value of an electrical signal is equivalent to its DC value. The average value of a pulse train is given by

$$V_{ave} = A\frac{PW}{T} = A\frac{duty\ cycle}{100}$$

where A is the pulse amplitude, PW is the pulse width, and T is the pulse train period.

(Remember, $duty\ cycle = 100\,PW/T$.) To control the speed of the DC motor, then, we can fix the pulse amplitude and period and vary the pulse width. This makes for an extremely stable controller.

FIGURE 30–14

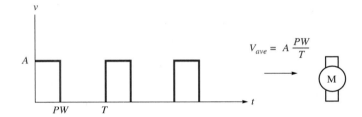

We can also control a DC motor with pulses of sinusoids. Figure 30–15A shows the very familiar half-wave rectified AC signal. The average value of the half-wave rectified signal was found in Chapter 2 to be

$$V_{ave} = \frac{V_p}{\pi}$$

where V_p is the peak value of the sine wave.

FIGURE 30–15

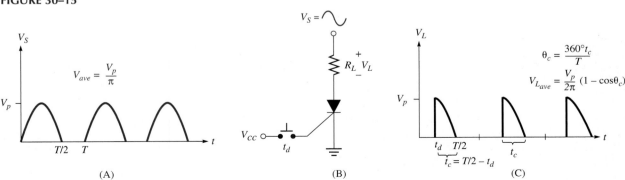

(A)   (B)   (C)

The circuit in Figure 30–15B is an SCR half-wave rectifier; the gate is pulsed t_d seconds after the sine wave begins. Triggered at t_d, the SCR conducts, and the load sees the remaining portion of the positive cycle of the input (Figure 30–15C). When the sine wave input reverses direction, current is forced into the SCR in the **OFF** direction. This is an example of forced commutation that turns the SCR **OFF.** The SCR is now ready for the next positive cycle.

As the delay time is increased, less of the positive cycle gets to the load. We also define a **conduction time** (t_c) as the time the SCR conducts. Because the SCR conducts from t_d to $T/2$, the conduction time is

$$t_c = \frac{T}{2} - t_d$$

Rather than use duty cycle to express the on-time of the pulse train, we use the **conduction angle,** which is defined as

$$\textbf{conduction angle: } \theta_c = \frac{2\pi}{T} t_c \text{ (radians)} = \frac{360}{T} t_c \text{ (degrees)}$$

If a half-wave rectifier conducts for a full half-cycle, it has a conduction angle of 180°. In terms of the conduction angle, the DC, or average, value of the load voltage is

$$V_{L_{ave}} = \frac{V_p}{2\pi} (1 - \cos \theta_c)$$

EXAMPLE 30–5
Conduction Angle and DC Value

Find the conduction angle and DC value of a rectified sinusoid with a peak value of $V_m = 170\text{V}$ and a frequency of 60Hz under these conditions (be careful to look at the subscripts):

(a) $t_d = 0\text{s}$ **(b)** $t_c = 4\text{ms}$

Solution   The period of this signal is

$$T = \frac{1}{f} = \frac{1}{60} = 16.7\text{ms}$$

(a) The delay time is 0s, so the sine wave here is on for the full half-cycle and

$$t_c = \frac{T}{2} - t_d = \frac{T}{2} = 8.33\text{ms}$$

Its conduction angle and average value are

$$\theta_c = \frac{360}{T} t_c = \frac{360}{16.7\text{ms}} 8.33 \text{ ms} = 180°$$

and

$$V_{L_{ave}} = \frac{V_p}{2\pi} (1 - \cos \theta_c) = \frac{170}{2\pi} (1 - \cos 180°) = 54.1\text{V}$$

(b) We are now given the conduction time directly, $t_c = 4\text{ms}$. The conduction angle, therefore, is

$$\theta_c = \frac{360}{T} t_c = \frac{360}{16.7\text{ms}} 4\text{ms} = 86.2°$$

The average value is

$$V_{L_{ave}} = \frac{V_p}{2\pi}(1 - \cos \theta_c) = \frac{170}{2\pi}(1 - \cos 86.2°) = 25.3\text{V}$$

Figure 30–16A shows a simple SCR gate control circuit that provides for a variable conduction angle. The gate voltage is found from the voltage divider:

$$v_g = \frac{R_2}{R_1 + R_2}v_s$$

As R_1 is varied, the gate voltage varies. Figure 30–16B shows that as R_1 is increased, the gate voltage decreases. Assume we need 2V at the gate to trigger this SCR. As you can see from the figure, as the gate voltage decreases, it takes a longer time to reach the necessary 2V. This time is the delay time.

FIGURE 30–16
SCR AC response

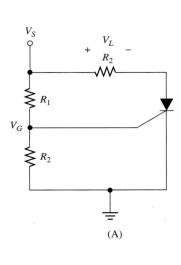

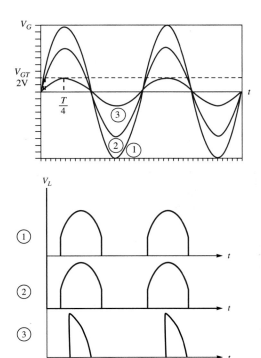

When R_1 is close to 0Ω, the gate voltage is the same as the supply voltage. In this case, the SCR will trigger almost immediately, resulting in zero delay time. With zero delay, the conduction angle is 180°.

When R_1 is maximum, the gate voltage drops to a value that reaches 2V at the peak of the sine wave ($t_d = T/4$). Any voltage less than this value will not trigger the SCR. The maximum delay time then is one-fourth of a cycle. This translates to a 90° conduction angle.

Figure 30–16C shows the load voltages for the three choices of R_1.

EXAMPLE 30–6
SCR with Four-Layer Diode

The four-layer diode in Figure 30–17A (p. 1022) has a breakover voltage of 30V. The SCR has a gate turn-on voltage of 2V. The driving voltage is a 170V 60Hz sine wave:

$$v_s = 170 \sin 2\pi60t$$

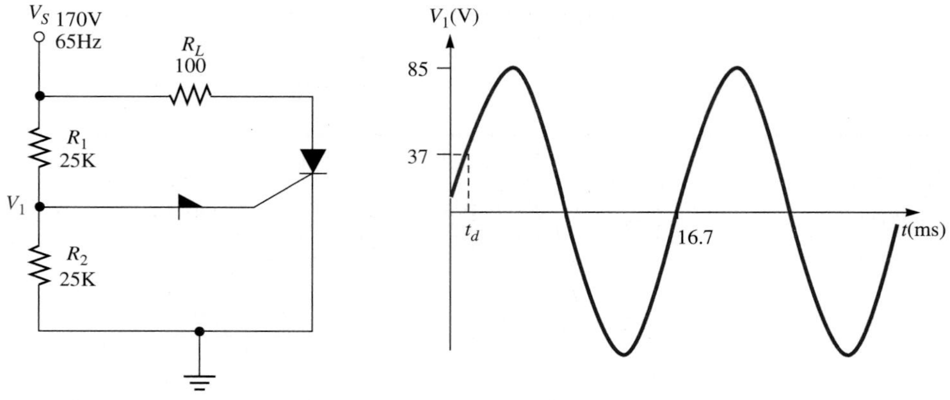

FIGURE 30–17 (A) (B)

(a) What voltage do you need at the anode of the four-layer diode to trigger the SCR (do not ignore the SCR turn-on voltage).

(b) Accurately sketch the voltage-divider voltage, and determine the delay time and conduction angle for this circuit.

(c) What is the DC value of the load voltage?

Solution

(a) Because we need enough voltage to trigger both the four-layer diode and the SCR gate, we need a total of

$$v_1 = 30 + 2 = 32\text{V}$$

(b) The voltage divider gives us

$$v_1 = \frac{R_2}{R_1 + R_2} v_S = \frac{25\text{K}}{25\text{K} + 25\text{K}} v_S = 0.5v_S = 85 \sin 2\pi 60t$$

In Figure 30–17B v_1 is plotted; it reaches 32V at the delay time:

$$85 \sin 2\pi 60t_d = 32$$

We find the delay time from (**Be sure your calculator is set to radian mode; $2\pi 60t$ is a radian angle.**)

$$t_d = \frac{\sin^{-1}(32/85)}{2\pi 60} = 1.02\text{ms}$$

The period of a 60Hz signal is

$$T = \frac{1}{f} = \frac{1}{60} = 16.7\text{ms}$$

The conduction time is

$$t_c = \frac{T}{2} - t_d = 8.33 - 1.02 = 7.31\text{ms}$$

The conduction angle is

$$\theta_c = \frac{360}{T} t_c = \frac{360}{16.7\text{ms}} 7.31\text{ms} = 158°$$

(c) The average voltage delivered to the load is found from

$$V_{L_{ave}} = \frac{V_p}{2\pi} (1 - \cos \theta_c) = \frac{170}{2\pi} (1 - \cos 158°) = 52.1\text{V}$$

Note: Because we use the conduction angle in degrees, you must be sure to reset your calculator to degree mode.

RC Time-Delay Circuit

We can get a larger range of conduction angles by using a capacitor in place of R_2 in the voltage divider (Figure 30–18). In Figure 30–18B we isolate the voltage divider and show the drive and capacitor voltages.

The capacitor voltage still has a smaller amplitude than the supply voltage. More importantly, the capacitor voltage is shifted to the right. That is, the capacitor introduces its own time delay; we can get time delays up to one quarter-cycle just from the capacitor. The RC circuit can provide total range of conduction angles from 0° to 180°.

The formula for the capacitor voltage is a bit involved:

$$v_c = A \sin(2\pi ft - \varphi)$$

where

$$A = \frac{V_S}{\sqrt{(2\pi fRC)^2 + 1}}$$

and

$$\varphi = \tan^{-1} 2\pi fRC$$

The time delay for this circuit is found by equating the capacitor voltage to the total voltage needed to trigger the SCR.

FIGURE 30–18

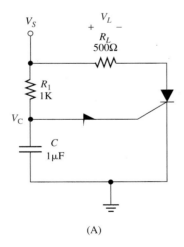

(A)

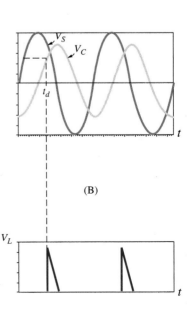

(B)

(C)

EXAMPLE 30–7
RC Triggering Circuit

The trigger circuit in Figure 30–18 (p. 1023) has these parameters: a 170V 60Hz supply,

$$V_{BR} = 30\text{V}, \ V_G = V_{GT} = 2\text{V}, \ R = 1\text{K}\Omega, \ C = 5\mu\text{F}.$$

(a) Find the delay time, conduction time, and conduction angle.
(b) Find the average voltage to the load.

Solution

(a) We find the capacitor voltage from

$$A = \frac{V_S}{\sqrt{(2\pi fRC)^2 + 1}} = \frac{170}{\sqrt{(2\pi 60 \times 1\text{KHz} \times 5\mu\text{F})^2 + 1}} = 79.7\text{V}$$

and

$$\varphi = \tan^{-1} 2\pi fRC = \tan^{-1} (2\pi 60 \times 1\text{K} \times 5\mu\text{F}) = 62.1° = 1.08\text{rad}$$

Note that we need the angle in radians to complete the math. The capacitor voltage is, therefore,

$$v_c = A \sin(2\pi ft - \varphi) = 79.7 \sin(2\pi 60t - 1.08)$$

The SCR will be triggered when the capacitor voltage reaches

$$v_c(t_d) = V_{BR} + V_{GT} = 32\text{V}$$

Therefore, we find the delay time from

$$79.7 \sin(2\pi 60 t_d - 1.08) = 32$$

which gives us

$$t_d = \frac{\sin^{-1} 32/79.7 + 1.08}{2\pi 60} = 3.96\text{ms}$$

The conduction time is

$$t_c = \frac{T}{2} - t_d = 8.33 - 3.96 = 4.34\text{ms}$$

The conduction angle is

$$\theta_c = \frac{360}{T} t_c = \frac{360}{16.7\text{ms}} 4.34\text{ms} = 93.6°$$

(b) The average load voltage is

$$V_{L_{ave}} = \frac{V_S}{2\pi} (1 - \cos \theta_c) = \frac{170}{2\pi} (1 - \cos 93.6°) = 28.8\text{V}$$

DRILL EXERCISE

Change the capacitor in the previous example to $C = 10\mu\text{F}$, and find A, φ, t_d, t_c, θ_c, and $V_{L_{ave}}$.

Answer $A = 43.6$, $\varphi = 1.31\text{rad}$, $t_d = 5\text{ms}$, $t_c = 3.33\text{ms}$, $\theta_c = 71.7°$, $V_{L_{ave}} = 18.6\text{V}$

1. How can AC pulses be used to control a DC motor?
2. What is the duty cycle of a pulse train?
3. What is the average value of a half-wave rectified signal?
4. Draw a 170V 60Hz sine wave. If a four-layer diode SCR circuit will trigger at 85V, indicate on your figure the delay time and the conduction time.
5. What is the conduction angle and how is it defined?
6. What is the relation between the conduction angle and the average, or DC, load voltage?
7. How does a voltage divider provide a time delay?
8. What is the range of conduction angles with a voltage divider trigger?
9. Draw an RC circuit. Draw the capacitor voltage if the input is a sine wave. How does the capacitor voltage compare to the voltage divider voltage?
10. What is the range of conduction angles for the RC trigger circuit?

30–5 ■ THE TRIAC

Figure 30–19 shows the TRIAC and its I-V characteristic. This device is essentially a bi-directional SCR. That is, it can pass current in either direction. For this reason we do not talk about an anode and a cathode in the TRIAC. Rather we label these terminals MT1 and MT2.

FIGURE 30–19
The TRIAC

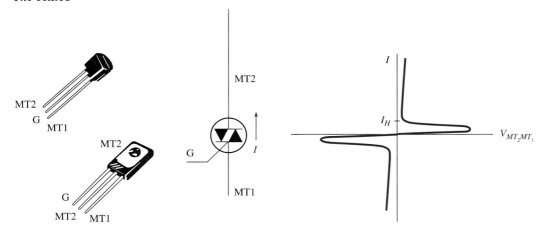

Interestingly, the TRIAC is triggered **ON** by either a positive or negative gate voltage. There may be a slight difference in the required gate voltage depending on whether MT2 is positive with respect to MT1 or vice versa.

Figure 30–20 (p. 1026) shows a data sheet that lists several commercial TRIACs. You can see the listings for the gate voltage and gate current for the four possible triggering arrangements:

V_{MT2} positive, V_G positive

V_{MT2} positive, V_G negative

V_{MT2} negative, V_G negative

V_{MT2} negative, V_G positive

Figure 30–21 (p. 1027) shows a TRIAC-DIAC voltage-divider triggering circuit. Note that because we use positive and negative gate voltages, we use a DIAC instead of a four-layer diode.

FIGURE 30–20

TRIACS

Case 29-02 TO-226AA (TO-92) Style 12			Case 77-04 Style 5			
					50V	
MAC97-3	MAC97A3	MAC97B3	T2322A	T2323A	100V	
MAC97-4	MAC97A4	MAC97B4	T2322B	T2323B	200V	
MAC97-5	MAC97A5	MAC97B5	T2322C	T2323C	300V	V_{DRM}
MAC97-6	MAC97A6	MAC97B6	T2322D	T2323D	400V	
MAC97-7	MAC97A7	MAC97B7	T2322E	T2323E	500V	
MAC97-8	MAC97A8	MAC97B8	T2322M	T2323M	600V	
					700V	
					800V	
8			25		I_{TSM} (Amps)	
10 10 10 10	5 5 5 7	3 3 3 5	10 10 10 10	25 40 25 40	I_{GT}@25°C (mA) MT2(+)G(+) MT2(+)G(−) MT2(−)G(−) MT2(−)G(+)	
2 2 2 2.5			2.2 2.2 2.2 2.2		V_{GT}@25°C (V) MT2(+)G(+) MT2(+)G(−) MT2(−)G(−) MT2(−)G(+)	
−40 to −110					T_J Operating Range (°C)	

On the positive half-cycle, a positive voltage is applied to MT2 of the TRIAC. A positive voltage is also applied to the DIAC. When the breakover voltage is reached, the DIAC fires, triggering the TRIAC.

On the negative half-cycle, a negative voltage is applied to MT2 of the TRIAC. A negative voltage is applied to the DIAC. When the breakover voltage is reached, the DIAC fires, triggering the TRIAC. We get conduction during both the positive and negative halves of the input cycle.

Because of the symmetry of the TRIAC output, the DC value of the load voltage in the TRIAC circuit is 0V. This doesn't make the TRIAC very useful for driving DC loads. AC loads, on the other hand, are driven by the RMS value of the signal. For example, the input sine wave has a DC value of 0V, but an RMS value of $V_p/\sqrt{2}$. As the conduction angle decreases, less of the sinusoid gets to the load, and the RMS value of the load voltage (and current) decreases.

FIGURE 30–21

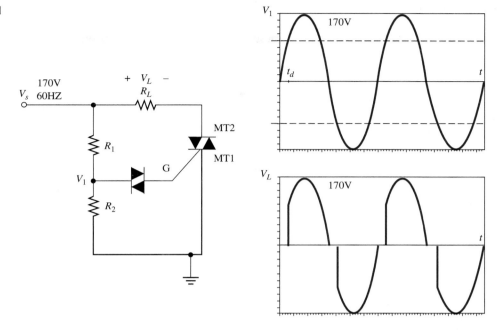

1. Besides the gate, how are the other two terminals of the TRIAC labeled?
2. Draw the TRIAC I-V characteristic. How does this curve compare to the SCR I-V characteristic?
3. For the SCR, the gate voltage must be positive. Is this true for the TRIAC?
4. Why do we usually use DIACs rather than four-layer diodes to trigger a TRIAC?
5. What is the DC value of a TRIAC output?
6. What other parameter is important in AC voltages and currents? How is this parameter affected by the conduction angle of a TRIAC circuit?

30–6 ■ THE UNIJUNCTION TRANSISTOR (UJT)

UJT Basics

Figure 30–22A (p. 1028) shows the **unijunction transistor** (UJT). There is an old saying that if it looks like a duck and sounds like a duck, it's a duck. In this case, however, the UJT looks like a transistor and sounds like a transistor, but it is not really a transistor. The UJT is a controlled switch, similar to the other devices in this chapter.

Also, the unijunction transistor is not technically a thyristor; however, it is often used to trigger the thyristors we have discussed. Note that the UJT is a PN device with three terminals; the P-type material forms the emitter, and the terminals connected to the N-type material form base 1 (B1) and base 2 (B2). The I-V characteristic for the UJT, shown in Figure 30–22B, (p. 1028), is the relation between the emitter current (I_E) and the emitter voltage (V_E). (Just to add a little confusion here, we note that most manufacturer's present the UJT characteristics curve by plotting V vs. I, as shown in Figure 30–22C, p. 1028.)

When the emitter voltage (V_E) of the UJT reaches V_p, the device is triggered **ON,** and there is a rush of emitter current that causes the emitter voltage to fall to near 0V. The peak voltage of the UJT is analogous to the breakover voltage of the four-layer diode.

An advantage of the UJT is that the trigger voltage (V_p) depends on the voltage between the two bases (V_{B2B1}), as shown in the family of V-I curves in Figure 30–22D (p. 1028).

FIGURE 30–22

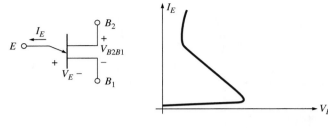

(A)

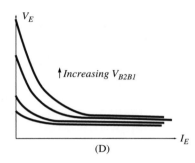

(B)   (C)

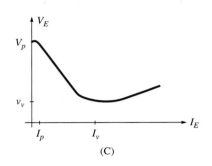

(D)

The circuit model for the UJT (Figure 30–23A) can help you understand the behavior and use of the UJT. The diode represents the PN junction. The total base resistance is split into two resistors: r_{B1} between B1 and $\hat{E}$, r_{B2} between $\hat{E}$ and B2. The total base 1–to–base 2 resistance is

$$r_{BB} = r_{B1} + r_{B2}$$

To turn the diode **ON** (to trigger the UJT), the emitter voltage must reach

$$V_E = 0.7 + V_{\hat{E}}$$

We find $V_{\hat{E}}$ from the voltage divider:

$$V_{\hat{E}} = \frac{r_{B1}}{r_{B1} + r_{B2}}V_{B_2B_1} = \frac{r_{B1}}{r_{BB}}V_{B_2B_1}$$

FIGURE 30–23

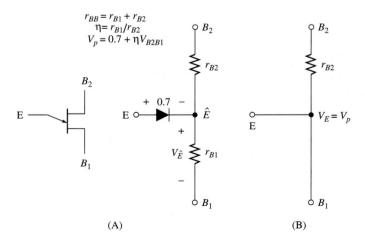

(A)   (B)

So, to trigger the UJT we need

$$V_E = V_p = 0.7 + \frac{r_{B1}}{r_{BB}} V_{B_2B_1}$$

You can see the dependence of V_p on the base 2–to–base 1 voltage. We define the UJT parameter known as the **intrinsic standoff ratio** (η) as

$$\eta = \frac{r_{B1}}{r_{B1} + r_{B2}} = \frac{r_{B1}}{r_{BB}}$$

We can then write the formula for V_p as

$$V_p = 0.7 + \eta V_{B_2B_1}$$

From a circuit's point of view, when the emitter voltage reaches V_p, the UJT triggers, and r_{B1} falls to near 0Ω (Figure 30–23B).

Figure 30–24 shows the data sheet for a commercial UJT. This device has typical intrinsic standoff ratio of 0.65 and a base 1–to–base 2 resistance of 7KΩ.

FIGURE 30–24
UJT data sheet

absolute maximum ratings: (25°C)

Power Dissipation	300 mw
RMS Emitter Current	50 mA
Peak Emitter Current	2 amperes
Emitter Reverse Voltage	30 volts
Interbase Voltage	35 volts
Operating Temperature Range	–65°C to +125°C
Storage Temperature Range	–65°C to +150°C

electrical characteristics: (25°C)

		Min.	Typ.	Max.
Intrinsic Standoff Ratio		0.56	0.65	
(V_{BB} = 10 V)	η	0.56	0.65	0.75
Interbase Resistance (kΩ)				
(V_{BB} = 3 V, I_E = 0)	R_{BB}	4.7	7	9.1
Emitter Saturation Voltage				
(V_{BB} = 10 V, I_E = 50 mA)	$V_{E(SAT)}$		2	
Emitter Reverse Current				
(V_{BB} = 30 V, I_{B1} = 0)	I_{EO}		0.05	12
Peak Point Emitter Current	$I_P(\mu A)$		0.04	5
(V_{BB} = 25 V)				
Valley Point Current				
(V_{BB} = 20 V)	I_V(mA)	4	6	

EXAMPLE 30–8
UJT V_p
Determination

Figure 30–25A (p. 1030) shows a common UJT circuit. If the UJT has the typical parameters given in Figure 30–24

(a) Find r_{B1} and r_{B2}.   **(b)** Find $V_{B_2B_1}$ and V_p.

Solution
(a) From the data given in Figure 30–24

$$\eta = 0.65 \quad \text{and} \quad r_{BB} = 7K\Omega$$

From the formula for η

$$\eta = \frac{r_{B1}}{r_{BB}}$$

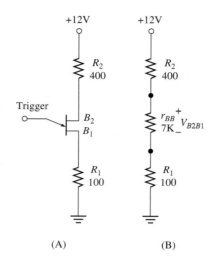

FIGURE 30–25   (A)   (B)

so

$$r_{B1} = \eta r_{BB} = 0.65 \times 7\text{K} = 4.55\text{K}\Omega$$

We find r_{B2} from

$$r_{BB} = r_{B1} + r_{B2} \rightarrow r_{B2} = r_{BB} - r_{B1} = 7\text{K} - 4.55\text{K} = 2.45\text{K}\Omega$$

Note: When the UJT fires, r_{B1} drops to approximately 100Ω.

(b) To find $V_{B_2B_1}$, we use the UJT circuit model in Figure 30.25B. Because the UJT is **OFF,** there is no emitter current. Therefore, $V_{B_2B_1}$ is found from the voltage divider:

$$V_{B_2B_1} = \frac{r_{BB}}{R_1 + R_2 + r_{BB}} V_s = \frac{7\text{K}}{0.1\text{K} + 0.4\text{K} + 7\text{K}} 12 = 11.2$$

The emitter trigger voltage is found from

$$V_p = 0.7 + \eta V_{B_2B_1} = 0.7 + 0.65 \times 11.2 = 7.98\text{V}$$

DRILL EXERCISE   Repeat the previous example if the UJT has $\eta = 0.75$.

Answer $r_{B1} = 5.25\text{K}\Omega$, $r_{B2} = 1.75\text{K}\Omega$, $V_{B_2B_1} = 11.2\text{V}$, $V_p = 9.1\text{V}$

The UJT is most often used as a relaxation oscillator (Figure 30–26). After V_p is set by V_S, R_1, and R_2, the circuit works as follows:

☐ The capacitor voltage charges towards V_S.

☐ When the capacitor voltage reaches V_p, the UJT triggers **ON.**

☐ When the UJT is triggered, the resistance between the emitter and base 1 drops to a low value, approximately 100Ω. This provides a low-resistance path through R_1 to discharge the capacitor.

☐ When the capacitor is discharged, the UJT turns **OFF.** Then the process repeats.

The sawtooth-like capacitor voltage is shown in Figure 30–26B. Also shown are the voltage waveforms V_{B1} and V_{B2}. When the UJT is triggered, there is a surge of emitter current. This creates a similar surge in the base 1–to–base 2 current. This second surge creates a positive-going voltage spike across R_1. Because V_{B1} is the voltage across this resistor, there is a positive-going spike in V_{B1}.

The surge current also creates a spike in the voltage across R_2. Since $V_{B2} = V_s - V_{R_2}$, when the voltage across R_2 increases, the voltage at V_{B2} decreases. This gives us the

FIGURE 30–26
UJT relaxation oscillator

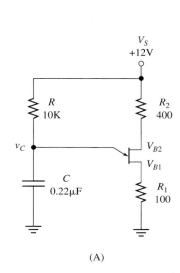

(A)

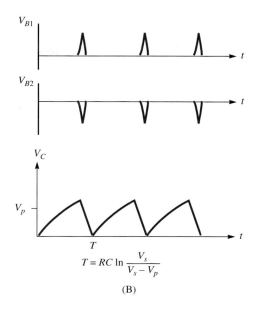

$$T = RC \ln \frac{V_s}{V_s - V_p}$$

(B)

negative-going spike in V_{B2} seen in Figure 30–25B. The voltage spikes at the two bases are often used to trigger other thyristors, such as the SCR and TRIAC.

The period of the pulse trains seen in Figure 30–26B is given by

$$T = RC\ln \frac{V_S}{V_S - V_P}$$

EXAMPLE 30–9
UJT Relaxation Oscillator

Assume that the UJT, the supply voltage, and R_1 and R_2 in Figure 30–26 have the same values and UJT parameters as the circuit in Example 30–8. Find the pulse train frequency if $R = 10K\Omega$ and $C = 0.22\mu F$.

Solution From Example 30–8 we know that the peak voltage for this circuit is

$$V_p = 8V$$

Therefore, the pulse train period is

$$T = RC \ln \frac{V_S}{V_S - V_p} = 10K \times 0.22\mu F \ln \frac{12}{12 - 8} = 2.42ms$$

The pulse train frequency is

$$f = \frac{1}{T} = \frac{1}{2.4ms} = 414Hz$$

The Programmable Unijunction Transistor (PUT)

A disadvantage of the UJT is that the standoff value is intrinsic to the device. You have to live with what the manufacturers supply. The programmable unijunction transistor (see Figure 30–27A, p. 1032) is a device that allows the user to set η.

Although the circuit symbol of the device might lead you to think that we trigger the PUT from the gate (G), this isn't so. We actually trigger the PUT from the anode. The voltage at the gate sets η.

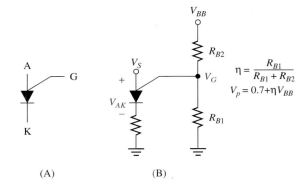

FIGURE 30–27
*Programmable
unijunction transistor
(PUT)*

$$\eta = \frac{R_{B1}}{R_{B1} + R_{B2}}$$

$$V_p = 0.7 + \eta V_{BB}$$

(A)   (B)

Figure 30–27B shows the basic PUT circuitry. The voltage at the gate is set by the voltage divider:

$$V_G = \frac{R_{B_1}}{R_{B_1} + R_{B_2}} V_{BB}$$

We define the standoff ratio (η) for the PUT as

$$\eta = \frac{R_{B_1}}{R_{B_1} + R_{B_2}}$$

The standoff ratio for the PUT is set by external resistors that are chosen by the designer.

The PUT will fire when the anode-to-cathode voltage (V_{AK}) equals the peak voltage (V_p), where

$$V_{AK} = V_p = 0.7 + \eta \, V_{BB}$$

While the PUT offers more triggering options than the UJT, the UJT can handle larger currents.

**REVIEW
QUESTIONS**

1. Draw the circuit symbol for the UJT. Draw the I-V curve for the UJT (both I vs. V and V vs. I).

2. When does the UJT fire?

3. What happens when the UJT fires?

4. On what does the emitter trigger voltage (V_p) depend?

5. What is the intrinsic standoff ratio, and how is it defined?

6. Draw the circuit model for the UJT.

7. If each base lead is connected to an external resistor, how do you find $V_{B_2B_1}$?

8. Draw the basic UJT relaxation oscillator and describe its operation.

9. What is the formula for the period of the relaxation oscillator output?

10. Draw the symbol of the PUT, and describe how it is triggered.

11. Compare the standoff ratios of the UJT and the PUT.

30–7 ■ ADDITIONAL THYRISTORS

There are a number of additional thyristors used in switching circuits. Rather than give any details or circuit examples for these devices (details that are available in thyristor handbooks) we will just show their circuit symbols and list their turn-on and turn-off properties.

The Gate Turn-Off Switch (GTO)

The gate turn-off switch (GTO) is shown in Figure 30–28. The GTO is similar to the SCR except that it can be turned on or off from the gate. The ability to turn off the GTO from the gate makes for simpler circuits than for the SCR. You do not need to provide for current diversion or forced commutation. A disadvantage of the GTO is that it requires much greater currents to turn the device on and off than the SCR.

FIGURE 30–28
The gate turn-off switch (GTO)

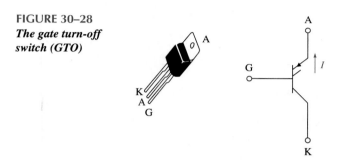

The Silicon Bilateral Switch (SBS)

The silicon bilateral switch (SBS) is shown in Figure 30–29A. It can be thought of as a gate-triggered DIAC. When the gate terminal is not used, the SBS behaves the same as the DIAC. The SBS turns on when the breakover voltage is exceeded; the breakover voltage of the SBS is usually less than 10V, smaller than the breakover voltage of DIACs.

The extra flexibility you get with the SBS is that you can change the breakover voltage with an external gate-to-anode voltage. If, for example, you create a gate-to-anode 1 voltage with a Zener diode, the breakover voltage for positive $V_{A_2 A_1}$ is given by $0.7 + V_Z$. The breakover voltage in the other direction does not change (Figure 30–29B).

FIGURE 30–29
Silicon bilateral switch (SBS)

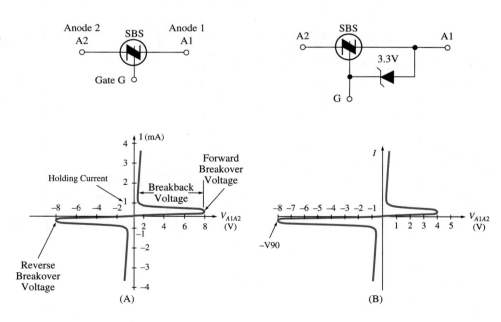

The Silicon-Controlled Switch (SCS)

The silicon-controlled switch (SCS), shown in Figure 30–30 (p. 1034), has two gates, the cathode gate and the anode gate. The SCS can be turned on or off from either gate. It is turned on either by a positive pulse at the cathode gate or by a negative pulse at the anode gate. The SCS is turned off by either a negative pulse at the cathode gate or by a positive pulse at the anode gate.

FIGURE 30–30
The silicon-controlled
switch (SCS)

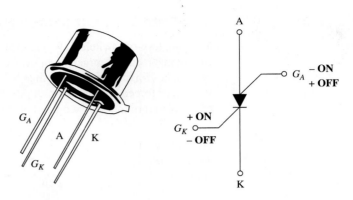

Thyristor Optoisolators

The SCR and the TRIAC are also available as optoisolators. In these devices, the gate drive is provided by an LED instead of a voltage (Figure 30–31).

FIGURE 30–31
The thyristor
optoisolator

 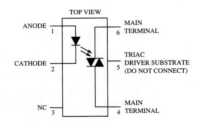

SUMMARY

We present this summary in tabular form. The table lists each device discussed in this chapter, its circuit symbol, and its on-off characteristics. The breakover voltage is indicated by V_{BR}.

DEVICE	SYMBOL	ON TRIGGER	OFF TRIGGER
Four-Layer Diode		$V_{AK} = V_{BR}$	Interrupt or divert current. Forced commutation.
DIAC		$V_{AK} = V_{BR}$ $V_{KA} = V_{BR}$	See four-layer diode.
SCR		$V_G = V_{GT}$	See four-layer diode.
TRIAC		$V_{MT2} > 0, V_G > 0$ $V_{MT2} > 0, V_G < 0$ $V_{MT2} < 0, V_G < 0$ $V_{MT2} < 0, V_G > 0$	See four-layer diode.
UJT		$V_E = 0.7 + \eta V_{B_2B_1}$	See four-layer diode.
PUT		$V_{AK} = 0.7 + \eta V_{BB}$	See four-layer diode.
GTO		$V_G > 0$	$V_G < 0$
SBS		$V_{A_1A_2} > 0, V_{GA_1} > 0$ $V_{A_2A_1} > 0, V_{GA_2} > 0$	See four-layer diode.
SCS		$V_{GC} > 0$ $V_{GA} < 0$	$V_{GC} < 0$ $V_{GA} > 0$
Optoisolators		*LED*	See four-layer diode.

| **PROBLEMS** | **SECTION 30–2 The Four-Layer Diode and the DIAC** |

1. Explain the on-off behavior of the four-layer diode.
2. Derive the pulse period formula given for the relaxation oscillator in Figure 30–4 (p. 1012).
3. For the circuit of Figure 30–4 (p. 1012): $V_S = 12V$, $V_{(BR)} = 7V$, $R = 2K\Omega$, $C = 2.2\mu F$. Find the pulse period and pulse train frequency.
4. Explain the behavior of the DIAC.

SECTION 30–3 The Silicon-Controlled Rectifier (SCR)

5. Explain the behavior of the SCR.
6. Describe the behavior of the SCR circuit in Figure 30–32.

FIGURE 30–32

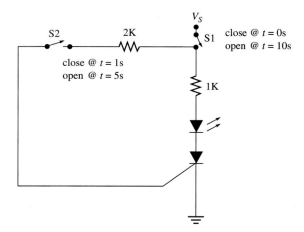

7. Figure 30–33 shows the gate voltage for the SCR of Figure 30–32 (Both switches are closed; the input is an AC voltage). If the voltage required to trigger the SCR is 10V
   (a) Find the delay time, the conduction time, and the conduction angle.
   (b) If the input is a 170V 60Hz sine wave, draw the load voltage, and find its average value.

FIGURE 30–33

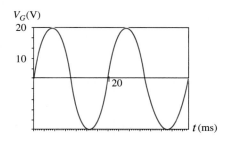

8. Find the delay time for the SCR circuit in Figure 30–34 (p. 1036).

$$V_{(BR)} = 15V \quad \text{and} \quad V_{GT} = 2V.$$

9. Find the delay time for the SCR circuit in Figure 30–35 (p. 1036). Use the data from the previous problem.

FIGURE 30–34

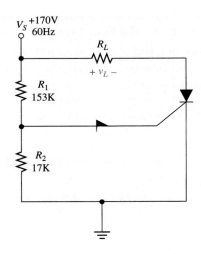

FIGURE 30–35

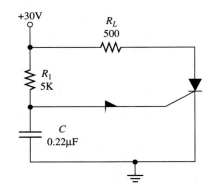

SECTION 30–4 AC SCR Applications

10. Draw a pulse train that has a magnitude of 50V, a period of 16.7ms, and a pulse width of 5ms.

   (a) What is the duty cycle of the pulse train?

   (b) What is the average voltage of the pulse train?

11. What pulse width would you need in the pulse train of Problem 10 to obtain an average voltage of 35V?

12. Figure 30–36 shows a sinusoidal pulse train that is the output of an SCR circuit.

   (a) What is the delay time?   **(b)** What is the conduction angle?

   (c) What is the average voltage?

FIGURE 30–36

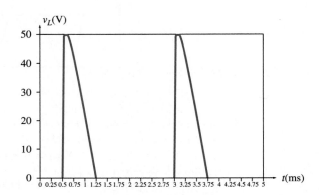

13. The SCR circuit in Figure 30–37 has a 100V 60Hz input signal. The breakover voltage for the four-layer diode is 20V. The gate turn-on voltage for the SCR is 2V.

(a) Find and sketch the voltage divider voltage (v_1).

(b) On the same plot, draw a line representing the total trigger voltage for this circuit.

(c) What is the delay time and conduction angle for this circuit?

(d) Sketch the load voltage.   (e) What is the average load voltage?

FIGURE 30–37

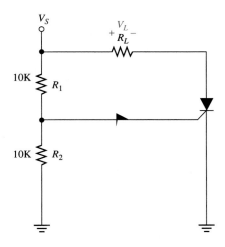

14. The SCR circuit in Figure 30–38 has a 50V 60Hz input signal. The breakover voltage for the four-layer diode is 20V. The gate turn-on voltage for the SCR is 2V.

(a) Find and sketch the capacitor voltage.

(b) On the same plot, draw a line representing the total trigger voltage for this circuit.

(c) What is the delay time and conduction angle for this circuit?

(d) Sketch the load voltage.   (e) What is the average load voltage?

FIGURE 30–38

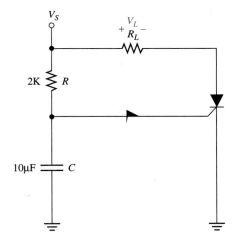

SECTION 30–5   The TRIAC

15. Explain the on-off behavior of the TRIAC.

16. The TRIAC circuit of Figure 30–39 (p. 1038) has an input signal of 20V; the gate turn-on voltage is 2V.

(a) Sketch the voltage divider voltage (v_1).

(b) On the same plot draw the line representing the trigger voltage for this circuit.

(c) What is the delay time and conduction angle for this circuit?

(d) Sketch the load voltage.

FIGURE 30–39

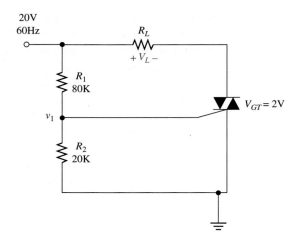

17. Redraw the TRIAC circuit of Figure 30–39, inserting a DIAC between the voltage divider and the TRIAC. Assume the DIAC has a breakover voltage of 10V, and repeat Problem 16.

SECTION 30–6 The Unijunction Transistor (UJT)

18. Describe the on-off behavior of the UJT.

19. A UJT has the following parameters: $\eta = 0.5$ and $r_{BB} = 5K\Omega$. Find r_{B1} and r_{B2}.

20. The UJT in the circuit in Figure 30–25 (p. 1030) has the parameters given in the previous problem.
 (a) Find V_{B2B1} and V_p. **(b)** Find the pulse train period and frequency.

21. The UJT in Figure 30–40 has the parameters $r_{BB} = 6K\Omega$ and $\eta = 0.6$.
 (a) Find r_{B1} and r_{B2}. **(b)** Find V_{B2B1} and V_p.
 (c) Find the pulse train period and frequency.

FIGURE 30–40

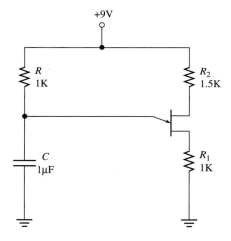

22. Explain the on-off behavior of the PUT.

23. For the PUT circuit of Figure 30–41, find η and V_p.

SECTION 30–7 Additional Thyristors

24. Figure 30–42 shows a number of thyristors and related devices. Match the device with the device name below:

(a) TRIAC	**(b)** Four-layer diode	**(c)** SCR	**(d)** DIAC	**(e)** PUT
(f) GTO	**(g)** UJT	**(h)** SCS	**(i)** Opto-TRIAC	**(j)** SBS

FIGURE 30–41

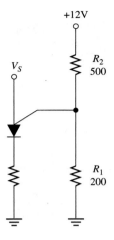

FIGURE 30–42

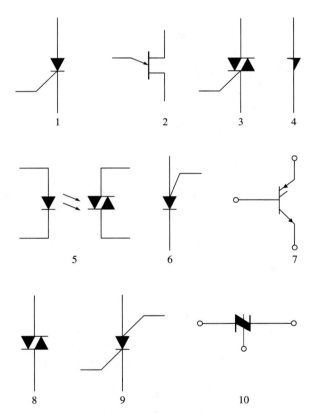

PN–JUNCTION PHYSICS

■ INTRINSIC SILICON

You remember from a physics or a chemistry class that an atom is composed of a positively-charged nucleus surrounded by several shells of negatively-charged electrons (Figure A1–A). Electrical and nuclear forces attempt to keep an atom's electrons tightly bound to the atom. These attractive forces have the weakest hold on the electrons in the outer, or **valence,** shell.

At room temperatures the electrons in the valence shell of conductors are bound very weakly to the nucleus. They break free easily and, so, are available to move and create current (Figure A1–B). In insulators, just the reverse is true. The valence electrons are so strongly bound to the nucleus that at normal energy levels they cannot break free.

FIGURE A–1

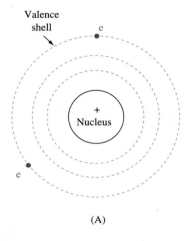

(A)

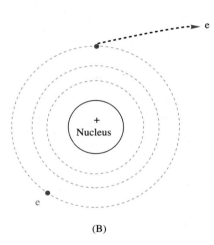

(B)

A **semiconductor** is a material that lies somewhere between the conductors and the insulators. At room temperature these materials have very few free electrons. However, a slight increase in the applied energy level creates a significant increase in the number of free electrons. This additional energy can come from an electrical source, from heat, from light, and even from mechanical stress.

The atoms that make the semiconductor materials are all located in the IVth column of the Periodic Table (Figure A2–A). That is, semiconductor atoms have four electrons in their valence shell (Figure A2–B). This is exactly half of the number of electrons needed to fill the shell. The most commonly used atom in this group is silicon (Si). Early semiconductor technology used germanium (Ge).

Semiconductor construction begins with the fabrication of a silicon wafer. We assume that the wafer is pure silicon. The silicon atoms in the wafer arrange themselves in a lattice structure, as the cross-section in Figure A–3 shows. Atoms like to completely fill their valence shells. Because each silicon atom has four valence electrons, four more electrons are needed to complete the valence shell. By sharing its electrons with four neighboring silicon atoms, each silicon atom can complete its shell. This sharing of electrons is known as **covalent bonding.**

FIGURE A–2

Periodic Table

III	IV	V
5 B	6 C	7 N
13 Al	14 Si	15 P
31 Ga	32 Ge	33 As
49 In	50 Sn	51 Sb

(A)

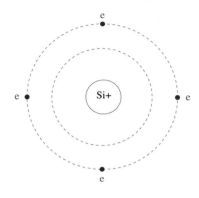

(B)

FIGURE A–3

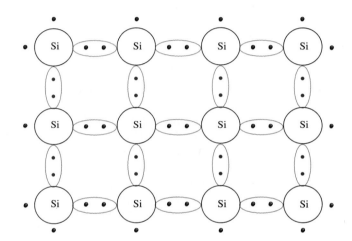

Covalent bonds are very strong and are hard to break. At low temperatures, therefore, most of the electrons in the silicon atom are tightly bound to the nucleus. There are very few free electrons, and the wafer is essentially a non-conductor.

At room temperature, a small number of the covalent bonds break. This happens because heat can give an electron enough energy to break away from its atom. Heat can create enough free electrons to turn the silicon into a weak conductor.

A silicon atom with four valence electrons is electrically neutral; that is, the atom has the same number of electrons and protons. Therefore, when an electron frees itself from a silicon atom, the atom is left with a net positive charge (Figure A4–A). The atom can—and often does—attract an electron from another atom. That is what is happening in Figure A4–B, in which the positive charge on the first atom is canceled by attracting an electron from the second atom. The movement of electrons from one atom to another creates a current of negative charges in the semiconductor.

Now the second atom has a positive charge. In effect, the positive charge has moved from the first atom to the second atom. Because the positive charge can move from one silicon atom to another, current in a semiconductor can also be created by mobile positive charges. These mobile positive charges are known as **holes** (h). It is the presence of both holes and electrons that give semiconductors their useful properties.

What are the electrical properties of a piece of silicon wafer sitting on the table top? If you shrink down to molecular size and enter the chip, you will see a very chaotic situation. A covalent bond breaks here, creating a free electron and a hole. Over there, a hole captures a free electron. The captured free electron returns a silicon atom to the neutral

FIGURE A–4

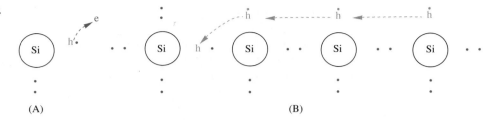

(A)   (B)

state, eliminating the hole. This gives the appearance of both electrons and holes moving randomly about the wafer. When an electron and a hole meet and cancel each other, we say they have **recombined.**

Your tiny self might also see that a small region of the chip is electrically negative, positive, or neutral. This is due to random movement of the free electrons and mobile holes. However, your normal-sized self, looking at the entire wafer, would see the wafer as electrically neutral. In intrinsic (pure) silicon, there are exactly as many free electrons as holes. This property has a very fancy name—**space charge neutrality.**

■ DOPING

Intrinsic silicon has the same number of mobile holes and electrons. We can add certain atoms to pure silicon to increase the number of mobile holes *or* electrons. This process is known as **doping** and the added atoms are known as **impurities.** Don't let the term *impurities* mislead you. The desired atoms are added in precisely controlled amounts.

The two most common impurity atoms are boron (B) and phosphorous (P). The most common method for doping is **diffusion.** In this process, a gas containing the desired impurity atom is passed over pure silicon in a heated chamber. The impurity atom diffuses from the gas into the silicon.

A second technique, known as **ion implantation,** is also used to dope silicon. In this method, a charged ion of the desired impurity atom is accelerated by an applied electric field and actually shot into the silicon.

N-Type Material

Let's see what happens when we diffuse phosphorous atoms into intrinsic silicon. Figure A–5 shows that the diffusion process results in the phosphorous atoms replacing some silicon atoms in the lattice structure. Now, phosphorous atoms have five electrons in the valence shell (Column 5, Periodic Table). Because only four of these electrons are needed to fill the covalent bonds with the four neighboring silicon atoms, there is one electron left over that is free to move.

FIGURE A–5

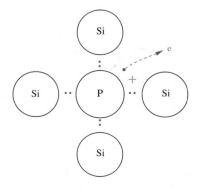

Note that when the free electron moves away from the phosphorous atom, a net positive charge, known as a positive **ion,** is left behind. This positive charge on the phosphorous ion remains with the ion and is not free to move about. Because a hole is a *mobile* positive charge, adding a phosphorous atom to silicon increases the density of free electrons but does not change the density of holes.

Our piece of silicon now has many more free electrons than free holes. For this reason, silicon doped with phosphorous (or any other Column 5 atom) is labeled **N-type material.** Because phosphorous donates a free electron, it is known as a **donor impurity.**

Because the number of free electrons in N-type material is much greater than the number of holes (due only to the intrinsic silicon), we say that electrons in N-type material are the **majority carriers.** Although small in number, the holes can still contribute to current and are known as the **minority carriers** in N-type material.

Usually, the concentration of electrons due to the donor atom is much greater than the concentration of electrons due to the intrinsic silicon. Likewise, the total positive charge is dominated by the immobile positive charge on the donor atom. Therefore, to get a basic understanding of semiconductor physics, we can ignore the minority carrier. We show this schematically in Figure A-6, where N-type material is represented with the immobile +charge and the free electron contributed by the donor atom.

FIGURE A–6

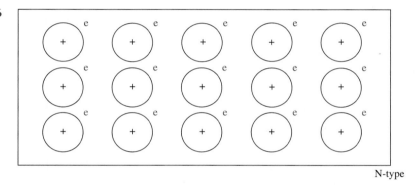

N-type

P-Type Material

Let's now diffuse boron (Column 3, Periodic Table) atoms into intrinsic silicon. The boron atoms will replace some silicon atoms in the lattice structure (Figure A7–A). Because boron atoms have only three electrons in the valence shell, it can only complete the covalent bonds of three of its neighboring silicon atoms. To complete the fourth covalent bond, an electron is stripped from a silicon atom (boron has a greater attraction to the electron than silicon).

FIGURE A–7

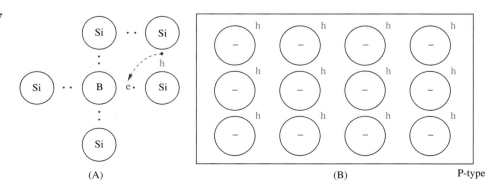

(A)   (B)   P-type

When the electron is stripped from the silicon atom, it leaves behind a positive hole. The electron attracted to the boron atom stays where it is, creating an immobile negative ion. Silicon doped with boron (or any other Column 3 atom) has many more holes than mobile electrons and is labeled **P-type material.** Because the boron receives an electron from a neighboring silicon atom, it is known as an **acceptor impurity.** In P-type material the holes are the majority carriers and the electrons are the minority carriers.

Usually, the concentration of holes due to the donor atom is much greater than the concentration of holes due to the intrinsic silicon. Likewise, the total negative charge is dominated by the immobile negative charge on the acceptor atom. We show this schematically in Figure A7–B, where P-type material is represented with the immobile −charges and the holes contributed by the acceptor atom.

■ THE PN JUNCTION

The semiconductor electronics industry was born when P-type material was joined to N-type material to form a **PN junction** (Figure A–8).

FIGURE A–8

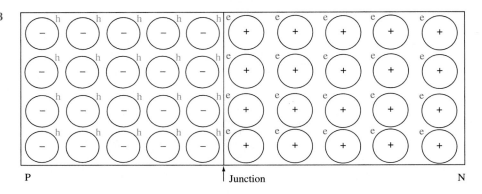

P   Junction   N

Let's first look at what happens to the electrons in the N-type material. There is a high concentration of electrons in the N-type material and a low concentration of electrons in the P-type material. Because of the concentration difference, electrons will begin to diffuse from the N-type into the P-type material (Figure A9–A).

At the same time, the concentration difference in the holes in the P- and N-type materials causes holes to begin to diffuse from the P-type into the N-type material. This initial migration of electrons and holes results in the situation shown in Figure A9–B, the small region surrounding the junction is now composed only of immobile ions.

Now, if electrons (or holes) were not electrically charged, they would keep moving until the electron (or hole) concentration was the same on both sides of the junction. However, electrons and holes have an electrical charge. As a result, when electrons move to the left, immobile positive ions are left behind so the N side of the junction becomes positively charged. As holes move to the right, immobile negative ions are left behind so the P side of the junction becomes negatively charged. This creates a voltage across the junction, known as the **junction potential** (V_γ) (Figure A9–B).

The positive charge on the N side acts to prevent more holes from crossing the junction. The negative charge on the P side acts to prevent more electrons from crossing the junction. An equilibrium is quickly reached in which only a limited number of holes and electrons cross the junction. This results in a finite region surrounding the junction where there are negative ions on P side and positive ions on the N side. This region has been depleted of mobile holes and electrons and is known as the **depletion region.**

In most silicon PN junctions, the equilibrium point is reached when the voltage across the PN junction is approximately $V_\gamma \approx 0.7V$. For germanium, the junction potential is lower ($V_\gamma \approx 0.2V$). Let's again shrink to molecular size and examine the junction potential with a tiny voltmeter. We place the −lead of the voltmeter at the left edge of the depletion region

FIGURE A–9

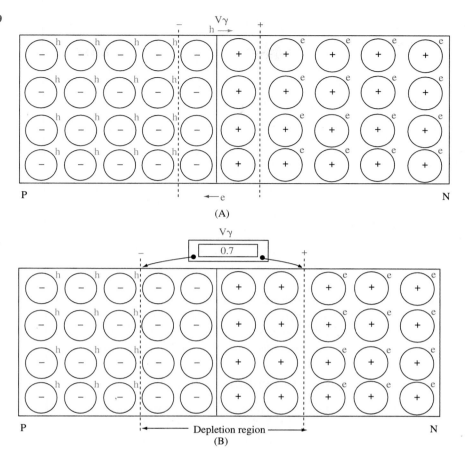

(A)

(B)

and use the +lead to measure the voltage along an axis that begins at the left edge of the depletion region and ends at the right edge. We find that the potential grows from 0V to approximately 0.7V. This potential difference is known as the **potential barrier** (Figure A–10).

Note that the width of the depletion region is also shown on the voltage plot.

FIGURE A–10

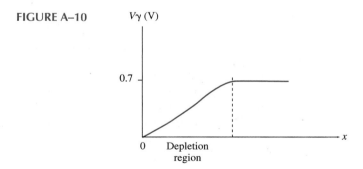

The Reverse-Biased PN Junction

In Figure A–11A we have connected a battery to the PN junction so that a positive voltage is applied to the N side and a negative voltage is applied to the P side. The positive voltage will draw electrons from the N material, creating more immobile positive ions at the junction. Likewise, the negative voltage on the N side will draw holes from the P material. Actually, electrons are injected into the P material where they recombine with holes. This has the same effect as drawing holes out of the wafer. As the reverse-bias potential increases, the depletion region grows and the potential barrier increases (Figure A11–B).

FIGURE A–11

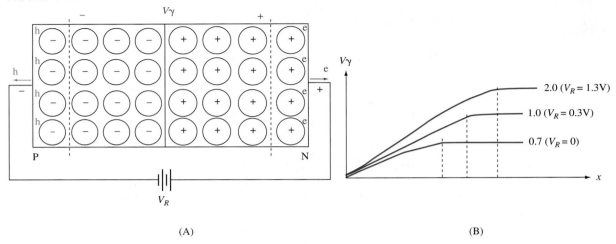

(A)   (B)

Is there any significant current in the reversed-biased PN junction? The answer is no. Because there are no mobile charges in the depletion region, it is impossible for a charge to leave the battery, cross the PN junction, and return to the battery. There is a small leakage current that is due to minority carriers finding their way across the depletion region.

The Forward-Biased PN Junction

We now apply a positive voltage to the P material and a negative voltage to the N material (Figure A–12A). The positive voltage injects holes into (actually, draws electrons from) the P material, where they migrate towards the junction. Likewise, electrons are injected into the N material and migrate to the junction. This results in a reduction in the number of immobile ions surrounding the junction. The depletion region narrows and the potential barrier is decreased (Figure A–12B).

FIGURE A–12

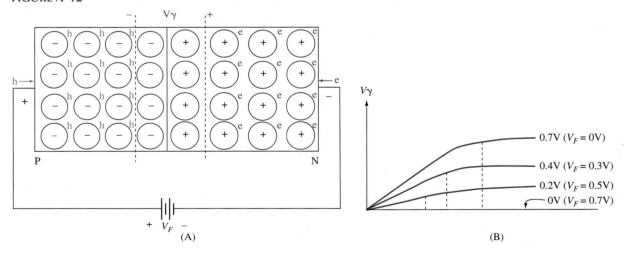

(A)   (B)

Increasing the voltage further reduces the depletion region and the potential barrier. When the applied voltage equals the junction potential (≈ 0.7V) the depletion region and the potential barrier disappear. At this applied voltage, the PN junction becomes a conductor and we get a measurable current.

BJT PHYSICS

Bipolar Junction Transistors (BJTs) are constructed by sandwiching P-type material between 2 N-type materials (NPN), or by sandwiching N-type material between 2 P-type materials (PNP). Because the PNP transistor exhibits the same physical behavior as the NPN, we will just discuss the behavior of the NPN transistor, shown schematically in Figure B–1.

Proper operation of the transistor requires that the base is lightly doped; that is, there is a low concentration of holes in the base. The emitter is heavily doped, so it has a high concentration of electrons. The collector is not as heavily doped as the emitter.

■ REVISITING THE REVERSE-BIASED PN JUNCTION

FIGURE B–1

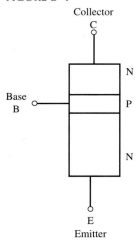

Let's first revisit the reverse-biased PN junction. The majority carrier in N-type material is the negatively-charged electron. The majority carrier in P-type material is the positively-charged hole. In reversed-biased PN junctions, the depletion region (or potential barrier) prevents majority current flow across the junction. That is, electrons stay in the N region and holes stay in the P region.

Consider now the minority carriers. The minority carriers in N-type material are the holes. The minority carrier in the P-type material are the electrons. In a forward-biased PN junction, we can ignore the minority carriers. Because the current due to majority carriers is zero in a reversed-biased PN junction, any current in the reversed-biased PN junction, must come from the minority carriers.

In Figure B–2 we show just the minority carriers in the PN junction. The positive voltage at the cathode (N) pulls electrons from the P-type material into the depletion region. Once in the depletion region, the electrons are attracted to the N side and pulled out at the cathode. The negative voltage at the anode (P) attracts holes from the N-type material into the depletion region. The holes are then attracted to the P side and pulled out at the anode.

These minority charges create a reverse-bias current in the PN junction. In a reverse-biased diode, this current is very small and can be ignored. However, the reverse-biased current is the key to the behavior of the transistor.

FIGURE B–2

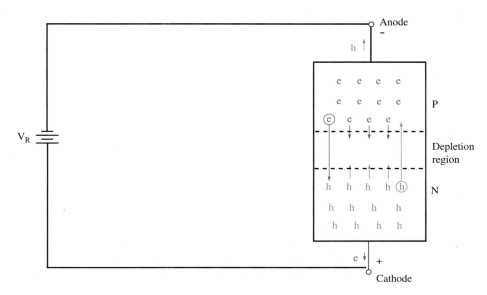

■ CURRENTS IN THE TRANSISTOR

Now we return to the transistor. In Figure B3–A we have connected the emitter to ground and connected the collector to +5V. We have also inserted ammeters to measure current at the three external terminals of the BJT.

FIGURE B–3

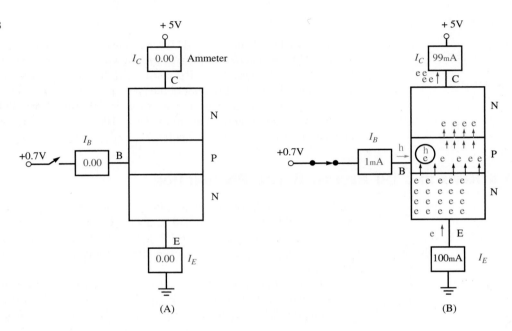

(A)   (B)

Because we have applied a positive voltage to the collector (N-type region), the base-collector voltage is negative, so the base-collector junction is reverse biased. Also, because the switch is open, the base-emitter junction is **OFF.** With both junctions **OFF,** there is no measurable current in the transistor.

We now close the switch to apply 0.7V to the base (Figure B3–B). The base-collector voltage is still negative, so the base-collector PN junction is still reverse biased. However, the applied base voltage forward-biases the base-emitter junction.

Let's assume that when the base-emitter junction is forward biased, enough electrons cross from the emitter into the base to create a 100mA current. The electrons that leave the emitter are replaced by electrons available from the ground connection. This creates an external emitter current of 100mA.

What happens to these emitter electrons when they are injected into the base is the key to the behavior of the transistor. Some of the injected electrons will meet and recombine with holes in the base. The electrons that recombine are effectively lost to us.

If the base is large and has a high concentration of holes, too many of the injected electrons will be lost to recombination. This is why the base is small and lightly doped. Typically, the base is doped so that less than 1% of the electrons injected into the base recombine with holes.

The holes in the base that recombine with the emitter electrons are replaced by holes supplied by the base voltage source (actually, electrons are pulled out of the base to create the required holes). The replacement of holes in the base creates the external base current—1mA in this example (1% of 100mA).

What happens to the remaining 99% of the electrons that have been injected into the base? Remember, electrons in the base (P-type material) are the minority carrier. Minority carriers produce a current in reverse-biased PN junctions, in this case, the base-collector junction. In the normal PN junction, minority carriers are very few and the reverse-biased current is very small.

Because the base is small and lightly doped, 99% or more of the electrons injected from the emitter do not recombine. This greatly increases the minority electron concentration in the base. Increasing the minority carrier concentration increases the reverse-biased PN junction current (the base-collector junction current in this case).

In fact, all of the emitter electrons that do not recombine in the base are swept across the base-collector junction and out of the collector. The collector current in our example is 99mA.

■ CURRENT AMPLIFICATION

What happens to the transistor currents if we increase the base voltage? As shown in Figure B–4, slight increases in base voltage create large increases in all transistor currents. Note that whatever the size of the transistor currents, they maintain a consistent relation. Base current is 1% of emitter current and collector current is 99% of emitter current.

FIGURE B–4

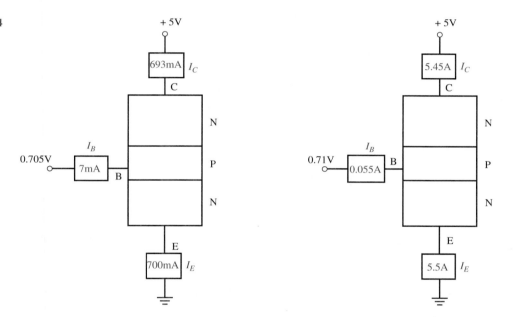

While we have described the current behavior of the transistor in terms of increases in the base voltage, we could also treat the base current as the input. That is, each time we inject a hole into the base, we draw many electrons from the emitter. Because 99% of these electrons are swept out at the collector, we can say that the collector current is dependent on the base current.

You can see from Figures B3–B and B-4 that the collector current in our example is 99 times greater than the base current. Because a small base current creates a large collector current, the transistor acts as a current amplifier.

GLOSSARY

■ A

acceptor impurity　An atom from column III of the Periodic Table that is inserted into intrinsic silicon to increase the density of holes.

ACTIVE　An operating state of a transistor in which significant DC current flows.

active filter　A filter built with an electronic amplifier.

AM demodulation　Retrieving the audio signal from an AM radio signal.

AM detection　See *AM demodulation.*

amplifier　An electronic device that produces a large output for a small input.

AND circuit　A logic circuit that produces a HIGH output only when all inputs are HIGH.

anode　The positive terminal of a device.

astable multivibrator　See *multivibrator, astable.*

automatic gain control　Circuitry that lowers the gain as the output increases.

■ B

bandwidth　The useful range of frequencies of an electronic device or filter.

band-pass filter　A filter designed to pass a limited band of frequencies.

bias current　The DC current required for proper operation of electronic devices.

Bipolar Junction Transistor (BJT)　A 3-terminal device in which the output current is controlled by the input current or voltage.

Bode gain plot　A plot of the gain, in decibels, of a system versus log frequency.

Box Model　For any amplifier, a general model composed of an input resistance, output resistance, and a gain that is represented with a controlled voltage source.

breadboard　A predrilled board with interconnected holes for building laboratory circuits.

breakdown　The sudden rush of current that occurs when a PN junction is sufficiently reverse biased.

breakdown voltage (BV)　Reverse voltage at which a PN junction goes into breakdown.

breakover voltage　The forward voltage at which thyristors turn on.

buffer amplifier　An amplifier with a gain close to 1, a very high input resistance, and a very low output resistance. Used to isolate loads from sources.

Butterworth Filter　A filter that is maximally flat in the pass band.

bypass capacitor　A capacitor used to bypass the emitter resistor (BJT) or source resistor (FET). Increases gain and decreases input resistance.

■ C

cascading amplifiers　Connecting together common-emitter and emitter-follower amplifiers.

cascode amplifier　A common-emitter amplifier connected to a common-base amplifier.

cathode　The negative terminal of a device.

center frequency   The frequency at which a band-pass filter has its maximum gain.

Chebychev filter   A filter with ripple in the pass band and a narrow transition region.

clamper   A circuit that prevents the output voltage from exceeding a set level without changing the input wave shape.

Class A amplifier   An amplifier that is biased near the middle of its ACTIVE region and always remains in the ACTIVE region.

Class AB amplifier   Similar to a Class B amplifier but biased slightly on.

Class B amplifier   An amplifier that is biased at CUT-OFF and turns on for half of the input cycle.

Class C amplifier   An amplifier that is biased deep into CUT-OFF and turns on only for a small portion of the input cycle.

clipper   A circuit that prevents the output from exceeding a set voltage by eliminating any part of a signal that is greater in magnitude than the set voltage.

closed-loop gain   The overall gain of a feedback system.

closed-loop system   See *feedback system.*

CMOS   A complementary pair of MOSFETs (matched N- and P-channel E-MOSFETs).

collector diode   The PN junction between the base and collector of a BJT.

Colpitts oscillator   An LC oscillator in which the feedback voltage is produced by a capacitive voltage divider.

common-base amplifier   A BJT amplifier in which the input is brought in at the emitter and the output is taken from the collector.

common-collector amplifier   See *emitter follower.*

common-emitter amplifier   A BJT amplifier in which the input is brought in at the base and the output is taken from the collector.

common mode   Both inputs are the same in a differential amplifier.

common mode rejection ratio   The ratio of the differential mode and common mode gains.

comparator   A circuit that produces a change in the output every time the input voltage exceeds a set reference voltage.

complementary pair   A pair of matched NPN and PNP BJTs or N- and P-channel FETs.

conductance   The inverse of resistance.

conduction angle   The portion of an input cycle (in degrees or radians) that is seen in the output of switching circuits.

conduction time   The portion of an input signal (in seconds) that is seen in the output of switching circuits.

conventional flow   Direction of current determined by the flow of positive charges.

coupling capacitor   A capacitor used to couple a signal source or a load to an amplifier.

cross-over distortion   The distortion in the output of a push-pull amplifier that is due to the delay in turning on the transistors.

current buffer   An amplifier that increases current available to a load.

current mirror   A two-transistor current source in which a reference current sets the output current.

current out–current feedback   Feedback in which the output variable is current and the feedback variable is also current.

current out–voltage feedback   Feedback in which the output variable is current and the feedback variable is voltage.

current source   A circuit that produces a current that is independent of the load.

current steering   Directing currents to various parts of a circuit.

curve tracer   An instrument that will show the family of I-V curves for a transistor.

CUT-OFF   The state of a transistor when all bias currents are 0A.

■ D

Darlington connection   Two transistors with their collectors tied together and the emitter of one connected to the base of the other.

DC bias levels   The DC voltages and currents in a transistor circuit.

DC restorer   See *clamper.*

DC Troubleshooting   A technique for locating problems in electronic circuits with DC measurements. See *Estimated Voltage* and *Measured Voltage.*

decade   A ten-fold increase in frequency.

decibel (dB)   A gain measure equal to 20 times the log of the magnitude.

depletion mode MOSFET (D-MOSFET)   A MOSFET with a conducting channel that is on under normal conditions.

depletion region   The region surrounding a reverse-biased PN junction in which there are no mobile charges.

derivative   The rate of change of a variable.

DIAC   A bi-directional thyristor that turns on when the voltage across it in either direction exceeds its breakover voltage.

difference amplifier   A device that amplifies the difference in two input signals.

differential amplifier   A difference amplifier in which the difference in inputs is limited to a few 100 millivolts.

differential mode   The two inputs to a differential amplifier are equal in magnitude and opposite in sign.

differentiation   A mathematical operation that finds the rate of change of one variable with respect to another.

diffusion   Process in which a high concentration of a chemical flows towards a low concentration.

diode   A two-terminal device that conducts current in one direction.

diode bridge   Four diodes connected to provide for full-wave rectification.

diode, ideal   A diode model in which the diode is either an open circuit or a short circuit.

donor impurity   An atom from column V of the Periodic Table that is inserted into intrinsic silicon to increase the density of electrons.

doping   The process of adding impurities to intrinsic silicon.

■ E

Early voltage   A BJT parameter that determines the output resistance.

efficiency   A rating of output stages found as a ratio of load power to supply power.

electron   Smallest negatively-charged particle.

electron flow   Direction of current determined by the flow of negative charges.

emitter diode   The BJT PN junction between the base and the emitter.

emitter follower   A common-collector amplifier having a voltage gain close to 1, a high input resistance, and a low output resistance See also *buffer.*

enhancement mode MOSFET (E-MOSFET)   A MOSFET with a conducting channel that is normally off.

envelope   The audio signal of an AM signal.

Estimated Voltage   The approximately-calculated DC bias voltage at the terminals of a transistor. See also *Measured Voltage.*

■ F

fall time   The time it takes a pulse to fall from 90% to 10% of its maximum pulsed value.

feedback factor   The ratio of the feedback signal to the output signal.

feedback fraction   See *feedback factor.*

field effect transistor (FET)   A transistor in which the gate voltage controls the output current. No current enters the gate of FETs.

filter   An electrical circuit designed to pass a limited range of frequencies.

flyback   The large voltage induced in an inductor when its current is suddenly interrupted.

forced commutation   A technique for turning off a thyristor by forcing current through it in the off direction.

forward bias   The voltage required to turn on a PN junction. Also, the region of operation in which the PN junction conducts.

forward gain   The gain in the forward path of a feedback system. Also known as the *open-loop gain.*

forward transfer admittance   The relation between a device's output current and input voltage.

four-layer diode   A one-way conducting thyristor that turns on when the voltage across it exceeds its breakover voltage.

free-running multivibrator   See *multivibrator, astable.*

frequency drift   Change in the frequency of an oscillator due to changes in component values.

full-wave rectifier   A rectifier that produces an output pulse for both the positive and negative halves of the input cycle.

■ G

gain-bandwidth product   The gain of an amplifier times its bandwidth.

gate threshold voltage $(V_{GS\,(th)})$ The voltage required in order to form a conducting channel in an E-MOSFET.

gate turn-off diode (GTO)   A three-terminal thyristor that can be turned off by a voltage applied to its gate.

gate-source cut-off voltage $(V_{GS\,(off)})$ The voltage required to turn off the channels of the JFET and D-MOSFETs.

ground bus   A large conductor that serves as the ground connection for a circuit.

■ H

half-power frequency   The frequency at which the gain drops to 0.707 of the maximum gain. See *3-dB frequency.*

half-wave rectifier   A rectifier that produces an output pulse only during one-half of the input cycle.

Hartley oscillator   An LC oscillator in which an inductive voltage divider provides the feedback voltage.

high-pass filter   A filter that passes only high frequencies.

hole   In semiconductor materials, the mobile positive charge.

hybrid-pi model   A model for the BJT based on its physical properties.

hysteresis   In a Schmitt trigger, the output goes positive for one reference voltage and negative for a second reference voltage.

h-parameter model   A BJT model based on input-output measurements.

■ I

I_B **control**   A BJT circuit in which the collector current is controlled by the base current.

ideal diode   A diode model in which the diode is either an open circuit or a short circuit.

impedance   The frequency dependent relation between voltage and current for a general two-terminal device.

impedance matching   Changing the apparent impedance of a load to optimize amplifier response.

impurity   An atom inserted into intrinsic silicon to change the concentration of holes or electrons. See *acceptor impurity* and *donor impurity*.

input offset voltage   The voltage required at the input of an Op-Amp to zero the output when all signal sources are turned off.

input resistance   The resistance seen at the input of an amplifier.

instrumentation amplifier   An integrated circuit amplifier that has a fixed gain or range of gains.

integration   A mathematical operation that finds the area under a curve. It is the inverse operation of differentiation.

intrinsic stand-off ratio   Parameter that determines the trigger voltage of a UJT.

inverter   A device in which the output goes positive when the input goes negative.

ion implantation   Doping a semiconductor by injecting ions of impurity atoms.

isolation amplifier   A buffer that isolates the load from the source.

I-V characteristic   A plot that shows device current versus device voltage.

I-V curve   See *I-V characteristic*.

■ **J**

junction   The boundary between N- and P-type material.

junction field effect transistor (JFET)   An FET that is controlled by varying the depletion region of a reverse-biased PN junction.

junction potential   The voltage across a PN junction.

■ **L**

leakage current   The current that flows in a device when it is off.

level detector   See *comparator*.

light emitting diode (LED)   A diode that will emit visible or infrared light when sufficient forward voltage is applied to it.

limiter   See *clipper*.

loop gain   In a feedback system, the product of the forward and feedback gains.

low-pass filter   A filter that passes only low frequencies.

■ **M**

majority carrier   The carrier with the highest concentration in doped silicon: holes in P-type and electrons in N-type material.

maximum reverse voltage (VR)   See *breakdown voltage*.

Measured Voltage   The actual voltage measured at a transistor terminal. See also *Estimated Voltage*.

metal oxide semiconductor FET (MOSFET)   A FET controlled by the field produced by a voltage applied to an insulated gate.

Miller capacitance   The large effective capacitance seen at the input of an amplifier when a capacitor is connected between the input and output.

minority carriers   The carrier with the lowest concentration in doped silicon: electrons in P-type and holes in N-type material.

multivibrator, astable   A free-running pulse generator.

multivibrator, monostable   A pulse generator that produces a single fixed-width pulse each time it is triggered.

■ N

N-type material   Semiconductor material doped to increase electron concentration.

NAND gate   A logic circuit that produces a LOW output only when all inputs are HIGH.

negative feedback   A feedback system in which a fraction of the output is subtracted from the input.

non-linear device   A device in which the output is not linearly dependent on the input.

NOR gate   A logic circuit that produces a HIGH output only when all inputs are LOW.

NPN BJT   A BJT built by sandwiching P-type material between two N-type materials.

■ O

octave   A doubling in frequency.

OHMIC region   The operating region of a FET when the drain-source voltage is very small. The FET acts as a resistor in this region.

one-shot multivibrator   See *multivibrator, monostable.*

open loop gain   See *forward gain.*

operational amplifier (Op-Amp)   An integrated circuit amplifier with very high gain, very high input resistance, and very low output resistance.

opto-coupler   A photo transistor driven by an LED.

opto-isolator   See *opto-coupler.*

Op-Amp   See *operational amplifier.*

oscillator   A signal generator that produces periodic signals, often sine waves.

oscilloscope   An instrument that records and displays time-varying voltages.

output resistance   The resistance seen at the output of an electronic device.

■ P

pass band   The range of frequencies that are passed by a filter. See *bandwidth.*

passive filter   A filter built without active devices (transistors, Op-Amps).

peak detector   A circuit that holds the peak value of the input.

peak inverse voltage (PIV)   See *breakdown voltage.*

peak sensitivity wavelength   The wavelength of optimal response of an optical device.

% efficiency   Efficiency expressed as a percentage. See *efficiency.*

phase angle   Angular measure of the delay between an output and an input signal.

phase shift   See *phase angle.*

phase shift oscillator   A negative-feedback resistor-capacitor oscillator.

phase splitter   An amplifier that produces two outputs that are 180° out of phase.

photo detector   See *photo diode.*

photo diode   A diode that will produce a current or a voltage in response to light.

photo transistor   A transistor that produces a current in response to light.

photoresistor   A device that has a resistance that depends on the light falling upon it.

pinch-off voltage   A FET parameter—the drain-source voltage at which current reaches its maximum value.

P-type material   Semiconductor material that is doped to increase the hole concentration.

PN junction   A junction of P-type and N-type material.

PNP BJT   A BJT built by sandwiching N-type material between two P-type materials.

PNPN diode   See *four-layer diode.*

positive feedback   A feedback system in which a portion of the output is added to the input.

potential barrier   The reverse-bias voltage on a PN junction that must be overcome for the PN junction to conduct current.

precision rectifier   A rectifier circuit that eliminates the diode voltage drop.

programmable unijunction transistor (PUT)   A unijunction transistor in which the intrinsic stand-off ratio can be set with external resistors.

push-pull amplifier   A power amplifier built with a complementary pair of Class B–biased transistors.

■ Q

quiescent levels   DC voltages and currents in a transistor circuit.

Q-point   In the BJT, DC collector current and collector-emitter voltage. In the FET, DC drain current and gate-source voltage.

■ R

ramp   A waveform that increases linearly with time.

reactance   For a capacitor or inductor, the frequency-dependent relation between voltage and current.

recombination   The process by which a hole and electron cancel each other.

rectification   The elimination of the positive or negative portion of a signal.

rectifier   A device that rectifies. See *diode*.

reflected resistance   The resistance seen at the primary of a transformer.

regeneration   Positive feedback used to increase the gain of an amplifier.

resonant frequency   The frequency at which an inductor-capacitor circuit oscillates.

reverse bias   The voltage applied to a PN junction that keeps the junction from conducting.

reverse blocking diode   See *four-layer diode*.

ripple   The variation in voltage seen in an AC-to-DC converter.

ripple (%)   Ripple expressed as a percent of the DC voltage.

rise time   The time it takes a pulse to rise from 10% to 90% of its pulsed value.

rms ripple factor   Ripple expressed in rms quantities.

root-mean square (rms)   The square root of the mean of a signal squared. rms values are used in power calculations.

■ S

Sallen-Key filter   An active filter built with a single Op-Amp.

SATURATION   In a BJT the state of operation in which the collector voltage equals the emitter voltage. Both the collector and emitter diodes are ON.

saturation current (I_{DSS})   In a JFET or D-MOSFET, the current when there is no gate-source voltage.

sawtooth generator   A signal generator that produces a sawtooth wave.

sawtooth wave   A signal consisting of a ramp that is periodically reset to zero.

schematic capture   Computer-aided design program in which the circuit is drawn on the screen.

Schmitt trigger   A comparator that has one reference voltage for a positive-going signal and another reference voltage for a negative-going signal.

Schockley diode   See *four-layer diode*.

Schottky diode   A metal semiconductor junction diode.

selectivity   The ability of an amplifier to respond only to a narrow range of frequencies.

semiconductor   A material that conducts poorly at room temperature but becomes a good conductor when energy is applied. The energy can be electrical, mechanical, thermal, or optical.

sensor   A device that converts other forms of energy into electrical signals.

signal generator   A device that can produce sine waves, square waves, and triangle waves.

signal tracing   A troubleshooting technique in which AC signals are traced through a circuit.

silicon-controlled rectifier (SCR)   A one-way conducting thyristor that is turned on by a gate-trigger voltage.

sine wave   A waveform that is mathematically described with a sine or cosine.

sine wave generator   A signal generator that produces sine waves.

slew rate   The maximum rate at which an Op-Amp output can change.

small signal   A signal that is small enough to allow a linear analysis.

solar cell   A photo diode designed to produce a fixed voltage in response to exposure to the sun.

solid state   Term used to describe semiconductor materials.

square wave   A wave shape that alternates between a positive and negative voltage of the same magnitude and has a pulse width that is one-half of the period.

square wave generator   A signal generator that produces square waves.

stable response   A system response which remains within finite limits.

stand-by power   The power used by an amplifier when there is no input.

step-down transformer   A transformer in which the secondary voltage is smaller than the primary voltage.

step-up transformer   A transformer in which the secondary voltage is larger than the primary voltage.

stop band   The range of frequencies that are blocked by a filter. These frequencies are not seen in the output.

system gain   The overall gain of an amplifier, including source and load resistances.

■ T

tank circuit   An inductor-capacitor circuit that resonates at a frequency of $1/\sqrt{LC}$.

temperature compensation   Prevention of variations in bias levels due to changes in temperature.

thermal voltage (V_T)   A BJT device parameter—26mV at room temperature.

Thevenin equivalent   A model for a two-port consisting of a voltage source in series with a resistance.

3dB-frequency   The frequency at which the gain is 3dB lower than the maximum gain, equivalent to the half-power frequency.

thyristor   An electronic switch.

transconductance   The relation between a device's output current and input voltage.

transducer   See *sensor.*

transformer   A pair of coupled inductors (coils) that transfer AC voltages from the primary to the secondary.

transformer reflected resistance   The apparent resistance seen at the primary of a transformer when there is a resistor in the secondary.

transition region   The region between a filter's pass-band and stop-band.

TRIAC   A bidirectional thyristor turned on by a gate trigger voltage.

triangle wave   A wave shape that is a periodic repetition of triangles.

triangle wave generator   A signal generator that produces triangle waves.

troubleshooting   The process of locating faults in a circuit.

tuned amplifier   An amplifier that will respond only to a set frequency.

tuned circuit   See *tank circuit.*

tuning capacitor diode   See *varactor.*

tunnel diode   A diode with a negative resistance region.

two-port   An electrical device that has an input side and an output side.

■ U

unijunction transistor (UJT)   A switch that is triggered when its emitter voltage exceeds a threshold set by its intrinsic stand-off ratio.

unity feedback   A feedback system in which the feedback signal equals the output signal.

unstable response   A response that grows uncontrollably.

■ V

valence shell   The outer electron shell of an atom.

varactor   A diode that acts as a capacitance that depends on the reverse-bias voltage.

variable frequency oscillator (VFO)   An oscillator in which the frequency can be varied.

varicap diode   See *varactor.*

virtual ground   A point in a circuit that is not wired to ground but still remains at 0V.

voltage amplifier   A device that produces a large output voltage from a small input voltage.

voltage-controlled oscillator (VCO)   An oscillator in which the frequency is set by a reference voltage.

voltage divider   A series resistor circuit that divides the supply voltage between the resistors.

voltage doubler   A diode circuit that provides a DC voltage that is twice the input voltage.

voltage gain   The magnitude of the ratio of output voltage to input voltage.

voltage out–current feedback   A feedback system in which the output is voltage and the feedback signal is current.

voltage out–voltage feedback   A feedback system in which the output is voltage and the feedback signal is also voltage.

voltage regulator   A device or circuit that provides a stable DC voltage.

voltage-variable resistor   A FET operated in its OHMIC region. The resistance of the FET depends on the gate-source voltage.

■ W

weighted summing amplifier   An amplifier that adds several inputs with a different weight assigned to each input.

Wien bridge oscillator   A sine wave generator that uses a bridge composed of one series RC and one parallel RC combination.

■ Z

Zener diode   A diode constructed to operate in the breakdown region in order to provide a constant voltage.

zero-crossing detector   A circuit that changes state every time the input signal crosses 0V.